Developments in Paleoenvironmental Research

For further volumes:
http://www.springer.com/series/5869

The loess-paleosol sections near Xian (photo1, by Li Li) and red clay section in Huating (photo 2, by Song Yougui), China. Loess-paleosol-red clay sequences have recorded the information about the south eastern humid summer monsoon and western inland aridification evolution. Unique archives reflecting the differentiation and coupling of Asian monsoon-arid environment, associated with the Tibetan Plateau growth.

Zhisheng An
Editor

Late Cenozoic Climate Change in Asia

Loess, Monsoon and Monsoon-arid Environment Evolution

Springer

Editor
Zhisheng An
Institute of Earth Environment
Chinese Academy of Sciences
Xi'an
China

ISSN 1571-5299
ISBN 978-94-007-7816-0 ISBN 978-94-007-7817-7 (eBook)
DOI 10.1007/978-94-007-7817-7
Springer Dordrecht Heidelberg New York London

Library of Congress Control Number: 2013958087

Printed on acid-free paper

Springer is part of Springer Science+Business Media (www.springer.com)

Preface

Climate change in Asia has long enjoyed tremendous interest and progress in the scientific community. The overall aims and scope of this volume are to capture this excitement and document these developments. Asia climate change study has now become a priority for the international paleoclimate community for two reasons. Firstly, the Asian continent, Tibetan Plateau and marginal seas have become a focus of paleomonsoon research. A series of large scale international field projects and drilling campaigns have been conducted to enhance our understanding of the Asian monsoon-arid environment history. Secondly, a sense of urgency has arisen in Asia as it continues to develop rapidly, and is already the most populated continent in the world. The goal of sustainable growth relies on a thorough understanding of climate change and the mechanisms involved. Our intention is to address this need by synthesizing our past findings to steer our future efforts in the most fruitful direction.

Preface

Acknowledgements

The present volume is a product of group efforts mainly by the State Key Laboratory of Loess and Quaternary Geology, Institute of Earth Environment, Chinese Academy of Sciences. This work was supported by the National Basic Research Program of China, the National Natural Science Foundation of China, the Chinese Academy of Sciences, State Key Laboratory of Loess and Quaternary Geology and Institute of Earth Environment.

The authors express their gratitude to Profs. Ray Bradley, Richard Arimoto, Alan Gillespie and George S. Burr for reviewing early versions of chapters. Their critical comments and constructive suggestions significantly improved the final presentation of the book. We thank the publishers and copyright holders for granting permissions to use published figures and data as cited in the book. We owe great debts to Yihe Tai, Yajuan Du, Ping Yao, Ming Li, and Yi Liu, Xiaohao Wen, Fengnian Wang, Shuhuan Du, Yuejun Si, Dongfeng Niu, Zhiwen Li, Yuanjun Guo, Yihua Guo for their kind help and assistance in the writing. Special thanks are due to Tamara, Judith and Li Li, whose continuous encouragement, organization and editorial works helped us pushing this book from the idea stage into reality.

Acknowledgments

Table of Contents

List of Abbreviations

AMS	Accelerator Mass Spectroscopy
ARM	Anhysteretic Remanent Magnetization
CGCM	Coupled General Circulation Models
CLIMAP	Climate: Long range Investigation, Mapping, and Prediction
CLP	Chinese Loess Plateau
COHMAP	Cooperative Holocene Mapping Project
DSDP	Deep Sea Drilling Project
EAM	East Asian Monsoon
EASM	East Asian Summer Monsoon
EBM	Energy-Balance Model
EC	Elemental Carbon
EMIC	Earth system models of intermediate complexity
FOAM	Fast Ocean Atmosphere Model
GIC	Global Iron Connections
GPTS	Geomagnetic Polarity Time Scale
IPWP	Indo-Pacific Warm Pool
IRSL	InFrared Stimulated Luminescence
ITCZ	Intertropcial Convergence Zone
LGM	Last Glacial Maximum
LIA	Little Ice Age
LGM	Last Glacial Maximum
MWP	Medieval Warm Period
NADW	North Atlantic Deep Water
NRM	Natural Remanent Magnetization
OC	Organic Carbon
OSL	Optically Stimulated Luminescence
PAHs	Polycyclic Aromatic Hydrocarbons
PCBs	Polychlorinated Biphenyls
PMIP	Paleoclimate Modeling Intercomparison Project
POPs	Persistent Organic Pollutants
QGS	Quartz Grain Size

SAR	Single-Aliquot Regenerative-dose
SCS	South China Sea
SIRM	Saturation Isothermal Remanent Magnetization
SMAR	Sensitivity-corrected Multiple Aliquot Regenerative dose
SST	Sea Surface Temperatures
TOC	Total Organic Carbon
TP	Tibetan Plateau
YD	Younger Dryas

List of Contributors

Zhisheng An State Key Laboratory of Loess and Quaternary Geology, Institute of Earth Environment, Chinese Academy of Sciences, Xi'an, China

Hong Ao State Key Laboratory of Loess and Quaternary Geology, Institute of Earth Environment, Chinese Academy of Sciences, Xi'an, China

George S. Burr University of , NSF- AMS Laboratory, Tucson, USA

Yanjun Cai State Key Laboratory of Loess and Quaternary Geology, Institute of Earth Environment, Chinese Academy of Sciences, Xi'an, China

Junji Cao State Key Laboratory of Loess and Quaternary Geology, Institute of Earth Environment, Chinese Academy of Sciences, Xi'an, China

Hong Chang State Key Laboratory of Loess and Quaternary Geology, Institute of Earth Environment, Chinese Academy of Sciences, Xi'an, China

Peng Cheng State Key Laboratory of Loess and Quaternary Geology, Institute of Earth Environment, Chinese Academy of Sciences, Xi'an, China

Yongming Han State Key Laboratory of Loess and Quaternary Geology, Institute of Earth Environment, Chinese Academy of Sciences, Xi'an, China

Zhangdong Jin State Key Laboratory of Loess and Quaternary Geology, Institute of Earth Environment, Chinese Academy of Sciences, Xi'an, China

Zhao Jin State Key Laboratory of Loess and Quaternary Geology, Institute of Earth Environment, Chinese Academy of Sciences, Xi'an, China

Baosheng Li School of Geographical Science, South China Normal University, Guangzhou, China

Li Li State Key Laboratory of Loess and Quaternary Geology, Institute of Earth Environment, Chinese Academy of Sciences, Xi'an, China

Xiaoqiang Li State Key Laboratory of Loess and Quaternary Geology, Institute of Earth Environment, Chinese Academy of Sciences, Xi'an, China

Yongxiang Li Department of Geology, Northwest University, Xi'an, China

Weiguo Liu State Key Laboratory of Loess and Quaternary Geology, Institute of Earth Environment, Chinese Academy of Sciences, Xi'an, China

Xiaodong Liu State Key Laboratory of Loess and Quaternary Geology, Institute of Earth Environment, Chinese Academy of Sciences, Xi'an, China

Yu Liu State Key Laboratory of Loess and Quaternary Geology, Institute of Earth Environment, Chinese Academy of Sciences, Xi'an, China

Zhonghui Liu Department of Earth Sciences, The University of Hong Kong, Hong Kong, China

Fengyan Lu State Key Laboratory of Loess and Quaternary Geology, Institute of Earth Environment, Chinese Academy of Sciences, Xi'an, China

Wei Pan Centre for Historical Environment and Socio-Economic Development in Northwest China; Shaanxi Normal University; Xi'an, China

Zicheng Peng University of Science and Technology of China, Hefei, China

Xiaoke Qiang State Key Laboratory of Loess and Quaternary Geology, Institute of Earth Environment, Chinese Academy of Sciences, Xi'an, China

Zhengguo Shi State Key Laboratory of Loess and Quaternary Geology, Institute of Earth Environment, Chinese Academy of Sciences, Xi'an, China

Shaohua Song State Key Laboratory of Loess and Quaternary Geology, Institute of Earth Environment, Chinese Academy of Sciences, Xi'an, China

Yougui Song State Key Laboratory of Loess and Quaternary Geology, Institute of Earth Environment, Chinese Academy of Sciences, Xi'an, China

Youbin Sun State Key Laboratory of Loess and Quaternary Geology, Institute of Earth Environment, Chinese Academy of Sciences, Xi'an, China

Liangcheng Tan State Key Laboratory of Loess and Quaternary Geology, Institute of Earth Environment, Chinese Academy of Sciences, Xi'an, China

Lide Tian Laboratory of Environment and Processes in the Tibetan Plateau, Institute of Tibetan Plateau Research, Chinese Academy of Sciences, Beijing, China

Xulong Wang State Key Laboratory of Loess and Quaternary Geology, Institute of Earth Environment, Chinese Academy of Sciences, Xi'an, China

Xinzhi Wu Institute of Vertebrate Paleontology and Paleoanthropology Chinese Academy of Sciences, Beijing, China

Feng Xian State Key Laboratory of Loess and Quaternary Geology, Institute of Earth Environment, Chinese Academy of Sciences, Xi'an, China

Hai Xu State Key Laboratory of Loess and Quaternary Geology, Institute of Earth Environment, Chinese Academy of Sciences, Xi'an, China

Libin Yan State Key Laboratory of Loess and Quaternary Geology, Institute of Earth Environment, Chinese Academy of Sciences, Xi'an, China

Xuefeng Yu State Key Laboratory of Loess and Quaternary Geology, Institute of Earth Environment, Chinese Academy of Sciences, Xi'an, China

Peizhen Zhang Institute of Geology, China Seismological Bureau, Beijing, China

Xiaoye Zhang Chinese Academy of Meteorological Sciences, Beijing, China

Xinbao Zhang Institute of Mountain Hazards and Environment, Chinese Academy of Sciences, Chengdu, China

Yunxiang Zhang Department of Geology, Northwest University, Xi'an China

Hongli Zhao State Key Laboratory of Loess and Quaternary Geology, Institute of Earth Environment, Chinese Academy of Sciences, Xi'an, China

Weijian Zhou State Key Laboratory of Loess and Quaternary Geology, Institute of Earth Environment, Chinese Academy of Sciences, Xi'an, China

Chapter 1
Introduction

Zhisheng An, Li Li, George S. Burr, Yougui Song, Libin Yan, Hong Chang, Youbin Sun, Yanjun Cai, Zhengguo Shi, Hai Xu, Hongli Zhao, and Weijian Zhou

Abstract This chapter gives a description of the research history and development of our current understanding of paleoclimate and paleoenvironments in China. It outlines major questions that have arisen after years of study of the Asian monsoon-arid climate system. Modern natural geography is introduced and this sets the stage for detailed reviews in the following chapters.

Keywords Asian monsoon · Tibetan plateau · Geomorphology · Loess · Glacial · Monsoon-arid zone · Monsoon variability · Global change

1.1 Introduction

This book is about climate change at the interface of the monsoon and arid regions of the Asian continent. The climate of the southern and eastern humid zones within this region are controlled by the Asian monsoon, while climate in the relatively arid zone to the northwest is controlled by the westerlies and the East Asian winter monsoon. The climate of the semi-arid zone in-between is influenced by both monsoons and the westerlies. The semi-arid zone marks the inland extent of the monsoon. This region changes with time, reflecting changing climate and monsoon frontal shifts. In recent years we have extended our view of climate change in China to include the western arid regions as an essential component of Asian climate. The current framework involves an eastern humid end-member and a western arid end-member, with the dynamic semi-arid transition zone in-between. Any comprehensive evaluation of Asian environmental change must recognize the coupled evolution of the Asian monsoon and monsoon-arid climate system as part of a whole, within a global framework.

Z. An (✉) · L. Li · Y. Song · L. Yan · H. Chang · Y. Sun· Y. Cai · Z. Shi · H. Xu· H. Zhao · W. Zhou
State Key Laboratory of Loess and Quaternary Geology,
Institute of Earth Environment, Chinese Academy of Sciences,
No. 10 Fenghui S. Rd, High-Tech Zone,
Xi'an PO Box 17, Xi'an, China
e-mail: anzs@loess.llqg.ac.cn

Z. An (ed.), *Late Cenozoic Climate Change in Asia*,
Developments in Paleoenvironmental Research 16,
DOI 10.1007/978-94-007-7817-7_1, © Springer Science+Business Media Dordrecht 2014

G. S. Burr
University of Arizona, NSF-Arizona AMS Laboratory,
85721 Tucson, Arizona, USA
e-mail: burr@email.arizona.edu

L. Li
e-mail: lili@ieecas.cn

Y. Song
e-mail: ygsong@loess.llqg.ac.cn

L. Yan
e-mail: yanlibin@ieecas.cn

H. Chang
e-mail: changh@loess.llqg.ac.cn

Y. Sun
e-mail: sunyb@ieecas.cn

Y. Cai
e-mail: caiyj@loess.llqg.ac.cn

Z. Shi
e-mail: shizg@ieecas.cn

H. Xu
e-mail: xuhai@ieecas.cn

H. Zhao
e-mail: zhaohl@ieecas.cn

W. Zhou
e-mail: weijian@loess.llqg.ac.cn

The entire region, from the monsoon-dominated zone to the arid zone, preserves unique records of past climate change. This is especially true of the semi-arid zone. Here the reader will find a comprehensive review of these records which provide evidence for spatial and temporal climatic and environmental change across the Asian continent since the Late Cenozoic. The dynamics underlying these changes are explored on the basis of these records, and an *Asian monsoon-arid environment system* is described that quantifies the controlling mechanisms of climate change and how they operate on different time scales.

The goal of the book is to present an integral framework for understanding natural environmental change and partly human-induced forcings in China that may be used to assist in the ongoing effort to manage the development of this region responsibly and sustainably. This goal extends far beyond the highly populated urban centers of East Asia because the Asian monsoon is an integral part of the global climate system, and monsoon-driven precipitation in particular, is vital to the maintenance and sustainable development of billions of people.

Monsoon circulation plays a very important role in the transportation of global water vapour and heat. The dynamics of the Asian summer monsoon is connected to the Equatorial Ocean and Southern Hemisphere, and the East Asian winter monsoon is closely associated with climatic conditions of northern high latitudes, including the North Atlantic Ocean and Arctic regions. Past Asian winter/summer monsoon variability has been shown to be influenced by changes in solar insolation, lower boundary conditions and greenhouse gases.

The Asian monsoon is the most powerful monsoon circulation system on Earth. It is driven by seasonal changes in solar radiation and the differential absorption of radiation by the oceans and continents. The Qinghai-Tibet Plateau, extending into the mid-troposphere, plays a major role in the regional energy balance and thus has a critical role in the dynamics of the Asian monsoon system. Asian monsoon dynamics interact with global climate through moisture and energy exchanges between high and low latitudes of the Northern Hemisphere, and between the Northern and Southern Hemispheres. During the summer, the Asian monsoon brings heavy rains to Asia and Africa, while monsoon winters are relatively dry. The lives of many people under the influence of the monsoon have adapted to these seasonal changes. While the wet monsoon brings much needed rain, it can also cause flooding and torrential storms that can bring crop and infrastructure damage with them. Vegetation, land use, hydrology, economy, culture, and society across the Asian monsoon region all depend on the evolution and variability of the Asian monsoon, to which this book is devoted.

Paleomonsoon research on the Asian monsoon zone began with the study of loess profiles. From the 1950's to the 1970's, scientists focused their attention on the distribution of loess on the Chinese Loess Plateau (CLP), its composition, structure and origin. In the 1970's to 1980's, Chinese loess studies explored correlations with other records of environmental change. This was the focus of the text, *Loess and Environment*, written by a Chinese group led and organized by Liu Tungsheng and An Zhisheng, that was published in 1985. This text became the Rosetta stone for Chinese loess studies. In the book, Liu and An successfully correlated Chinese loess-paleosol sequences with climatic records of deep sea marine cores. At that point, the scientific community began to realize that Chinese loess records represented a unique and valuable archive of continental environmental change that should be studied carefully alongside marine sediment and ice core records. In the late 1980's, the research group led by An Zhisheng continued to explore the dynamics of Chinese environmental change, demonstrating that loess-paleosol sequences provide sufficient resolution to study monsoon variability. The scope of the work was expanded to include research on a variety of continental archives. That research revealed that the 600,000 km^2 of loess-paleosol sequences distributed along the middle reaches of the Yellow River are formed under the influence of monsoonal climate changes, with alternating dominance between the East Asian winter and summer monsoons. Furthermore, the eolian depositional flux recorded on the Loess Plateau reflects aridity changes in the dust source region. Therefore, the eolian loess-paleosol sequences can be used as high resolution records of past monsoons and associated climate changes of arid inland Asia. This discovery solved the puzzle of how such massive and widely-distributed sediments came to be deposited. The hypothesis that the monsoon is the controlling force for paleoenvironmental change in East Asia was proposed in the late 1990s, as correlations between loess, Quaternary sediment records, and historical records made this connection increasingly clear. The monsoon control hypothesis has since been acknowledged by the international community, as it explains a series of natural phenomena, like the shifts of desert margins, lake levels, snow/forest lines, vegetative zones, mammalian migration and sea surface temperature changes in Asia.

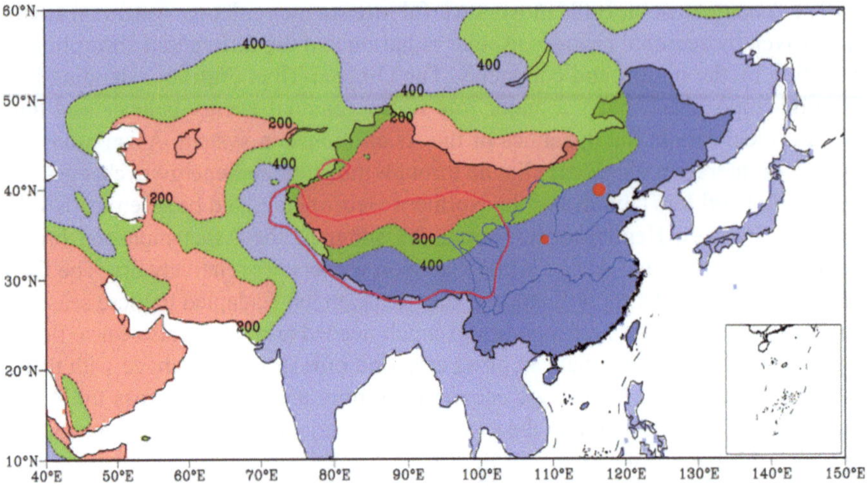

Fig. 1.1 The spatial distribution of annual mean precipitation (Unit: mm) based on the CRU dataset (New et al. 2002) from 1971 to 2000. The elevation of the pink contour is 3000 m. The red, green, and blue shaded areas indicate the regions where precipitation is less than 200 mm, between 200 and 400 mm, and more than 400 mm, respectively. The deeper red, green and blue shaded areas are located in China. The big and small red solid circles indicate the position of Beijing and Xi'an, respectively

1.2 Geography of the Asian Monsoon-arid zone

1.2.1 Climatic Setting

The continent of Asia is traversed from southwest to northeast by the Asian monsoon belt, the largest monsoon system on Earth. The Asian monsoon extends from the Arabian Sea in the west, to northeastern China and Korea in the east. What sets the Asian monsoon apart from other monsoon systems is the presence of the Tibetan Plateau (TP), which forms a massive wall along its southern margin (Wang 2006). The total annual precipitation across the monsoon belt ranges from more than 1200 mm in coastal regions, to approximately 200 mm in the Asian interior. Annual precipitation provides a convenient metric of monsoon influence and we can subdivide the Asian continent into three distinct environmental regions accordingly: 1) a relatively wet and humid southeastern region with greater than 400 mm annual precipitation; 2) an interior arid northwest region with less than 200 mm annual precipitation; and 3) an intermediate semi-arid belt with 200–400 mm annual precipitation (Fig. 1.1). These three regions are the basic components of the *Asian monsoon-arid environment system* which will be discussed throughout this text.

The Asian monsoon is mainly composed of two meteorologically distinct, but overlapping, subsystems: the Indian (South Asian) monsoon and the East Asian monsoon (Fig. 1.2). The overall Asian monsoon region is characterized by its large area, strong intensity and distinct seasonality. During summer, the Asian monsoon brings heavy rainfall to Asia, while in winter it provides for relatively colder, drier climate.

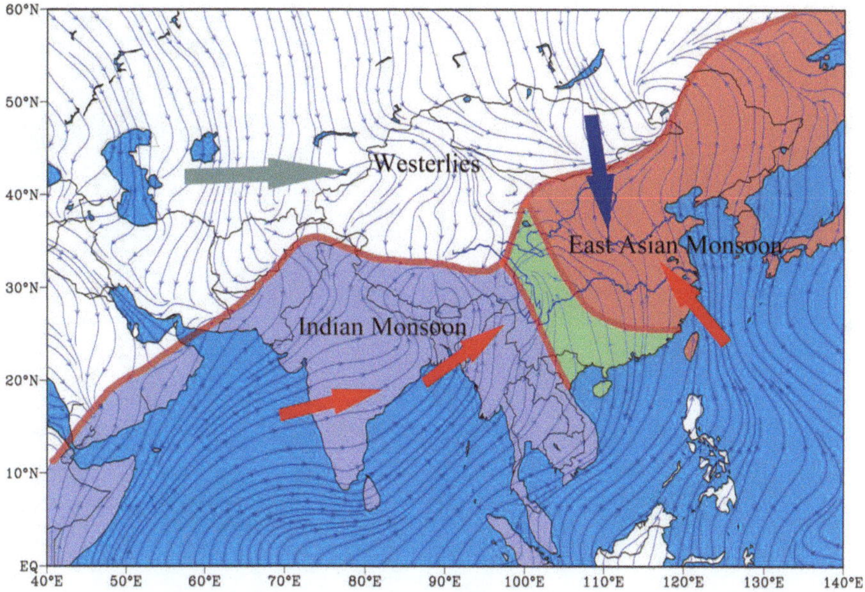

Fig. 1.2 A diagram illustrating the Asian summer monsoon, winter monsoon and westerlies. The *red line*, which is the same as the northern boundary of quasi-southern wind, 10 m above the surface (blue streamline) in July based on the ERA-40 dataset (Uppala et al. 2005), indicates the northern boundary of the monsoonal area. The *red*, *blue* and *grey* arrows indicate the frontal motions of the summer monsoon, winter monsoon and prevailing winds of the westerlies, respectively. The *light blue* and *red* regions are impacted by the Indian Monsoon and East Asian Monsoon, respectively, while the *light green* region is influenced variably by both the Indian Monsoon and East Asian Monsoon. The *left* and *right* limits of this transition region are suggested by Wang (2006) and Liao Ke et al. (1999), respectively

The Indian monsoon is a typical tropical monsoon system that is driven by the seasonal inter-hemispheric thermal contrast between the Asian continent and the Indian Ocean. The system is anchored by the Indian Low (Fig. 1.3a) and southwesterly winds dominate during the summer. The TP is the source of a latent heat anomaly within the mid-troposphere that drives the onset and development of the Indian monsoon (Yanai et al. 1992). The moisture carried by the Indian monsoon originates in the Indian Ocean. It produces copious rainfall every year which is distributed over southern Asia, southeastern Asia and southwestern China.

The East Asian monsoon is distinct from the Indian monsoon due to the very different orography bounding the two systems along their northern margins and because of zonal land-ocean configuration differences. The East Asian monsoon is a shallow system, primarily dominating the lower troposphere. It is regarded as one of the most important global climatic bridges between low and high latitudes and is the only subtropical/temperate monsoon on Earth. It delivers moisture from the tropical Pacific, which is the primary source of water vapor for northern China. Cross-equatorial air streams from Australia also contribute to the strength of the East Asian monsoon. Especially in winter, a Siberian-Mongolian high pressure cell

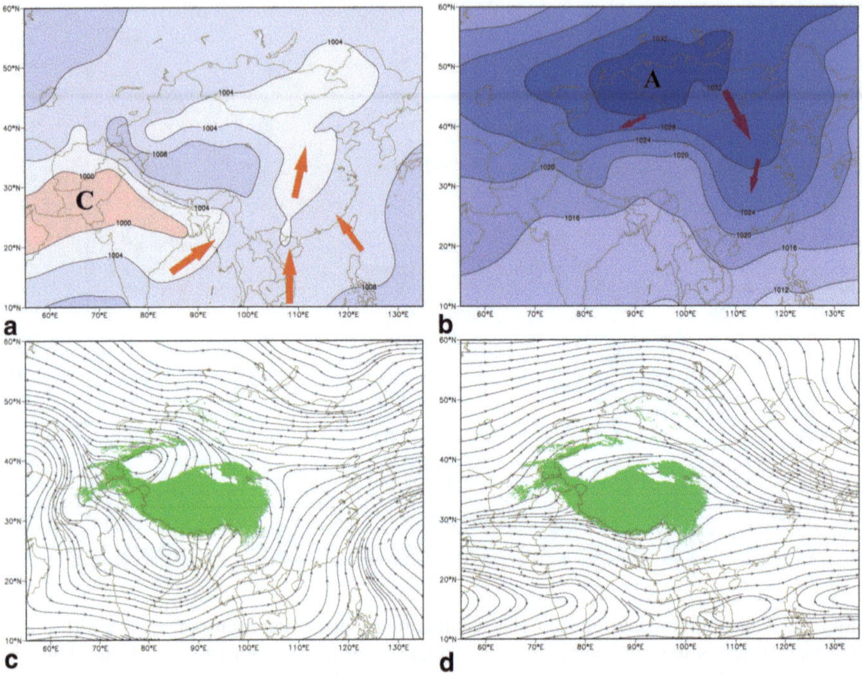

Fig. 1.3 Modern mean sea level pressure and 700 hPa flow fields in July (a, c) and January (b, d) from 1971 to 2000 based on NCEP/NCAR reanalysis (Kalnay et al. 1996). "C" in a and "A" in b indicate the center of the Indian Low and the center of the Siberian–Mongolian High. The *green* shaded areas in (c, d) indicate the TP where the elevation exceeds 3000 m

develops over the continent (Fig. 1.3b), facilitating the injection of cold air masses into Mongolia and China. The net result is the world's most intense winter monsoon.

The Asian interior is a vast arid region primarily due to the geographic barrier imposed by the TP. It is very difficult for moisture to reach this region and annual precipitation is generally less than 200 mm. The westerlies prevail for the whole year across the plateau, especially in the middle and high troposphere (Fig. 1.3c, d), providing a climatic teleconnection between the North Atlantic and Asia (Porter and An 1995). A northeasterly branch of the winter monsoon system can also impact arid areas through the lower atmosphere. Asian deserts are the main source regions for airborne dust in East Asia, and are among the three largest dust sources in the world. This dust is transported by the westerlies across the Asian continent and out to sea. Much of the dust ends up in the North Pacific Ocean, but it can travel as far as Greenland, as discussed in Chap. 4.

The phased growth of the elevated terrain of the TP has significantly intensified the Asian monsoon (An et al. 2001) as well as aridity inland (Manabe et al. 1990) since the Late Cenozoic. It also serves to divide the westerlies into two branches when they encounter the plateau, with each branch following the northern or southern margins (Fig. 1.3c, d). This contributes to enhanced divergence-driven subsidence downstream, and hence to the arid climate in the Asian interior (Manabe et al. 1990).

Under glacial conditions (e.g., the last glacial maximum, LGM), the monsoon-arid climate pattern over Asia changed. The impact of ice sheets in the Northern Hemisphere likely weakened the Asian monsoon during summer, while strengthening it during winter. As sea level decreased and the continental margin grew, the East Asian summer monsoon retreated southeastward, leaving less rainfall and more severe aridity in the interior. In winter, the intensified Siberian High increased the advection of cold air into lower-latitudes, resulting in colder temperatures in northern China (Jiang et al. 2010). The westerlies were also strengthened and shifted southward during the LGM.

1.2.2 Vegetation

Vegetation differences can be used to divide China into two regions along the line—Daxinganling-Lüliangshan-Liupanshan-eastern Tibetan Plateau, and roughly corresponding to the zonation defined by annual precipitation. The southeast region is dominated by monsoonal climate and features a variety of forest types. The northwest region is weakly affected by the summer monsoon and is comprised of dry grasslands and deserts.

In the southeastern region the vegetation zonal distribution changes significantly along a north-south temperature gradient, as reflected by forest types. From north to south, cold temperate coniferous forests give way to temperate broad-leaved forests, warm temperate deciduous broad-leaved forests, subtropical evergreen broad-leaved forests, tropical monsoon/rain forests and finally an equatorial rainforest belt. In northwest China, zonal vegetational divisions are less distinct, with only a north and south differentiation observed in the Xinjiang temperate desert region, bounded by the Tianshan. Here a temperate desert zone is found to the north and a warm temperate desert zone is found to the south.

In addition to zonal vegetation changes affected by the Asian monsoon in summer, longitudinal vegetation changes occur north of the Kunlunshan-Qinling-Huaihe line that separates temperate and warm temperate regions. From southeast to northwest, deciduous broad-leaved forests and mixed forests give way to grasslands (steppe meadow—steppe—desert steppe) and deserts (steppe desert – typical desert). In the southern subtropical and tropical forested areas, longitudinal vegetation differences are negligible. These are summarized in Fig. 1.4.

1.2.3 Hydrology

The Asian continent contains a large number of rivers and lakes, with some of the world's longest rivers, including the Yangtze, Yellow River, Lena River, and Mekong River; and large lakes like the Caspian Sea, Aral Sea, Lake Baikal, Lake

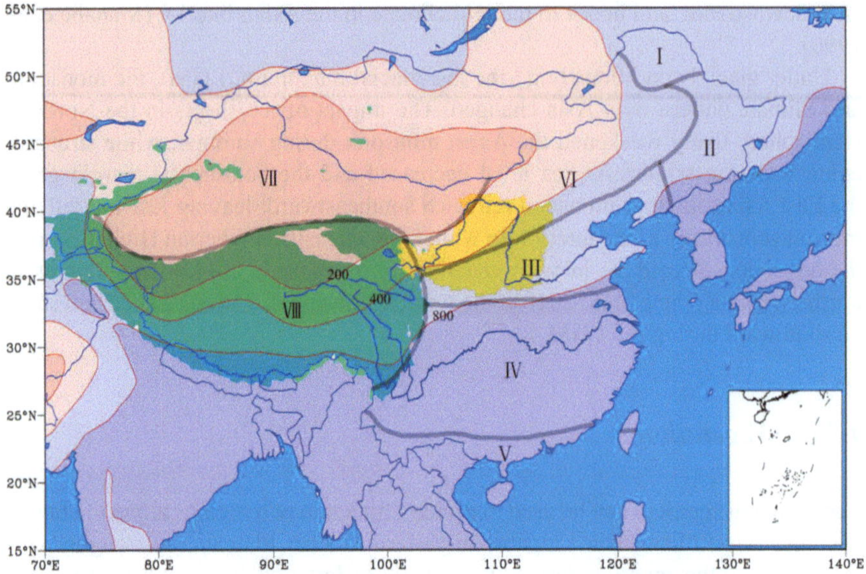

Fig. 1.4 Vegetation map of Asia. *Red thin lines* show 200, 400 and 800 mm isohyet precipitation contours. Arid areas fall within the 200 mm isohyet contours. Semi-arid areas are bounded by the 200–400 mm isohyet contours. The semi-humid zone lies within the 400–800 mm isohyet contours and the humid zone is bounded by the 800 mm isohyet contour. The *green* and *yellow* shaded areas are the Tibetan Plateau and Loess Plateau, respectively. Coarse gray lines indicate vegetation type zones. They are: I cold temperate coniferous forest; II temperate conifer/broad-leaved mixed forest zone; III warm temperate deciduous broad-leaved forest; IV subtropical evergreen broad-leaved forest; V tropical rain forest; VI temperate grassland; VII temperate desert zone; VIII Tibetan Plateau Alpine vegetation

Balkhash, and Lake Qinghai (Fig. 1.5). Most of the Asian rivers originate from high inland plateaus, and ultimately flow into the Pacific Ocean, Indian Ocean or Arctic Ocean. The distribution of these rivers is uneven, and closely related to tectonic features and climate. East Asia, Southeast Asia and South Asia have the largest river network because of abundant monsoon precipitation and large annual runoff, while central to western Asia has sparse rivers due to low precipitation and high evaporation. The Amur River, Yellow River, Yangtze River, Zhujiang River, Red River and Mekong River are mainly supplied by monsoon precipitation and flow into the Pacific Ocean. The Salween River, Irrawaddy River, Ganges River, Brahmaputra River and Indus River are tropical monsoonal rivers. They originate from the southern Tibetan Plateau and flow into the Indian Ocean. The Ob River, Yenisei River, and Lena River are supplied mainly by meltwater, and originate on the Mongolian Plateau and the north slope of the mountains in southern Siberia. These flow into the Arctic Ocean. Dry western Asia has few large rivers, with much of the runoff terminating in inland basins as playa lakes (endorheic systems). Runoff supplied by meltwater and rain is small, with the highest water levels in spring and lowest in summer.

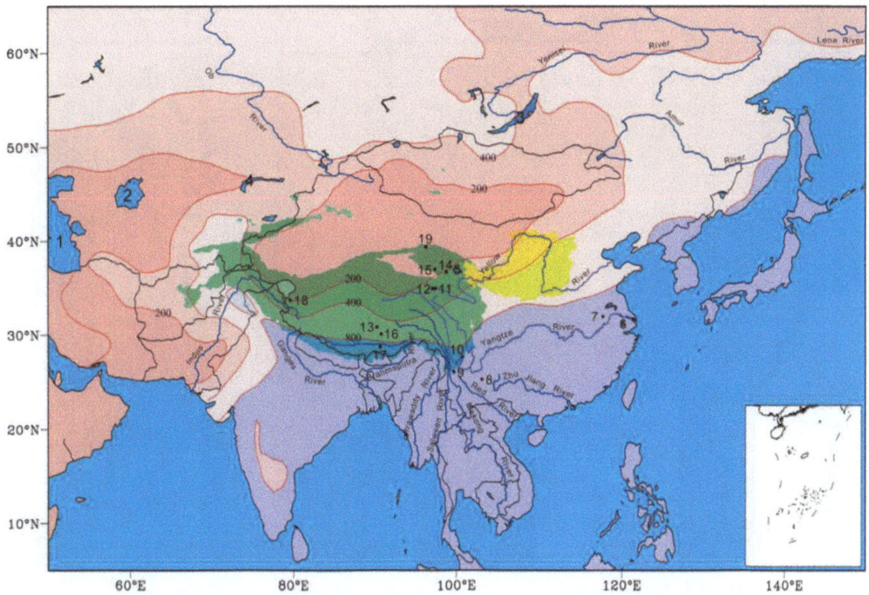

Fig. 1.5 Location of Asian rivers and lakes. Rivers are labeled on the map and lakes mentioned in the text are numbered. These include: 1) Caspian Sea, 2) Aral Sea, 3) Lake Baikal, 4) Lake Balkhash, 5) Lake Qinghai, 6) Lake Taihu, 7) Lake Chaohu, 8) Lake Dianchi, 9) Lake Erhai, 10) Lake Lugu, 11) Lake Eling, 12) Lake Gyaring, 13) Lake Selincuo, 14) Chaka salt Lake, 15) Qarhan salt Lake, 16) Lake Namtso, 17) Lake Yamdrok-tso, 18) Lake Pangong tso, 19) Lake Sugan

There are a large number of lakes widely distributed across China. The eastern plains feature freshwater lakes because of ample monsoon precipitation and relatively low evaporation. These lakes (e.g. Lake Taihu and Lake Chaohu) generally have low total dissolved solids (TDS) and high HCO_3 concentrations, and belong to the bicarbonate water type. Lakes in northeastern China have variable TDS, with freshwater, saltwater, and saline examples. The water types of these lakes are variably influenced by different bedrocks. On the Yunnan-Guizhou Plateau, monsoon precipitation is plentiful while evaporation is relatively weak. The TDS of lakes here are generally higher than those of eastern China because of stronger weathering in a karst environment. These lakes (e.g. Lake Dianchi, Lake Erhai and Lake Lugu) have relatively high HCO_3 concentrations and belong to the bicarbonate water type. On the Tibetan plateau, water sources of lakes are mainly supplied by the Indian summer monsoon on long time scales. Variable monsoon precipitation, evaporation, and melt water supply produce a wide range of TDS values within these lakes. For example, Lake Eling and Lake Gyaring are freshwater lakes, while Lake Selincuo and Lake Qinghai are saltwater lakes. In addition, there are a large number of saline lakes located in Qinghai Province (e.g. Chaka salt lake and Qarhan salt lake). These lakes belong to different water types, e.g. bicarbonate water (Lake Namtso, Lake Eling), sulfate type (Lake Selincuo, Lake Yamdrok-tso), and chloride type (Lake Qinghai, Lake Pangong Tso, Lake Sugan). Over the Xinjiang-Inner Mongolia

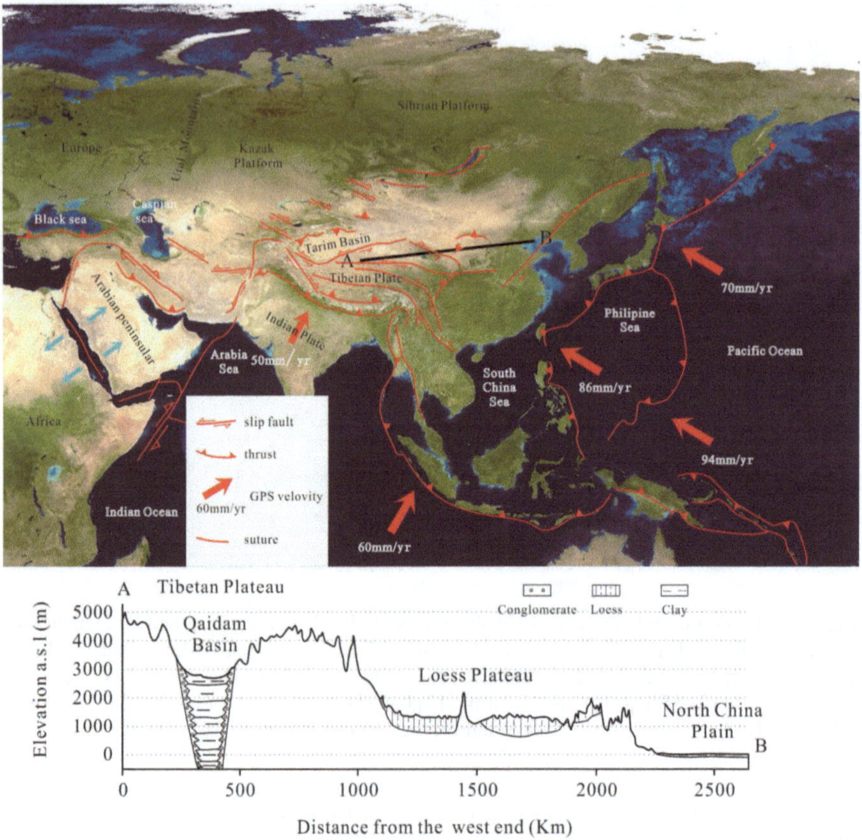

Fig. 1.6 *Upper panel* shows DEM map with simplified active tectonic map of Asia and surrounding regions. Main faults compiled from the literature. (Armijo et al. 1986, 1989; Avouac and Tapponnier 1993; Burchfield et al. 1995; Fu et al. 2004; Peltzer and Tapponnier 1988; Tapponnier and Molnar 1977; Tapponnier et al. 2001; Taylor and Yin 2009; Yin 2010). GPS data showing relative plate motions with respect to the Eurasian Plate. (Wang Q et al. 2001; Sella et al. 2002; Zhang et al. 2004; Allmendinger et al. 2007; Vergnolle et al. 2007). *Lower panel* is a profile from the Kunlun Mountains to the Bohai Gulf

region, because of strong evaporation and small precipitation, lakes generally have high TDS values, and saltwater and saline lakes are common. Water types of these lakes are also variable, and include bicarbonate, sulfate, and chloride types.

1.3 Geomorphology and tectonic background

Asia is the earth's largest continent occupying an area of 44,180,000 km^2. By virtue of its enormous size it contains many geomorphological extremes. A characteristic and dominant geomorphological feature in China is the stepped terrain that rises from east to west across the country, with three conspicuous geographic zones

(Fig. 1.6). The lowest zone is represented by coastal plains. The second zone is represented by Inner Mongolia and the Loess Plateau and the highest zone is the Tibetan Plateau.

There are many active faults in Asia, characterized by a full spectrum of styles and structural orientations over a vast region (Fig. 1.6) (Tapponnier and Molnar 1977; Armijo et al. 1986, 1989; Peltzer and Tapponnier 1988; Avouac and Tapponnier 1993; Burchfield et al. 1995; Tapponnier et al. 2001; Wang Q et al. 2001; Sella et al. 2002; Fu et al. 2004; Zhang et al. 2004; Allmendinger et al. 2007; Vergnolle et al. 2007; Taylor and Yin 2009; Yin 2010). Although Asia has had a long and complex geological history, its present morphology is largely the product of the youngest episode in its history. The collision between India and the Eurasian plate since the Eocene brought about drastic changes in the continent's evolution (Molnar and Tapponnier 1975; Besse et al. 1984; Patriat et al. 1984; Tapponnier et al. 2001; Rowley et al. 2006; Wang E et al. 2008). This magnificent event marks the beginning of the neotectonic period in Asia since the Cenozoic.

The large scale Cenozoic sutures around Asia mark the initiation of a new episode of the continent's neotectonic evolution. These can be divided into three major groups: 1) peri-Arabian, 2) Tibetan Plateau, and 3) peri-Australian. The Tibetan Plateau was the first major piece of the former Gondwanaland to collide with Asia. The collision apparently occurred diachronously from west to east during the Eocene, as a consequence of which the subcontinent rotated in a counterclockwise direction. The most dramatic effect of this collision was the formation of the Tibetan Plateau, which is the largest concentration of continental material on Earth. This event was characterized by significant lithospheric thickening accompanied by the expulsion, east- and southeastwards, of vast regions of Asia along enormous intracontinental strike-slip faults. The high seismicity associated with these mammoth structures attests to their current activity. Various lines of indirect evidence suggest an offset on the order of hundreds of kilometers for both the Altyn Tagh and Red River faults since the beginning of Cenozoic (Leloup et al. 2001; Yin et al. 2002; Yin 2010).

The gross outlines of Asia's landform are the product of neotectonic evolution, where active tectonics dominate. Second-order features in the young collisional mountain ranges and highlands of central Asia are shaped by climate. The high ranges that flank the periphery of the continent are the products of Cenozoic continental collisions, as is the highland region of central Asia. The lowland in north Asia (Siberia) lies in a region that has been essentially stable throughout the Cenozoic (Van der Voo et al. 1999).

A conspicuous aspect of Asia's Quaternary geology is the relatively restricted area affected by Pleistocene glaciation, as compared with Europe and North America. This is likely a function of Asia's enormous continental area and the aridity of the northern interior regions. During the Pleistocene, most of the Asian ice cover formed in western Siberia, where the ice sheet was confluent with northern European ice across the Urals. Elsewhere in Asia glaciers were confined to high mountains. The southernmost Quaternary glaciers in Asia were found in the Himalaya, and in the Sanjiang area of China.

1.4 Historical Perspective of Environmental Change Research

Extensive studies have been carried out over the last half-century on the monsoon-arid regions in Asia, starting with studies of the Quaternary. In the 1980s, Liu Tungsheng (Liu et al. 1985), An Zhisheng and Lu Yanchou (An and Lu 1984) were among the first to rigorously correlate Chinese loess-paleosol sequences to marine sediment records. Later, An et al. (1990, 1991a) established a paleoclimatic link between loess-paleosol records and marine records using a multi-proxy approach that led to a paleomonsoon theory which regards East Asian environmental changes as the result of monsoon variations during the past 2.6 Ma. Since the beginning of the 21st century, Chinese scholars have further extended their research to include paleoenvironments from throughout the monsoon-arid regions. These studies form the basis for an understanding of regional environmental change and the focus has now shifted to the dynamics of Asian environmental change within a larger temporal and spatial context. Rapid development and wide application of new high-resolution dating technologies during the last 20 years have contributed to more reliable time series records. Improvements to semi-quantitative and quantitative climate proxy analyses have significantly enhanced the resolution and precision of paleoenvironmental reconstructions. These improvements have brought us to the current state of understanding of the mechanisms that control changes across the Asian monsoon-arid environmental gradient.

1.4.1 Climatic Implications of Loess-paleosol Sequences and Glacial-interglacial Cycles

Large continental ice sheets in the Northern Hemisphere have grown and retreated many times in the past. Glacial periods are colder, dustier and generally drier than interglacial periods and these differences are preserved in a variety of geologic archives. Glacial-interglacial cycles are apparent in many marine and terrestrial sediments from middle to high latitude regions, characteristic of Quaternary climate. As early as the 10th century (North Song Dynasty in China), Shen Kuo formulated the notion of cold-warm climate change that experienced geographic shifts with time, based upon paleontological evidence, such as fossil shells and petrified bamboo.

In 1840, Agassiz was the first to propose that there were glacial periods in Earth's climatic history. In 1909, A. Penck and E. Brückner published *Die Alpen im Eiszeitalter*, which divided glacial periods into four stages: *Donau*, *Gunz*, *Mindel Riss* and *Würm*. Relatively warm periods in-between were recognized as interglacials. Afterwards, corresponding glacials were located in North Europe, North America and Asia. Besides the Donau glacial period, the sequential Nebraska, Kansas, Illinios, and Wisconsin Glacial stages were recognized in North America. In the

1930–1940s, Lee recognized that the Poyang, Dagu, Lushan and Dali glacial events corresponded to the glacial sequence in the Alps (Lee et al. 1947), a conclusion which is still questioned by many scholars (Shi et al. 1989).

In the 1950s, Emiliani seized an opportunity to use oxygen isotope analyses of carbonates from the Caribbean Ocean as a means to study paleoclimate in Quaternary deposits. His work laid the groundwork for the study of sediment core Vema 28–238, from the Pacific Ocean. That record yielded excellent oxygen isotope and magnetic time series records for the past 870,000 years and provided a global chronological framework for future studies of climate change (Shackleton and Opdyke 1973). The boundaries of 22 marine isotope stages (MIS) representing alternating periods of high and low Northern Hemisphere ice volume were recognized and dated. The next step was to correlate these with continental loess records (Kukla 1977).

Loess-paleosol and red clay sequences widely distributed on the LP, are now well-known for their rich store of high resolution paleoenvironmental information. Chinese scholars have undertaken their exploration since the 1970s, initially by comparison with corresponding deep sea sediment records. This constituted a big step from the classic 4-stage glacial-interglacial framework, to a detailed multi-cyclic climate framework. It was learned that loess formed during cold dry glacial periods, while plaeosols formed during warm, wet interglacial periods. A correlation between marine sediments and alternating loess-paleosol layers was pioneered by Lu and An (Lu and An 1979). Rigorous comparisons between marine sediments and loess-paleosol sequences from the LP were made by An and Lu (1984), and Sasajima and Wang (1984), and compiled in the book *Loess and Environment* (Liu et al. 1985). A magnetostratigraphic record from the last 2.5 Ma was established through paleomagnetism studies from loess cored at Luochuan (Heller and Liu 1982). Afterwards, a good correlation was found between Chinese loess-paleosol sequences and global glacial cycles, recorded by $\delta^{18}O$ records from deep sea sediments. More detailed correlations were made by Kukla (Kukla and Cilek 1996), Rutter (Rutter et al. 1991) and Ding (Ding et al. 1998). Ding contributed a detailed correlation of loess and marine records in the early Pleistocene, especially from 1.1 ~ 2.6 Ma. The proposed theory of glacial-interglacial cycles preserved in loess marked a new stage of Chinese loess study, integrated with global environmental research (Liu et al. 1985). Continental loess-paleosol sequences have since become one of the three pillars that comprise the foundation of global climate change research, along with marine sediments and ice cores.

1.4.2 History and Variability of Paleomonsoon Activity and its Connection to Global Change

Chinese geologists have found that paleomonsoon variations are closely related to environmental evolution on geologic time scales. In the 1980s, many scholars noted the importance of paleomonsoons and their distinct influence on loess accumulation (Yang et al. 1985; Li et al. 1988). In the early 1990s, An et al. (1990, 1991b) proposed that loess accumulated as dust, transported by the northerly winter monsoon, and interbedded paleosols derived their moisture from the southerly summer monsoon. This conclusion was based on a systematic analysis of bio-geological evidence, including: loess-paleosol sequences, desert advances and retreats, lake level changes, mammal faunal migrations, SST changes in the South China Sea, and tree line and snow line changes. This finding laid the groundwork for the study of loess-paleosol sequences as archives of East Asian monsoon variations over the past 2.5 Ma, through the analysis of the alternating dominance of winter and summer paleomonsoons.

The next step in our understanding of Asian monsoon history came with loess studies that covered the past 20 and 130 ka. The improved resolution of these time series allowed for increased scrutiny with regard to the variability and behaviour of monsoon climate on different time scales. This led to the formulation of the Asian monsoon control hypothesis (An et al. 1990, 1991 a, b), that interprets environmental changes in East Asia as a consequence of adjustments of lower boundary conditions, including land-sea distribution, SST changes in the South China Sea, ice volume changes, and insolation effects. This theory provided a framework for further research on the Asian paleomonsoon and its connections to past global change (Porter and An 1995; Ding et al. 1995; An and Porter 1997).

Asian climate experienced continuous change in both time and frequency domains since the onset of the Asian monsoon (An 2000; An et al. 2006; Sun and Wang 2005; Ding et al. 1995; Guo et al. 2002; Qiang et al. 2011). These changes have been studied by placing emphasis on high resolution paleomonsoon records from the last glacial-deglacial cycle, the Holocene and especially the last 2 ka. This includes records from loess (An et al. 1990), lacustrine sediments (Zhou et al. 1996), marine sediments (Wang and Wang 1989) tree rings (Liu et al. 2006), ice cores (Yao et al. 1996; Thompson et al. 2000), speleothems (Wang Y et al. 2001; Yuan et al. 2004; Cheng et al. 2009; Cai et al. 2010; Tan et al. 2009) and historical documents (Wang and Gong 2000; Zhang and Lu 2007). East Asian monsoon climate variability, instability, periodicity, asynchroneity and the possible mechanisms underlying these variations have been explored (An et al. 2000). Fast flickering and abrupt events at millennial time scales are superimposed on the orbital winter and summer monsoon changes during the last glacial cycle. These varied in amplitude and spatial distribution, indicating relatively large variability and instability in the East Asian monsoon, and they correspond to modern day frequent cold air mass movements and distinct precipitation variability. Monsoon events are also affected by the interaction of climatic events between high and low latitudes that may have feedbacks on the monsoon at millennial time scales (Zhou et al. 2001). The influence of the Northern Atlantic meridional overturning circulation on the East Asian monsoon on millennial time scales has also been suggested (Sun et al. 2012). Cross equatorial currents from the Southern Hemisphere also impact the Asian monsoon at glacial/interglacial and millennial time scales (An et al. 2011; An 2000; Cai et al. 2006). This occurs through a pressure push and thermal pull induced by the cross equatorial pressure gradient (Tomas and Webster, 1997; An 2000; Webster and Fassulo 2003; An et al. 2011).

The relationship between East Asian monsoonal changes and the Tibetan Plateau uplift/growth have been studied extensively. A simultaneous increase in winter and summer monsoons, intensified aridification of the Asian interior, and the enhanced global ice volume may be indicative of plateau growth from 3.6–2.6 Ma (An et al. 2001). It has also been shown that monsoon variations are linked to dust accumulation and Northern Hemisphere glaciation, a conclusion that is supported by geological evidence and climate model simulations.

1.4.3 The Concept of the Asian Monsoon-arid zone as an Integrated Environmental System and its global Significance

The Asian continent is a key component of the global climate system. Climate on the continent is influenced by both the Asian monsoon and the westerlies. The interaction of cold continental high latitude air masses with warm, moist marine air masses occurs on a large scale, and as such, significant feedbacks exist in land-sea-atmosphere interactions in Asia and global climate. These have operated in tandem to produce environmental changes on the Asian continent since the Late Cenozoic.

An et al. (1991a) established the loess flux curve which is a function of aridity in the Asian interior over the past 130 ka. Later Sun and An (2002) reconstructed the aridity history of inland Asia for the last 7 Ma years. A recent study from Qinghai Lake shows that the content of >25 μm eolian dust component may reflect variations in the westerlies through time. An anti-phase relationship between the westerlies and the summer monsoon has also been identified on glacial-interglacial and glacial millennial time scales (An et al. 2012).

The existence of the Tibetan Plateau enhances Asian monsoon circulation and also obstructs the inland transport of warm moist air from the Indian Ocean (Ruddiman et al. 1989, Kutzbach et al. 1993). Descending air flow along the northern flank of the TP affects aridification in the Asian interior (Ye and Qian 1974; RGES 1978; Manabe et al. 1974; Liu et al. 2001) and so induces environmental changes. The coupling between the phased growth of the Tibetan Plateau and paleoclimate in the monsoon-arid system since the Miocene has been demonstrated within a global context, and it is responsible for changes in the paleomonsoon (An et al. 2006). This is especially evident in the northeastern expansion of the Tibetan Plateau, which has led to a simultaneous intensification of the East Asian monsoon and increased aridity in the Asian interior. Asian dust has also been shown to impact global climate (An et al. 2001; Shi et al. 2011). Iron-bearing dust that is deposited in the North Pacific Ocean increases ocean productivity which consumes atmospheric CO_2 (Jickells et al. 2005).

1.5 Scientific Significance and Major Questions of Asian Monsoon-arid Zones

1.5.1 Major Scientific Questions and Significance on a Tectonic Time Scale

The process, governing law and dynamics of climate change on a tectonic time scale refers to a period of more than 10^5 years. Marine studies show that during the Cenozoic the Earth has experienced both relatively cold and warm extremes (Zachos et al. 2001). Tectonic movements, Northern and Southern Hemisphere ice volumes and atmospheric CO_2 concentration changes are considered to be major factors that drive climate change on a tectonic time scale. Mountain and plateau growth, and the opening and closure of oceanic straits in particular, alter the land-sea configuration, and coupling of the atmosphere and ocean, leading to regional and global climate change (Broccoli and Manabe 1990; Ramstein et al. 1997). As an important component of lower boundary conditions, global ice volume changes exert a significant effect on polar-equatorial energy transfer, and so alter the global climate system (Zachos et al. 2001). As it is clear from the work described above, geologic records from the Asian monsoon-arid zone are ideal candidates for understanding long term climatic and environmental change, at least since the Oligocene. In order to explore the dynamic mechanisms of climate change on tectonic time scales in the Asian monsoon-arid zone, the following scientific questions should be addressed:

1. When and why did the Asian monsoon climate and inland aridification start?
2. What stages has the Asian monsoon-arid zone gone through and what were the dynamic drivers controlling climate?
3. How have the uplift and growth of the Tibetan Plateau, change in global ice volume, and CO_2 concentration changes affected the paleomonsoon and westerlies?
4. How did the retreat of the Tethys and subsequent land-sea redistribution affect the Asian monsoon-arid environment?

1.5.2 Major Scientific Questions and Significance on Orbital and Millennial Time scales

Orbital-scale monsoon-arid environmental changes are driven primarily by orbitally-induced changes in solar irradiation (An et al. 1991b; Wang Y et al. 2008), and are strongly modulated by changing lower boundary conditions (i.e. ice volume) (An et al. 1991b; Ding et al. 1995; Liu et al. 1999). Loess-based proxies (e.g. magnetic susceptibility and grain size) display distinct glacial-interglacial variations that are similar to marine benthic $\delta^{18}O$ records (a proxy for global ice volume) (An et al. 1991a,b; Ding et al. 1995; Bloemendal et al. 1995; Liu et al. 1999; Ding et al. 2002), suggesting a strong coupling between Northern Hemisphere ice volume and East Asian monsoon variability (Liu and Ding 1993; Ding et al. 1995). By contrast, high-resolution, absolutely-dated speleothem $\delta^{18}O$ records with a dominant precessional period are nearly in-phase, in response to Northern Hemisphere summer insolation (Wang Y et al. 2008). To fully understand the dynamics driving orbital-scale East Asian monsoon variability, the following scientific questions should be further investigated:

1. Why do the dominant periods differ between Chinese loess and speleothems?
2. What are the climatic implications of different phases for monsoon proxies at the obliquity and precession bands?
3. How can we evaluate the relative roles of external (solar radiation) and internal (i.e. ice volume and greenhouse gas) factors that may influence the orbital-scale monsoon variability?

In recent years we have seen the expansion of spatial coverage and temporal resolution of paleoclimate records from China's interior. Since environmental changes in the Asian interior are influenced by the westerlies, winter monsoon, and high-latitude climate (An et al. 2011), these should be distinctly different from those of the monsoon-affected areas. By studying the whole of the Asian arid-monsoon system we can now explore the following:

1. How did the westerlies, Asian summer monsoon and winter monsoon interact at millennial timescales?
2. How do the monsoon-arid environmental systems respond to high-latitude climate change?
   Since orbital-scale changes in the Indian monsoon differ from the East Asian monsoon in terms of frequency and phase (Clemens and Prell 2003; we may also ask:
3. How do the Asian monsoon systems interact with the Southern Hemisphere climate system, especially with regard to latent heat transport, and along cross-equatorial pressure gradients?
4. What is the different influence of changing lower boundary conditions (i.e. Northern Hemisphere ice sheets, SSTs and Antarctic temperatures) on the East Asian and Indian monsoon circulations?

An integral set of orbital-scale paleo-proxy records from both regions could greatly enhance our understanding of the dynamical mechanisms involved in these two monsoon systems.

1.5.3 Forcing of Abrupt Events: Ocean Circulation, Solar Activities, Sea Ice, Ocean-atmosphere Interactions (super ENSO), and the Interaction Between High and Low Latitude Climates

Abrupt events on millennial or centennial time scales are more immediate concerns for the well-being of the human race as compared to relatively long-term changes on tectonic and orbital time scales. Greenland ice cores preserve a detailed record of abrupt climate change on millennial and centennial time scales, and over the past twenty years the study of abrupt events in the monsoon-arid environment system has become a priority, especially in East Asian monsoon areas. Abrupt events on a millennial time scale generally coincide with global events as recorded in Greenland ice cores and North Atlantic Ocean sediments. These abrupt events from the Late Pleistocene to the Holocene have been identified in a variety of archives in monsoon areas. Several key questions have arisen in their study, including:

1. What are the history, variability and origins of these abrupt events, especially in the inland arid zone where well-dated high-resolution records are scarce?
2. How are these abrupt events and related environmental changes related to global climate change?
3. How can we separate high and low latitude climatic influences on the mid-latitude monsoon-arid zone in Asia?
4. Are there any teleconnections between abrupt events in the Asian monsoon-arid zone and the Southern Hemisphere?
5. What forcings were responsible for the abrupt events, i.e. ocean circulation changes (including fresh water discharge), solar activity, sea ice extent and ocean-atmosphere interactions (super ENSO)?

Reference

Allmendinger RW, Reilinger R, Loveless J (2007) Strain and rotation rate from GPS in Tibet, Anatolia, and the Altiplano. Tectonics 26:TC3013

An ZS (2000) The history and variability of the East Asian paleomonsoon climate. Quatern Sci Rev 19:171–187

An ZS, Kukla GJ, Porter SC, Xiao JL (1991a) Magnetic susceptibility evidence of monsoon variation on the Loess Plateau of central China during the last 130,000 years. Quatern Res 36:29–36

An ZS, Lu YC (1984) A climatostratigraphic subdivision of late Pleistocene strata named Malan formation in north China. Kexue Tongbao(Chinese Science Bulletin), 29(9): 1239–1242

An ZS, Liu Tungsheng, Lu Yanchou, Porter SC, Kukla G, Wu Xihao, Hua Yingming (1990) The long-term paleomonsoon variation recorded by the loess-paleosol sequence in central China. Quatern Int 7:91–95

An ZS, Wu XH, Wang PS, Wang SM, Dong GR, Sun XJ, Zhang DE, Lu YC, Zheng SH, Zhao SL (1991b) Paleomonsoons of China over the last 130,000 years - paleomonsoon variation. Science in China, series B 34:1016–1024

An ZS, Clemens SC, Shen J, Qiang XK, Jin ZD, Sun YB, Prell WL, Luo JJ, Wang SM, Xu H, Cai YJ, Zhou WJ, Liu XD, Liu WG, Shi ZG, Yan LB, Xiao XY, Chang H, Wu F, Ai L, Lu FY (2011) Glacial-Interglacial indian summer monsoon dynamics. Science 333:719–723

An ZS, Colman SM, Zhou WJ et al (2012) Interplay between the Westerlies and Asian summer monsoon recorded in Lake Qinghai sediments since 32 ka. Scientific Reports 2:619, doi:10.1038/srep00619

An ZS, Kutzbach JE, Prell WL, Porter SC (2001) Evolution of Asian monsoons and phased uplift of the Himalayan Tibetan plateau since Late Miocene times. Nature 411:62–66

An ZS, Porter SC (1997) Millennial-scale climatic oscillations during the last interglaciation in central China. Geology 25:603–606

An ZS, Porter SC, Kutzbach JE, Wu XH, Wang SM, Liu XD, Li XQ, Zhou WJ (2000) Asynchronous Holocene optimum of the East Asian monsoon. Quatern Sci Rev 19:743–762

An ZS, Zhang PZ, Wang EQ, Wang SM, Qiang XK, Li L, Song YG, Chang H, Liu XD, Zhou WJ, Liu WG, Cao JJ, Li XQ, Shen J, Liu Y, Ai L (2006) Changes of the monsoon-arid environment in China and growth of the Tibetan Plateau since the Miocene. Quatern Sci 26:678–693 (in Chinese)

Armijo R, Tapponnier P, Mercier J, Tong-Lin H (1986) Quaternary extension in southern Tibet: Field observations and tectonic implications. J Geophys Res 91: 13803–13813, 13872

Armijo R, Tapponnier P, Tonglin H (1989) Late Cenozoic right-lateral strike-slip faulting in southern Tibet. J Geophys Res 94:2787–2838

Avouac JP, Tapponnier P (1993) Kinematic model of active deformation in central Asia. Geophys Res Lett 20:895–898

Besse J, Courtillot V, Pozzi JP, Westphal M, Zhou YX (1984) Palaeomagnetic estimates of crustal shortening in the Himalayan thrusts and Zangbo suture. Nature 311:621–626

Bloemendal J, Liu X, Rolph T (1995) Correlation of the magnetic susceptibility stratigraphy of Chinese loess and the marine oxygen isotope record: chronological and palaeoclimatic implications. Earth Planet Sci Lett 131:371–380

Broccoli A, Manabe S (1990) Can existing climate models be used to study anthropogenic changes in tropical cyclone climate? Geophys Res Lett 17:1917–1920

Burchfield B, Chen Z, Liu Y, Royden L (1995) Tectonics of the Longmen Shan and adjacent regions, central China. International Geology Review 37:661–735

Cai YG, An ZS, Cheng H, Edwards RL, Kelly MJ, Liu WG, Wang XF, Shen CC (2006) High-resolution absolute-dated Indian Monsoon record between 53 and 36 ka from Xiaobailong Cave, southwestern China. Geology 34:621–624

Cai YJ, Tan LC, Cheng H, An ZS, Edwards RL, Kelly MJ, Kong XG, Wang XF (2010) The variation of summer monsoon precipitation in central China since the last deglaciation. Earth Planet Sci Lett 291:21–31

Cheng H, Edwards RL, Broecker WS, Denton GH, Kong XG, Wang YJ, Zhang R, Wang XF (2009) Ice Age Terminations. Science 326:248–252

Clemens SC, Prell WL (2003) A 350,000 year summer-monsoon multi-proxy stack from the Owen Ridge, Northern Arabian Sea. Mar Geol 201:35–51

Ding ZL, Derbyshire E, Yang SL, Yu ZW, Xiong SF, Liu TS (2002) Stacked 2.6–Ma grain size record from the Chinese loess based on five sections and correlation with the deep-sea $\delta^{18}O$ record. Paleoceanography 17(3): 5-1–5-21

Ding ZL, Liu TS, Rutter NW, Yu ZW, Guo ZT, Zhu RX (1995) Ice-volume forcing of East Asian winter monsoon variations in the past 800,000 years. Quatern Res 44:149–159

Ding ZL, Sun JM, Liu TS, Zhu RX, Yang SL, Guo B (1998) Wind-blown origin of the Pliocene red clay formation in the central Loess Plateau, China. Earth Planet Sci Lett 161:135–143

Fu B, Ninomiya Y, Lei X, Toda S, Awata Y (2004) Mapping active fault associated with the 2003 Mw 6.6 Bam (SE Iran) earthquake with ASTER 3D images. Remote Sensing of. Environment 92:153–157

Guo ZT, Ruddiman WF, Hao QZ, Wu HB, Qiao YS, Zhu RX, Peng SZ, Wei JJ, Yuan BY, Liu TS (2002) Onset of Asian desertification by 22 Myr ago inferred from loess deposits in China. Nature 416:159–163

Heller F, Liu TS (1982) Magnetostratigraphical dating of loess deposits in China. Nature 300:431–433

Jiang DB, Lang XM (2010) Last Glacial Maximum East Asian Monsoon: Results of PMIP Simulations. J Clim 23:5030–5038

Jickells TD, An ZS, Andersen KK, Baker AR, Bergametti G, Brooks N, Cao JJ, Boyd PW, Duce RA, Hunter KA, Kawahata H, Kubilay N, laRoche J, Liss PS, Mahowald N, Prospero JM, Ridgwell AJ, Tegen I, Torres R (2005) Global Iron Connections Between Desert Dust, Ocean Biogeochemistry, and Climate. Science 308:67–71

Kalnay E, Kanamitsu M, Kistler R, Collins W, Deaven D, Gandin L, Iredell M, Saha S, White G, Woollen J, Zhu Y, Chelliah M, Ebisuzaki W, Higgins W, Janowiak J, Mo KC, Ropelewski C, Wang J, Leetmaa A, Reynolds R, Jenne R, Joseph D (1996) The NCEP/NCAR 40-year reanalysis project. Bull Am Meteorol Soc 77:437–471

Kukla G, Cilek V (1996) Plio-Pleistocene megacycles: record of climate and tectonics. Palaeogeogr Palaeoclimatol Palaeoecol 120:171–194

Kukla GJ (1977) Pleistocene land–sea correlations I. Europe. Earth-Science Reviews 13:307–374

Kutzbach J, Prell W, Ruddiman WF (1993) Sensitivity of Eurasian climate to surface uplift of the Tibetan Plateau. The Journal of Geology:177–190

Lee J S (1947) Quaternary glaciations in the Lushan area, central China. Memoirs of the Institute of Geology, Academia Sinica 2:1–70

Leloup PH, Arnaud N, Lacassin R, Kienast J, Harrison T, Trong TTP, Replumaz A, Tapponnier P (2001) New constraints on the structure, thermochronology, and timing of the Ailao Shan-Red River shear zone, SE Asia. J Geophys Res 106:6683–6732

Liao Ke, Institute of Geography of Chinese Academy of Sciences, National Bureau of Surveying and Mapping (China) (1999) The national physical atlas of China. China Cartographic Publishing House Beijing, China

Li J, Feng Z, Tang L (1988) Late quaternary monsoon patterns on the Loess Plateau of China. Earth Surf Proc Land 13(2):125–135

Liu TS, Ding ZL (1993) Stepwise coupling of monsoon circulations to global ice volume variations during the late Cenozoic. Global and Planetary Change, 7(1):119–130.

Liu TS, Ding ZL, Rutter N (1999) Comparison of Milankovitch periods between continental loess and deep sea records over the last 2.5 Ma. Quatern Sci Rev 18:1205–1212

Liu TS et al (1985) Loess and the Environment. China Ocean Press, Beijing

Liu Y, An ZS, Ma HZ, Cai QF, Liu ZY, Kutzbach JK, Shi JF, Song HM, Sun JY, Yi L (2006) Precipitation variation in the northeastern Tibetan Plateau recorded by the tree rings since 850 AD and its relevance to the Northern Hemisphere temperature. Science in China Series D: Earth Sciences 49:408–420

Liu X D, Li L, An Z S (2001) Tibetan Plateau uplift and drying in Eurasian interior and Northern Arfica (in Chinese). Quat Sci, 21:114–122

Lu Y C, An ZS (1979) The quest for series of nature environmental changes during the Brunhes Epoch. Chin Sci Bull 24:221–224 (in Chinese)

Manabe S, Broccoli A (1990) Mountains and arid climates of middle latitudes. Science 247:192–195

Manabe S, Terpstra TB (1974) The effects of mountains on the general circulation of the atmosphere as identified by numerical experiments. J Atmos Sci 31:3–42

Molnar P, Tapponnier P (1975) Cenozoic tectonics of Asia: effects of a continental collision. Science 189:419–426

New M, Lister D, Hulme M, Makin I (2002) A high-resolution data set of surface climate over global land areas. Climate Res 21:1–25

Patriat P, Achache J (1984) India-Eurasia collision chronology has implications for crustal shortening and driving mechanism of plates. Nature 311:615–621

Peltzer G, Tapponnier P (1988) Formation and evolution of strike-slip faults, rifts, and basins during the India-Asia collision: An experimental approach. J Geophys Res 93: 15085–15015, 15117

Porter SC, An ZS (1995) Correlation between climate events in the North Atlantic and China during the last glaciation. Nature 375:305–308

Qiang XK, An ZS, Song YG, Chang H, Sun YB, Liu WG, Ao H, Dong JB, Fu CF, Wu F, Lu FY, Cai YY, Zhou WJ, Cao JJ, Xu XW, Ai L (2011) New eolian red clay sequence on the western Chinese Loess Plateau linked to onset of Asian desertification about 25 Ma ago. Science China Earth Sciences, 54(1):136–144

Ramstein G, Fluteau F, Besse J, Joussaume S (1997) Effect of orogeny, plate motion and land-sea distribution on Eurasian climate change over the past 30 million years. Nature 386:788–795

RGES (Research Group on Experimental Simulation) (1978) The annulus simulation of summer large-scale convective systems over Qinghai Tibetan Plateau Scientia Sinica 21:347–364

Rowley DB, Currie BS (2006) Palaeo-altimetry of the late Eocene to Miocene Lunpola basin, central Tibet. Nature 439:677–681

Ruddiman WF, Raymo ME, Martinson DG, Clement BM, Backman J (1989) Pleistocene evolution: northern hemisphere ice sheets and North Atlantic Ocean. Paleoceanography 4:353–412
Rutter N, Zhongli D, Evans M, Tungsheng L (1991) Baoji-type pedostratigraphic section, Loess Plateau, north-central China. Quatern Sci Rev 10:1–22
Sasajima S, Wang YY (1984) The recent research of loess in China - stratigraphy, magnetostratigraphy, chronology, pedology, paleotology and paleoclimatology, Kyoto Institute of Natural History, Kyoto, pp. 183
Sasajima S, Wang YY (1984) Recent research on loess in China. Kyoto University, Kyoto
Sella GF, Dixon TH, Mao A (2002) REVEL: A model for recent plate velocities from space geodesy. J Geophys Res 107:209–226
Shackleton NJ, Opdyke ND (1973) Oxygen isotope and palaeomagnetic stratigraphy of equatorial Pacific core V28-238: oxygen isotope temperatures and ice volumes on a 105 year and 106 year scale. Quatern Res 3:39–55
Shi YF, Cui ZJ, Li JJ (1989) Problems on Quaternary Glaciation and Environments in Eastern China, Science Press, Beijing (in Chinese)
Shi ZG, Liu XD, An ZS, Yi BQ, Yang P, Mahowald N, 2011: Simulated variations of eolian dust from inner Asian deserts during late Pliocene-Pleistocene periods, Climate Dynamics, 37: 2289–2301
Sun XJ, Wang PX (2005) How old is the Asian monsoon system?–Palaeobotanical records from China. Palaeogeography, Palaeoclimatology, Palaeoecology, 222(3):181–222
Sun YB, An ZS (2002) History and variability of Asian interior aridity recorded by eolian flux in the Chinese Loess Plateau during the past 7 Ma. Science in China Series D: Earth Sciences 45:420–429
Sun YB, Clemens SC, Morrill C, Lin XP, Wang XL, An ZS (2012) Influence of Atlantic meridional overturning circulation on the East Asian winter monsoon. Nature Geosci 5:46–49
Tan LC, Cai YJ, Cheng H, An ZS, Edwards RL (2009) Summer monsoon precipitation variations in central China over the past 750 years derived from a high-resolution absolute-dated stalagmite, Palaeogeography, Palaeoclimatology, Palaeoecology, 280:432–439
Tapponnier P, Molnar P (1977) Active faulting and tectonics in China. J Geophys Res 82:2905–2930
Tapponnier P, Xu ZQ, Roger F, Meyer B, Arnaud N, Wittlinger G, Yang JS (2001) Oblique stepwise rise and growth of the Tibet Plateau. Science 294:1671–1677
Taylor M, Yin A (2009) Active structures of the Himalayan-Tibetan orogen and their relationships to earthquake distribution, contemporary strain field, and Cenozoic volcanism. Geosphere 5:199–214
Thompson LG, Yao TD, Mosley-Thompson E, Davis ME, Henderson KA, Lin PN (2000) A High-Resolution Millennial Record of the South Asian Monsoon from Himalayan Ice Cores. Science 289:1916–1919
Tomas RA, Webster PJ (1997) The role of inertial instability in determining the location and strength of near-equatorial convection. Quart J Royal Meteoroll Soc 123(542):1445–1482
Uppala SM, Kållberg P, Simmons A, Andrae U, Bechtold V, Fiorino M, Gibson J, Haseler J, Hernandez A, Kelly G (2005) The ERA-40 re-analysis. Quart J Royal Meteoroll Soc 131:2961–3012
Van der Voo R, Spakman W, Bijwaard H (1999) Mesozoic subducted slabs under Siberia. Nature, 397(6716):246–249
Vergnolle M, Calais E, Dong L (2007) Dynamic of continentaldeformationin Asia. J Geophys Res 112:B11403. doi:10.1029/2006JB004807
Wang B (2006) The Asian Monsoon. Springer Berlin Heidelberg
Wang CS, Zhao XX, Liu ZF, Peter CL, Stephan AG, Robert SC, Yi HS, Zhu LD, Liu S, Li YL (2008) Constraints on the early uplift history of the Tibetan Plateau. Proceedings of the National Academy of Sciences 105:4987–4992
Wang EG, Wang L, Lan LF, Luo C, Zhuang WL, Peng JB, Cao Y (2008) High-performance polymer heterojunction solar cells of a polysilafluorene derivative. Appl Phys Lett 92:033307

Wang Pingxian, Zhang Jijun, Zhao Quanhong (eds) (1988) Foraminifera and Ostracoda in Sediments of the East China Sea. China Ocean Press, Beijing

Wang Q, Zhang PZ, Freymueller JT, Bilham R, Larson KM, Lai XA, You XZ, Niu ZJ, Wu JC, Li YX, Liu JN, Yang ZQ, Chen QZ (2001) Present-day crustal deformation in China constrained by global positioning system measurements. Science 294:574–577

Wang SW, Gong DY (2000) Enhancement of the warming trend in China. Geophys Res Lett 27:2581–2584

Wang YJ, Cheng H, Edwards RL, Kong XG, Shao XH, Chen ST, Wu JL, Jiang XY, Wang XF, An ZS (2008) Millennial-and orbital-scale changes in the East Asian monsoon over the past 224,000 years. Nature 451:1090–1093

Wang YJ, Cheng H, Edwards RL, An ZS, Wu JY, Shen CC, Dorale JA (2001) A high-resolution absolute-dated late Pleistocene monsoon record from Hulu Cave, China. Science 294:2345–2348

Wang LJ, Wang PX (1989) The transfer function method was used to calculate the temperature of the South China Sea Trial. Chin Sci Bull 34(1):53–56

Webster PJ, Fasullo J (2003) in: Holton J, Curry JA (ed) Encyclopedia of Atmospheric Sciences. Academic Press, London, p 1370–1385

Yanai M, Li CF, Song ZS (1992) Seasonal heating of the Tibetan Plateau and its effects on the evolution of the Asian summer monsoon. J Meteorol Soc Jpn 70:319–351

Yang HR, Xu X (1985) Quaternary cliamte and environment in China. In: Yang HR (ed) Paper collections on Quaternary glaciation and Quaternary geology vol 2. Geological House, Beijing, p 89–103

Yao TD, Thompson LG, Mosley-Thompson E, Yang ZH, Zhang XP, Lin PN (1996a) Climatological significance of $\delta^{18}O$ in north Tibetan ice cores. J Geophys Res 101:29531–29537

Yao TD, Jiao KQ, Tian LD, Yang ZH, Shi WL, Thompson LG (1996b) Climatic variations since the Little Ice Age recorded in the Guliya Ice Core. Science in China Series D-Earth Sciences 39:587–596

Ye DZ, Qian ZJ (1974) The preliminary annulus simulation of the influence of the heating effect of the Tibetan Plateau on the atmospheric general circulation over Eastern Asia in summer. Scientific Sinica 3:301–320

Yin A (2010) Cenozoic tectonic evolution of Asia: A preliminary synthesis. Tectonophysics 488:293–325

Yin A, Rumelhart P, Butler R, Cowgill E, Harrison T, Foster D, Ingersoll R, Zhang Q, Zhou X Q, Wang X F, Hanson A, Raza A (2002) Tectonic history of the Altyn Tagh fault system in northern Tibet inferred from Cenozoic sedimentation. Geol Soc Am Bull 114:1257–1295

Yuan DX, Cheng H, Edwards RL, Dykoski CA, Kelly MJ, Zhang ML, Qing JM, Lin YS, Wang YJ, Wu JY, Dorale JA, An ZS, Cai YJ (2004) Timing, Duration, and Transitions of the Last Interglacial Asian Monsoon. Science 304:575–578

Zachos J, Pagani M, Sloan L, Thomas E, Billups K (2001) Trends, Rhythms, and Aberrations in Global Climate 65 Ma to Present. Science 292:686–693

Zhang DE, Lu LH (2007) Anti-correlation of summer/winter monsoons? Nature 450:E7–E8

Zhang P Z, Shen ZK, Wang M, Gan WJ, Bürgmann R, Molnar P, Wang Q, Niu ZJ, Sun JZ, and Wu C, Sun HR, You XZ (2004) Continuous deformation of the Tibetan Plateau from global positioning system data. Geology 32:809–812

Zhou WJ, Donahue DJ, Porter SC, Jull TA, Li XQ, Stuiver M, An ZS, Matsumoto E, Dong GR (1996) Variability of Monsoon Climate in East Asia at the End of the Last Glaciation. Quaternary Research 46(3):219–229

Zhou WJ, Head MJ, An ZS, De Deckker P, Liu ZY, Liu XD, Lu XF, Donahue D, Jull AJT, Beck JW (2001) Terrestrial evidence for a spatial structure of tropical-polar interconnections during the Younger Dryas episode. Earth Planet Sci Lett 191:231–239

Chapter 2
Chinese Loess and the East Asian Monsoon

Zhisheng An, Youbin Sun, Weijian Zhou, Weiguo Liu, Xiaoke Qiang, Xulong Wang, Feng Xian, Peng Cheng, and George S. Burr

Abstract This chapter offers a comprehensive review of loess studies on the Loess Plateau, and how they provide evidence for past monsoon variability. The chronological framework that has been established for the LP with the use of multiple chronometers is explained in detail. Sedimentological, paleomagnetic and geochemical climate proxies are also discussed, and the history, variability and dynamics of the paleomonsoon is described at tectonic, orbital and millennial time scales.

Keywords Chinese Loess Plateau · Loess · Paleosol · Red clay · Climate proxy · Grain size · Magnetic susceptibility · Color reflectance · Isotopic geochemistry · $\delta^{18}O$ · ^{10}Be · Geomagnetic field · C4 plant · C3 plant · Tectonic time scale · Orbital time scale · Oligocene · Holocene

Z. An (✉) · Y. Sun · W. Zhou · W. Liu · X. Qiang · X. Wang · F. Xian · P. Cheng
State Key Laboratory of Loess and Quaternary Geology, Institute of Earth Environment, Chinese Academy of Sciences, No. 10 Fenghui S. Rd, High-Tech Zone, Xi'an PO Box 17, Xi'an, China
e-mail: anzs@loess.llqg.ac.cn

Y. Sun
e-mail: sunyb@ieecas.cn

W. Zhou
e-mail: weijian@loess.llqg.ac.cn

W. Liu
e-mail: liuwg@loess.llqg.ac.cn

X. Qiang
e-mail: qiangxk@loess.llqg.ac.cn

X. Wang
e-mail: wxl@loess.llqg.ac.cn

F. Xian
e-mail: xianf@ieecas.cn

P. Cheng
e-mail: chp@ieecas.cn

G. S. Burr
AMS Laboratory, Physics Department, University of Arizona 85721 Tucson, Arizona, USA
e-mail: burr@email.arizona.edu

Z. An (ed.), *Late Cenozoic Climate Change in Asia,*
Developments in Paleoenvironmental Research 16,
DOI 10.1007/978-94-007-7817-7_2, © Springer Science+Business Media Dordrecht 2014

2.1 Introduction

East Asian monsoon variation, characterized by alternating dominance between cold-dry winter monsoon and warm-humid summer monsoon at the glacial-interglacial time scale, was directly driven by external solar radiation and strongly modulated by changing lower boundary conditions through interactions among atmosphere, ocean, land and the ice system. Eolian sequences on the Chinese LP indicate the onset of the East Asian paleomonsoon at least by the late Oligocene, and significant monsoon variability on tectonic to millennial time scales afterwards. Long-term evolution of the East Asian monsoon was closely related with the uplift of the Tibetan Plateau and the expansion of the northern hemisphere ice sheet, as revealed by geological evidence and numerous modeling results. High-resolution loess records demonstrate that millennial-scale monsoon variability during the last glaciation was closely related with the cold-air activities of high latitudes in the Northern Hemisphere, which was dynamically linked to rapid climate change in the North Atlantic. Future research on Chinese loess should focus on reliable chronology and sensitive monsoon proxies for both the high-resolution loess and the largely underlying red clay sequences. Furthermore, temporal-spatial variability of the East Asian monsoon system should be further investigated to understand its linkage to high- and low-latitude processes and their forcing mechanisms within the global context.

2.1.1 Research History of Chinese Loess

Widespread loess-paleosol and the underlying red clay sequences on the Chinese Loess Plateau are well known for their great thickness, high resolution and richness of paleoenviromental information. Since 19th century, western researchers represented by Richthofen (1877) started to conduct loess surveys in China, and for the first time brought forward the preliminary understanding of its eolian origin based on sedimentary characteristics. In the early 20th century, Andersson (1923) studied the Hipparion fossil of Baode in Shanxi, differentiating Chinese loess into the Pliocene Baode red clay with Hipparion fossils and the Late Pleistocene Malan Loess. Then, based on vertebrate fossils, Teilhard de Chardin and Young (1930) further divided the underlying loess beneath the Malan Loess into the Early Pleistocene and Middle Pleistocene red clay. In the 1960s, on the basis of loess-paleosol stratigraphy and fossil mammal features, Liu and Chang (1962) divided the Chinese loess into the Early Pleistocene Wucheng Loess, Middle Pleistocene Lishi Loess, and Late Pleistocene Malan Loess, which laid the foundation for loess chronology studies, and further proved their eolian origin on the basis of numerous geological and biological evidence.

Since the late 1970s, based on the characteristic alternating loess and paleosol strata in loess sections, Chinese researchers have proposed the concept of the Chinese loess-paleosol sequence, and systematically discussed Quaternary environmental variation (Lu and An 1979; Liu et al. 1980). Accordingly they have created a detailed stratigraphic division and nomenclature of the loess-paleosol sequences of the Loess Plateau (Lu and An 1979; Liu 1985; Kukla and An 1989; Ding and Liu 1989). On the basis of the paleomagnetic study of the Luochuan loess drilling hole, Heller and Liu (1982) established the magnetostratigraphy of Chinese loess-paleosol sequence back to 2.5 Ma. Afterwards, researchers around the world found that the Chinese loess-paleosol sequence can be well-correlated with the global ice-volume change recorded by $\delta^{18}O$ of deep-sea sediments (Heller and Liu 1984; An and Lu 1984; Liu 1985; Kukla 1987; Kukla et al. 1988; Rutter et al. 1991; Rutter 1992; Ding et al. 1995; Bloemendal et al. 1995; Liu et al. 1999a; Ding et al. 2002). This realization brought Chinese loess studies to a new stage as they became a basis for larger global environmental change research (Liu 1985) together with deep-sea sediments and polar ice cores as three pillars of global climate-change studies.

The next phase of study arose from an analysis of the temporal-spatial distribution, stratigraphy, and biological fossil features of Chinese loess-paleosol sequences, combined with modern climatology studies. In the early 1990s An et al. (1990, 1991a) suggested that the Chinese loess-paleosol sequence was a good record of the East Asian monsoon variation. On the 10^4-yr time scale the East Asian monsoon is alternately dominated by glacial winter and interglacial summer monsoons. An et al.'s paleomonsoon theory pushed the study of Chinese loess toward a new stage that combined it with the study of atmospheric dynamics.

In the most recent phase of study, research has been extended to the underlying red clay formation, especially the Mio-Pliocene red clay sequences (Sun et al. 1998b; Ding et al. 1999b; Guo et al. 2002). Research on the Loess Plateau revealed that the basal age of the red clay in the east of the Liupan Mountain is 6–8 Ma (Sun et al. 1997; Sun et al. 1998a; 1998b; Ding et al. 1998b; An et al. 1999; Qiang et al. 2001), and firmly established its eolian origin (Liu et al. 1988; An et al. 1991a; Ding et al. 1997; Ding et al. 1998b; Lu et al. 2001; Yang and Ding 2004; Liu et al. 2006a; Qiao et al. 2006; Li et al. 2006). However, the distribution of sources may have differed from those for the later loess. Recently, the basal age of the red-clay sequence at Shilou has been pushed back to >11 Ma (Xu et al. 2009). The oldest red-clay deposits have been found mainly on the western Loess Plateau (west of Liupan Mountain), for example, the red clay sequences of Qin'an (22–6.2 Ma) (Guo et al. 2002) and Zhuanlang (25.6–4.8 Ma) (Qiang et al. 2011).

Like the Quaternary loess-paleosol sequence, the red clay sequence appears to contain a good record of East Asian monsoon variation (Sun et al. 1997; 1998a; 1998b; Ding et al. 1999b; 2001b; Guo et al. 2002). In addition, the eolian dust-flux variation of the loess-red clay sequence can record the onset of aridification of inland Asia (An et al. 1991b). The aridification variation history of the last 7 Ma has been revised accordingly (Sun and An 2002, 2005). The discovery of the Qin'an red clay suggested that the inland Asia aridification started in the Miocene, and even back to the Late Oligocene (Guo et al. 2002, 2008; Qiang et al., 2011).

Thus, the eolian deposits on the Chinese Loess Plateau have recorded the history and variability of Asian Monsoon climate and the history of aridification of the inland Asia since the late Oligocene (An 2000; An et al. 2006; Sun and An 2002, 2005; Guo et al. 2002). Both the Monsoon and interior aridification relate not only to late Cenozoic global cooling, but also are influenced by regional tectonic events, especially the phased uplift of the Tibetan Plateau (An et al. 2001; An et al. 2006). Further studies on the eolian sequences on the Chinese Loess Plateau, are expected to reveal not only the coupling evolution of monsoon-aridity system of East Asia, but also shed light on the dynamic mechanisms linking regional and global environmental variation on different time scales.

2.1.2 East Asian Monsoon Evolution Recorded by Chinese loess

The East Asian monsoon is an important part of global atmospheric circulation. Unlike other monsoon systems (African monsoon and Indian monsoon), the East Asian monsoon has a unique and strong mid-latitude winter circulation ('winter monsoon'). In winter and spring, cold air masses formed in the high-latitudes of central Asia move southward along the east side of the Tibetan Plateau, resulting in cold, windy and dry conditions on the Chinesse Loess Plateau. The winter monsoon can extend as far south as the tropical areas in the South China Sea, and even cross the equator. As summer begins, warm and humid air originating from the tropical ocean moves northward, ultimately reaching the China-Mongolia border area (~40°N). This is the highest latitude reached by any monsoon system in the world, and provides indispensable precipitation for this area. The East Asian monsoon not only has a profound influence on climate in East Asia, with its clear seasonal contrast between warm-humid summer and cold-dry winters, but also dynamically links to the global climate system.

Although Chinese Quaternary geologists suspected that the environment of eastern China has likely fluctuated along with the strength of the East Asian monsoon variation on long (10^3–$10^{4\text{-yr}}$ time scales), it wasn't until the early 1990s that thorough, systematic research on the loess-paleosol sequence in central China revealed the strong correlation between the alternating layers of loess and paleosols in the Chinese Loess Plateau and fluctuations in the anti-phased East Asian summer and winter monsoons. An et al. (1991a) proposed that most of the loess was deposited during times when the East Asian summer monsoon was weakened, and did not reach as far north as the northern margin of the Loess Plateau. The dust was transported to the Loess Plateau by the northerly winds of the winter monsoon during the relatively cool and dry seasons. In contrast, the interbedded paleosols formed under the warmer, more humid summers associated with strong summer monsoons. An et al. (1991a) therefore suggested that the East Asian paleomonsoon variation has been the principal and direct driving force controlling environmental conditions in East Asia. Subsequent studies on monsoon-climate records provided by loess, speleothems, and lake and marine sediment cores, unveiled the framework and mechanisms of the East Asian monsoon climate variation during different geological periods. These lines of evidence have propelled the East Asian paleomonsoon studies to general awareness and placed the loess studies in a global context (e.g., An et al. 1990; An 2000; Liu and Ding 1998).

In recent years, studies have attempted to obtain information on the formation and evolution of the East Asian monsoon, to discover the relationship between East Asian monsoon variation and global ice volume and the uplift of the Tibetan Plateau, and to explore the global significance of the East Asian monsoon and Tibetan Plateau uplift (An et al. 2001). Paleoclimate research has already been extended spatially from the inland Loess Plateau to the marginal seas of southern China and the Warm Pool in the western Pacific (Wang et al., 2003, 2005). The range of study also has expanded to include different geological records of the northwestern inland arid area, revealing the relationship between the evolution of the East Asian monsoon and climate change at both high and low latitudes. Only through research on the response of the East Asian monsoon to the overall behavior of various factors of the global climate system, viewed through the framework of the global change, we can fully understand the mechanisms of the East Asian monsoon variability.

Loess research has revealed that the East Asian winter monsoon plays a leading role in the transportation of Asian dust and deposition of the loess-paleosol sequence on the Loess Plateau (An et al. 1991b). In winter, dust is carried by the East Asian winter monsoon blowing from inland Asia onto the Loess Plateau and deposited there, forming the loess. However, the monsoon is not the only factor driving deposition in the Loess Plateau. Recent studies on the grain-size composition of the loess have shown that high-altitude westerly circulation also influences the transportation and deposition of loess (Porter and An 1995; Sun et al. 2004; Sun et al. 2008b). The warm and humid air masses of the summer monsoon originating in the equatorial ocean episodically have promoted soil development. Based on magnetostratigraphic results and direct dating methods such as ^{14}C and luminescence dating, the basic chronology of the Chinese loess/red clay sequence has been established. By using proxy indices sensitive to monsoon variation in the loess/red clay sequence, the evolution history of the East Asian paleomonsoon has been reconstructed back to 22–25 Ma (Guo et al. 2002; Qiang et al. 2011).

The following text will review the progress of recent studies of Chinese loess and the East Asian paleomonsoon evolution from four aspects: temporal and spatial distributions of the loess, chronology, proxy indices, and the East Asian monsoon evolution on different time scales. A brief summary of the evolution characteristics and dynamic mechanisms of the East Asian paleomonsoon variability will be presented.

2.2 Temporal and Spatial Distributions of Chinese Loess

Chinese loess is deposited over an area of ~630 000 km^2, or 6.6% of China (Liu et al. 1985). It is mainly located between 33 and 47°N, and 75 and 127°E, north of the Qinling, Qilian and Kunlun mountains, from Xinjiang through Gansu, Shaanxi, Shanxi, Henan, western Liaoning, to the Songliao Plain. The loess area is arc-shaped, protruding toward the south. The sandy deserts of Horqin and Otindag in northeastern China, Mu Us, Tengger and Badain Juran deserts in Inner Mongolia, Taklimakan and Gurbantunggut deserts in northwestern China border it on the north

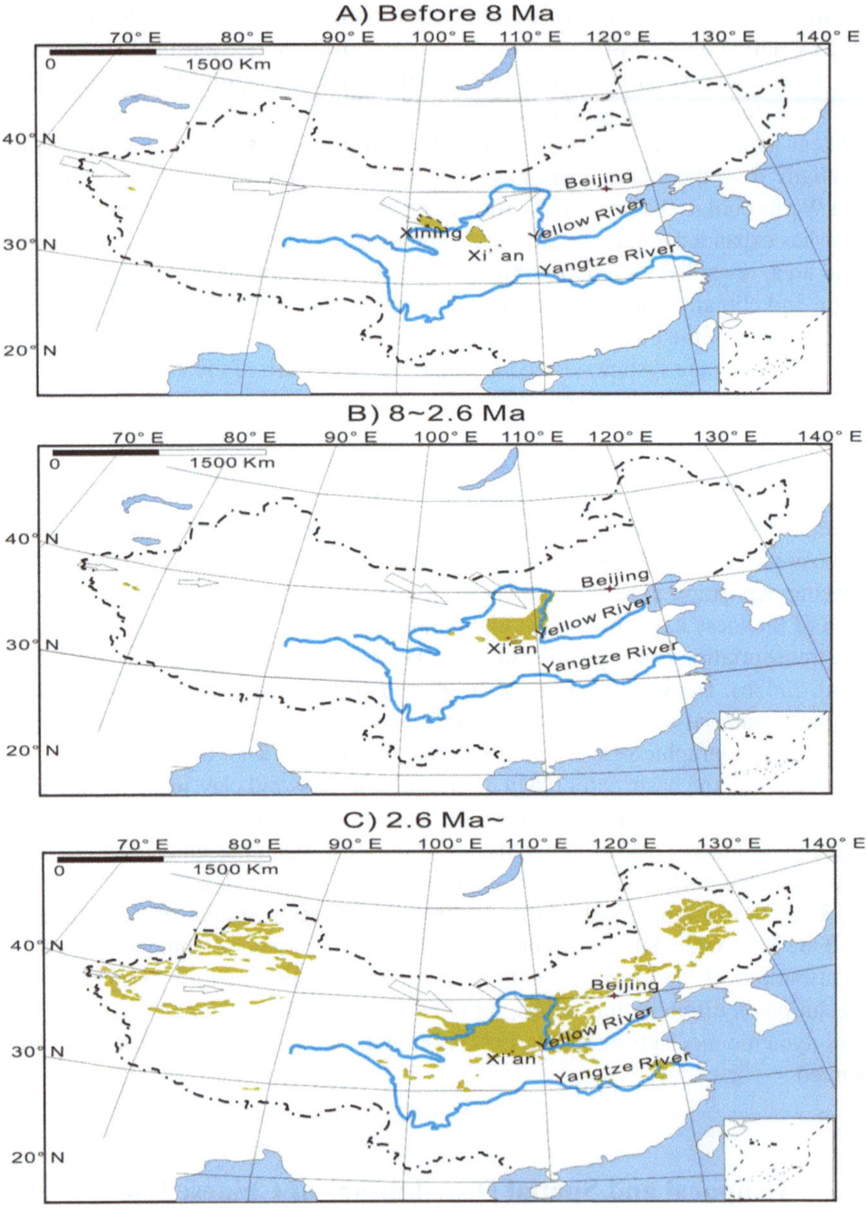

Fig. 2.1 Distribution of loess-paleosol and red clay sequences over northern China

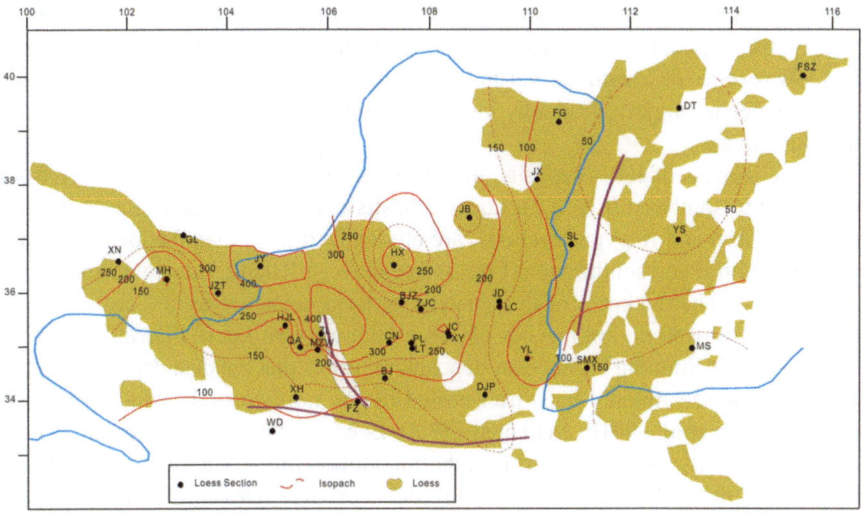

Fig. 2.2 Spatial variability of thickness of loess deposits on the Chinese Loess Plateau

(Fig. 2.1c). This spatial distribution of the loess is clearly related to the anticyclonic winter monsoon formed in the stable high-pressure cells centred in Lake Baikal and Mongolia. Outside this zone, the dust flux has been lower but loess nevertheless is dispersed over Nanjing, Jiujiang, Wuhan, Sanxia, Sichuan Basin, Chuanxi Basin, Qaidam Basin, and eastern Tibet. Elevations at which the loess is found vary greatly depending on the locality. In the Turpan Basin of Xinjiang, loess is found on the surface at an elevation of 150 m below sea level. On the continental shelf of the East China Sea and the Yellow Sea, loess is located in seawater at depths of 50–60 m during the global glacial maxima. Loess deposits are 4500 m above sea level the Pamir Plateau, 5340 m in the western Kunlun mountains, and 4300 m on the Chuanxi Plateau (Liu et al. 1985; Sun 2005b). On the central Loess Plateau, elevations of the loess tablelands are 1000–1800 m, whereas in the western Loess Plateau the loess deposits accumulated around 2000–2500 m above sea level.

Spatial distribution of the red clay is not identical to that of the later Quaternary loess-paleosol sequences (Fig. 2.1). Miocene red clay is mainly located in Tianshui–Qin'an area, to the west of the Liupan Mountains (Guo et al. 2002; Qiang et al. 2011). Red clay is found only sporadically in the Xining Basin (Lu et al. 2004b) and on the eastern Loess Plateau near Shilou in Shanxi (Xu et al. 2009). Late Miocene to Pliocene red clay sequences, however, are extensively distributed on the central Loess Plateau (e.g. Zheng et al. 1992; Sun et al. 1998a; 1998b; Ding et al. 1999b; Ding et al. 2001b). Early Pleistocene Wucheng loess is situated north of 35°N and east of 110°E, including the northern slope of the western Kunlun mountains. Middle Pleistocene Lishi loess extends to southeastern China, having a larger distribution area than the Wucheng loess. For example, Xiashu loess in the mid-low reaches of the Yangtze River is the counterpart of the Lishi and Malan loess.

Late Pleistocene Malan loess covering different geomorphologic units is even more widely distributed, extending to the Miaodao Islands in the east, and to the middle and low reaches of Yangtze River in the south. The zonal expansion of loess since the Pleistocene suggests gradually increasing aridity at the dust sources throughout the Quaternary.

The thickness of loess deposits on the Loess Plateau decreases gradually from northwest to southeast (Fig. 2.2). Loess deposits in the northwestern Loess Plateau (e.g., Lanzhou and Jingyuan sections) are obviously thicker than the loess sequences in the central and southeastern Loess Plateau (e.g., Xifeng, Luochuan and Weinan sections). The thickness of the Quaternary loess sequence on the central Loess Plateau is 100–180 m, but reaches 450 m on the western Loess Plateau (Fig. 2.2). Southeastward thinning of the loess deposits indicates that the dust deposited on the Loess Plateau has mainly been transported by the northwesterly winter monsoon (Liu 1985; An et al. 1991b; Xiao et al. 1992). Loess sections from the central Loess Plateau are fairly complete, and usually encompass the whole of the Quaternary (Liu et al. 1985). More recently, a 670-m loess core was retrieved from the western Kunlun by drilling, includes loess from the Quaternary and late Pliocene (Fang Xiaomin, personal communication 2010).

The loess sedimentation rate is normally ~200 m/Ma near Lanzhou in the western Loess Plateau, decreasing to 50–100 m/Ma in the central Loess Plateau. Loess sedimentation rates in Xinjiang and the Kunlun are all > 100 m/Ma. Spatial variations in the thickness and sedimentation rate of loess deposits over northern China indicate that the loess deposits originated from inland Asia and were transported dominantly by the winter monsoon.

2.2.1 Quaternary Loess-paleosol Sequences

Before the 1990s, numerous studies on loess grain size, mineral components and dust-transport mechanisms had solidified the understanding that on the Loess Plateau, loess thickness and grain size both decrease gradually from northwest to southeast, whereas the degree of soil development increases (Liu et al. 1985). In contrast, because of the limited exposure and evident confinement of the underlying red clay deposits in localized depressions in the paleo-topographic surface, the scientific understanding of their thickness, spatial distribution, origin and chronology of the red clay was greatly hampered. In the past two decades, dozens of typical loess/red-clay sections have been analyzed (Fig. 2.3). Of these, the Luochuan, Xifeng, Duanjiapo, Jingbian, Weinan, Lingtai, Fugu, Shilou, Jiaxian, and Jingchuan sections in the central Loess Plateau (east of the Luipan Mountains and west of Lvliang Mountains) all share the same time duration for the loess (2.6 Ma), and basal ages ranging from 5 to 11 Ma for the red clay. The Lanzhou, Jingyuan loess, Qin'an, and Zhuanglang sections in the western Loess Plateau (west of the Liupan Mountains) have basal ages for ranging from 1.4 to 1.8 Ma for the Quaternary loess,

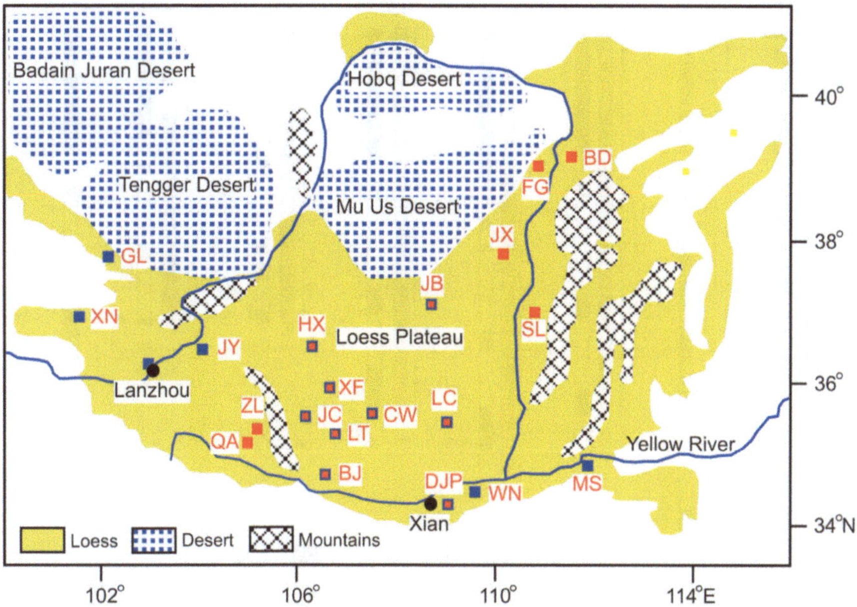

Fig. 2.3 Typical loess and red clay sections on the Chinese Loess Plateau. Sections are located at Baoji (BJ), Baode (BD), Changwu (CW), Duanjiapo (DJP), Fugu (FG), Gulang (GL), Huanxian (HX), Jiaxian (JX), Jingbian (JB), Jingchuan (JC), Jingyuan (JY), Lantian (LT), Lanzhou, Lingtai (LT), Luochuan (LC), Mangshan (MS), Qin'an (QA), Shilou (SL), Weinan (WN), Xifeng (XF), Xining(XN) and Zhuanglang (ZL)

and 22 to 25 Ma for the red clay, respectively (Burbank and Li 1985; Liu 1985; Yue et al. 1991; Guo et al. 2002; Qiang et al. 2011).

The following is a brief comparison of typical loess/red-clay sequences on the Loess Plateau. Quaternary loess-paleosol sequences on the Chinese Loess Plateau share many similarities (Fig. 2.4), with typical profiles exposed at Luochuan (Liu et al. 1985), Xifeng (Kukla 1987), Baoji (Rutter et al. 1991; Ding et al. 1993), Lingtai (Sun et al. 1998b; Ding et al. 1999b), and Lanzhou (Burbank and Li 1985). It is clear that the thickness of the loess exceeds 300 m in the Lanzhou section, near the dust sources, but decreases to 170 m in the Xifeng and Lingtai sections of the middle Loess Plateau, and is only 130–140 m in the Baoji and Luochuan sections. The pedogenic intensity of paleosol layers is significantly lower in the northerly Luochuan and Xifeng sections than in the more southerly Lingtai and Baoji sections. Paleosol development in the Wucheng Loess best shows this spatial discrepancy. In the Luochuan and Xifeng sections, the Wucheng Loess can be divided into three paleosol complex (WS 1–3) interbedded with three loess complexes (WL1–3). In Baoji and Weinan sections, however, Wucheng Loess can be further subdivided into 17 paleosol layers (S15–32) intercalated by 18 loess layers (L15–L33), suggesting that in the southern Loess Plateau paleosol units were well-developed resulting from the stronger influence of the the East Asian summer monsoon. The depth of

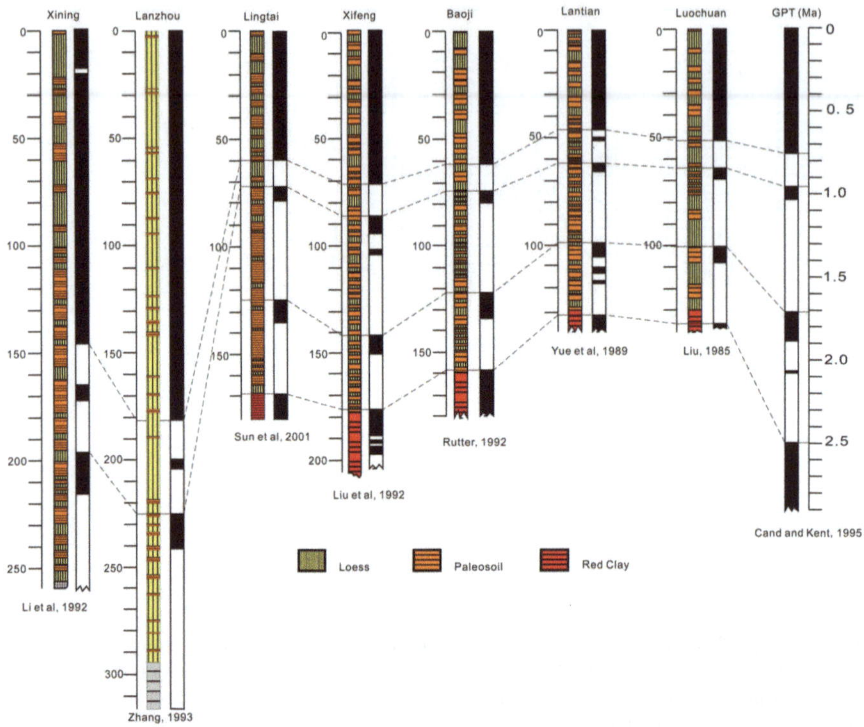

Fig. 2.4 Comparison of Quaternary loess-paleosol sequences on the Chinese Loess Plateau. For locations of sections, see fig. 2.3 GPTS—Geomagnetic Polarity Time Scale *black* sections have normal polarity, *white* sections show reversed

the typical marker layers (S1, S5, L9, L15, etc.) in different sections shows that loess strata are normally thicker in the northwestern Loess Plateau, while paleosol strata are commonly compacted in the southeastern Loess Plateau. As descriebd above, this is interpreted to result from the alternating dominance of the winter and summer monsoons over the past 2.5 Ma.

2.2.2 Red Clay Deposits on the Central CLP

In the past two decades, investigation of the dominantly Tertiary red clay sequence has become a hot topic in Chinese paleoclimate research. There have been two motivations: to determine the onset and evolution of Asian aridification and cast light on the paleoclimatic history. Through systematic field geological surveys, many late Cenozoic eolian sequences including the Tertiary red clay on the central Loess Plateau were studied intensively to conduct stratigraphy and paleomagnetism studies. These sections include the Lingtai, Xifeng and Bajiazui sections on the middle

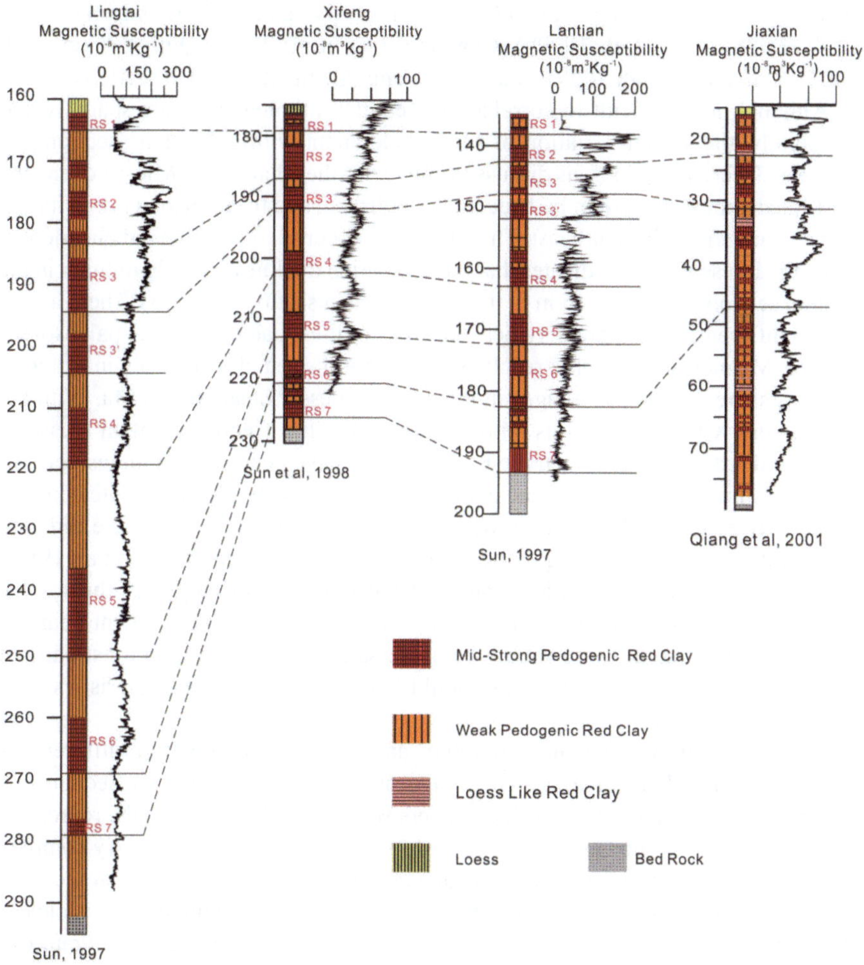

Fig. 2.5 Late Miocene to Pliocene red clay sequences on the central Chinese Loess Plateau

Loess Plateau, Jiaxian section in the northern Loess Plateau, and the Duanjiapo section in the southern Loess Plateau (Fig. 2.5).

The Tertiary red clay deposits are easily recognized because of their eminent reddish color and stronger pedogenic degree than the overlying loess-paleosol sequence, with nearly horizontal layers of carbonate nodules distributed within. Field observation suggested that there was no depositional hiatus between the red clay and the overlying loess-paleosol sequence. Unlike the loess-paleosol sequences, no marker layers can be found within the upper, middle and lower parts of the red clay deposits for stratigraphic correlation. Currently, the more reliable red clay sections for magnetostratigraphic study on the Loess Plateau include the Duanjiapo (Zheng et al. 1992), Bajiazui and Zhaojiachuan sections (Sun et al. 1997; 1998a), Jiaxian

(Ding et al. 1998b; Qiang et al. 2001), Jingchuan (Ding et al. 2001a), Lingtai (Sun et al. 1998b; Ding et al. 1999b), Chaona (Song et al. 2001), Qin'an (Guo et al. 2002), Shilou (Xu et al. 2009), and Zhuanglang (Qiang et al. 2011) sections.

Although it is difficult to correlate the red clay lithological strata spatially and temporally, the spatial variation of soil development within the same section and between different sections, as discussed above, implies that by the Mio-Pliocene the East Asian monsoon system had become established, and the monsoon precipitation mainly came from the southeast. The stratigraphic correlation of the red clay deposits on the Loess Plateau indicates that the deposition thickness is ~120 m thick in the Lingtai section, but only ~55 m in the Zhaojiachuan section (Fig. 2.5). There are no spatial differentiating characteristics like those of the Quaternary loess-paleosol sequence. Magnetostratigraphic results showed that the red clay deposits in different sections share similar initial ages in the central Loess Plateau, 6.6–8 Ma. With the Lingtai and Xifeng sections as representatives, the late Cenozoic eolian loess-paleosol and red clay deposits on the Loess Plateau have been studied intensively for understanding continental climate evolution and its links to global climate change.

In early research, the red clay was commonly divided into the Jingle red clay and Baode red clay, according to the vertebrate fauna in the strata (Liu et al. 1985). This division is not pragmatic because of the limited distribution of the fossils. Soil-strata division is the new method used in red clay studies in recent years. A proposed soil-strata division should be universally applicable and in most parts of the Loess Plateau. The strata units should have recognizable characteristics on a coarse regional scale.

Lithostratigraphy, magnetostratigraphy, and climate stratigraphy in different areas on the Loess Plateau consistently suggest that the whole red-clay sequence is composed of multiple reddened loess layers with intercalated paleosols, representing the late Tertiary monsoon variation history on the Loess Plateau. Every complex is comprised of several depositional-pedogenic strata units. In the southern sections, for example at Duanjiapo and Lingtai, every depositional-pedogenic unit is clearly discernible. In the middle area, for example at Xifeng and Luochuan, strata can be difficult to identify, but the whole combination of paleosol and reddish loess is evident. Therefore, as a model for regional correlation, it may be possible to utilize the paleosol and reddish loess alternations within the red clay. Consistent with the nomenclature of the Quaternary loess-paleosol sequence, the paleosol complex in the red clay is divided into seven units: from top to bottom, RS1–RS7 (Fig. 2.5).

2.2.3 Late Oligocene to Miocene Loess and Red Clay Sequences on the Western Loess Plateau

The most complete Miocene red clay and loess sequence (QA-I) (105E, 34N) is 253.1 m thick and located 27 km northwest of Qin'an city (Guo et al. 2002). A second similar section (QA-II, 220.6 m thick) has been found 2 km from QA-I. QA-I contains 231 visually definable reddish layers interbedded with yellow-brown

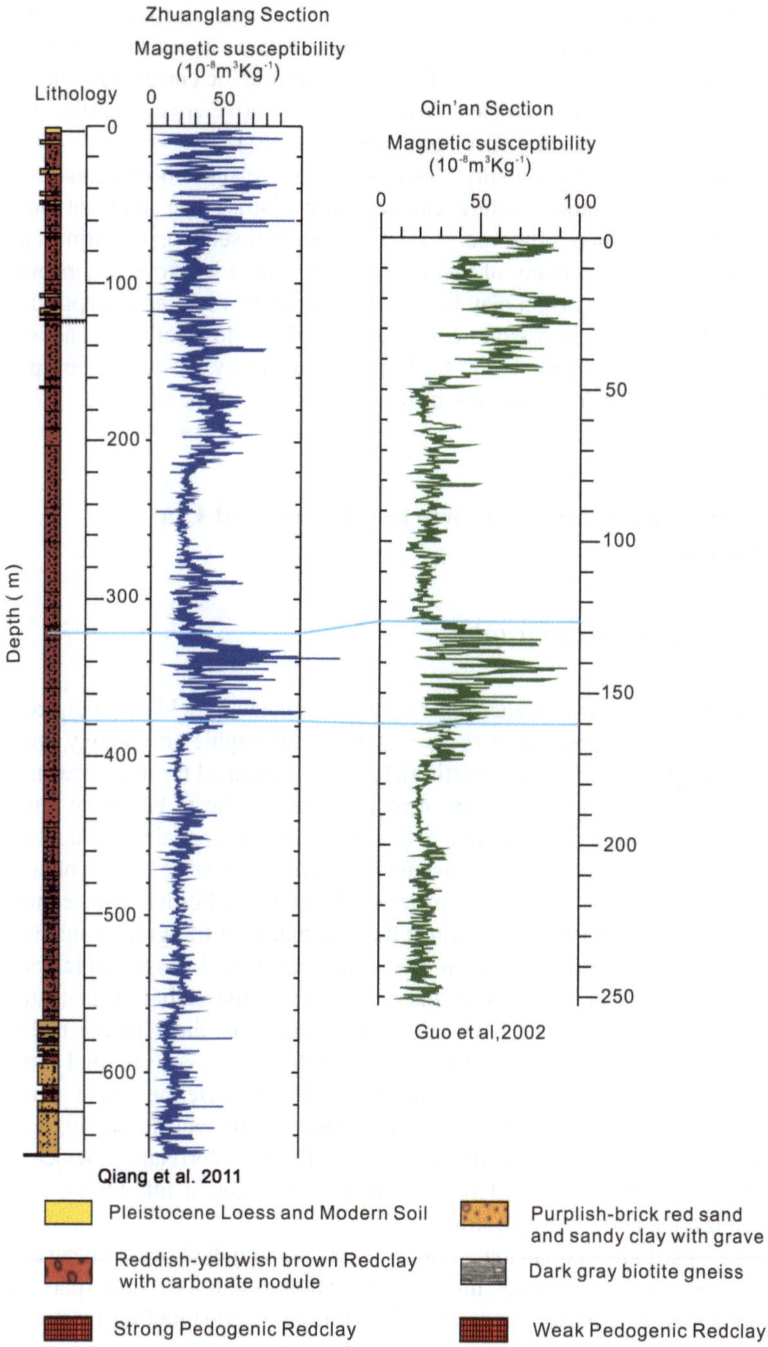

Fig. 2.6 Miocene loess-red clay sequences at Qin'an and Zhuanglang

or brown silty layers, all nearly horizontal. This layering is reflected in variations of magnetic susceptibility (Fig. 2.6), with higher values in red layers than in surrounding white layers. According to field lithostratigraphic correlation, the top part of QA-II was truncated by erosion. A spatially correlative erosion surface was observed at a depth of 59.8 m in QA-I and 20.0 m in QA-II.

The reddish layers have a silty clay or clay texture and moderate to strong angular blocky or prismatic structure, characteristic of soils. The abundant biological pores and clay coatings observed in microscopic thin sections confirm that these red beds are paleosols, many of them composite soils. Fine-grained ferrimagnetic minerals of pedogenic origin may be responsible for their higher susceptibility values, as in Quaternary loess-soil sequences in China. Most soils are underlain by carbonate nodule horizons similar to those in Quaternary loess. The non-paleosol silty layers have a yellow-brown or brown color and massive structure.

2.3 Chronological Framework of Loess/Red Clay Sequences

2.3.1 Magnetostratigraphy

Magnetostratigraphy is a science utilizing the geomagnetic field variations recorded by the remanent magnetism of rocks to solve stratigraphic chronology problems. Polarity variation of the geomagnetic field can be obtained through measurements of the natural remanent magnetism of rocks, normal polarity being the same with current geomagnetic polarity, and reversed being the opposite. Throughout the whole of geological history, a geomagnetic polarity time scale can be obtained by arranging normal and reversed polarities in chronological order. The geomagnetic polarity time scale (GPTS), determined through magnetostratigraphic methods and magnetic anomaly stripes, has been extended to the Late Jurassic, which provides a solid foundation for the division and correlation of the global magnetostratigraphy (Cande and Kent 1995). Major polarity chrons since the Miocene include the Brunhes normal chron, the Matsuyama reversed chron, the Gauss normal chron, and the Gilbert reversed chron (Cande and Kent 1995). Because the dipole component of the geomagnetic field reverses simultaneously throughout the world, magnetostratigraphy is well-suited for global-scale stratigraphic correlation, which can be made between strata from different continents, of different lithologies, and even between marine and terrestrial strata.

Widely distributed eolian loess/red clay sequences on the Loess Plateau are characterized by good consistency, high sedimentation rate, a long time span, and enriched paleoclimate information. They therefore are well-suited for magnetostratigraphic study (Liu et al. 1985) that can establish its time frame, variation history, and characteristics of the fluctuating geomagnetic field. Since the 1970s, numerous studies have been conducted to construct magnetostratigraphic profiles of multiple

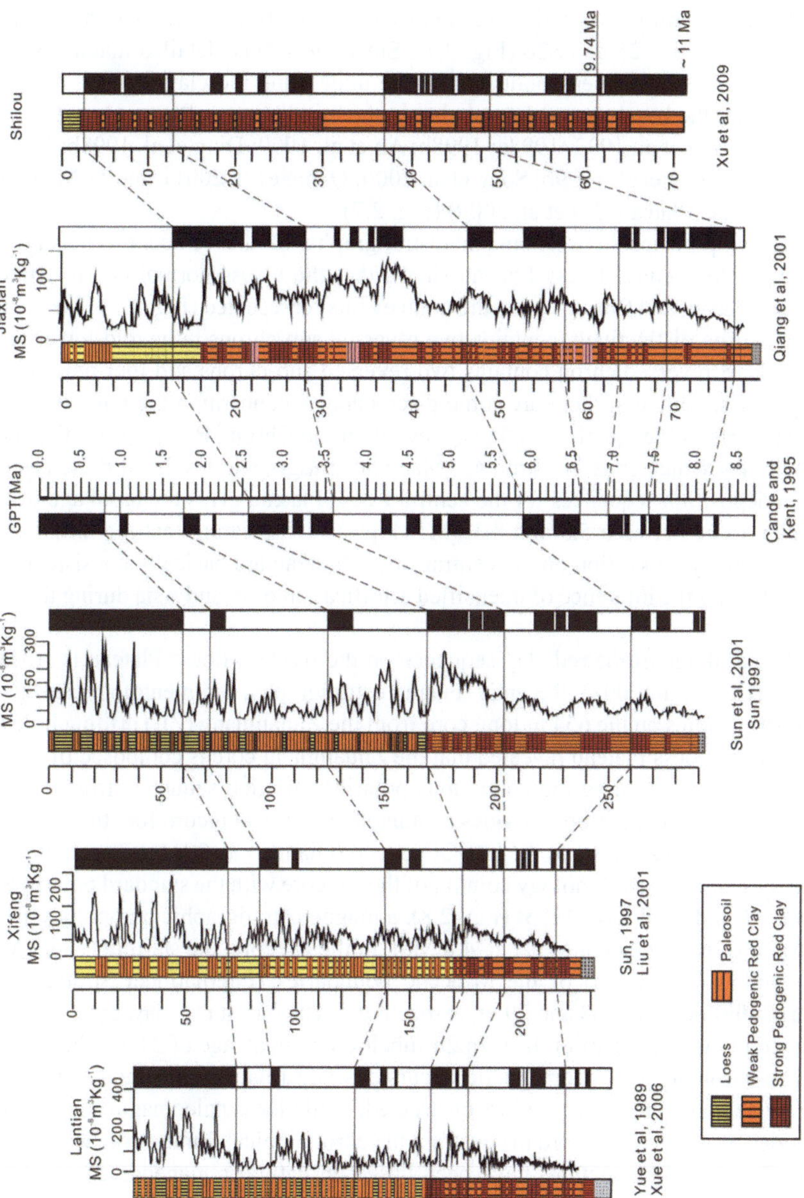

Fig. 2.7 Pedostratigraphy and magnetostratigraphy in central Loess Plateau

loess sections (An et al. 1977; Heller and Liu 1982; Liu 1985; Zhu et al. 1994a). Paleomagnetic results for the loess indicate that the Brunhes/Matuyama boundary lies in the lower part of L8, or the upper part of S8, the Matuyama/Gauss boundary in L33, the Jaramillo magnetic reversal lies between L10 and L13, and the Olduvai reversal between L25 and S26 (Fig. 2.7). Since the 1990s, detailed magnetostratigraphic studies have been conducted on the underlying red-clay deposits. Results suggest that the basal age of the red clay in the middle Loess Plateau is between 6 and 8 Ma (Sun et al. 1997; 1998a; 1998b; An et al. 1999; Ding et al. 1998b; Yu and Ding 1998; Ding et al. 1999b; Song et al. 2000; Qiang et al. 2001), and 11 Ma in the eastern Loess Plateau (Xu et al. 2009) (Fig. 2.7).

Intercomparisons between magnetostratigraphy of the Pliocene red clay sequence of the central Loess Plateau shows that the Gauss normal chron, Gilbert reversed chron, and Chron 3A are the main events represented (Fig. 2.7). The Gauss normal chron distinctively contains two reversed sub-chrons in its mid-lower part. The Gilbert reversed chron contains two reversed sub-chrons and four normal polarity zones. Although there are some discrepancies concerning the polarity in the bottom parts of the red clay sections, they all end in Chron 3B, except for the Jiaxian red clay which ends in Chron 4n. Magnetostratigraphic results of these typical eolian deposition sequences for the central Loess Plateau gives the basal age of the red clay to be between 5.3 and 8.35 Ma (Fig. 2.7), indicating that the initial ages of the red clay deposition on the central Loess Plateau are basically consistent and strengthening the inference of intensified aridification of inland Asia during the late Miocene.

The basal age of the red clay sequence on the western Loess Plateau is at least 22 Ma (Guo et al. 2002). Recently, magnetostratigraphy, sedimentology, and geochemstry studies on the 654 m long core from the Zhuanglang (ZL) drilling hole on the western Loess Plateau revealed that the Zhuanglang core is composed of eolian red clay strata recording the continuous polarity variation sequence from 25.6 to 4.8 Ma. This core therefore provides a valuable geological record for studies on inland Asia aridification since the late Oligocene (Qiang et al. 2011). Correlating the polarity sequence and lithology column of the ZL core with the standard polarity sequence (Cande and Kent 1995) (Fig. 2.8), a magnetostratigraphic chronology from C7An to C3n.3n is obtained (25.6–4.8 Ma). In the ZL core, the starting (23.03 Ma) and ending (5.332 Ma) of the Miocene boundaries (International Stratigraphic Chart 2009) occur at 568 and 29 m, respectively. The oldest red clay, at 624.8 m, is situated in the upper part of the C7n.2n subchron, with an age of 24.85 Ma.

The chronology of geomagnetic excursions and relative paleointensities from marine and ice cores were well reconstructed, while loess paleomagnetism studies provided a continental record to constrain the chronological framework. Most of the geomagnetic reconstructions have been based on natural remanence analysis and recently supplemented by cosmogenic ^{10}Be studies. Representative stacked relative geomagnetic paleointensity records for marine cores such as SINT800 (Guyodo and Valet 1999), NAPIS 75 (Laj et al. 2000) and PISO-1500 (Channell et al. 2009) have been well studied. Therefore, it is feasible for us to use these records to synthesize various records, with the goal of refining our knowledge of the past changes in

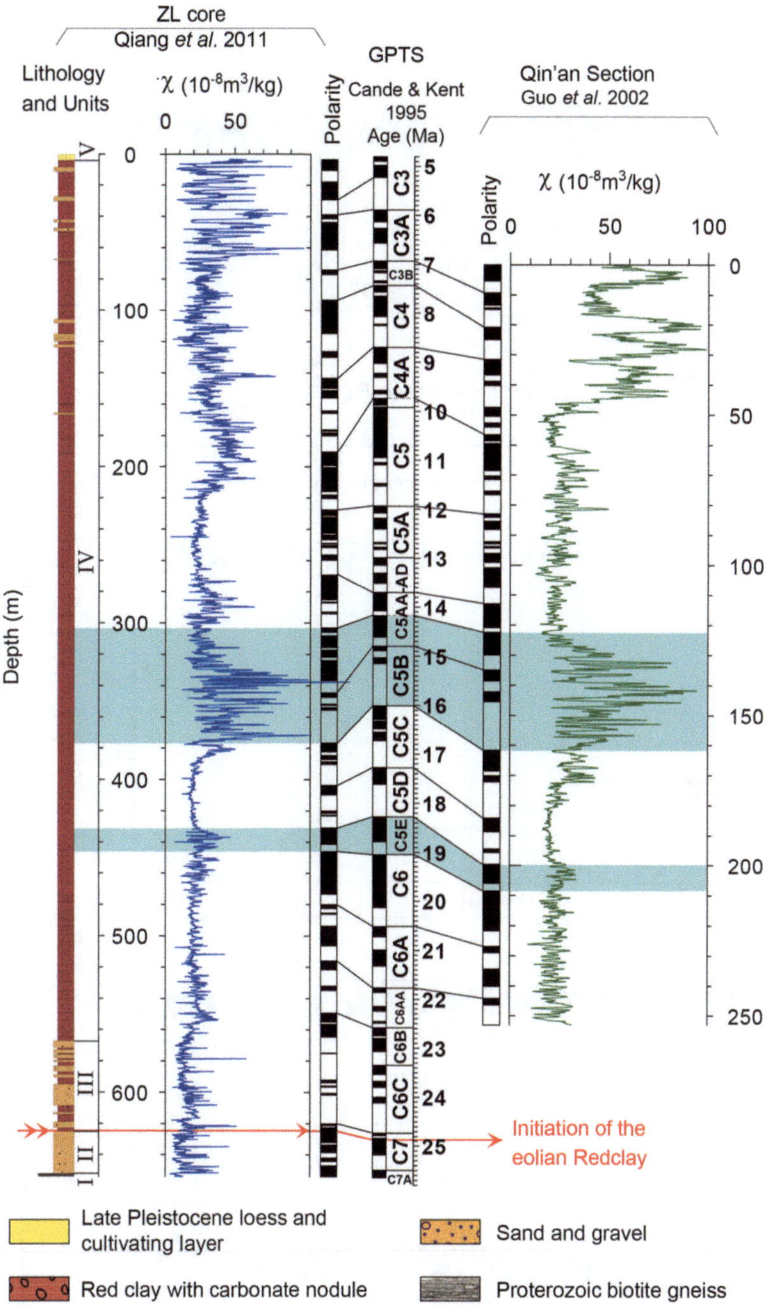

Fig. 2.8 Pedostratigraphy and magnetostratigraphy on the western Loess Plateau (Guo et al. 2002; Qiang et al. 2011). GPTS = geomagnetic polarity time scale

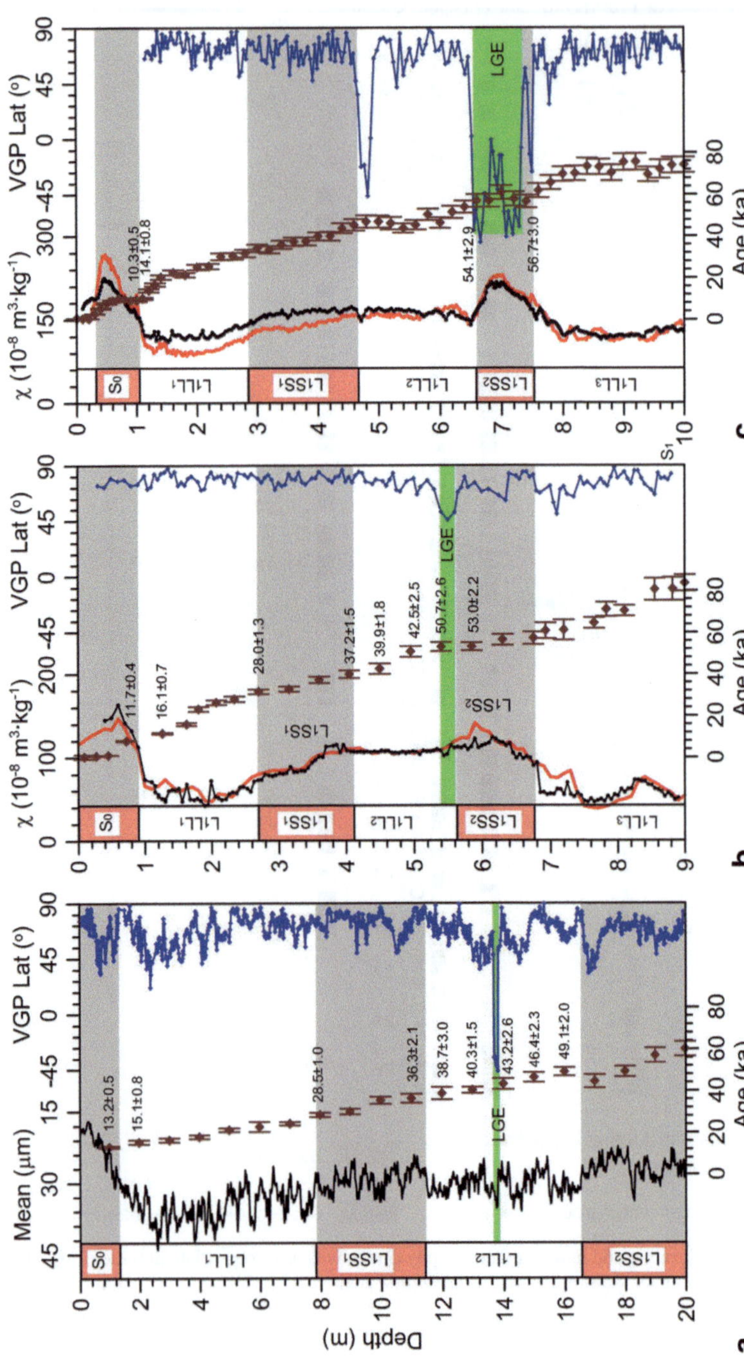

Fig. 2.9 Comparison of (A) Gulang mean grain size and virtual geomagnetic pole (VGP) with magnetic susceptibility and VGP results of (B) Luochuan and (C) Weinan sections (Zhu et al. 2007). Also shown are the OSL dates of three loess section: Gulang (Sun et al. 2012), Luochuan (Lu et al. 2007), and Weinan (Kang et al. 2013)

the geomagnetic field and paleoclimate system. Similar geomagnetic studies from Chinese loess have been based mainly on natural remanence analysis. Features of short-lived excursions including the Laschamp and Blake events, thought to have affected the entire globe, have all been identified in Chinese loess (Zhu et al. 1994b; Zhu et al. 1998; Zhu et al. 2000; Zhu et al. 2006; Fang et al. 1997; Wang et al. 2010; Sun et al. 2013) (Fig. 2.9). Using the normalized natural-remanence parameter in loess, Zheng et al. (1995) and Pan et al. (2001) also tried to recover the relative paleointensity changes. The resulting records exhibit obvious dissimilarity with the globally stacked paleointensity curves. Nevertheless, the existence of these excursions provides more time controls for the loess chronology construction (Zhu et al. 2007; Sun et al. 2013).

2.3.2 Chronological Control Using Optically Stimulated Luminescence (OSL) Dating

2.3.2.1 Introduction to OSL Dating of Loess

Chinese loess is typically composed of fine-grained particles (<50 μm) that have been transported great distances before it was deposited. This is a particularly favorable circumstance from the standpoint of OSL, as the ample sunlight exposure resets the internal luminescence clock in the loess grains (Prescott and Robertson 1997), thus satisfying an essential prerequisite for luminescence dating (Lu et al. 1988; Prescott and Robertson 1997). Typically, more than 80% of Chinese loess grains are less than 50 μm and 10–15% are less than 10 μm (Liu et al. 1985; Wintle 1990), and thus are suitable for direct analysis. Much work on Chinese loess has been carried out with fine grains of 4–11 μm after removal of carbonate. Other OSL studies of Chinese loess have dated fine- and medium-grained quartz and etched feldspar (Stevens et al. 2006; Lai 2010).

For the past 40 years, Chinese researchers have used luminescence dating to establish chronological control in loess profiles. The OSL dating technique has become a fundamental tool for quantifying time scales in Holocene soils, last-glacial loess and interglacial paleosols. In the 1970s, Li et al. (1977) first introduced Thermoluminescence (TL) dating into Chinese Quaternary studies, and Li (1982) and Li and Sun (1986) reported the first set of TL ages from Chinese loess. In the late 1980s, TL was used extensively to establish the chronology of the Malan loess (An and Lu 1984). TL also provided the foundation for time scales for Holocene soils, last-glacial loess and last-interglacial palsosol sequences (Lu et al. 1987; Lu et al. 1988; Forman 1991).

Here we focus on the measurement of grains in the size range 4–11 μm. Within this material there are three signals available for OSL dating: quartz OSL, polymineral infrared stimulated luminescence (IRSL) and a post-IR OSL signal. The quantities measured are equivalent doses (ED or De) corresponding to a quartz OSL De value, two De values for polymineral IRSL and a post-IR OSL signal, respectively.

In the following section we review the methodological improvement in ED determination and their application in dating Chinese loess.

2.3.2.2 ED Determination in Quartz

Single-Aliquot Regenerative-dose (SAR) protocol OSL signals from quartz grains are frequently used for dating late Quaternary sediments. The reliability of OSL dating and its increasing popularity, can be largely attributed to the introduction of the SAR protocol. This reliable method provides a means to use the test dose OSL response to correct for sensitivity changes during OSL measurements (Wintle and Murray 2006). The SAR protocol is complemented by some basic tests (e.g. repeated ratio, recuperation ratio) that estimate whether the protocol is appropriate for the samples to be dated (Wintle and Murray 2006).

Detailed descriptions of the principal features and main assumptions of an SAR protocol are available in the literature (e.g. Table 1:(Murray and Olley 2002)). For the protocol to be useful and reliable, sensitivity changes of the measured regenerative dose cycles must be accurately corrected. This is done using the OSL response to the test dose. Murray and Olley (2002) have reviewed the performance of the SAR dating technique applied to a variety of materials, including modern to late Pleistocene sediments, and older samples. Their results summarized 53 OSL dates from a variety of sediment types with independent age control. The observed correspondence between SAR OSL ages and independent dates over the past 130 ka support the view that the SAR protocol can be regarded as a robust luminescence dating method.

Sensitivity-corrected Multiple Aliquot Regenerative dose (SMAR) protocol Lu et al. (2007) developed an OSL protocol optimized for dating quartz grains from the abundant 4–11 μm size fraction in loess. Their method is known as the SMAR protocol. The SMAR protocol also uses an OSL response as a test dose to correct for sensitivity changes. Normally, more than three aliquots are measured to obtain natural OSL intensities, and at least five aliquots are used to construct a dose-response curve that brackets natural OSL intensity.

A methodological study points to three advantages of the SMAR protocol: 1) good performance in respect to sensitivity, 2) reduction in the scatter of experimental results, and 3) elimination of buildup of OSL signals (Lu et al. 2007). To further validate the reliability of the SMAR protocol, OSL ages from the Luochuan section were dated (Fig. 2.10). Two OSL ages from this profile accurately reflect the known age of the L_1/S_1 transition (corresponding to MIS 5/4) at (71.1±3.4) ka and (70.4±2.6) ka. This correspondence demonstrates the reliability of OSL ages determined by SMAR.

Comparison between the SAR and SMAR protocols In a comparative study, the SAR and SMAR protocols were applied to 23 samples from the Luochuan section, using the same preheating and experimental conditions. It is clear that ED values determined by the SAR and SMAR protocols are in good agreement for samples

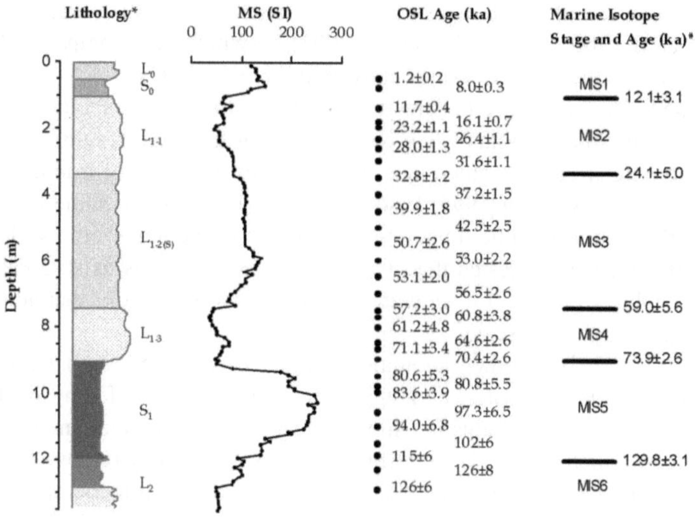

Fig. 2.10 OSL chronology of the Luochuan loess profile (Lu et al. 2007)

from the last-glacial loess, with ED values less than 120 Ga. But the SAR ED values show unexpected underestimation for older samples, with an average underestimation of about 8 % for the last-interglacial paleosol. This is caused by the use of blue-light stimulation in the SAR protocol for bleaching, rather than the broader and more powerful spectrum of the SOL 2 sunlamp. This conclusion is also supported by the dose recovery tests of Wang et al. (2006).

Murray et al. (2007) used the SAR protocol to date well-described marine sediments, which were likely deposited during the early Eemian interglacial at about 130 ka on the river Sula in northern Russia. Various quality control checks are used to show that the OSL behavior is satisfactory; with a mean for 16 ages of (112±2) ka. This represents an underestimate of ~14 % compared to the expected age, a discrepancy similar to that reported elsewhere. This underestimation rate is consistent with the SAR underestimation rate reported by (Lu et al. 2007). Lowick et al. (2010) and Lowick and Preusser (2011) also found age underestimation in the OSL high-dose region when analyzing fine-grained quartz that dated to the Eemian in sediments from the Alps.

2.3.2.3 IRSL and Post-IR OSL Dating Using Polymineral Grains

IRSL dating The IRSL signal in polymineral grains mainly comes from feldspar, and previous studies have found that IRSL signals from polymineral grains display anomalous fading during laboratory storage (Wintle 1973; Banerjee et al. 2001; Watanuki et al. 2003). De values measured on fine quartz from loess, using the SAR protocol, produced comparable results to the IRSL ED protocol for a set of nine Holocene samples with ages < 25 Ga. An analogous finding was obtained by Banerjee et al. (2001) and Stokes et al. (2003), in which they compared IRSL and post-IR OSL results with De values corresponding to < 40 Ga.

The IRSL De values are underestimated by up to ~34 % for older samples in Luochuan polymineral grains when compared to quartz OSL results. This may be due to anomalous fading, which can affect the depopulation of trapped electrons (Visocekas 1985; Spooner 1994; Visocekas et al. 1994), since some traps release electrons through a possible tunneling mechanism on geologic time scales. Wang et al. (2006) further suggested that two types of IRSL signals exist: a natural and relatively stable signal, and regenerated signals in the laboratory that are unstable over a long enough time scale.

Kang et al. (2010) studied anomalous fading of the IRSL signal for the past 170 ka in Luochuan samples, and suggested that anomalous fading of the IRSL signal from polymineral grains in Chinese loess has both thermal and non thermal dependences, which proved that preheating cannot completely remove this component. So at this stage we conclude that IRSL ages from polymineral grains are subject to anomalous fading and may be expected to underestimate ages for last glacial and interglacial loess-paleosol sequences.

Post-IR OSL dating For polymineral grains, the total OSL emission originates from fine-grained quartz and feldspar grains, and their relative contribution depends on the previous IR stimulation. Wang et al. (2006) applied a linear modulation technique to identify the parameters for the post-IR OSL peak-shaped curve, and found that the effect of prior IR exposure on the post-IR OSL signal is significant and after a certain amount of time (e.g. 300 s) the fast component post-IR OSL signal approaches that of the quartz OSL signal, and after even longer IR bleaching times, the post-IR OSL signal still maintains a significant 'slow component' signal, which can be removed by subtraction for dating purposes. It is therefore necessary to determine plateau region for a plot of post-IR OSL ED values versus IR exposure to determinate the appropriate IR exposure time prior to measurement.

2.3.2.4 OSL Dating Applications

Holocene soils Roberts and Wintle (2001) used a double SAR protocol on fine-grained polymineral samples (4–11 μm) for a high-resolution OSL dating study of a terraced loess section (Duowa, Qinghai Province, 35°39′N, 102°38′E) on the western margin of the Loess Plateau (LP). This site preserves a continuous and varying accumulation of dust deposited throughout the Holocene. The authors found that from 12 ka to 2.5 ka the accumulation rate stayed around 0.2 mm/yr. After 2.5 ka it increased to ~0.8 mm/yr and then to 3.4 mm/yr for the past 0.68 ka. Lai and Wintle (2006) used the accumulation rate changes determined by OSL ages at the Yunbao section to locate the boundary between the Pleistocene and Holocene at (13.49 ± 1.15) ka. Zhao et al. (2007) also used OSL dating to obtain a chronology for a loess-paleosol sequence in the Guanzhong Basin on the LP. Two relatively dry episodes in the Holocene (3.7–2.4 ka and 1.2–0.8 ka) were found at the site, and changes in the sedimentation rate through the section suggest that agricultural expansion may have accelerated dust deposition. Stevens and Lu (2010) dated five loess-paleosol sections on the LP, and observed that magnetic susceptibility increased abruptly and dramatically between 9 and 6 ka ago. The authors suggested that the Holocene 'optimum' in the region may be a more recent phenomenon than previously thought and observed that this increase in summer-monsoon intensity significantly post-dated the insolation peak at 11.5 ka.

Last Glacial loess An advantage of the OSL technique dating loess is that age estimates are related to the most recent sunlight exposure which, in the context of the Asian interior, means dust deposition from the last glacial period. OSL results from the Luochuan loess section show that dust accumulation was apparently continuous for the period from 74 to 12 ka (Lu et al. 2007). Based upon the OSL dates, they found that there was a high average accumulation rate of loess during the creation of the last interstadial paleosol (L1–2(S)), corresponding to MIS 3, not only in L1-1 and L1-3 in cold-dry glacial conditions (MIS 2 and 4) (Lu et al. 2007). This study serves as an example of how OSL dating can provide an independent time scale for the study of loess deposition from the last glacial cycle.

Moveover, Stevens et al. (2006), Lu et al. (2006) and Stevens et al. (2007) claimed to identify significant hiatuses in Chinese loess sections based on high-resolution OSL dating, but their claims lacked verification in the field. Lai et al. (2007) applied OSL dating to the study of the Yuanbao section from 51 ka to 17 ka, on the western edge of the LP and found no evidence of hiatuses at the scale of their investigation. Lai (2010) also dated Luochuan loess and published OSL ages using coarse silt-sized (45–63 μm) quartz, and also proved that this loess section is continuous over the last glaciation. More recently, OSL-dated loess sections from Gulang and Jingyuan on the northwest LP demonstrated that loess deposition was uninterrupted during the last glaciation, and the OSL dates compared well to other independent age estimates (Sun et al. 2010b, 2012).

Buylaert et al. (2008) dated the Zhongjiacai, Tuxiangdao and Ledu sections and found continuous sedimentation through the last glaciation in the Zhongjiacai loess at Baicaoyuan town, and also for the Ledu section. The Tuxiangdao section proved to be an exception however, with a hiatus at about 20–30 ka. This is understandable as the position of the Tuxiangdao profile is peripheral to the central CLP.

2.3.3 Chronological Control Using ^{14}C dating

2.3.3.1 Introduction

For more than 60 years, radiocarbon (^{14}C) dating has been a valuable tool for determining chronological frameworks in geologic deposits from the late Quaternary. Numerical age control using radiocarbon has made it possible to determine sedimentation rates in dust and lake deposits; and quantify magnitudes of environmental and geomorphological change through time. The 'absolute' numerical chronologies that radiocarbon dating provides also make it possible to study climate variability through time, by correlation and comparison of multiple proxy records. Radiocarbon dating of the CLP began using decay-counting techniques with proportional or liquid scintillation counters in samples containing grams of carbon. The decay-counting method remains a viable technique, but for sediments its application has always been limited, due to the small amounts of carbon in sediments (typically <2%). This limitation was lifted with the advent of accelerator mass spectroscopy (AMS), which can be used to date sub-milligram quantities of carbon. Before the 1990s, Chinese scientists sent their samples overseas for AMS radiocarbon measurement, as there was no AMS facility in China. This began to change, first with the establishment of a specialized small sample (100 mg) processing line in Xi'an in the early 1990s (Zhou et al. 1994b), and later with the acquisition of a dedicated AMS facility, installed at the Xi'an AMS Center (Zhou et al. 2006).

The establishment of accurate ^{14}C chronologies for well-defined loess-paleosol sequences and limnetic facies deposits is critical to our efforts to understand global climate change during the late Quaternary. An initial ^{14}C chronological framework for the loess plateau in China was established in the mid-1980s (Zheng 1984; Qiao et al. 1987). Many ^{14}C results indicated possible younger contamination. The approach taken at the Xi'an Laboratory of Loess and Quaternary Geology has been first to investigate all of the available potential materials for radiocarbon dating from each site, and then to evaluate their suitability for a particular chronology. In practice this means comparing radiocarbon dates from wood, charcoal, plant matter, shell, carbonate, sediment organic matter, pollen, peat for both loess-paleosol sequences and limnetic facies. Over these years this site-specific approach has led to the development of a variety of specialized pre-treatment methods described here. In addition, we present some key findings from sites in the arid and semi-arid regions of China.

2.3.3.2 Datable Material from Loess-paleosols and Other Sediments

The reliability of a ^{14}C date depends on the material being analyzed. Radiocarbon age determinations rely on the assumption that the proportion of radiocarbon to stable carbon in a sample can be related (directly or indirectly) to the known proportion of atmospheric radiocarbon at the time the sample was deposited. It was recognized from the beginning that some degree of post-depositional alteration can occur (Libby 1955) and by studying each sample type we can better understand this process and how to correct it. The goal is to determine which sample types and sample pretreatments produce the most reliable dates—that is, to isolate a carbon fraction that is most resistant to post-depositional alteration.

Wood, charcoal and plant matter Wood, charcoal and plant matter are excellent candidates for radiocarbon dating because they derive their ^{14}C directly from the atmosphere, and they are frequently well preserved in sediments. Wood, charcoal and plant matter can be picked under a microscope and treated using acid, base and acid (AAA pretreatment), in order to remove calcareous material and mobile humic acids (Olsson 1980; Zhou et al. 1990b; Donahue 1993). Head (1979) concluded that cellulose is the most inert and reliable chemical component for ^{14}C dating. Given a sufficient amount of wood or plant matter, it is also possible to extract cellulose by leaching the sample with sodium chlorite ($NaClO_2$) following the AAA pretreatment (e.g. Zhou et al. 1990b)

Shell A common mollusc found throughout the LP in China is the drought-enduring and cold-resistant genus, *Cathaica* Möllendorff. Zhou et al. (1999) have tested the suitability of dating such molluscs, collected from semi-arid and arid regions in China. Most of the snail species studied produce aragonite shells which are intrinsically unstable under earth-surface conditions (Jamieson 1953). This instability leads to post-depositional re-crystallization that can incorporate older limestone-derived bicarbonate into the shells, and hence skew their radiocarbon ages. X-ray diffraction provides an effective screening tool for ^{14}C dating, and this is routinely used to detect the presence of calcite in recrystallized samples from both paleosol-loess and limnetic deposits (Zhou et al. 1999).

Carbonate On both the Tibetan Plateau and LP, it is common for evapotranspiration to exceed precipitation which facilitates the formation of pedogenic carbonates in the sediments. Zhou et al. (1992) have found a good correspondence between the age of pedogenic carbonate and the humin fraction from the same stratigraphic horizon in the Baxie loess profile. In that study, primary and secondary carbonates were collected by successive CO_2 evolution, using 50% acetic acid and 10% HCl solutions, respectively to derive the reliable ^{14}C dating results for the loess-paleosol deposits.

Sediment organic matter In the absence of wood or charcoal, organic matter from paleosols is often used for ^{14}C dating. Owing to the presence of younger soluble organics introduced through agricultural activity or from other modern sources, organic matter in soil may only provide a minimum age (Head et al. 1989; Zhou et al. 1990b). The AAA pretreatment method can be used to chemically differentiate sediments into two primary fractions: humic acids and humin. Humic acids are operationally defined as the organic fraction that is soluble in the alkali step of the AAA pretreatment. For silt and clay-rich sediments this fraction may produce reasonable ages (e.g. Head et al. 1989; Zhou et al. 1992). Organic solvent extraction can be also used to remove the modern contamination and obtain a reliable fraction for dating. Head et al. (1989) found that the NaOH-insoluble fraction of paleosols, following solvent extraction, yields the most reliable results.

Pollen Pollen concentrated from sediments is potentially an excellent material for radiocarbon dating, for the following reasons: 1) pollen is produced annually and deposited *in situ* in sediments; 2) its exine, or outer layer, is composed of sporopollenin, which is very stable and does not exchange carbon with sediments; 3) pollen undergoes limited translocation in loess and paleosols, <10 cm (Wu et al. 1992), because within the loess/desert belt (in the arid steppe environment) soil development is generally very weak, although limited mixing and percolation do occur during pedogenesis (Dimbleby 1985); 4) pollen is an abundant plant fossil, well preserved in sediments, and AMS allows for the analysis of 100 µg of carbon or less; 5) pollen is predominantly terrestrial, and its age should not be influenced by reservoir effects (Regnéll 1992). Most studies employing pollen dating focus on lake sediments and peat, where pollen concentrations are high. Zhou et al. (1997) extended this work to eolian sediments and established an extraction method for pollen from loess. Although pollen concentrates have excellent chemical properties as a radiocarbon archive, they come with the proviso that the physical ease by which they may be transported must be considered. Pollen dates are not uniformly reliable where the winter monsoon is strong.

Peat Peats occur in relatively moist and isolated biomes, scattered throughout the semi-arid regions of China, especially along the semi-arid to semi-humid environmental boundary. Peats are often produced by annual growth of sedges common to central China, and may be deposited continuously over long time periods. They can provide excellent age control for paleoclimatic studies, owing to the significant amount of carbon that they contain, as for example, at the Midiwan site (Zhou et al. 1996). Peats require careful pretreatment to produce consistently reliable ages. These questions were carefully discussed in Zhou et al. (1996).

2.3.3.3 Holocene Stratigraphic Divisions Using ^{14}C Dating in Loess-paleosol and Eolian Sand-paleosol Sequences in China

Using the techniques described above, a growing number of loess-paleosol sequences spanning the last 10 ka have now been dated. These profiles provide valuable records of global climate change, including the effects of human occupation. Six examples of these records from across the LP are shown in Fig. 2.11. The sections are: 1) the Baxie profile (35°45' N, 103°35' E) (An et al. 2000a) at the western margin of the plateau; 2) the Yangtaomao profile (38°48' N, 110°27' E) (Zhou et al. 1996) on the northern margin of the plateau; 3) the Luochuan profile (35°45' N, 109°25' E) (Zhou et al. 2007b) from the central plateau, 4) the Qishan profile (34°27', N107°45' E) (Huang et al. 2002a) from the southern plateau; 5) the Baimapo profile (34°10' N, 109°19' E) (Zhou et al. 1994a), southern plateau; and 6) the Beizhuangcun profile (34°30' N, 109°30' E) (Zhou et al. 1990b) from the southern plateau. The susceptibility curves for the six profiles in Fig. 2.11 reflect the common tendency for higher SI values between 10 and 5 cal ka BP, when paleosol complexes developing on the LP, and lower SI values after 5 cal ka BP, when loess deposition increased. All of the profiles studied show that the Holocene optimum occurred between 10 and 5 cal ka BP (Fig. 2.11), and that more arid conditions prevailed afterwards. Superimposed on this broad trend are distinct geographic differences expressed at different sites, reflecting temporal-spatial variability in climate conditions across the plateau during the Holocene.

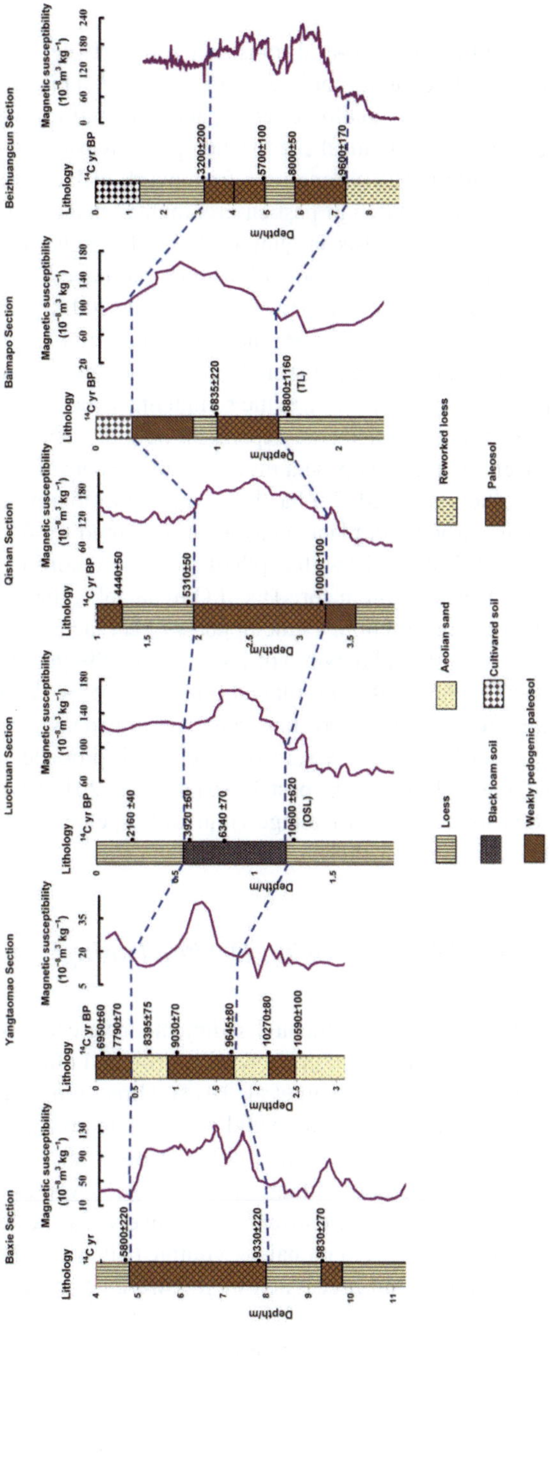

Fig. 2.11 ^{14}C dating results of six sections on the CLP

2.3.3.4 Case Study for Correcting the Old Carbon Effects for Qinghai Lake Sediments

Qinghai Lake is situated in the semi-arid transition zone between inland arid and humid monsoon areas, and lies at the confluence of the East Asian monsoon and the Westerlies. Because of its unique geologic setting and potential for high-resolution records, it makes Qinghai Lake an ideal site to study past environmental changes in an inland continental setting. The chronologic framework of lacustrine sediments is a primary component for research of past environmental changes. Unfortunately however, the old carbon effect exists in Qinghai Lake. It results in the ^{14}C dates that are much older than the true ages and various methods have been employed to correct for this effect and to get reliable ages (Yu and Kelts 2002; Shen et al. 2005; Wang et al. 2007a; Yu et al. 2007) The results from these studies are discrepant. Moreover, their established chronologies were based on too few ^{14}C dates and without considering the variation of lithology in the sedimentary sequence. Because the radiocarbon ages from Qinghai Lake core depend on both the decay of ^{14}C and on the local hard water effect and thus pose a multivariable geophysical problem. An analytical approach (Zhou et al. 2007a) has been applied to this problem at Lake Qinghai. Total organic carbon and plant seeds were measured by AMS from core sequence (1Fs, 36° 48' N, 100° 08' E) with depth of ca 18 m. According to the differences in lithology and TOC content in core 1Fs of Qinghai lake, three sections were identified for age-model construction: 1) the upper-499 cm lacustrine silty clay to clay; 2) the middle unit of silty clay with silt layers from 499–901 cm; and 3) the lower 901–1861 cm silty clay, loess-like silt and fine sand layers. We applied linear regressions separately for these three portions, and determined an average hard water effect of about 135, 1143 and 2523 yrs, respectively. This produced a reliable ^{14}C chronology for the Qinghai Lake for the past 32 ka (An et al. 2012), which is highly consistent with the absolutely dated Dongge-Hulu cave speleothem $\delta^{18}O$ records (Dykoski et al. 2005; Wang et al. 2008).

2.3.4 Chronological Control Using Orbital Tuning

Numerous studies of oceanic and continental sediments have established a cause-and-effect relationship between variations in Earth's orbit and climate (e.g. Imbrie et al. 1984; Shackleton et al. 1990; Ding et al. 1994). Direct tuning proxy indicators of the EAM to orbital parameters is a valid approach to construct loess chronostratigraphy (Ding et al., 1994; Lu et al. 1999; Heslop et al. 2000; Ding et al. 2002). Previous studies have proposed that winter monsoon variability is tightly linked to variation in the Northern Hemisphere ice sheets via the Siberian High and thus closely related to changes in global ice volume (Liu and Ding 1993; Ding et al. 1995). Orbital obliquity and precession targets were generated from the orbital solution of Laskar (1990) employing the SPECMAP-defined lags of −69° and −78° for obliquity and precession, respectively (Imbrie et al. 1984). Using these

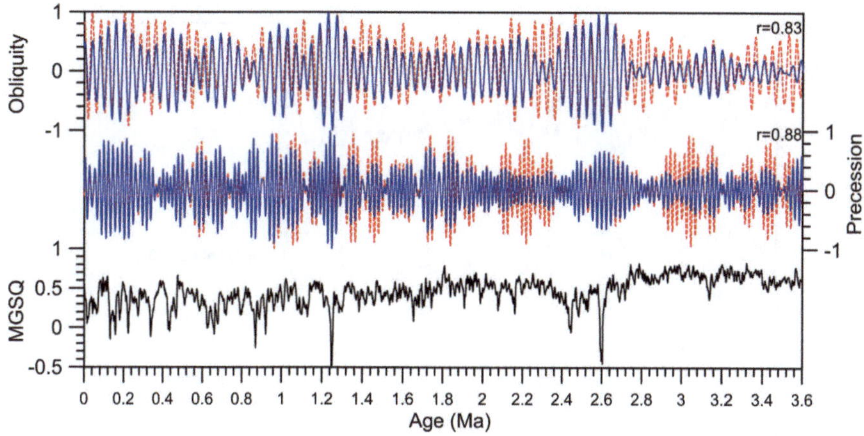

Fig. 2.12 Mean grain size of Quartz (MGSQ), precesion and obliquity variations used in the astronomical chronology (Sun et al. 2006a)

targets, the tuning method of Yu and Ding (1998) was utilized to generate an astronomical chronology for the loess-based proxy records. This approach assumes that winter monsoon strength maxima (MGSQ) are in phase with large global ice volume throughout the past 3.6 Ma, a reasonable assumption given our current understanding of the dynamics involved (Ding et al. 1995). The orbital-tuning procedure involves simultaneously tuning the 41-ka and 21-ka components of the grain size (MGSQ maxima) to the lagged obliquity minima and precession maxima (for a detailed description of the tuning method see Yu and Ding 1998 and Ding et al. 2002).

Evaluation of the astronomical chronology is based on two independent approaches. One way is to compare the filtered components of MGSQ with the lagged obliquity and precession curves (Fig. 2.12). The filtered 41-ka and 21-ka components of the LT and ZJC grain size records are in phase with the obliquity and precession curves, and exhibit similar amplitude modulation. The correlation coefficient r between the filtered 41-ka component of MGSQ and obliquity record is 0.85 for LT and 0.83 for ZJC. The correlation coefficient between the filtered grain-size and orbital curves at the precession frequency band is 0.83 for LT and 0.79 for ZJC. High correlation coefficients and similar amplitude modulation suggest that the orbitally tuned chronologies are well-constrained by the obliquity and precession records. Whereas the good phase match is a function of the tuning, the amplitude matches are independent of the tuning, indicating a mechanistic link between orbital insolation forcing and loess-paleosol variance as recorded by MGSQ.

Comparing the grain-size time series with orbital obliquity and precession curves provides additional support for the chronological model. Cross-spectral analyses of MGSQ with the orbital curve (ETP) were performed separately for six time intervals: 0–0.6 Ma, 0.6–1.2 Ma, 1.2–1.8 Ma, 1.8–2.4 Ma, 2.4–3 Ma and 3–3.6 Ma. Cross-spectral results show that for the last 3.6 Ma, the MGSQ time series for both the LT and ZJC profiles closely match the orbital signals over the obliquity (41-ka)

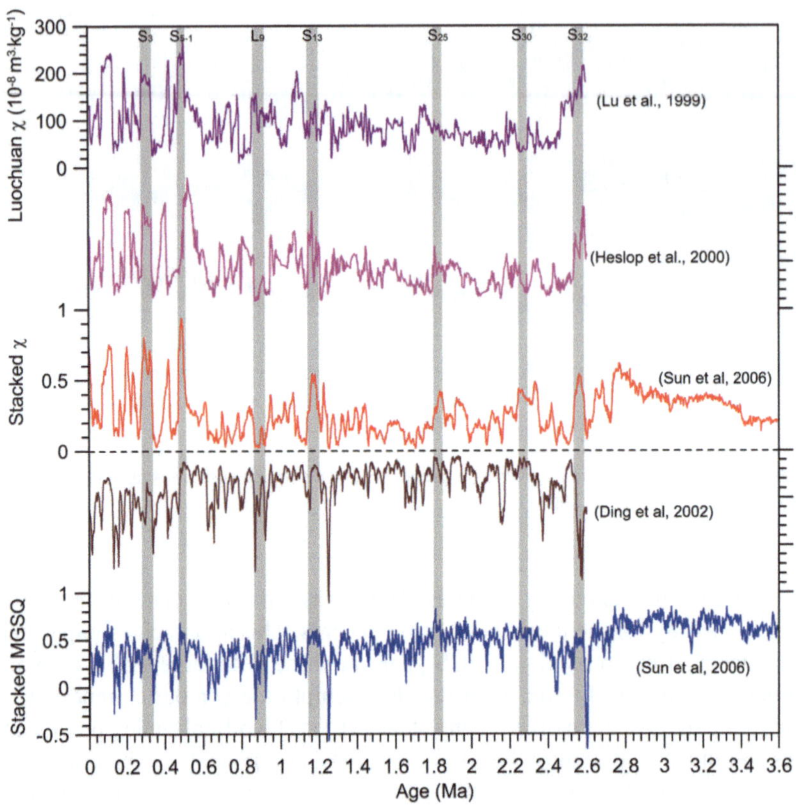

Fig. 2.13 Comparison of different astronomical chronologies. (Lu et al. 1999; Heslop et al. 2000; Ding et al. 2002; Sun et al. 2006)

and precession (23- and 19-ka) bands. The coherence in the obliquity and precession periodicities is well above the 95% significance level during most time intervals, further suggesting the validity of our orbital-tuned time scales. The coherence over the eccentricity (100-ka) band is more variable, dropping below the 95% significance level within some time intervals (e.g., 2.4–3 Ma).

The astronomical timescale yields several characteristics indicating improvements relative to recently published age models spanning the last 2.6 Ma (Lu et al. 1999, Heslop et al. 2000; Ding et al. 2002). Comparison of the age models developed for LC, LT and ZJC sections with the composite marine oxygen isotope record Shackleton et al. (1990, 1995) indicates that the boundary ages of S_3, S_5, L_9, S_{13}, S_{22}, S_{25}, S_{28}, S_{30-31} and S_{32} are different among these age models (shown as shaded areas in Fig. 2.13). Based upon the comparison of the records with the $\delta^{18}O$ curve and the sedimentation rate (SR) derived from these age models, we find that the bottom of L_9 was extended to 0.95 Ma, whereas S_{13}, S_{22}, S_{25} and S_{30-31} were compressed to varying extents in the Heslop2000 age model (Heslop et al. 2000). Furthermore, S_{5-1} is assigned to the lower MIS 13 and MIS 14. In the Ding2002 age model (Ding et al. 2002), S_3, S_{25} and $S_{30–31}$ were compressed into a small amount of time, and the top of S_{28} was extended to 2.05 Ma. Moreover, the bottom silty loess layer (L_{33}) was wrongly correlated to interglacial oxygen isotope stages (MIS 101–103). In contrast, the Sun2006 age model better correlates paleosol units to one or more interglacial isotope stages, while loess layers are better correlated to glacial isotope stages, as expected from a paleoclimatic basis. Significant improvement of Sun2006 age model relative to Heslop2000 and Ding2002 age models can be clearly identified within the time interval 1.6–2.6 Ma, as indicated by the less variable sedimentation rates and the better correlation of MGSQ with the marine oxygen isotope record (Sun and An 2004).

Although orbital tuning is a valid approach to construct an astronomical chronology for the Chinese loess, a proper target curve is the key to assuring its reliability. The target curve is dependent on the forcing-response relationship between the orbital parameters and the loess-based monsoon proxies, i.e., the phase relation between the loess proxies and the solar insolation at the obliquity and precession frequencies. The current orbital-tuning approach adopts the same stable phase lags (8 ka for obliquity and 5 ka for precession). This application might be revised in the following two ways: (1) the phase lag might be decreased in the early Pleistocene due to limited ice-sheet cover in high-latitude of northern Hemisphere (Lisiecki and Raymo 2005), and (2) the phase lag also might be decreased in the latest Pleistocene as suggested by the high-resolution, 'absolutely' dated speleothem records (Wang et al. 2008; Cheng et al. 2009).

Besides orbital tuning, different models were also applied to construct the loess chronology. Magnetic susceptibility was firstly used to approximate the timescale of loess sequences (Kukla et al. 1988), based on the assumption that magnetic susceptibility values are influenced by the dilution of dust input. The magnetic susceptibility model was challenged by largely pedogenenic-induced magnetic enhancement of Chinese loess-paleosoil sequences (e.g. Zhou et al. 1990a; Evans and Heller 1994). A grain size model was also employed to construct an independent chronology of Chinese loess (Porter and An 1995; Vandenberghe et al. 1997), due to the close relationship between grain size (e.g., coarse fraction content) and sedimentation rate over the last two glacial cycles (Ding et al. 2001c). However, correlation coefficients between dust flux and grain size are not stationary (Sun and An 2005), and thus additional time controls are needed to generate a reliable chronology back to the Pliocene.

2.4 Proxies Indices of the Paleomonsoon

Geological, biological and chemical evidence derived from loess-paleosol sequences of the Loess Plateau for the last 2.6 Ma indicate that loess results from the deposition of dust transported by the northerly winter monsoon. Particle-size, aeolian flux and detrital quartz composition can therefore be used as proxy indices of the winter monsoon (Liu 1985; Kukla 1987; Kukla et al. 1988; Kukla and An 1989; An et al. 1990; An et al. 1991a; Ding et al. 1994; Xiao et al. 1995). In contrast, the interbedded paleosols were formed by pedogenesis under summer monsoon conditions. Magnetic susceptibility, organic carbon content, stable carbon isotope ratio, ^{10}Be content, chemical weathering index and carbonate content of paleosols can therefore be used as proxy indices of the summer monsoon (An et al. 1991b; Shen et al. 1992; Lin and Liu 1992; Jia et al. 1995; Balsam et al. 2004; Ning et al. 2008; Zhou et al. 2007b).

2.4.1 *Sedimentology*

2.4.1.1 Grain Size

Grain-size composition of eolian deposits is determined by three factors: (1) the property of dust sources, (2) the wind strength, and (3) the sink-to-source distance (Pye 1987). The grain size of last glacial loess deposits decreases systematically from northwest to southeast over the Chinese Loess Plateau, implying that the loess mainly originated from the sandy deserts of the Gobi in northwestern China and transported by the near-surface northwesterly wind which can be regarded as the winter monsoon (Liu 1985; An et al. 1991b). Historical records of dust influx together with observations of modern dust storms and meteorological satellite images suggest that the winter monsoon and Northern Hemisphere westerlies are the main agents of dust transportation (Zhang 1982; Liu 1985). Stronger winds are associated with more dust storms, coarser particle size and larger dust input to the Loess Plateau. Therefore, the grain size of loess deposits is widely employed as a proxy indicator of wind strength to reconstruct Quaternary variations of the East Asian winter monsoon. By utilizing mathematical methods, different grain-size populations have been decomposed from the bulk grain-size data to infer the dust transported by westerly jet over a long-range and by winter monsoon from a proximal source (Sun et al. 2004; Qin et al. 2005; Prins et al. 2007).

Various grain-size parameters, such as median/mean grain size (An et al. 1991a; Xiao et al. 1992; An 2000; Ding et al. 2002; Sun and An 2005), grain-size ratios (Vandenberghe et al. 1997; 2004) and coarse-fraction content (Porter and An 1995; Lu et al. 1999; Zhang et al. 1999), are widely employed for the reconstruction of winter monsoon variations. Moreover, grain-size parameters are also used to determine variations in the extent of the deserts (Ding et al. 1999b) and to ascertain the origin of the late Tertiary red-clay deposits (Ding et al. 1998b; An et al. 2000b; Lu et al. 2001; Yang and Ding 2004; Wen et al. 2005). By comparing the grain-size distributions of the loess, paleosol and red clay deposits between the paired bulk and quartz samples, post-depostional pedogenic modification can be explored (Sun et al. 2006b). Grain-size parameters derived from the bulk and the quartz samples, such as mean grain size (MGS), coarse-fraction content and U-ratio, all exhibit similar variations in the loess-paleosol sequence (Quaternary, 0–2.6 Ma), whereas these same grain-size parameters of quartz samples show different variability from those of bulk samples in the red clay sequence (late Miocene to Pliocene, 2.6–7 Ma).

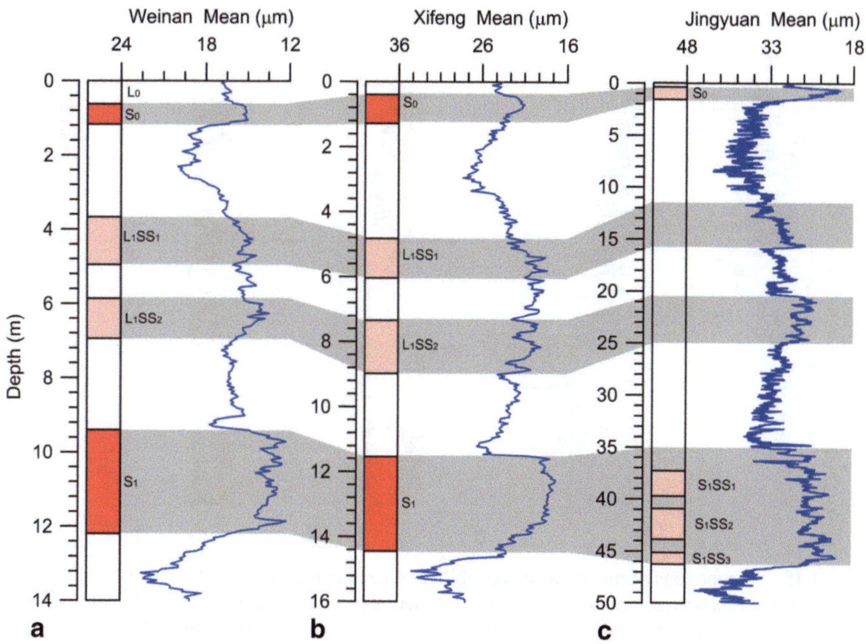

Fig. 2.14 Mean grain-size variations of three loess sections (Sun et al. 2010b)

MGS variations can be used to determine the relative intensity and gradient of the East Asian winter monsoon across the LP. The glacial-interglacial difference in the MGS variation is larger at the JY section (~30 μm) than the WN (~14 μm) and XF (~18 μm) sections (Fig. 2.14). Moreover, the average MGS value shows a twofold increase from the central LP to the northwest LP during glacial (~14 μm) relative to interglacial periods (~7 μm). These spatial patterns are consistent with investigations on the spatial variations of grain size for the last two glacial cycles (Yang and Ding 2004; Nugteren and Vandenberghe 2004), indicating that: 1) glacial-interglacial contrast in the winter monsoon fluctuations is higher in the northwest LP than in the central LP, and 2) southeast-northwest gradient of the winter monsoon variation is larger during glacial periods than interglacial periods.

In contrast to the grain-size distributions of quartz samples, grain-size distributions of bulk samples are skewed toward the fine portion and are greater in the fractions of clay and fine silt. Stepwise increases in the similarity of the grain-size distributions between paired bulk and quartz samples from late Tertiary red clay sequence to the Quaternary loess-paleosol sequence indicate that the relative intensity of pedogenic modification is stronger in the redclay sequence than in the overlying loess–paleosol sequences (Sun et al. 2006b). The effect of pedogenic modification on the paleomonsoon reconstruction is addressed by comparing the grain-size variations of the bulk and quartz samples (Fig. 2.15). For the loess–paleosol sequence,

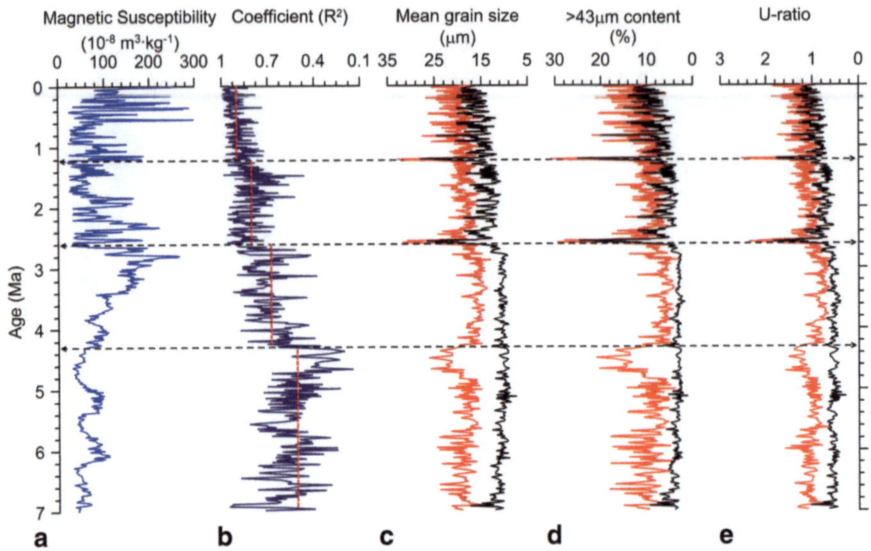

Fig. 2.15 Comparisons of magnetic susceptibility (A) and correlation coefficient (B) of the grain-size distributions between paired bulk and quartz samples with mean grain size (C), N43 μm content (D) and U-ratio (E) of bulk (*black lines*) and quartz samples (*red lines*) from the Lingtai section. Dashed arrows indicate shifts in the relative intensity of podegenic alteration at the Lingtai section. (Sun et al. 2006b)

mean grain size, >43 μm content and U-ratio of the bulk and quartz samples exhibit similar variations at the glacial-interglacial scales. However, changes in the grain-size parameters for the bulk redclay samples are less distinctive than for the quartz samples. Due to the strongly pedogenic modification to the grain-size distributions of the red-clay deposits, grain size of bulk red clay samples is an imperfect indicator of changes in the dust-transport dynamics of strongly weathered paleosol layers during the late Miocene and the early Pliocene.

To ascertain the advantage of quartz grain size as a reliable indicator of the wind strength (Xiao et al. 1995), we compare mean grain-size variations of the bulk and quartz samples through time (Fig. 2.15). In the loess-paleosol sequences, the mean grain size of both bulk and quartz samples exhibits similar orbital-scale variability, although the mean grain size of quartz particles is relatively coarse compared to the bulk samples. For the red clay, the mean grain size of bulk samples shows little variability, whereas the mean grain size of quartz particles exhibits oscillations very similar to those of the overlying loess-paleosol sequences. Moreover, the mean grain-size difference between the bulk and quartz samples is larger in the red clay than in the overlying loess-paleosol sequences. These differences suggest that pedogenic modification of the non-quartz components is stronger in the red clay than in the overlying loess-paleosol deposits (Sun et al. 2006b).

Although the particle size of eolian deposits might be influenced by factors such as the extent of the source areas and/or aridity, source-sink distance and paleowind intensity (An et al. 1991a; Xiao et al. 1992; Rea 1994; Ding et al. 2001c), comparison of the quartz mean grain size (QGS) with dust flux suggests that the aridity in inland Asia and paleowind intensity were weakly related prior to 2.7 Ma, but became gradually coupled over the last 2.7 Ma (Sun and An 2005). There has been no significant change in the provenance of the red clay (Sun 2005a; Wang et al. 2007b; Sun 2010), and therefore the Pleistocene interpretation of the QGS as an indicator of the wind strength (Xiao et al. 1995) can be extended into the Pliocene and Miocene (Sun et al. 2006a; Sun et al. 2006b).

2.4.1.2 Color Reflectance

Changing color, one of the most striking features of Chinese loess-paleosol sequences, is closely linked to the degree of soil development (Liu 1985; Porter 2000; Ji et al. 2001; Chen et al. 2002). The light-yellowish loess layers (weakly weathered soils) were deposited during cold and dry periods, whereas the interbedded brown-reddish paleosols units (strongly weathered soils) were formed under warm and wet conditions (Kukla and An 1989; An et al. 1990). On the Loess Plateau, alternations of light-yellowish loess and brown-reddish paleosol units can well document distinctive variations of the East Asian paleomonsoon (An et al. 1990; An et al. 1991a; Ding et al. 1993; Gao et al. 1992). At typical loess sections such as Luochuan, Xifeng, Baoji, and Lingtai, 33 loess layers (L_1–L_{33}) intercalated by palesol units (S_0–S_{32}) have been investigated intensively to reconstruct cyclic oscillations of the winter and summer monsoons over the last 2.6 Ma (Kukla and An 1989; An et al. 1990; Ding et al. 1993; 1999a). Moreover, several strongly weathered paleosol layers (e.g., S_1 and S_5), distinguished by an intense reddish color, have been widely employed as markers for stratigraphic characterization and correlation over the Loess Plateau (An and Wei 1980; Liu 1985; Han et al. 1998).

Previous studies suggested that color difference between loess and paleosol is influenced by multiple factors including mineralogy (Porter 2000), grain size (Chen et al. 2002), and the formation of pedogenic minerals (e.g., iron oxides, organic matter and carbonate) (Gao et al. 1992; Fang et al. 1999; Porter 2000; Yang and Ding 2001, 2003; Chen et al. 2002). For example, during interglacial times, relatively warm and humid climate enhances the production of magnetic minerals (Zhou et al. 1990a; Maher 1998), strong leaching of carbonate (An et al. 1991b), and production of organic matter (Lin and Liu 1992) in soils. All these post-depositional processes can redden and darken the loess color (Chen et al. 2002; Shen et al. 2006). Porter (2000) employed a gray-scale proxy from digital photographic images and suggested that dominant light minerals in eolian dust (i.e., quartz and carbonate) are likely a key factor influencing the loess color. Based on diffuse reflectance spectrophotometry, Ji et al. (2001) suggested that mineralogic changes (i.e., the relative proportion of hematite and goethite) are responsible for the color changes in the loess sequences. However, Chen et al. (2002) used whiteness to characterize the loess-paleosol alternations and confirmed that fine-grained pedogenic components, particularly iron-oxide minerals, are the most conspicuous contributors to whiteness changes. Although previous studies realized that organic matter and carbonate might influence the loess color, their relative impacts were never evaluated quantitatively.

By comparing the color difference between original samples and pretreated samples from two loess-paleosol sequences, impacts of three major geochemical factors (organic matter, carbonate, and iron oxides) on the changing color of Chinese loess were quantitatively evaluated by Sun et al. (2011). The results indicate that iron oxides and organic matter would darken the loess, whereas carbonate could lighten it. Their impacts on the loess lightness decrease gradually from iron oxides and organic matter to calcium carbonate. Changes in the redness, however, are controlled dominantly by the types and concentrations of iron oxides (especially hematite). Thus, loess redness can be regarded as a more straightforward proxy of the summer monsoon intensity than the lightness.

Though Ji et al. (2001) pointed out that neither susceptibility nor redness may indicate the complete extent of soil development and suggested that the combination of redness and susceptibility may provide a reasonable estimate of pedogenesis, a direct combination of these two proxies has not been adopted by the loess research community so far. Since these two proxies are associated with the concentrations of different iron oxides formed during pedogenesis under different climatic conditions, a new summer monsoon index was generated by stacking these two proxies in order to obtain a thorough understanding of the summer monsoon variability (Sun et al. 2011). A good comparison of the stacked summer monsoon index with a first-derivation ratio estimated from the color reflectance suggests that both proxies can be used to infer the precipitation variability. Unlike the magnetic susceptibility and redness variations, the stacked summer monsoon index clearly demonstrates a two-phase enhancement of the summer monsoon variability during the penultimate deglaciation (Sun et al. 2011).

A good correlation between color parameters (e.g., gray scale, whiteness, redness) and magnetic susceptibility (a proxy for the summer monsoon intensity) suggests that color of Chinese loess can be used as an alternative and convenient proxy to reflect temporal and spatial variability of the summer monsoon at orbital to millennial time scales (Fang et al. 1999; Porter 2000; Chen et al. 2002; Yang and Ding 2003; Ji et al. 2004) (Fig. 2.16). However, apparent discrepancies are also evident among different summer monsoon-related proxies. For example, increase of the hematite content (mostly related to redness) lead to enhancement of magnetic susceptibility during the L_2/S_1 transition and the most developed paleosol complex (S_5) (Guo et al. 2000; Ji et al. 2001; Torrent et al. 2007). Long-term trends and glacial-interglacial variability between magnetic susceptibility and the concentration of spectrally identified hematite are also different (Ji et al. 2001; Balsam et al. 2004). In addition, redness and the ratio of free to total Fe_2O_3 for the late Tertiary red clay sequence display significant variability similar to that of the overlying loess-paleosol sequences; whereas magnetic susceptibility of the red clay sequence showes less variability, differing significantly from the large amplitude fluctuations occurred in the Quaternary loess-paleosol sequences (Ding et al. 2001b; Ji et al. 2006).

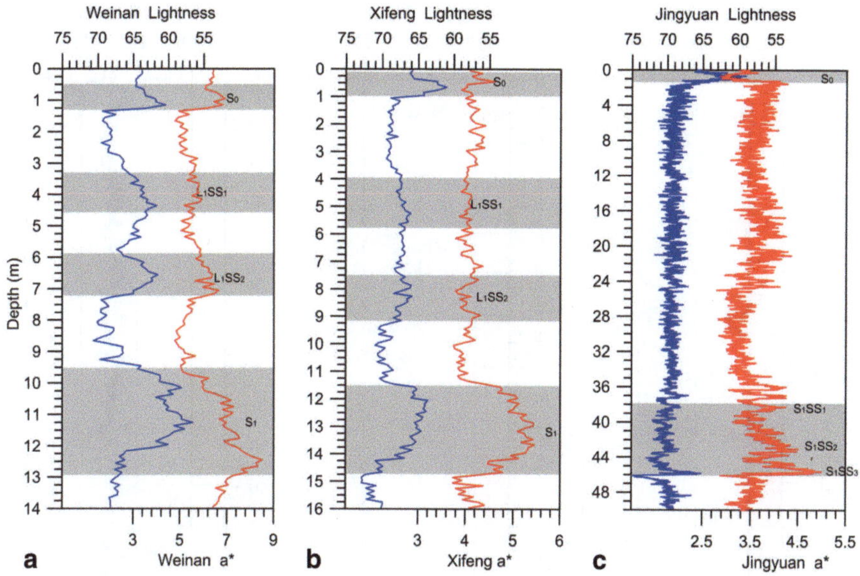

Fig. 2.16 Variations of magnetic susceptibility, lightness and redness of three loess sections

2.4.1.3 Mineralogy

Carbonate content Carbonate content is a function of the soil-forming environment, such as the balance between precipitation-induced leaching intensity and temperature-related evapotranspiration effectiveness (Reheis 1987). A seasonally wet/dry climate rather than the amount of rainfall, with a critical balance between rainfall and evaporation, is considered to be a key factor in driving carbonate dissolution and re-precipitation (Rossinsky Jr and Swart 1993). In the monsoon-affected area (e.g., the Chinese Loess Plateau), variation in carbonate content is controlled mainly by the effective precipitation (i.e., the balance between precipitation-induced leaching and temperature-induced evaporation) (Lu 1981; Liu 1985; An et al. 1991b). Carbonate content has been employed as a precipitation proxy to reconstruct rapid summer-monsoon oscillations as well as the long-term summer monsoon evolution. (Chen et al. 1997; 1999a; Fang et al. 1999; Chen et al. 2007)

At the Lingtai section, the carbonate content exhibits high amplitude variability (0.7%–25%) in the red clay formation, slightly lower (0.2%–21.8%) variability from L_{33} to L_{15}, and low variability (0–19.2%) above L_{15} (Fig. 2.17). The primary difference among the three stages is the gradual decrease in the average carbonate content and the glacial-interglacial variability, suggesting stepwise strengthening of the effective precipitation, particularly during the warm periods. The decreased glacial-interglacial variability of carbonate content in the upper loess-paleosol sequence (S_0–S_{14}) was probably caused by the relatively heavy precipitation during interglacial intervals, resulting in strong dissolution of carbonate not only in the

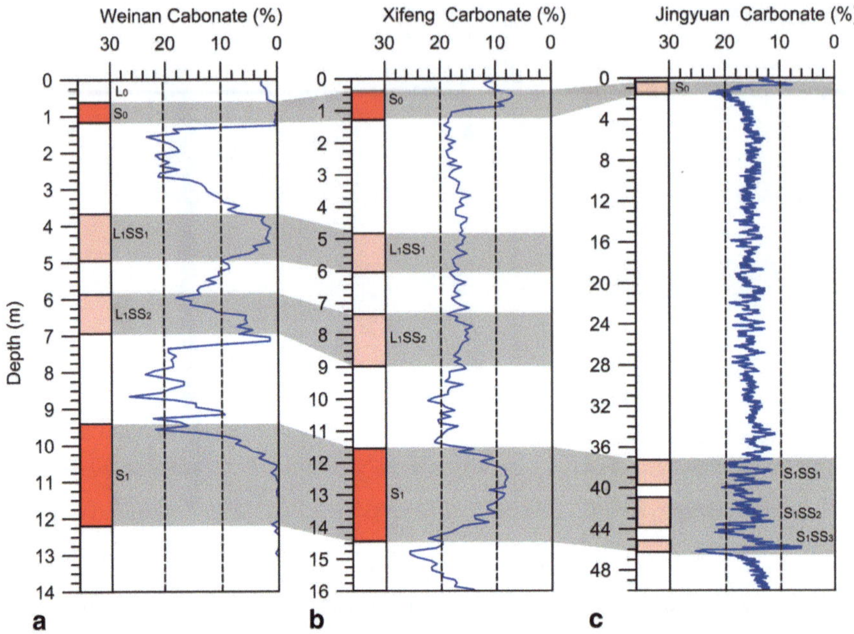

Fig. 2.17 Carbonate content variations of three loess sections

paleosol layers but also in the uppermost part of the underlying loess. In addition, U-series dating and stable isotope analysis of carbonate nodules in Chinese loess suggest that eluviation and reprecipitation of soil carbonates also occurred during glacial stages (Rowe and Maher 2000).

Fed(free iron)/Fet(total iron) ratio The free iron to total iron ratio (Fed/Fet) is a measurement of iron liberated from iron-bearing silicate minerals by chemical weathering relative to total iron. This parameter is widely used by European pedologists (Duchaufour 1983). It has been recognized that the ratio of citrate-bicarbonate-dithionite (CBD)-extractable iron (or free iron) to total iron concentrations is controlled basically by pedogenetic weathering degrees of the loess-soil sequence, and soil weathering is in turn closely associated with monsoonal precipitation (Guo et al. 1998; Ding et al. 2001b). The so-called free iron extracted by the CBD method consists of finely divided hematite and goethite, amorphous inorganic Fe oxides and organic-complexed Fe (Mehra and Jackson 1960). It is possible that the finely divided hematite and goethite may be largely deposited with eolian particles; however, the amorphous inorganic Fe oxides and organic-complexed Fe are produced essentially by chemical weathering. Free-iron contents can also be partly determined by total iron contents of the loess-soil sequence. Therefore, the influence of depositional processes has to be eliminated or largely reduced in developing chemical weathering indices associated with iron contents.

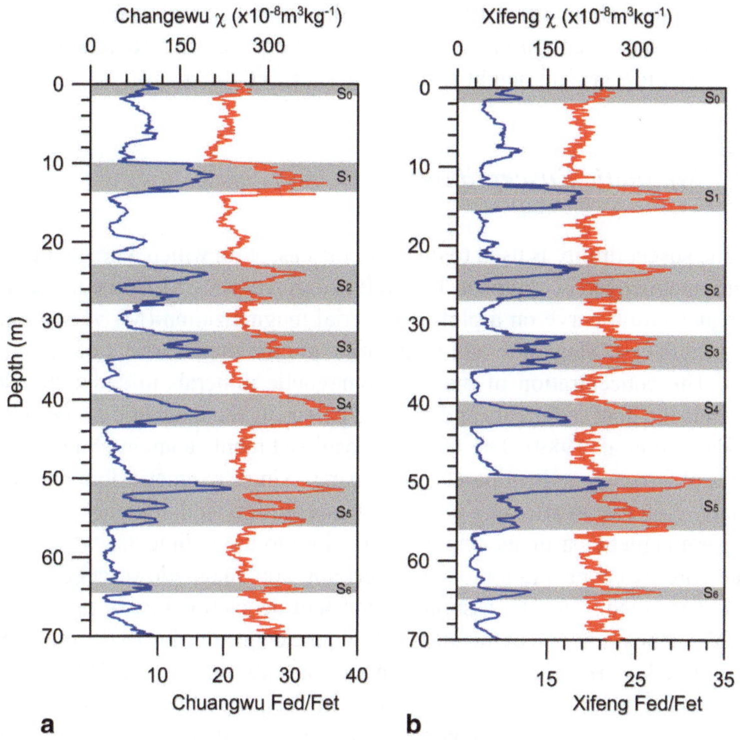

Fig. 2.18 Fed/Fet and magnetic susceptibility variations of two loess sections. (Guo et al. 2009)

The Fed/Fet reveals a series of spatially correlative intervals of lower weathering intensity (LW) in the last-glacial loess, indicating significantly cooler or/and drier conditions. The timings of these LW events in the last glacial period are close to those of the Heinrich layers in the North Atlantic Ocean, indicating that the climate in the area affected by the East Asian monsoon has experienced high-frequency changes more or less synchronous with the Heinrich events (Guo et al. 1996). The Fed/Fet ratio of four loess sections along a north-south transect of the Loess Plateau shows higher values in soil units than in loess horizons and a southward increase. The tempo-spatial variability pattern of the Fed/Fet ratio is in good agreement with the variations in the East Asia summer monsoon intensity identified by other monsoon proxies and soil-development observations. The Fed/Fet ratio thus is a/good proxy indicator of the summer-monsoon evolution (Ding et al. 2001b; Guo et al. 2009) (Fig. 2.18).

The Lingtai Fe_2O_3 ratio had high values during three time intervals: 4.8–4.1 Ma, 3.4–2.6 Ma, and the interglacial periods of the past 0.8 Ma. The increase in summer-monsoon intensity over these three intervals coincides with the well-developed soil characteristics. It is therefore concluded that the East Asia summer monsoon has experienced a non-linear evolution since the late Miocene. The occurrence of a monsoon peak at about 4.8–4.1 Ma implies that the Tibetan Plateau, one of the most important boundary conditions in maintaining the monsoon circulation, must have been uplifted to a critical height at that time (Ding et al. 2001b).

2.4.2 Magnetic Susceptibility

Magnetic susceptibility is used to measure the ease with which a magnetic material becomes magnetized (Thompson 1986). It is defined as $\kappa = M/H$, the initial slope of the magnetization curve on a plot of material magnetization (M) versus magnetic field strength (H). Magnetite and maghemite are the main carriers of the magnetic signals. The concentration of these ferromagnetic minerals in sediments such as loess lends the host sediments and derived paleosols the magnetic properties of the minerals (Liu et al. 1989). The loess-paleosol sediments contain abundant $< 1\ \mu m$ particles of ferromagnetic minerals comprising a single magnetic domain. These are especially effective in affecting magnetic susceptibility and, together with ultrafine super-paramagnetic particles, impart high values to the sediments.

There are two conceptual models describing how magnetic susceptibility of the loess is controlled. The first assumes a constant deposition rate of magnetic particles. If the deposition rate of the loess is high, the concentration of the magnetic particles will be low, resulting in low magnetic susceptibility values. If the deposition rate of loess is lowered, the susceptibility will rise proportionately. Second, eluviation and weathering during pedogenesis leads to the formation of a large number of ultrafine particles and increases the concentration of magnetic particles in the soil, again resulting in higher magnetic susceptibilities in paleosols (Zhou et al. 1990a; An et al. 1991b; Zheng et al. 1991; Han et al. 1991; Jia et al. 1996; Liu et al. 1999b).

Although the mechanism increasing the magnetic susceptibility requires further study, it is widely accepted that the fine ferromagnetic particles formed during pedogenesis will increase magnetic susceptibility in soils and paleosols. It is for this reason that magnetic susceptibility is regarded as a proxy of summer-monsoon intensity.

The earliest susceptibility measurement on Chinese loess was carried out by An et al. (1977). They noticed that the loess and paleosol susceptibility values differed significantly. Heller and Liu (1982) later measured the magnetic susceptibility κ of Chinese loess and paleosols systematically. They compared the κ curve of Chinese loess sequences with the marine benthic oxygen isotope curve for the first time, and pointed out that variations in κ contained information on global environmental change (Heller and Liu 1982, 1984; Heller and Liu 1986). They further noted that κ of the Chinese loess-paleosoil sequence varied with the strength of the warm and humid East Asian summer monsoon, due to pedogenesis, and can thus be viewed as a proxy record of Asian monsoon variability (An et al. 1991a). The κ curve of the Luochuan and Jixian loess-paleosol sequences, which are 100 km from each other, record completely synchronized changes, validating the value of κ as a proxy for environmental change (Han et al. 1991).

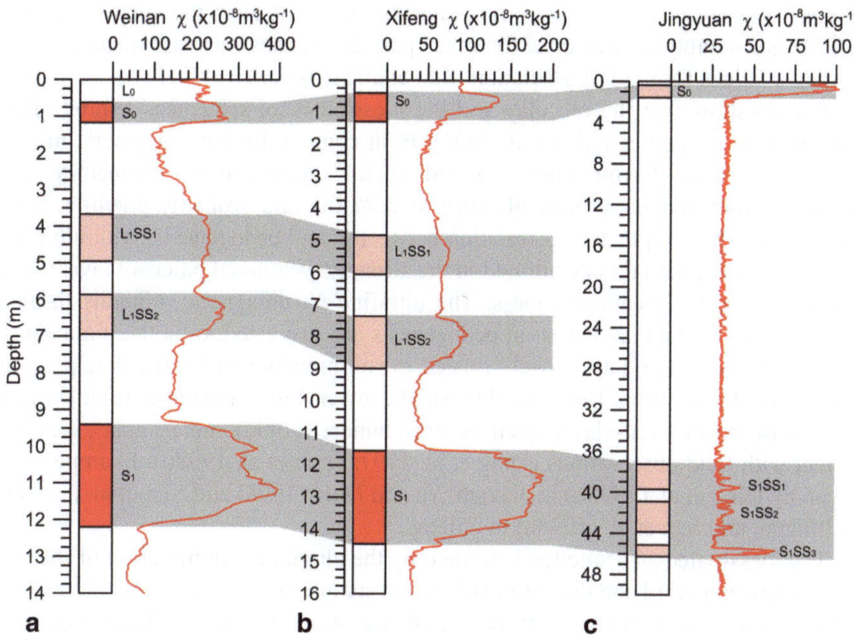

Fig. 2.19 Magnetic susceptibility variations of three loess sections

The above efforts laid the foundation for the study of κ in loess. Magnetic susceptibility as a proxy of paleoenvironment change was widely applied in Quaternary environmental research. Because the magnetic susceptibility of loess and paleosols acts as a proxy index of Asian summer monsoon variation, much research has focused on relating κ and climatic parameters, especially precipitation (An et al. 1991a; Heller et al. 1993; Liu et al. 1995a, 1995b; Porter et al 2001). Lu et al. (1994) investigated the κ values of topsoil samples to study the correlation between the κ and climatic conditions (such as temperature, precipitation), to estimate the average annual precipitation characteristics of different regions, and to make a meaningful attempt at paleoenvironment reconstruction.

Comparing magnetic susceptibility curves of different sections from the Loess Plateau suggests that: 1) these curves have similar glacial-interglacial variations, being higher during warm periods and lower during cool periods; 2) values of κ of loess sections from the northwestern Loess Plateau are lower those than from the southern and eastern Loess Plateau; and 3) κ shows more distinct precessional cycles during the last interglaciation in the loess sequence from the northwestern LP than elsewhere (Fig. 2.19). Variations of climate during the last 600 ka are recorded more obviously in κ curves of sections from central, eastern, and southern parts of the LP than from northwestern parts of the LP (Hao and Guo, 2005). This may be explained as follows. In sections from the central and southeastern LP, values of κ during the interglacial periods are notably higher than during those from cool-dry glacial periods, due to relatively strong pedogenisis associated with warm-humid climate. However, in sections from the northwestern LP, even though κ values for MIS 5 are notably higher than for other periods, no obvious discrepancy exists among κ values for the last glaciation (Sun et al. 2010b).

Studies show that the red clay and the loess-paleosol sequences exhibit similar magnetic properties (Liu et al. 2001). Both contain the same magnetic minerals: low-coercivity ferromagnetic minerals such as magnetite and maghemite, and high-coercivity magnetic minerals such as hematite and probably goethite. Both sequences exhibit a positive correlation between κ and pedogenesis, with ultrafine ferromagnetic material concentrated in the strongly weathered paleosol layers, and not in the weakly weathered loess. The ultrafine ferromagnetic minerals formed by bacterial or chemical action in pedogenesis are suspected to be the main cause of the κ variations and are closely related to soil moisture and paleo-precipitation (Heller and Evans 1995; Liu et al. 1995b). The lower (low κ and upper-middle (high κ) parts of most LP red-clay sequences, even hundreds of kilometers apart, are consistent with field observations (Ding et al. 1997; Ding et al. 1998b). Examples are given by Ding et al. (1998b) in Jiaxian, Yu and Ding (1998) and Sun et al. (1998b) in Lingtai, and Sun et al. (1998a) in Xifeng.

The loess-paleosol sequence is formed by the alternating domination of the dry and cold winter wind and the warm and humid summer wind. The magnetic susceptibility of loess and paleosol can be regarded as a proxy index of Asian monsoon summer variation (An et al. 1991; Liu et al. 1995a). Our data suggest that the red clay has been through similar processes and its magnetic susceptibility has a similar interpretation. The eolian red clay extends the loess-paleosol proxy record of paleomonsoon evolution back into Late Miocene times.

The three above conclusions consistently show that magnetic susceptibility of the loess deposits was influenced by the variation of the East Asian summer monsoon. From the southeastern to northwestern LP, East Asian summer monsoon influences on pedogenesis of loess deposits decrease. Thus, values of magnetic susceptibility of sections from the northwestern LP are much lower than that of sections from the southeastern LP, and magnetic-susceptibility curves of sections from the southeastern LP record more obviously the variation of climate changes than magnetic susceptibility curves for the northern and western parts of the LP (Hao and Guo, 2005).

Spatial variations of the magnetic susceptibility of red clay sections on the Chinese Loess Plateau are mostly consistent (Fig. 2.20). Magnetic-susceptibility values of the four sections from Fig. 2.20 all generally increase from bottom to top, with periodic but consistent excursions. Before ~3.4 Ma, the excursions lasted ~1 Ma. After ~3.4 Ma, κ values increased strikingly, especially in the Duanjiapo section; and after ~2.7 Ma, the excursion durations decreased. MS values of sections from the northern LP are relatively lower than those from the south. Before 3.5 Ma, κ values in the Bajiaju and Zhaojiachuan sections are $<50\times10^{-8}\cdot m^3\cdot kg^{-1}$, while in the Lingtai and Duanjiapo sections they are $>50\times10^{-8}\cdot m^3\cdot kg^{-1}$ In the Bajiaju and Zhaojiachuan sections at 2.5–3.5 Ma, $50\times10^{-8}\cdot m^3\cdot kg^{-1}<\kappa<100\times10^{-8}\cdot m^3\cdot kg^{-1}$, while in the Lingtai and Duanjiapo sections at the same time $100\times10^{-8}\cdot m^3\cdot kg^{-1}<\kappa<200\times10^{-8}\cdot m^3\cdot kg^{-1}$.

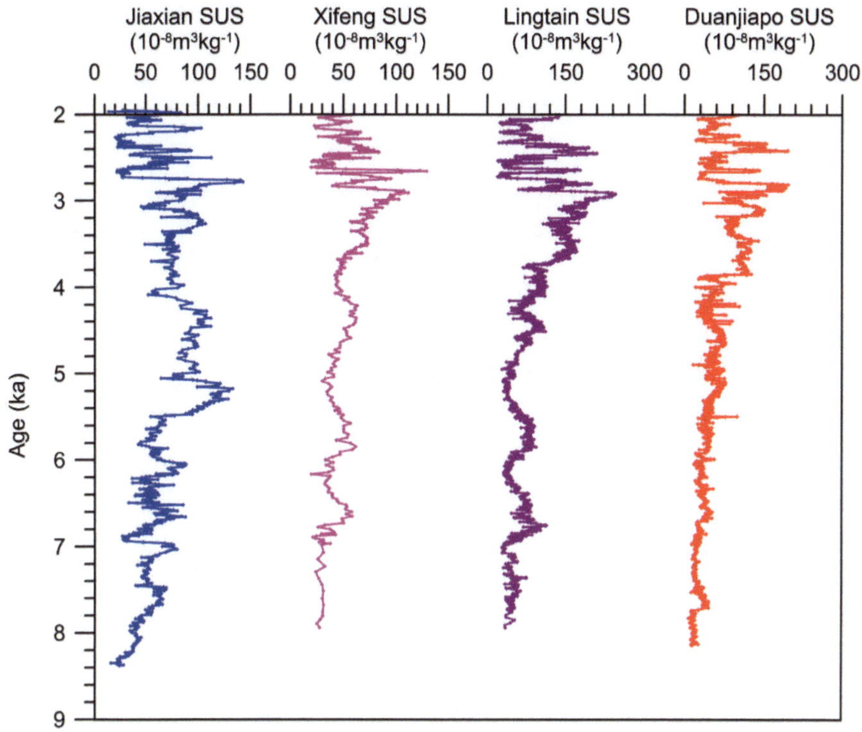

Fig. 2.20 Magnetic susceptibility variations of four red-clay sections

2.4.3 *Elemental Geochemistry*

Geochemical compositions of eolian depositions on the LP were generally controlled by three factors, including dust-provenance change, grain-size sorting and post-depositional weathering (Gallet et al. 1996; Chen et al. 1999b; Chen et al. 2006). Based on discrepancies in the geochemical behavior of individual elements under different climatic conditions, a number of geochemical ratios were established to indicate paleoclimate and provenance change (Gallet et al. 1996; Zhang et al. 1996; Ding et al. 2001b; Jahn et al. 2001; Sun 2002). For example, SiO_2/Ti_2O_3 and Zr/Rb ratios have been used to monitor wind strength of the East Asian winter monsoon (Peng and Guo 2001; Liu et al. 2004; Chen et al. 2006). In contrast, Rb/Sr, Fed/Fet, and Ba/Sr ratios have been employed to estimate relative pedogenic intensity, and hence to infer the intensity of the East Asian summer monsoon (Guo et al. 1996; Guo et al. 1998; Chen et al. 1999b; Pang et al. 2007).

The Rb/Sr ratio is considered as a proxy indicator of chemical weathering intensity because Rb mainly occurs in K-bearing minerals which are relatively resistant to weathering, and Sr remains in Ca-bearing minerals which are sensitive to weathering (Chen et al. 1999b). Although the geochemical composition and grain-

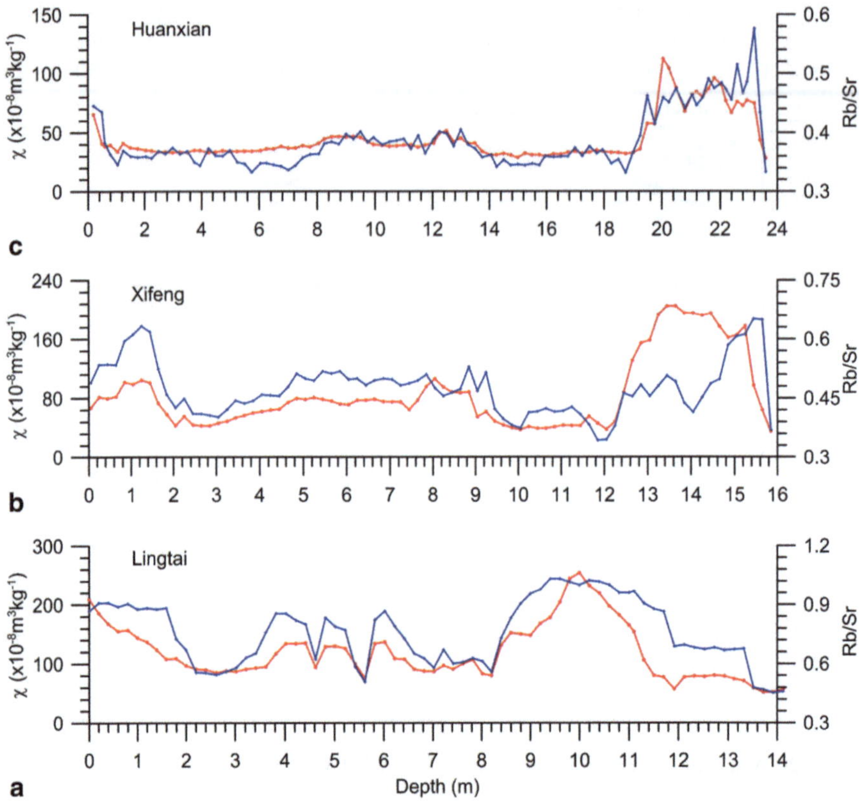

Fig. 2.21 Rb/Sr variations of three loess sections. (Chen et al. 1999b)

size distribution of dust deposits during different geological times are dissimilar, the Rb/Sr ratio is largely dependent on the post-depostional chemical weathering. Comparison of the Rb/Sr ratio of surface soils with modern climate variables suggests that high Rb/Sr ratios corresponds well to high mean annual temperature and precipitation, suggesting that variations of Rb/Sr are controlled mainly by regional climatic conditions (Chen et al. 2003). Moreover, variations in Rb/Sr of three loess-paleosol sequences along a north-south transect are consistent with the alternations of loess and paleosol layers, with higher ratios in interglacial periods and vice versa (Fig. 2.21). Spatially, the Rb/Sr ratios decrease gradually from the southeast to the northwest for both loess and paleosol layers, probably in response to the strength of the East Asian summer monsoon.

Variations of the Zr/Rb ratio in three loess sections encompassing the last 130 ka display a generally similar pattern to that of mean grain size of bulk samples (Fig. 2.22). Although a positive correlation exists between the Zr/Rb ratio and MGS, the correlation coefficient decreases in the southern and eastern sections of the LP, where intense pedogenesis occurred in relatively warm and humid climates.

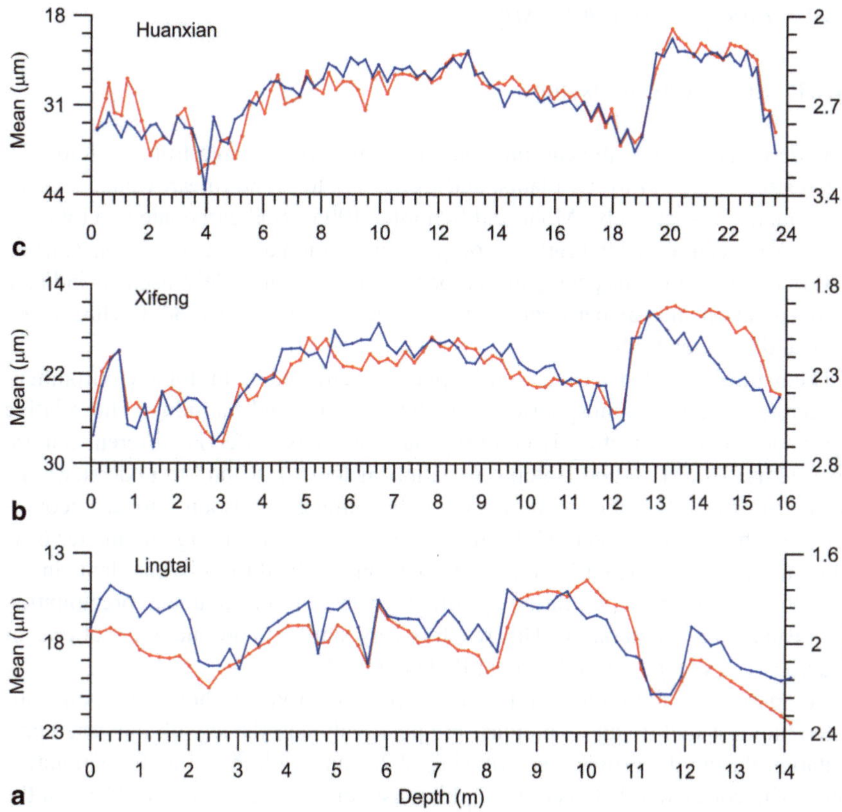

Fig. 2.22 Zr/Rb variations of three loess sections. (Chen et al. 2006)

These results demonstrate that the Zr/Rb ratio reflects original eolian grain size and may serve as a reliable index of the strength of East Asian winter-monsoon winds (Chen et al. 2006). Zr/Rb ratios in loess-paleosol sequences are sensitive to loess–paleosol alternations and are positively correlated to grain size. Because Zr and Rb are derived from an eolian source region and remain stable during the post-depositional weathering and pedogenesis, the Zr/Rb ratio is a good measure of initial dust properties which are closely related to winter-monsoon strength. The Zr/Rb ratio thus provides a proxy indicator of East Asian winter-monsoon circulation. In addition, SiO_2/TiO_2 and SiO_2/Al_2O_3 ratios have also been used to infer grain size characteristics and therefore the winter-monsoon intensity (Liu et al. 1995a).

2.4.4 *Isotopic Geochemistry*

2.4.4.1 Stable Carbon Isotope

In modern ecosystems, the carbon isotopic compositions range from −32 to −20 (mean value: −27‰) for C_3 plants, and from −15 to −9‰ (mean value: −13‰) for C_4 plants, respectively (Mohr and Schopfer 1995). Soil carbonate and total organic carbon can record the carbon isotopic composition of land vegetation. Carbon isotope compositions in pedogenic carbonate have been widely used to indicate paleovegetation and paleoenvironmental changes (e.g. Cerling 1984; Cerling 1999; Wang and Follmer 1998).

The natural abundance of C_3 and C_4 plants is affected by multiple environmental factors including atmospheric pCO_2, temperature and humidity. The relative importance of these factors is controversial, and may differ in different natural ecosystems. To test carbon isotopic variation of loess TOC and to understand the mechanism of C_4/C_3 ratios as a proxy for the summer monsoon, modern ecosystems have been investigated. C_4/C_3 ratios estimated from soil organic matter $\delta^{13}C$ values are consistent with C_4 biomass proportion obtained from living plants in the field (Fig. 2.23). The C_4/C_3 ratio is positively related to mean annual precipitation and mean annual temperature. This result is very similar to productivity patterns of modern C_4/C_3 vegetation in the Great Plains in the USA.

Comparison of these data with various climatic variables shows that both estimated biomass and living C_4 biomass are positively correlated with increased precipitation during the growing season (Fig. 2.23) and with the annual precipitation range. The correlation between the C_4 biomass (or soil organic matter $\delta^{13}C$) in the modern ecosystem and the annual precipitation amount is significant. The abundance of C_4 plant biomass is also positively correlated with temperature, although weaker than with precipitation. A higher ratio of C_4/C_3 plant biomass is favored by increased warm-season rainfall in semiarid regions with high seasonal contrast because of the high water-use efficiency of C_4 plants, enabling them to out-compete C_3 plants in hot and dry environments with warm-season precipitation (Huang 2001). These findings are similar to those from other semiarid regions (Huang et al. 2001; Johnson et al. 1999).

Carbon isotope of TOC as a summer monsoon proxy In the past 20 years, carbon isotopes of the loess-paleosol sequence have been studied to shed light on the variation of the paleovegetation and corresponding paleoclimate and paleoenvironment (Lin et al. 1991; Lin and Liu 1992; Gu et al. 2003; Zhang et al. 2003; Li and Liu 2003; An et al. 2005; Liu et al. 2005b). Previous studies in isotope geochemistry and paleoenvironment suggested that $\delta^{13}C$ signals must be free from primary detritus carbonate, and only $\delta^{13}C$ values of secondary carbonate in paleosols can be used for paleoclimatic and paleo-environmental studies (Cerling 1999; Frakes and Sun 1994; Han et al. 1996a; Gu et al. 2003; Li and Liu 2003; Ning et al. 2008).

Carbon isotope ratios of TOC record past changes in the relative production of C_3 and C_4 plants (Cerling 1989; Quade and Cerling 1995; Hatte et al. 1998; Wang

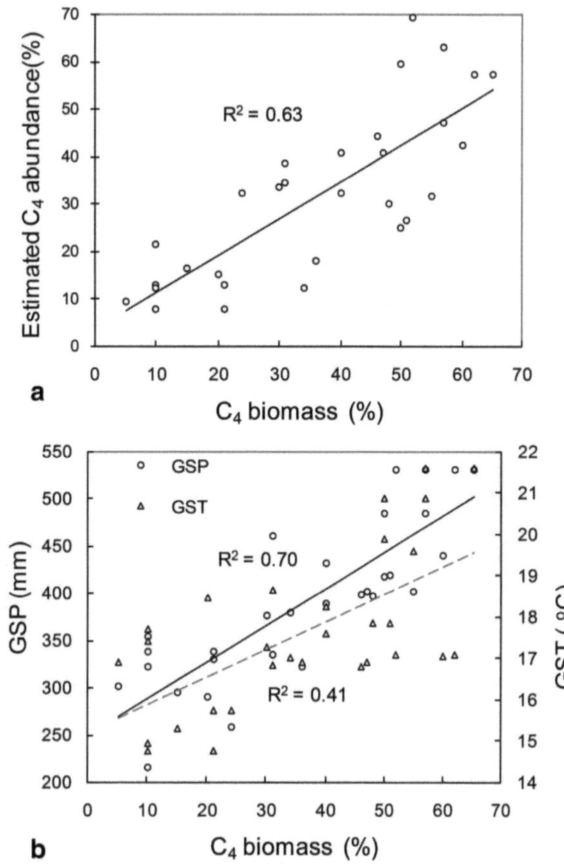

Fig. 2.23 Correlation of modern C_4 plant biomass and soil organic matter $\delta^{13}C$ values with climate parameters. a) Linear correlation of C_4 plant biomass: weight ratio of C_4 plants/(C_3 plants + C_4 plants) with mean growing season precipitation. b) Correlation of modern C_4 plant biomass with climate parameters: growing seasonal precipitation (GSP) and growing seasonal temperature (GST) (An et al. 2005)

et al. 1997; Nordt et al. 2002). Unlike lake or marine sediments where organic matter may derive from both terrestrial and marine organisms (Street Perrott et al. 1997; Huang et al. 1999), the loess organic matter contains little material from lower organisms such as algae and can be directly used to estimate changes in C_3 and C_4 plants. A number of studies have shown that carbon isotope ratios of loess TOC display major changes that are associated with glacial-interglacial transitions (Lin et al. 1991; Wang et al. 1997; Zhang et al. 2003; Gu et al. 2003; Vidic and Montanez 2004). In addition, because leaf waxes are only produced by high plants, their isotope ratios unambiguously record information of land plants responding to climatic change (Goni and Eglinton 1996; Huang et al. 1999; Huang et al. 2001; Freeman and Colarusso 2001; Zhang et al. 2003). In recent decades, compound-specific carbon isotopes of loess sediment from the LP have been studied intensively for reconstructing climatic change in the LP.

Because soil isotope variations had been related to changes in the relative proportions of C_3 and C_4 plants, An et al. (1991a) measured the C isotopes of TOC from the LP. Their results shown that an increase in the C_4/C_3 ratio occurred in the paleo-

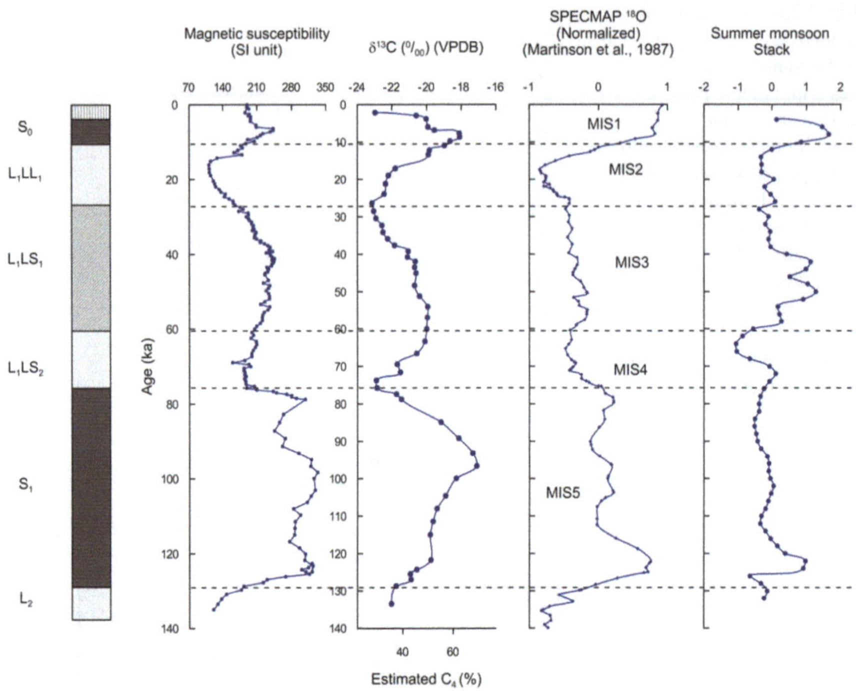

Fig. 2.24 Variation of $\delta^{13}C$ values in organic matter and magnetic susceptibility in the Lantian profile (The paleosol horizons have higher $\delta^{13}C$ values than that in the adjacent loess layers. Three high $\delta^{13}C$ peaks are present in paleosol stages (S0, L1LS1, and S1)

sols, consistent with increasing summer monsoon the κ proxy of summer monsoon change (Lin et al. 1991). Later, other investigations of loess sequences confirmed that carbon isotopes offer a reliable proxy of the East Asian summer monsoon for the LP.

The carbon isotopes of loess TOC from the Lantian profile in the southeastern LP show that higher $\delta^{13}C$ values of soil organic material are found in the weak paleosol (L_1LS_1) and especially in paleosol horizons (stages S_0 and S_1) (Fig. 2.24). Isotopic and magnetic values reach their maxima in the middle of the paleosol horizons. The loess stages (L_1LL_1, L_1LL_2 and L_2) in all the profiles have similar low $\delta^{13}C$ values, about −23‰. Loess in the studied areas is covered by natural vegetation. Dust or aerosol from source areas, generally more arid and less vegetated, may contribute detrital carbon to soil in the LP. However, aeolian dust contains very little soil organic matter. We suggest that in the arid northwesternern LP the organic matter in loess is derived from regionally mixed plant matter, expecting that the wind-transported organic matter in the loess would be greater than in the paleosol. The most common and well-developed proxy for the East Asia summer-monsoon strength is κ (An et al. 1990, 1991a). In paleosol stages (S_0, L_1LS_1 and S_1) of the Lantian profile, a more positive carbon isotope ratio and the corresponding C_4/C_3

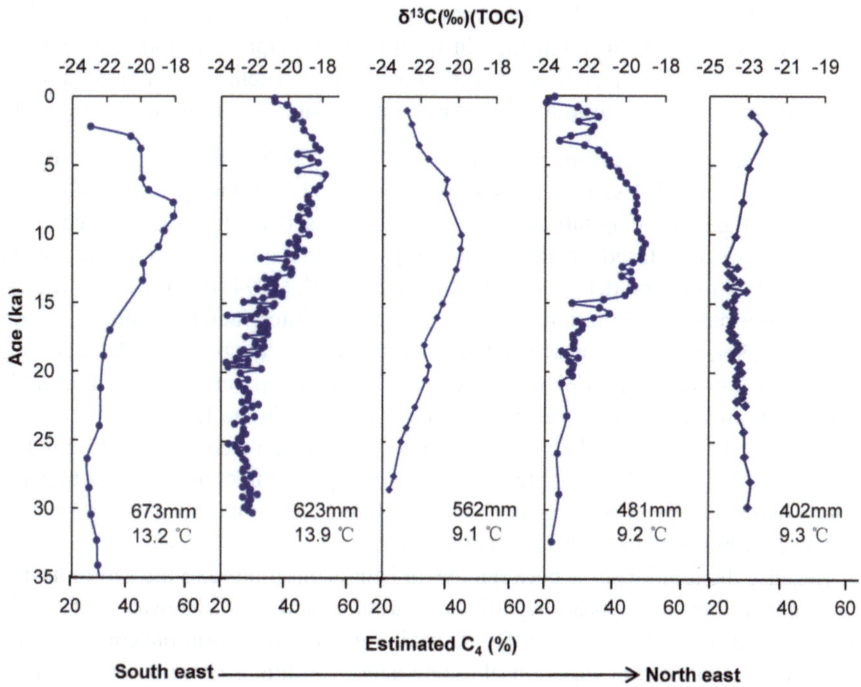

Fig. 2.25 Change in $\delta^{13}C$ values of total organic carbon (TOC) of the five loess—paleosol sections along the transect at Lantian, Weinan, Xunyi, Xifeng and Huanxian. (The C_4/C_3 ratios are evidently increase with increasing summer intensity (magnetic susceptibility proxy) from Southeast to Northwest

ratio increase, is consistent with the higher κ values, indicating a stronger summer monsoon.

Carbon isotopes from five loess-paleosol sequences (Lantian (34°04′N, 107°02′E), Weinan (34°20′N, 109°29′E), Xunyi (34°35′N, 108°20′E), Xifeng (35°42′N, 107°38′E) and Huanxian (36°37′N, 107°19′E)) spanning the last 35 ka across a climatic gradient, with precipitation and temperature decreasing from northwest to the southeast (Fig. 2.25). At present, the mean annual precipitation at Huanxian is ~402 mm and mean annual temperature about 9.3 °C, whereas the mean annual precipitation and temperature at Lantian are 673 mm and 13.2 °C, respectively. The $\delta^{13}C$ values of TOC in the five profiles show strong similarity with the variations in κ for the last 35 ka. Our data suggest that changes in $\delta^{13}C$ record the changes in summer-monsoon strength. Along an increasing summer-monsoon transect from northwest to southeast, $\delta^{13}C$ values of TOC also increased. The Huanxian profile is located at a site with the lowest precipitation (nearest to the Tenggerli and MuUs deserts) and temperature. As expected from our surface transect, for the Holocene the $\delta^{13}C$ values throughout the Huanxian profile are as much as 5–7‰ lower than the other four profiles.

The results of both modern ecosystem and the loess profiles indicate the factors controlling C_4 plant abundance in the LP are precipitation and temperature. C_4 plants are favored over C_3 plants at times of higher temperature and summer precipitation, or during periods of increased East Asian summer-monsoon strength.

Reconstructing paleo-precipitation using soil TOC carbon isotope The study of climatology in the Loess Plateau has been concerned with the quantitative reconstruction of paleo-precipitation. Research has focused on finding practical precipitation indices. Based on the relationships between rainfall and pedogenes is recorded by κ of the modern soil in the LP, Maher et al. (1994) quantitatively reconstructed the paleoprecipitation since the last glacial-interglacial transition. Maher and Thompson (1995) reconstructed the precipitation for the past 1.1 Ma in Xifeng using κ of the loess-paleosol sequences. The relationship between κ of the modern topsoil and the annual precipitation are significant in the LP (Lu et al. 1994). Also, researchers have studied the variation of the annual paleo-precipitation in the Weinan section of the Loess Plateau using the opal phytolith climatic factor transfer functions or geochemical transfer functions (free Fe_2O_3, total Fe and carbonates) (Wu et al. 1994; Lu et al. 1994; Sun et al. 1994).

Because the relationships between the moisture conditions and the carbon isotopic compositions of plants are significant (Farquhar et al. 1982; Stewart et al. 1995; Schulze et al. 1996; Liu et al. 2005b), the soil organic carbon isotopic compositions could have recorded the variation of the moisture conditions (precipitation) during the terrestrial plant growth process. Hatte et al. (1998) quantitatively reconstructed the variation of the paleo-precipitation of the last glacial stage in Nussloch using the $\delta^{13}C$ values of soil organic matter in the loess sediments, where only C_3 plants grew because of the cold climate and high latitudes. However, in the Chinese Loess Plateau, there has been a mixed vegetation system of both C_4 and C_3 plants at least since the last interglacial stage. Therefore, the influence of both C_4/C_3 ratios and precipitations on soil organic carbon isotopes should be considered in reconstructing paleo-precipitation in LP loess-paleosol sediments.

Much research has been devoted to studying the relations between plant carbon isotopes and rainfall. In eastern Australia, the $\delta^{13}C$ values of the vegetation have been found to respond to precipitation within the full rainfall range of 300–1700 mm, with a coefficient of −0.33‰ per 100 mm increase in annual precipitation (Stewart et al. 1995). In contrast, the response of plant $\delta^{13}C$ value to precipitation in northern Australia is limited to the precipitation range 200–450 mm, with $\delta^{13}C$ values remaining constant over the range of 450–1800 mm (Schulze et al. 1996). Wang et al. (2003a) concluded that, in the LP, mean $\delta^{13}C$ values decrease 0.49‰ if the annual precipitation increases 100 mm. To eliminate the influences of other species, Liu et al. (2005a) investigated the relationship between the $\delta^{13}C$ values of the dominant C_3 (*Stipa bungeana,* and *Heteropappus less*) and C_4 plants (*Bothriochloa ischaemum*), and the amount of precipitation in northwest China along a monsoon rainfall gradient. Their results showed that the sensitivity of the $\delta^{13}C$ response to precipitation is −1.1‰/100 mm for the C_3 plants and −0.61‰/100 mm for the C_4 plant. An et al. (2005) demonstrated the linear relation between the C_4/C_3 ratios

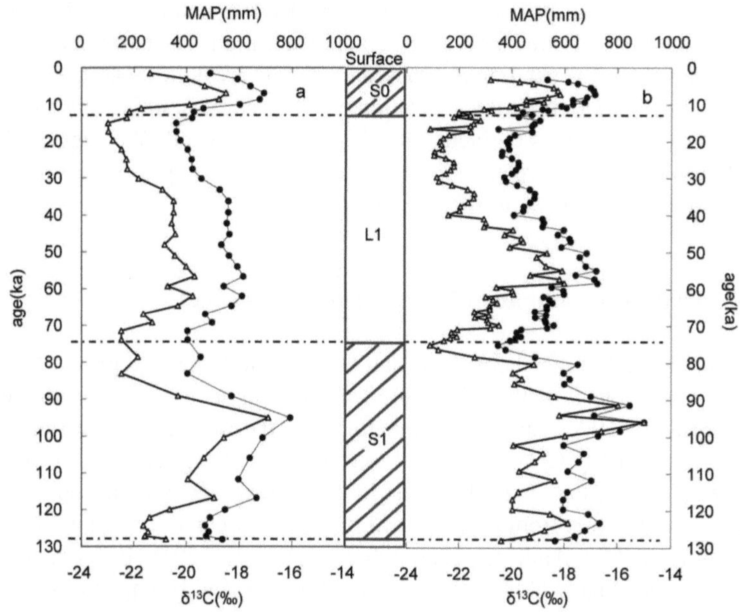

Fig. 2.26 Variation of soil $\delta^{13}C$ TOC and mean annual precipitation (MAP) in the Lantian and Weinan sections during the past 130 ka. (a is the Lantian section, b is the Weinan section; solid dots and hollow triangles indicate the MAP and soil $\delta^{13}C$ TOC value, respectively)

of soils and annual precipitation in modern ecosystems. In accordance with data from modern ecosystems (Liu et al. 2005b) and utilizing its linear relationship with C_4/C_3 ratios, annual precipitation in the LP can be estimated using the $\delta^{13}C$ values of the soil organic matter.

The $\delta^{13}C_{TOC}$ values from the loess-paleosol sequences show that the annual precipitation has varied between 360 mm and 792 mm at the Lantian site since the last interglacial stage, with the maximum occurring in S1 and the minimum in L1. Rainfall was more abundant in S1 than in S0 in the Lantian area, with maximum annual precipitation of 792 mm and 693 mm, respectively (Fig. 2.26). The mean annual precipitation at the Weinan site has varied between 348 mm and 845 mm since the last interglacial stage, and the maximum annual precipitation for this area is greater than that of the Lantian. Nonetheless, in common with the Lantian site, the rainfall at the Weinan site was more abundant in S1 than in S0, with maximum annual precipitation of 845 mm and 714 mm, respectively.

Many climatic studies involving the loess-paleosol sequence have been applied toward discerning variations in the East Asian summer monsoons or paleo-precipitation (Maher et al. 1994; Maher and Thompson 1995; Lu et al. 1996; Sun et al. 1999; Wu et al. 1994). Although Sun et al. (1999) have suggested that the annual precipitation in the Weinan area has varied from less than 200 mm up to 700 mm since the last interglacial stage, the amount of precipitation calculated in the present study exceed their results. A variety of geological materials can be used to reconstruct paleo-precipitation, and the results from diverse methods can differ significantly. Which methods are best remains an open question. The correlations between soil carbon isotopes, vegetation biomass, and climate are quite clear: the soil carbon isotopes reflect variations in the vegetation type and corresponding climate because the soil organic matter is derived from the vegetation itself. We suggest that the reconstruction of paleo-precipitation using the carbon isotopes of the soil organic matter in the Loess Plateau is a viable method.

2.4.4.2 Hydrogen Isotopes of Loess-paleosol Sequences

The large mass difference between D and H results in large isotopic fractionations and high variability in nature. Development of compound-specific hydrogen isotope analysis (CSHIA) using gas chromatography/thermal conversion/isotope ratio mass spectrometry (GC/TC/ IRMS) in recent years (Burgoyne and Hayes 1998; Hilkert et al. 1999) has spurred interest in obtaining paleoclimatic and paleoenvironmental information from D/H ratios of individual organic compounds (Xie et al. 2000; Sessions 1999; Andersen et al. 2001; Sauer et al. 2001; Huang et al. 2002b, 2004; Yang and Huang 2003; Dawson et al. 2004; Liu and Huang 2005).

Although hydrogen isotope composition (δD) of plant leaf wax is expected to vary systematically at a global scale (Smith and Ziegler 1990; Yakir 1992), D/H ratios of homologous compounds from different plants growing on the same site exhibit significant variability (Bi et al. 2005; Liu et al. 2006b; Hou et al. 2007). In addition, hydrogen isotope composition from a single plant utilizing source water with a constant δD value varies seasonally (Sessions 2006). Efforts have been made to characterize and explain the significance of the δD variability in leaf *n*-alkanes and the factors that control their behavior (Chikaraishi and Naraoka 2001, 2003; Yang and Huang 2003; Sessions and Hayes 2005; Smith and Freeman 2006; Liu et al. 2006b; Sachse et al. 2006; Hou et al. 2007). These studies, although using limited regional data, have suggested that δD in plant leaf wax is related to precipitation δD and humidity/aridity changes, as well as plant physiology and life forms.

Hydrogen isotopic compositions of individual *n*-alkanes from the Xifeng profile are shown in Fig. 2.27. δD values of *n*-alkanes range from −140 to −191 ‰ for C_{27}, from −146 to −188 ‰ for C_{29}, and from −161 to −202 ‰ for C_{31}, respectively. The δD values of C_{29} and C_{27} have a similar range (−140 to −190 ‰), but the δD values of C_{31} were lower than both C_{29} and C_{27} *n*-alkanes. The δD values of C_{27}, C_{29} and C_{31} *n*-alkanes show similar trends through time. These results indicate that hydrogen isotope ratios of individual C_{27}, C_{29} and C_{31} *n*-alkanes respond similarly to climatic and environmental factors, which is consistent with their common origin from higher-plant leaf waxes.

The δD values of individual *n*-alkanes vary systematically with successions of loess paleosol layers over the last 130 ka. In general, the δD values of *n*-alkanes are lower in the paleosol stages (S_0, L_1LS_1 and S_1) than those in loess stages, L_1LL_1, L_1LL_2 and L_2. For example, the long-chain *n*-alkanes in L_1LL_1 during last glacial maximum (LGM) have the highest δD values of ca. −140 ‰ for C_{27}, −147 ‰ for

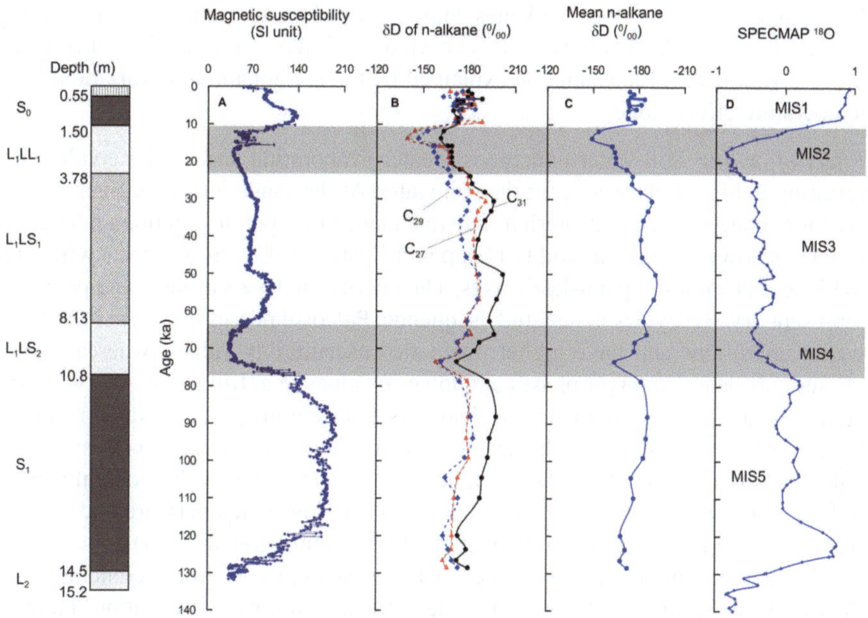

Fig. 2.27 Hydrogen isotopic compositions of individual *n*-alkanes from the Xifeng profile

C_{29} and −140‰ for C_{31} *n*-alkanes, whereas corresponding values during interglacial times are typically ca −180‰ for C_{27}, −180‰ for C_{29} and −190‰ for C_{31}. We also observe that the molecular ratios of C_{31}/C_{29} *n*-alkanes in general, show a negative correlation with δD values of the C_{27}, C_{29} and C_{31} *n*-alkanes in loess and paleosol samples.

Multiple climatic and environmental factors can affect leaf-wax δD values through their effects on plant source waters (e.g., temperature, aridity, monsoon intensity, origin of air masses), on evapotranspiration (e.g., aridity) and on vegetation types (e.g., Chenopodiaceae vs. Gramineae). Therefore, variations in δD values of individual leaf-wax *n*-alkanes must be interpreted in terms of combined effects of changes in climate and vegetation.

Precipitation δD values and temperature effect The Global Network of Isotopes in Precipitation (GNIP), sponsored by the International Atomic Energy Agency and the World Meteorological Organization (IAEA/WMO 2000), has recorded the monthly isotopic compositions in precipitation in China. The δD values of the precipitation from three meteorological stations located in the Loess Plateau are higher during warm summer season (May to October) than in cold seasons. The full range of variation is almost 70‰. Based on the seasonal isotopic trend, we infer that precipitation during warmer periods (e.g., the Holocene) would have higher δD values than during colder periods (e.g., the last glaciation). There are no published reconstructions on the amplitude of glacial-interglacial temperature changes for the Chinese Loess Plateau to allow accurate estimation of precipitation δD variations.

Nevertheless, our leaf-wax *n*-alkanes show *lower* δD values during the Holocene than during 21–13 ka, contrary to the expected trend in precipitation δD. Therefore, the leaf-wax δD profile cannot be explained by the precipitation δD variations and temperature effect alone.

Effect of aridity Soil water undergoes greater evaporation under drier conditions, resulting in higher δD values for the soil water. At the same time, plant-leaf waters are more concentrated in deuterium due to enhanced evapotranspiration under conditions of lower relative humidity (Yapp and Epstein 1982). Both effects will lead to higher δD values of plant-leaf waxes. The alternating layers of loess and paleosol represent dry/wet cycles in our study sequence. Paleosol phases represent relatively wetter conditions and have higher magnetic susceptibility values, whereas loess phases correlate with dryer phases and lower κ values (e.g. Kukla and An 1989, An et al. 1991a). The high δD values of leaf waxes during the past 130 ka correspond to times of loess deposition, whereas the low δD values of leaf waxes correspond to paleosol horizons (Fig. 2.27). Notably, however, the lowest δD values of leaf waxes did not occur during the LGM at 21 ka, when the lowest temperature and aridity were recorded in ice cores (e.g. Petit et al. 1999; Members et al. 2004).

The fact that the highest δD values of leaf-wax *n*-alkanes do not coincide with the LGM can be attributed to the combined effects of aridity and variations on precipitation δD values. The low temperature at the LGM may have lowered the δD values of the precipitation so much that aridity-induced isotopic enrichment in leaf waxes was significantly reduced. However, D/H fractionations due to higher aridity were large while those due to lower temperature were small 15–12 ka, leading to overall higher δD values for the leaf waxes. The highest δD values of *n*-alkanes were reached at around 15–12 ka, when temperature had risen, but it was still very dry (consistent with the deposition of loess, rather than the creation of soil). Dry conditions combined with the temperature effect could also explain the D/H excursions around 73 ka, where the maximum δD values are also somewhat displaced from the lowest κ values.

Effect of vegetation Vegetation history in the Chinese Loess Plateau for the past glacial-interglacial cycles is poorly understood because of the poor preservation of pollen. Leaf-wax D/H ratios and C_{31}/C_{27} and C_{31}/C_{29} *n*-alkane ratios, however, offer new insights into the vegetation distributions for the region. Phases of high *n*-alkane δD values and low C_{31}/C_{29} and C_{31}/C_{27} ratios (15–12 ka; and 75–73 ka) can be inferred to result from more abundant Chenopodiaceae shrubs. The study on the vegetation of Mongolia shows that the abundance of Chenopodiaceae species (either C_4 or C_3) increases with increasing aridity (Pyankov et al. 2000). In arid and semi-arid southern Australia, the Chenopod shrubland community lies also essentially within the dry, winter rainfall belt (O'Neill 1995).

The δD values of C_{27}, C_{29} and C_{31} *n*-alkanes trend consistent with molecular distributions in the entire Xifeng loess sequence. For example, the distribution of *n*-alkanes of loess stages in the last glacial stage (22–12 ka) is dominated by C_{29} and C_{27}, with peak values 0.88 for C_{31}/C_{29} and 1.3 for C_{31}/C_{27}, respectively. High δD values of leaf waxes characterize this period. In contrast, the C_{31} component is the dominant *n*-alkane, with a peak value 1.99 for C_{31}/C_{29} and 4.2 for C_{31}/C_{27}, respectively in the Holocene, early glacial and last interglacial periods. During the same periods, the δD values of leaf waxes show relatively low values. There is a general inverse relationship between *n*-alkane δD values and the C_{31}/C_{27} and C_{31}/C_{29} ratios for the last 130 ka in the Xifeng profile.

2.5 Monsoon Precipitation and Geomagnetic Field Changes Traced by Cosmogenic ^{10}Be

2.5.1 Study History of ^{10}Be in Chinese loess

The cosmogenic radionuclide ^{10}Be is produced primarily by nuclear interactions between secondary cosmic-ray particles with oxygen and nitrogen in the atmosphere (Lal and Peters 1967). Its production rate is mainly regulated by the geomagnetic field intensity on time scales of thousands to hundreds of thousands years (e.g. Frank 2000), so that its accumulated concentration in sediments can, in principle, be used to derive high-resolution records of geomagnetic field changes. An increasing number of studies of the paleo-geomagnetic intensity from ^{10}Be in marine and ice-core sediments have been reported (Frank et al. 1997; Wagner et al. 2000; Muscheler et al. 2005; Christl et al. 2010). The results have validated the ^{10}Be technique using AMS, as a viable tool to trace past paleo-geomagnetic intensity changes. In Chinese loess, Shen et al. (1989, 1992) first measured ^{10}Be concentrations and tried to derive an alternative time scale for loess over the past 730 ka by matching the ^{10}Be curve with SPECMAP $\delta^{18}O$ records (Imbrie et al. 1984). Beer et al. (1993) and Shen et al. (2000) developed a simple model to separate the detrital and pedogenic component in the magnetic susceptibility using the ^{10}Be flux as a proxy with which to estimate monsoonal precipitation changes. Gu et al. (1996) studied ^{10}Be in bulk samples, in different size fractions and in selective fractions chemically leached from several representative loess-paleosol and red clay sequences. The results indicated that about 72 and 89% of the ^{10}Be resides in the <4 μm grain-size fractions of the loess and paleosol samples, respectively. Gu et al. (2006) also discussed the potential application of ^{10}Be as a dating method for loess-paleosol chronology.

Broadly speaking, studies on ^{10}Be in Chinese loess before 2006 were primarily concerned with paleoclimate (Shen et al. 1992; Shen et al. 2000; Beer et al. 1993; Gu et al. 1996; Gu et al. 1997), where the effects of the geomagnetic field were not accounted for. This situation arose because the relationship between ^{10}Be and the geomagnetic field in loess archives is complicated by multiple sources and changing sedimentological conditions (Fig. 2.28). If we could disentangle the mixed signals of the ^{10}Be preserved in loess then this question might be resolved. Recently, a new methodology and conceptual model was developed for geomagnetic tracing from Chinese loess ^{10}Be. Records of global geomagnetic intensity covering the past 80 ka and 130 ka have now been generated (e.g. Zhou et al. 2007b; Zhou et al. 2010;

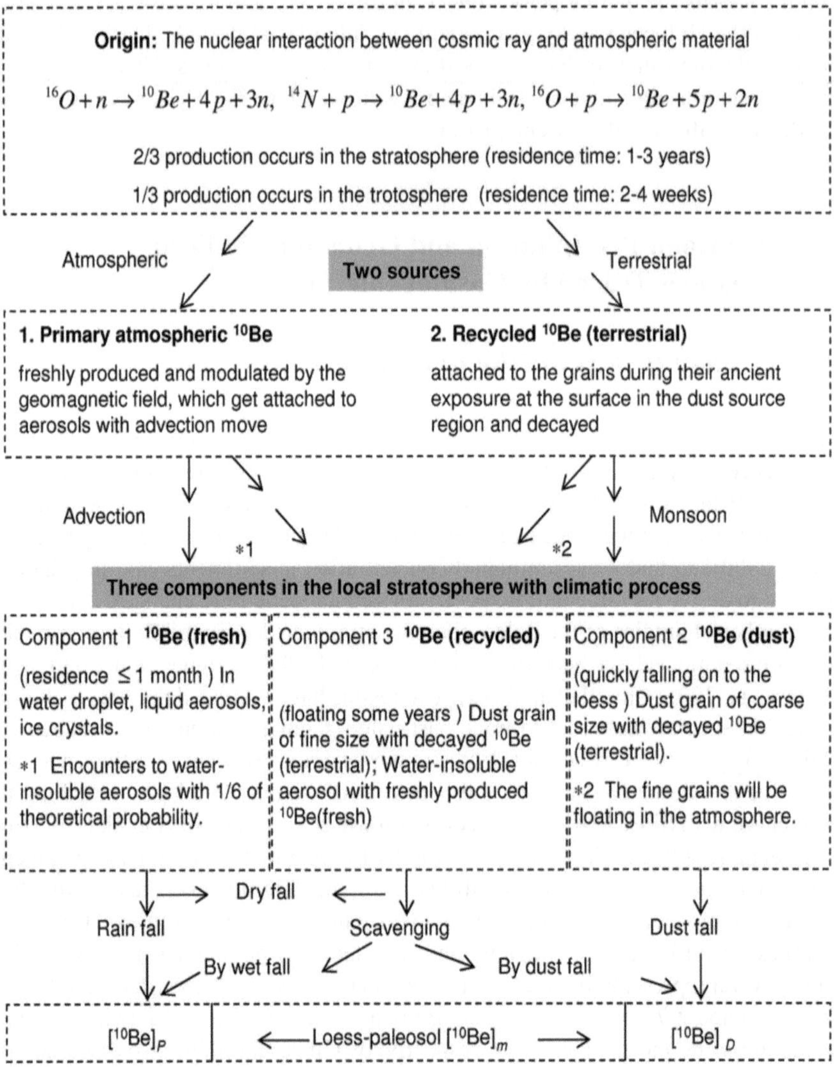

Fig. 2.28 The origin, transportation and accumulation process of ^{10}Be in loess. $[^{10}Be]_m$—measured ^{10}Be concentration in loess; $[^{10}Be]_D$—dust-derived ^{10}Be concentration component; $[^{10}Be]_P$—^{10}Be concentration component from rainfall

Xian et al. 2008), including a quantitative paleoprecipitation reconstruction for the CLP (Zhou et al. 2007b). These results support an expanded effort to extend ^{10}Be records to the whole Quaternary and even longer. In this chapter, we will introduce the new progress of geomagnetic tracing studies from Chinese loess ^{10}Be in detail.

To better understand the methods for reconstructing the geomagnetic field and wet precipitation changes from loess ^{10}Be, we make use of some nomenclature as follows in the next section.

2.5.2 Nomenclature

Concentration

$[^{10}Be]_m$	measured ^{10}Be concentration in loess
$[^{10}Be]_D$	dust-derived ^{10}Be concentration component
$[^{10}Be]_P$	^{10}Be concentration component from rainfall
$[^{10}Be]_{m,e}$	^{10}Be concentration for a particular time or magnetic field intensity estimated by linear regression analysis
$\Delta[^{10}Be]_m$	the residual ^{10}Be concentration for a particular time or magnetic field calculated as the difference between $[^{10}Be]_m$ and $[^{10}Be]_{m,e}$
$<\Phi_{D,}P>$	the climate correction factor
$\overline{(\Phi_{D,}P)}$	the average climate status
$\overline{[^{10}Be]_{m,e}}$	the average estimated ^{10}Be concentration for a particular time or magnetic field intensity by linear regression analysis
$<[^{10}Be]_m>$	the deserved ^{10}Be concentration after the measured $[^{10}Be]_m$ accumulated during the climate condition $(\Phi_{D,}P)$ is scaled to the value under the average climate condition $\overline{(\Phi_{D,}P)}$.

Flux

Φ_D	dust flux (=*loess accumulation rate* × *dry bulk density*)
$\Phi_{\Delta[^{10}Be]_m}$	residual ^{10}Be flux (=$\Delta[^{10}\mathrm{Be}]_\mathrm{m}$ × dust flux)

Precipitation

P	precipitation rate (rainfall)

Magnetic susceptibility

SUS_m	measured magnetic susceptibility in loess
SUS_D	magnetic susceptibility component carried in dust
SUS_P	magnetic susceptibility component related to the wet precipitation

Magnetic intensity units

M	magnetic intensity of the earth at a certain time

M_0 magnetic intensity of the earth at present
ΔM geomagnetic field variation

Production rate terms

Q_M global ^{10}Be production rate at magnetic field M
Q_{Mo} global ^{10}Be production rate with the present magnetic field
Pr local ^{10}Be production rate on the CLP
Pr_{avg} average local ^{10}Be production rate on the CLP over the past 130 ka
ΔPr variation in local atmospheric ^{10}Be production rate

2.5.3 Paleogeomagnetic and ^{10}Be Time Series Records From Loess

2.5.3.1 Paleo-geomagnetic Time-series Records from Loess

One reason for the interest in tracing magnetic-field changes from the ^{10}Be concentrations in the loess lies in the fact that the paleo-geomagnetic features of the loess are still controversial. Studies of relative paleo-geomagnetic intensity variations through time provide important information about the nature of geomagnetic field changes and even the workings of the Earth's geodynamo (Merrill and Mcfadden 1995). The relative paleointensity records have commonly been determined using paleomagnetic stratigraphy or by time series cosmogenic nuclide concentrations (e.g. ^{10}Be). In the first method, natural remanent magnetization (NRM) is determined. This parameter is related to both the alignment of magnetic grains by the geomagnetic field and the concentration of magnetic particles in a sample. Anhysteretic Remanent Magnetization (ARM), Saturation Isothermal Remanent Magnetization (SIRM) and magnetic susceptibility are commonly used as normalizing factors to estimate relative paleointensity (Pan et al. 2001). Numerous individual relative paleointensity records from marine sediments have been published over the past two decades (Tric et al. 1992; Valet and Meynadier 1993; Weeks et al. 1995; Roberts et al. 1997; Yamazaki 1999; Channell et al. 2000), as well as synthetic curves derived from stacked records, such as the Sint-200 (Guyodo and Valet 1996), Sint-800 (Guyodo and Valet 1999), NAPIS75 (Laj et al. 2000) and the recent PISO-1500 records (Channell et al. 2009).

Despite the fact that several studies have reconstructed magnetostratigraphic records from Chinese loess (e.g. An et al. 1977; Heller and Liu 1982; Yu and Ding 1998), only a few studies have documented long-term relative paleo-intensities. A complicating factor is a possible delay in remanence "lock-in" in loess. Magnetostratigraphic analyses and model simulations on Chinese loess suggest that remanence "lock-in" might have a delay of as much as 20,000 years (e.g. Zhou and Shackleton 1999; Evans and Heller 2001). A second complicating factor involves the dissimilarities between relative paleo-intensity records from Chinese loess and marine records (Pan et al. 2001). These factors show that the determination of relative paleogeomagnetic intensity in loess-paleosol profiles is complex, possibly due to sedimentological differences associated with changes in the monsoon intensity over the LP and its effect on NRM. Another good example is the long-standing uncertainty surrounding the difference in timing of the Brunhes-Matuyama (B/M) polarity reversal (~780 ka) in marine sediments, as compared to loess deposits (Zhu et al. 1993; Zhu et al. 1994a; Sun et al. 1993; Zhou and Shackleton 1999; Spassov et al. 2001). Hence, uncertainties remain concerning the acquisition process of post-depositional remanence in Chinese loess (e.g. Liu et al. 2008). To remove these, it is necessary to develop a method such as ^{10}Be-tracing to decouple the climatic and geomagnetic influences on the paleomagnetic signal.

2.5.3.2 ^{10}Be Time Series Records from loess

Cosmogenic ^{10}Be is rapidly removed from the atmosphere and deposited on Earth mainly through wet precipitation. Since beryllium has very low solubility, it tends to remain where it is deposited, whether in sediments, on the surface of soil particles, or in ice cores. Unlike ^{14}C, which actively exchanges with atmospheric and sediment carbon, sedimentary ^{10}Be atoms reliably reflect atmospheric production changes with time in sediment archives. Therefore ^{10}Be can be considered as a good candidate for paleoenvironmental tracing studies (Frank 2000).

The ^{10}Be deposition rate in sediments is closely related to atmospheric ^{10}Be production and associated wet deposition. The atmospheric ^{10}Be production rate is primarily related to changes in cosmic-ray flux, solar activity and the intensity of the earth's geomagnetic field (Masarik and Beer 1999). On time-scales of thousands to hundreds of thousands years, the ^{10}Be production rate is known to be modulated by the earth's magnetic field. The production of ^{10}Be follows the general relation (Elsasser et al. 1956; Frank 2000):

$$\frac{Q_M}{Q_{M_0}} = const. \times \frac{1}{\sqrt{\frac{M}{M_0}}} \tag{1}$$

where Q_M is the global nuclear disintegration rate for a corresponding field strength, M. In this equation, M_0 is a reference point taken as the present field strength. This relationship makes it possible to infer paleomagnetic intensity changes from past production rates of cosmogenic ^{10}Be, accumulated in sediment profiles.

Wind-blown loess sediments have been shown to be a storehouse of very long and potentially high-resolution records of both variations in the geomagnetic field, and of variability in the East Asian monsoon (Liu and Ding 1998). While there have been numerous efforts to extract climate or magnetic intensity records from Chinese loess, the exercise has proved challenging because the climatic and geomagnetic signals contained in these sediments are tightly intertwined (e.g. An 2000; Evans and Heller 2001; Porter et al. 2001). This is true for both magnetic remanence as well as for records of cosmogenic isotopes found in loess, such as ^{10}Be, both of which have been used in attempts to reconstruct climate and geomagnetic field variations (Zhou et al. 2007b; Zhou et al. 2010).

For paleomagnetic reconstructions based on ^{10}Be, some ^{10}Be comes from remobilized dust that fell to Earth at some time in the past, whereas a second component is from atmospheric fallout of ^{10}Be recently generated by cosmic-ray interactions in the atmosphere. It is the flux of this latter fallout component that is regulated by the geomagnetic field intensity. The interpretation of fallout ^{10}Be is somewhat complicated by the fact that some of it makes its way to the ground via adsorption onto dry particulates, and the rest is deposited by wet precipitation. Although the dry fallout fraction is usually less than 10% (Wallbrink and Murray 1994), the local ^{10}Be fallout flux to the ground depends strongly on the wet precipitation amount. To further complicate matters, solar magnetic field variations also influence atmospheric production (Masarik and Beer 1999), though solar activity mainly acts on shorter periods than are being considered here. Therefore, the loess ^{10}Be record is complexly intertwined with variations in geomagnetic field and climate, which results in the fact that both loess ^{10}Be concentration and flux curves are not as straightforward as marine, lake and ice records for reflecting geomagnetic field changes (Zhou et al. 2007a; 2007b; 2010).

In order to derive a robust geomagnetic paleointensity from Chinese loess, the high-resolution ^{10}Be analyses were first performed on the 1216-cm and 1488-cm cores taken from the Luochuan (35°45′N, 109°25′E) and Xifeng (35°42′N, 107°38′E) loess sections, both located east of the Liupan mountain region in the central LP (Fig. 2.29). The two cores were sampled at approximately 80–100 year intervals (1 cm) for magnetic susceptibility measurements at the Xi'an AMS Center. The Luochuan core was sampled at ~4 cm intervals for ^{10}Be measurements, except during the interval representing the LGM, where it was sampled at 1 cm intervals. The Xifeng core was sampled at 8–10 cm intervals, except the interval representing MIS 3 and part of the last interglacial paleosol S_1, where it was sampled at 2–4 cm intervals for ^{10}Be analysis. The total 306 and 230 BeO samples from Luochuan and Xifeng were chemically prepared at the Xi'an AMS Center and measured on the 3MV accelerator mass spectrometer of Xi'an AMS Centre, VERA (Vienna Environmental Research Accelerator) and NSF-Arizona AMS labs. Measured precisions were better than 3% in about 99% of the samples.

Calendar-age models for these two loess sequences were derived from a grain-size model (Porter and An 1995) based on the correlation between the proxies in loess and the SPECMAP marine oxygen isotope stages, applying MIS 2/1, 5/4 and 6/5 transitions to the inferred L_1/S_0 (12.3 ka) and S_1/L_1 (79 ± 1 ka) (Johnsen et al. 2001) and S_1/L_2 (129.3 ± 0.9 ka) (Yuan et al. 2004) as stratigraphic boundaries for age interpolation. For the Luochuan section, the age scale was also cross-checked by combination of optically stimulated luminescence (OSL) and ^{14}C ages (Fig. 2.30). The Xifeng section was carefully correlated to Luochuan, other well-documented sections on the CLP, and to the Dongge-Hulu cave $\delta^{18}O$ record (Wang et al. 2001; Dykoski et al. 2005) for the past 130 ka.

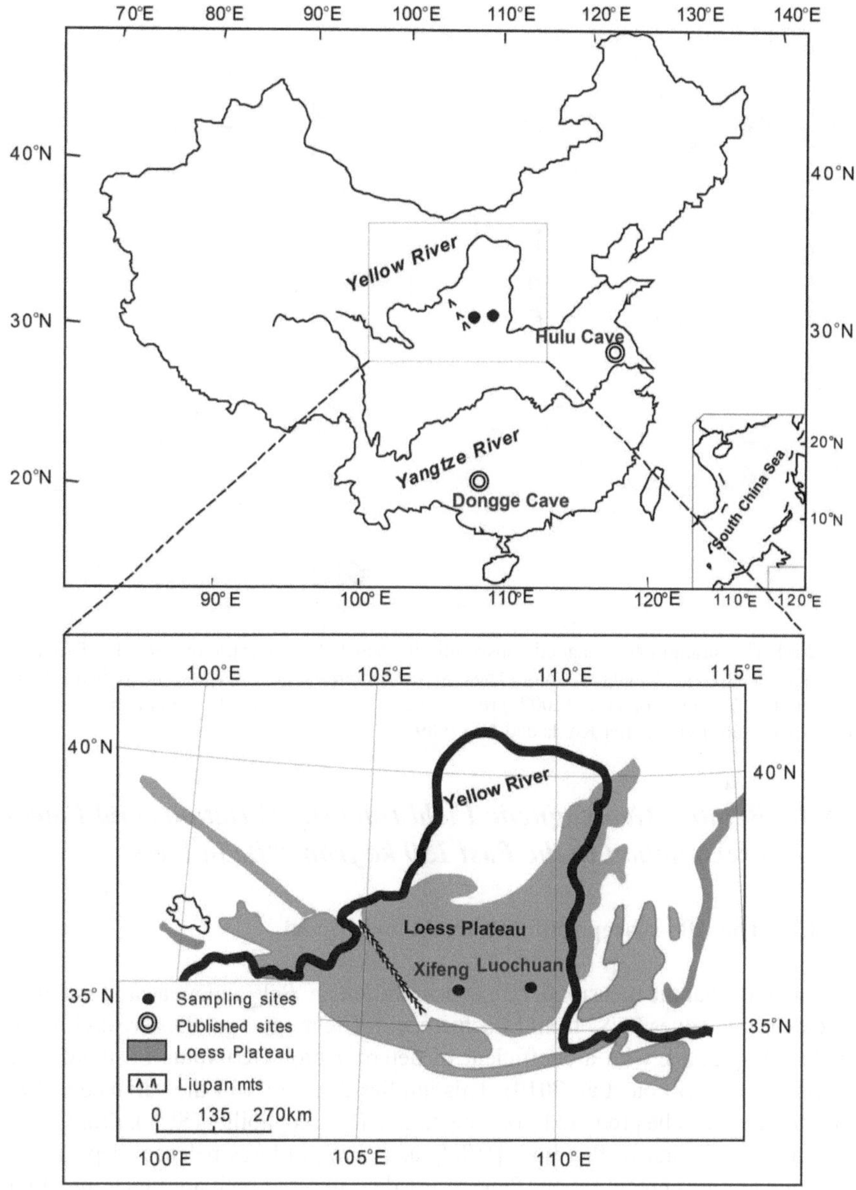

Fig. 2.29 Map showing the CLP (shaded areas), Liupan mountains, two sites for ^{10}Be tracing studies (*solid circles*) and the locations of Hulu Cave and Dongge Cave (*hollow circles*)

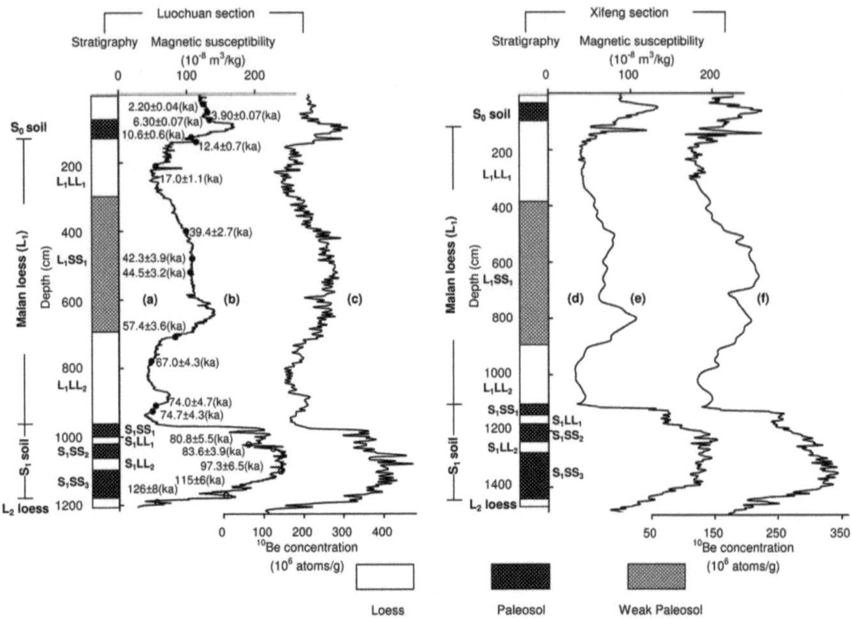

Fig. 2.30 The stratigraphy, magnetic susceptibility and ^{10}Be concentration records of (a–c) in Luochuan and (d–f) in Xifeng sections. (The OSL (*black circles*) and ^{14}C (*black dots*) data are after Lu et al. (2007) and Zhou et al. (2007a) respectively. The Malan loess (L_1) and Last interglacial soil (S_1) are sub-divided after Kukla and An (1989)

2.5.4 Relative Geomagnetic Field Intensity Variations and Paleo-precipitation in the Past 130 ka from ^{10}Be in loess

2.5.4.1 The ^{10}Be-susceptibility Resemblance Model

It is obvious that the measured ^{10}Be concentrations $[^{10}Be]_m$ and measured magnetic susceptibility values SUS_m from Luochuan and Xifeng loess profiles are highly correlated (Fig. 2.30), with a coefficient of determination (R^2) value of ~0.90, over the past 130 ka (Zhou et al. 2010). This implies that ~90 % of the variance in ^{10}Be concentration can be predicted from the magnetic susceptibility (SUS). From previous studies it is known that both $[^{10}Be]_m$ and SUS_m in loess-paleosol deposits are controlled by wet precipitation, P, and dust flux, Φ_D, and both are functions of climate. Here we describe a new "^{10}Be-susceptibility resemblance model" as a simple approach to understand the relationship between loess ^{10}Be and paleoenvironmental changes. This new model is fundamentally different from earlier studies (e.g. Beer et al. 1993; Heller et al. 1993; Shen et al. 2000), as it defines the measured (meteoric) ^{10}Be concentration, $[^{10}Be]_m$, and the measured susceptibility, SUS_m as functions of three variables, with $[^{10}Be]_m = f(P, Pr, \Phi_D)$ and $SUS_m = f(P, \Phi_D, X)$. Pr is the local production rate of ^{10}Be and X is a soil condition factor related to pedogenesis.

This model accounts for geomagnetic modulation in the time-dependent production term *Pr* rather than assuming a constant geomagnetic field as in the afore mentioned studies (e.g. Beer et al. 1993; Shen et al. 2000). The influence of non-wet precipitation factors (X and Φ_D) on pedogenic susceptibility, illustrate the fact that pedogenic susceptibility itself can not be taken as a pure wet precipitation proxy (e.g. Heller et al. 1993; Beer et al. 1993).

There exist three sources of ^{10}Be in the atmosphere over the LP (Fig. 2.28). One is the freshly produced cosmogenic ^{10}Be, which was mainly attached on the water soluble aerosols and contemporarily deposited to loess mainly through wet precipitation (rainfall) (Baumgartner et al. 1997; Gu et al. 1996). The second is the intrinsic ^{10}Be (Gu et al. 1996), produced at an earlier time and remobilized by wind erosion. Among the remobilized dust particles from distant sources, the coarser grains are directly deposited on the loess surface (dust fall), while the finer grains remain suspended in the atmosphere, on which a portion of the freshly produced cosmogenic ^{10}Be atoms might be adsorbed, forming a third component, recycled ^{10}Be (Baumgartner et al. 1997). The recycled ^{10}Be component was either locally scavenged by dust fall or rainfall at some later time (Gu et al. 1996) or recycled from some distant source (Gu et al. 1996; Lal 2007)· Since the second and third ^{10}Be signals are both related to dust fall process, we then simply define the sum concentration $[^{10}Be]_D$ which includs both fractions for the sake of discussion.

Similarly, SUS_m from loess-paleosols also consists of two components (e.g. An and Sun 1995; Shen et al. 2000): 1) SUS_D, from the same reworked dust that carries the $[^{10}Be]_D$ signal; and 2) SUS_P, which forms *in situ* as pedogenic ferromagnetic mineral grains (Maher et al. 1994; An and Sun 1995). The dust-fall ferromagnetic mineral grain flux and ^{10}Be dust-fall flux share a positive linear correlation with Φ_D (e.g. Beer et al. 1993; Zhou et al. 2007b). Pedogenic ferromagnetic mineral grains form beneath the surface of the loess through a complex chemical/biochemical pedogenic process that is related to soil-forming factors (e.g. Zhou et al. 1990a; An et al. 1991b; Maher and Thompson 1995). We account for this process in loess-paleosols in the same fashion as rainfall ^{10}Be accumulation. We assume that there is a pedogenic ferromagnetic mineral grain 'flux', analogous to the rainfall ^{10}Be flux as both quantities are positively correlated with wet precipitation P (e.g. Zhou et al. 2007b, 2010).

The rainfall ^{10}Be flux is linearly related to the ^{10}Be production rate *Pr*, which depends on the geomagnetic field strength, whereas the pedogenic ferromagnetic mineral grain 'flux' is linearly related to the number of ferromagnetic mineral grains in the loess-paleosols, or to their formation rate in relation to soil conditions, expressed as X. Hence, the wet-determined ^{10}Be flux is a function of the two variables P and *Pr*, whereas the pedogenic ferromagnetic mineral grain 'flux' is a function of the two variables P and X.

The rainfall ^{10}Be atoms in loess possess their concentration meaning only when they are mixed with loess particles, and pedogenic ferromagnetic mineral grains, as dust accumulation was still continuing during pedogenesis (e.g. An et al. 1991a; Porter et al. 2001). Both ^{10}Be concentration and 'concentration' SUS_P 'concentration' are anti-correlated to dust flux Φ_D in the same manner, which is a dilution effect of the dust flux Φ_D (e.g. Porter et al. 2001), while carrying the dust components $[^{10}Be]_D$ and SUS_D to the loess. Hence, the wet-determined concentration $[^{10}Be]_P$ depends on three variables (P, Pr, Φ_D) and SUS_P also depends on three variables (P, X, Φ_D). $[^{10}Be]_P$ and SUS_P follow a direct linear correlation to one another (Zhou et al. 2007b; Zhou et al. 2010), because the effect of the geomagnetic modulation on $[^{10}Be]_P$ and the influence of the soil condition variations on SUS_P are much weaker than that of the climate factors (P, Φ_D), as discussed below. Consequently, it can be concluded from the new "^{10}Be-Susceptibility resemblance model" and the high similarity between $[^{10}Be]_m$ and SUS_m in Fig. 2.30, that

$$\left[{}^{10}Be\right]_m = \left[{}^{10}Be\right]_D + \left[{}^{10}Be\right]_P \tag{2}$$

Where $[^{10}Be]_P = f(P, Pr, \Phi_D)$ and $[^{10}Be]_m = f(\Phi_D, P, Pr)$;

$$SUS_m = SUS_D + SUS_P \tag{3}$$

Where $SUS_P = f(P, X, \Phi_D)$ and $SUS_m = f(\Phi_D, P, X)$;

$$[^{10}Be]_{m,e} = aSUS_m + b \tag{4}$$

$$SUS_m = c[^{10}Be]_m + d, r^2 \approx 0.9 \tag{5}$$

Where the subscript e refers to the estimated value from the regression equation. The residual signals from this linear regression analysis are essentially due to the influence of the geomagnetic field modulation on $[^{10}Be]_P$ and possible soil-condition variations on SUS_P, besides the incompletely identical correlation of $[^{10}Be]_m$ and SUS_m to climate factors (Φ_D, P) and other unmentioned factors as well as their uncertainties.

2.5.4.2 Separating Geomagnetic-field and Climatic Signals in ^{10}Be from Loess

It will be suggested later that SUS_P is also determined by the soil condition factor, X, in addition to the monsoon climate condition, but the soil condition variation in loess is so small that the effect of the ΔX on the pedogenic SUS_P change is negligible. Thus, the soil condition in loess can be considered to be constant over long time scales (e.g. Maher et al. 1994; Han et al. 1996b; Zhou et al. 2007a) and the pedogenic SUS_P, or the total SUS_m can then be taken as a climate proxy (e.g. East Asian Monsoon controlled dustfall Φ_D and rainfall P),

$$SUS_m \approx SUS_P \tag{6}$$

With this definition, and based on the high similarity between the $[^{10}Be]_m$ curve and the SUS_m curve (Fig. 2.30), Zhou et al. (2007b) proposed the idea of taking magnetic susceptibility as a proxy for ^{10}Be components affected by climate factors ($\Phi_{D,}P$), to separate the geomagnetically modulated ^{10}Be signal from loess to reconstruct a curve representing the changes of the earth's geomagnetic magnetic field through time. To better understand the essence of such a separation for geomagnetic intensity reconstruction from loess ^{10}Be, Zhou et al. (2007a) further introduced an important Mean Value Concept (MVC) for regression data analysis in geosciences. The concept and method are essentially to trace the independent variable x_2 changes from the known dual-dependent variable $y(x_1, x_2)$ and another variable x_1 (or its proxy). Details about the MVC method are presented in Zhou et al. (2007a).

The main step of the data analysis under the MVC is first to extract the climate-affected ^{10}Be composition through the regression between $[^{10}Be]_m$ and SUS_m (Fig. 2.30) as shown in equations 7 and 8,

Luochuan:

$$[^{10}Be]_{m,e} = 1.2431 \times SUS_m + 105.14, \mathrm{r} = 0.94, \mathrm{n} = 306, \mathrm{p} < 0.0001 \tag{7}$$

Xifeng:

$$[^{10}Be]_{m,e} = 1.2276 \times SUS_m + 79.69, \mathrm{r} = 0.94, \ \mathrm{n} = 230, \mathrm{p} < 0.0001 \tag{8}$$

Where the estimated $[^{10}Be]_{m,e}$ are determined by the varying climate factors (proxy) assuming a constant atmospheric average ^{10}Be production rate Pr_{avg} over the past 130 ka, as determined using the MVC. Then the residual signal is given by equation (9),

$$\Delta[^{10}Be]_m = [^{10}Be]_m - [^{10}Be]_{m,e} \tag{9}$$

Where $[^{10}Be]_{m,e} = f(\Phi_D, P, Pr_{\mathrm{avg}})$;

The concentrations $\Delta[^{10}Be]_m$ (Fig. 2.31) are then determined by subtracting the estimated values from the measured values for the past 130 ka.

It should be pointed out that here we use total $[^{10}Be]_m$ vs SUS_m instead of the wet-determined $[^{10}Be]_P$ vs SUS_P for the Pr reconstruction, because it is difficult to extract the pure wet-determined $[^{10}Be]_P$ or SUS_P from the measured $[^{10}Be]_m$ or SUS_m. Furthermore, the dust fall $[^{10}Be]_D$ and SUS_D are in dependent of the ^{10}Be production rate changes, so that both the residuals represent change in rainfall ^{10}Be, caused by the changes in Pr as shown in equations 10.

$$\Delta[^{10}Be]_m = f(\Phi_D, P, \Delta Pr) \tag{10}$$

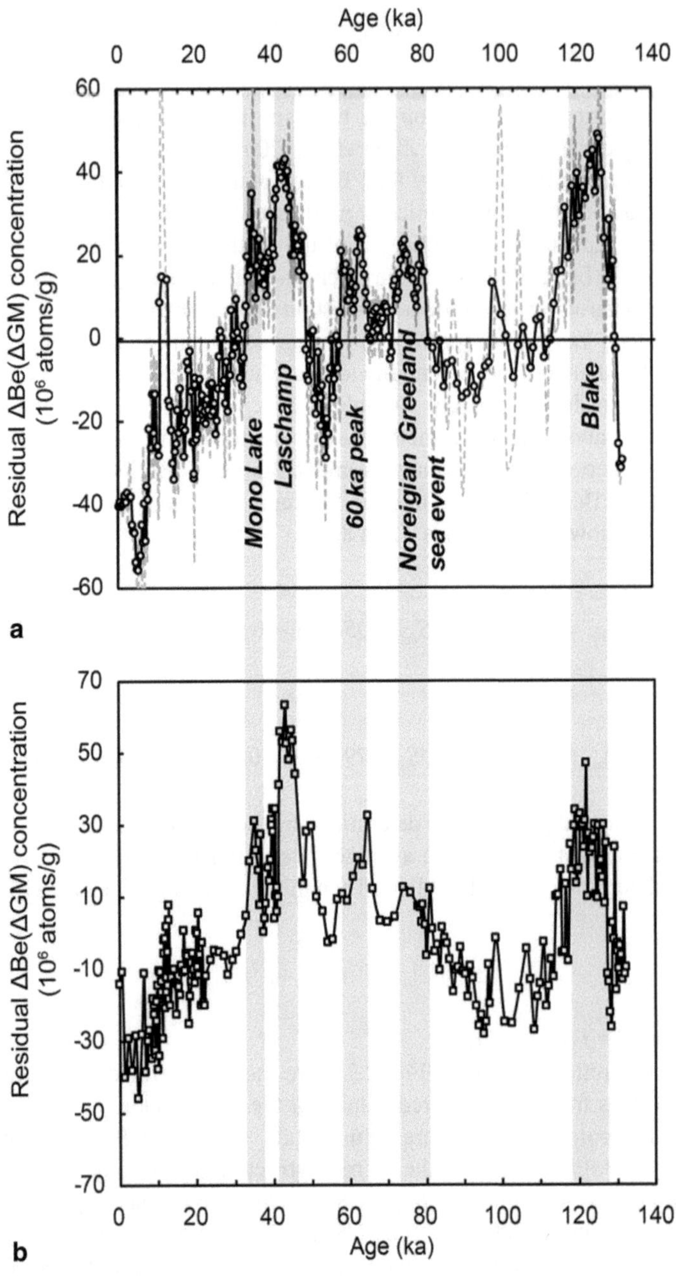

Fig. 2.31 The separated residual change-amount of the ^{10}Be concentration (ΔBe) caused by the atmospheric ^{10}Be production rate variation and corresponding geomagnetic excursion events. (a) The Luochuan section, (b) The Xifeng section

From the afore mentioned analysis, it suggests that the key point for separating the geomagnetic modulated ^{10}Be signals in loess is how to remove the most climatic contamination effects. The MVC method (Zhou et al. 2007a) is based on the linear regression between $[^{10}Be]_m$ and SUS_m regarding this purpose. Since there was no perfect method to separate the climatic signal completely in loess, we then develop the alternative way by correcting and scaling the ^{10}Be concentration itself (MVCS) under the same climate condition, to derive the geomagnetic modulated ^{10}Be signal in addition to the MVC method. The MVCS method can be described briefly as follows.

According the regression between $[^{10}Be]_m$ and SUS_m as shown in equations 7 and 8, we can firstly define the climate correct factor $<\Phi_{D,} P>$ from equation 11.

$$<\Phi_D, P> = \frac{[^{10}Be]_{m,e}}{\overline{[^{10}Be]_{m,e}}} = \frac{aSUS_m + b}{\overline{aSUS_m + b}} \tag{11}$$

which represents the ratio of the estimated $[^{10}Be]_{m,e}$ under varying climate condition ($\Phi_{D,} P$) to the average estimated value under average climate status.

Then the scaled $<[^{10}Be]_m>$ is given by equation (12)

$$<[^{10}Be]_m > \frac{[^{10}Be]_m}{<\Phi_D, P>} \tag{12}$$

which means that the measured $[^{10}Be]_m$ accumulated under the time-varying climate condition ($\Phi_{D,} P$) is scaled to the value under the average climate condition $\overline{(\Phi_{D,} P)}$.

From this dividing calculations, the ^{10}Be concentration component affected by the climate variations relative to the average $\overline{(\Phi_{D,} P)}$ conditions have been firstly removed from the $[^{10}Be]_m$. It only leaves the ^{10}Be concentration signals mainly modulated by the geomagnetic intensity or ^{10}Be production rate variations ΔPr under the certain climatic conditions. Similar as the $\Delta[^{10}Be]_m$ calculated by MVC method, this finally scaled $<[^{10}Be]_m>$ signal also can be used to trace the past geomagnetic field changes. In our present chapter, we just show the ^{10}Be tracing studies results from the simple MVC analysis method.

2.5.4.3 Relative Geomagnetic Field Intensity Sequence from ^{10}Be Production Rate and its Comparison with Global Signals in the Past 130 ka

A geomagnetic field variation (ΔM) would affect ^{10}Be accumulation in loess only through wet precipitation rather than directly. Therefore, the flux of the variation $\Phi_{\Delta[^{10}Be]_m}$ is then related not only to atmospheric ^{10}Be production rate variation ΔPr, but also to the contemporary wet precipitation P, whereas the concentration variation $\Delta[^{10}Be]_m$ is additionally related to the dust flux Φ_D by its dilution effect. Therefore, $[^{10}Be]_m$, SUS_m, P and Φ_D and the production rate Pr constitute a multivari-

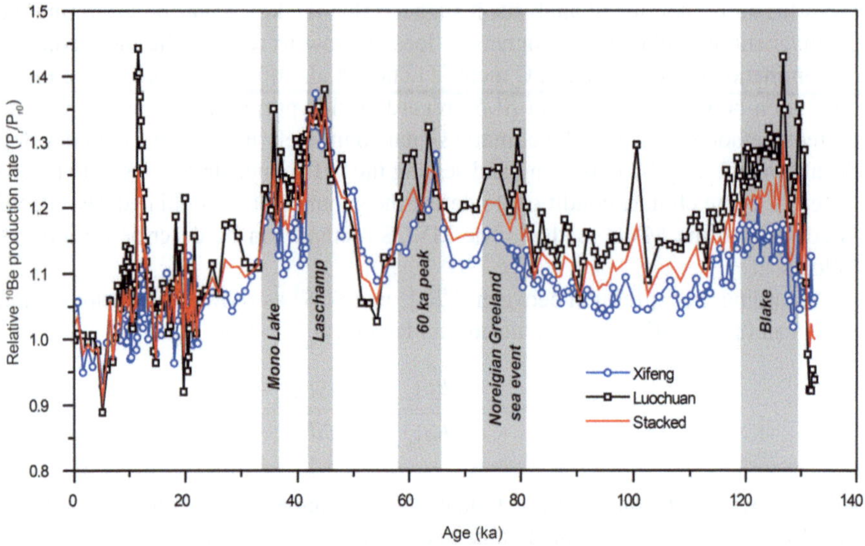

Fig. 2.32 The derived relative ^{10}Be production rate records from Luochuan (*black*), Xifeng loess section (*blue*) and the stacked curve (in *red*) over the past 130 ka

able geophysical system with non-independent variables, where the ΔPr effect on $\Delta[^{10}Be]_m$ depends upon the other independent variables P and Φ_D. Hence, the calculated geomagnetically modulated ^{10}Be variation $\Delta[^{10}Be]_m$ from equations 7, 8 and 9 will contain information on climatic factors (wet precipitation P and Φ_D) as shown in equation (10).Of course we need to develop new methods to eliminate the effects of the potential climatic factors P and Φ_D contained in equation (12) in the future.

Figure 2.31 shows the 130 ka ^{10}Be production rates derived for the Luochuan and Xifeng sections respectively. They clearly reveal the Mono Lake, Laschamp, Blake and other related excursion events, as the abnormal peaks in production rate due to the low paleointensity during typical excursions. However, there also exist minor differences between the two records, which might be due to the errors from the imperfect correlation of $[^{10}Be]_m$ and SUS_m to climate factors (P and Φ_D) and other unrecognized factors as well as their uncertainties. We also present the synthesized ^{10}Be production rate record (red line in Fig. 2.32) to avoid the random errors. Although the stacked record is similar to the two individual Pr curves with a relative standard deviation of about 3 %, the synthesized curve is relatively robust.

Since the fifth-order polynomial function describing the dependence of ^{10}Be production rate on geomagnetic field intensity is not easy to invert (Wagner et al. 2000), we developed an empirical formula (Xian et al. 2008) as shown in (13) to convert the ^{10}Be production rate signal to paleointensity changes.

$$\frac{M}{M_0} = 0.9864 \times \left(\frac{Pr}{Pr_0}\right)^{-3.1555} - \left(0.4806 \times \left(\frac{Pr}{Pr_0}\right)^2 - 1.2893\left(\frac{Pr}{Pr_0}\right) + 0.8239\right) \quad (13)$$

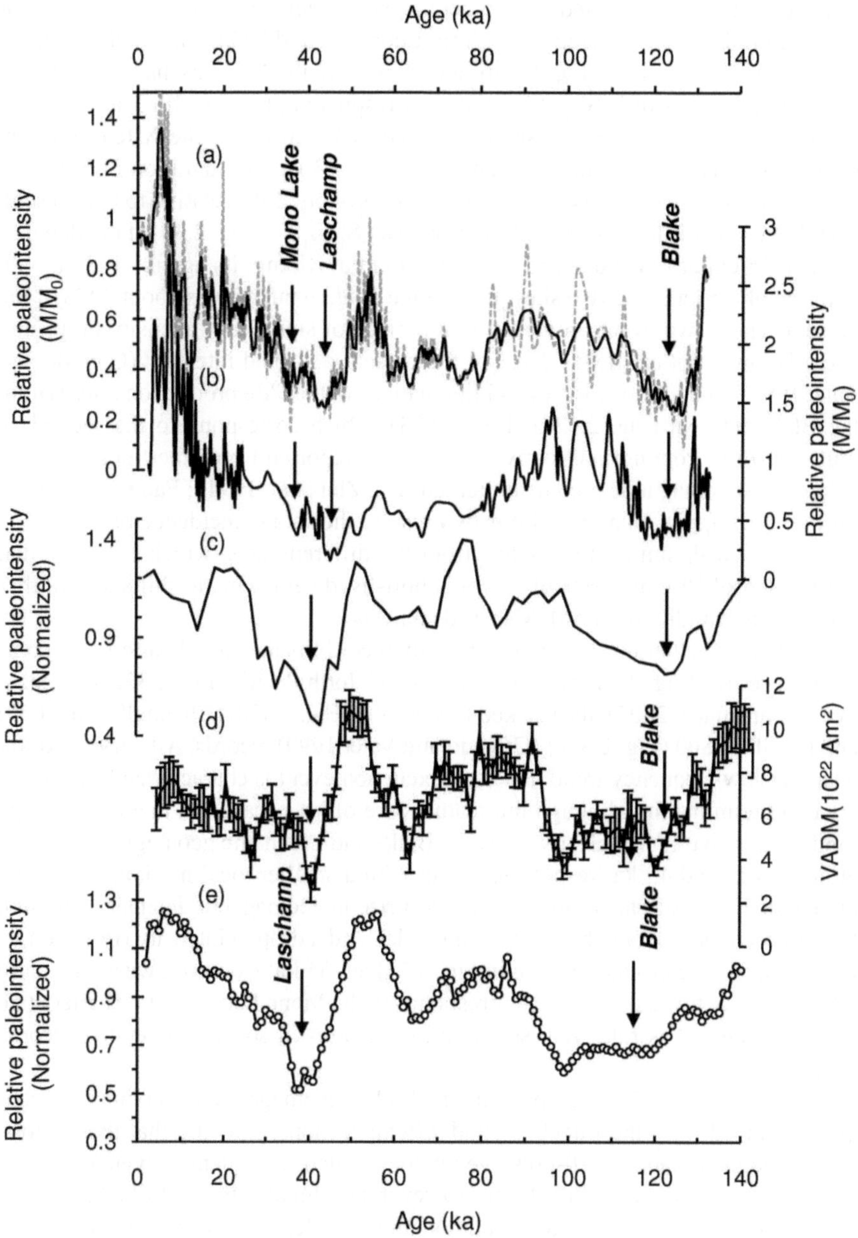

Fig. 2.33 The comparison of ^{10}Be resulting geomagnetic intensity records (a) Luochuan, (b) Xifeng with the individual records of (c) North Pacific (Yamazaki et al. 2007), (d) PISO-1500 (Channell et al. 2009) and (e) SINT800 result (Guyodo and Valet 1999)

The plot of the reconstructed 130 ka geomagnetic intensities (Fig. 2.33 a and b) or the ^{10}Be production rates (Fig. 2.32) from Luochuan and Xifeng and their synthesis record clearly reveal the Laschamp and Blake events. The Laschamp event lies in the middle layer of L_1SS_1 (Fig. 2.30), corresponding to mid MIS 3 at the depth of 492–524 cm in the Luochuan section and 584–620 cm in the Xifeng section, respectively. The age of the Laschamp event is ~43 ka with duration of 3.9 ka in the two sections. Its average relative intensity is about 20 % relative to the present. The Blake event is located at the lower paleosol S_1 (S_1SS_3, Fig. 2.30) at the depth of 1124–1176 cm at Luochuan and 1320–1424 cm at Xifeng. Its age is ~123 ka with duration of 8.5 ka in the two sites; its average relative intensity is about 30 % of the present. These two events coincide with periods of significant intensity drops and occur below a critical field <50 % of the present-day field intensity (Guyodo and Valet 1999). There also exist several fluctuations of the ^{10}Be production rates within the Blake interval (Fig. 2.32 and Fig. 2.33), which correspond to the secondary multiple shifts from normal to reversed polarity reported from other loess sections using the paleomagnetic remanent method (e.g. Zhu et al. 1994b; Fang et al. 1997). Further overlapping analysis of our two results shows a coincidence of the excursion events with similar time spans from the different sites, which indicates that loess-paleosol ^{10}Be may reliably record short-lived geomagnetic shifts and further supports the validity of our MVC-based analysis.

To check whether our ^{10}Be paleointensity record fits the global pattern, we compare our results (Fig. 2.33a and b) with those of North Pacific (Fig. 2.33c) (Yamazaki and Kanamatsu 2007), the stacked PISO-1500 (Fig. 2.33d) (Channell et al. 2009) and the SINT-800 (Fig. 2.33e) (Guyodo and Valet 1999) records. All curves exhibit a similar low-frequency trend and each excursion event is characterized by a rapid start and termination, although the features are offset because of the different age models. The five curves clearly show the Blake and Laschamp geomagnetic minima at about 123 and 43 ka, respectively, and with a smaller local minimum at about 60–65 ka. Our result also show a long-term rise in geomagnetic intensity common to most records observed between 40 and 5 ka, and a drop in intensity (reflected by an increased ^{10}Be production rate in Fig. 2.33) at ~35 ka. However, we are not sure whether or not this dip at ~35 ka corresponds to the Mono Lake event (Wagner et al. 2000), because it is relatively small and unobvious in some comparable records (Fig. 2.33).

The Mono Lake, Laschamp, Blake and other geomagnetic excursions are reliably extracted from the Luochuan and Xifeng sections, showing that these short-lived excursions can be well-preserved in loess-paleosol sequences even as far east as the Liupan Mountains (Fig. 2.29). However, the climatically controlled pedogenic, lock-in, smoothing, dust-accumulation and post-depositional remanence effects all have influenced the physical and chemical processes of ferromagnetic mineral formation in different paleosols and loess units. These processes result in the ambiguity of the geomagnetic-excursion signal in some stratigraphic levels, which is hard to detect by paleomagnetic measurement (e.g. Zhu et al. 1994b). As a result, whether the paleosol is capable of recording the Blake event has been debated (e.g. Zheng et al. 1995; Zhu et al. 1998). In contrast, the atmospheric ^{10}Be production

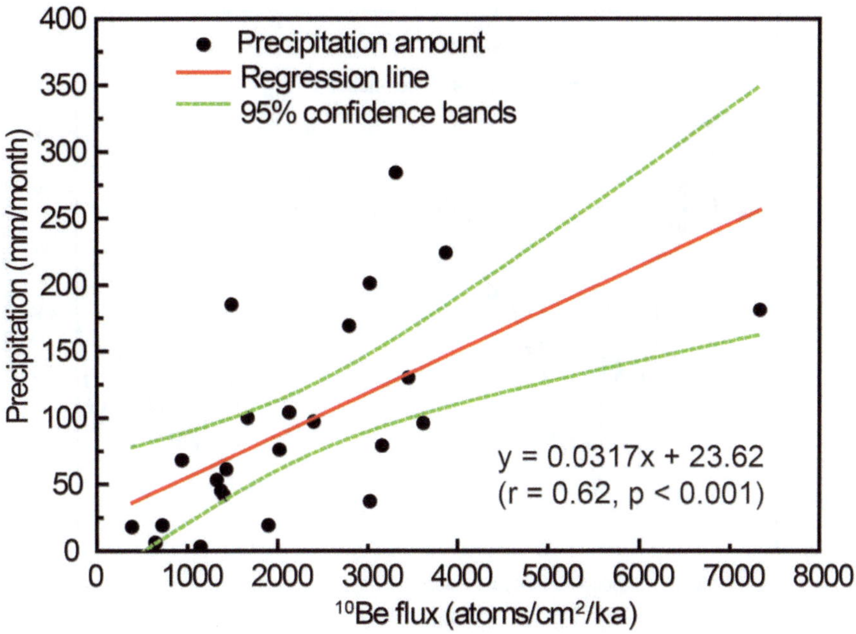

Fig. 2.34 The derived function between Monsoon precipitation amount and $[^{10}Be]_P$ flux. (Ishikawa et al. 1995)

rate signals derived here reflect the Blake geomagnetic excursion and are independent of lithology. The tracing approach of loess-paleosol ^{10}Be provides an effective means of generating an extremely long record of geomagnetic excursion and paleointensities over the last 2.6 Ma. The comprehensive combination analysis on typical excursion events with independent ages from radionuclide and OSL dating will potentially improve chronology studies on Chinese loess at the 10^3-yr scale.

2.5.4.4 Reconstruction of Paleoprecipitation Variations by Using ^{10}Be and ^7Be in the Past 80 ka

The atmospheric fallout $[^{10}Be]_P$ flux is regulated by the geomagnetic field intensity and strongly depends on the amount of monsoonal wet precipitation (Zhou et al. 2007a). This suggested a way to derive a high-resolution wet-precipitation record from the newly produced $[^{10}Be]_P$ flux.

Studies of cosmogenic ^7Be, which has a similar atmospheric production pathway to ^{10}Be, have proved that the dry-fallout fraction is generally second-order in magnitude. The rest, ~92–97 % of total ^7Be, arrives by wet precipitation (Wallbrink and Murray 1994). The correlation between the amount of precipitation and the ^7Be flux is well-fitted by a linear regression function.

In three different ^7Be *vs* precipitation studies from Austria, Germany and Switzerland, a relatively uniform $^{10}Be/^7Be$ flux ratio in modern tropospheric precipita-

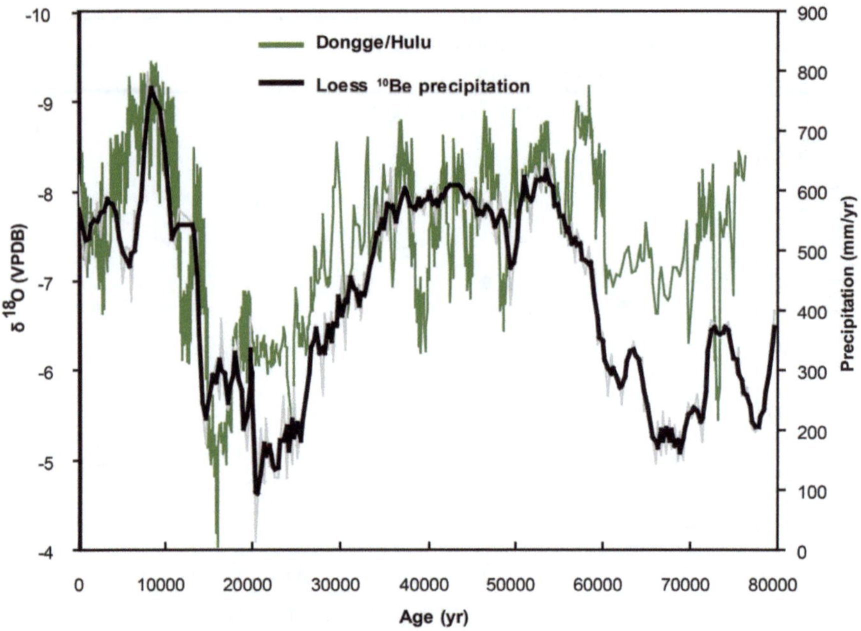

Fig. 2.35 The comparison of our ^{10}Be resulting paleoprecipitation curve (*black*) with the speleothem $\delta^{18}O$ records (*green*) over the past 80 ka. The $\delta^{18}O$ is a combination from Dongge (Dykoski et al. 2005) and Hulu (Wang et al. 2001) caves

tion was found (Priller et al. 2004). Since ^{7}Be has a very short half-life (~53 days), this means that there is no recycled dust component to confuse its relationship with rainfall amount. Therefore, we can use ^{7}Be as an analog for ^{10}Be in modern rainfall (Ishikawa et al. 1995) (Fig. 2.34). However, we must adjust the intercept (corresponding to the dry-fallout fraction) to match the precipitation observed at Luochuan today, in order to derive the function relating Monsoon precipitation amount and the precipitation-related ^{10}Be flux (Zhou et al. 2007b). In this way, a quantitative 80-ka precipitation record can be extracted from the ^{10}Be signal from the loess.

2.5.4.5 Comparison of Paleoprecipitation Variations on the Loess Plateau with Monsoon Proxy and Insolation Variations Over the Last 80 ka

Figure 2.35 shows the resulting paleo-precipitation record derived from the loess ^{10}Be record after removal of the effects of dust and geomagnetic modulation (in black). The general shape of this curve appears well correlated with the speleothem $\delta^{18}O$ records from Dongge (Dykoski et al. 2005) and Hulu (Wang et al. 2001) caves in SE China (in green), which are widely regarded as a robust record of Asian monsoon intensity. Our record indicates extremely low rainfall during the LGM

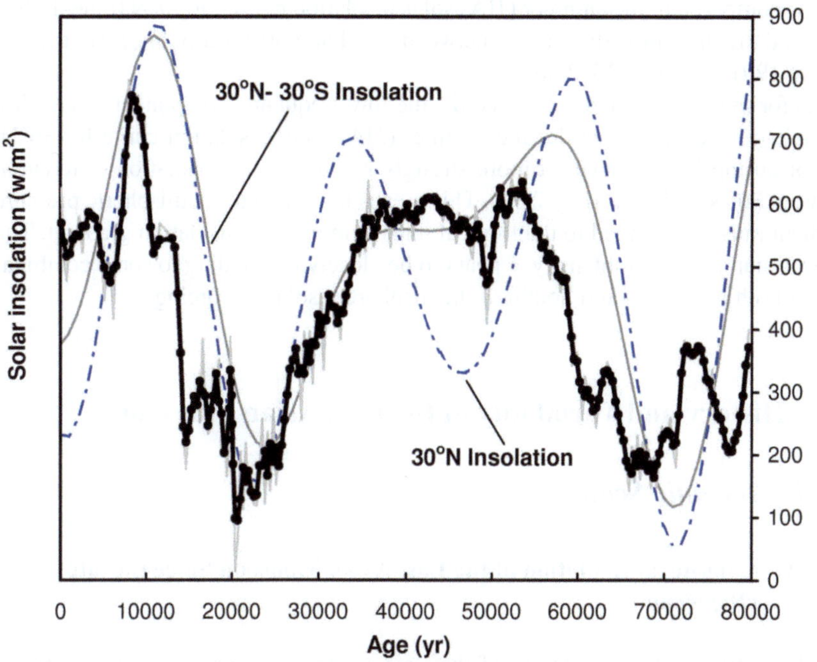

Fig. 2.36 Loess ^{10}Be precipitation record compared to records of solar insolation changes (Berger and Loutre 1991). The blue dashed line is the summer (JJA) solar insolation curve for 30°N, and the grey line is the insolation differential between 30°N and 30°S

and MIS 4, with mean amounts of ~100 mm and ~200 mm yr^{-1}, respectively. Monsoonal precipitation rose gradually after MIS 4 to near-modern levels during MIS 3, with the mean amount about 600 mm yr^{-1}. After ~35 ka, precipitation began a long decline, culminating in a 5 ka-long minimum during the LGM, ending at ~20 ka. Precipitation then increased stepwise in dramatic fashion at ~20 ka, and again at 14.3 ka. It was constant or decreased slightly between 13 and 11 ka, followed by another abrupt precipitation increase culminating in a Holocene maximum at 9 ka. Precipitation then dropped rapidly to a Holocene minimum at 5.8 ka followed by a modest rise to 4 ka and subsequent gentle decay and another small rise to the present during the late Holocene. The ^{10}Be precipitation record also suggests that the Holocene Optimum in East Asia is between about 6 and 10 ka, as presented by Zhou et al. (2004).

Both the ^{10}Be precipitation record and the Dongge and Hulu $\delta^{18}O$ curves show the significant strengthening of the East Asian Monsoon during MIS 3. This is similar to the significantly damped events reflected by Guliya ice-core $\delta^{18}O$ and lake records from the NW Tibetan Plateau (Thompson et al. 1997; Shi et al. 2001). However, this kind of feature does not appear in the marine and ice-core records, suggesting the local specificity of climate changes in East Asia. To explain this surprising feature of Chinese climatic changes during MIS 3, the ^{10}Be precipitation record is compared to the summer (JJA) solar insolation curve for 30°N (blue dashed line) and the insolation differential between 30°N and 30°S (gray line) (Berger and Loutre 1991) in Fig. 2.35, 2.36.

As for the Dongge and Hulu records, the low-frequency component of the ^{10}Be precipitation record resembles the summer (JJA) solar insolation curve for 30°N, except during MIS 3, when it more strongly resembles the insolation differential between 30°N and 30°S (Fig. 2.36). This means that the inter-hemispheric pressure gradient must be coupled to the tropical inter-hemispheric insolation gradient. The paleo-Asian monsoon intensity appears to be closely responding to some combination of both Northern- and Southern- hemisphere insolation forcing.

2.6 History and Variability of the East Asian Monsoon

2.6.1 Tectonic Scale

2.6.1.1 Long-term Evolution of the East Asian Monsoon Since the late Oligocene

The formation and development of the East Asian monsoon circulation were affected significantly by the uplift of the Tibetan Plateau (An et al. 2001; Liu and Chen 2000). Studies on the loess-paleosol and red clay sequence on the LP, which is close to the Tibetan Plateau and the northwestern inland deserts, show that the loess results from dust-deposition transported mainly by northerly winter monsoon, while the alternating paleosol strata relate closely to the moisture-laden southerly summer monsoon.

Miocene dust deposits in Qin'an to the west of the Liupan Mountains might suggest the initial aridification of western China by early Miocene. The magnetic susceptibility in the Qin'an red clay sequence is relatively low, with two exceptional high values around 16–14 Ma and 8.2–6.2 Ma (Guo et al. 2002), implying two episodes of intensified pedogenesis and summer monsoons. High susceptibility is also found around 7 Ma in the Dongwan section (Hao and Guo 2004). Multiple eolian dust-deposition sequences on the LP east of the Liupan Mountains share similar basal ages of 7–8 Ma (Sun et al. 1998a, 1998b; Ding et al. 1999b, 2001a). No high susceptibilities are found in these sections, possibly due to hydrous mineral alteration processes. The lower strata from 21.8 to 19.2 Ma in the Linxia Basin contain forest-steppe animal fossils in large quantities. Pollen studies there show an environmental transition from steppe to forest at ~21.8 Ma, reflecting increased precipitation and a warm-humid climate then. These lines of evidence suggest the initiation of the East Asian monsoon at least by early Miocene (Li et al. 1997; Ma et al. 1998).

Based on the loess magnetostratigraphy and stable-isotope records, An et al. (2005) proposed that the East Asian monsoon intensified at 3.4, 2.5, 1.2, 0.9, 0.5, and 0.15 Ma, linked to phased uplift of the Tibetan Plateau at 1.4–1.2, 0.5, and

0.1 Ma. This tectonic-monsoon linkage was highlighted by significant intensification of the East Asian summer monsoon around 1.2 Ma and 0.5 Ma, respectively (An et al. 1990; Xiao and An 1999).

The evolution of the East Asian monsoon recorded by loess and red clay sequences were extended back to the Miocene (An et al. 1999, 2001; An 2000; Guo et al. 2002). A combination of geological and biological evidence from the land and Ocean with and the numerical simulations indicates the coupling between the uplift of the Tibetan Plateau and changes in the East Asian monsoon, and also the onset of the Northern Hemisphere Glaciation (An et al. 2001). The importance of the plateau uplift from 3.6 to 2.6 Ma shows in the sharp increase in the κ of the red clay exceeding $200 \times 10^{-8} \cdot m^3 \cdot kg^{-1}$ and simultaneous increase of coarse grains transported by winter monsoons. The intensified East Asian winter and summer monsoons are consistent with the pollen records in the Jiuxi Basin implying frequent alternation of forest and steppe vegetation, and larger ^{13}C values of the red clay implying expansion of C_4 vegetation (An et al. 2005). The abundance of the zooplankton *N. dutertrei* (%) in the 1146 sediment core from the South China Sea increased at 3.2 Ma, also suggesting monsoon intensification then (Wang et al. 2003).

From 1.3 to 1.0 Ma, the susceptibility of the loess-paleosol sequence increased again, exceeding $150 \times 10^{-8} \cdot m^3 \cdot kg^{-1}$. Simultaneously the C_4 vegetation expansion implied by the C isotope data and increase in *N. dutertrei* percentage from the 1146 core suggest the intensification of the East Asian summer monsoon around 1.2 Ma. The high coarse-grain content of the corresponding L_{15} silt layer on the LP suggested intense winter monsoons during this time. After 0.6 Ma, the most developed paleosol (S_5), with $\kappa \leq (250\text{–}300) \times 10^{-8} \cdot m^3 \cdot kg^{-1}$, corresponds to another expansion of C_4 vegetation. At this time, the planktonic foraminifer content in the South China Sea also increased significantly, and loess strata show increased coarse-grain content, suggesting intensification of the winter monsoon.

The loess-paleosol and red clay sequences, the South China Sea records, and their correlation with global climate records shown in Fig. 2.37. Five important stages in the evolution of the coupled monsoon and arid environmental system since the Miocene are summarized in the following:

- 25–22 Ma—the climate system controlling Chinese terrestrial environment changed from a planetary to a monsoon system. The inland arid area and the area affected by the East Asian monsoon started to differentiate, and eolian dust deposits first appeared on the LP. With the development of the summer monsoon, the eastern humid area extended to the north. The South China Sea expanded, and the Himalayas and Tibetan Plateau were uplifted.
- 16–14 Ma—the inland arid area further expanded; monsoon intensification recorded a major uplift episode in the Tibetan Plateau.
- 10–7 Ma—major uplifting of the Tibetan Plateau and Inland aridification intensified. The Chinese monsoon/arid-environment system had become establiushed and the Indian Ocean summer monsoon intensified. Eolian red clay deposits were widely developed on the Chinese Loess Plateau.

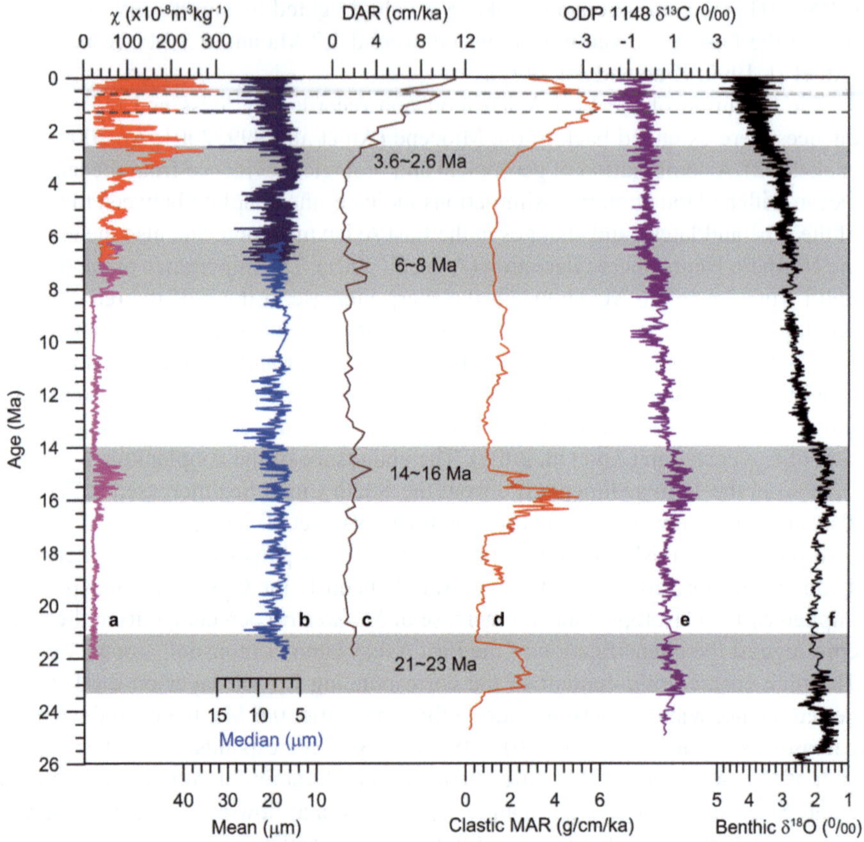

Fig. 2.37 According to monsoon sequence, aridification sequence, the South China Sea records, and the correlation with the global records described above, important evolution stages of the Chinese terrestrial monsoon-arid environment differentiation and coupling since the Miocene

- 3.6–2.6 Ma—the Tibetan Plateau expanded toward north and northeast, and the onset of large-scale glaciations resulted in the establishment of the modern Chinese monsoon/arid-environment pattern.
- After 2.6 Ma—The Tibetan Plateau continued to expand toward the northeast in episodes at 1.3–1.0 Ma and after 0.6 Ma, reflected in the significant intensification of the summer monsoon and inland aridifaction.

It is noteworthy that from 8 to 7 Ma, at 3 Ma and 1.2 Ma, and after 0.6 Ma, C_4 vegetation expanded episodically (An et al. 2005) (Fig. 2.38). The irregular expansion of the Tibetan Plateau, the resulting intensification of the monsoon and inland aridification were the dynamic factors causing this, since C_4 vegetation requires a semi-arid to semi-humid steppe environment and summer precipitation. The growth of the Tibetan Plateau had profound effects on the ecological environment in East Asia, and caused feedbacks from the East Asian monsoon system. To further

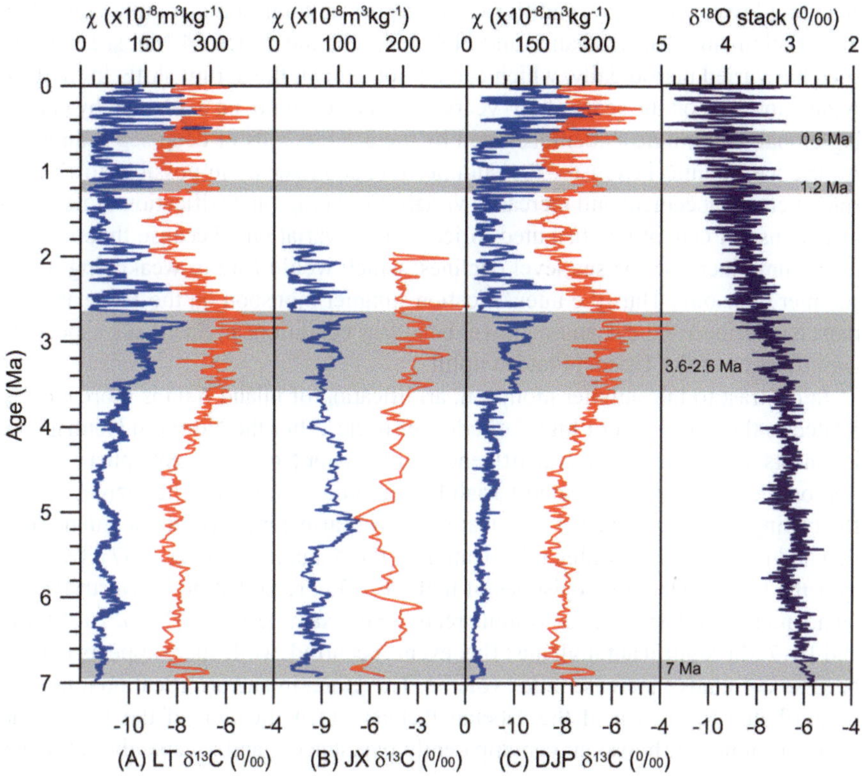

Fig. 2.38 MS curve in the past 7 Ma and it is noteworthy that from 8 to 7 Ma, at 3 Ma and 1.2 Ma, and after 0.6 Ma, a number of C_4 vegetation expansion events occurred. (An et al. 2005)

demonstrate the tectonic-climate linkage, numerical tests have been made to check the impacts of the plateau growth on environmental change. In these tests, a Global Circulation Model (GCM) was used in a series of sensitivity experiments in which the plateau elevation was altered. Plateau elevation was found to have a strong influence on the formation and development of the monsoon and arid environment, and the uplift increased the temporal variability of the East and South Asian monsoons on the ~10^4 yr-scale. Another set of numerical experiments, obtained by changing the topographic elevation and area of the Tibetan Plateau using the Community Climate Model (CCM3) of the National Center for Atmospheric Research, show that along with the uplift of the Tibetan Plateau, Central Asian aridification intensified, reflected by the precipitation rate in summer, and the summer monsoon strengthened (An et al. 2001). These numerical experiments illustrate the causal relationship between the growth of the Tibetan Plateau and the development of the Asian monsoon and inland aridification.

To understand the growth of the Tibetan Plateau, it is important to investigate the relationship between regional climate in East Asia and global climate. Except for changing lower boundary conditions, the East Asian monsoon is mainly controlled by variations in solar irradiance and global ice volume. Since 24 Ma, solar irradiance has varied periodically, which cannot not explain the abrupt shifts in the East Asian monsoon on the scale of 10^6 years, which also cannot be explained by changing global ice volume as demonstrated by the $\delta^{18}O$ records of deep-sea sediments. As the ice volume increases, the summer monsoon seems impossible to display staggered enhancement and increased variability. Abrupt intensification of the summer monsoon cannot be attributed to ice volume variations, because theoretically ice-volume increases as sea level declines, which would lead to weakening of the summer monsoon. Thus the intensification summer monsoon on the 10^6 year scale must be ascribed to the changing lower boundary conditions in East Asia, especially the influence of the Tibetan Plateau uplift.

In contrast to the summer monsoon, aridification of inland Asia is more closely related to the expansion of high-latitude ice sheets. After the Northern Hemisphere ice sheets formed at ~3 Ma, and influence of the major Northern Hemisphere glaciation on the East Asian monsoon variability became prominent. The increasing but fluctuating ice volume during the Quaternary was also conducive to the intensification and increasingly variable aridification of the Asian inland (Fig. 2.37). Even so, over the last 25 Ma a few episodes of high dust flux recorded in the North Pacific do not match well with the ice volume record. For example at 24–22 Ma, 16–14 Ma, and 10–7 Ma, significant high dust flux events occurred, while there appears to have been no correlative increase in ice volume. Therefore, inland aridification must still be attributed to growth of the Tibetan Plateau. Intensification of the East Asian summer monsoon, however, can apparently overshadow and reverse the influence of the Tibetan Plateau.

2.6.1.2 C_4 Expansion Over the Past 7 Ma Linked to the Summer Monsoon Variation

The relative influence of climatic (precipitation, temperature) and atmospheric (pCO_2) conditions on the natural abundance of C_4 and C_3 plants is an important but widely debated topic in earth-system history (i.e. Quade et al. 1989; Cerling et al. 1997; Pagani et al. 1999; Huang et al. 2001; Schefuß et al. 2003). Because C_4 plants concentrate CO_2 and have very low levels of photorespiratory inhibition of photosynthesis (Farquhar et al. 1982), they have an important advantage over C_3 plants at times of low pCO_2/O_2 ratios (Sage 2001). The widespread expansion of C_4 plants in the late Miocene has been attributed to a significant reduction of atmospheric pCO_2 levels (Cerling et al. 1997). However, recent evidence suggests that aridity and temperature are also important factors in controlling the relative ratios of C_4/C_3 plants (Pagani et al. 1999; Huang et al. 2001; Gu et al. 2003, Schefuß et al. 2003, Zhang et al. 2003). C_3 plants can increase in abundance relative to C_4 plants at low atmospheric pCO_2 levels (e.g., during the last glaciation) if precipitation is sufficiently high (Schefuß et al. 2003). The sensitivities of C_3 and C_4 plants to environmental factors are also likely to differ in different natural ecosystems.

In the central Chinese Loess Plateau, the well-known loess-paleosol/red clay eolian sequence provides continuous records of East Asian climate and vegetation change since the late Miocene. The high-resolution carbon-isotope time series derived from pedogenic carbonate (nodules) and soil TOC covering the last 7 Ma represent the history of C_4 vegetation in the Loess Plateau (An et al. 2005).

The proportion of C_4/C_3 vegetation is estimated based on the average end-member organic $\delta^{13}C$ values of C_3 and C_4 plants in the LP, −27.3‰ and −12.6‰, respectively (Liu et al. 2003a). For the carbonate nodules, the proportion of C_4/C_3 vegetation is estimated by the average $\delta^{13}C$ values of C_3 and C_4 plants, which are 12‰ and 2‰, respectively (Quade et al. 1989), using the following equation:

$$\delta^{13}C_{soil} = \delta^{13}C_{C4\_}sc\, f + \delta^{13}C_{C3\_}sc\,(1-f) \tag{14}$$

where f is relative abundance of C_4 plants and $(1-f)$ is relative abundance of C_3 plants; and $^{13}C_{C4\_}sc$ and $^{13}C_{C3\_}sc$ are $\delta^{13}C$ values of the pure C_4 carbonate and pure C_3 carbonate, respectively.

Carbon isotopic compositions of 367 and 528 pedogenic carbonate samples from the Lantian and Lingtai sections comprise two high-resolution carbon isotope time series covering the last 7 Ma (Fig. 2.38). These C_4 expansions covary with strengthened East Asian summer-monsoon circulation.

At the glacial-interglacial scale, both C_3 and C_4 soil carbonate have higher $\delta^{13}C$ values in the interglacial soils than in the glacial loess, indicating more C_4 plant biomass during soil-forming intervals (Zhang et al. 2003). Over the last 1.6 Ma, the maximum soil organic matter $\delta^{13}C$ values during interglacial periods at Lantian averaged −18‰, and the glacial minimum soil organic matter $\delta^{13}C$ values averaged −24‰, reflecting an increase of C_4 plant biomass from ~22% in glacial to ~63% in interglacial. In corresponding paleosol layers, isotope data of soil organic matter and soil carbonate are highly consistent with magnetic susceptibility (Fig. 2.38), a widely used proxy for changes in East Asian summer-monsoon circulation (An et al. 1991a).

Such C_4/C_3 shifts were unlikely caused by glacial-interglacial variations in atmospheric pCO_2, because there were more C_3 plants in the LP during glacial periods when CO_2 concentrations were low, as shown by ice core records (Petit et al. 1999). This suggests that low pCO_2 alone is insufficient to drive an expansion of C_4 plants in the absence of other favorable climatic conditions (Huang et al. 2001; Schefuß et al. 2003). Because monsoonal influence on the LP was strengthened during interglacial relative to glacial periods (An 2000; Guo et al. 2000), the greater proportion of C_4 biomass during interglacial intervals agrees well with our observation from modern vegetation regimes that the C_4/C_3 ratio increases with enhanced monsoonal influence (Fig. 2.38). This vegetation pattern in northern China is consistent with the vegetation changes reported from other semiarid areas (Johnson et al. 1999), but differs from the pattern in regions with enhanced winter precipitation (Huang et al. 2001). This again indicates that C_4 biomass is favored by increased warm-seasonal rainfall and temperatures in semiarid regions.

At 7 Ma, soil carbonate $\delta^{13}C$ values at Lingtai averaged −6.7‰, suggesting a ~38% C_4 biomass. This is consistent with the observation for southern Asia (Cerling et al. 1997; Morgan et al. 1994; Quade et al. 1989) that C_4 plants existed in the Asian continent since ~8 Ma. From ~6.8 to 4.5 Ma, soil carbonate $\delta^{13}C$ values at Lingtai oscillated between −7.5‰ and −8.9‰ and averaged −8‰, indicating the vegetation type dominated by C_3 plants. The first shift of soil carbonate $\delta^{13}C$ values at Lingtai, toward higher $\delta^{13}C$ values, occurred at ~4.5 Ma, reaching peak values of ~ −3.5‰ at ~2.8 Ma. This indicates a substantial expansion of the C_4 plant biomass to an estimated proportion of 60%. This shift coincided with an increasing degree of soil development, as shown by the enhanced magnetic susceptibility at both Lingtai and Lantian, and by the Rb/Sr ratio at Xifeng (An et al. 2001), an index of the chemical weathering intensity of the loess. Because chemical weathering in the LP mainly depends upon summer precipitation and temperature (An et al. 2001; Guo et al. 2000), the increase in C_4 biomass must have resulted from increased warm-season rainfall and temperature, and hence strengthened summer-monsoon circulation.

From 2.7 Ma to 1.6 Ma, carbon isotope values from both sites gradually decreased to their lowest values (Fig. 2.38), suggesting a corresponding decrease in C_4 plant biomass. Magnetic susceptibility of the soil units exhibits a similar change. This corresponds to an interval with weak soil development (Guo et al. 2000), an indication of generally dry and cold conditions. A second shift to higher carbon isotope values is observed for 1.6–1.0 Ma, suggesting another significant expansion of C_4 plants that approached ~38%. This interval also was characterized by higher susceptibility values with a series of overlapped and developed soil layers (Guo et al. 2000), indicating increased summer monsoon strength. This ^{13}C-enriched peak at ~1.2 Ma is followed by a lower one around 0.9 Ma, corresponding to coarse-grained loess unit L_9, an indication of very dry and cold conditions in the LP then (Kukla and An 1989).

Since 0.6 Ma, $\delta^{13}C$ values of soil carbonate at both sites, particularly in paleosols, has shifted continuously to higher values, indicating expansion of C4 biomass during interglacial periods. The maximum $\delta^{13}C$ values at Lantian (−3.4‰ for soil carbonate) and Lingtai (−4.9‰) suggest a proportion of C_4 plant biomass of > 60% at Lantian and ~51% at Lingtai. Similarly, the magnetic susceptibility of the soil units gradually increased to the highest values in the sequences. In the LP, paleosols younger than 0.7 Ma are the most-developed (Guo et al. 2000), characterized by clay-leached features and deeper red coloring. These features indicate much more humid and warmer conditions, attributable to strengthened East Asian summer monsoons.

Comparison of the carbon isotope data with other climate indices clearly shows that C_4 plant biomass was favored by warm-season precipitation and temperature in the LP. This observation is valid at glacial-interglacial and longer time scales, and is also confirmed by the modern distribution of plants and soil organic matter $\delta^{13}C$ values in this region. The abundance of C_4 biomass decreases from southeast to northwest, in the same direction as the seasonal temperature and precipitation gradients associated with summer-monsoon penetration. At geological time scales, covariance between the magnetic susceptibility and $\delta^{13}C$ records indicates that C_4 biomass expansions were driven by strengthened summer-monsoon circulation at both glacial-interglacial and longer time scales. Integrated studies on paleoclimatic evidence and modeling imply that expansion of the northeastern Tibetan Plateau enhances the East Asian summer monsoon but leads to drying in central Asia (An et al. 2001). This is reflected in the increased eolian flux to the LP and the North Pacific Ocean (Rea et al. 1998) and is consistent with C_4 vegetation expansion under enhanced monsoonal climates within semiarid regions (Johnson et al. 1999).

Although the multiple C_4 plant expansions in East Asia appear to be tightly coupled with strengthened summer-monsoon circulation, we cannot rule out other possible influences, such as the impact of changing atmospheric pCO_2 on long-term expansions of C_4 biomass (Cerling et al. 1997; Raymo and Ruddiman 1992) and the long-term increase of global ice volume with associated lowered temperature contributing to the decline of C_4 plant biomass. Both regional and global factors were likely important in driving biomass change at these time scales (Fox and Koch 2003; Schefuß et al. 2003).

2.6.2 Modes, Periodicity and Amplitude Variations of Summer and Winter Monsoons Since the late Miocene

The alternating loess-paleosol sequences, characterized by large-amplitude fluctuations of both magnetic susceptibility and grain size, clearly document an enhanced summer monsoon during Pleistocene interglaciations and a strengthened winter monsoon during Pleistocene glaciations (Kukla and An 1989; An et al. 1990; Heller and Evans 1995; Liu and Ding 1998; Xiao and An 1999; An 2000). The underlying Pliocene red clays, however, exhibit relatively low-amplitude orbital-scale oscillations of both magnetic susceptibility and bulk grain size (An et al. 1999; Ding et al. 1999b; Ding et al. 2000; An et al. 2000b). Previous studies have tried to decipher the paleoclimatic system during the period of the red clay formation (An et al. 1999; Ding et al. 1999b; Guo et al. 2002; Vandenberghe et al. 2004); however the trend and variability of East Asian monsoon circulation during the Mio-Pliocene are still not well-understood, mainly due to inconsistent interpretations of these proxy indicators. For example, bulk grain size of the red clay has been interpreted as an indicator of westerly circulation (Ding et al. 1998b), winter monsoon intensity (An et al. 2000b; An et al. 2001) or combinations thereof (Vandenberghe et al. 2004; Sun et al. 2008a). Similarly, interpretation of red clay magnetic susceptibility as an indicator of summer monsoon strength remains controversial (Sun et al. 1998a; An et al. 2001; Ding et al. 2001b; Liu et al. 2003b).

Based on magnetic susceptibility and depositional rate changes of the loess/red clay sequences over the last 7.2 Ma from the LP, together with a comparison of marine records from the North Pacific Ocean, the evolutionary process of the Late Cenozoic glaciation in the Northern Hemisphere can be divided into three stages: an incipient stage around 7.2–3.4 Ma, initiation at about 3.4–2.6 Ma, and culmination

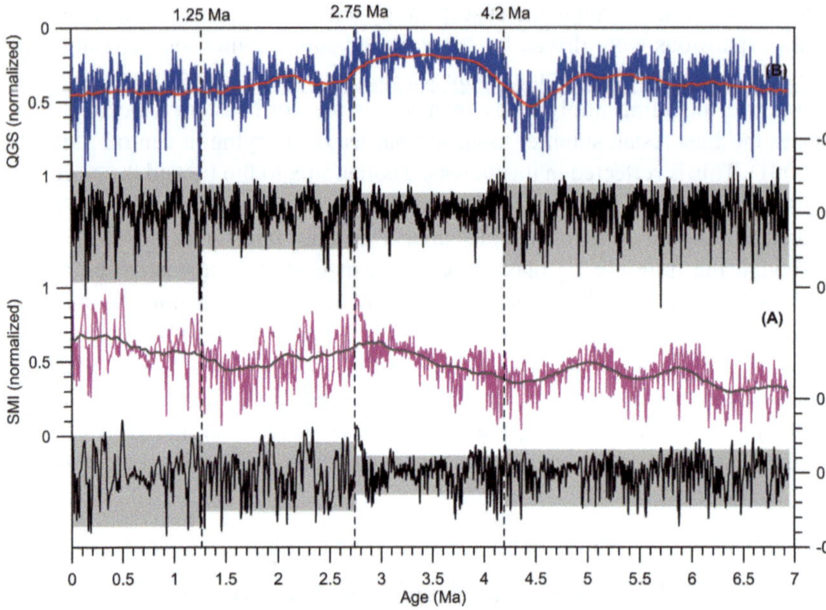

Fig. 2.39 East Asian monsoon oscillations over the past 7 Ma were reconstructed by using magnetic susceptibility and carbonate content as summer monsoon (precipitation) proxies (SMI), and quartz grain size (QGS) as a winter monsoon (wind-intensity) index (Sun et al. 2010a)

since 2.6 Ma (An et al. 1999). The evolution of the East Asian monsoon is characterized by paired fluctuations of winter and summer monsoons, and is basically composed of an initial stage of weak winter and summer monsoons, a transitional stage of simultaneous increase in both, and the prevailing stage of strong winter and weak summer monsoons during glacials, or weak winter and strong summer monsoons during interglacials. Grain-size variations of red clay and loess-paleosol sequences on the central LP confirm the re arrangement of past atmospheric circulation around 2.7 Ma and suggest several shifts afterward that linked the interaction of regional climate to tectonic events (Ding et al. 2000; Ding et al. 2005). Recently, East Asian monsoon oscillations over the past 7 Ma were reconstructed by using stacked magnetic susceptibility and carbonate content as a new summer monsoon index (SMI), and quartz grain size (QGS) as a winter monsoon (wind intensity) index (Fig. 2.39). The results suggest that precipitation and wind intensity exhibited significant orbital-scale variations prior to 4.2 Ma, followed by slightly damped variability between 4.2 and 2.75 Ma. Subsequently, East Asian monsoon circulation experienced two large shifts, at about 2.75 and 1.25 Ma, characterized by stepwise strengthening of glacial wind and interglacial precipitation (Sun et al. 2010a).

It is well-known that during the Pleistocene a strong winter monsoon (wind) and summer monsoon (precipitation) were generally 180° out of phase at glacial-interglacial scales, characterized by alternating dominance between a strong summer monsoon during interglacials and a strong winter monsoon during glaciations (An

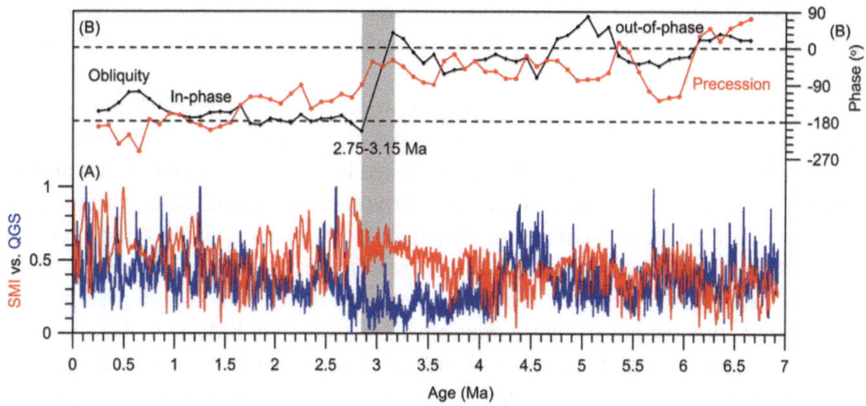

Fig. 2.40 The time-varying correlation coefficient between the QGS and SMI of the red clay deposits. (Sun et al. 2010a)

2000). Phase relations for the Pliocene interval, however, indicate that strong winter and summer monsoons were nearly in-phase (Clemens et al. 2008). Since magnetic susceptibility of the red clay is less sensitive to orbital-scale monsoon variations than that of the loess, the implications of the in-phase relationship between the summer and winter monsoons need further investigation. The phase relationship between the precipitation (SMI) and wind (QGS) was evaluated using two approaches. First, the time-varying correlation coefficient between the QGS and SMI was calculated using a 200 ka window with 20 ka time steps (e.g., 0–200 ka, 20–220 ka, etc.). Second, evolutive cross-spectral analysis of the QGS with SMI was performed using a 500 ka window with 100 ka time steps (e.g., 0–500 ka, 100–600 ka, etc.) to ascertain the phase relationship between these two proxies.

The time-varying correlation coefficient between the QGS and SMI of the red clay deposits differs significantly from that of the overlying loess-paleosol sequence. A rampfit function regression to the correlation coefficient reveals a gradual shift in the relationship between QGS and SMI from 3.15 to 2.75 Ma. Prior to 3.15 Ma, positive coefficient values indicate an almost in-phase relationship between wind strength and precipitation, whereas after 2.75 Ma negative values indicate that strong wind strength is associated with weak precipitation (Fig. 2.40). For the interval 3.15 to 2.75 Ma, the coefficient values reveal a transitional pattern of the coupling mode between wind strength and precipitation. Cross-spectral analysis reveals that the power spectrum and coherence between these two records are identical after 2.75 Ma, characterized by dominant peaks and strong coherency at the 100-, 41-, 23- and 19-ka periods. Prior to 3.15 Ma, spectral peaks and high coherence are identified at the obliquity and precession periodicities. Evolutive cross-spectral results between these two proxies show that phase relationships changed from nearly in-phase prior to 3.15 Ma to 180° out-of-phase after 2.75 Ma at the obliquity and precession periods. Both rampfit regression and cross-spectral results suggest that a transition of orbital-scale coupling between winter wind strength

(QGS) and summer precipitation (SMI) occurred around 3.15–2.75 Ma. Below, the paleoclimatic implications of this remarkable shift and the possible driving mechanisms are explored.

Since grain size reflects mainly surface processes (An et al. 1991a; Xiao et al. 1992; Sun et al. 2006b), it is best to consider QGS as a weathering-resistant and reliable indicator of wind strength. As suggested by previous studies (Porter and An 1995; Xiao et al. 1995; Sun et al. 2006a), orbital- to millennial-scale variations in QGS are closely linked to changes in Northern Hemisphere high latitude climate. In contrast, the SMI is strongly related to more complex post-depositional processes such as leaching, evaporation, and pedogenesis; all dependant on both precipitation and temperature. An out-of-phase relationship between the SMI and QGS since the Pleistocene suggests that strong wind and heavy precipitation occurred at opposite times within an orbital cycle. During warm and wet interglacial periods, a high SMI indicates strong leaching and pedogenesis induced by heavy precipitation, while low QGS corresponds to relatively weak wind strength. During cold and dry periods, leaching and pedogenic intensities would be significantly weakened due to limited precipitation, whereas wind intensity would be significantly strengthened.

In contrast, during the late Miocene and Pliocene high SMI (driven predominantly by the carbonate content change) coincided with coarser QGS, indicating that more effective precipitation occurred in the cold times. As mentioned above, the SMI variation in red clay (mainly induced by carbonate content) is controlled by the competing effects of precipitation-induced leaching and temperature-induced evaporation. If the SMI in the red clay formation was controlled mainly by leaching, similar to the overlying loess-paleosol sequence, we can infer an in-phase relationship between strong wind and heavy precipitation prior to 2.75 Ma. This inference confirms previous findings based on the evolutive phase results for the magnetic susceptibility and QGS records, implying that heavy precipitation occurred during the winter at obliquity minima (Clemens et al. 2008). Alternatively, if the SMI was related to temperature-controlled processes such as evaporation, higher SMI would indicate less production of secondary carbonate due to weak evaporation under cold conditions. We would then attribute the in-phase relationship between these two proxies prior to 3.15 Ma to the seasonal temperature difference. In either case, opposite-phase relationships between the QGS and SMI indicate that the wind and precipitation coupling during the late Miocene to Pliocene was fundamentally different than during the Pleistocene glacial–interglacial cycles.

To ascertain the long-term evolution of the summer monsoon during the Pliocene to early Miocene, much research has focused on the types and concentrations of magnetic minerals that are strongly linked with post-depositional weathering and pedogenic processes. The red clay is characterized by a lower magnetic susceptibility and by relatively higher super-paramagnetic grain content than in the Quaternary loess-paleosol sequence, suggesting that magnetic minerals within the two units are different (Liu et al. 2003b). Free Fe_2O_3 concentrations (mainly hematite and goethite) in the red clay exhibit large-amplitude oscillations with relatively high values compared to the overlying loess-paleosol sequences (Ding et al. 2001b), consistent with the relatively high hematite content and the ratio of hematite to goethite in the red-clay formation relative to the overlying loess and paleosol layers (Ji et al. 2006).

These magnetic mineral evidence suggests that the Neogene red clay formation was formed under relatively warmer and drier climatic conditions relative to those of the Pleistocene interglacial periods.

The 2.75 Ma climate shift was a global phenomenon, although it was particularly dramatic in the monsoonal regions. For example, marine records of African climate document a shift toward more arid conditions and increased variability after 2.8 Ma (Tiedemann et al. 1994; deMenocal 1995), though this shift was not supported by a statistical reassessment of previously published dust flux data (Trauth et al. 2009). The phase relationship between the Asian monsoon proxies and orbital forcing shifts significantly at both obliquity and precession frequency bands around 2.7 Ma (Clemens et al. 1996; Clemens et al. 2008). The SMI variation reveals a relatively weak summer monsoon prior to 4.2 Ma, followed by continued intensification of the East Asian summer monsoon until 2.75 Ma, consistent with monsoon precipitation variability inferred from the carbonate isotope record of the loess and red clay sequences (An et al. 2005) and the hematite/goethite ratio of South China Sea sediments (Zhang et al. 2009). The benthic $\delta^{18}O$ record reveals that the global temperature was about 3°C warmer during the Pliocene than during the Pleistocene (Zachos et al. 2001; Ravelo et al. 2004). Considering the effects of increasing precipitation and decreasing temperature together, we propose that prior to 2.75 Ma, monsoonal circulation was characterized by alternating dominance of cold-wet winter season and warm-dry summer season. After 2.75 Ma, when the effects of NH ice volume came to dominate the system, glacial intervals (cold and dry) were characterized by strong winter monsoons and weak summer monsoons whereas interglacial intervals (warm and humid) were characterized by the opposite, weak winter monsoons and strong summer monsoons.

In the past decade, the most striking advance of our knowledge about East Asian monsoon variability was based on high-precision numerically dated speleothem records, which can provide robust timing of abrupt monsoon events as benchmarks for correlating and calibrating different paleoclimate records (Wang et al. 2001, 2008; Cheng et al. 2009). However, the East Asian summer monsoon variations revealed by the speleothem records are different from the loess-based proxies, in terms of both frequency and phase. The spleothem $\delta^{18}O$ is characterized by dominant precession cycles over the last three glacial cycles, nearly in-phase with Northern Hemisphere summer insolation. By contrast, the loess magnetic susceptibility exhibits spectral peaks at the three primary orbital periods (eccentricity, obliquity and precession), corresponding well to the ice-volume change. These apparent discrepancies need further investigation, particularly in improving the chronology and in ascertaining the paleoclimatic interpretation of the monsoon proxies.

2.6.3 Land-ocean Comparison Over the last 7 Ma

Recent progresses both in studies of Chinese loess and deep-sea sediments have provided robust and longer records of winter monsoon variation and ice-volume change back to the late Miocene. However, when and how the winter monsoon

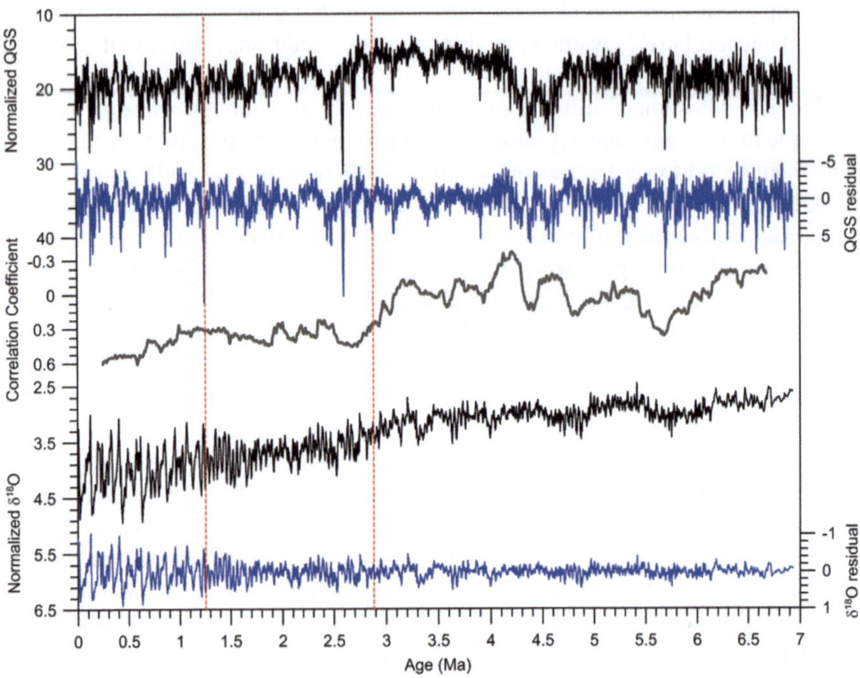

Fig. 2.41 Comparison of quartz grain size with bethic $\delta^{18}O$ record over the past 7 Ma. Also shown are the residuals correlation coefficient of the quartz grain size and bethic $\delta^{18}O$ records after removing >500 ka components

became coupled with global ice-volume change remains uncertain. Comparison of quartz grain size (a reliable winter monsoon proxy) generated from two loess-paleosol and red-clay sequences with a stacked benthic $\delta^{18}O$ record (a global ice-volume proxy) indicates that at longer (>500 ka) time scales, the winter monsoon became strongly coupled with global ice-volume change at 2.1 Ma, while at orbital time scales the winter monsoon variations were significantly influenced by global ice-volume change after ~3.3 Ma (Sun et al. 2010c). Correlation coefficients between these two records further indicate that winter monsoon intensity was strongly coupled with global ice-volume change at 0–1.1 Ma, 1.4–1.8 Ma, and 2–2.85 Ma (Fig. 2.41). In spite of these close connections, amplitude mismatches between these two records are evident in terms of both long-term trend and glacial-interglacial fluctuations. The mismatches suggest that additional processes might have played a role in modulating the response of the winter monsoon variation to ice-volume forcing.

East Asian monsoon evolution is dynamically linked to changes in solar insolation and internal boundary conditions (e.g., Prell and Kutzbach 1997; An 2000; Wang et al. 2008), and is closely affected by the integrated behaviors of the global climatic system (i.e., interactions between the atmosphere, ocean, land and ice systems) (Webster et al. 1998; Wang 2006). Both numerical climate-modeling studies and geological evidence suggest that uplift of the Tibetan Plateau and growth of the Northern Hemisphere ice sheet might have played important roles in enhancing monsoon variability (Prell and Kutzbach 1997; An et al. 2001; Liu and Yin 2002; Clift et al. 2008; Kutzbach et al. 1993; Ruddiman 1997). At orbital time scales, variations of magnetic susceptibility measured in Chinese loess suggest that changes in the intensity of the summer monsoon show both similarities in climatic cycles and differences in amplitudes of climatic fluctuations relative to global ice volume (Kukla and An 1989; An et al. 1990; Bloemendal et al. 1995). In contrast, variation of the winter monsoon as revealed by loess grain size indicates stepwise coupling to global ice volume change (Liu and Ding 1993; Ding et al. 1995; Liu et al. 1999a; Sun et al. 2010c).

Magnetic susceptibility and grain size of Chinese loess exhibit a dominant 100 ka cycle over the last 0.8 Ma, consistent with the periodicity revealed by the deep-sea $\delta^{18}O$ record (Ding et al. 1995; Liu et al. 1999a; Lu et al. 2004a; Sun et al. 2010c). However, absolute-dated oxygen isotope records (a proxy for monsoon precipitation) from cave speleothems reveal a dominant 23 ka cycle over the past 224 ka that is almost synchronous with summer insolation at 65°N (Wang et al. 2008). Such a difference in the periodicity suggests that the forcing mechanisms of the summer and winter monsoons are not necessarily similar. Most likely, the winter monsoon (northwesterly wind) is tightly coupled with the high latitude climate (Ding et al. 1995), whereas the summer monsoon (rainfall) is strongly related to the low latitude climate as well as cross-equatorial transport from the southern Hemisphere (An 2000; Wang et al. 2008). To assess the forcing mechanisms of the East Asian monsoon better, it is critical to differentiate the impacts of regional tectonic events and global climate change on the long-term trend and glacial-interglacial variability of the winter and summer monsoons.

Comparison of Milankovitch periods between continental loess and deep-sea records indicates that at about 1.6 Ma the ice sheets in the Northern Hemisphere may have reached a critical size, sufficient to modulate changes in the global climate system (Liu et al. 1999a). Further comparison of a stacked loess grain-size record with a composite marine $\delta^{18}O$ record shows that for the past 1.8 Ma, the loess-paleosol record can be correlated almost cycle-to-cycle with the marine record (Ding et al. 2002). Comparisons of quartz grain size and the Zr/Rb ratio with deep-sea $\delta^{18}O$ records reveal that good land-ocean correlation can be extended back to 2.6 Ma (Sun and An 2004; Liu et al. 2004). Since previous studies were based mainly upon visual inspection and didn't separately evaluate the impacts of global ice volume on the long-term trend and glacial-interglacial variability of the winter monsoon, when (1.6 Ma or earlier?) and how (stepwise or gradually?) the winter monsoon became coupled with global ice-volume change remains unclear.

The results of a comparative time-domain analysis of proxies of the East Asian winter monsoon and global ice volume for the past 7 Ma are presented in Fig. 2.41. Unlike previous studies, straightforward statistical analyses (i.e., filtering and correlation analysis) of the data sets are employed to define intervals of strengthened and weakened relationships between the two signals, on tectonic and orbital time scales. The objectives are to discern similarities and differences between these two records, particularly in terms of the long-term trend and glacial-interglacial variability, and then to evaluate when and how the East Asian winter monsoon became coupled with global ice-volume variability. Finally, possible roles of global ice volume and regional tectonic events in driving the long-term strengthening trend and increasing glacial-interglacial amplitude of the winter monsoon variation since the late Miocene are assessed.

Comparison of QGS with stacked $\delta^{18}O$ reveals notable mismatches around 1.25 Ma and 2.6 Ma (Fig. 2.41). Transient coarsening of the QGS indicates that during these two intervals, either the winter monsoon intensity was exceptionally strong or aridity and the geographic range of dust source areas were significantly expanded (Xiao and An 1999; Ding et al. 2002). Since global ice volume did not exhibit corresponding increases at these two time intervals, the marked QGS coarsening is likely related to changes in regional climatic/tectonic conditions, e.g. the stepwise uplift in the northeast margin of the Tibet Plateau (Sun and Liu 2000; An et al. 2001). To compare the QGS with the stacked $\delta^{18}O$ record better, the QGS values around these two short intervals are truncated. Since chronology of these two records was generated by orbital tuning for the last 3.6 Ma, the error was estimated to be less than 10 ka (Lisiecki and Raymo 2005; Sun et al. 2006a). For the convenience of comparison, these two data sets were then normalized and interpolated linearly at 2-ka intervals.

Both the QGS and stacked $\delta^{18}O$ records exhibit distinct orbital-scale variations superimposed on long-term trends. Since the long-term trend and glacial-interglacial variation might be related to different forcing mechanisms, it is necessary to separate these two components from the original data sets. Previous studies reveal that loess grain size and deep-sea $\delta^{18}O$ records are characterized by cyclic fluctuations at earth-orbital eccentricity (400 and 100 ka), obliquity (41 ka), and precession bands (23 and 19 ka) (Liu et al. 1999a; Lu et al. 2004b; Lisiecki and Raymo 2005; Sun et al. 2006a). Thus, a low-pass filter (>500 ka) is employed to separate the long-term trend from these two data sets, and the residual can be regarded as an expression of the orbital-induced variability.

Visual comparison of loess grain-size and benthic $\delta^{18}O$ records suggests that although the winter monsoon and global ice volume exhibit similar variability at glacial-interglacial time scales, the coupling relationship varies over the last 7 Ma (Fig. 2.41). To assess coupling between loess grain size and benthic $\delta^{18}O$ better, we calculated the time-varying correlation coefficient between QGS and stacked $\delta^{18}O$ using a 500 ka window with 20 ka time steps (e.g., 0–500 ka, 20–520 ka, etc.). Correlation coefficients indicate that the East Asian winter-monsoon strength is variably coupled with global ice-volume change: prior to 2.75 Ma, the winter monsoon was decoupled with global ice-volume change; during 2.75–1.25 Ma, the winter monsoon was moderately coupled with global ice-volume change; since 1.25 Ma, the winter monsoon became strongly coupled with global ice-volume change.

Prior to 2.75 Ma, the winter monsoon intensity was characterized by an obvious change in the amplitude around 4.2 Ma, whereas the $\delta^{18}O$ record exhibits a gradually increasing trend with glacial-interglacial variations superimposed (Fig. 2.41). From 2.75 to 1.25 Ma, correlation coefficients between these two records are between 0.3 and 0.45, indicating that winter-monsoon intensity started to be influenced by global ice-volume change at glacial-interglacial time scales. The QGS exhibits distinct glacial-interglacial variations with uniform amplitude, while the $\delta^{18}O$ record displays similar glacial-interglacial variations but superimposed on a gradually increasing trend. After 1.25 Ma, the QGS exhibits large-amplitude variations, corresponding well with glacial-interglacial fluctuations of the $\delta^{18}O$ record (Fig. 2.41). Correlation coefficients increase from 0.3 to 0.6, indicating that winter monsoon intensity became strongly coupled to global ice-volume change.

Comparison of the long-term trends revealed by these two records is evaluated separately to further evaluate the relationship between the winter monsoon and global ice volume change. The long-term trend of global ice volume increased gradually over the interval of 7–0.9 Ma and became stable over the last 0.9 Ma. In contrast, winter monsoon evolution is characterized by strengthening before 4.2 Ma, slight weakening during 4.2–2.75 Ma and then gradual strengthening from 2.75 to 0.9 Ma. It is obvious that the long-term trends of these two records are significantly different before 2.75 Ma and match very well thereafter.

In general, both the quartz grain size and magnetic susceptibility records of the eolian sediments from the Lingtai section can be correlated cycle-by-cycle with the marine $\delta^{18}O$ record over the past 7 Ma despite of a few mismatches (Fig. 2.42). Smaller (larger) grain size and higher (lower) susceptibility units correspond to lower interglacial (higher glacial) $\delta^{18}O$ units. This correlation is much better during the past 2.6 Ma than during the 7–2.6 Ma interval. This indicates that the global ice volume may have played an important role in modulating the orbital variability of the East Asian monsoon. This land-ocean climate linkage is significantly typical in the Quaternary. Generally, growth of ice cover during glacial periods would increase the continental albedo and subsequently decrease sensible heating of the Asian continent. This would result in decreased temperature gradient between the Asian mainland and ocean areas during summer seasons and increased temperature gradient during winter seasons, which in turn weakened summer monsoon and strengthened winter monsoon. In contrast, the decay of ice cover during interglacial periods would result in intensified summer monsoon and weakened winter monsoon along similar lines of reasoning. Therefore, the cycle-by-cycle correlation of the quartz grain size and magnetic susceptibility of the eolian sediments to the marine $\delta^{18}O$ record indicates that the coupling between East Asian monsoon and global ice volume at orbital time scales appears to be traced back to the Pliocene. In addition, frequency-dependent magnetic susceptibility from the Jiaxian red clay (Qiang et al. 2005) shows small values that vary with large-amplitude oscillations at orbital scales during the late Miocene. During the Pliocene, however, it is characterized by increased values that vary with quite small-amplitude orbital-scale oscillations.

Mechanisms driving the monsoon variation include both external and internal forcing (e.g., changing boundary conditions and interactions among climate components) (An et al. 1990; Liu and Ding 1993; Ding et al. 1995; Guo et al. 1996). At tectonic time scales, it has been suggested that global climate change and regional tectonic events might have different impacts on the East Asian monsoon evolution (Xiao and An 1999; An et al. 2000b; Sun and Liu 2000). For example, increase in the global ice volume will result in strengthening of the winter monsoon and

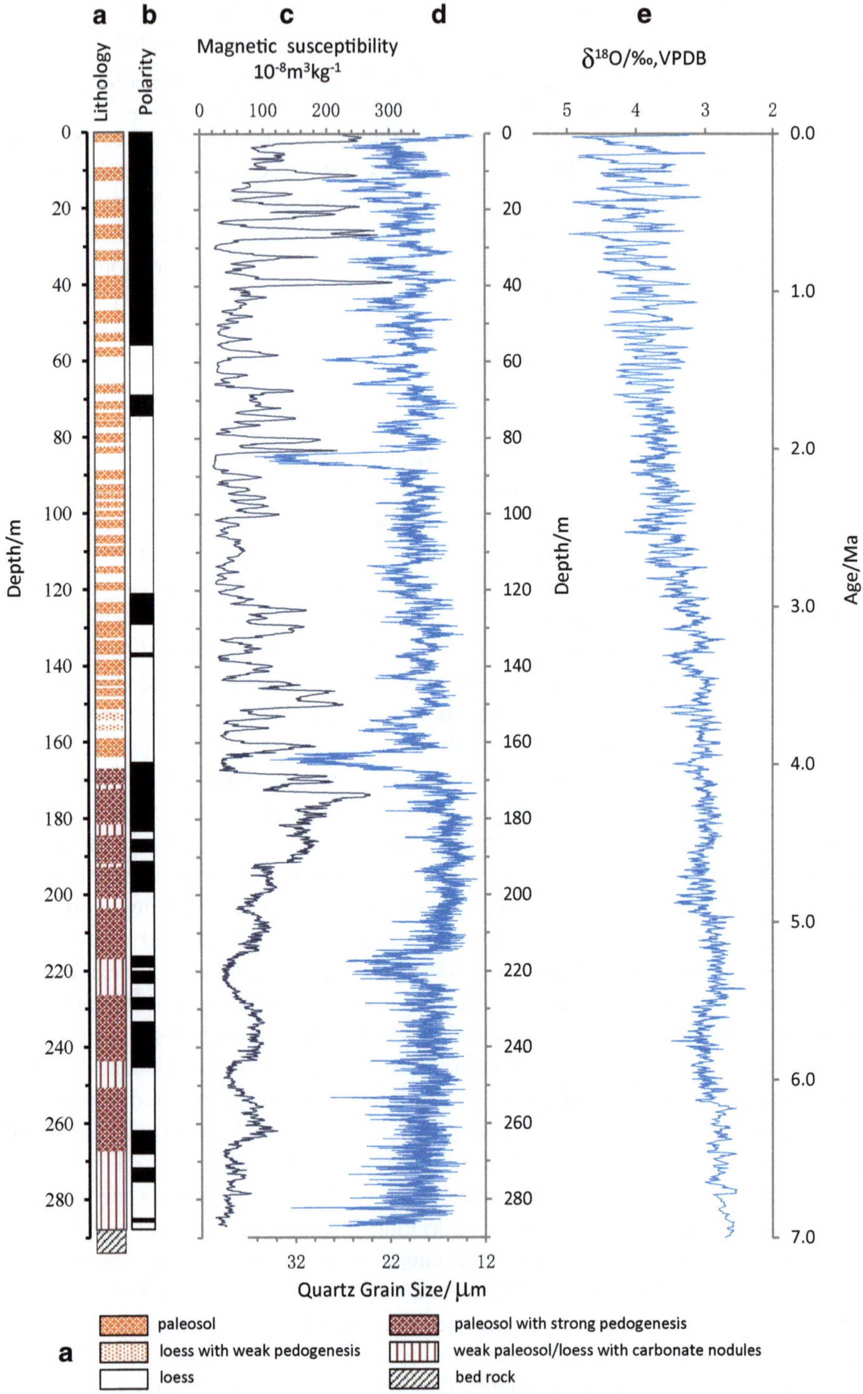

Fig. 2.42 Comparisons of the quartz grain size and magnetic susceptibility records of the eolian sediments from the Lingtai section with marine $\delta^{18}O$ record over the past 7 Ma

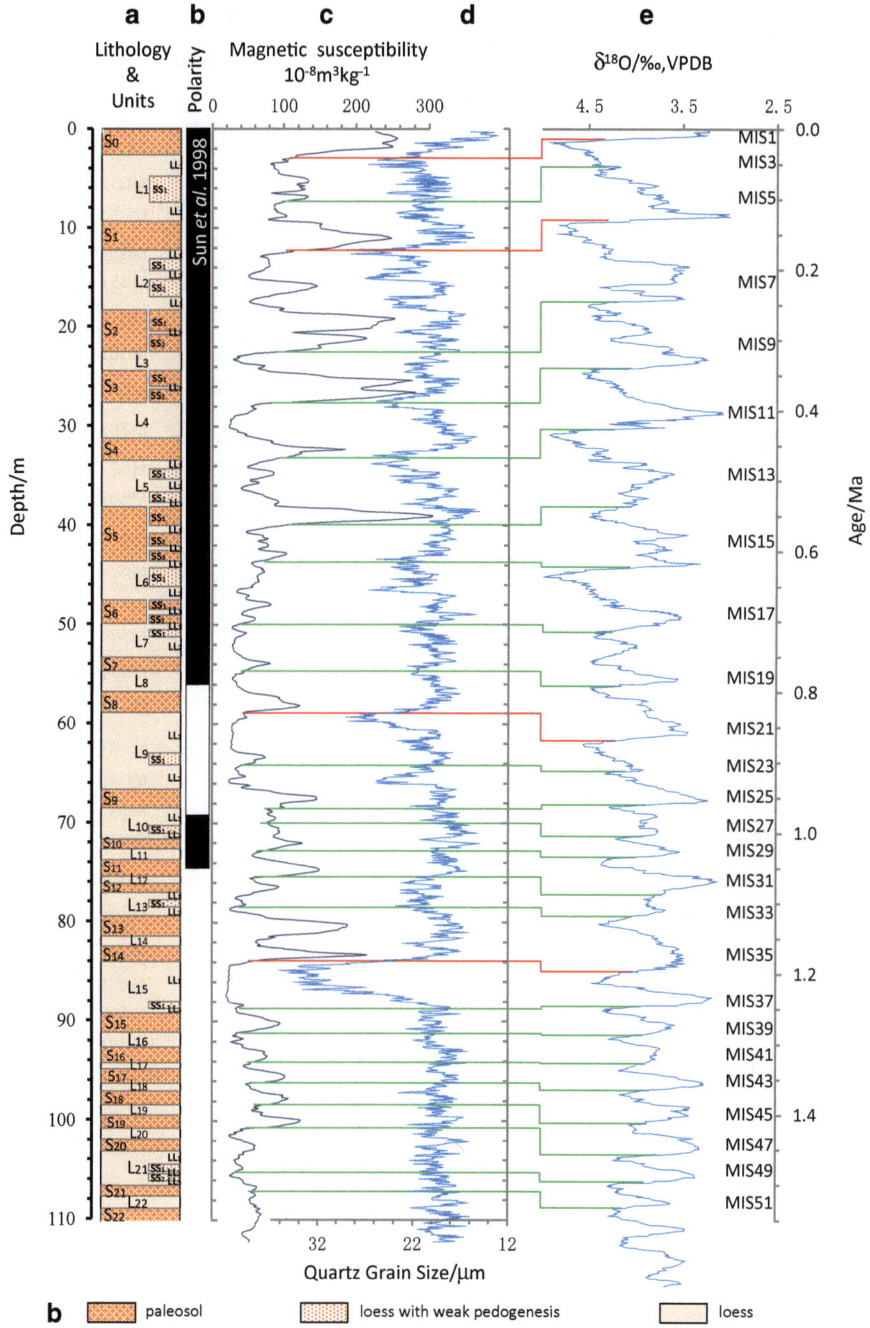

Fig. 2.42 (continued)

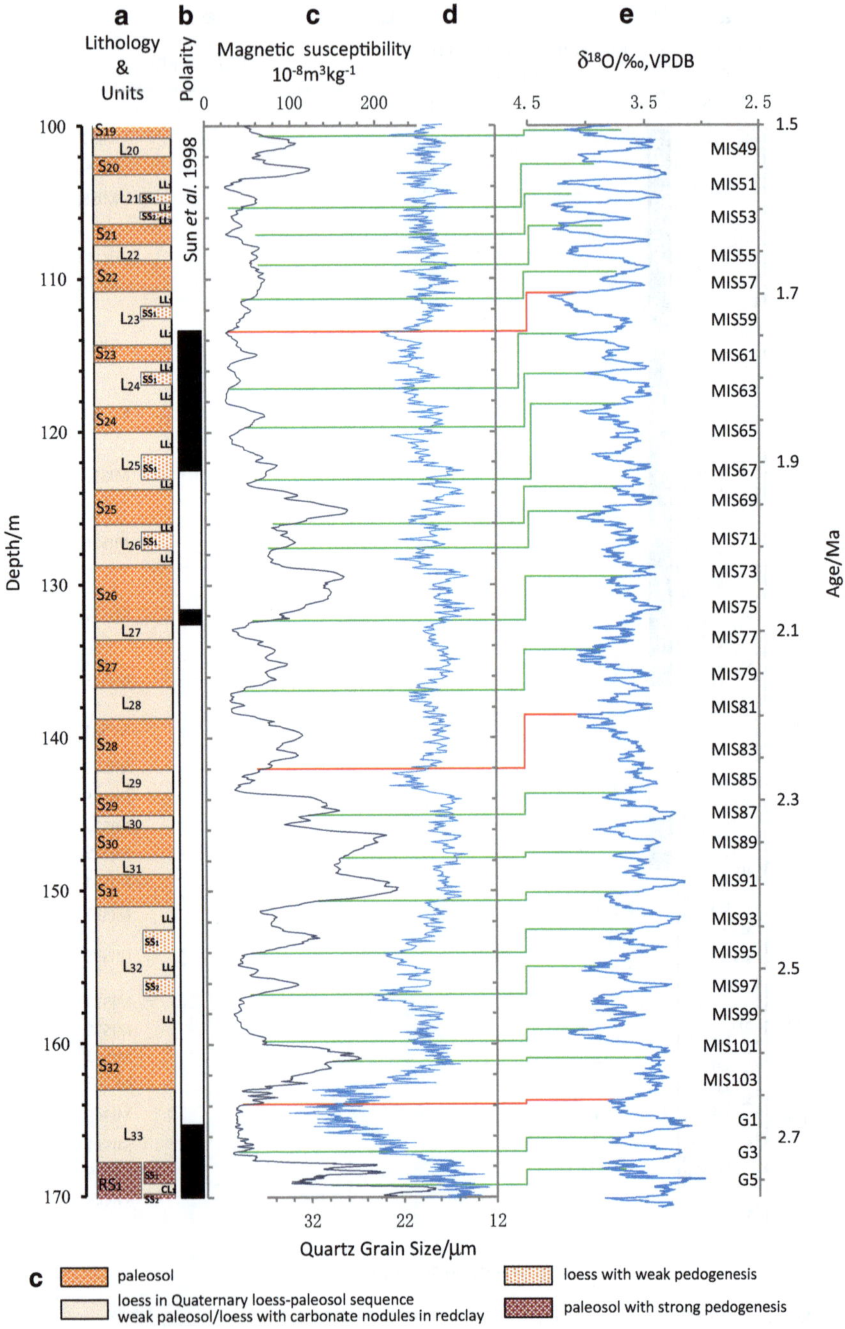

Fig. 2.42 (continued)

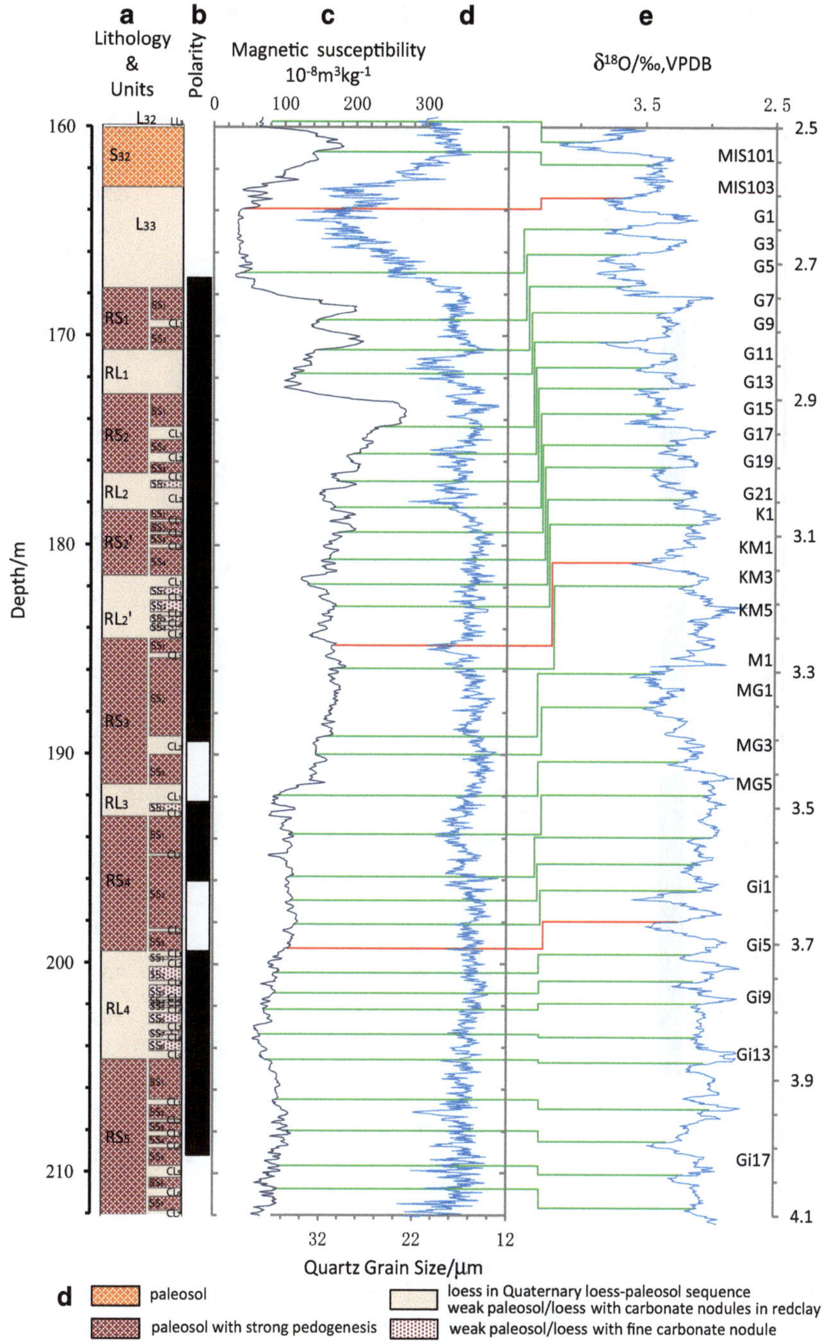

Fig. 2.42 (continued)

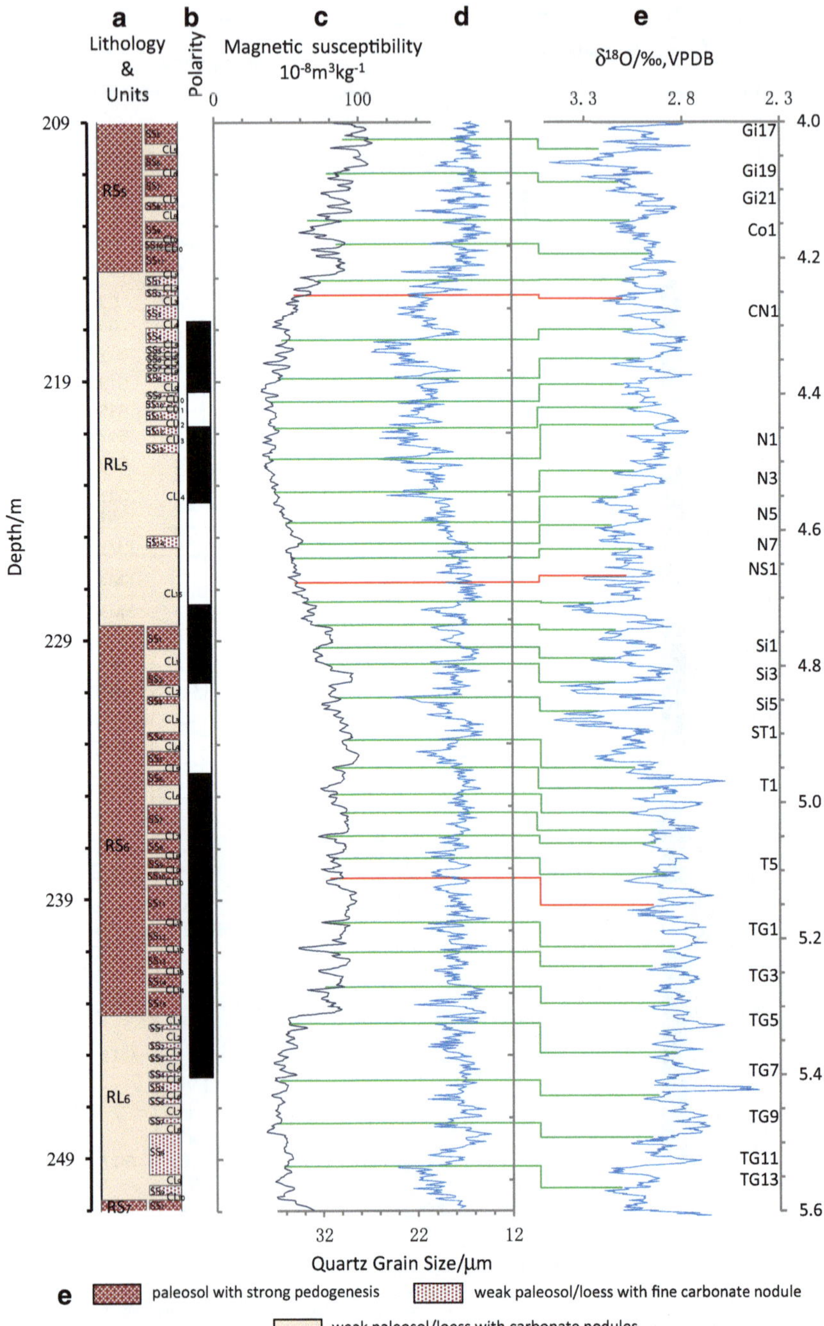

Fig. 2.42 (continued)

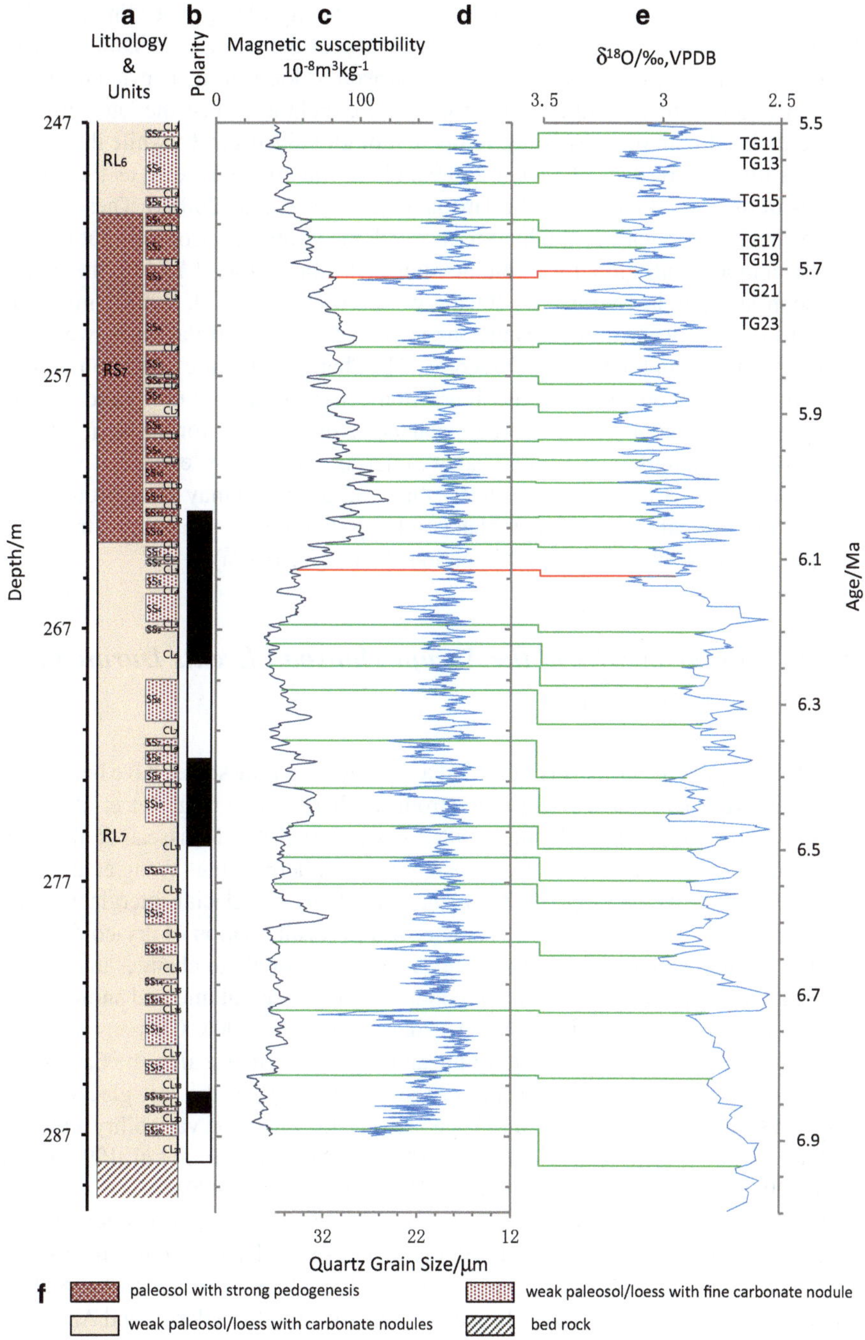

Fig. 2.42 (continued)

weakening of the summer monsoon, whereas tectonic uplift (e.g., the Qinghai-Tibet Plateau) will lead to simultaneously strengthening of the winter and summer monsoons (An et al. 2001). More recently, comparison of loess grain size and deep-sea records suggests that strengthening monsoonal circulation since the late Miocene was associated with progressive development of global ice sheets, while the abrupt changes that occurred at 7.6, 3.4, 1.2–0.9 and 0.6 Ma are interpreted as the consequence of the episodic uplift of the Tibetan Plateau (Sun et al. 2008a). These results suggest that with regard to the long-term trend, the winter-monsoon evolution was clearly coupled with global ice volume after 2.75 Ma. Before 2.75 Ma, the winter monsoon and perhaps westerlies (Qiang et al. 2010; Sun et al. 2008a) might be related to global ice volume changes as indicated by the correlation between the quartz grain size and marine $\delta^{18}O$ records. The changed climatic condition around 2.75 Ma might be attributed to the interaction between tectonic uplift and climate change factors. Tectonic changes (e.g., enhanced uplift of the northeastern Tibetan Plateau) and strengthened continental weathering induced by enhanced summer monsoon around 3 Ma (An et al. 2001; Zhang et al. 2009) may have brought ice volume to a critical threshold, at which point expansion in the ice volume triggered more extensive winter monsoon fluctuations (Liu et al. 1999a).

2.6.4 Characteristics of the Abrupt Monsoon Events During the last Glacial Cycle

East Asian monsoon variability at orbital and millennial time scales have been studied extensively using loess sequences from northern China (e.g. An et al. 1990; Porter and An 1995; Liu and Ding 1998; Xiao and An 1999; Rousseau et al. 2000; An 2000) and cave stalagmites from eastern and southern China (Wang et al. 2001; Yuan et al. 2004; Wang et al. 2008; Cheng et al. 2009). On glacial-interglacial time scales, the East Asian monsoon is characterized by an alternation of dry-cold winter and warm-humid summer monsoons and is coupled strongly to changes in external solar radiation and internal boundary conditions (e.g., ice volume and snow cover, latent heat sources) (Ding et al. 1995; An 2000; Wang et al. 2008; Clemens and Prell 2003; Clemens et al. 2008). At suborbital and millennial time scales, the East Asian monsoon has undergone significant changes at least during the last glacial cycle (Fig. 2.43), implying a dynamical linkage between the monsoon variability and high latitude Northern Hemisphere climate (Porter and An 1995; Guo et al. 1996; Chen et al. 1997; Ding et al. 1998a; Ding et al. 1999a; Fang et al. 1999).

Unlike the well-dated cave stalagmites, accurate dating of Chinese loess remains problematic. Previous assessments of rapid monsoon changes relied mainly on indirect chronologies; for example, tuning the loess-based proxies to the SPECMAP time scale or interpolation based on grain-size models (Porter and An 1995; Guo et al. 1996; Chen et al. 1997; Vandenberghe et al. 1997). Recently, optically stimulated luminescence (OSL) dating has been successfully applied to Chinese loess-paleosl sequences. These OSL results generated considerable debate regard-

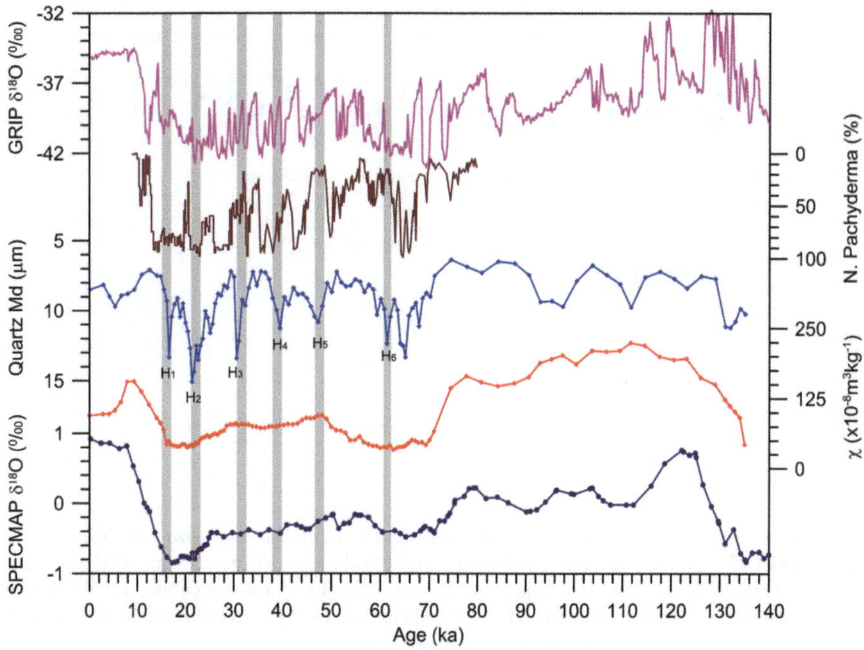

Fig. 2.43 Comparison of loess proxies (magnetic susceptibility and quartz median size) and speleothem $\delta^{18}O$ with GRIP, North Atlantic and SPECMAP curves

ing the overall continuity of Chinese loess-paleosol sequences. For example, an investigation of three late Quaternary loess sections in the central Loess Plateau revealed a sedimentary hiatus within the last deglaciation (Stevens et al. 2006), whereas the deglacial loess sequence from the western Loess Plateau were found to be continuous (Lai and Wintle 2006). On the basis of fine-grained quartz OSL-age estimates, loess accumulation at the classical Luochuan section was found to be nearly continuous during the last glaciation (Lu et al. 2007). Similarly, OSL dates estimated from the luminescence signal of coarse silt-sized quartz indicates that during the interval of 20–50 ka, no obvious hiatus was observed at a loess sequence in the western Loess Plateau (Lai et al. 2007). In summary, as is the case with all large sedimentary deposits, there are varying degrees of continuity among sections, depending on extant depositional and post-depositional processes.

Paleoclimate signals recorded in Chinese loess may be modified to various degrees by the post-depositional processes such as surface mixing and pedogenesis. For example, field observations have suggested that the surface layer of uncompacted eolian deposits can be disturbed by low-level winds, particularly during cold and dry times when vegetation cover is sparse (e.g., Liu 1985). Pedogenesis, a post-depositional weathering process, occurs during warm and humid conditions, and is characterized by carbonate leaching, elemental mobilization, and in-situ formation of magnetic and clay minerals (Liu 1985; Bronger and Heinkele 1989; Zhou et al. 1990a). Pedogenesis in the LP occurs during interglacial or interstadial periods

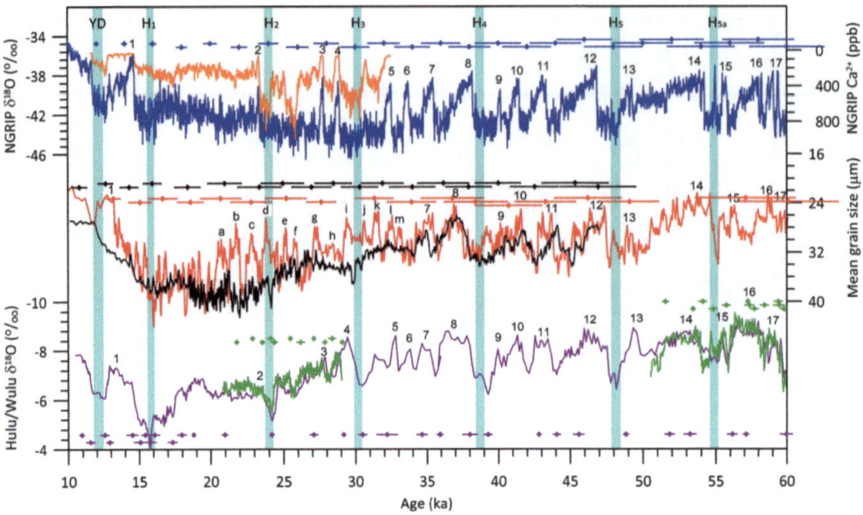

Fig. 2.44 Comparison of Gulang (*red*) and Jingyuan (*black*) mean grain size with NGRIP $\delta^{18}O$ (*blue*) and Ca^{2+} (*orange*) records and Hulu (*purple*)/Wulu (*green*) speleothems. OSL ages and errors (1σ error bars) of two loess sequences are marked on the grain-size curves. The ^{230}Th ages and errors (2σ error bars) are shown on the speleothem records. Age errors for the NGRIP ice core are the maximum counting errors on the GICC05 time scale

when sediment supply and transport are limited and the climate is warmer and/or wetter (Kemp 2001). At these times, surface mixing acts on the weathering product profiles as a low-pass filter attenuating high-frequency variability of loess proxies (e.g., magnetic susceptibility), particularly in the central LP which has exceptionally low dust influx rates (Feng et al. 2004). Further investigation of these processes will improve our understanding and interpretation of climate proxies generated from Chinese loess sequences.

Recently, a detailed OSL dating of two thick loess sequences from the western LP is presented, offering a reliable chronology for last-glacial deposits (Sun et al. 2010b; Sun et al. 2012). Magnetic susceptibility and mean grain size records from three loess-paleosol sequences along a northwest-southeast transect are investigated to evaluate impacts of post-depositional processes on these loess-based proxy records. In the central LP, rapid monsoon signals have been attenuated to various degrees depending on the sedimentation rate and pedogenic intensity (Porter and An 1995; Xiao et al. 1995; Guo et al. 1996; Sun et al. 2010b). In the northwestern Loess Plateau, however, due to high sedimentation rate and relatively weak pedogenesis, high-resolution grain-size oscillations reliably record rapid monsoon changes and can be well correlated to the rapid climate changes recorded in the Greenland ice-core and Hulu cave speleothems. These results indicate that: 1) loess sequences developed within the flat tableland of the central and western LP are ~ continuously deposited during the last glaciation; 2) post-depositional processes have distinct impacts on rapid monsoon signals recorded in loess sequences from different regions;

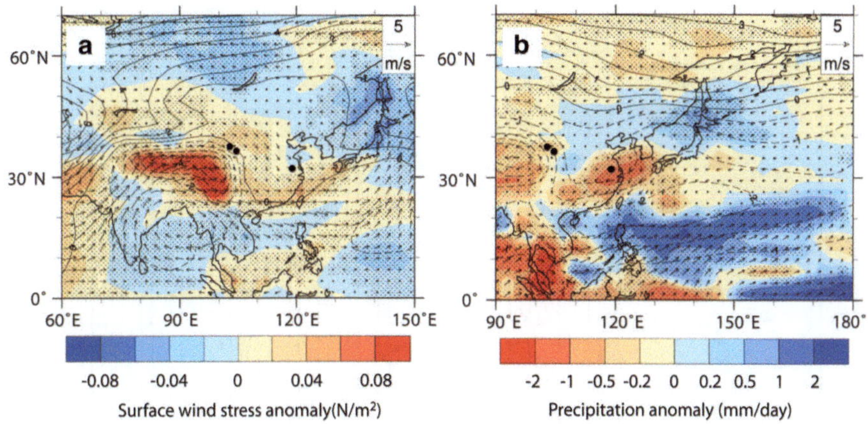

Fig. 2.45 CCSM3 simulated surface wind stress and precipitation anomalies between the water-hosing experiment and the control simulation. (a) Boreal winter (DJF) surface wind stress (N/m^2, color), sea level pressure (hPa, lines) and 750 mb wind vectors (m/s). (b) Boreal summer (JJA) precipitation (mm/day, color), sea level pressure (hPa, lines) and wind vectors from the lowest atmospheric model level (m/s)

and 3) reconstructed millennial-scale variations of the East Asian winter monsoon over the past 60 ka are broadly correlated with temperature variations over Greenland precipitation changes in Eastern Asia (Fig. 2.44).

Detailed comparison of multiple proxies generated from different archives is often compromised by a lack of high-precision absolute-dated chronologies. High-resolution OSL-dated grain-size records of two loess sequences developed in the northwestern LP allow robust comparison of loess grain size with Chinese speleothem and Greenland ice-core records. Over the past 60 ka, loess grain-size and speleothem $\delta^{18}O$ variations match well in both magnitude and timing, implying a strong coupling of millennial-scale winter and summer monsoon variability. To ascertain rapid monsoon changes recorded in Chinese loess, the key issue is to obtain a robust chronology for the loess sequences. Based on the synchronicity of abrupt changes in speleothem $\delta^{18}O$ and loess grain size during the last glaciation, it seem possible to generate a new and reliable loess chronology beyond the maximum range of OSL dating can be generated by matching loess grain size to absolutely dated speleothem $\delta^{18}O$ records, rather than to the orbitally tuned benthic $\delta^{18}O$.

Integration of terrestrial, marine and ice-core records allows better understanding of hemispheric coupling mechanisms of abrupt climate changes. North Atlantic "water-hosing" experiments suggest a dynamic response of monsoonal wind and precipitation in East Asia to a reduction of the Atlantic meridional overturning circulation (AMOC). Freshwater forcing results in an increased latitudinal temperature gradient and strong winter winds throughout much of the atmosphere in the Northern Hemisphere mid-latitudes (Fig. 2.45a). Meanwhile, the North Pacific subtropical high weakens following the hosing, and the resulting strong cyclonic circulation anomaly leads to reduced convergence along the Meiyu front in East Asia and

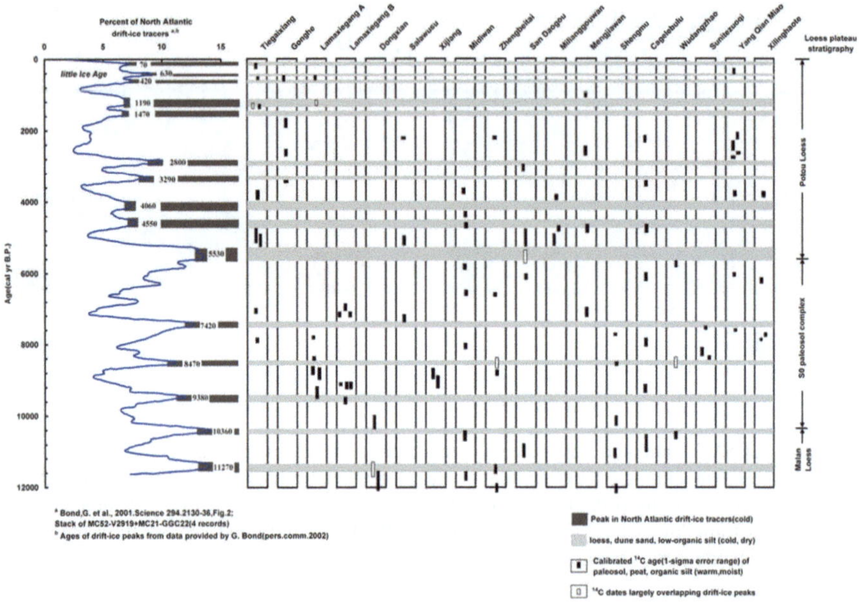

Fig. 2.46 Temporal distribution of calibrated radiocarbon dates from paleosols and peat units in 18 stratigraphic sections (Porter and Zhou 2006), compared with time series of variations in drift-ice tracers in the North Atlantic Ocean (Bond et al. 2001)

decreased summer precipitation near Hulu Cave (Fig. 2.45b). This combination of proxy data and modeling results indicates that the AMOC plays a key role in driving abrupt climate change in both the winter and summer monsoon systems, and that the westerlies transmit the winter-season North Atlantic signal to monsoonal regions of Asia.

The close correlation of Chinese millennial-scale climate variations and ice-rafting events in the North Atlantic has also been extended through the Holocene. The eolian record from north-central China supports the likelihood of an atmospheric teleconnection between these two regions throughout the last 12 ka. It also is consistent with earlier suggestion that changes in the physical oceanography of the North Atlantic Ocean affect the climate downwind, including the strength and path of storm tracks within the westerly wind system that passes over the Chinese desert and loess regions (Fig. 2.46) (Porter and Zhou 2006). Peaks in drift-ice tracers match times when eolian sand and loess were deposited in north-central China under cold, dry climatic conditions. Intervening times of diminished drift ice are synchronous with episodes of soil formation and peat accumulation under a mild, moist summer monsoon-dominated climate.

2.7 Summary

East Asian monsoon climate is the manifestation of the inter-coupling of the global atmosphere, land, ocean, and ice system within the specific geographic setting of East Asia. Researches on the magnetostratigraphy of the high-resolution loess-paleosol/red clay sequence and the proxy indices of monsoon climate show that the East Asian monsoon was initiated at least by the late Oligocene. The East Asian monsoon climate has been characterized by alternating periods of cold-dry winter monsoons and warm-humid summer monsoons. Strong winter monsoon and summer precipitation demonstrate relatively large temporal-spatial variability. Abrupt monsoon events indicate that the East Asian paleomonsoon fluctuated on a millennial or shorter scales, and was teleconnected with the global climate system. The evolution of the East Asian monsoon is controlled by the feedback of East Asian atmosphere circulation to the solar irradiance changes induced by orbital variation, and the changing lower boundary conditions. The long-term variation of the East Asian monsoon, which is closely related with the overall behavior of the Earth climate system, not only correlates closely with global ice-volume variations, but also with the variation of the underlying topographic surface, such as the sea-land distribution, uplift of the Tibetan Plateau, sea-temperature anomalies, and land desertification.

Multiple loess-based proxies show that the intensive uplift of the Tibetan Plateau increased the variability of the East Asian monsoon climate, and had a great effect on global cooling and the onset of Quaternary glaciation. Interaction and coupling of different factors within the Earth climate system also clearly influenced the variability of the East Asian monsoon climate. Therefore, to study the East Asian paleomonsoon, we should first emphasize surveys of high-resolution proxy sequences of the East Asian monsoon and obtain accurate chronologies of the proxy indices. This will allow us to further understand the origin and evolution of the East Asian monsoon, and in the meantime, to study the instability of the East Asian paleomonsoon and its temporal-spatial variation on tectonic to centennial time scales. This in turn would provide a background and a mechanism for understanding of the modern monsoon and future climate changes. Meanwhile, it is necessary to consider the land, ocean, atmosphere, and ice system in East Asia as an integrated monsoonal-arid environment system within the global change framework. Assessment of how factors within the East Asian monsoon environment system interact with each other, and their relationship with global climate factors, with special attention to unique factors in the East Asia environment such as the Tibetan Plateau uplift, will provide a more thorough understanding of the dynamic mechanism driving East Asian monsoon variability on different time scales.

References

An Z, Lu Y (1984) A climatostratigraphic subdivision of Late Pleistocene strata named by Malan formation in North China. Kexue Tongbao 29:1239–1241

An ZS (2000) The history and variability of the East Asian paleomonsoon climate. Quat Sci Rev 19(1-5):171–187

An ZS, Colman SM, Zhou WJ, Li XQ, Brown ET, Jull AJT, Cai YJ, Huang YS, Lu XF, Chang H, Song YG, Sun YB, Xu H, Liu WG, Jin ZD, Liu XD, Cheng P, Liu Y, Ai L, Li XZ, Liu XJ, Yan LB, Shi ZG, Wang XL, Wu F, Qiang XK, Dong JB, Lu FY, Xu XW (2012) Interplay between the Westerlies and Asian monsoon recorded in Lake Qinghai sediments since 32 ka. Sci Rep-Uk 2:619. doi:10.1038/Srep00619

An ZS, Huang YS, Liu WG, Guo ZT, Clemens S, Li L, Prell W, Ning YF, Cai YJ, Zhou WJ, Lin BH, Zhang QL, Cao YN, Qiang XK, Chang H, Wu ZK (2005) Multiple expansions of C4 plant biomass in East Asia since 7 Ma coupled with strengthened monsoon circulation. Geology 33:705–708

An ZS, Kukla G, Porter SC, Xiao JL (1991b) Late Quaternary Dust Flow on the Chinese Loess Plateau. Catena 18(2):125–132

An ZS, Kukla GJ, Porter SC, Xiao JL (1991a) Magnetic-susceptibility evidence of monsoon variation on the Loess Plateau of central China during the Last 130,000 Years. Quat Res 36 (1):29–36. doi:10.1016/0033-5894(91)90015-W

An ZS, Kutzbach JE, Prell WL, Porter SC (2001) Evolution of Asian monsoons and phased uplift of the Himalaya-Tibetan plateau since Late Miocene times. Nature 411(6833):62–66. doi:10.1038/35075035

An ZS, Liu TS, Lu YC, Porter SC, Kukla G HWX, Hua YM (1990) The long-term paleomonsoon variation recorded by the loess-paleosol sequence in central China. Quat Int 718:91–95

An ZS, Porter SC, Kutzbach JE, Wu XH, Wang SM, Liu XD, Li XQ, Zhou WJ (2000a) asynchronous holocene optimum of the east Asian monsoon. Quat Sci Rev 19(8):743–762

An ZS, Sun D, Chen M, Sun Y, Li L, Chen B (2000b) Red clay sequences in Chinese Loess Plateau and recorded paleoclimate events of the Late Tertiary. Quat Sci 20(5):435–446 (in Chinese)

An ZS, Sun DH (1995) Discussion on the Monsoon variation over the loess plateau in the last glacial cycle. In: China contribution to global change. Beijing Science Press, Beijing (in Chinese)

An ZS, Wang JD, Li HM (1977) Paleomagnetic research of the Luochuan loess section. Geochimica 4:239–249 (in Chinese)

An ZS, Wang SM, Wu XH, Chen MY, Sun DH, Liu XM, Wang FB, Li L, Sun YB, Zhou WJ, Zhou J, Liu XD, Lu HY, Zhang YX, Dong GG, Qiang XK (1999) Eolian evidence from the Chinese Loess Plateau: the onset of the Late Cenozoic Great Glaciation in the Northern Hemisphere and Qinghai-Xizang Plateau uplift forcing. Sci China Ser D 42(3):258–271

An ZS, Wei LY (1980) The fifth paleosol in the Lishi Loess and its paleoclimatic significance. Acta Pedol Sin 1:1–12 (in Chinese)

An ZS, Zhang PZ, Wang EQ, Wang SM, Qiang XK, Li L, Song YG, Chang H, Liu XD, Zhou WJ, Liu WG, Cao JJ, Li XQ, Shen J, Liu Y, Ai L (2006) Changes of the monsoon-arid environment in China and geowth of the Tibetan Plateau since the Miocene. Quat Sci 26:678–693 (in Chinese)

Andersen N, Paul HA, Bernasconi SM, McKenzie JA, Behrens A, Schaeffer P, Albrecht P (2001) Large and rapid climate variability during the Messinian salinity crisis: Evidence from deuterium concentrations of individual biomarkers. Geology 29(9):799–802

Andersson JG (1923) Essays on the Cenozoic of northern China. Mem Geol Surv China. Ser A 3:1–152

Balsam W, Ji J, Chen J (2004) Climatic interpretation of the Luochuan and Lingtai loess sections, China, based on changing iron oxide mineralogy and magnetic susceptibility. Earth Planet Sci Lett 223(3-4):335–348

Banerjee D, Murray AS, Bøtter-Jensen L, Lang A (2001) Equivalent dose estimation using a single aliquot of polymineral fine grains. Radiat Meas 33 (1):73–94. doi:10.1016/s1350-4487(00)00101-3

Baumgartner S, Beer J, Wagner G, Kubik P, Suter M, Raisbeck GM, Yiou F (1997) ^{10}Be and dust. Nucl Instrum Meth B 123 (1-4):296–301. doi:10.1016/s0168-583x(96)00751-3

Beer J, Shen C, Heller F, Liu T, Bonani G, Beate D, Suter M, Kubik PW (1993) ^{10}Be and magnetic susceptibility in Chinese loess. Geophys Res Lett 20(1):57–60. doi:10.1029/92gl02676

Berger A, Loutre MF (1991) Insolation values for the climate of the last 10 million years. Quat Sci Rev 10 (4):297–317. doi:10.1016/0277-3791(91)90033-q

Bi XH, Sheng GY, Liu XH, Li C, Fu JM (2005) Molecular and carbon and hydrogen isotopic composition of n-alkanes in plant leaf waxes. Org Geochem 36(10):1405–1417. doi:10.1016/j.orggeochem.2005.06.001

Bloemendal J, Liu XM, Rolph TC (1995) Correlation of the magnetic susceptibility stratigraphy of Chinese loess and the marine oxygen isotope record: chronological and palaeoclimatic implications. Earth Planet Sci Lett 131 (3-4):371–380. doi:10.1016/0012-821x(95)00016-6

Bond G, Kromer B, Beer J, Muscheler R, Evans MN, Showers W, Hoffmann S, Lotti-Bond R, Hajdas I, Bonani G (2001) Persistent solar influence on North Atlantic climate during the Holocene. Science 294(5549):2130

Bronger A, Heinkele T (1989) Micromorphology and genesis of paleosols in the luochuan Loess Section, China - pedostratigraphic and environmental implications. Geoderma 45 (2):123–143. doi:10.1016/0016-7061(89)90046-3

Burbank DW, Li JJ (1985) Age and paleoclimatic significance of the loess of lanzhou, North China. Nature 316(6027):429–431. doi:10.1038/316429a0

Burgoyne TW, Hayes JM (1998) Quantitative production of H2 by pyrolysis of gas chromatographic effluents. Anal Chem 70(24):5136–5141. doi:10.1021/ac980248v

Buylaert JP, Murray AS, Vandenberghe D, Vriend M, De Corte F, Van den haute P (2008) Optical dating of Chinese loess using sand-sized quartz: Establishing a time frame for Late Pleistocene climate changes in the western part of the Chinese Loess Plateau. Quat Geochronol 3(1-2):99–113. doi:10.1016/j.quageo.2007.05.003

Cande SC, Kent DV (1995) Revised calibration of the geomagnetic polarity timescale for the late Cretaceous and Cenozoic. J Geophys Res 100:6093–6095

Cerling T, Quade J, Yang W, Bowman J (1989) soil and paleosol as ecologic and paleoecologic indicators. Nature 341:138–139

Cerling TE (1984) The stable isotopic composition of modern soil carbonate and its relationship to climate. Earth Planet Sci Lett 71(2):229–240

Cerling TE (1999) Stable carbon isotopes in palaeosol carbonates. In Special publication of the international association of sedimentologists 27:pp 43–60

Cerling TE, Harris JM, MacFadden BJ, Leakey MG, Quade J, Eisenmann V, Ehleringer JR (1997) Global vegetation change through the Miocene/Pliocene boundary. Nature 389(6647):153–158. doi:10.1038/38229

Channell JET, Stoner JS, Hodell DA, Charles CD (2000) Geomagnetic paleointensity for the last 100 kyr from the sub-antarctic South Atlantic: a tool for inter-hemispheric correlation. Earth Planet Sci Lett 175 (1–2):145–160. doi:10.1016/s0012-821x(99)00285-x

Channell JET, Xuan C, Hodell DA (2009) Stacking paleointensity and oxygen isotope data for the last 1.5 Myr (PISO-1500). Earth Planet Sci Lett 283(1–4):14–23. doi:10.1016/j.epsl.2009.03.012

Chen FH, Bloemendal J, Feng ZD, Wang JM, Parker E, Guo ZT (1999a) East Asian monsoon variations during Oxygen Isotope Stage 5: evidence from the northwestern margin of the Chinese loess plateau. Quat Sci Rev 18(8-9):1127–1135

Chen FH, Bloemendal J, Wang JM, Li JJ, Oldfield F (1997) High-resolution multi-proxy climate records from Chinese loess, evidence for rapid climatic changes over the last 75 kyr. Palaeogeogr Palaeoclimatol Palaeocl 130(1–4):323–335

Chen J, An Z, Head J (1999b) Variation of Rb/Sr ratios in the Loess-Paleosol sequences of central China during the Last 130,000 Years and their implications for monsoon paleoclimatology. Quat Res 51(3):215–219. doi:10.1006/qres.1999.2038

Chen J, Chen Y, Liu LW, Ji JF, Balsam W, Sun YB, Lu HY (2006) Zr/Rb ratio in the Chinese loess sequences and its implication for changes in the East Asian winter monsoon strength. Geochim Cosmochim Acta 70(6):1471–1482. doi:10.1016/j.gca.2005.11.029

Chen J, Ji JF, Balsam W, Chen Y, Liu L, An ZS (2002) Characterization of the Chinese loess-paleosol stratigraphy by whiteness measurement. Palaeogeogr Palaeoclimatol Palaeocl 183 (287–297)

Chen XL, Fang XM, An ZS, Han WX, Wang X, Bai Y, Hong Y (2007) An 8.1 Ma calcite record of Asian summer monsoon evolution on the Chinese central Loess Plateau. Sci China Ser D 50(3):392–403. doi:10.1007/s11430-007-0016-x

Chen Y, Chen J, Liu L, Ji J, Zhang J (2003) Spatial and temporal changes of summer monsoon on the Loess Plateau of Central China during the last 130 ka inferred from Rb/Sr ratios. Sci China Ser D 46(10):1022–1030. doi:10.1007/BF02959397

Cheng H, Edwards RL, Broecker WS, Denton GH, Kong XG, Wang YJ, Zhang R, Wang XF (2009) Ice age terminations. Science 326(5950):248–252. doi:10.1126/science.1177840

Chikaraishi Y, Naraoka H (2001) Organic hydrogen-carbon isotope signatures of terrestrial higher plants during biosynthesis for distinctive photosynthetic pathways. Geochemical J 35(6):451–458

Chikaraishi Y, Naraoka H (2003) Compound-specific delta D-delta C-13 analyses of n-alkanes extracted from terrestrial and aquatic plants. Phytochemistry 63(3):361–371. doi:10.1016/S0031-9422(02)00749-5

Christl M, Lippold J, Steinhilber F, Bernsdorff F, Mangini A (2010) Reconstruction of global ^{10}Be production over the past 250ka from highly accumulating Atlantic drift sediments. Quat Sci Rev 29(19-20):2663–2672. doi:10.1016/j.quascirev.2010.06.017

Clemens SC, Murray DW, Prell WL (1996) Nonstationary phase of the plio-pleistocene Asian monsoon. Science 274(5289):943–948

Clemens SC, Prell WL (2003) A 350,000 year summer-monsoon multi-proxy stack from the Owen ridge, Northern Arabian sea. Mar Geol 201 (1-3):35–51. doi:10.1016/S0025-3227(03)00207-X

Clemens SC, Prell WL, Sun Y, Liu Z, Chen G (2008) Southern hemisphere forcing of Pliocene δ^{18}O and the evolution of Indo-Asian monsoons. Paleoceanography 23 (10.1029)

Clift PD, Hodges KV, Heslop D, Hannigan R, Van Long H, Calves G (2008) Correlation of Himalayan exhumation rates and Asian monsoon intensity. Nat Geosci 1(12):875–880. doi:10.1038/Ngeo351

Dawson D, Grice K, Wang SX, Alexander R, Radke J (2004) Stable hydrogen isotopic composition of hydrocarbons in torbanites (Late Carboniferous to Late Permian) deposited under various climatic conditions. Org Geochem 35(2):189–197. doi:10.1016/j.orggeochem.2003.09.004

deMenocal PB (1995) Pliopleistocene African climate. Science 270(5233):53–59

Dimbleby G (1985) The palynology of archaeological sites. Academic Press, New York, pp 176

Ding ZL, Derbyshire E, Yang SL, Sun JM, Liu TS (2005) Stepwise expansion of desert environment across northern China in the past 3.5 Ma and implications for monsoon evolution. Earth Planet Sci Lett 237(1-2):45–55. doi:10.1016/j.epsl.2005.06.036

Ding ZL, Derbyshire E, Yang SL, Yu ZW, Xiong SF, Liu TS (2002) Stacked 2.6-Ma grain size record from the Chinese loess based on five sections and correlation with the deep-sea delta O-18 record. Paleoceanography 17(3):1033. doi:10.1029/2001pa000725

Ding ZL, Liu T, Rutter NW, Yu Z, Guo Z, Zhu R (1995) Ice-Volume Forcing of East Asian Winter Monsoon Variations in the Past 800,000 Years. Quat Res 44(2):149–159. doi:10.1006/qres.1995.1059

Ding ZL, Liu TS (1989) Progresses of loess research in China (part 1): Loess stratigraphy. Quat Sci (1):24–35 (in Chinese)

Ding ZL, Ren JZ, Yang SL, Liu TS (1999a) Climate instability during the penultimate glaciation: Evidence from two high-resolution loess records. China J Geophys Res 104 (B9):20123-20132. doi:10.1029/1999jb900183

Ding ZL, Rutter NW, Liu TS (1993) Pedostratigraphy of China loess deposits and climatic cycle in the last 2.5 Myr. Catena 20(73):91

Ding ZL, Rutter NW, Liu TS (1997) The onset of extensive loess deposition around the G/M boundary in China and its palaeoclimatic implications. Quat Int 40:53–60. doi:10.1016/S1040-6182(96)00061-4

Ding ZL, Rutter NW, Liu TS, Sun JM, Ren JZ, Rokosh D, Xiong SF (1998a) Correlation of Dansgaard-Oeschger cycles between Greenland ice and Chinese loess. Paleoclimates 2:281–291

Ding ZL, Rutter NW, Sun JM, Yang SL, Liu TS (2000) Re-arrangement of atmospheric circulation at about 2.6Ma over northern China: evidence from grain size records of loess-palaeosol and red clay sequences. Quat Sci Rev 19 (6):547–558. doi:10.1016/s0277-3791(99)00017-7

Ding ZL, Sun JM, Liu TS, Zhu RX, Yang SL, Guo B (1998b) Wind-blown origin of the Pliocene red clay formation in the central Loess Plateau, China. Earth Planet Sci Lett 161 (1-4):135–143. doi:10.1016/s0012-821x(98)00145-9

Ding ZL, Xiong SF, Sun JM, Yang SL, Gu ZY, Liu TS (1999b) Pedostratigraphy and paleomagnetism of a similar to 7.0 Ma eolian loess-red clay sequence at Lingtai, Loess Plateau, north-central China and the implications for paleomonsoon evolution. Palaeogeogr Palaeoclimatol Palaeocl 152(1-2):49–66

Ding ZL, Yang SL, Hou SS, Wang X, Chen Z, Liu TS (2001a) Magnetostratigraphy and sedimentology of the Jingchuan red clay section and correlation of the Tertiary eolian red clay sediments of the Chinese Loess Plateau. J Geophys Res 106 (B4):6399-6407. doi:10.1029/2000jb900445

Ding ZL, Yang SL, Sun JM, Liu TS (2001b) Iron geochemistry of loess and red clay deposits in the Chinese Loess Plateau and implications for long-term Asian monsoon evolution in the last 7.0 Ma. Earth Planet Sci Lett 185 (1-2):99–109. doi:10.1016/s0012-821x(00)00366-6

Ding ZL, Yu ZW, Rutter NW, Liu TS (1994) Towards an orbital time scale for Chinese loess deposits. Quat Sci Rev 13 (39–70)

Ding ZL, Yu ZW, Yang SL, Sun JM, Xiong SF, Liu TS (2001c) Coeval changes in grain size and sedimentation rate of eolian loess, the Chinese Loess Plateau. Geophys Res Lett 28(10):2097–2100

Donahue D (1993) Measurements of radiocarbon ages at the University of Arizona accelerator mass spectrometer facility. Collected Oceanic Works 16(1):56–69

Duchaufour P (1983) Pédologie. Tome I. Pédogenèse et Classification, 2 edn. Masson, Paris

Dykoski CA, Edwards RL, Cheng H, Yuan DX, Cai YJ, Zhang ML, Lin YS, Qing JM, An ZS, Revenaugh J (2005) A high-resolution, absolute-dated Holocene and deglacial Asian monsoon record from Dongge Cave. China. Earth Planet Sci Lett 233(1–2):71–86. doi:10.1016/j.epsl.2005.01.036

Elsasser W, Ney EP, Winckler JR (1956) Cosmic-ray intensity and geomagnetism. Nature 178(4544):1226–1227

EPICA Members (2004) Eight glacial cycles from an Antarctic ice core. Nature 429(6992):623–628. doi:10.1038/Nature02599

Evans ME, Heller F (1994) Magnetic enhancement and paleolimate: Study of a loess/paleosol couplet across the Loess Plateau of China. Geophys J Int 117:257–264

Evans ME, Heller F (2001) Magnetism of loess/palaeosol sequences: recent developments. Earth-Sci Rev 54(1-3):129–144

Fang XM, Li JJ, VanderVoo R, MacNiocaill C, Dai XR, Kemp RA, Derbyshire E, Cao JX, Wang JM, Wang G (1997) A record of the Blake Event during the last interglacial paleosol in the western Loess Plateau of China. Earth Planet Sci Lett 146 (1-2):73–82. doi:10.1016/S0012-821x(96)00222-1

Fang XM, Ono Y, Fukuksawa H, Pan BT, Li JL, Guan DH, Oi KC, Tsukamoto S, Torii M, Mishima T (1999) Asian summer monsoon instability during the past 60,000 years: magnetic susceptibility and pedogenic evidence from the western Chinese Loess Plateau. Earth Planet Sci Lett 168(3-4):219–232

Farquhar G, O'Leary M, Berry J (1982) On the relationship between carbon isotope discrimination and the intercellular carbon dioxide concentration in leaves. Functl Plant Biol 9 (2):121–137. doi:http://dx.doi.org/10.1071/PP9820121

Feng ZD, Wang HB, Olson C, Pope GA, Chen FH, Zhang JW, An CB (2004) Chronological discord between the last interglacial paleosol (S1) and its parent material in the Chinese Loess Plateau. Quat Int 117:17–26. doi:10.1016/S1040-6182(03)00112-5

Forman SL (1991) Late Pleistocene chronology of loess deposition near Luochuan, China. Quat Res 36 (1):19–28. doi:10.1016/0033-5894(91)90014-v

Fox DL, Koch PL (2003) Tertiary history of C4 biomass in the Great Plains, USA. Geology 31(9):809–812. doi:10.1130/g19580.1

Frakes LA, Sun JZ (1994) A Carbon-Isotope Record of the Upper Chinese Loess Sequence - Estimates of Plant Types during Stadials and Interstadials. Palaeogeogr Palaeoclimatol Palaeocl 108(1–2):183–189

Frank M (2000) Comparison of cosmogenic radionuclide production and geomagnetic field intensity over the last 200 000 years. Philos T Roy Soc A 358(1768):1089–1107. doi:10.1098/rsta.2000.0575

Frank M, Schwarz B, Baumann S, Kubik PW, Suter M, Mangini A (1997) A 200 kyr record of cosmogenic radionuclide production rate and geomagnetic field intensity from ^{10}Be in globally stacked deep-sea sediments. Earth Planet Sci Lett 149 (1-4):121–129. doi:10.1016/s0012-821x(97)00070-8

Freeman KH, Colarusso LA (2001) Molecular and isotopic records of C4 grassland expansion in the late miocene. Geochim Cosmochim Acta 65 (9):1439–1454. doi:10.1016/s0016-7037(00)00573-1

Gallet S, Jahn BM, Torii M (1996) Geochemical characterization of the Luochuan loess-paleosol sequence, China, and paleoclimatic implications. Chem Geol 133 (1–4):67–88. doi:10.1016/S0009-2541(96)00070-8

Gao WY, Lin BH, Sun DH (1992) Catoptrics studies for loess and climatic variations. In: Liu TS, An ZS (eds) Loess, Quaternary Geology, Global Change. Science Press, Beijing, pp 92–100 (in Chinese)

Goni MA, Eglinton TI (1996) Stable carbon isotopic analyses of lignin-derived CuO oxidation products by isotope ratio monitoring gas chromatography mass spectrometry (irm-GC-MS). Org Geochem 24 (6–7):601–615. doi:10.1016/0146-6380(96)00052-6

Gu ZY, Guo Z, Lal D (2006) ^{10}Be Concentration relation to chemical compositions of Chinese loess and red clay as a potential dating method. Quat Sci 26(2):244–249 (in Chinese)

Gu ZY, Lal D, Liu TS, Guo ZT, Southon J, Caffee MW (1997) Weathering histories of Chinese loess deposits based on uranium and thorium series nuclides and cosmogenic Be-10. Geochim Cosmochim Acta 61 (24):5221–5231. doi:10.1016/S0016-7037(97)00313-X

Gu ZY, Lal D, Liu TS, Southon J, Caffee MW, Guo ZT, Chen MY (1996) Five million year ^{10}Be record in Chinese loess and red-clay: climate and weathering relationships. Earth Planet Sci Lett 144 (1-2):273–287. doi:10.1016/0012-821x(96)00156-2

Gu ZY, Liu Q, Xu B, Han J, Yang S, Ding Z, Liu T (2003) Climate as the dominant control on C3 and C4 plant abundance in the Loess Plateau: Organic carbon isotope evidence from the last glacial-interglacial loess-soil sequences. Chinese Sci Bull 48(12):1271–1276. doi:10.1007/bf03183950

Guo ZT, Berger A, Yin QZ, Qin L (2009) Strong asymmetry of hemispheric climates during MIS-13 inferred from correlating China loess and Antarctica ice records. Clim Past 5(1):21–31

Guo ZT, Biscaye P, Wei LY, Chen XH, Peng SZ, Liu TS (2000) Summer monsoon variations over the last 1.2 Ma from the weathering of loess-soil sequences in China. Geophys Res Lett 27(12):1751–1754. doi:10.1029/1999gl008419

Guo ZT, Liu TS, Fedoroff N, Wei LY, Ding ZL, Wu NQ, Lu HY, Jiang WY, An ZS (1998) Climate extremes in Loess of China coupled with the strength of deep-water formation in the North Atlantic. Global Planet Change 18 (3-4):113–128. doi:10.1016/S0921-8181(98)00010-1

Guo ZT, Liu TS, Guiot J, Wu NQ, Lu H, Han J, Liu J, Gu Z (1996) High frequency pulses of East Asian monsoon climate in the last two glaciations: Link with the North Atlantic. Clim Dynam 12(10):701–709. doi:10.1007/s003820050137

Guo ZT, Sun B, Zhang ZS, Peng SZ, Xiao GQ, Ge JY, Hao QZ, Qiao YS, Liang MY, Liu JF, Yin QZ, Wei JJ (2008) A major reorganization of Asian climate by the early Miocene. Clim Past 4(3): 153–174.

Guo ZT, Ruddiman WF, Hao QZ, Wu HB, Qiao YS, Zhu RX, Peng SZ, Wei JJ, Yuan BY, Liu TS (2002) Onset of Asian desertification by 22 Myr ago inferred from loess deposits in China. Nature 416(6877):159–163

Guyodo Y, Valet JP (1996) Relative variations in geomagnetic intensity from sedimentary records: The past 200,000 years. Earth Planet Sci Lett 143 (1-4):23–36. doi:10.1016/0012-821x(96)00121-5

Guyodo Y, Valet JP (1999) Global changes in intensity of the Earth's magnetic field during the past 800 kyr. Nature 399(6733):249–252

Han J, Hus JJ, Liu T, Paepe R, Vandenberghe RE (1991) Magnetic properties of the Malan and Lshi formations. Quaternary Sciences 11(4):310–325 (in Chinese)

Han J, Jiang W, Liu T (1996a) Carbon isotopic records of the paleoclimatic changes in loess. Sci China Ser D 39 (5):460–467

Han JM, Lv HY, Wu NQ, Guo ZT (1996b) The magnetic susceptibility of modern soils in China and its use for paleoclimate reconstruction. Stud Geophys Geod 40(3):262–275

Han JT, Fyfe WS, Longstaffe FJ (1998) Climatic implications of the S5 paleosol complex on the southernmost Chinese Loess Plateau. Quat Res 50:21–33

Hao Q, Guo Z (2004) Magnetostratigraphy of a late Miocene-Pliocene loess-soil sequence in the western Loess Plateau in China. Geophys Res Lett 31(9):L09209

Hao QZ, Guo ZT (2005) Spatial variations of magnetic susceptibility of Chinese loess for the last 600 ka: implications for monsoon evolution. J Geophys Res 110: B12101.

Hatte C, Fontugne M, Rousseau DD, Antoine P, Zoller L, Tisnerat-Laborde N, Bentaleb I (1998) delta C-13 variations of loess organic matter as a record of the vegetation response to climatic changes during the Weichselian. Geology 26 (7):583–586. doi:10.1130/0091-7613 (1998) 026 <0583:Cvolom>2.3.Co;2

Hays JD, Imbrie J, Shackleton NJ (1976) Variations in the Earth's Orbit: Pacemaker of the Ice Ages. Science 194(4270):1121–1132. doi:10.1126/science.194.4270.1121

Head M (1979) Structure and chemical properties of fresh and degraded wood: their effects on radiocarbon dating measurements. MSc thesis

Head M, Zhou W, Zhou M (1989) Evaluation of 14C ages of organic fractions of paleosols from loess-paleosol sequences near Xian, China. Radiocarbon 31(3):680–696

Heller F, Evans ME (1995) Loess Magnetism. Rev Geophys 33(2):211–240

Heller F, Liu T (1986) Palaeoclimatic and sedimentary history from magnetic susceptibility of loess in China. Geophys Res Lett 13(11):1169–1172. doi:10.1029/GL013i011p01169

Heller F, Liu TS (1982) Magnetostratigraphical Dating of Loess Deposits in China. Nature 300(5891):431–433

Heller F, Liu TS (1984) Magnetism of Chinese Loess Deposits. Geophys J Roy Astr S. 77(1):125–141

Heller F, Shen CD, Beer J, Liu XM, Liu TS, Bronger A, Suter M, Bonani G (1993) Quantitative estimates of pedogenic ferromagnetic mineral formation in Chinese loess and palaeoclimatic implications. Earth Planet Sci Lett 114 (2–3):385–390. doi:10.1016/0012-821x(93)90038-b

Heslop D, Langereis CG, Dekkers MJ (2000) A new astronomical timescale for the loess deposits of northern China. Earth Planet Sci Lett 184 (125–139)

Hilkert AW, Douthitt CB, Schluter HJ, Brand WA (1999) Isotope ratio monitoring gas chromatography mass spectrometry of D H by high temperature conversion isotope ratio mass spectrometry. Rapid Commun Mass Sp 13(13):1226–1230

Hou JZ, D'Andrea WJ, MacDonald D, Huang YS (2007) Hydrogen isotopic variability in leaf waxes among terrestrial and aquatic plants around Blood Pond, Massachusetts (USA). Org Geochem 38(6):977–984. doi:10.1016/j.orggeochem.2006.12.009

Huang CC, Pang JL, Huang P, Hou CH, Han YP (2002a) Holocene climatic events on the loess tableland in the western Guanzhong basin, China. Arid Land Geography 25(1):10–15 (in Chinese)

Huang YS, Shuman B, Wang Y, Webb T (2002b) Hydrogen isotope ratios of palmitic acid in lacustrine sediments record late Quaternary climate variations. Geology 30(12):1103–1106

Huang YS, Shuman B, Wang Y, Webb T (2004) Hydrogen isotope ratios of individual lipids in lake sediments as novel tracers of climatic and environmental change: a surface sediment test. J Paleolimnol 31(3):363–375

Huang YS, Street-Perrott FA, Metcalfe SE, Brenner M, Moreland M, Freeman KH (2001) Climate Change as the Dominant Control on Glacial-Interglacial Variations in C3 and C4 Plant Abundance. Science 293(5535):1647–1651. doi:10.1126/science.1060143

Huang YS, Street-Perrott FA, Perrott RA, Metzger P, Eglinton G (1999) Glacial-interglacial environmental changes inferred from molecular and compound-specific δ13C analyses of sediments from Sacred Lake, Mt. Kenya. Geochim Cosmochim Acta 63 (9):1383–1404. doi:10.1016/s0016-7037(99)00074-5

IAEA/WMO (2000) Global Network for Isotopes in Precipitation

Imbrie J, Hays J, Martinson D, McIntyre A, Mix AC, Morley JJ, Pisias NG, Prell WL, Shackleton NJ (1984) The orbital theory of Pleistocene climate: support from revised chronology of the marine ^{18}O record. In: Berger A (ed) Milankovitch and Climate, Part 1. Springer, New York, pp 269–305

Ishikawa Y, Murakami H, Sekine T, Yoshihara K (1995) Precipitation scavenging studies of radionuclides in air using cosmogenic ^{7}Be. J Environ Radioact 26 (1):19–36. doi:10.1016/0265-931x(95)91630-m

Jahn BM, Gallet S, Han JM (2001) Geochemistry of the Xining, Xifeng and Jixian sections, Loess Plateau of China: eolian dust provenance and paleosol evolution during the last 140 ka. Chem Geol 178 (1–4):71–94. doi:10.1016/S0009-2541(00)00430-7

Jamieson JC (1953) Phase equilibrium in the system calcite-aragonite. J Chem Phys 21(8):1385–1390. doi:10.1063/1.1699228

Ji JF, Balsam WJ, Chen J (2001) Mineralogic, climatic interpretations of the Luochuan loess section(China)based on diffuse reflectance spectrophotometry. Quatern Research 56(1):23–30

Ji JF, Chen J, Balsam WL, Lu HY, Xu HF (2004) High resolution hematite/goethite records from Chinese loess sequences for the last glacial-interglacial cycle: rapid climatic response of the East Asian Monsoon to the tropical Pacific. Geophys Research Lett 31:L03207

Ji JF, Chen J, Zhou W, Zhao L, Wang HT (2006) Mineralogical evidences of a warm and dry summer monsoon for the late Tertiary red clay formation, China. Geochim Cosmochim Acta 70 (18):A292–A292. doi:10.1016/j.gca.2006.06.593

Jia RF, Xin JL, Zhao L (1995) Organic matter changes in the Weinan loess section and implications for humidity variations. Geochimica 24:66–74 (in Chinese)

Jia RF, Yan BZ, Li RS, Fan GC, Lin BH (1996) Characteristics of magnetotactic bacteria in Duanjiapo loess section, Shaanxi province and their environment significance. Sci China Ser D 39(5):478–485

Johnsen SJ, Dahl-Jensen D, Gundestrup N, Steffensen JP, Clausen HB, Miller H, Masson-Delmotte V, Sveinbjörnsdottir AE, White J (2001) Oxygen isotope and palaeotemperature records from six Greenland ice-core stations: Camp Century, Dye-3, GRIP, GISP2, Renland and NorthGRIP. J Quat Sci 16(4):299–307. doi:10.1002/jqs.622

Johnson BJ, Miller GH, Fogel ML, Magee JW, Gagan MK, Chivas AR (1999) 65,000 years of vegetation change in central Australia and the Australian summer monsoon. Science 284(5417):1150–1152

Kang SG, Wang XL, Li XN, Lu YC (2010) Anomalous fading of the IRSL signal of polymineral grains in Chinese loess. Radiat Meas 45(1):22–28. doi:10.1016/j.radmeas.2009.08.006

Kang SG, Wang XL, Lu YC (2013) Quartz OSL chronology and dust accumulation rate changes since the Last Glacial at Weinan on the southeastern Chinese Loess Plateau. Boreas 42:815–829

Kemp RA (2001) Pedogenic modification of loess: significance for palaeoclimatic reconstructions. Earth-Sci Rev 54 (1–3):145–156. doi:10.1016/S0012-8252(01)00045-9

Kukla G (1987) Loess Stratigraphy in Central China. Quat Sci Rev 6 (3-4):191–219. doi:10.1016/0277-3791(87)90004-7

Kukla G, An ZS (1989) Loess Stratigraphy in Central China. Palaeogeogr Palaeoclimatol Palaeocl 72(1–2):203–225

Kukla G, Heller F, Ming LX, Chun XT, Sheng LT, Sheng AZ (1988) Pleistocene Climates in China Dated by Magnetic-Susceptibility. Geology 16 (9):811-814. doi:10.1130/0091-7613

Kutzbach JE, Prell WL, Ruddiman WF (1993) Sensitivity of Eurasian Climate to Surface Uplift of the Tibetan Plateau. J Geol 101(2):177–190

Lai ZP (2010) Chronology and the upper dating limit for loess samples from Luochuan section in the Chinese Loess Plateau using quartz OSL SAR protocol. J Asian Earth Sci 37(2):176–185. doi:10.1016/j.jseaes.2009.08.003

Lai ZP, Wintle AG (2006) Locating the boundary between the Pleistocene and the Holocene in Chinese loess using luminescence. Holocene 16(6):893–899

Lai ZP, Wintle AG, Thomas DSG (2007) Rates of dust deposition between 50 ka and 20 ka revealed by OSL dating at Yuanbao on the Chinese Loess Plateau. Palaeogeogr Palaeoclatol Palaeoecol 248(3–4):431–439. doi:10.1016/j.palaeo.2006.12.013

Laj C, Kissel C, Mazaud A, Channell JET, Beer J (2000) North Atlantic palaeointensity stack since 75 ka (NAPIS-75) and the duration of the Laschamp event. Philos T Roy Soc A 358(1768):1009–1025

Lal D (2007) Recycling of cosmogenic nuclides after their removal from the atmosphere; special case of appreciable transport of Be-10 to polar regions by aeolian dust. Earth Planet Sci Lett 264(1–2):177–187. doi:10.1016/j.epsl.2007.09.030

Lal D, Peters B (1967) Cosmic ray produced radioactivity on the earth. In: Handbuch der Physik. Springer-Verlag, Berlin

Laskar J (1990) The chaotic motion of the solar system: a numerical estimate of the size of the chaotic zones. Icarus (88):226–291

Li FJ, Wu NQ, Pei YP, Hao QZ, Rousseau DD (2006) Wind-blown origin of Dongwan late Miocene-Pliocene dust sequence documented by land snail record in western Chinese Loess Plateau. Geology 34(5):405–408. doi:10.1130/G22232.1

Li HH (1982) Thermoluminescence age of loess. Geochimica 1:82–84 (in Chinese)

Li HH, Sun J (1986) Formation age of Malan loess dated by thermoluminescence. Sci China Series B 29(11):1212–1220

Li J, Pei Z, Lu Y (1977) A preliminary study of both the thermoluminescence of the quartz silts in loess and the age determination of the loess layers. Kexue Tongbao 7:498–502

Li JJ, Fang XM, Van der Voo R, Zhu JJ, Mac Niocaill C, Cao JX, Zhong W, Chen HL, Wang JL, Wang JM, Zhang YC (1997) Late Cenozoic magnetostratigraphy (11-0 Ma) of the Dongshanding and Wangjiashan sections in the Longzhong Basin, western China. Geol Mijnbouw 76(1–2):121–134

Li YM, Liu DS (2003) delta C-13(SC) values of the loesspaleosol sequence in Dali area, Shaanxi Province, China, and paleoenvironmental implications. Chin Sci Bull 48(9):902–907. doi:10.1360/02wd0224

Libby W (1955) Radiocarbon Dating, 2nd Ed. University of Chicago Press, Chicago

Lin BH, Liu R (1992) The stable isotope evidence of the variance of the Summer Monsoon in the Loess Plateau from recent 800 ka. Chin Sci Bull 37 (18):1691–1693

Lin BH, Liu RM, An ZS (1991) Preliminary research on stable isotopic compositions of Chinese loess. In: Liu T (ed) Environment and Global Ghange. Science Press, Beijing, pp 124–131 (in Chinese)

Lisiecki LE, Raymo ME (2005) A Pliocene-Pleistocene stack of 57 globally distributed benthic delta O-18 records. Paleoceanography 20 (1):Pa1003. doi:10.1029/2004pa001071

Liu JF, Guo ZT, Qiao YS, Hao QZ, Yuan BY (2006a) Eolian origin of the Miocene loess-soil sequence at Qin'an, China: Evidence of quartz morphology and quartz grain-size. Chin Sci Bull 51(1):117–120. doi:10.1007/s11434-005-0811-8

Liu L, Chen J, Ji J, Chen Y (2004) Comparison of paleoclimatic change from Zr/Rb ratios in Chinese loess with marine isotope records over the 2.6-1.2 Ma BP interval. Geophys Res Lett 31 (15):L15204. doi:10.1029/2004GL019693

Liu QS, Roberts AP, Rohling EJ, Zhu R, Sun Y (2008) Post-depositional remanent magnetization lock-in and the location of the Matuyama-Brunhes geomagnetic reversal boundary in marine and Chinese loess sequences. Earth Planet Sci Lett 275(1–2):102–110. doi:10.1016/j.epsl.2008.08.004
Liu T, Wen Q, An Z, Zheng H (1980) Loess deposits and the palaeoclimatic evolution since the Pleistocene in China. Paper presented at the International Geology Congress, Paris
Liu TS, et al (1985) Loess and the Environment. China Ocean Press, Beijing, pp 1–251
Liu TS, Chang TH (1962) The Huangtu (loess) of China. Acta Geol Sin 42(42):1–14 (in Chinese)
Liu TS, Ding ZL (1993) Stepwise Coupling of Monsoon Circulations to Global Ice Volume Variations during the Late Cenozoic. Global Planet Change 7(1-3):119–130
Liu TS, Ding ZL (1998) Chinese loess and the paleomonsoon. Annu Rev Earth Pl Sc 26:111–145
Liu TS, Ding ZL, Rutter N (1999a) Comparison of Milankovitch periods between continental loess and deep sea records over the last 2.5 Ma. Quat Sci Rev 18(10–11):1205–1212
Liu TS, Guo ZT, Liu JQ, Han JM, Ding ZL, Gu ZY, Wu NQ (1995a) Variations of Eastern Asian Monsoon over the Last 140,000 Years. B Soc Geol Fr 166(2):221–229
Liu WG, An ZS, Zhou WJ, Head MJ, Cai DL (2003a) Carbon isotope and C/N ratios of suspended matter in rivers: an indicator of seasonal change in C-4/C-3 vegetation. Appl Geochem 18 (8):1241–1249. doi:10.1016/S0883-2927(02)00249-4
Liu WG, Feng XH, Ning YF, Zhang QL, Cao YN, An ZS (2005a) delta C-13 variation of C-3 and C-4 plants across an Asian monsoon rainfall gradient in arid northwestern China. Global Change Biol 11(7):1094–1100. doi:10.1111/j.1365-2486.2005.00969.x
Liu WG, Huang Y (2005) Compound specific D/H ratios and molecular distributions of higher plant leaf waxes as novel paleoenvironmental indicators in the Chinese Loess Plateau. Org Geochem 36 (6):851–860. doi:http://dx.doi.org/10.1016/j.orggeochem.2005.01.006
Liu WG, Ning YF, An ZS, Wu ZH, Lu HY, Cao YN (2005b) Carbon isotopic composition of modern soil and paleosol as a response to vegetation change on the Chinese Loess Plateau. Sci China Ser D 48(1):93–99. doi:10.1360/02yd0148
Liu WG, Yang H, Li LW (2006b) Hydrogen isotopic compositions of n-alkanes from terrestrial plants correlate with their ecological life forms. Oecologia 150(2):330–338. doi:10.1007/s00442-006-0494-0
Liu X, Yin ZY (2002) Sensitivity of East Asian monsoon climate to the uplift of the Tibetan Plateau. Palaeogeogr Palaeoclimatol Palaeoecol 183(3):223–245
Liu XD, Chen BD (2000) Climatic warming in the Tibetan Plateau during recent decades. Int J Climatol 20 (14):1729–1742
Liu XM, An Z, Rolph T, Qiang X, Hesse P, Lu H, Zhou J, Cai Y (2001) Magnetic properties of the Tertiary red clay from Gansu. Sci China Ser D 44(7):635–651. doi:10.1007/BF02875337
Liu XM, Hesse P, Rolph T, Beget JE (1999b) Properties of magnetic mineralogy of Alaskan loess: evidence for pedogenesis. Quat Int 62:93–102. doi:10.1016/S1040-6182(99)00027-0
Liu XM, Liu TS, Xu TC, Liu C, Chen MY (1988) The Chinese Loess in Xifeng .1. The Primary Study on Magnetostratigraphy of a Loess Profile in Xifeng Area, Gansu Province. Geophys J Int 92(2):345–348
Liu XM, Rolph T, An ZS, Hesse P (2003b) Paleoclimatic significance of magnetic properties on the Red Clay underlying the loess and paleosols in China. Palaeogeogr Palaeoclimatol Palaeocl 199 (1–2):153–166. doi:10.1016/S0031-0182(03)00504-2
Liu XM, Rolph T, Bloemendal J, Shaw J, Liu TS (1995b) Quantitative Estimates of Palaeoprecipitation at Xifeng, in the Loess Plateau of China. Palaeogeogr Palaeoclimatol Palaeocl 113(2–4):243–248
Liu XM, Xu T, Liu T (1989) The anisotropy of magnetic susceptibility and the genesis of Chinese loess: Implication to Quaternary research. Sci China Series B (4):441–448
Liu Z, Wu G, Shao L, Wang J, Huang B (2003b) Evolution of the South China Sea and monsoon history revealed in deep-sea records. Chin Sci Bull 48(23):2549–2561
Lowick SE, Preusser F (2011) Investigating age underestimation in the high dose region of optically stimulated luminescence using fine grain quartz. Quat Geochron 6(1):33–41. doi:10.1016/j.quageo.2010.08.001

Lowick SE, Preusser F, Pini R, Ravazzi C (2010) Underestimation of fine grain quartz OSL dating towards the Eemian: Comparison with palynostratigraphy from Azzano Decimo, northeastern Italy. Quat Geochron 5(5):583–590. doi:10.1016/j.quageo.2009.12.003

Lu H, Zhang FQ, Liu X, Duce RA (2004a) Periodicities of palaeoclimatic variations recorded by loess-paleosol sequences in China. Quat Sci Rev 23(18–19):1891–1900. doi:10.1016/j.quascirev.2004.06.005

Lu HY, Han JM, Wu NQ, Guo ZT (1994) Determination of modern soil in China and its paleoclimatic significance. Sci China series B 24 (12):1290–1297

Lu HY, Liu XD, Zhang FQ, An ZS, Dodson J (1999) Astronomical calibration of loess-paleosol deposits at Luochuan, central Chinese loess plateau. Palaeogeogr Palaeoclimatol Palaeoecol 154:237–246

Lu HY, Stevens T, Yi S, Sun X (2006) An erosional hiatus in Chinese loess sequences revealed by closely spaced optical dating. Chin Sci Bull 51(18):2253–2259. doi:10.1007/s11434-006-2097-x

Lu HY, Vandenberghe J, An ZS (2001) Aeolian origin and palaeoclimatic implications of the 'Red Clay' (north China) as evidenced by grain-size distribution. J Quat Sci 16 (1):89–97

Lu H, Wu N, Liu T, Han J, Qin X, Sun X, Wang Y (1996) Seasonal climatic variation recorded by phytolith assembages from the Baoji loess sequence in central China over the last 150 000 a. Sci China Ser D 26(6):2013–2016

Lu HY, Wang XY, An ZS, Miao XD, Zhu RX, Ma HZ, Li Z, Tan HB, Wang XY (2004b) Geomorphologic evidence of phased uplift of the northeastern Qinghai-Tibet Plateau since 14 million years ago. Sci China Ser D 47(9):822–833. doi:10.1360/03yd0315

Lu YC (1981) Pleistocene climatic cycles and variation of CaCO3 content in a loess profile. Chin J Geol 16:122–131 (in Chinese)

Lu YC, An ZS (1979) Quest for series of national environmental chagnes in loess plateau during Brunhes epoch. Kexue Tongbao 24:221–224

Lu YC, Prescott JR, Robertson GB, Hutton JT (1987) Thermoluminescence dating of the Malan loess at Zhaitang, China. Geology 15 (7):603–605

Lu YC, Wang XL, Wintle AG (2007) A new OSL chronology for dust accumulation in the last 130,000 yr for the Chinese Loess Plateau. Quat Res 67(1):152–160. doi:10.1016/j.yqres.2006.08.003

Lu YC, Zhang J, Xie J (1988) Thermoluminescence dating of loess and palaeosols from the Lantian section, Shaanxi province, China. Quat Sci Rev 7 (3–4):245–250. doi:10.1016/0277-3791(88)90011-x

Ma Y, Li J, Fan X (1998) Pollen-based vegetational and climaticrecords during 30.6 to 5.0 My from Linxia area, Gansu. Chin Sci Bull 43(3):301–304

Maher BA (1998) Magnetic properties of modern soils and Quaternary loessic paleosols: paleoclimatic implications. Palaeogeogr Palaeoclimatol Palaeocl 137(1–2):25–54

Maher BA, Thompson R (1995) Paleorainfall reconstructions from pedogenic magnetic susceptibility variations in the Chinese Loess and Paleosols. Quat Res 44(3):383–391. doi:10.1006/qres.1995.1083

Maher BA, Thompson R, Zhou LP (1994) Spatial and Temporal Reconstructions of Changes in the Asian Palaeomonsoon—a new mineral magnetic approach. Earth Planet Sci Lett 125(1–4):461–471

Masarik J, Beer J (1999) Simulation of particle fluxes and cosmogenic nuclide production in the Earth's atmosphere. J Geophys Res 104 (D10):12099-12111. doi:10.1029/1998jd200091

Mehra OP, Jackson ML (1960) Iron oxide removal from soils and clays by a Dithionite-Citrate System Buffered with Sodium Bicarbonate. Clays and Minerals 5:317–327

Merrill RT, Mcfadden PL (1995) Dynamo Theory and Paleomagnetism. J Geophys Res 100 (B1):317–326. doi:10.1029/94jb02361

Milankovitch M (1930) Mathematische Klimalehre und Astronomische Theorie der Klimaschwankungen. In: Köppen (ed) Handbuch der Klimalogie Band 1. Teil A Borntrager, Berlin

Mohr H, Schopfer P (1995) C4 Plants and CAM Plants. In: Plant Physiology. Springer Berlin Heidelberg, pp. 245–257. doi:10.1007/978-3-642-97570-7_15

Morgan ME, Kingston JD, Marino BD (1994) Carbon Isotopic Evidence for the Emergence of C4 Plants in the Neogene from Pakistan and Kenya. Nature 367(6459):162–165. doi:10.1038/367162a0

Murray A, Olley J (2002) Precision and accuracy in the optically stimulated luminescence dating of sedimentary quartz: a status review. Geochronometria 21:1–16

Murray AS, Svendsen JI, Mangerud J, Astakhov VI (2007) Testing the accuracy of quartz OSL dating using a known-age Eemian site on the river Sula, northern Russia. Quat Geochron 2(1-4):102–109. doi:10.1016/j.quageo.2006.04.004

Muscheler R, Beer R, Kubik PW, Synal HA (2005) Geomagnetic field intensity during the last 60,000 years based on ^{10}Be and ^{36}Cl from the Summit ice cores and ^{14}C. Quat Sci Rev 24(16-17):1849–1860. doi:10.1016/j.quascirev.2005.01.012

Ning YF, Liu WG, An ZS (2008) A 130-ka reconstruction of precipitation on the Chinese Loess Plateau from organic carbon isotopes. Palaeogeogr Palaeoclimatol Palaeocl 270(1-2):59–63. doi:10.1016/j.palaeo.2008.08.015

Nordt LC, Boutton TW, Jacob JS, Mandel RD (2002) C-4 plant productivity and Climate-CO2 variations in south-central Texas during the late Quaternary. Quat Res 58(2):182–188. doi:10.1006/qres.2002.2344

Nugteren G, Vandenberghe J (2004) Spatial climatic variability on the Central Loess Plateau (China) as recorded by grain size for the last 250 kyr. Global Planet Change 41(3–4):185–206. doi:10.1016/j.gloplacha.2004.01.005

O'Neill AL (1995) Spectral monitoring of climate change. http://Ltpwww.gsfc.nasa.gov / ISSSR–95

Olsson IU (1980) 14C extractives from wood. Radiocarbon 22 (2):515–524

Pagani M, Freeman KH, Arthur MA (1999) Late Miocene Atmospheric CO2 Concentrations and the Expansion of C4 Grasses. Science 285(5429):876–879. doi:10.1126/science.285.5429.876

Pan YX, Zhu RX, Shaw J, Liu QS, Guo B (2001) Can relative paleointensities be determined from the normalized magnetization of the wind-blown loess of China? J Geophys Res 106(B9):19221–19232

Pang JL, Huang CC, Liu A, Wang LJ (2007) Ba/Sr and Rb/Sr ratio of holocene loess-palaeosol sequences and its significiance in south Loess Plateau. Quat Sci 27:357–364 (in Chinese)

Peng S, Guo Z (2001) Geochemical indicator of original eolian grain size and implications on winter monsoon evolution. Sci China Ser D 44(1):261–266. doi:10.1007/BF02911995

Petit JR, Jouzel J, Raynaud D, Barkov NI, Barnola JM, Basile I, Bender M, Chappellaz J, Davis M, Delaygue G, Delmotte M, Kotlyakov VM, Legrand M, Lipenkov VY, Lorius C, Pepin L, Ritz C, Saltzman E, Stievenard M (1999) Climate and atmospheric history of the past 420,000 years from the Vostok ice core, Antarctica. Nature 399(6735):429–436

Porter SC (2000) High-resolution paleoclimatic information from Chinese eolian sediments based on grayscale intensity profiles. Quat Res 53:70–77

Porter SC, An ZS (1995) Correlation between climate events in the North Atlantic and China during the last glaciation. Nature 375(6529):305–308. doi:10.1038/375305a0

Porter SC, Hallet B, Wu XH, An ZH (2001) Dependence of near-surface magnetic susceptibility on dust accumulation rate and precipitation on the Chinese Loess Plateau. Quat Res 55(3):271–283

Porter SC, Zhou WJ (2006) Synchronism of Holocene East Asian monsoon variations and North Atlantic drift-ice tracers. Quat Res 65(3):443–449. doi:10.1016/j.yqres.2006.01.007

Prell W, Kutzbach J (1997) The impact of Tibet-Himalayan elevation on the sensitivity of the monsoon climate system to changes in solar radiation. In: Ruddiman WF (ed) Tectonic Uplift and Climate Change. Plenum Publishing Corporation, New York, pp. 171–201

Prescott JR, Robertson GB (1997) Sediment dating by luminescence: a review. Radiat Meas 27 (5–6):893–922. doi:10.1016/s1350-4487(97)00204-7

Priller A, Berger M, Gaggeler HW, Gerasopoulos E, Kubik PW, Schnabel C, Tobler L, Wild EM, Zanis P, Zerefos C (2004) Accelerator mass spectrometry of particle-bound ^{10}Be. Nucl Instrum Meth B 223:601–607. doi:10.1016/j.nimb.2004.04.111

Prins MA, Vriend M, Nugteren G, Vandenberghe J, Lu HY, Zheng HB, Weltje GJ (2007) Late Quaternary aeolian dust input variability on the Chinese Loess Plateau: inferences from unmixing of loess grain-size records. Quat Sci Rev 26(1–2):230–242. doi:10.1016/j.quascirev.2006.07.002
Pyankov VI, Gunin PD, Tsoog S, Black CC (2000) C-4 plants in the vegetation of Mongolia: their natural occurrence and geographical distribution in relation to climate. Oecologia 123(1):15–31
Pye K (1987) Aeolian dust and dust deposits. Academic Press, London, pp 1–334
Qiang MR, Chen FH, Wang ZT, Niu GM, Song L (2010) Aeolian deposits at the southeastern margin of the Tengger Desert (China): Implications for surface wind strength in the Asian dust source area over the past 20,000 years. Palaeogeogr Palaeoclimatol Palaeocl 286(1–2):66–80. doi:10.1016/j.palaeo.2009.12.005
Qiang XK, An ZS, Li HM, Chang H, Song YG (2005) Magnetic properties of Jiaxian red clay sequences from northern Chinese Loess Plateau and its paleoclimatic significance. Sci China Ser D 48(8):1234–1245. doi:10.1360/03yd0251
Qiang XK, An ZS, Song YG, Chang H, Sun YB, Liu WG, Ao H, Dong JB, Fu CF, Wu F, Lu FY, Cai YJ, Zhou WJ, Cao JJ, Xu XW, Ai L (2011) New eolian red clay sequence on the western Chinese Loess Plateau linked to onset of Asian desertification about 25 Ma ago. Sci China Earth Sci 54(1):136–144. doi:10.1007/s11430-010-4126-5
Qiang XK, Li ZX, Powell CM, Zheng HB (2001) Magnetostratigraphic record of the Late Miocene onset of the East Asian monsoon, and Pliocene uplift of northern Tibet. Earth Planet Sci Lett 187(1-2):83–93
Qiao Y, Huang B, Zhou M (1987) 14C Dating of loess. In: T.S L (ed) Loess and the environment. China Ocean Press, Beijing, pp 81–92 (in Chinese)
Qiao YS, Guo ZT, Hao QZ, Yin QZ, Yuan BY, Liu TS (2006) Grain-size features of a Miocene loess-soil sequence at Qinan: Implications on its origin. Sci China Ser D 49(7):731–738. doi:10.1007/s11430-006-0731-8
Qin X, Cai B, Liu T (2005) Loess record of the aerodynamic environment in the east Asia monsoon area since 60,000 years before present. J Geophys Res 110 (B1):B01204. doi:10.1029/2004JB003131
Quade J, Cerling TE (1995) Expansion of C-4 Grasses in the Late Miocene of Northern Pakistan - Evidence from Stable Isotopes in Paleosols. Palaeogeogr Palaeoclimatol Palaeocl 115(1-4):91–116
Quade J, Cerling TE, Bowman JR (1989) Development of Asian monsoon revealed by marked ecological shift during the latest Miocene in Northern Pakistan. Nature 342(6246):163–166
Ravelo AC, Andreasen DH, Lyle M, Lyle AO, Wara MW (2004) Regional climate shifts caused by gradual global cooling in the Pliocene epoch. Nature 429(6989):263–267. doi:10.1038/Nature02567
Raymo ME, Ruddiman WF (1992) Tectonic Forcing of Late Cenozoic Climate. Nature 359(6391):117–122
Rea DK (1994) The paleoclimatic record provided by eolian deposition in the deep-sea—the geologic history of wind. Rev Geophys 32(2):159–195
Rea DK, Snoeckx H, Joseph LH (1998) Late Cenozoic eolian deposition in the North Pacific: Asian drying, Tibetan uplift, and cooling of the northern hemisphere. Paleoceanography 13(3):215–224
Regnéll J (1992) Preparing pollen concentrates for AMS dating—a methodological study from a hard-water lake in southern Sweden. Boreas 21(4):373–377. doi:10.1111/j.1502-3885.1992.tb00042.x
Reheis MC (1987) Climatic implications of alternating clay and carbonate formation in semiarid soils of South-Central Montana. Quat Res 27(3):270–282
Richthofen F (1877) China, vol 1. Verlag von Dietrich Reimer, Berlin
Roberts AP, Lehman B, Weeks RJ, Verosub KL, Laj C (1997) Relative paleointensity of the geomagnetic field over the last 200,000 years from ODP Sites 883 and 884, North Pacific Ocean. Earth Planet Sci Lett 152(1–4):11–23

Roberts HM, Wintle AG (2001) Equivalent dose determinations for polymineralic fine-grains using the SAR protocol: application to a Holocene sequence of the Chinese Loess Plateau. Quat Sci Rev 20(5-9):859–863

Rossinsky Jr V, Swart PK (1993) Influence of climate on the formation and isotopic composition of calcretes. In: Swart PK, Lohmann, K.C., McKenzie, J., Savin, S (eds.) Climate change in continental isotopic records, vol 78. vol Geophys. Monogr. Ser. AGU, Washington, D. C. doi:10.1029/GM078p0067

Rousseau DD, Wu NQ, Guo ZT (2000) The terrestrial mollusks as new indices of the Asian paleomonsoons in the Chinese loess Plateau. Global Planet Change 26:199–206

Rowe PJ, Maher BA (2000) 'Cold' stage formation of calcrete nodules in the Chinese Loess Plateau: evidence from U-series dating and stable isotope analysis. Palaeogeogr Palaeoclimatol Palaeocl 157 (1-2):109–125. doi:10.1016/S0031-0182(99)00157-1

Ruddiman WF (1997) Tectonic uplift and climate change. Plenum Press, New York, London

Rutter N (1992) Xiii Inqua Congress 1991—Chinese Loess and global change—Presidential-Address. Quat Sci Rev 11 (3):275–281. doi:10.1016/0277-3791(92)90001-O

Rutter N, Ding ZL, Liu TS (1991) Comparison of isotope Stages 1–61 with the Baoji-Type pedostratigraphic section of North-Central China. Can J Earth Sci 28(7):985–990. doi:10.1139/E91-089

Sachse D, Radke J, Gleixner G (2006) delta D values of individual n-alkanes from terrestrial plants along a climatic gradient—Implications for the sedimentary biomarker record. Org Geochem 37(4):469–483. doi:10.1016/j.orggeochem.2005.12.003

Sage RF (2001) Environmental and evolutionary preconditions for the origin and diversification of the C4 photosynthetic syndrome. Plant Biol 3:202–203

Sauer P, Eglinton TI, Hayes JM, Schimmelmann A, Sessions AL (2001) Compound-specific D/H ratios of lipid biomarkers from sediments as a proxy for environmental and climatic conditions. Geochim Cosmochim Acta 65:213–222

Schefuß E, Schouten S, Jansen JHF, Sinninghe DJS (2003) Africa vegetation controlled by tropical sea surface temperatures in the mid-Pleistocene period. Nature 422:418–421

Schulze E, Ellis R, Schulze W, Trimborn P, Ziegler H (1996) Diversity, metabolic types and δ13C carbon isotope ratios in the grass flora of Namibia in relation to growth form, precipitation and habitat conditions. Oecologia 106(352):369

Sessions AL (2006) Seasonal changes in D/H fractionation accompanying lipid biosynthesis in Spartina alternflora. Geochim Cosmochim Acta 70(9):2153–2162. doi:10.1016/j.gca.2006.02.003

Sessions AL, Burgoyne TW, Schimmelmann A, Hayes JM (1999) ractionation of hydrogen isotopes in lipid biosynthesis. Org Geochem 30:1193–1200

Sessions AL, Hayes JM (2005) Calculation of hydrogen isotopic fractionations in biogeochemical systems. Geochim Cosmochim Acta 69(3):593–597

Shackleton NJ, Berger A, Peltier WR (1990) An alternative astronomical calibration of the lower Pleistocene timescale based on ODP Site 677. Earth Env Sci T R So 81 (04):251–261. doi:10.1017/S0263593300020782

Shackleton NJ, Hall MA, Pate D (1995) Pliocene stable isotope stratigraphy of Site 864. In: Pisias N, Mayer L, Janecek T, Palmer-Julson A, van Andel T (eds) Proceedings of the Ocean Drilling Program, Scientific Results, vol 138. Ocean Drilling Program, College Station, TX pp 337–355. doi:10.2973/odp.proc.sr.138.117.1995

Shen C, Beer J, Heller F (2000) ^{10}Be-susceptibility model and quantitative estimates of pedogenic ferromagnetic material flux in Chinese loess. Nucl Instr Meth Phys Res B 172:551–554

Shen C, Beer J, Liu T (1992) ^{10}Be in Chinese loess. Earth Planet Sci Lett 109(1–2):169–177

Shen C, Liu T, Beer J (1989) ^{10}Be and the accumulating process of Chinese Loess. Sci China Series B 7:744–751

Shen J, Liu XQ, Wang SM, Matsumoto R (2005) Palaeoclimatic changes in the Qinghai Lake area during the last 18,000 years. Quat Int 136:131–140. doi:10.1016/j.quaint.2004.11.014

Shen ZX, Cao JJ, Zhang XY, Arimoto R, Ji JF, Balsam WL, Wang YQ, Zhang RJ, Li XX (2006) Spectroscopic analysis of iron-oxide minerals in aerosol particles from northern China. Sci Total Environ 367(2-3):899–907

Shi Y, Yu G, Liu X, Li B, Yao T (2001) Reconstruction of the 30–40 kabp enhanced Indian monsoon climate based on geological records from the Tibetan Plateau. Palaeogeogr Palaeoclimatol Palaeoecol 169(1):69–83

Smith B, Ziegler H (1990) Isotopic fractionation of hydrogen in plants. Botanica Acta 103:335–342

Smith FA, Freeman KH (2006) Influence of physiology and climate on δD of leaf wax n-alkanes from C3 and C4 grasses. Geochim Cosmochim Acta 70:1172–1187

Song YG, Fang XM, Li JJ, An ZS, Miao XD (2001) The Late Cenozoic uplift of the Liupan Shan, China. Sci China Ser D 44:176–184

Song YG, Fang XM, Li JJ, An ZS, Yang D, Lu LQ (2000) Age of red clay at Chaona section near Liupan Mountain and its tectonic significance. Quat Sci 20:457–463 (in Chinese)

Spassov S, Heller F, Evans M (2001) The Matuyama/Brunhes Geomagnetic Polarity Transition at Lingtai and Baoji, Chinese Loess Plateau. Phys Chemi Earth (A) 26 (11-12):899–904

Spooner NA (1994) The anomalous fading of infrared-stimulated luminescence from feldspars. Radiat Meas 23(2-3):625–632

Stevens T, Armitage SJ, Lu HY, Thomas DSG (2006) Sedimentation and diagenesis of Chinese loess: Implications for the preservation of continuous, high-resolution climate records. Geology 34(10):849–852. doi:10.1130/G22472.1

Stevens T, Lu H (2010) Radiometric dating of the late Quaternary summer monsoon on the Loess Plateau, China. Geological Society, London. Special Publications 342(1):87–108. doi:10.1144/sp342.8

Stevens T, Thomas DSG, Armitage SJ, Lunn HR, Lu HY (2007) Reinterpreting climate proxy records from late Quaternary Chinese loess: A detailed OSL investigation. Earth-Sci Rev 80(1–):111–136. doi:10.1016/j.earscirev.2006.09.001

Stewart G, Turnbull M, Schmidt S, Erskine P (1995) 13C natural abundance in plant communities along a rainfall gradient: A biological integrator of water availability. Aus J Plant Physiol 22:51–55

Stokes S, Hetzel R, Bailey RM, Tao MX (2003) Combined IRSL-OSL single aliquot regeneration (SAR) equivalent dose (D-e) estimates from source proximal Chinese loess. Quat Sci Rev 22 (10–13):975–983. doi:10.1016/S0277-3791(03)00044-1

StreetPerrott FA, Huang YS, Perrott RA, Eglinton G, Barker P, BenKhelifa L, Harkness DD, Olago DO (1997) Impact of lower atmospheric carbon dioxide on tropical mountain ecosystems. Science 278(5342):1422–1426. doi:10.1126/science.278.5342.1422

Sun DH, An Z, Shaw J, Bloemendal J, Sun Y (1998a) Magnetostratigraphy and palaeoclimatic significance of Late Tertiary aeolian sequences in the Chinese Loess Plateau. Geophys J Int 134(1):207–212. doi:10.1046/j.1365-246x.1998.00553.x

Sun DH, Bloemendal J, Rea DK, An ZS, Vandenberghe J, Lu HY, Su RX, Liu TS (2004) Bimodal grain-size distribution of Chinese loess, and its palaeoclimatic implications. Catena 55:325–340

Sun DH, Liu DS, Chen MY, An ZS, Shaw J (1997) Magnetostratigraphy and palaeoclimate of Red Clay sequences from Chinese Loess Plateau. Sci China Ser D 40(4):337–343. doi:10.1007/Bf02877564

Sun DH, Shaw J, An Z (1993) Matuyama/Brunhes (M/B) transition recorded in Chinese Loess. J Geomagn Geoelectr 45:319–330

Sun DH, Shaw J, An ZS, Cheng MY, Yue LP (1998b) Magnetostratigraphy and paleoclimatic interpretation of a continuous 7.2Ma Late Cenozoic eolian sediments from the Chinese Loess Plateau. Geophys Res Lett 25(1):85–88

Sun DH, Su RX, Bloemendal J, Lu HY (2008a) Grain-size and accumulation rate records from Late Cenozoic aeolian sequences in northern China: Implications for variations in the East Asian winter monsoon and westerly atmospheric circulation. Palaeogeogr Palaeoclatol Palaeoecol 264(1–2):39–53. doi:10.1016/j.palaeo.2008.03.011

Sun JM (2002) Provenance of loess material and formation of loess deposits on the Chinese Loess Plateau. Earth Planet Sci Lett 203 (3–4):845–859. doi:10.1016/S0012-821x(02)00921-4

Sun JM (2005a) Nd and Sr isotopic variations in Chinese eolian deposits during the past 8 Ma: Implications for provenance change. Earth Planet Sci Lett 240(454):466

Sun JM, Liu TS (2000) Stratigraphic evidence for the uplift of the Tibetan Plateau between similar to 1.1 and similar to 0.9 myr ago. Quat Res 54(3):309–320. doi:10.1006/qres.2000.2170

Sun J, Diao G, Wen Q, Zhou H (1999) A preliminary study on quantitative estimate of Palaeoclimate by using geochemical transfer function in the Loess Plateau. Geochimica 28(3):265–272 (in Chinese with English Abstr.)

Sun JM, Zhu ZK (2010) Temporal variations in Pb isotopes and trace element concentrations within Chinese eolian deposits during the past 8 Ma: Implications for provenance change. Earth Planet Sci Lett 290:438–447

Sun JZ (2005b) Loess. Hong Kong Archaeological Society, Hong Kong

Sun X, Du N, Weng C, Lin R, Wei K (1994) Paleovegetation and paleoenvironment of Nanaisi lake, Xinjiang, N.W. China during the last 14,000 years. Quat Sci 3:239–248 (in Chinese)

Sun Y, He L, Liang L, An Z (2011) Changing color of Chinese loess: Geochemical constraint and paleoclimatic significance. J Asian Earth Sci 40(6):1131–1138

Sun YB, An ZS (2002) History and variability of Asian interior aridity recorded by eolian flux in the Chinese Loess Plateau during the past 7 Ma. Sci China Ser D 45(5):420–429

Sun YB, An ZS (2004) An improved comparison of Chinese loess with deep-sea 18O record over the interval 1.6–2.6 Ma. Geophys Res Lett 31:L13210

Sun YB, An ZS (2005) Late Pliocene-Pleistocene changes in mass accumulation rates of eolian deposits on the central Chinese Loess Plateau. J Geophys Res 110:D23101. doi:10.1029/2005jd006064

Sun YB, An ZS, Clemens SC, Bloemendal J, Vandenberghe J (2010a) Seven million years of wind and precipitation variability on the Chinese Loess Plateau. Earth Planet Sci Lett 297(3-4):525–535. doi:10.1016/j.epsl.2010.07.004

Sun YB, An ZS, Clemens SC (2010c) Non-stationary response of Plio-Pleistocene East Asian winter monsoon variation to ice volume forcing. Geological Society, London, Special Publications, SP342:79–86

Sun YB, Clemens SC, An ZS, Yu ZW (2006a) Astronomical timescale and palaeoclimatic implication of stacked 3.6-Myr monsoon records from the Chinese Loess Plateau. Quat Sci Rev 25(1–2):33–48. doi:10.1016/j.quascirev.2005.07.005

Sun YB, Clemens SC, Morrill C, Lin X, Wang X, An Z (2012) Influence of Atlantic meridional overturning circulation on the East Asian winter monsoon. Nat Geosci 5(1):46–49. doi:10.1038/ngeo1326

Sun YB, Lu HY, An ZS (2006b) Grain size of loess-laeosol and Red Clay deposits on the Chinese Loess Plateau: Significance for understanding pedogenic alteration and palaeomonsoon evolution. Palaeogeogr Palaeoclimatol Palaeocl 241:129–138

Sun Y, Qiang, X, Liu, Q, Bloemendal, J, Wang X (2013) Timing and lock-in effect of the Laschamp geomagnetic excursion in Chinese Loess, Geochem. Geophys. Geosyst., doi:10.1002/2013GC004828

Sun YB, Tada RJ, Chen JC, Liu QS, Toyoda S, Tani A, Ji JF, Isozaki Y (2008) Tracing the provenance of fine-grained dust deposited on the central Chinese Loess Plateau. Geophys Res Lett 35 (1):L01804. doi:10.1029/2007gl031672

Sun YB, Wang XL, Liu QS, Clemens SC (2010b) Impacts of post-depositional processes on rapid monsoon signals recorded by the last glacial loess deposits of northern China. Earth Planet Sci Lett 289(1-2):171–179. doi:10.1016/j.epsl.2009.10.038

Teilhard de Chardin P, Yang CC (1930) Preliminary observations on the pre-Loessic and postPontian formations in W. Shansi and N. Shensi. Geol Mem, Ser A (8):32–35

Thompson L, Yao T, Davis M (1997) Tropical climate instability: the last glacial cycle from a Qinghai-Tibetan ice core. Science 276:1821–1825

Thompson R (1986) Modelling magnetization data using SIMPLEX. Phys Earth Planet Ints 42:113–127

Tiedemann R, Sarnthein M, Shackleton NJ (1994) Astronomical timescale for the Pliocene Atlantic 18O and dust flux records of Ocean Drilling Program site 659. Paleoceanography 9:619–638

Torrent J, Liu QS, Bloemendal J, Barro´n V (2007) Magnetic enhancement and iron oxides in the upper luochuan loess-paleosol sequence, Chinese Loess Plateau. Soil Sci Soc Am J 71:1570–1578. doi:10.2136/sssaj2006.0328

Trauth MH, Larrasoan JC, Mudelsee M (2009) Trends, rhythms and events in Plio-Pleistocene African climate. Quat Sci Rev 28:399–411

Tric E, Valet JP, Tucholka P, Paterne M, Labeyrie L, Guichard F, Tauxe L, Fontugne M (1992) Paleointensity of the Geomagnetic-Field during the Last 80,000 Years. J Geophys Res 97 (B6):9337–9351. doi:10.1029/91jb01620

Valet J, Meynadier L (1993) Geomagnetic field intensity and reversal during the past four million years. Nature 366:234–238

Vandenberghe J, An ZS, Nugteren G, Lu HY, VanHuissteden K (1997) New absolute time scale for the Quaternary climate in the Chinese loess region by grain-size analysis. Geology 25 (1):35–38. doi:10.1130/0091-7613 (1997) 025<0035:Natsft>2.3.Co;2

Vandenberghe J, Lu HY, Sun D, Huissteden Jv, Konert M (2004) The late Miocene and Pliocene climate in East Asia as recorded by grain size and magnetic susceptibility of the Red Clay deposits (Chinese Loess Plateau). Palaeogeogr Palaeoclimatol Palaeocl 204:239–255

Vidic NJ, Montanez IP (2004) Climatically driven glacial-interglacial variations in C-3 and C-4 plant proportions on the Chinese Loess Plateau. Geology 32(4):337–340. doi:10.1130/G20222.1

Visocekas R (1985) Tunnelling radiative recombination in labradorite: Its association with anomalous fading of thermoluminescence. Nuclear Tracks and Radiation Measurements (1982) 10 (4–6):521–529. doi:10.1016/0735-245x(85)90053-5

Visocekas R, Spooner NA, Zink A, Blanc P (1994) Tunnel afterglow, fading and infrared emission in thermoluminescence of feldspars. Radiat Meas 23 (2–3):377–385. doi:10.1016/1350-4487(94)90067-1

Wagner G, Masarik J, Beer J, Baumgartner S, Imboden D, Kubik PW, Synal HA, Suter M (2000) Reconstruction of the geomagnetic field between 20 and 60 kyr BP from cosmogenic radionuclides in the GRIP ice core. Nucl Instrum Meth B 172 (1–4):597–604. doi:10.1016/s0168-583x(00)00285-8

Wallbrink P, Murray A (1994) Fallout of ^{7}Be in South Eastern Australia. J Environ Radioact 25:213–228

Wang B (2006) The Asian Monsoon. Springer Verlag, Berlin Heidelberg New York

Wang DJ, Wang YC, Han JT, Duan MG, Shan JZ, Liu TS (2010) Geomagnetic anomalies recorded in L9 of the Songjiadian loess section in southeastern Chinese Loess Plateau. Chin Sci Bull 55(6):520–529

Wang G, Han J, Liu D (2003a) The carbon isotope composition of C3 herbaceous plants in loess area of northern China. Sci China Ser D 46(10):1069–1076. doi:10.1007/BF02959402

Wang H, Ambrose SH, Liu CLJ, Follmer LR (1997) Paleosol stable isotope evidence for early hominid occupation of east Asian temperate environments. Quat Res 48(2):228–238. doi:10.1006/qres.1997.1921

Wang H, Follmer LR (1998) Proxy of monsoon seasonality in carbon isotopes from paleosols of the southern Chinese Loess Plateau. Geology 26 (11):987–990. doi:10.1130/0091-7613

Wang PX, Clemens S, Beaufort L, Braconnot P, Ganssen G, Jian ZM, Kershaw P, Sarnthein M (2005) Evolution and variability of the Asian monsoon system: state of the art and outstanding issues. Quat Sci Rev 24(5-6):595–629

Wang PX, Jian Z, Zhao Q, Li Q, RJ Wang, ZF Liu, GX Wu,L Shao,J Wang,B Huang,D Fang,J Tian, J Li, X Li, G Wei, Xi Sun, Y Luo, X Su, S Mao, Chen M (2003) Evolution of the South China Sea and monsoon history revealed in deep-sea records. Chin Sci Bull, 48(23): 2549–2561

Wang P, Jian Z, Zhao Q, Li Q, Wang R, et al. (2003b) Evolution of the South China Sea and monsoon history revealed in deep-sea records. Chinese Sci Bull 48: 2549–2561.

Wang X, Lu Y, Zhao H (2006) On the performances of the single-aliquot regenerative-dose (SAR) protocol for Chinese loess: fine quartz and polymineral grains. Radiat Meas 41(1):1–8. doi:10.1016/j.radmeas.2005.02.010

Wang Y, Cheng H, Edwards RL, Kong X, Shao X, Chen S, Wu J, Jiang X, Wang X, An Z (2008) Millennial- and orbital-scale changes in the East Asian monsoon over the past 224,000 years. Nature 451(7182):1090–1093. doi:10.1038/nature06692

Wang Y, Shen J, Wu J, Liu X, Zhang E, Liu E (2007a) Hard-water effect correction of lacustrine sediment ages using the relationship between 14C levels in lake waters and in the atmosphere: the case of Lake Qinghai. J Lake Sci 19(5):504–508 (in Chinese)

Wang YJ, Cheng H, Edwards RL, An ZS, Wu JY, Shen CC, Dorale JA (2001) A high-resolution absolute-dated Late Pleistocene monsoon record from Hulu Cave, China. Science 294(5550):2345–2348

Wang YX, Yang JD, Chen J, Zhang KJ, Rao WB (2007b) The Sr and Nd isotopic variations of the Chinese Loess Plateau during the past 7 Ma: Implications for the East Asian winter monsoon and source areas of loess. Palaeogeogr Palaeoclimatol Palaeoecol 249(3-4):351–361. doi:10.1016/j.palaeo.2007.02.010

Watanuki T, Murray AS, Tsukamoto S (2003) A comparison of OSL ages derived from silt-sized quartz and polymineral grains from Chinese loess. Quat Sci Rev 22 (10-13):991–997. doi:10.1016/S0277-3791(03)00053-2

Webster PJ, Magana VO, Palmer TN, Shukla J, Tomas RA, Yanai M, Yasunari T (1998) Monsoons: Processes, predictability, and the prospects for prediction. J Geophys Res 103 (C7):14451–14510. doi:10.1029/97jc02719

Weeks RJ, Laj C, Endignoux L, Mazaud A, Labeyrie L, Roberts AP, Kissel C, Blanchard E (1995) Normalized Natural Remanent Magnetization Intensity during the Last 240000 Years in Piston Cores from the Central North-Atlantic Ocean - Geomagnetic-Field Intensity or Environmental Signal. Phys Earth Planet In 87(3–4):213–229

Wen L, Lu H, Qiang X (2005) Changes in grain-size and sedimentation rate of the Neogene Red Clay deposits along the Chinese Loess Plateau and implications for the palaeowind system. Sci China Ser D 48(9):1452–1462. doi:10.1360/01yd0558

Wintle AG (1973) Anomalous fading of thermoluminescence in mineral samples. Nature 245(5421):143–144

Wintle AG (1990) A Review of Current Research on Tl Dating of Loess. Quat Sci Rev 9(4):385–397

Wintle AG, Murray AS (2006) A review of quartz optically stimulated luminescence characteristics and their relevance in single-aliquot regeneration dating protocols. Radiat Meas 41(4):369–391

Wu N, Lu H, Sun X, Guo Z, Liu J, Han J (1994) Climate transfer function from opal phytolith and its application in paleoclimate reconstrution of China loess-paleosol sequence. Quat Sci 3:270–279 (in Chinese)

Xian F, An Z, Wu Z, Beck J, Yu H, Kang Z, Cheng P (2008) A simple model for reconstructing geomagnetic field intensity with ^{10}Be production rate and its application in Loess studies. Sci China Ser D 51(6):855–861. doi:10.1007/s11430-008-0054-z

Xiao JL, An Z (1999) Three large shifts in East Asian monsoon circulation indicated by loess-paleosol sequences in China and late Cenozoic deposits in Japan. Palaeogeogr Palaeoclimatol Palaeoecol 154(3):179–189

Xiao JL, Porter SC, An ZS, Kumai H, Yoshikawa S (1995) Grain-Size of Quartz as an Indicator of Winter Monsoon Strength on the Loess Plateau of Central China during the Last 130,000-Yr. Quat Res 43(1):22–29

Xiao JL, Zheng HB, Zhao H (1992) Variation of winter monsoon intensity on the Loess Plateau, central China during the last 130,000 years: evidences from grain size distribution. The Quat Res 31:13–19

Xie S, Nott CJ, Avsejs LA, Volders F, Maddy D, Chambers FM, Gledhill A, Carter JF, Evershed RP (2000) Palaeoclimate records in compound-specific delta D values of a lipid biomarker in ombrotrophic peat. Org Geochem 31(10):1053–1057

Xu Y, Yue LP, Li JX, Sun L, Sun B, Zhang JY, Ma J, Wang JQ (2009) An 11-Ma-old red clay sequence on the Eastern Chinese Loess Plateau. Palaeogeogr Palaeoclimatol Palaeoecol 284(3–4):383–391. doi:10.1016/j.palaeo.2009.10.023
Yakir D (1992) Variations in the natural abundance of oxygen-18 and deuterium in plant carbohydrates. Plant Cell Environ 15:1005–1020
Yamazaki T (1999) Relative paleointensity of the geomagnetic field during Brunhes Chron recorded in North Pacific deep-sea sediment cores: orbital influence? Earth Planet Sci Lett 169 (1–2):23–35. doi:10.1016/s0012-821x(99)00064-3
Yamazaki T, Kanamatsu T (2007) A relative paleointensity record of the geomagnetic field since 1.6 Ma from the North Pacific. Earth Planet Sci Lett 59:785–794
Yang H, Huang Y (2003) Preservation of lipid hydrogen isotope ratios in Miocene lacustrine sediments and plant fossils at Clarkia, northern Idaho, USA. Org. Geochem 34:413–423
Yang SL, Ding ZL (2001) Seven million-year iron geochemistry record from a thick eolian red clay-loess sequence in Chinese Loess Plateau and the implications for paleomonsoon evolution. Chin Sci Bull 46(4):337–341
Yang SL, Ding ZL (2003) Color reflectance of Chinese loess and its implications for climate gradient changes during the last two glacial-interglacial cycles. Geophys Res Lett 30(20):2058. doi:10.1029/2003GL018346
Yang SL, Ding ZL (2004) Comparison of particle size characteristics of the Tertiary 'red clay' and Pleistocene loess in the Chinese Loess Plateau: implications for origin and sources of the 'red clay'. Sedimentology 51(1):77–93. doi:10.1046/j.1365-3091.2003.00612.x
Yapp C, Epstein S (1982) A re-examination of cellulose carbon-bound hydrogen D measurement and some factors affecting plant-water D/H relationships. Geochim Cosmochim Acta 46:955–965
Yu J, Kelts K (2002) Abrupt changes in climatic conditions across the late-glacial/Holocene transition on the N. E. Tibet-Qinghai Plateau: evidence from Lake Qinghai, China. J Paleolimnol 28(2):195–206
Yu SY, Shen J, Colman SM (2007) Modeling the radiocarbon reservoir effect in lacustrine systems. Radiocarbon 49(3):1241–1254
Yu ZW, Ding ZL (1998) An automatic orbital tuning method for the Paleoclimate records. Geophys Res Lett 25:4525–4528
Yuan DX, Cheng H, Edwards RL, Dykoski CA, Kelly MJ, Zhang ML, Qing JM, Lin YS, Wang YJ, Wu JY, Dorale JA, An ZS, Cai YJ (2004) Timing, duration, and transitions of the Last Interglacial Asian Monsoon. Science 304(5670):575–578
Yue L, Lei X, Qu H (1991) A Magnetostratigraphic study on hte Jingyuan loess section, Gansu, China. Quat Sci (4):349–352 (in Chinese)
Zachos J, Pagani M, Sloan L, Thomas E, Billups K (2001) Trends, rhythms, and aberrations in global climate 65 Ma to present. Science 292(5517):686–693
Zhang DE (1982) Analysis of dust deposition phenomena in historical times. Kexue Tongbao 27(5):294–297
Zhang XY, Arimoto R, An ZS (1999) Glacial and interglacial patterns for Asian dust transport. Quat Sci Rev 18:811–819
Zhang XY, Zhang GY, Zhu GH, Zhang D, An ZS, Chen T, Huang XP (1996) Elemental tracers for Chinese source dust. Sci China Ser D 39(5):512–521
Zhang YG, Ji JF, Balsam W, Liu LW, Chen J (2009) Mid-Pliocene Asian monsoon intensification and the onset of Northern Hemisphere glaciation. Geology 37(7):599–602. doi:10.1130/G25670a.1
Zhang Z, Zhao M, Lu H (2003) Lower temperature as the main cause of C4 plant declines during the glacial periods on the Chinese Loess Plateau. Earth Planet Sc Lett 214:467–481
Zhao H, Chen FH, Li SH, Wintle AG, Fan YX, Xia DS (2007) A record of Holocene climate change in the Guanzhong Basin, China, based on optical dating of a loess-palaeosol sequence. Holocene 17(7):1015–1022. doi:10.1177/0959683607080530
Zheng H (1984) Holocene loess in the middle reaches of Yellow River. Geochemistry 3:237–346

Zheng HB, An ZS, Shaw J (1992) New contributions to Chinese Plio-Pleistocene magnetostratigraphy. Phys Earth Planet In 70:146–153

Zheng HB, Oldfield F, Yu LH, Shaw J, An ZS (1991) The Magnetic-Properties of Particle-Sized Samples from the Luo-Chuan Loess Section—Evidence for Pedogenesis. Phys Earth Planet In 68(3-4):250–258

Zheng HB, Rolph T, Shaw J, An ZS (1995) A Detailed Paleomagnetic Record for the Last Interglacial Period. Earth Planet Sci Lett 133 (3-4):339–351. doi:10.1016/0012-821x(95)00089-U

Zhou LP, Oldfield F, Wintle AG, Robinson SG, Wang JT (1990a) Partly pedogenic origin of magnetic variations in Chinese loess. Nature 346(6286):737–739

Zhou LP, Shackleton NJ (1999) Misleading positions of geomagnetic reversal boundaries in Eurasian loess and implications for correlation between continental and marine sedimentary sequences. Earth Planet Sci Lett 168 (1–2):117–130. doi:10.1016/s0012-821x(99)00052-7

Zhou WJ, An ZS, Head MJ (1994a) Stratigraphic division of holocene loess in China. Radiocarbon 36(1):37–45

Zhou WJ, An ZS, Lin BH, Xiao JL, Zhang JZ, Xie J, Zhou MF, Porter SC, Head MJ, Donahue DJ (1992) Chronology of the Baxie Loess profile and the history of monsoon climates in China between 17,000 and 6000 years Bp. Radiocarbon 34(3):818–825

Zhou WJ, Chen M, Xian F, Song S, Wu Z, Jull AJT, Liu W (2007a) The mean value concept in mono-linear regression of multi-variables and its application to trace studies in geosciences. Sci China Ser D 50(12):1828–1834. doi:10.1007/s11430-007-0123-8

Zhou WJ, Donahue D, Jull AJT (1997) Radiocarbon AMS dating of pollen concentrated from eolian sediments: Implications for monsoon climate change since the late Quaternary. Radiocarbon 39(1):19–26

Zhou WJ, Donahue DJ, Porter SC, Jull TA, Li XQ, Stuiver M, An ZS, Matsumoto E, Dong GG (1996) Variability of monsoon climate in East Asia at the end of the last glaciation. Quat Res 46(3):219–229

Zhou WJ, Head M, Kaihoa L (1994b) Small sample dating in China. Radiocarbon 36(1):47–49

Zhou WJ, Head MJ, Wang FB, Donahue DJ, Jull AJT (1999) The reliability of AMS radiocarbon dating of shells from China. Radiocarbon 41(1):17–24

Zhou WJ, Priller A, Beck JW, Wu ZK, Chen MB, An ZS, Kutschera W, Feng X, Yu HG, Lin L (2007b) Disentangling geomagnetic and precipitation signals in an 80-Kyr Chinese loess record of ^{10}Be. Radiocarbon 49(1):139–160

Zhou WJ, Xian F, Beck JW, Jull AJT, An ZS, Wu ZK, Liu M, Chen MB, Priller A, Kutschera W, Burr GS, Yu HG, Song SH, Cheng P, Kong XH (2010) Reconstruction of 130-Kyr Relative Geomagnetic Intensities from ^{10}Be in Two Chinese Loess Sections. Radiocarbon 52(1):129–147

Zhou WJ, Yu XF, Jull AJT, Burr GS, Xiao JY, Lu XF, Xian F (2004) High-resolution evidence from southern China of an early Holocene optimum and a mid-Holocene dry event during the past 18,000 years. Quat Res 62(1):39–48. doi:10.1016/j.yqres.2004.05.004

Zhou WJ, Zhao X, Lu X, Lin L, Wu Z, Peng C, Zhao W, Huang C (2006) The 3MV multi-element AMS in Xi'an, China: Unique features and preliminary tests. Radiocarbon 48(2):285–293

Zhou WJ, Zhou MF, Head J (1990b) ^{14}C chronology of Beizhuang Cun sedimentation sequence since 30,000 years BP. Chin Sci Bull 35(7):567–572

Zhu RX, Coe RS, Guo B, Anderson R, Zhao XX (1998) Inconsistent palaeomagnetic recording of the Blake event in Chinese loess related to sedimentary environment. Geophys J Int 134(3):867–875

Zhu RX, Ding ZL, Wu HN, Huang BC, Jiang L (1993) Details of magnetic polarity transition recorded in Chinese Loess. J Geomagn Geoelectr 45(4):289–299

Zhu RX, Laj C, Mazaud A (1994a) The Matuyama-Brunhes and upper jaramillo transitions recorded in a Loess Section at Weinan, North-Central China. Earth Planet Sci Lett 125(1-4):143–158

Zhu RX, Liu QS, Pan YX, Deng CL, Sun JM (2006) Identifying the origin of the magnetic directional anomalies recorded in the Datong loess profile, northeastern Chinese loess plateau. Geophys J Int 164(2):312–318. doi:10.1111/j.1365-246X.2006.02848.x

Zhu RX, Pan YX, Coe RS (2000) Paleointensity studies of a lava succession from Jilin Province, northeastern China: Evidence for the Blake event. J Geophys Res 105 (B4):8305-8317. doi:10.1029/1999jb900448

Zhu RX, Zhang R, Deng CL, Pan YX, Liu QS, Sun YB (2007) Are Chinese loess deposits essentially continuous? Geophys Res Lett 34 (17):L17306. doi:10.1029/2007gl030591

Zhu RX, Zhou LP, Laj C, Mazaud A, Ding ZL (1994b) The blake geomagnetic polarity episode recorded in Chinese Loess. Geophys Res Lett 21(8):697–700

Chapter 3
Asian Monsoon Variability Recorded in Other Archives

Yanjun Cai, Zhangdong Jin, Weijian Zhou, Yu Liu, Zhonghui Liu, Baosheng Li, Xuefeng Yu, Liangcheng Tan, Lide Tian, Zicheng Peng, Shaohua Song, Xiaoqiang Li, Hongli Zhao, Fengyan Lu, Wei Pan, and Zhisheng An

Abstract This chapter provides an extensive review of paleomonsoon variability on millennial- to decadal time scales in various archives other than loess, including speleothems, lake sediments, peat deposits, desert deposits, marine sediments, corals, ice cores, tree rings, pollen archives and historical documents. Case studies will be presented to introduce the major proxy indexes associated with these archives, and the reconstructed history of Asian monsoon variability. Advances in the study of monsoon variability and the peculiarity of different archives will also be introduced in this chapter.

Keywords Speleothems · Lake sediments · Peat · Desert · Marine sediments · Coral · Ice cores · Tree ring · Pollen · Historical documents

3.1 Overview of Various Paleoclimate Archives

As described in the previous chapter, the eolian loess sequence, a unique geological archive in the world, plays an important role in studying the climatic and environmental changes in China as it documents the variability of both Asian monsoon and the aridification of inland Asia. It has the advantages of being relatively continuous and long-lasting for studying the monsoon changes and inland aridifications on

Y. Cai (✉) · Z. Jin · W. Zhou · Y. Liu · X. Yu · L. Tan · S. Song · X. Li · H. Zhao · F. Lu · Z. An
State Key Laboratory of Loess and Quaternary Geology, Institute of Earth Environment, Chinese Academy of Sciences, No. 10 Fenghui S. Rd, High-Tech Zone,
PO Box 17, 710075 Xi'an, China
e-mail: caiyj@loess.llqg.ac.cn

Z. Jin
e-mail: zhdjin@ieecas.cn

W. Zhou
e-mail: weijian@loess.llqg.ac.cn

Z. An (ed.), *Late Cenozoic Climate Change in Asia,*
Developments in Paleoenvironmental Research 16,
DOI 10.1007/978-94-007-7817-7_3,

Y. Liu
e-mail: liuyu@loess.lqg.ac.cn

X. Yu
e-mail: xfyu@loess.llqg.ac.cn

L. Tan
e-mail: tanlch@ieecas.cn

S. Song
e-mail: sshh@ieecas.cn

X. Li
e-mail: lixiaoqiang@ivpp.ac.cn

H. Zhao
e-mail: zhaohl@ieecas.cn

F. Lu
e-mail: lfy@ieecas.cn

Z. An
e-mail: anzs@loess.llqg.ac.cn

Z. Liu
Department of Earth Sciences, The University of Hong Kong, Hong Kong, China
e-mail: zhliu@hku.hk

B. Li
School of Geographical Science, South China Normal University, 510631 Guangzhou, China
e-mail: libsh@scnu.edu.cn

L. Tian
Laboratory of Environment and Processes in the Tibetan Plateau, Institute of Tibetan Plateau Research, Chinese Academy of Sciences, 100085 Beijing, China
e-mail: ldt@itpcas.ac.cn

Z. Peng
University of Science and Technology of China, No.96, JinZhai Road Baohe District, 230026 Hefei, China
e-mail: pzc@ustc.edu.cn

W. Pan
Centre for Historical Environment and Socio-Economic Development in Northwest China, Shaanxi Normal University, 710062 Xi'an, China
e-mail: panwei@snnu.edu.cn

tectonic and orbital time scales, and even has the potential for studies of monsoon variations on millennial time scales at some sites. However, other geo-biological archives and historical records are also essential in studying the Asian monsoon changes, as they provide indispensable information for fully understanding Asian monsoon changes and its dynamic forces and environmental effects, especially on the millennial to decadal time scales. For example, taking advantage of high-precision absolute U-series dating, speleothems from karst caves, provide the ages of the abrupt events that occurred within the Asian monsoon regions and make it possible to correlate these events with global climate changes, enhancing greatly our understanding of the linkage between Asian monsoon and global climate change. With multiple proxies, lake sediments are good at providing the integrated information on hydrology, weathering processes, biomass and other aspects within the drainage systems. Coral records may indicate the changes in moisture supply from the tropical ocean, ocean circulation, sea surface temperature and other parameters linked closely with monsoon climate. Tree-ring records provide annual changes of regional temperature, precipitation and the hydrological environment, and are the basis to fully describe the present global warming and the anthropogenic contribution. As the direct index of vegetation, pollen assemblages indicate the succession of vegetation associated with monsoon changes, revealing variations in moisture availability and temperature. Therefore, studies of these various archives, as well as peatbogs, ice cores, marine and desert deposits and historical records, greatly expanded our knowledge of the variability of Asian monsoons and their driving forces.

This chapter will focus on various archives used in the study of Asian monsoon changes, including speleothems, lake sediments, peat bogs, corals, tree rings, ice cores, marine deposits, desert deposits and historical records. Case studies will be presented to introduce the major proxy indices associated with these archives, and the reconstructed history of Asian monsoon variability. Meanwhile, advances in the study of monsoon variability and the peculiarity of different archives will also be covered in this chapter.

3.2 High Resolution Speleothem Records and Their Climatic Significance

3.2.1 Introduction

Speleothems, including stalagmites, stalactites and flowstones, are secondary carbonate deposits in karst caves (Moore 1952). They are formed by the loss of carbon dioxide and subsequent calcium carbonate precipitation from supersaturated groundwater that enters a cave. The carbon dioxide carried in groundwater is generated largely by root respiration and microbial oxidation of soil organic matter derived from vegetation at the surface. The geochemical processes that participate in carbonate formation in the cave are affected chiefly by ambient temperature and humidity, factors that are ultimately controlled by climatic and environmental changes. Speleothem growth rate and composition, including stable isotopes, trace elements and organic materials, have been shown to provide a reliable record of climatic and environmental change at a given cave site.

Determining palaeoclimatic conditions by using speleothems began in the 1960s (Galimov and Grinenko 1965). In 1968, Hendy & Wilson proposed that the stable oxygen isotope composition of speleothems could be used to reconstruct high-resolution palaeo-temperature records to distinguish regional and global temperature changes that result from the advance and retreat of ice sheets. With this motivation, they investigated the regional temperature changes in New Zealand. Soon after, Hendy (1971) suggested two criteria to assess whether isotopic equilibrium existed during the time of calcite deposition. Meanwhile, Duplessy et al. (1970) reconstructed temperature changes in France from 130 to 90 ka BP and compared them to marine foraminifera records. Thompson and Schwarcz (1976) reconstructed temperature changes during the last 200 ka BP in West Virginia, which was supported in a study by Harmon et al. (1978). However, the speleothem work didn't capture the attention of the paleoclimatological community and went largely unnoticed. Consistent improvements in U/Th dating techniques (Edwards et al. 1987; Shen et al. 2002; Hellstrom 2003) and construction of exceptionally high-resolution $\delta^{18}O$ time series from speleothems, comparable to ice core records (Wang et al. 2001, 2005a, 2008a; Spötl and Mangini 2002; Yuan et al. 2004; Cai et al. 2006; Spötl et al. 2006) have stimulated a surge of renewed effort to use speleothems as high-resolution paleoclimatic archives. Speleothems can provide chronologies for climate change studies with very high resolution because their depositional sequences are unambiguous and they can be dated directly. In addition, they are widely distributed from high to low latitudes, offering global coverage. These factors account for why they have become one of the most important natural archives for the study of global climate change.

Gascoyne (1992) summarized the variety of ways that cave calcite deposits can be studied as paleoclimate archives, including: 1) depositional abundance within a time period, 2) variations in ^{13}C and ^{18}O content in the calcite, and 3) ^{2}H and ^{18}O in fluid inclusions, and 4) trace element concentrations. More recently we can add to this list, pollen assemblages trapped in the calcite. All of these proxies are potentially available as synchronous palaeoclimatic indicators. Previous work had tended to concentrate mainly on the timing and abundance of deposits as a palaeoclimatic indicator for the last 300,000 years. However, recent research has shown that stable isotope composition, especially that of oxygen, has proved the most fruitful up to now.

3.2.2 The Speleothem Records of Asian Monsoon

In the Asian monsoon region, speleothems have been instrumental in the study of climatic and environmental change. Based on dating results of cave calcite deposits, Zhang et al. (1985) and Zhao et al. (1989) used abundance of the deposits to investigate climate conditions in southern China. Higher abundance was interpreted as evidence for warm intervals, and periods with relatively less deposition were interpreted as cold intervals. They found three hiatuses, at 130, 210 and 220–230 ka BP, which they ascribed to cold climate conditions. With ^{14}C dates and stable isotope measurements of two stalagmites from Maomaotou Cave, Guilin city, Wang (1985) discussed the cold and warm fluctuations at the site. At the same time, Chen et al. (1986) reconstructed temperature changes during the period from 80-50 ka BP, using carbon and oxygen isotope data of speleothems associated with human remains from the Zhoukoudian site in Beijing.

Driven by the development of global change sciences and the previous achievements of speleothem studies, the study of speleothems started to flourish in China in the 1990s. Chinese scientists began to use speleothem $\delta^{18}O$ records to calculate paleotemperatures. Huang (1992) discuss temperature changes in Hubei Province based on oxygen isotope records, and a chronology constrained by ^{14}C dates. Zhu and Zhang (1992) used the O'Neil equation to calculate temperature and then compared their curve with deep-sea isotope records and produced a speleothem $\delta^{18}O$ record with 100 ka BP glacial-interglacial cycles. Using the modern precipitation $\delta^{18}O$ values as that of paleo-precipitation, Li et al. (1994) reconstructed temperature changes in Guilin city for the last 350 ka BP. Tan and Lin (1995) dated a speleothem from Jiguan Cave, Henan Province by using the AMS ^{14}C method and measured stable oxygen and carbon isotope compositions. These were interpreted in terms of Holocene temperature changes by comparing the record with other climatic and environmental records. Hong et al. (1995) calculated temperature changes by substituting the $\delta^{18}O$ of paleo-precipitation with that of modern precipitation and compared a reconstructed temperature record with the deep-sea records.

It is worth noting that Prof. Yuan had the insight to recognize the sensitivity of karstification to environmental change and pointed out that speleothems were vital to understanding climate change in southern China, where loess, ice and lake sediments are rare or absent (Yuan 1995). Subsequently, he and his group carried out systematic speleothem studies in the Guilin area. By using multiple dating methods, e.g., α counting, TIMS-U-series, AMS ^{14}C and ^{210}Pb, combined with carbon and oxygen stable isotope analyses, they reconstructed the history of climate change in the Guilin region during the last 200 ka BP and identified some millennial and centennial time scale events, including the YD event (Yuan et al. 1999). At the same time, Prof. Liu Tungsheng, Prof. Tan and others also studied speleothems for palaeoclimatological and palaeoenvironmental investigations (Tan and Liu 1995; Tan et al. 1998).

A pioneering paper was subsequently published by Wang et al. (2001), which showed clear orbital-timescale climatic signals in an oxygen isotopic speleothem record from Hulu Cave. The record was tightly correlated with ice core results from the Greenland Ice Sheet Project (GISP2) (Meese et al. 1997) and the publication of this paper brought the whole field of speleothem research to a new level. The Hulu Cave study strongly supports the hypothesis that East Asian monsoon climate is driven by insolation changes and that the East Asian monsoon is an important component of the global atmosphere–ocean circulation system on millennial timescales. It confirmed that millennial timescale climatic events observed in GISP2 reflected changes in the entire Northern Hemisphere (NH) and that during glacial times, the East Asian monsoon climate was unstable. The study also confirmed the dust layer count chronology of GISP2, which agreed with the absolutely-dated speleothems. This in turn, explained the apparent age discrepancies between the GISP2 ages and ice-flow model ages used in the European Greenland Ice Core Project (GRIP) record (Dansgaard et al. 1993), providing a means to correct the GRIP data. Yuan et al. (2004) extended the Asian monsoon record to the last interglacial. Recently, the speleothem Asian monsoon record has been extended even further, and now includes the last four glacial-interglacial periods (Wang et al. 2008a; Cheng et al. 2009). These ultrahigh resolution and absolutely-dated records indicate that the Asian monsoon responded directly to NH insolation and show clearly the precessional cycles (Wang et al. 2008a, Cheng et al. 2009). A focus of the recent work has been on millennial timescale events during the last two glacial-interglacial cycles. During glacial times, the frequency of these events increased while their duration shortened, implying a possible connection between glacial ice sheets and the East Asian monsoon, in connection with North Atlantic Ocean circulation (Wang et al. 2008a). All of these studies (Wang et al. 2001, 2005a, 2008a; Tan et al. 2003; Yuan et al. 2004; Cai et al. 2006, 2010a, b; Hu et al. 2008; Zhang et al. 2008; Cheng et al. 2009) refine and extend our understanding of the history, variability and forcing of the Asian monsoon-arid climate system, and its connection with global climate.

3.2.3 The Interpretation of Speleothem $\delta^{18}O$ Values

Speleothem $\delta^{18}O$ records from southern China indicate that large amplitude changes of precipitation occur on orbital to decadal timescales. However, what these changes represent is still being debated. Different mechanisms have been suggested to interpret $\delta^{18}O$ variations in calcite. For example, Wang et al. (2001), Cheng et al. (2006), and Cheng et al. (2009) interpret $\delta^{18}O$ to represent a "ratio of the amount of summer to winter precipitation" or "relative summer to winter monsoon intensities." Yuan et al. (2004) interpret cave $\delta^{18}O$ as "the fraction of water vapor removed from air masses between the tropical Indo-Pacific and southeastern China." Both of these two interpretations link the cave $\delta^{18}O$ with the East Asian summer monsoon intensity and have been frequently used to interpret speleothem records from Asian monsoon regions (Wang et al. 2005a; Zhang et al. 2008; Dykoski et al. 2005; Kelly et al. 2006; Cosford et al. 2008; Cai et al. 2006). However, a few researchers argue that this signal could reflect changes in moisture sources, such as the Pacific Ocean, South China Sea, and Indian Ocean (Maher 2008; Tan 2009; Dayem et al. 2010; Clemens et al. 2010). Answering this question is crucial for us to understand the variation of the Asian summer monsoon and its connection with global climate change, and is a key to reconciling inconsistencies in phase relationships between Asian summer monsoon records from speleothems and their marine counterparts, on orbital timescales (Clemens et al. 2010).

Under equilibrium conditions, both cave temperature and drip water $\delta^{18}O$ affect the calcite $\delta^{18}O$ (Hendy, 1971). The speleothem records from China indicate that the amplitude of calcite $\delta^{18}O$ changes is about 3.0‰ within the Holocene and about 5–7‰ during glacial–interglacial cycles (Yuan et al. 2004; Wang et al. 2001, 2005a, 2008a; Hu et al. 2008; Cheng et al. 2009; Cai et al. 2010a). For example, speleothem $\delta^{18}O$ records from Jiuxian Cave show Holocene amplitudes of 3‰ and glacial–interglacial amplitudes of 5–6‰. Due to the small temperature-dependent fractionation between calcite and water ($-0.23‰\cdot{}^{\circ}C^{-1}$, O'Neil et al. 1969), the observed $\delta^{18}O$ variations must be dominated by changes in drip-water $\delta^{18}O$, i.e., the amount-weighted $\delta^{18}O$ of the precipitation, because a shift of only 2 ‰ would be the upper bound of the temperature effect for calcite–water fractionation from the LGM to the present, in central China (Johnson et al. 2006, references therein).

Precipitation $\delta^{18}O$ shows a clear positive correlation with temperature in high latitude regions, while amount effects (i.e., negative correlation between precipitation amount and $\delta^{18}O$) dominate in tropical regions, where there is no significant relationship between $\delta^{18}O$ and temperature (Dansgaard 1964; Rozanski et al. 1993). By analyzing the data from 10 sites within the Global Network for Isotopes in Precipitation (GNIP) using a multiple regression method, Johnson and Ingram (2004) suggested that in South China, where precipitation is dominated by the summer monsoon, precipitation $\delta^{18}O$ is anti-correlated with rainfall amount, so that the amount effect prevails. In northwest China, which is beyond the reach of the summer monsoon, precipitation $\delta^{18}O$ is positively correlated with temperature, so that the temperature effect prevails. At the confluence between the southern monsoon region and northwestern inland arid region, precipitation $\delta^{18}O$ is influenced by both temperature and monsoonal precipitation amount: in winter, the temperature effect is dominant, and in summer the amount effect is dominant.

Although temperature dominates the oxygen isotope composition of precipitation during the winter season, it has a small effect on the amount-weighted $\delta^{18}O$ values of precipitation, i.e., the $\delta^{18}O$ values of the drip water in the cave, because winter precipitation accounts for only 30–35 % of the total annual precipitation in this region. Furthermore, the temperature-induced fractionation between the drip water and speleothem calcite, i.e., $-0.23\,‰ \cdot °C^{-1}$ (O'Neil et al. 1969), can offset the effect of the temperature on precipitation $\delta^{18}O$, which is not likely to be more than about $0.24\,‰ \cdot °C^{-1}$ in central China (Johnson et al. 2006). Therefore, temperature would have a minor effect on the speleothem calcite $\delta^{18}O$, and other factors that control the $\delta^{18}O$ of drip water are expected to dominate, such as changes in summer monsoon precipitation amount, changes in the *summer/winter* precipitation ratio and *precipitation/evaporation* (P/E) ratio, changes in moisture recycling that occurred along the transport pathway and changes in the source composition, due to global ice volume and circulation changes.

On different timescales, the contribution of different factors to speleothem calcite $\delta^{18}O$ may change. For example, in the Holocene, as the ice volume and ocean circulation did not change significantly, the moisture source isotope composition will experience only minor changes (Hu et al. 2008). At the same time, boundary conditions of the East Asian continent that could change the distance from the moisture source to the study region (An et al. 1992) did not change substantially either, in the Holocene. Although it is difficult to quantify the effects of changes in the P/E ratio, from moisture recycling and circulation, their effects are likely to be minor under relatively stable boundary conditions (i.e., in the Holocene) (Hu et al. 2008). Therefore, the $\delta^{18}O$ of speleothem calcite can be regarded primarily as an indicator of the amount of summer monsoon precipitation during the Holocene; where lighter values indicate an intensified summer monsoon with increased monsoon precipitation, and heavier values indicate a weakened summer monsoon and decreased monsoon precipitation. This general relationship also holds for glacial times, in central China.

However, when we interpret changes in the absolute values between the glacial and interglacial, and the $\delta^{18}O$ variation during the deglaciation, additional factors need to be taken into consideration, such as changes in atmospheric circulation and moisture sources (Hoffmann and Heimann 1997; Maher 2008; Dayem et al. 2010). Additionally, ice volume changes should increase the $\delta^{18}O$ of the moisture source by about 1.0 ‰ (Schrag et al. 1996); the ratio of the summer and winter precipitation would also change as the summer monsoon would start later and retreat earlier (An et al. 1992). Finally, decreased humidity and temperature would lower the degree of continental recycling (Rozanski et al. 1993).

It is well know that two monsoon systems affect climate change in China, i.e., in the southeast, the East Asian monsoon dominates while in southwestern China, the Indian monsoon dominates. In southern China, where the climate is controlled both by the East Asian and Indian summer monsoon systems, moisture from the Pacific (including the South China Sea) and Indian Oceans contribute to the precipitation. According to the modern precipitation $\delta^{18}O$ pattern in monsoon regions of China, Cai et al. (2001) proposed that the large variation of speleothem $\delta^{18}O$ from Qixing Cave, Guizhou Province, may reflect changes in moisture source. Recently, Tan (2009, 2011) suggested that the short-term variation of speleothem $\delta^{18}O$ may indicate circulation changes. As the fraction of moisture from the Indian Ocean increases, the precipitation and speleothem $\delta^{18}O$ become depleted, because of the long transport distance. On the contrary, when the fraction of Pacific Ocean moisture increases, speleothem $\delta^{18}O$ becomes enriched because of the reduced transport distance. Using a numerical climate model with an embedded oxygen-isotope model, Pausata et al. (2011) proposed that the speleothem $\delta^{18}O$ changes mainly reflect the intensity of Indian rather than East Asian monsoon precipitation, as the water vapor exported from the Indian basin to East Asia is isotopically enriched during Heinrich events. However, if the short-term variation of precipitation $\delta^{18}O$ is dominated by circulation changes, the precipitation $\delta^{18}O$ should display similar variations within the monsoon regime. The fact is that the "amount effect" is significant, while the rainfall amount changes in different patterns. This observation cannot be interpreted as circulation changes. Hence, the amount effect should dominate on the short term, although it is unrealistic to categorically exclude any contribution from circulation changes.

In summary, under different boundary conditions, changes of speleothem calcite $\delta^{18}O$ may be dominated by different factors (Cai et al. 2010a) in Asian monsoon regions. On orbital and sub-orbital timescales the calcite $\delta^{18}O$ of speleothems from the Asian monsoon regions can be used as a proxy of monsoon intensity, while on decadal to centennial timescales, it can be used as a proxy for summer monsoon precipitation. It is worth noting that temperature remains a very important factor to be considered when interpreting the speleothem $\delta^{18}O$ changes, as speleothem $\delta^{18}O$ is controlled by both the temperature and drip water $\delta^{18}O$ and past temperature changes are not yet well constrained enough to accurately evaluate its contribution to the speleothem $\delta^{18}O$ through the temperature-dependent fractionation relationship. Moreover, the influence of temperature on the drip water $\delta^{18}O$ is also not well assessed, especially in northern China where the temperature effect is much more significant than in southern China. In conclusion, the interpretation of speleothem $\delta^{18}O$ is not straightforward, as multiple factors contribute to the changes of the speleothem $\delta^{18}O$. Furthermore, the dominant factor may change with the time-scales of interest, the boundary conditions and perhaps the location of the cave.

3.2.4 The Variability of the Asian Summer Monsoon Recorded by Speleothems

Only after recognizing abrupt warming events from Greenland ice cores and ice-raft events from the North Atlantic, did millennial timescale abrupt events associated with the Asian monsoon attract the attention of researchers. In 1993, Sirocko et al. discussed centennial timescale climatic events of the Indian summer monsoon during the last 24 ka BP, and suggested that these events reflected non-linear responses to insolation changes. Porter and An (1995) discovered the teleconnection between the East Asian winter monsoon and North Atlantic climate by comparing large grain size events in loess with the Heinrich Events found in the North Atlantic Ocean. After that, the variability of Asian monsoon climate has been a focus of monsoonal climate studies. Climatic events on millennial timescales have been found from various kinds of records, such as loess, lake and ocean sediments (Guo et al. 1999; Xiao et al. 1995; An and Poter, 1997; Altabet et al. 2002). However, the age models associated with these various archives are insufficient to assess the fine details of phasing among the monsoon records themselves, and in the details of monsoon phasing relative to high latitude climate change, with sufficient resolution to determine which system initiated the abrupt transitions.

The exceptionally well-dated monsoon records of Hulu cave (Wang et al. 2001) shed light on the study of millennial timescale events. Since then, abrupt events on millennial timescales have been confirmed by speleothem records from around the globe, including caves in tropical regions. For example, the speleothem records from Dongge Cave replicate observations from Hulu records during deglaciation (Dykoski et al. 2005, Fig. 3.1) and a speleothem record from Brazil, South America, mirror the changes in Asian monsoon climate (Wang et al. 2006, Fig. 3.2). The highly correlated relationships among different records within the tropical and monsoonal regions demonstrate the advantage of the high-precision absolute U-series dating method in the study of abrupt climatic events, and imply that the speleothem $\delta^{18}O$ time series mainly record climatic and environmental changes and can be used to understand the interaction and connection of regional and global climatic changes.

High-resolution speleothem records not only benchmark the age of the abrupt events, but also link together these abrupt events and North Atlantic climate changes with absolute ages. The speleothem growing in northern Brazil consistently corresponded to weak Asian monsoon events, indicating a close connection and inter-

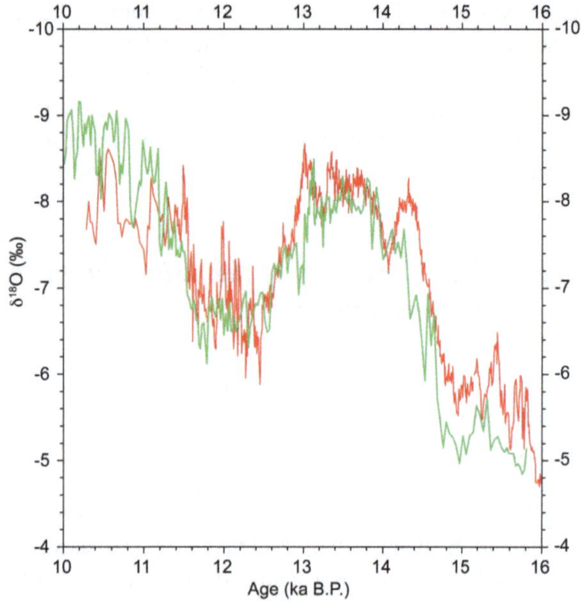

Fig. 3.1 Comparison of the deglacial sequence of Dongge stalagmite D4 (*Red? or green?*) and Hulu stalagmite H82 (*grey*) from 10 ka to 16 ka. Though, 1200 km apart, both D4 and H82 show a remarkably similar $\delta^{18}O$ pattern suggesting that regional speleothem records can be correlated using an absolute U-series chronology. (after Dykoski et al. 2005)

action between the Northern and Southern Hemispheres, and implying that the shift of the Intertropical Convergence Zone (ITCZ) may be the linkage for the mirroring of abrupt events in both hemispheres (Wang et al. 2004a; Wang et al. 2006). Modeling results show that ice-rafted events in the North Atlantic may result in the weakening of North Atlantic Deep Water (NADW) production, and an increase of sea ice during the winter season, hence leading to a decrease in temperature in the Northern Hemisphere (Zhang and Delworth 2005). Temperature decreases at high latitudes of the Northern Hemisphere can increase the temperature gradient over the Northern Hemisphere, resulting in a southward shift of the ITCZ and weakening of the Asian summer monsoon (Chiang and Bitz 2005).

The abrupt climatic events found in the Northern Hemisphere during the glacial were widespread around the world, implying instability of glacial climate. The Asian monsoon speleothem record during the last two glacial-interglacial cycles shows that millennial climatic events not only occurred in the last glacial, but also during the penultimate glacial. The duration of these events decreases, and their frequency increases during the glacial build-up in both the last and penultimate glacial periods. This suggests that ice sheet size affects their character and pace (Wang et al. 2008a; Fig. 3.3), implying an effect of mean climate condition on the abrupt events (Sima et al. 2004; Wang and Mysak 2006).

The East Asian and Indian Monsoons are two major components of the Asian monsoon system. It has been argued that the depleted cave $\delta^{18}O$ may be produced by two different sources at two different times within the East Asian monsoon regions. A light $\delta^{18}O$ response is produced via summer monsoon dynamics associated with the long transport path from the Indian Ocean into Asia during summer. Another light $\delta^{18}O$ response is produced by cooler winter season atmospheric precipitation. Thus, light cave $\delta^{18}O$ may reflect a combination of both summer and winter

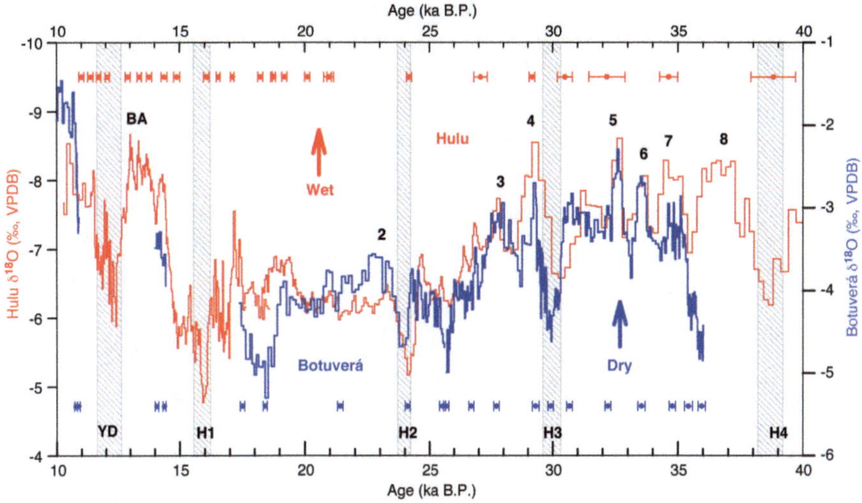

Fig. 3.2 Comparison of speleothem δ18O records from Botuvera Cave, southern Brazil (*Blue curve*), and Hulu Cave, eastern China (*red curve*, Wang et al. 2001). Here, inverse scales were used for the two records. Wet periods at the Hulu site are correlated with dry periods at Botuvera, and vice versa. The pattern demonstrates a low-latitude atmospheric precipitation see-saw mechanism. (from Wang et al. 2006)

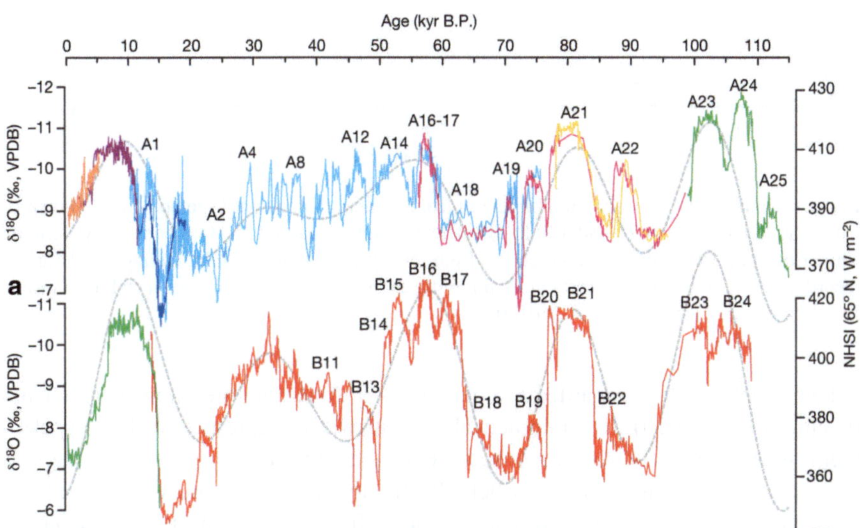

Fig. 3.3 Comparison of millennial-timescale oscillations of the Asian monsoon between the last and penultimate interglacial-glacial cycles. The duration and frequency of millennial-timescale events changed systematically throughout the course of a glacial period, and the general nature of these changes is reproducible for the last two glacial periods. NHSI is Northern Hemisphere Solar Insolation. (from Wang et al. 2008a)

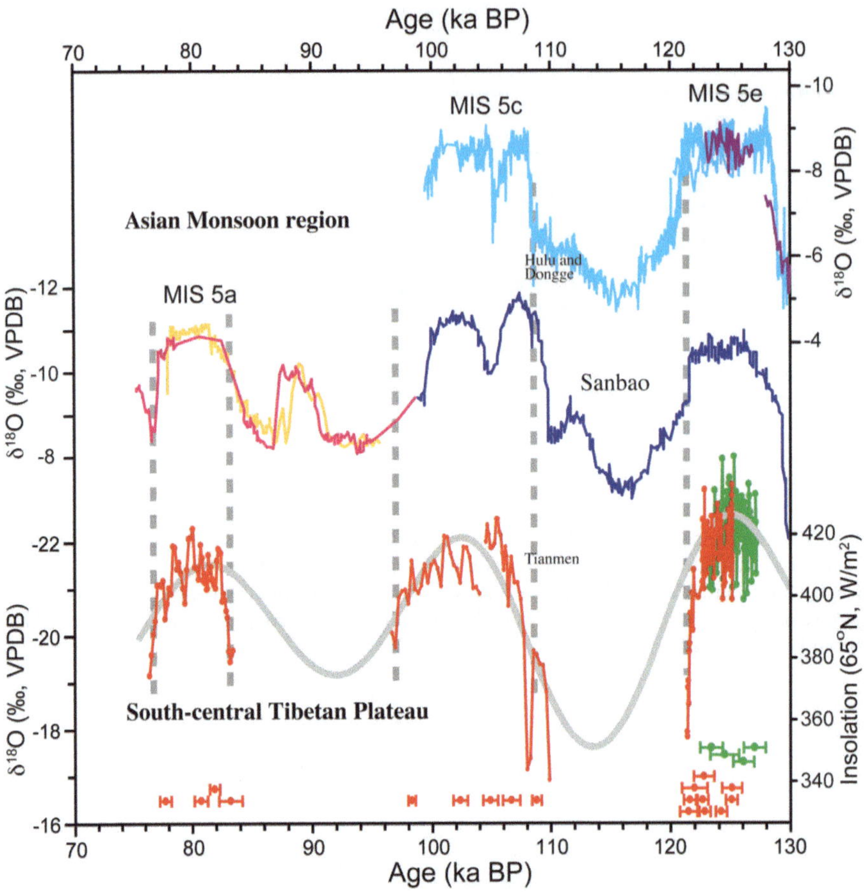

Fig. 3.4 Comparison of Tianmen $\delta^{18}O$ record from Tibetan Plateau with speleothem $\delta^{18}O$ records from East Asia between 77 and 130 ka BP. The Tianmen stalagmite $\delta^{18}O$ record shows striking similarities to East Asian Monsoon records from low altitude regions (Hulu, Dongge, and Sanbao records) within their dating uncertainties, implying an in-phase relationship between East Asia and Indian monsoons. (from Cai et al. 2010b)

dynamics and can't be regarded as a pure index of Asian summer monsoon intensity (Clemens et al. 2010). It is impracticable to answer this question with the speleothem record from East Asian monsoon regions alone. So, the record from regions affected by the Indian monsoon, where the precipitation is dominated by the Indian summer monsoon and moisture mainly comes from the Indian Ocean and Bay of Bengal, is essential to clarify this issue.

Modern observation indicates that the precipitation over southern Tibet is dominated by the Indian summer monsoon, and variations in rainfall $\delta^{18}O$ values are closely associated with the Indian monsoon changes. The speleothem record from Tianmen Cave, southern Tibet, the first cave record from the Tibetan Plateau (TP), characterizes a precipitation $\delta^{18}O$ history and demonstrates the variability of Indian summer monsoon on both orbital and millennial time-scales during Marine Isotope

Stage 5 (i.e., 5a, 5c and 5e). It shows remarkable similarities to the East Asian Monsoon records from the adjacent regions, providing essential evidence that the Asian Monsoon system, including the East Asian and Indian Monsoon sub-systems, responds largely and directly to changes in NH summer insolation (Fig. 3.4, Cai et al. 2010b). This record reveals that speleothem $\delta^{18}O$ records from East Asian monsoon regions, where the changes of moisture sources and winter precipitation may have significant influences on the precipitation $\delta^{18}O$, reflect mainly the variation of Asian summer intensity on orbital to millennial time-scales. At the same time, extremely low $\delta^{18}O$ values during Marine Isotope Stages 5a, 5c, and 5e suggest that precipitation, rather than temperature, was a major factor controlling $\delta^{18}O$ in precipitation on orbital time scales in south-central TP. This finding may help in the interpretation of regional ice core $\delta^{18}O$ records. In addition, the large range of orbital-scale shifts in meteoric $\delta^{18}O$ (>9‰) raises important considerations related to reconstructing the uplift history of the plateau (Cai et al. 2010b).

3.2.5 Centennial to Decadal Cycles Recorded in the Speleothems

High-resolution speleothem records show clearly the centennial to decadal timescale variation of Asian monsoon climate. Speleothem records from Oman demonstrate that the strength of the Asian summer monsoon circulation has gradually decreased since the early Holocene, in response to an orbitally-forced reduction in summer insolation. Abrupt changes in monsoon intensity superimposed on this trend appear to be linked to North Atlantic cooling cycles and, ultimately, to solar variability (Neff et al. 2001; Burns et al. 2002; Fleitmann et al. 2003, 2007). The Dongge record further confirms the contribution of solar activity to the Asian summer monsoon, indicating a close connection between Asian summer monsoon and temperature changes in northern polar regions (i.e., Greenland). A speleothem record with a temporal resolution of 2–3 years from central China indicates that the strengthening and weakening of the Asian summer monsoon is consistent with temperature changes over the Northern Hemisphere, and regional climate appears to have had an impact on the stability of the dynastic governments in ancient China (Zhang et al. 2008). However, it is worth noting that the speleothem $\delta^{18}O$ records do not show uniform changes on centennial to decadal timescales (Fig. 3.5, Tan et al. 2009). This may be due to factors not connected with climate or it might reflect different responses of regional precipitation to monsoon climate change. Thus, the spatial variation between speleothem records on a large scale, at centennial to decadal timescales, is still open to debate.

3.2.6 Comparison of Millennial Events Between Northern and Southern Hemispheres

Comparison of Antarctic and Greenland ice cores reveal seesaw variability in millennial timescale events, i.e., cold events in the northern high-latitude region

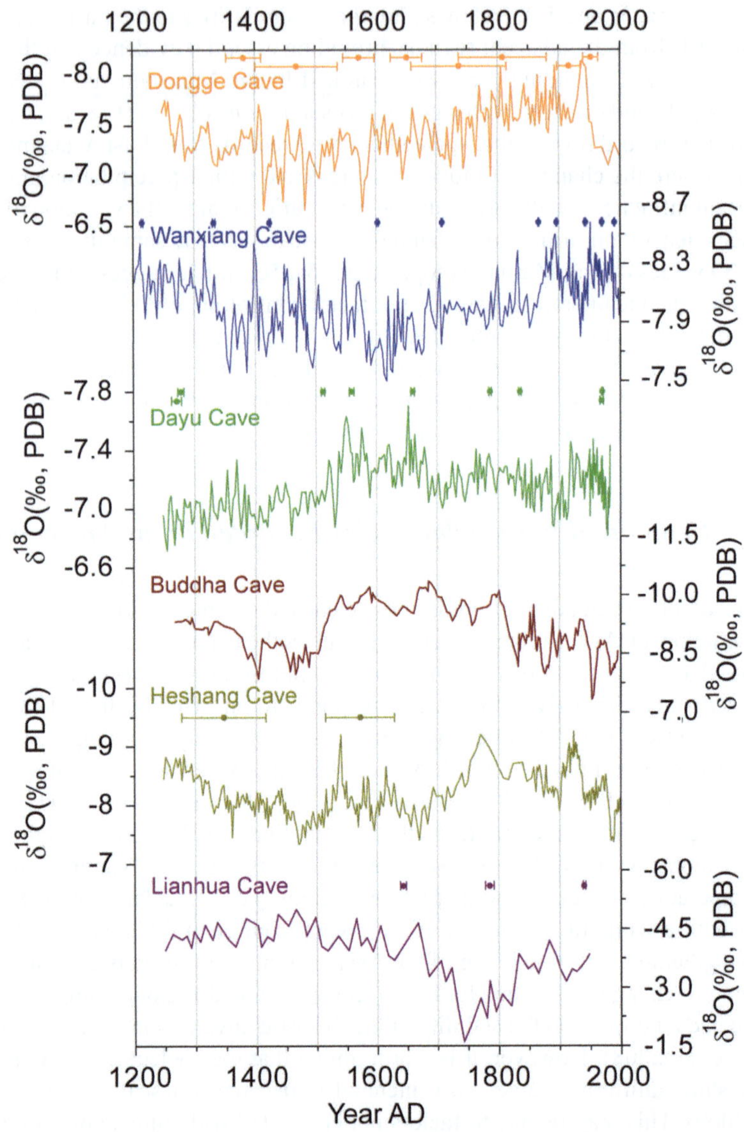

Fig. 3.5 Comparison of the speleothem $\delta^{18}O$ records from the Dayu Cave, Buddha Cave (Paulsen et al. 2003), Wanxiang Cave (Zhang et al. 2008), Dongge Cave (Wang et al. 2005a), Heshang Cave (Hu et al. 2008), and Lianhua Cave (Cosford et al. 2008). The comparison demonstrates obvious regional differences from south to north in central China during the last 750 years. (from Tan et al. 2009)

correlated with warm events in the Antarctic (Blunier and Brook 2001). NADW circulation changes have been suggested as a means of producing such a correlation (Broecker 1998). In Asian monsoon regions, the summer monsoon intensified during Dansgaard-Oeschger (D-O) warming periods while it weakened during the cold Heinrich events. It has been proposed that an enhanced East Asian summer monsoon is related to intensified cross-equatorial flow associated with decreased Antarctic temperatures (An 2000), as well as increased temperatures over the Eurasian continent. A comparison of the Hulu Cave record with polar ice core records shows that both high-northern latitude and Antarctic temperature changes influence the East Asia summer monsoon (Rohling et al. 2009).

The parallel study of ocean sediments and cave calcite indicate that climate changes over the North Atlantic affect the Indian summer monsoon significantly. Weak Indian summer monsoon events are always linked to cold events in the North Atlantic. The forcing mechanism for this linkage may be related to an increase in snow cover on the Eurasian continent, especially on the TP. This imparts a change in albedo resulting in slow warming over the Eurasian continent in spring. In this case, the establishment of the Indian low comes later in the year than normal, leading to a weakened Indian summer monsoon. It is also plausible that cooling in the North Atlantic pushes the westerly circulation southward, retarding the northward progression of the monsoon front in the beginning of summer and accelerating its retreat in the fall, resulting in a shorter and weaker monsoon.

The Indian summer monsoon is affected not only by NH climate, but also by Southern Hemisphere climate, because the Mascarene High (in the southern Indian Ocean) is also an indispensible component of the Indian monsoon system. The Bond cycle, which includes both Heinrich events and D-O oscillations, is the predominant feature of climate change in the regions around North Atlantic and Greenland. However, more and more evidence indicates that these events are not limited to the Northern Hemisphere, but are also present in South America and New Zealand (Lowell et al. 1995), the Antarctic (Blunier et al. 1998), and the Southern Ocean (Charles et al. 1996). Forcing by Heinrich events and D/O oscillations drive climate change in both hemispheres. However, the changing pattern of climate at the two hemispheres is not the same on the millennial time scale (Leuschner and Sirocko 2000). Millennial timescale variability in GISP2 and GRIP are characterized by abrupt temperature increases, followed by gradual decreases and then abrupt returns to baseline glacial conditions. In contrast, warming and cooling was gradual in the Antarctic (Blunier and Brook 2001).

The oxygen isotopic record of stalagmite XBL-1 from southwestern China reveals millennial timescale variability of the Indian Monsoon between 53 and 36 ka, synchronous with changes in the East Asian Monsoon recorded at Hulu Cave and similar to D-O cycles recorded in Greenland ice (Cai et al. 2006). This record, in general, confirms the chronology of Hulu Cave and demonstrates the variability of the Indian Monsoon on millennial timescales. The XBL-1 record correlates with high-latitude ice core records from the Northern Hemisphere. However, some Indian Monsoon characteristics in this record more closely resemble features in the Antarctic record, following an anticorrelation. This suggests a climatic link between the high southern latitudes, in addition to the established link to the climate of the high northern latitudes (Cai et al. 2006). Therefore, the climate of the high southern latitudes must contribute to variations in the Indian summer monsoon and affect the Indian summer monsoon system on a millennial timescale.

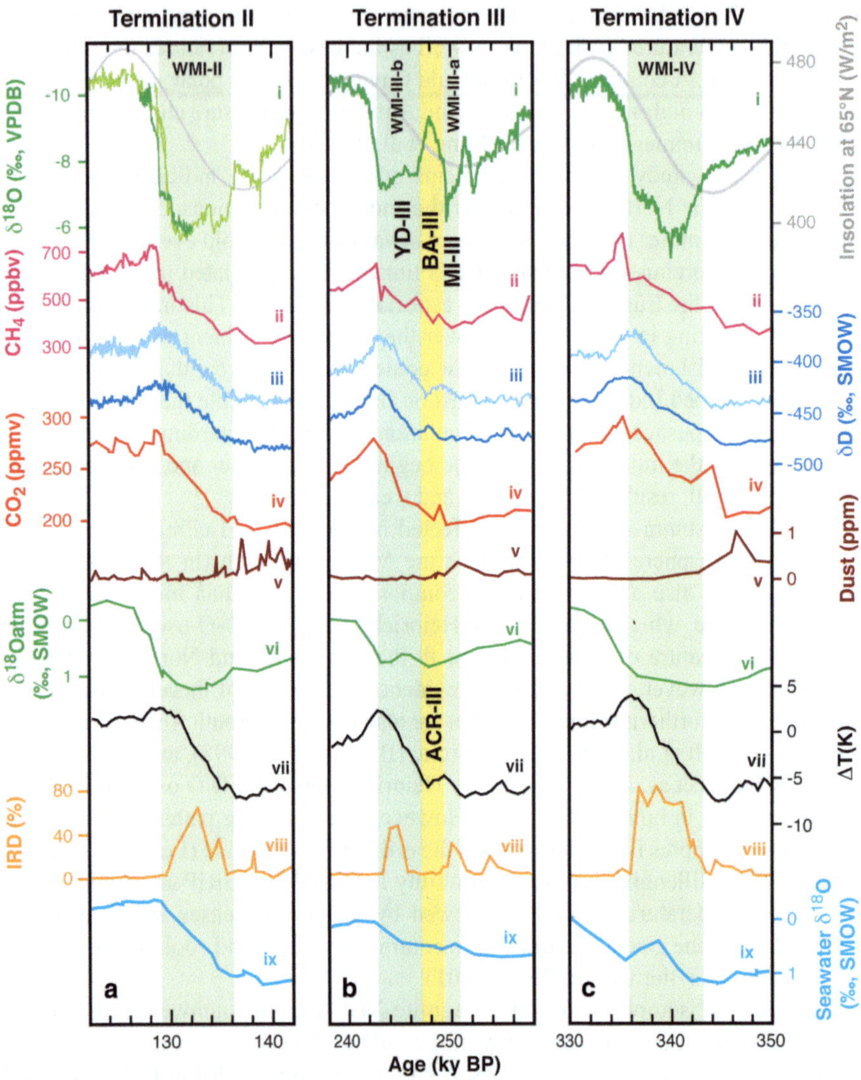

Fig. 3.6 ^{230}Th-dated oxygen isotope records of stalagmites from Sanbao Cave, China, can be precisely correlated with ice cores and marine sediments, establishing the timing and sequence of major events. (from Cheng et al. 2009)

3.2.7 Millennial Events: The Key to Establishing Accurate Chronologies in Ocean Sediments and Polar Ice Cores

The most significant advantage of speleothem records is that they can provide high-precision absolute ages covering the past 500 ka BP. The comparison of Greenland ice core records dated by layer counting and U-series dated speleothem records serve as a model for utilizing speleothem records. This method can be extrapolated into the past and absolute dates from speleothems can be used to calibrate ocean sediment records and older ice core records. Methane concentrations in polar ice cores show a close correspondence with Asian monsoon events, and this connection is useful in studies of relatively old ice. Speoleothem studies provide the necessary chronological framework. This approach was adopted by Cheng et al. (2009), to determine an absolute chronology for ocean sediments and ice cores during the last four terminations and to discuss the mechanisms/processes during the ice age (three terminations in Fig. 3.6). In another study, Sun et al. (2010) transferred a speleothem chronology to loess-paleosol sequences by correlating abrupt events in both records. This made it possible to correlate the loess records with other absolutely dated records.

3.2.8 Perspective of Speleothem Study in the Asian Monsoon Region

An increasing number of studies indicate that changes in the Asian monsoon-arid system are closely linked to climate in the high northern latitudes, Southern Hemisphere climate, due to shifts of the ITCZ. However, these records are far from enough for us to recognize and understand the variability of the Asian monsoon-arid climate system on millennial to decadal timescales, and their driving forces. Although most researchers agree that abrupt Asian monsoon events respond to North Atlantic climatic changes, some propose that the Asian monsoon itself may play a key role in magnifying small changes in tropical climate, resulting in abrupt millennial timescale events. To thoroughly understand the Asian monsoon-arid system, we need more absolutely-dated high-resolution speleothem records from both high and low latitudes. Only through the spatial and temporal comparison and testing with model simulations, can we fully understand the mechanisms that control abrupt climatic changes on millennial timescales. Speleothems are among the most valuable natural archives for us to understand these abrupt climatic and environmental changes in the past.

3.3 Asian Monsoon Variability from Lake Sediment Records

3.3.1 East Asian Summer Monsoon Variability

Lacustrine sediments provide a rich archive for the study of East Asian monsoon variability through time. This is true in part because of the relatively large number of lakes in China and adjacent countries that have experienced continuous sedimentation throughout the Holocene and beyond. These lakes are also geographically distributed across a large area within the domain of the monsoon climatic system

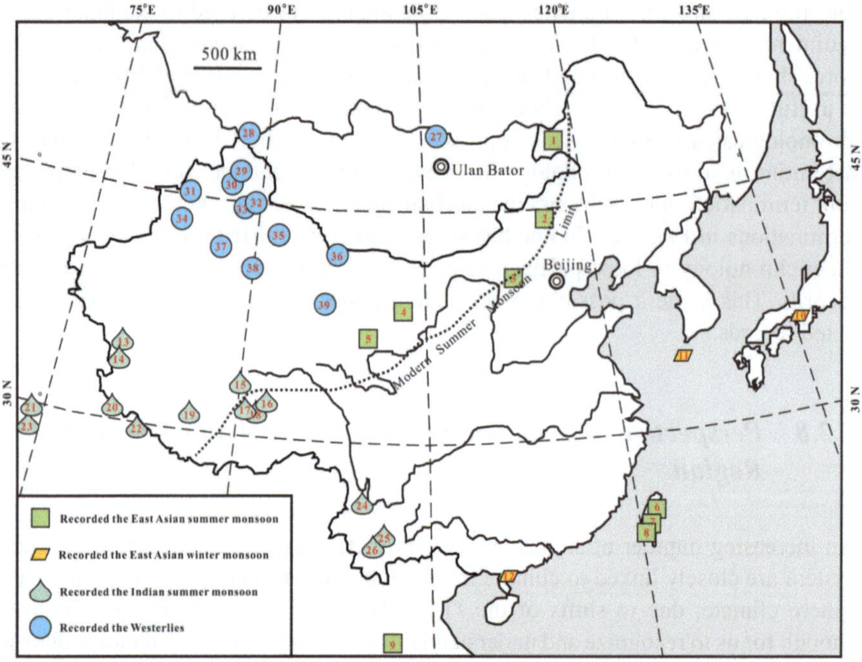

Fig. 3.7 Spatial distribution of lacustrine records (numbers indicated refer to the records, see Table 3.1) from Asia. The records are numbered according to the dominant climate system and from north to south

which translates into spatial control from a paleo perspective. Most studies of the East Asian summer monsoon that are based on lacustrine sediments are from hydrologically closed basins located in transitional positions of humid areas which are sensitive to monsoonal climate. These lakes include Lake Qinghai, Dali Lake, Daihai Lake, Hulun Lake, and Zhuyeze Lake in northern China; and Lake Biwa in central Japan (Fig. 3.7). Currently the longest lacustrine record of East Asian monsoon variability is from Lake Biwa, which extends back to 300 ka before present (BP) (Iwamoto and Inouchi 2007). Most studies from other records have focused on late glacial to Holocene time periods.

Numerous proxies of lacustrine sediments have been used to understand the influence of the East Asian summer monsoon. These proxies reflect sedimentological, palynological, geochemical and paleomagnetic characteristics of the lake deposits (Table 3.1). Among them, total organic carbon (TOC) has proved to be one of most effective paleomonsoon proxies because TOC in lake sediments responds sensitively to changes in biomass (Meyers and Lallier-Vergès 1999; Hodell et al. 1999), which are affected by precipitation and temperature variations in monsoonal regions. In boreal summer, the East Asian summer monsoon is characterized by high precipitation and temperature. These favorable hydrological circumstances lead to increase in biomass within the catchment area and productivity in the lake. This results in increasing sedimentary TOC, which integrates the contributions form different sources of organic matter (Meyers and Lallier-Vergès 1999). In this review,

Table 3.1 List of lake sites with paleomonsoon and Westerlies records from north to south shown in Fig. 3.7

Site	Lake name	Core location	Elevation (m)	Proxies	Time duration (ka)	References
AM-1	Hulun Lake	49°07.62′N, 117°30.36′E	545.3	grain size, sediment components, pollen	10.0–0	Wen et al. 2010a,b
AM-2	Dali Lake	43°15.68′N, 116°36.26′E	1226	TOC, TIC, grain size	11.5–0	Xiao et al. 2008, 2009
AM-3	Daihai Lake	40°35′11″N, 112°40′3″E	1221	Rb/Sr, $CaCO_3$, grain size, pollen	10.8–0	Xiao et al. 2004; Jin et al. 2006
AM-4	Zhuyeze (Sanjiaocheng)	39°00′38″N, 103°20′25″E	1320	$CaCO_3$, pollen, TOC	11.8–0	Chen et al. 1999, 2006
	(Qingtu Lake)	39°03′N, 103°40′E		grain size, $CaCO_3$, TOC, C/N, $\delta^{13}C_{org}$		Long et al. 2010
AM-5	Lake Qinghai	36°54′N, 100°10.8′E	3196	$CaCO_3$, ostracod $\delta^{13}C$, $\delta^{18}O$	17.5–0	Lister et al. 1991; Shen et al. 2005a
AM-6	Retreat Lake	24°29′30″N, 121°26′15″E	2230	TOC, C/N, $\delta^{13}C_{org}$	10.0–0	Selvaraj et al. 2011
AM-7	Chia-min Lake	23°17′42″N, 121°01′34″E	1673	TOC, TN, C/N	4.0–0	Lou and Chen 1997
AM-8	Great Ghost Lake	22°52′15″N, 120°51′15″E	2150	TOC, TN, C/N, C/S	2.4–0	Lou et al. 1997
AM-9	Kara Lake	13°33′N, 107°08′E	150	LOI, pollen, charcoal	9.3–0	Maxwell 2001
WM-10	Lake Biwa	35°14′42″N, 136°3′14″E	85	quartz grain size, flux	145–0	Xiao et al. 1997
WM-11	Tongsu maar	33°21′16″N, 126°37′32″E	692	quartz flux	6.5–0	Lim et al. 2005
WM-12	Lake Huguang Marr	21°9′N, 110°17′E	87	magnetic properties, Ti content	16–0	Yancheva et al. 2007
IM-13	Sumxi Co	34°32′N, 80°50′E	5058	shell content, $CaCO_3$, TOC	14.9–0	Gasse et al. 1991
IM-14	Bangong Co	33°40′N, 79°00′E	4241	$CaCO_3$, pollen, biogenic remains	11.3–0	Gasse et al. 1996
IM-15	Lake Zigetang	32°00′N, 90°54′E	4560	pollen	10.8–0	Herzschuh et al. 2006
IM-16	Ahung Co	31°37.2′N, 92°03.6′E	4575	$CaCO_3$, % dolomite, TOC, C/N, $\delta^{13}C_{org}$, $\delta^{13}C_{car}$ and $\delta^{18}O_{car}$	7.5–0	Morrill et al. 2006
IM-17	Seling Co	31°34′N, 89°00′E	4530	mineralogy, $\delta^{13}C_{TIC}$ and $\delta^{18}O_{TIC}$	13.8–0	Gu et al. 1993
IM-18	Co Ngoin	31°31′18″N, 91°32′55″E	4515	ostracod $\delta^{18}O$, $\delta^{13}C$, Mg/Ca, Sr/Ca	2010–840	Jin et al. 2009, 2011
		31°30′01″N, 91°31′41″E		Rb/Sr, TOC, lightness	13.5–4.5	Jin et al. 2005
IM-19	Zabuye Salt Lake	31°21′00″N, 84°4′12″E	4421	TOC, $CaCO_3$, $\delta^{13}C_{TIC}$ and $\delta^{18}O_{TIC}$	34.8–0	Wang et al. 2002
IM-20	Goting Basin	30°49′N, 79°49′E	3600	mineral magnetism, oxides, TOC, C/N	25–13	Juyal et al. 2009
IM-21	Lunkaransar Lake	28°30′N, 73°45′E	–	fossil pollen, lake level	11.5–0	Bryson and Swain 1981

Table 3.1 (continued)

Site	Lake name	Core location	Elevation (m)	Proxies	Time duration (ka)	References
IM-22	Kathmandu Basin	27°49′N, 85°29′E	1340	bio-Si, diatom assemblages	600–0	Hayashi et al. 2009
IM-23	Didwana Lake	27°20′N, 74°35′E	–	fossil pollen, lake level	23.5–0	Swain et al. 1983
IM-24	Heqing Basin	26°33′43″N, 100°10′14″E	2190	TOC, Rb/Sr, Tsuga	2600–130	An et al. 2011
IM-25	Xingyun Hu	24°20′N, 102°47′E	1723	$CaCO_3$, $\delta^{18}O_{TIC}$	25.1–0	Hodell et al. 1999
IM-26	Qilu Hu	24°10′N, 102°45′E	1797	$CaCO_3$, grain size, TOC, TN	50.0–0	Hodell et al. 1999
W-27	Gun Nuur	50°15′N, 106°36′E	600	magnetism, TOC, $\delta^{13}C_{org}$	11.5–0	Wang et al. 2004b
W-28	Hoton-Nur Lake	48°37′18″N, 88°20′45″E	2083	pollen, diatom	11.5–0	Rudaya et al. 2009
W-29	Ulungur Lake	47°14′N, 87°08′E	479	pollen, grain size, $CaCO_3$, TOC, TN, $\delta^{13}C_{org}$, ostracod $\delta^{13}C$, $\delta^{18}O$	10.0–0	Jiang et al. 2007; Liu et al. 2008a
W-30	Manas Lake	45°45′N, 86°00′E	280	$\delta^{18}O$, $\delta^{13}C$ of carbonate, pollen	13.8–0	Rhodes et al. 1996
W-31	Aibi Lake	44°52.2′N, 82°44.1′E	190	pollen	2.5–0	Feng et al. 2006
		44°48′N, 82°27′E		$\delta^{13}C$, $\delta^{18}O$ of ostracod and TOC	11.5–8.0, 1.5–0	Wu et al. 2004a, 2005
W-32	Dongdaohaizi Lake	44°41.7′N, 89°33.5′E	430	pollen, LOI, susceptibility, grain size	4.5–0	Feng et al. 2006
W-33	Sichang Lake	44°18.6′N, 89°8.6′E	589	pollen	1.2–0	Feng et al. 2006
W-34	Yili Valley	43°51′25.7″N, 81°57′54.3″E	928	pollen, A/C	15.0–3.3	Li et al. 2011a
W-35	Barkol Lake	43°36′57″N, 92°46′27″E	1585	fossil pollen, A/C	16.7–0	Tao et al. 2010
		43°42′N, 92°50′E		$CaCO_3$ and its $\delta^{13}C$ and $\delta^{18}O$, grain size	9.4–0	Zhong et al. 2010
W-36	Juyanhai Lake	42°20.13′N, 101°15.35′E	904	mineralogy, elemental concentration	4.1–0	Chen et al. 2010a
W-37	Bosten Lake	41°56.9′N, 86°46.5′E	1048	grain size, pollen, ostracod $\delta^{13}C$, $\delta^{18}O$, Mg/Ca, Sr/Ca	16–1.5	Huang et al. 2009; Mischke and Wünnemann 2006
W-38	Lop Nur	40°28′29″N, 90°20′83″E	780	grain size, microfossils, pollen, elements	2.2–0	Ma et al. 2008
W-39	Yanchi playa	39°45′N, 99°18′E	2840	magnetism, lightness, $CaCO_3$	16.4–0	Yu et al. 2006a

AM East Asian summer monsoon, *WM* Asian winter monsoon, *IM* Indian summer monsoon, *W* westerly climate we use TOC to compare between lacustrine records and understand the variation of the East Asian summer monsoon.

3.3.1.1 Lake Biwa

Lake Biwa (34°58′–35°31′N, 135°52′–136°17′E) is the largest lake in Japan, with an area of 674 km^2, a maximum water depth of 104 m and an altitude of 84 m above sea level (a.s.l.). The lake has a catchment of 3848 km^2. Mean annual temperature and precipitation at Lake Biwa are 14.4 °C and 1618 mm, respectively. The rainy seasons at the lake are directly related to the East Asian summer monsoon.

Lacustrine sedimentation at Lake Biwa has been continuous over the last ca. 400 ka. Sediments reach a maximum thickness of approximately 250 m, with a sedimentation rate in excess of 30 cm ka^{-1}. During the past decades, several sediment cores have been obtained near the centre of the southern depression in the North Basin of the lake and provided high-resolution and continuous records of East Asian summer monsoon activity through time.

The total carbon (TC) content of Lake Biwa sediments serves as a proxy for the East Asian summer monsoon, as it reflects primary productivity changes that are controlled by precipitation in the catchment area (Yamada 2004; Iwamoto and Inouchi 2007). The TC content analysis for a 150 m sediment core from Lake Biwa with a time resolution of approximately 150 years has been used to reconstruct millennial-scale East Asian monsoonal changes over the past 300 ka (Iwamoto and Inouchi 2007). The TC contents within the Takashima-oki core over the past 300 ka are high during interglacial stages and low during glacial stages (Fig. 3.8a), corresponding to trends in the benthic $\delta^{18}O$ curve (Lisiecki and Raymo 2005), a proxy of global ice volume change. Variation of the TC content suggests that the summer monsoon, which controls the primary production of the lake, varied at a millennial scale over the past 300 ka BP. During the Holocene, TC contents have increased since ~8 ka BP, and were highest during 2–6 ka BP. It is therefore apparent that the East Asian summer monsoon was strongest during the period 2–6 ka BP. Relatively low TC content indicates that the East Asian summer monsoon during the last glacial stage was not as strong as during the Holocene. Figure 3.8b shows variations in the East Asian summer monsoon in Japan during the last glacial stage that correspond to D–O cycles recognized in the GISP2 ice core. During the last glacial stage (MISs 2–4), TC contents vary directly with D–O cycles, except for those in Interstadials 3 and 4. Variations in TC contents show decreasing trends that correspond to cold events that were contemporaneous with Heinrich events (HEs). These findings suggest that enhancement of the summer monsoon in Japan is synchronized with warming in Greenland, and with increased primary productivity in Lake Biwa at such times. However, enhancement of the East Asian monsoon associated with warming D–O cycles may lag behind the rise in temperature seen in the D–O cycles. During the Holocene and MIS 5, the East Asian summer monsoon was strong, while during MIS 7 the monsoon experienced large amplitude variations, with a ~20 ka cycle (precession cycle). Frequency analysis of TC content indicate that the East

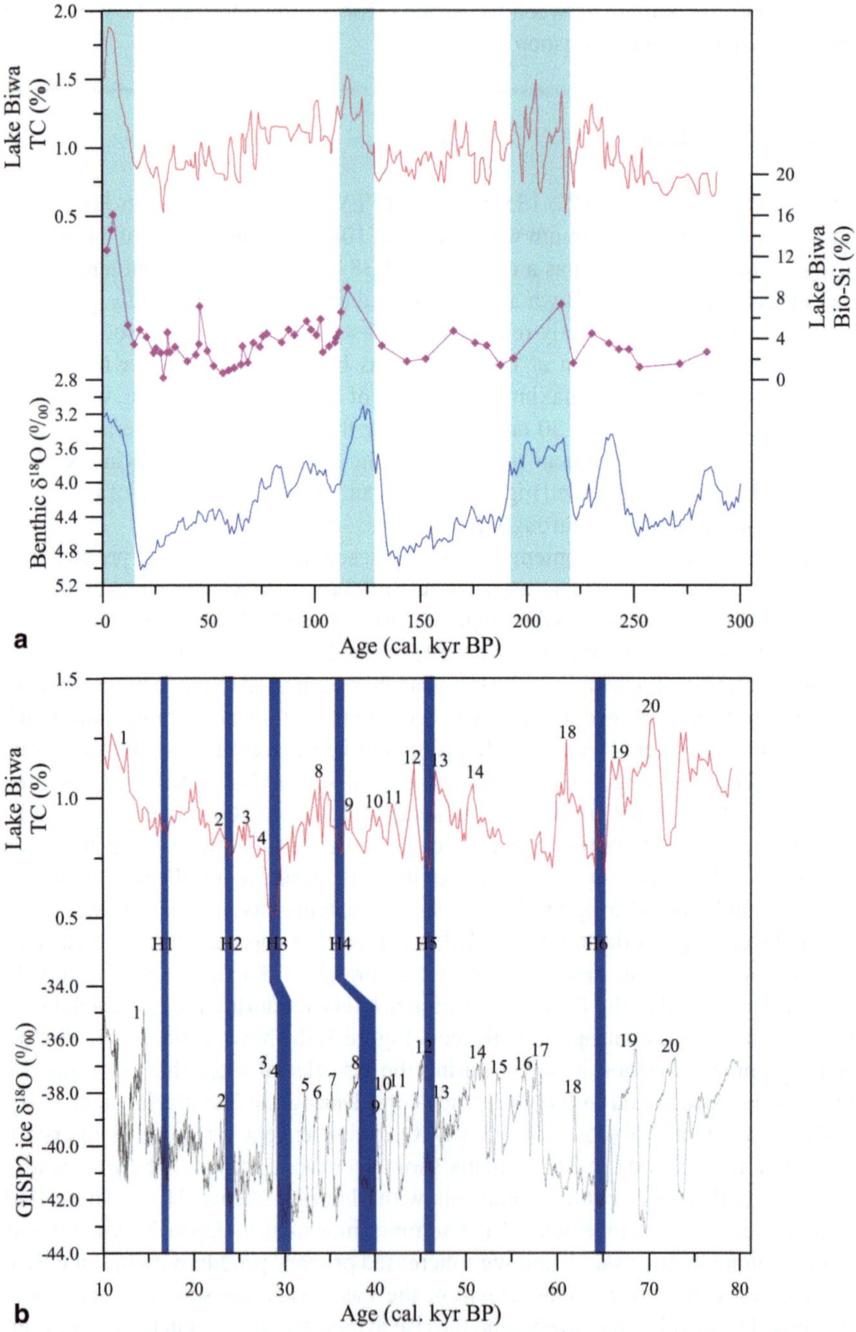

Fig. 3.8 (a) Variations in total carbon (TC) and biogenic silica content of the Takashima-oki core from Lake Biwa (after Iwamoto and Inouchi 2007), indicators of Asian summer monsoon intensity; and a marine oxygen isotope record for the period MIS 2–4 (Lisiecki and Raymo 2005). Shaded bars mark the intervals of strong Asian monsoons. (b) Correlation between the TC contents of the Takashima-oki core and the GISP2 oxygen isotope record over the past 80 ka (Meese et al. 1997). Shaded bars indicate decreased TC content associated with Heinrich Events (H1–H6) in the North Atlantic. Numbers indicate interstadial numbers of the DO cycles

Asian summer monsoon had a cycle of 1500 years over the past 300 ka, likely associated with iceberg discharge events in the North Atlantic and D–O cycles. This result serves as another example of how the East Asian summer monsoon underwent abrupt millennial-scale changes during interglacial stages, such as MISs 5 and 7.

3.3.1.2 Retreat Lake

Retreat Lake (24°29′30″N, 121°26′15″E) is a hydrologically closed subalpine lake in northeastern Taiwan with an elevation of 2230 m a.s.l. The lake is situated directly along the route of moisture-laden, southeasterly East Asian summer monsoon winds and close to the main axis of the Kuroshio Current. This shallow freshwater lake (maximum depth 1.5 m) experiences a mild mean annual temperature of 12 °C and high mean annual precipitation of ~3000 mm.

TOC records in two sediment cores from Retreat Lake document five centennial-scale periods (~7.3–6.9, 6.35–5.9, 5.8–5.1, 4.5–~2.1, and 2.0–1.6 ka BP) of relatively reduced East Asian summer monsoon precipitation, superimposed on a long-term decreasing monsoon trend from ~8 ka BP onwards (Fig. 3.9b), suggesting an unstable optimum climate in Taiwan (Selvaraj et al. 2011). The first three intervals of weak East Asian summer monsoon indicated by low TOC content within the Holocene optimum (~8.6–4.5 ka BP) correlate with solar activity, revealing that the centennial-scale summer monsoon variations in a warm climate were related to weak solar irradiance. The desiccation of the lake (4.5–~2.1 ka BP), followed by a weak monsoon event at ~2.0–1.6 ka BP, may coincide with decreased sea surface temperatures in the tropical Pacific. Centennial-scale reorganizations in the dynamics controlling circulation in the tropical Pacific may have played an important role in the East Asian summer monsoon instability, particularly during the late Holocene (Selvaraj et al. 2011).

3.3.1.3 Dali Lake

Dali Lake (43°13′-43°23′N, 116°29′-116°45′E) in central-eastern Inner Mongolia is a closed lake basin situated in a transitional zone at the northern margin of the East Asian monsoon. It has an area of 238 km^2, a maximum water depth of 11 m, and an elevation of 1226 m a.s.l. Mean annual temperature is 1–2 °C with a July average of 16–18 °C and a January average of −17 to −24 °C. Mean annual precipitation is 350–400 mm and mean annual evaporation reaches 1287 mm.

TOC concentrations were characterized in an 8.5 m core recovered from the depocenter of the lake (Fig. 3.9e, Xiao et al. 2008, 2009). The data indicate that Dali Lake reached its highest level during the early Holocene (11.5–7.6 cal ka BP). The middle Holocene (7.6–3.45 cal ka BP) was characterized by dramatic fluctuations in lake level with three intervals of low stands occurring 6.60–5.85, 5.10–4.85 and 4.45–3.75 cal ka BP. The rise in lake level since ca. 7.6 cal ka BP has been shown to be related to increased monsoonal precipitation over the lake region, concurrent with increased temperatures. The summer monsoon was relatively stronger during the middle Holocene between ca. 7.6 and 3.45 cal ka BP, and assumed to follow a

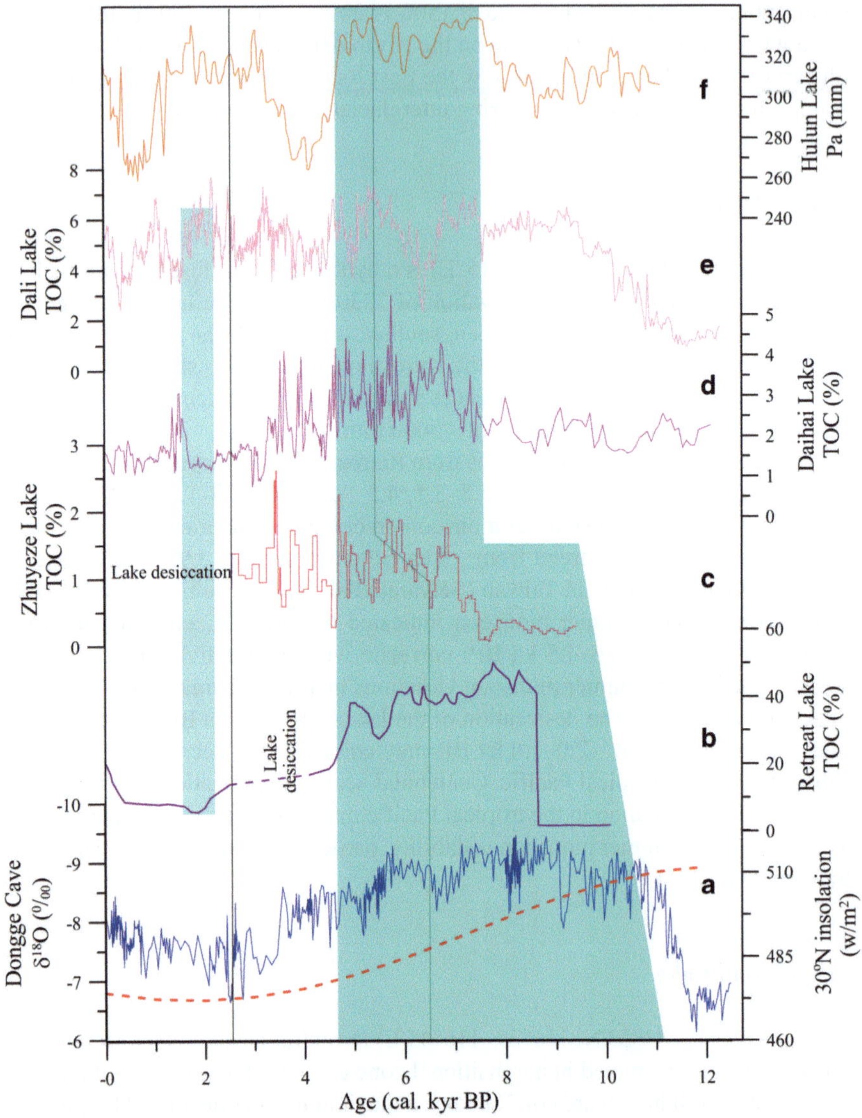

Fig. 3.9 Comparison of total organic carbon (TOC) records from (b) Retreat Lake (after Selvaraj et al. 2011), (c) Zhuyeze paleolake (after Long et al. 2010), (d) Daihai Lake (after Xiao et al. 2006), (e) Dali Lake (after Xiao et al. 2009) and (f) mean annual precipitation (Pa) reconstructed by pollen record from Hulun Lake (after Wen et al. 2010b) with stalagmite $\delta^{18}O$ record from (a) Dongge Cave (Dykoski et al. 2005). Black lines show weak monsoons, and the red dashed line represents summer insolation at 30°N

weakening trend after ca. 3.45 cal ka BP. High variability of the East Asian monsoon climate during the middle and late Holocene might have been directly associated with variations in the intensity and frequency of El Niño-Southern Oscillation (ENSO) events.

3.3.1.4 Daihai Lake

Daihai Lake (40°28′-40°39′N, 112°32′-112°48′E) lies to the southwest of Dali Lake and is also located along a transitional zone sensitive to the East Asian summer monsoon. The hydrologically closed lake has a surface area of 133.46 km^2 with a total catchment area of 2289 km^2, and a maximum depth of 16.05 m. Annual average air temperature is ~3.5 °C. Mean annual precipitation is 400 mm, and evapotranspiration is high, at 1938 mm yr^{-1}. A 12.08 m long sediment core (DH99-A) was recovered below a water depth of 15.50 m in the central part of Daihai Lake in August 1999. The sediment core contains irregularly laminated, grey-black silty clay. The chronology is based on six accelerator mass spectrometry (AMS) radiocarbon dates on terrestrial plant macrofossils or bulk organic fractions, and on two aragonite U-Th dates analyzed by Thermal Ionization Mass Spectrometry (Xiao et al. 2004).

Xiao and his colleagues also used TOC concentrations to assess relative precipitation (Fig. 3.9d), and produced a detailed Holocene history of changes in precipitation over the region (Xiao et al. 2006). From ~10.25 to 7.90 cal ka BP, TOC contents were relatively low and constant, indicating mild and dry climatic conditions. The middle Holocene between ~7.90 and 3.10 cal ka BP was marked by intensified and highly variable monsoon conditions, as indicated by high and variable TOC values in the lake sediments. The period between ~7.90 and 4.45 cal ka BP exhibits a warm and humid climate, with distinct fluctuations, i.e., cool and humid ~7.90–7.25 cal ka BP, warm and slightly humid ~7.25–6.05 cal ka BP, warm and humid between ~6.05 and 5.10 cal ka BP, mild and slightly humid ~5.10–4.80 cal ka BP, and mild and humid ~4.80–4.45 cal ka BP. The period corresponds to the Holocene optimum. During the period of ~4.45–2.90 cal ka BP, woody plants declined, and climate generally became cooler and drier than in the preceding period. This period is characterized by a cold, dry episode from ~4.45 to 3.95 cal ka BP, a warm, slightly humid interval between ~3.95 and 3.50 cal ka BP and a mild, slightly dry episode from ~3.50 to 2.90 cal yr BP, and appears to be a transition from warm and humid, to cold and dry climatic conditions. Since ~2.90 cal ka ago, forests disappeared and the vegetation density decreased, and TOC decreased to its lowest values reflecting a cool and dry climate. However an interval of somewhat more favorable growth conditions was observed between ~1.70 and 1.35 cal.ka BP, when woody plants recovered, along with relatively higher TOC values.

3.3.1.5 Hulun Lake

Hulun Lake (48°30′40″–49°20′40″N, 117°00′10″–117°41′40″E) in Inner Mongolia is located along the northeastern margin of the East Asian summer monsoon. The lake has an area of 2339 km^2, with a catchment area of 37,214 km^2 within the borders of China. Mean annual temperature is 0.3 °C and mean annual precipitation is 280 mm, which falls in July–August when the monsoon rainfall belt migrates to its northern limit. Mean annual evaporation reaches 1400–1900 mm, which is five to six times the annual precipitation. The water balance of Hulun Lake is closely tied to the strength of the East Asian summer monsoon.

A quantitative mean annual precipitation (Pa) record for the Holocene has been reconstructed for Hulun Lake (Wen et al. 2010a, 2010b). The results are shown in Fig. 3.9f. A warm and dry climate prevailed over the lake region until ~8.0 cal. ka BP. During the period ~8.0–4.4 cal. ka BP, precipitation increased by ~10% and temperature gradually declined. The interval between ~4.4 and 3.35 cal. ka BP was marked by extremely dry and relatively cold conditions. Precipitation recovered from ~3.35 to 1.0 cal. ka BP, with temperatures rising during the period ~3.35–2.05 cal. ka BP, and dropping from ~2.05 to 1.0 cal. ka BP. The lack of precipitation during the early Holocene could be the result of a weakened summer monsoon at that time due to the influence of remnant ice sheets in the Northern Hemisphere (Wen et al. 2010a). The observed increases in monsoonal precipitation during the middle to late Holocene were likely driven by ocean–atmosphere dynamics in the western tropical Pacific at that time.

3.3.1.6 Zhuyeze Lake

Zhuyeze Lake is situated in the Minqin Basin between the Tengger and Badain Jaran deserts. It is a large paleolake that was fed by the Shiyang River, flowing out of the Qilian Mountains since the late Pleistocene (Pachur et al. 1995). It lies along the northwestern margin of the modern Asian summer monsoon belt. The major runoff in this drainage is from monsoonal precipitation, with only 7% from glacial melt water from the Qilian Mountains. This region is dominated by the summer monsoon precipitation from July to September, which accounts for the bulk of the moisture for the whole year. In winter, cold and dry continental air masses prevail in the area. Mean annual temperature and precipitation at Zhuyeze Lake are 7.8 °C and 110 mm, respectively, while evaporation is over 2600 mm. The lake reached approximately its maximum area during the late Pleistocene, and has shrunk since the 1950s with a present day average area of 580 km^2.

TOC data, along with grain size, carbonate, C/N and $\delta^{13}C_{org}$ from Zhuyeze paleolake show that the climate was warm and dry in the early Holocene (9.5–7.0 cal ka BP), cool and humid in the mid-Holocene (7.0–4.8 cal ka BP), and increasingly drier in the late Holocene (since 4.8 cal ka BP) (Long et al. 2010; Fig. 3.9c). The pattern of the effective moisture changes was synchronous in the East Asian monsoon marginal zone, while the moisture optimum during the Holocene was out of phase between the Asian monsoonal marginal region and the Asian region dominated by the monsoon (that is, a wet monsoon Asia corresponding to a dry Zhuyeze during the early Holocene, and a decreased monsoon and wet Zhuyeze during the mid-Holocene). The discrepancy was more likely due to the greatly enhanced rate of evaporation (E) than to the higher monsoonal precipitation (P) in arid lands, resulting from high temperatures during the early Holocene within the Asian monsoon marginal area.

3.3.1.7 Lake Qinghai

Lake Qinghai (36°32′–37°15′N, 99°36′–100°47′E), the largest, hydrologically closed lake with brackish to saline water (salinity = 15.5 g l^{-1}) on the TP, is developed within an intermontane depression at the tectonically active margin of the plateau. Today, Lake Qinghai, an altitude of 3194 m a.s.l., has a surface area of about 4,260 km^2 and a catchment area of about 29,660 km^2. In the Lake Qinghai area, average annual air temperature (1951–2007) is ~1.2 °C. Mean annual precipitation (1951–2000) is 357 mm, but evaporation is 3–4 times higher than precipitation. The water is fed mainly by direct rainfall and runoff through several perennial rivers and about 50 seasonal waterways. The lake output is dominated by evaporation. The surface water area of Lake Qinghai has shrunk from 4980 km^2 to about 4260 km^2 during the last century (Jin et al. 2010).

The lake lies in a transitional zone where climate is sensitive to the East Asian and Indian monsoons and the Westerlies. At present, the East Asian summer monsoon circulation reaches this region in summer and induces clear seasonality in precipitation, while the Westerlies climate is dominant in winter. Lake Qinghai has experienced large fluctuations in water level, salinity and stable isotope contents in response to climate change. Changes in environmental conditions associated with these climate systems are preserved in the sediments. Lake Qinghai sediments mainly consist of detrital materials of riverine and eolian origins, combined with authigenic carbonates. Thus, Lake Qinghai is an ideal site to study the competing influence of both the Westerlies and East Asian summer monsoon on northeastern TP. Previously published results of paleolimnological studies of Lake Qinghai, reviewed by Colman et al. (2007), show that the importance of Lake Qinghai sediment records for understanding past global changes has been increasingly recognized since the 1870s, due to its unique geographical location. However, limnological studies using modern techniques did not start until the 1960s, when fundamental data on lake biology, water chemistry, and hydrology were first obtained during a multi-disciplinary expedition. Interest in paleoclimatic change from lake sediments continued to increase in the 1980s. A variety of proxy analyses of lake sediments, including pollen, carbon, elemental and isotope geochemistry have revealed late glacial to Holocene changes in lake hydrology and catchment environment associated with the advance and retreat of the Asian summer monsoon front.

In 1988, a 5.6 m core, taken in the deepest part of the basin, was analyzed in detail for its sedimentological characteristics, and ostracod $\delta^{18}O$ values. The oxygen isotope record of the ostracods has been shown to be related to the precipitation/evaporation (P/E) budget at the lake, and by extension, past lake levels (Lister et al. 1991). The chronology for the Qinghai study is based on 3 AMS radiocarbon dates measured on vegetal remains of aquatic origin (*Ruppia*) that ranged between 10.9 and 8.4 ka BP. No measurement of ^{14}C activity was conducted on modern waters and the precise timing of environmental events younger than 8.4 ka BP could not be ascertained. The age of the core base is estimated at 14.3 ka BP by extrapolation. The record suggests that the regional climate remained arid until at least 12.5 ka BP, when an incipient monsoon stage occurred, inducing an initial, minor amplitude increase in lake level.

Sedimentary evidence documents a further period of aridity after 10.8 ka BP, synchronous with the Younger Dryas event, which is tentatively attributed to a monsoonal weakening. At 10.2 ka BP, the basin began filling in response to a permanent strengthening of the Asian monsoon, as shown by a rapid decrease in $\delta^{18}O$ values. The overall positive shift in $\delta^{18}O$ values through the early-middle Holocene is interpreted as being due to steady evaporative enrichment of ^{18}O in the lake-water reservoir, essentially independent of lake level, and reflecting long-term stable climatic and hydrological conditions. Negative deviations from that trend suggest however that lake level rose around 10.0 ka BP, from ~9.5 to 8.5, and 8.2 to 7.2 ka BP. Brief positive deviations (falling lake levels) are observed around 8.2 and 6.8 ka BP

A stranded shoreline terrace at +12 m may represent the highest Holocene lake level and was dated at about 7–6 ka BP. After ~3 ka BP, a possible overall weak negative $\delta^{18}O$ shift may represent a naturally changing isotopic steady state (Lister et al. 1991). However, the highest $\delta^{18}O$ values for the whole Holocene period are observed around 3.7 ka BP, and from about 2.4 to 1.5 ka BP.

In July 2000, a 7.95 m long core (QH-2000) was recovered from the southeastern part of Lake Qinghai in water depth of 22.3 m (Shen et al. 2005a). Detailed records from the core QH-2000 also covered the last 18 ka. Pollen analyses and total organic carbon (TOC), total nitrogen (TN), grain size, and carbonate measurements were conducted on the core QH2000 (Shen et al. 2005a) to provide complementary information on late glacial to Holocene climate change. Five distinct stages of regional climate changes can be defined. Minor increases in TOC, TN, and carbonate contents after ca 17.0 cal ka BP indicate the termination of LGM. But the lake level was still low. A decrease in tree pollen percentages between 12.5 and 11.5 cal ka BP may correlate with the Younger Dryas stadial, although chronologies are uncertain. Steady warming began at 11.0 cal ka BP, along with higher primary productivity as marked by significant increases in TOC, TN, and carbonate content. This trend was frequently punctuated by a number of centennial-scale cooling episodes, possibly including the "8.2-ka event", implying unstable climate conditions during the mid-Holocene climate optimum (Chen et al. 2001a; Wünnemann et al. 2003). A significant shift in climate conditions occurred at ~4.0 cal ka BP. The $\delta^{13}C$ values continued to become more negative, possibly related to an increasing input of terrigenous organic matter as lake level continuously lowered. However, pollen records indicate that the C_3/C_4 ratio of the vegetation did not change significantly during the Holocene (Shen et al. 2005a).

Both records indicated that maximum warming occurred about 10.0 cal ka BP, accompanied by increased monsoon precipitation in the eastern TP. Superimposed on this general pattern is small-amplitude, centennial-scale oscillations during the Holocene. However, uncertainties in the core chronology make it difficult to correlate these abrupt changes with the short-term dry events deduced from other proxy records. Radiocarbon dates of sediment cores from Lake Qinghai using bulk organic matter are problematic because organic matter may have been reworked and delivered to the lake from various pre-aged sources by river and groundwater discharges. Linearly fitting the calibrated radiocarbon ages of organic matter to depth in cores gives a sediment–surface intercept of 1100 years for Core QH85-14B (Kelts et al. 1989), 1039 years for Core QH2000 (Shen et al. 2005a), and 439 years for QH85-16A (Zhang et al. 1994a). These varied reservoir effects with depth imply an increasing proportion of organic matter containing old carbon of terrestrial origin, and perhaps increasing groundwater influence.

3.3.1.8 Other Lake Records

Limited lacustrine records exist from mainland Southeast Asia. One example is Kara Lake. It is a permanent, closed-basin, volcanic lake in Ratanakiri Province, northeastern Cambodia. Maxwell (2001) reconstructed a 9.3 ka monsoon record at Kara Lake from pollen, organic and inorganic carbon using a 15 m sediment core. This first well-dated palynological record from Cambodia shows that late glacial conditions ended about 8.5 ^{14}C ka BP, more than 1000 years later than sites in southwest China. Summer monsoon intensity increased over the period ca. 8.4–5.3 ^{14}C ka BP, similar to most other sites in the Asian monsoon region. Subsequent conditions were drier, and perhaps warmer, than present.

In another study, a lake sediment record from Yanchi playa in the central Hexi Corridor, northwestern China, shows millennial- to centennial-scale climate variability during the Holocene that has been regulated by both the westerlies and the East Asian summer monsoon in a roughly synchronous manner (Yu et al. 2006a). Cold–dry intervals in the central Hexi Corridor generally correspond to cooling phases in high latitudes of the Northern Hemisphere, and cold, warm and wet intervals largely correlate with relatively warm phases in other northern high latitude sites. A possible mechanism involving NAO/AO and ENSO dynamics is tentatively proposed to account for this synchronicity. The climatic imprint on sites from the NW China dry lands may be spatially variable, in response to the alternating climate control by either the westerlies or East Asian Summer Monsoon. Greater spatial control and improved chronologies should shed light on this question.

From the synthesis above of lacustrine records of the East Asian monsoon (Table 3.1, Fig. 3.9b–f), it is clear that the resolution, temporal and spatial control of these studies is providing an expanded understanding of the East Asian monsoon through time. Although lacustrine records from different lakes may impart a local, basin-specific signal (from factors such as size, hydrology, sediment preservation, etc), a consistent history of the spatial and temporal variability of the East Asian summer monsoon can be determined through correlation of the various records. Two observations are obvious from Fig. 3.9, with regard to Holocene monsoonal precipitation: 1) variability in the proxies increase from south to north and from east to west, with decreasing but fluctuating TOC contents in the sediments, and 2) the inferred duration of the Holocene optimum is shorter in the north as compared to the south. These observations indicate that lake sediments in the north are sensitive to strong monsoons, but do not respond to weak monsoons. This could be because the lakes become desiccated intermittently, during weak monsoon periods with corresponding low precipitation.

The increase in the number of proxy records is not entirely consistent, however, creating a conflicting picture of Holocene precipitation changes across Asia. In particular, a debate has arisen with regard to East Asian monsoonal precipitation during the early and middle Holocene. Two viewpoints have emerged: 1) that monsoonal precipitation was weak during the early Holocene but strengthened during the middle Holocene, as documented by sediment cores from Daihai Lake (Xiao et al. 2004, 2006; Jin et al. 2006) and Qingtu Lake (Long et al. 2010), and 2) that monsoonal precipitation strengthened during the early Holocene but weakened during the middle Holocene, as suggested by results from Zhuyeze Lake (Chen et al. 2001a, 2006) and Lake Qinghai (Lister et al. 1991; Shen et al. 2005a). Such a discrepant view may be due to: 1) ambiguities in the ages and proxies obtained from different sedimentary sequences, 2) hydrological differences in lakes from different geomorphic settings, and 3) regional differences related to uplift and atmospheric circulation (Xiao et al. 2009; Long et al. 2010).

Additional explanation shed light on the early Holocene hydroclimatological puzzle as shown in Fig. 3.9. The PMIP2 850 hPa wind fields in summer (JJA) and winter (DJF) during mid-Holocene (6 ka) and last glacial maximum (LGM, 21 ka) periods (Fig. 3.10, Braconnot et al. 2007) show that 1) Retreat Lake, Dali Lake and Daihai Lake were mainly affected by East Asian monsoon, 2) the East Asian Summer Monsoon circulation reached both Zhuyeze Lake and Lake Qinghai in summer as at present, and 3) Hulun Lake might be mainly controlled by the Westerlies even during the mid-Holocene. During the LGM, the climates of all these lakes except Retreat Lake in Taiwan were mainly controlled by the Westerlies. Therefore, we suggested that weak precipitation during the early Holocene recorded in Hulun Lake, Daihai Lake, Dali Lake and Zhuyeze reflect not only the weak East Asian monsoon, but also the imprint of the Westerlies (Fig. 3.9). This is also supported by the inference by Xiao et al. (2008). They indicated that the expansion of Dali Lake during the early Holocene resulted from the input of the snow/ice melt, rather than the monsoonal precipitation, in response to the increase in summer solar radiation in the Northern Hemisphere (Fig. 3.9, Xiao et al. 2008). If Westerly atmospheric circulation was dominant as suggested above, precipitation records from the lakes located in transitional positions of humid areas would reflect the moisture derived from both the Westerlies and/or the East Asian monsoon (Fig. 3.10), rather than the latter alone. If so, care should be taken in using the proxies in the northern and western lakes to infer weak monsoons.

3.3.2 Asian Winter Monsoon Variability

Eolian dust flux in central Asia has been correlated with variations in the strength of the winter monsoon and the aridity of the source areas at glacial-interglacial time scales (An et al. 1991a). This eolian dust-climate relationship was also suggested

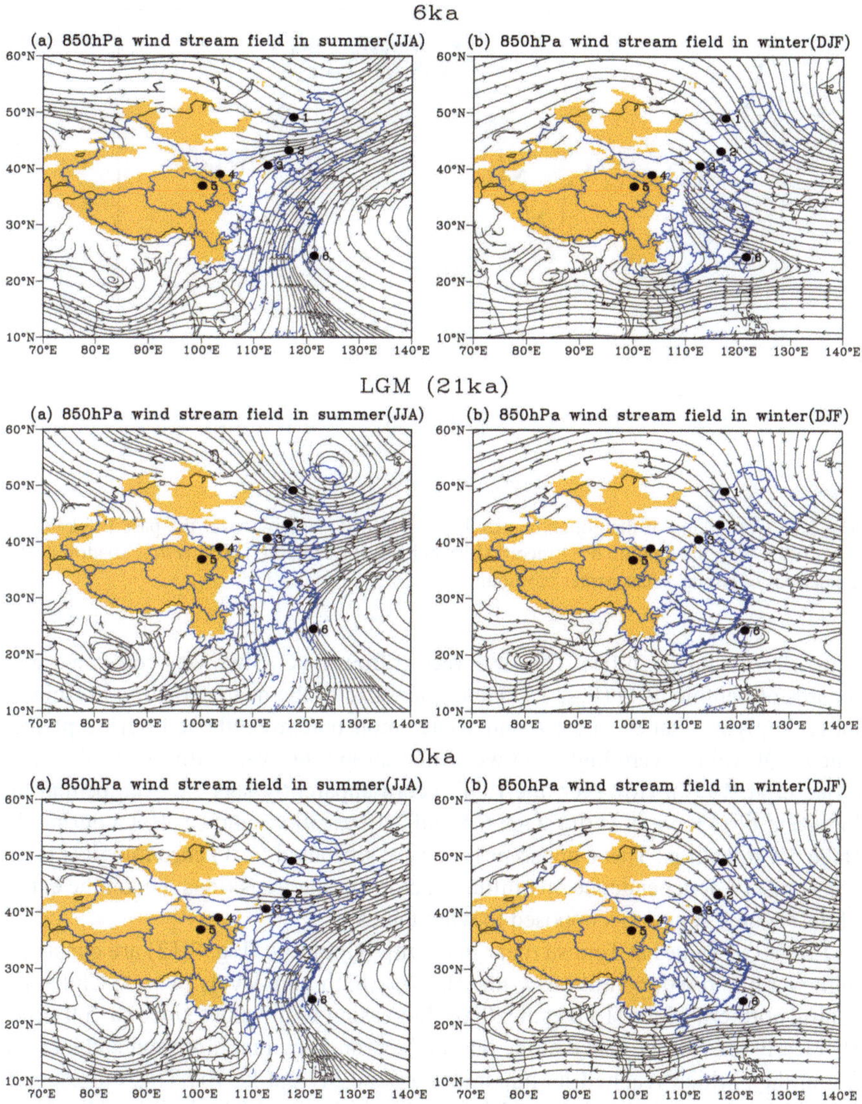

Fig. 3.10 850 hPa wind fields in summer (JJA) and winter (DJF) during pre-industrial (PI, 0 ka), mid-Holocene (6 ka) and last glacial maximum (LGM, 21 ka) periods based on the CCSM3 modeling output in the frame of PMIP2 (Braconnot et al. 2007). The elevations of yellow areas exceed 1500 m. Black solid circles represent the location of lake records: 1. Hulun Lake, 2. Dali Lake, 3. Daihai Lake, 4. Zhuyeze Lake, 5. Lake Qinghai, 6. Retreat Lake

by eolian quartz flux (EQF) results from Lake Biwa and by Ti content of sediments from Lake Huguang Maar in coastal southeast China.

Through the identification of atmospherically transported quartz grains in the Lake Biwa sediments, EQF was used to infer variations in the intensity of the East Asian winter monsoon and/or the westerlies, and the aridity of inland Asia (Xiao

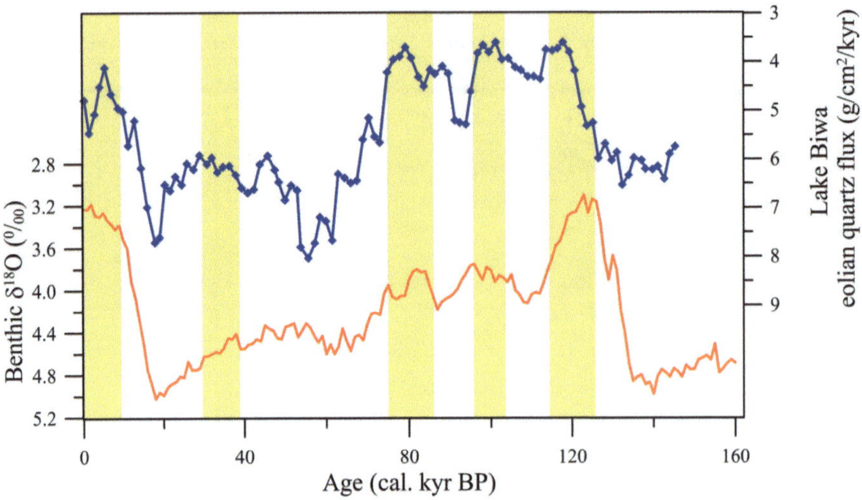

Fig. 3.11 Correlation of the eolian quartz flux (EQF) of Lake Biwa over the past 145 ka (Xiao et al. 1997) with the benthic $\delta^{18}O$ record (Lisiecki and Raymo 2005). The *yellow* shadows mark the periods of strong Asian winter monsoon

et al. 1997; Yamada 2004). The EQF record from Lake Biwa was correlated with records of benthic $\delta^{18}O$ (Fig. 3.11) and grain-size of the quartz fraction in Chinese loess, implying that the East Asian winter monsoon strengthened during periods when EQF values were high, and weakened during intervals with low EQF values (Xiao et al. 1997). However, the EQF record apparently lags ca. 5 ka behind the loess and $\delta^{18}O$ records at the stage 6/5 and 2/1 transitions and ca. 10 ka during the stage 5/4 transition. These apparent lags were attributed to: 1) chronological problems or 2) EQF dependence on multiple factors during transitions, i.e. on the extent of dust source regions as opposed to wind intensity.

Low EQF values at Tongsu maar in the mid-Holocene (Fig. 3.12) are consistent with warm and wet conditions with relatively high lake levels in west China and central northern China (Lim et al. 2005). Increasing EQF values since ca. 5.0 cal ka BP clearly provide evidence of the onset of a stronger winter monsoon and increasing aridity, which reached its maximum around 3.1 cal ka BP. This corresponds to the lowest lake level in Lake Yiema, a typical closed lake basin in the desert of arid northwestern China (Chen et al. 1999) and the desiccation around 4.2 ka BP of Co Ngoin, central TP (Jin et al. 2005). Lower EQF values from 2.0–1.0 cal ka BP imply less aridity and/or a weaker winter monsoon than during the late-Holocene, further supported by increasing evergreen and broadleaved deciduous tree pollen between ca. 1.7 and 1.3 cal ka BP in central China, and strong summer circulation at ca. 1.3 cal ka BP and increased winter monsoon at ca. 1.0 cal ka BP in the desert/loess transition areas. Thus, the Cheju EQF record suggests that long-term variability in eolian dust deposition in downwind areas has been controlled by the aridity of source areas and winter monsoon strength during the mid- to late

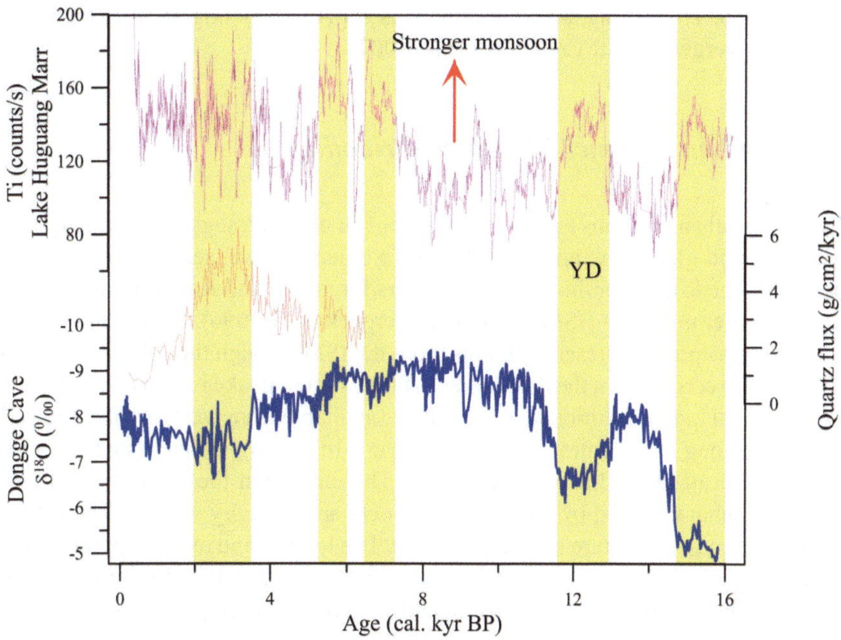

Fig. 3.12 Comparison of the records of eolian quartz flux from Tongsu Maar in Cheju Island, Korea (From Lim et al. 2005), Ti content from Lake Huguang Maar (Yancheva et al. 2007) with Dongge Cave $\delta^{18}O$ record (Dykoski et al. 2005). Both quartz flux and Ti content from the Marr were used to reflect the intensity of the Asian winter monsoon. The yellow shadows mark the periods of strong Asian winter monsoon along with weak Asian summer monsoon

Holocene. Furthermore, Lim et al. (2005) argued that centennial-scale periodicities in EQF in Cheju Island can be explained as a response to short-term changes in solar activity.

By using magnetic properties and Ti contents in the sediments from Lake Huguang Maar, the strength of the winter monsoon winds in coastal southeast China was reconstructed over the past 16 ka (Yancheva et al. 2007). Lake Huguang Maar receives a minimal quantity of material by runoff and thus acts as a natural sediment trap for dust delivered to the site by northerly winds of the winter monsoon. The Ti content of the sediments reflects this eolian input to the lake (Fig. 3.12). High Ti contents indicate stronger winter monsoon winds before the Bølling–Allerød warming, during the Younger Dryas episode, and during the middle and late Holocene. However, higher Ti contents during the middle and late Holocene than during the Late Glacial might indicate that the signal of Ti content in the sediments does not reflect the Asian winter monsoon alone. The Lake Huguang Maar record of winter monsoon strength shows a remarkable anticorrelation with radiometrically dated stalagmite $\delta^{18}O$ records from Dongge cave, East Asia (Fig. 3.12). Strong winter monsoon phases reconstructed from Huguang Maar correlate with higher $\delta^{18}O$ at Dongge cave, which indicates lower summer precipitation rates and a weaker summer monsoon. This strong anticorrelation was explained by migration of the intertropical convergence zone (Yancheva et al. 2007).

3.3.3 Indian Summer Monsoon Variability

Climate model simulations and available paleoclimatic data suggest that changes in precession and glacial boundary conditions (e.g. ice volume, sea surface temperature (SST, albedo) are dominant forcing factors for millennial-scale variations in the Indian summer monsoon (ISM) strength (Overpeck et al. 1996). Some central Asia lake sediments proxies preserve an imprint of the ISM through time. These complement marine records from the Arabian Sea and continental lake records from Africa. Reconstructed high resolution lake records are available from western Tibet (Sumxi Co and Bangong Co), central Tibet (Co Ngoin, Ahung Co), Heqing (southwestern China), and Rajasthan (Lake Lunkaransar). The climate at these areas is entirely monsoonal, characterized by warm wet summers and cool dry winters. The Indian Ocean is the major moisture source at present. The longest and most comprehensive of these records is from Heqing paleolake, which yields detailed information on past atmospheric circulation patterns, in addition to winter temperatures.

3.3.3.1 Heqing Paleo-Lake

A 666 m long sediment core was retrieved from the central part of the Heqing paleolake basin in summer 2002, with 97% recovery (An et al. 2011). The lake sediments mainly consist of horizontally laminated greyish-green calcareous clay and silty clay with thin-bedded silt and fine sand layers, except for two intervals of sand layers with fine gravels. Comparison between the Heqing core magnetic polarity sequence and the Geomagnetic Polarity Time Scale (GPTS) indicate that the Brunhes/Matuyama (B/M) and Matuyama/Gauss (M/G) boundaries are located at 152.00 m and 614.47 m respectively. The Jaramillo (232.45–212.62 m) and Olduvai (435.16–403.40 m) subchrons were identified in the Matuyama chron. In addition, there are four possible short geomagnetic excursions during the Matuyama chron at core depths of 188.20–185.27 m, 350.94–348.22 m, 373.05–366.70 m and 512.75–509.89 m, probably corresponding to the Santa Rosa, Gardar, Gilsa, and Reunion events, respectively. The magnetic polarity ends at the first normal subchron (*C2An.1n*) of the Gauss normal chron. Based on an average sedimentation rate of the interval between lower Olduvai and M/G boundaries, the bottom age of the Heqing core can be extrapolated to 2.78 Myr.

Samples at 10–20 cm intervals over the upper 200 m depth (~900 yr resolution) and 50 cm intervals below 200 m (~2000-yr resolution) were analyzed for proxies including total organic carbon (TOC), Rb/Sr, and pollen. Given the in-phase relationship and strong correlation between Rb/Sr and TOC, the two records were combined by normalizing each to unit variance and averaging to produce a stacked ISM

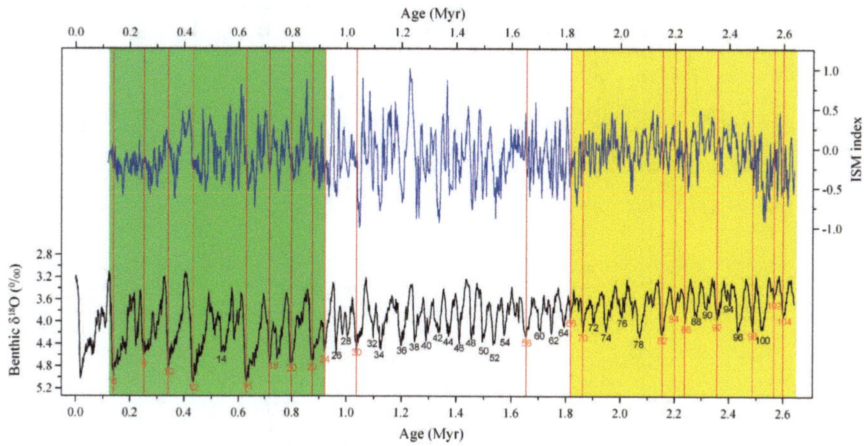

Fig. 3.13 The timing of the Heqing ISM minima relative to global ice volume maxima (benthic $\delta^{18}O$, LR04 age scale) (Lisiecki and Raymo 2005) during the glacials within three intervals over the 2.60 Myr. The red numbers and lines indicate the times when the ISM minima led the ice volume maxima while the black numbers indicate in phase relationships between the two. Note that in the middle interval (1.82–0.92 Myr) most of ISM minima (17 of 19) are coincident with ice volume maxima, while in the young and old intervals most and half of ISM minima led ice volume maxima respectively. (from An et al., 2011)

index that captures the full amplitude of variability with high values indicating enhanced ISM circulation. Based on the phase relationship between ISM minima and $\delta^{18}O$ maxima in the benthic foram record (An et al. 2011), the ISM record is separated into three intervals, i.e. 0.92–0.13, 1.82–0.92 and 2.60–1.82 Myr (Fig. 3.13). Visual inspection reveals that there is much more structure in the Heqing ISM record than in either the benthic $\delta^{18}O$ record or the Antarctic δD record. The Heqing ISM spectrum contains spectral peaks at 73-, 55-, and 29-ka periods, in addition to primary orbital periods associated with eccentricity (100 ka) and precession (23 and 19 ka). Climatically, we attribute the relative complexity of the ISM spectrum to the combined influence of NH ice volume and SH temperature.

During strong interglacials, ISM maxima coincide with NH ice volume minima and Antarctic temperature maxima (Fig. 3.14). From peak interglacials into glacials, the ISM weakens as ice volume increases (sky blue arrows, Fig. 3.14). However, the weakening of the ISM into glacials is not smooth or linear; a number of strong 'secondary' ISM peaks exist that have no correlative structure in the $\delta^{18}O$ record (e.g. peaks labeled a through g in Fig. 3.14). Transitioning from interglacials into glacials, ISM strength reaches a minimum as benthic $\delta^{18}O$ reaches ~4.5 ‰, corresponding to low, often minimum Antarctic temperatures (δD minima). These ISM minima occur 14 to 35 ka earlier than the timing of NH glaciation maxima. Similar leads are found in the 350-ka Summer Monsoon Stack from the northern Arabian Sea (Clemens and Prell 2003). Subsequently, the ISM strengthens gradually (black arrows, Fig. 3.14) even as NH ice volume continues to grow. Finally, the ISM intensifies rapidly along with the glacial termination. However, we note that ISM

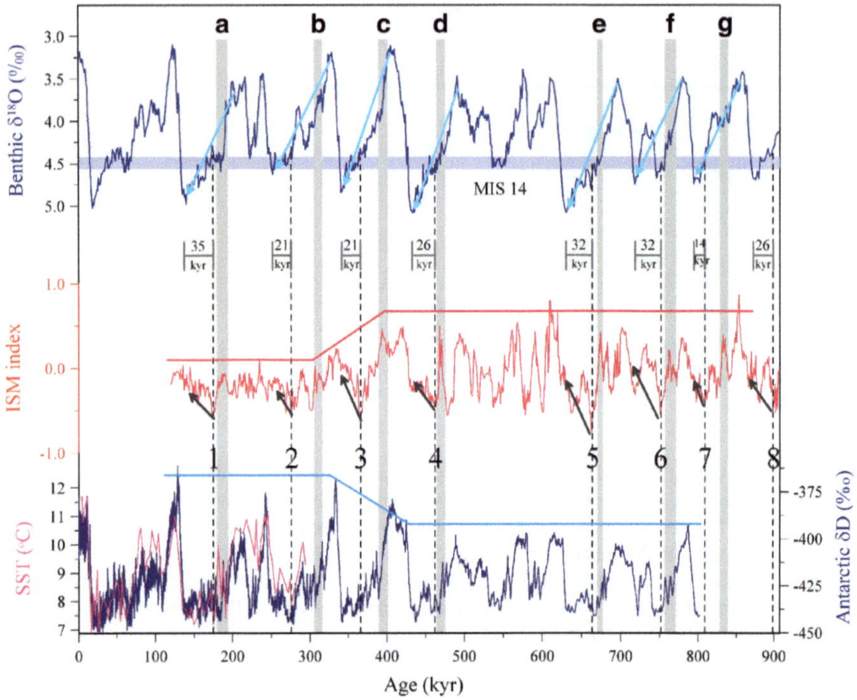

Fig. 3.14 Comparison between stacked index of the Indian summer monsoon (red, ISM), global ice volume (Lisiecki and Raymo 2005) (blue, benthic $\delta^{18}O$, LR04 age scale), and Antarctic deuterium (Jouzel et al. 2007) (dark blue, EDC δD, EDC3 age scale) during the past 900 ka. The mid-latitude southern hemispheric sea surface temperature (SST) (Mashiotta et al. 1999) (pink, RC11-120) varies simultaneously with Antarctic temperature. The horizontal grey bar in the benthic $\delta^{18}O$ shows a ~4.5‰ threshold. Note that ISM minima lead global ice volume maxima by 14 to 35 ka. The black dashed vertical lines show the minima of the ISM, labeled from 1 to 8. The sky blue arrows show the transitions from interglacial to glacial, and the black arrows show strengthening ISM prior to global ice volume maxima. Secondary ISM peaks (grey shadings, a through g) occur during the slow transition from interglacials into glacials. Thick solid red and sky-blue lines show the long-term amplitude relationship between ISM and Antarctic temperature. (from An et al., 2011)

minima do not occur early during moderate glacial intervals. For example, during MIS 14 the ISM minimum is in phase with ice volume maximum.

At the glacial-interglacial timescale, the ISM often reaches its minima, and then begins to strengthen, prior to ice volume maxima within the 2.60–1.82 Ma and 0.92–0.13 Ma intervals (Fig. 3.13). These two intervals are also characterized by low-amplitude ISM variability. We interpret the early ISM minima and subsequent strengthening in glacial times as a response to low temperatures and increasing ice volume in the Antarctic which tend to enhance atmospheric meridional circulation and the Mascarene High in the southern Indian Ocean. We propose that this 'pressure push' mechanism more than compensates for the effects of increased NH glaciation which serves to weaken the Indian low. In contrast, within the 1.82–0.92 Ma interval, the ISM minima are much more in phase with ice volume maxima and the ISM is characterized by high amplitude variability. We interpret this as dominance of the northern thermal (pull) forcing, with a subdued southern pressure (push) influence.

3.3.3.2 Lake Records from the Tibetan Plateau

An ISM record is also available from a 206.5 m long sediment core from Co Ngoin, central Tibet, but it covers only the early to mid-Pleistocene period (Jin et al. 2009, 2011). High-resolution limnic records of the abundance of ostracod shells and their geochemical compositions ($\delta^{18}O$, Mg/Ca and Sr/Ca) from Co Ngoin show distinct variations in lake water chemistry in association with changes in the ISM over the period 2.01 to 0.84 Ma BP (Fig. 3.15). ISM climate conditions during the early Pleistocene were characterized by frequent oscillations which are preserved in the ostracod geochemical records. A major climate change occurred in the Co Ngoin area at about 0.93 Ma BP. A pronounced fluctuation in the ostracod geochemical proxies during the mid-Pleistocene transition indicates increased glaciation and a strong monsoon seasonal contrast over the plateau, coincident with increases in both Chinese loess grain size and Arabian Sea lithogenic flux.

Hodell et al. (1999) produced a reconstructed paleoclimate record for southwestern China covering the past 25 ka, on the basis of $CaCO_3$ records from sediment cores from two Yunnan lakes – Xingyun Hu and Qilu Hu. The 25 ka record is marked by alternating warm, interglacial or interstadial stages and cold, glacial or stadial periods. From 25 to 12 cal ka BP, low $CaCO_3$ and TOC concentrations in sediments suggest cold temperatures and a weak ISM. Abrupt changes are recorded in most sediment proxies at 12 cal ka BP, suggesting major deglacial intensification of the ISM associated with the end of the last deglaciation. The ISM was strongest from 12 to 8 cal ka BP, but it weakened gradually afterwards. Spectral analysis of the carbonate and quartz signals reveals significant power concentrated at periods of 7.2 and 8.9 cal ka BP, respectively, which may reflect a nonlinear climate response to precessional forcing at a time of reduced eccentricity modulation (Hodell et al. 1999). A TOC record from proglacial lake sediments at Goting in the Higher Central Himalaya registered millennial-to centennial-scale fluctuations of the ISM during the Last Glacial (Juyal et al. 2009). Optically stimulated luminescence dating indicates that the high resolution stratigraphic record of ISM variability at Goting extends back through the Last Glacial Stage (*ca.* 25-13 ka BP). Millennial-scale ISM variability is attributed to climatic instability in higher northern latitudes, modulated by the position of the westerlies. However, centennial-scale abrupt changes are attributed to changes in albedo on the Himalaya and TP.

Terrestrial records of past change from Tibetan lake sediments since the last glacial indicate that the ISM was significantly weaker than present during glacial times, much stronger than present during the early to mid-Holocene, and relatively weaker since then (Fig. 3.16, Gasse et al. 1991; van Campo and Gasse 1993; Wang et al. 2002; Jin et al. 2005; Herzschuh et al. 2006). These lacustrine records suggest

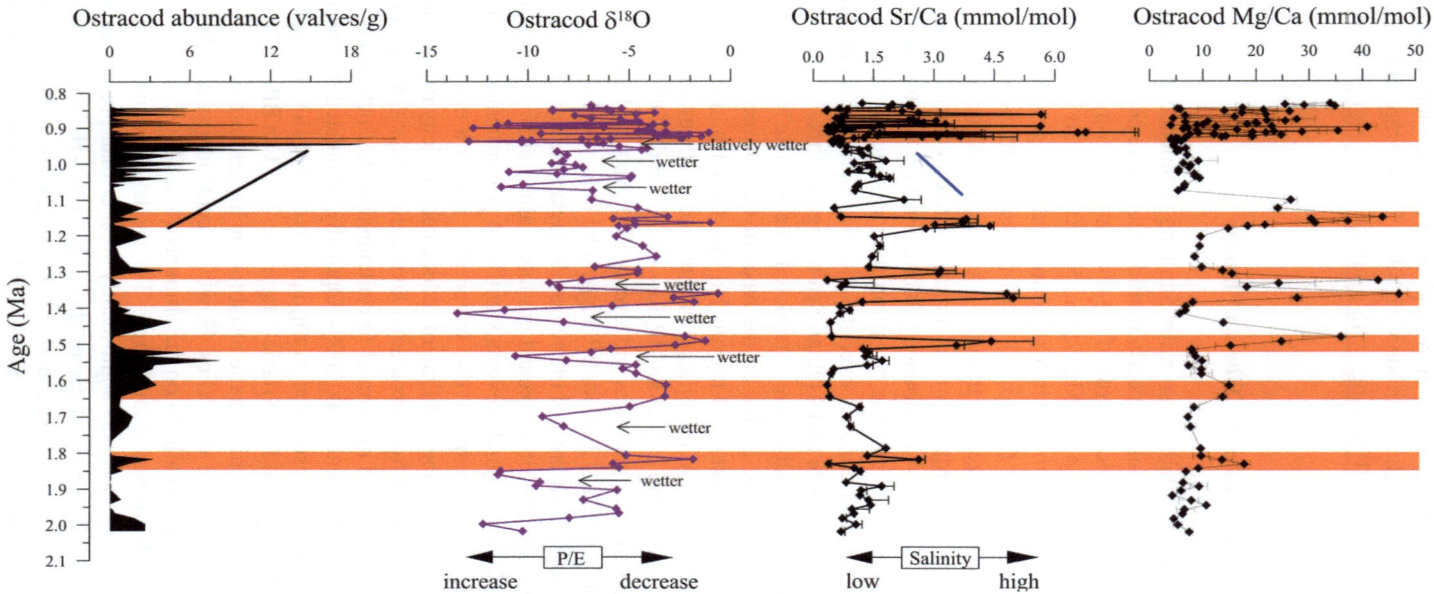

Fig. 3.15 Variations in ostracod oxygen-isotope and trace element (Mg, Sr) geochemistry and abundance of the adult valves of *Qinghaicypris* Huang 1979 over the period of 2.01 to 0.84 Myr BP from Co Ngoin sediments in central Tibet. (after Jin et al. 2011)

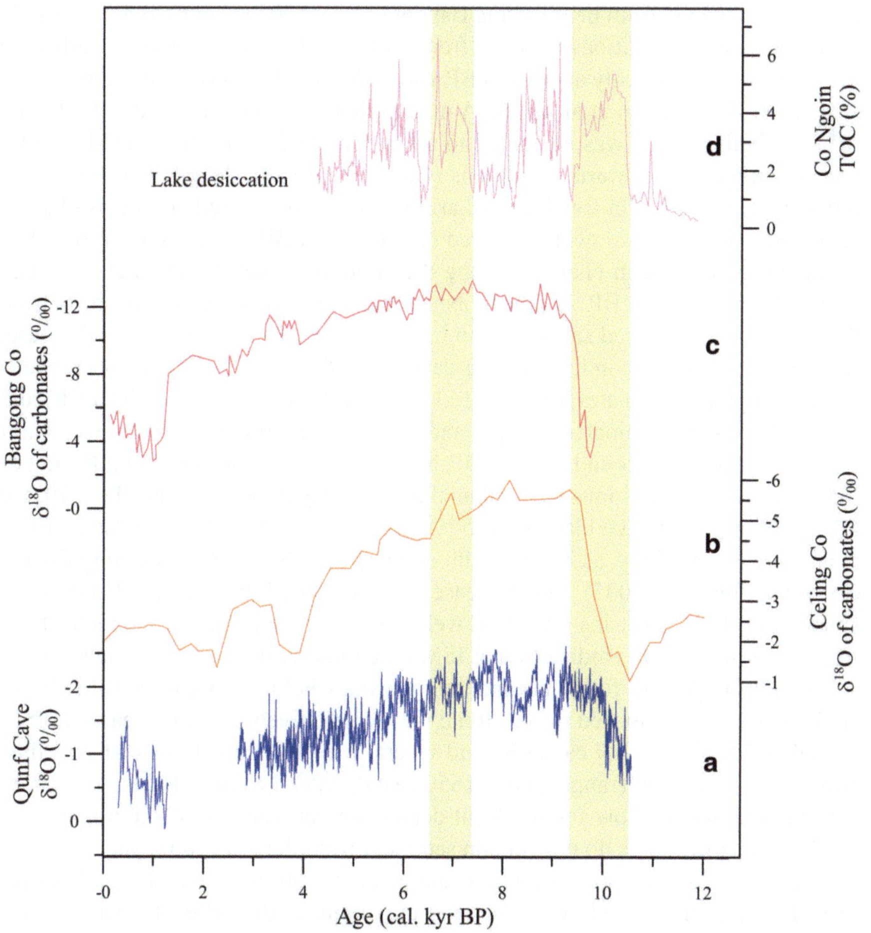

Fig. 3.16 Correlation of the ISM records from lacustrine sediments with (a) stalagmite $\delta^{18}O$ from Qunf Cave (southern Oman) (Fleitmann et al. 2003). The data of from (b) Celing Co (Gu et al. 1993), (c) Bangong Co (Gasse et al. 1996), and (d) Co Ngoin (Jin et al. 2005). Yellow shaded bars mark the intervals of strong ISM during the Holocene

that the driest period occurred approximately 13 to 12.5 ka BP, followed by a two step increase in ISM strength in the early Holocene. Peak wetness at the Tibetan sites generally occurred between 10 and 5 ka BP, after which it became gradually drier, punctuated by centuries long dry events (van Campo and Gasse 1993; Gasse and Van Campo 1994; Wang et al. 2002; Jin et al. 2005). For example, a 13 ka record from Sumxi Co (western Tibet) showed that conditions in the early-middle Holocene were warmer and wetter than present (Gasse et al. 1991). Importantly, they found evidence for warm, humid pulses at ~ 12.5 and ~ 10.0 ka BP and a return to cold, dry conditions at ~ 11–10 ka BP, with a 2–3 ka lag behind precessional forcing, none of which can be explained by orbital variations. A remarkably similar scenario was indicated from neighboring Bangong Co, ~120 km south of Sumxi Co, in which wet/warm conditions deduced from pollen and stable isotopes of carbonate were established suddenly at ~9.6 ka BP, and ended at ~1.2 ka BP (Fig. 3.16, Fontes et al. 1996; van Campo et al. 1996). A major environmental change at ~10–9.5 ka BP from both records was attributed to a rapid strengthening of the ISM circulation which led to wet-warm conditions (Gasse et al. 1991, 1996). This event was followed by a long-term trend toward aridity which culminated around 4-3 ka BP, punctuated by a reversal event centered on 8.0-7.7 ka BP. Bangong Co proxy data also indicate short-term changes during the Holocene, such as dry spells at ~6.2, ~3.9-3.2, and ~1.3 ka BP. Another Tibetan lacustrine paleoclimate record is from Seling Co, a large closed saline lake in Central Tibet (Gu et al. 1993). In this case, similar paleoclimatic conditions were derived on the basis of $\delta^{18}O$ measurements from authigenic carbonates, over a ~12 ka period. From ~12.2 to 10.0 ka BP, cold dry conditions predominated. A rapid and large change from dry cold to wet warm conditions occurred around 10.0 ka BP, and was attributed to the strengthening of ISM circulation. Conditions wetter than those of today prevailed from about 10.0 to 4.2 ka BP. A progressive increase in $\delta^{18}O$ values from 5.5 to 4.2 ka BP is ascribed to an increase in aridity, associated with cooling, as at Sumxi Co, Bangong Co and Co Ngoin (Jin et al. 2005). The decrease in ISM strength through the Holocene is apparent in other lacustrine records as well. A geochemical study of sediments from Ahung Co (not shown) indicates that ISM precipitation decreased through the late and middle Holocene following maximum values before 7.5 cal ka BP (Morrill et al. 2006). Superimposed on this trend are two relatively rapid decreases in ISM rainfall at 7.0–7.5 and 4.7 cal ka BP and several wet–dry oscillations with a median spacing of ~500 years (range: 200 to 1650 years). After 4.0 cal ka BP, the lake probably became too shallow for sediment deposition, similar to Co Ngoin (Jin et al. 2005), so that the cores do not contain sediment from the last ~4000 years.

The most characteristic features of these records are the sudden establishment of wet conditions around 10 ka BP, and maximum aridity after 4-3 ka BP. This implies that climate evolution at these sites has been driven by the same factors. In all cases, enhanced supplies of rainfall during the early Holocene were attributed to the strengthening of the ISM. After ~6.3 ka BP, a general trend toward aridity was due to a change in the precipitation regime, from monsoon precipitation to convective rains. The reconstructed results from central and western Tibet also agree with climatic reconstructions from northwestern India. Pollen records, with relatively low-resolution, from lake cores from Lundaransar and Didwana Lakes provide histories of the Holocene ISM rainfall in Rajasthan, north India (Bryson and Swain 1981; Swain et al. 1983). The data shows that the ISM was weak or absent in latest glacial time, an enhanced ISM circulation during the early Holocene, and a decrease to the present. The records indicate that effective summer precipitation during the early Holocene was considerably higher than modern values. Some shorter term variations were explained in terms of the behavior of the present day interannual variability. There is thus good consistency in the evolution of sites influenced by the ISM. However, the short-term dry periods, in phase with rapid climatic changes observed at northwestern India, cannot be accounted for by the smooth variations of orbital parameters.

These records indicate that it is still important to recognize the central role of glacial boundary conditions in modifying the response of the ISM to astronomical forcing. At the glacial-interglacial timescale, the ISM variability most likely responds to mechanisms modulating the strengths of both the Indian low and the Mascarene high. During interglacials ISM maxima are induced by an enhanced Indian low associated with global ice volume minima (northern thermal pull), whereas during glacials the lead ISM minima are strongly modulated by an intensified Mascarene high associated with Southern Hemisphere high latitude cooling (southern pressure push) (An et al. 2011). For nonlinear periods superimposed on an otherwise linear trend, Overpeck et al. (1996) hypothesized that ISM strength slowly increased in response to insolational forcing until the declining influence of glacial boundary conditions triggered strong summer warming on the TP generating the pressure gradient needed for a strong ISM.

After the early Holocene the glaciers retreated and millennial-scale variations in the ISM followed a more or less linear response to insolational forcing. This accounts for the gradually weakening monsoon observed after 5 ka BP. Available data suggest that the TP may have warmed abruptly at the same time that the ISM increased in strength, particularly between 10 and 9.5 ka BP, and that these abrupt changes were coincident with rapid decreases in snow and glacial coverage (Fang 1991; Gasse et al. 1991; Gasse and Van Campo 1994; Wang et al. 2002). This is a response to maximum summer (July) solar radiation in the Northern Hemisphere, resulting in an amplified seasonal contrast expressed in the summer monsoon index (An et al. 2000). The times of the Holocene optimum in ISM and East Asian summer monsoon are asynchronous, which is a response to the Earth's changing solar radiation and glacial boundary conditions (An et al. 2000). The appearance of a precipitation peak in southwestern China ca. 12.0 cal ka BP was probably associated with the tropical ISM which peaked ca. 12.0 cal ka BP, some 3000 years before the East Asian monsoon maximum. Possibly the remnant NH ice sheets delayed the development of the East Asian monsoon, but had less influence on the ISM. Therefore, more well-dated paleoclimatic records from the ISM-affected region are needed to fully test this hypothesis. Growing attention has also focused on the abrupt changes in monsoon strength that apparently took place during the Holocene, in the absence of significant changes in glacial boundary conditions (Gasse and Van Campo 1994).

3.3.4 Non-monsoonal Climatic Variability Dominated by the Westerlies

Limited lacustrine records are available for non-monsoonal regions of arid and semi-arid central Asia, where climate is dominated by the westerlies, and Holocene climate change remains controversial. Earlier studies indicated that Holocene westerly climate change at Xinjiang were similar to those for east China. Examples in-

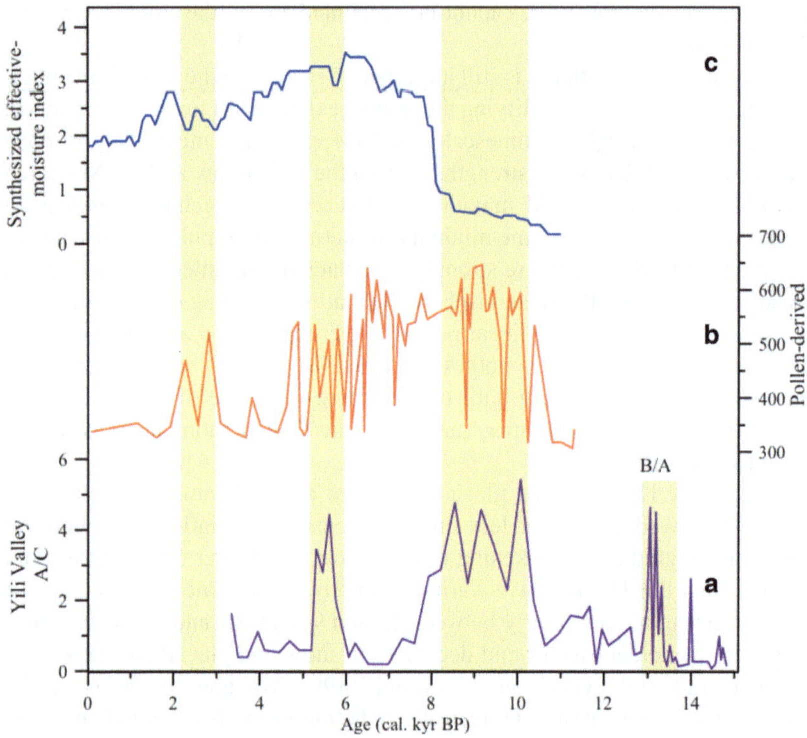

Fig. 3.17 Correlation of typical lacustrine records from central Asia with synthesized effective-moisture index, both reflecting the variation of the Westerlies climate. The data are (a) *Artemisia*/Chenopodiaceae (A/C) ratio from Yili Valley (Li et al. 2011a), (b) pollen-derived annual precipitation from Hoton-Nur (Rudaya et al. 2009), and (c) synthesized effective-moisture index (Chen et al. 2008a). Yellow shaded bars mark the intervals of relatively strong Westerlies reflected by lacustrine records since the Late Glacial, which is not consistent with the synthesized effective-moisture index. B/A, Bølling-Allerød

clude the carbonate oxygen isotope study and pollen records from Manas Lake (Sun et al. 1994; Rhodes et al. 1996; Wei and Gasse 1999), as well as pollen records from Lake Tashkol and Kendegelukol in the Altay Mountains (Blyakharchuk et al. 2004; Fig. 3.7). These results have been related to the moisture-bearing summer monsoon which extended its influence westward to Xinjiang and northward to northern Mongolia during the early mid-Holocene. However, in a recent review by Chen et al. (2008a), the authors dismissed the interpretation of a traditional insolation forcing mechanism after an evaluation of the Holocene westerlies and moisture history in arid central Asia. They suggested that moisture changes in arid central Asia during the Holocene were synchronous, including a dry early Holocene (11–8.0 ka BP), a wetter (less dry) early to mid-Holocene (8.0–4.0 ka BP), and a moderately wet late Holocene (since 4.0 ka BP) (Fig. 3.17c), but the effective-moisture history in arid central Asia was out-of-phase with that sequence in monsoonal Asia. Furthermore, recent studies also indicate that the climate was dry during the late glacial and early

Holocene period, while the mid-Holocene was humid and the late Holocene was relatively dry, suggesting that climate was mainly controlled by the Westerlies (Jiang et al. 2007; Liu et al. 2008a; Huang et al. 2009; Tao et al. 2010).

A high-resolution fossil pollen record from sedimentary cores at Barkol Lake, eastern Xinjiang (Tao et al. 2010) indicates that the site was dominated by desert since 16.7 cal ka BP. During the late glacial and early Holocene the climate was arid and during the mid-late Holocene it was relatively wet at Barkol Lake. The desert had extremely arid climate and lower effective moisture in the period 16.7–7.9 cal ka BP, especially from 16.7 to 8.9 cal ka BP when the lake may have dried up completely. This was followed by a climatic optimum from 7.9 to 4.3 cal ka BP, when the effective moisture increased. After that, a short but extremely arid climatic event occurred between 4.3 and 3.8 cal ka BP. From 3.8 to 0.53 cal ka BP moisture again increased, and for the last 0.53 cal ka the climate deteriorated. A more detailed Holocene paleoclimatic history was reconstructed from a 2.50 m sediment section also from Barkol Lake (Zhong et al. 2010). Holocene climate was reconstructed based on sedimentology, mineralogy and stable isotopes of authigenic phases, indicating the following sequence of conditions: cold and dry (9.4-7.4 cal ka BP), wet-warm (7.4-5.9 cal ka BP), dry-temperate (5.9-3.1 cal ka BP), wet-cold (3.1-1.1 cal ka BP), dry-warm (1.1-0.5 cal ka BP) and dry (0.5 cal ka BP to the present). A similar Holocene climatic pattern was established for Ulungur Lake on the basis of multiproxy data (pollen assemblages, $CaCO_3$, TOC, TN, $\delta^{13}C_{org}$, and ostracod $\delta^{13}C$ and $\delta^{18}O$) from a 2.25 m sediment core. Notable climatic fluctuations included a temperate and dry period from 10.0 to 7.6 cal. ka BP, a warm and wet period from 7.6 to 5.3 cal. ka BP, warm and moist between 5.3 and 3.6 cal. ka BP, temperate and dry from 3.6 to 2.1 cal. ka BP, temperate and moist between 2.1 and 1.3 cal. ka BP, and cool and dry after 1.3 cal. ka BP (Jiang et al. 2007; Liu et al. 2008a). Variations in carbonate content indicate that temperature was generally high between 9.6 and 7.7 cal ka BP, low between 7.7 and 6.7 cal ka BP, intermediate between 6.7 and 0.56 cal ka BP, and low during the last 560 years (Liu et al. 2008a).

The interpretation of cold and dry climate during the early Holocene is supported by ice and other lacustrine core records in central Asia during this period, such as Guliya ice core, and sediment cores from Aibi and Manas Lakes in China, Lake Issyk-Kul of Kyrgyzstan, and the Valley of Lakes in Mongolia. Gun Nuur in northern Mongolia was completely dessicated during the early Holocene (Wang et al. 2004b). In Guliya ice core on the TP two cold events occurred between 9.6 and 9.2 ka BP and between 8.4 and 8.0 ka BP. The latter period possibly reached a temperature decrease of 10 °C (Wang et al. 2002). Lake studies indicate low lake levels during the early Holocene, with cold (or cool) and dry conditions. Meanwhile, studies of the Aral Sea and the Caspian Sea to the west also report low lake levels during the early Holocene, when compared to the present (Boomer et al. 2000; Kazanci et al. 2004). Arid climate and/or low lake levels may also have occurred in central Asia during the late glacial and early Holocene, implying a different pattern of Holocene environmental change than observed in the Asian monsoonal area. The largest inland fresh water lake in China, Bosten Lake also experienced dry conditions during the late glacial and early Holocene (Huang et al. 2009). Within the early Holocene, an abrupt cold-wet event between 8.2 and 7.6 cal. ka BP was accompanied by a peat layer, reduced weathering, low $\delta^{13}C$ and $\delta^{18}O$ of authigenic carbonate within the Aibi Lake catchment (Jin et al. 2004; Wu et al. 2005), generally in agreement with the widespread cold event of 8.2 cal. ka BP. However, a much different picture of effective moisture changes for the last 15 ka in central Asia was reconstructed on the basis of palynological investigations of the continuous lacustrine sediments in Yili Valley, Xinjiang (Li et al. 2011a). The Bølling-Allerød (15-12.9 cal ka BP) was relatively warm and humid, with some fluctuations. Then, the increasing trend of effective moisture was interrupted by the cold and relatively dry Younger Dryas. A sharp increase in effective moisture marked the beginning of the Holocene in the Yili Valley. The early Holocene (10.6-7.6 cal ka BP) was the wettest time, allowing a temperate steppe to develop (Fig. 3.17a).

Warm and wet/moist conditions during the middle Holocene are observed in records from Manas (Rhodes et al. 1996), Aibi (Wu et al. 2005), Bosten (Huang et al. 2009) and Barkol Lakes (see Table 3.1, Zhong et al. 2010). Although a dry climate with desert vegetation arose in the early mid-Holocene (7.6-6.5 cal ka BP), spanning 1100 years, a humid phase emerged between 6.5 and 5.2 cal ka BP (Li et al. 2011a). Proxy data from Eastern Juyan (Alashan Plateau) also indicate that a period of favorable climate conditions occurred between 5.4 and 1.7 cal ka BP except for a short-term dry episode between 3.2 and 3.0 cal ka BP (Mischke et al. 2002). High available moisture with cooler temperature in Central Asia was most likely caused by increased cyclonic westerlies from the North Atlantic (Magny et al. 2003; Huang et al. 2009).

Different climatic conditions for the late Holocene were suggested by lacustrine records from arid central Asia, although an evaluation of spatial and temporal patterns of effective moisture variations in arid central Asia shows a relatively dry Medieval Period, a wet Little Ice Age and increasing moisture during recent decades (Chen et al. 2010b). For example, the data of pollen, $\delta^{13}C$ and $\delta^{18}O$ of carbonate and TOC, sedimentary facies, loss on ignition, magnetic susceptibility and grain size from three plain lakes (Sichang, Aibi, and Dongdaohaizi Lakes) in northern Xijiang show a pattern of temperature and moisture dominated by cold-wet and warm-dry conditions during the late Holocene (Wu et al. 2004a; Feng et al. 2006). Wei et al. (1999) assumed that the differences of the climatic changes after 3.5 cal ka BP possibly resulted from re-domination by the Westerly jet and the retreat of the Asian monsoon in the northern Xinjiang. Furthermore, multi-proxy analysis (grain size, microfossils, plant seeds, and elements) of sediments in the central West Lake of Lop Nur indicated favorable environmental conditions resulting from increased rainfall in Medieval times (Ma et al. 2008). However, a multiproxy record of a 4.24 m-long lake core from East Juyanhai Lake in western Inner Mongolia shows that the climate was relatively dry from ~AD 800–1100, with the ensuing humid environment at the end of this period gradually changing to extremely dry (AD 1100–1550) at the Little Ice Age maximum (Chen et al. 2010a).

Although lacustrine records with various proxies from the Westerlies-dominated region are limited, the climatic conditions are quite different from those of the Asian monsoonal region. Especially, dry climates during the early Holocene in the regions of central Asia are indicated by $\delta^{18}O_{carbonate}$ records of the lakes, which cannot be explained in terms of the invasion of the Asian summer monsoon into the interior. However, Rudaya et al. (2009) attributed a noticeable increase in precipitation and lake levels between 11 and 8 ka BP derived from pollen data of Hoton-Nur Lake sediment in Mongolian Altai to strengthening of the summer monsoon (Fig. 3.17b); while Mischke and Wünnemann (2006) attributed the higher moisture availability in the early mid-Holocene at Bosten Lake to a strong climatic influence of the ISM and its interplay with westerly winds. On the other hand, the moisture changes in Yili Valley were thought to be mainly influenced by the North Atlantic Ocean SSTs through the Westerlies. The mean position of the Siberian High Pressure cell probably made a great contribution to the drought between 7.6 and 6.5 cal ka BP. Therefore, we do not think that a simple pattern of the climatic evolution could effectively explain the complexity of the Holocene hydro-thermal combination patterns (such as cold and dry, cold and wet, warm and dry and warm and wet) in central Asia. Well-dated paleoclimatic records from central Asia are needed to better understand the controlling mechanisms of regional variability and to further resolve the controversies concerning the timing, moisture source and driving circulation systems of the Westerlies.

3.4 Peat Records of Asian Monsoon Variability since 20 ka

3.4.1 Peat Localities and Paleoclimatic Implications

Peatlands, the areas of landscape that have a naturally accumulated layer of organic material at the surface (Chambers and Charman 2004), cover around 4×10^6 km^2 of the earth's surface with wide distribution thoughout the world (Charman 2002; Joosten and Clarke 2002; Lappalainen 1996; Lars 1994; Strack 2008). In the high latitude regions of the Northern Hemisphere, low temperature is beneficial to the accumulation of organic material, resulting in extensive areas with peat coverage in North America, North Europe and Russia, for example (Strack 2008).

Peat sequences are good paleoenvironmental archives because they possess several advantages for environmental change research: 1) their accumulation rate is relatively higher than other archives (e.g. loess), 2) they provide higher time resolution than other archives; 3) they contain a variety of potential climate proxies, providing opportunity for multi-proxy climatic reconstructions; 4) they are easily sampled and dated to establish a chronological framework (Chambers and Charman 2004).

China is a country with rich peat resources. The total peatlands in China cover an area of about 1×10^6 km^2, with wide distribution from Hainan Province in the south to Heilongjiang Province in the north, and from the region of Aibi Lake in the west to the Sanjiang Plain in the east (Fig. 3.18). There are six regions with abundant peat resources: south China: (I), the area of lower-middle reaches of the Yangtze River, (II), northeast China, (III) southwest China (IV), north China (V), and northwest

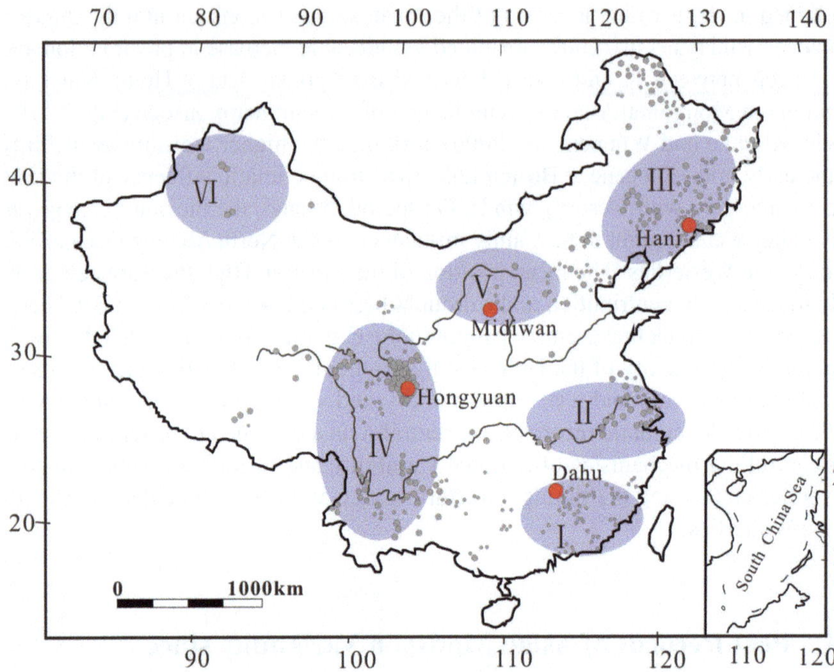

Fig. 3.18 Distribution of peat resources in China. (redrawing after Yin 1999)

China (VI). The largest alpine peatland in China is located in the Zoigê-Hongyuan area in the northeastern TP (Chai 1990; Yin 1999).

Although the first description of peat was in the "Diary of the Travels of Xu Xiake" more than 400 years ago, the study of peat records of paleoclimate in China has lagged until recently, developing fast over the last 20 years (Xie et al. 2000; Zhou et al. 2001, 2002a, b, 2004, 2005, 2007, 2010a, b; Hong et al. 2000, 2001, 2003, 2005, 2009; Hong et al. 2010; Yu et al. 2006b, 2011; Wang et al. 2004c; Zheng et al. 2009a, b, 2011a, b; Zhu et al. 2009; Ma et al. 2009; Yu and Liu 2010; Zhong et al. 2010a, b). Paleoclimate reconstructions have focused on the monsoonal region. Peat sequences from Northeast China and Southwest China preserve continuous records of the history and variability of monsoonal climate in these two regions for the entire Holocene (Zhou et al. 2002a, 2010a, b; Hong et al. 2000, 2001, 2003, 2005, 2009; Hong et al. 2010; Yu et al. 2006b, 2011; Wang et al. 2004c; Xu et al. 2002, 2006; Zheng et al. 2009a, 2011a; Seki et al. 2009; Yamamoto et al. 2010a, b), and records of the relationship between human activities and natural process (Hong et al. 2000; Yu et al. 2010a). Many sections from south China and the desert-loess transition zone are mainly composed of sand and interbeded peat, that record information from multiple climatic events (Zhou et al. 1997, 1999, 2002a, 2004, 2005, 2007; Zheng et al. 2009a, b; Zhong et al. 2010a, b).

3.4.2 Monsoon Variability of the Last Deglaciation Recorded in Peat Deposits of the Semi-Arid Zone

As discussed in Chap. 1, the semi-arid zone, or desert-loess transition zone, lies between the isohyet levels of 200 and 400 mm. It is a climatically sensitive zone that responds to temporal advances and retreats of the monsoon front. Water in this region is a critical ecological factor. In periods with enhanced precipitation, the desert retreats to the north, and paleosols or peat layers form. In times of reduced precipitation, the desert advances southward, and loess or sand layers form in the semi-arid zone. Thus, alternating peat and sand record climate changes related to the changing position of the boundary between the semi-arid zone and the arid zone.

An et al. (1993) first presented evidence for an episode of intensified summer monsoon in central China during the Younger Dryas (YD) using paleoclimatic proxies from the Baxie loess section. This work was subsequently reinforced by the well-dated Midiwan peat-sand sequence from the semi-arid zone (Zhou et al. 1996). The YD event in the Midiwan profile begins with the Heiheze silt, which is followed by silty peat, and then eolain sand (Fig. 3.19). Variations of organic carbon isotopes show climatic shifts from cold-dry (11,200-10,600 ^{14}C yr before present (BP)) to cool-humid (10,600-10,200 ^{14}C yr BP) and back to cold-dry (10,200-10,000 ^{14}C yr BP). The dry-humid-dry climate pattern during the YD period can be found in many records from monsoonal China; for example: Jingbian, Shenmu and Dongxiang (Zhou et al. 2001), and even in peat records from northeast China (Zhou et al. 2010b; Hong et al. 2010). During the middle period of the YD, the summer monsoon was intensified as observed in both the Midiwan record and Hulu Cave speleothem record (Wang et al. 2001). This increase was a response to elevated sea surface temperatures (SST) in the Southwest Pacific (Fig. 3.19). This kind of spatial pattern during the YD period is consistent with a sustained influence from El Niño Southern Oscillation (ENSO), which is to say that palaeo-ENSO events might have affected the spatial pattern of Asian monsoons during the YD, as a dynamic link between the ocean, land and atmosphere. The interaction of cold air from high latitudes and moist air from the low latitude ocean, over the semi-arid zone is likely to produce enhanced precipitation in this region (Zhou et al. 2001).

3.4.3 Holocene Monsoon Variability Recorded in East Asian Peat Deposits

3.4.3.1 Comparison of Peat Records with North Atlantic Ice Rafted Detritus Records

The key to correlating paleoclimatic archives is a reliable chronological framework. Pretreatment protocols on different deposits from the Chinese Loess Plateau (LP) and adjacent regions for radiocarbon dating were established by Weijian Zhou and her research group in State Key Laboratory of Loess and Quaternary Geology

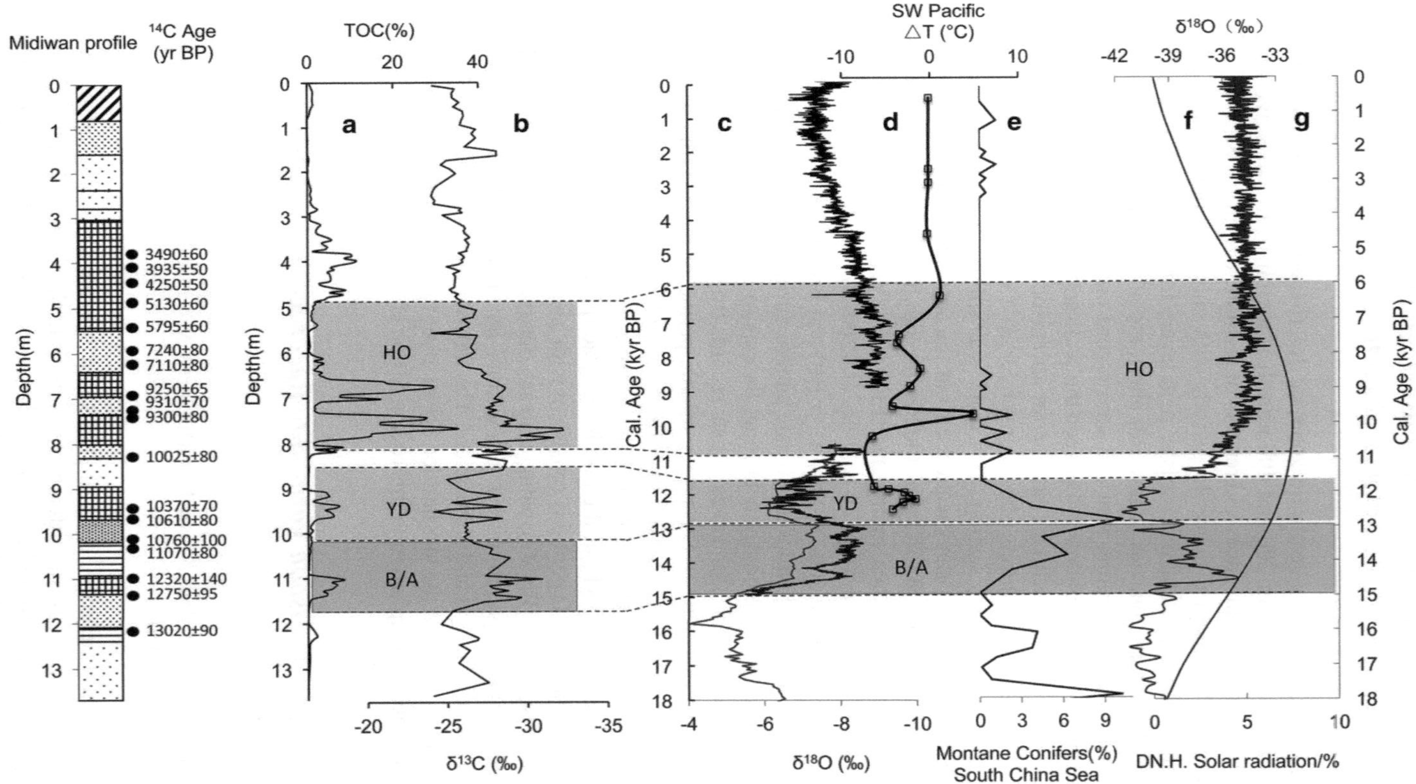

Fig. 3.19 Comparison of peat climatic record in Midiwan with other archives from different regions. a and b are the total organic carbon (TOC %) and organic carbon isotope values for the Midiwan section, respectively. c is the speleothem oxygen isotope records from Hulu Cave (Wang et al. 2001) and Dongge Cave (Wang et al. 2005a). d is the a sea surface temperature record from the southwest Pacific (Gagan et al. 2000). e is the percentage of montane conifer pollen in a marine record from the South China Sea (Sun and Li 1999). f is a GISP2 ice core oxygen isotope record (Stuiver et al. 1995). g is the variation of solar radiation in the Northern Hemisphere. (Berger 1978)

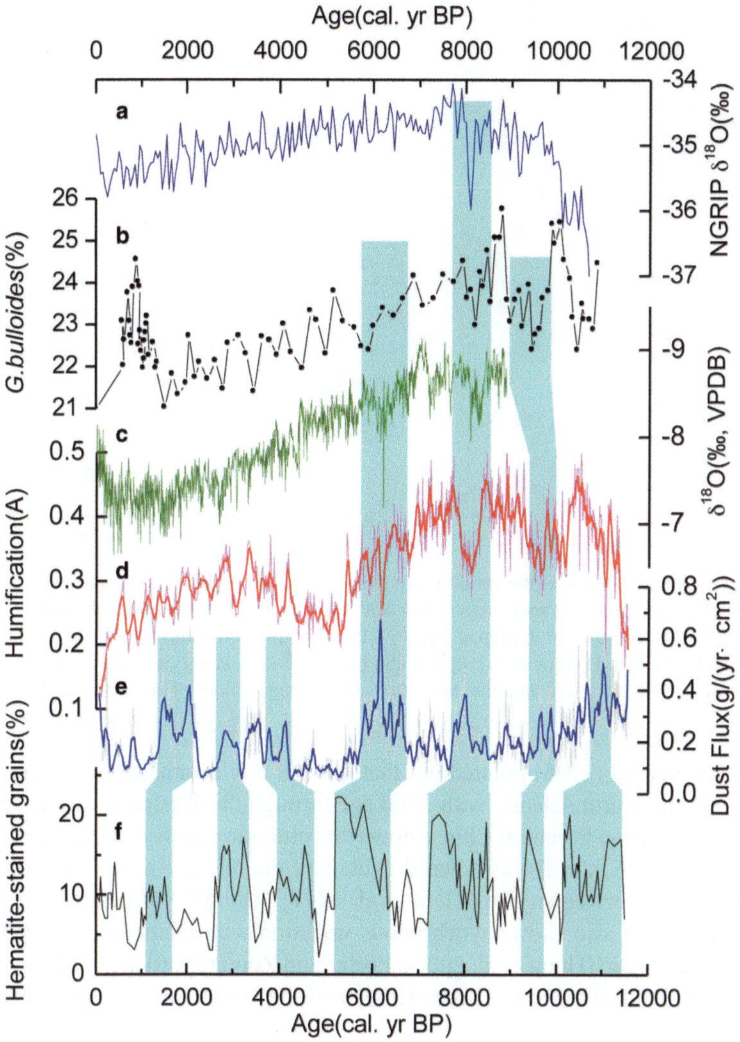

Fig. 3.20 A comparison of monsoon proxies from Hongyuan peat with other records. a) is the $\delta^{18}O$ of the NGRIP ice core (NGRIP Members 2004); b) is the percentage of *G. bulloides* in the Indian Ocean (Gupta et al. 2003); c) is the $\delta^{18}O$ of Dongge Cave (Wang et al. 2005a); d) and e) are humification index and dust flux in Hongyuan peat (Yu et al. 2011); and f) is the percent hematite stained grains in core MC52-VM29-191 from the North Atlantic (Bond et al. 2001)

(SKLLQG) to obtain reliable dating results (Head et al. 1989; Zhou et al. 1991a, b, 1992, 1994a, 1996). The comparison of well-dated peat records from China with other paleoclimate archives have allowed for close comparisons necessary to understand past climate dynamics.

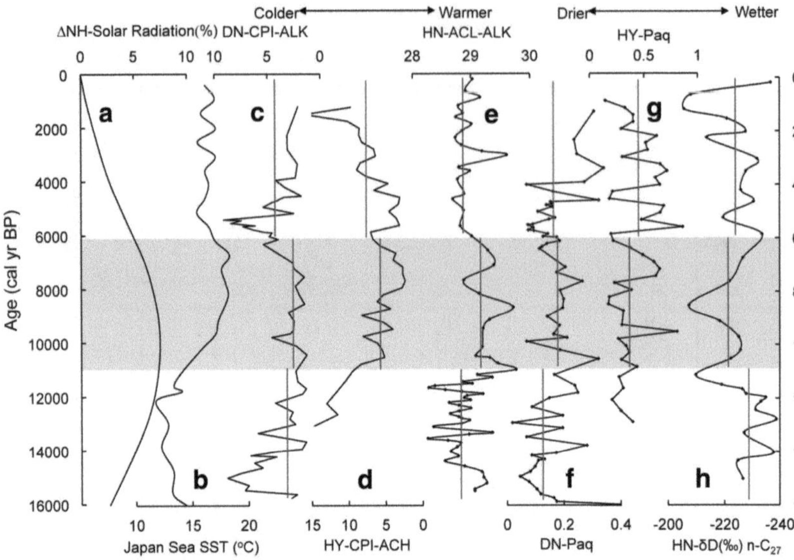

Fig. 3.21 Spatial patterns of the variations in temperature and precipitation since 16 ka as inferred by biomarkers in three peat sequences (Zhou et al. 2010b). a) solar radiation in the Northern Hemisphere, b) SST record from the Japan Sea, c) CPI-ALK in Dahu peat, d) CPI-ACH in Hongyuan peat, e) CPI-ALK in Hani peat, f) Paq in Dahu peat, g) Paq in Hongyuan peat, and h) δD of C27 n-alkanes in Hani peat. Red and blue lines show the average values of proxies in each stage

Porter and An (1995) pointed out that there is a teleconnection between LP records and the North Atlantic with a linkage through the Westerlies. Evidence from the semi-arid zone in central China shows that climatic events in this region, including the Bølling-Older Dryas-Allerød cycle and the YD event, are consistent with events in the Norwegian Sea (Zhou et al. 1997). The $\delta^{18}O$ of peat cellulose from northeast China also shows synchronous variation with temperature records from the North Atlantic (Hong et al. 2009). Porter and Zhou (2006) also pointed out that the climate teleconnection between China and the North Atlantic is also present in Holocene records.

The total organic carbon (TOC) and grayscale records (Zhou et al. 2002a) from the well-dated Hongyuan peat sequence can be placed into a one-to-one correspondence with ice rafted detritus (IRD) records from the North Atlantic Ocean (Bond et al. 2001). A power spectrum analysis of the Hongyuan TOC sequence yields a 1463-yr periodicity that corresponds to the 1450 year Bond cycle, strengthening the view that climate in central China shares a teleconnection with climate in the North Atlantic. Other proxies from the same section, such as carbon isotopes (Hong et al. 2003), humification (Wang et al. 2004c; Yu et al. 2006b) and dust flux (Yu et al. 2011) also demonstrate a tight correlatioin with IRD events (Fig. 3.20). The IRD events correspond to times of a weak Indian Monsoon (Hong et al. 2003) and a relatively strong winter monsoon (Yu et al. 2011).

3.4.3.2 Implications of the Spatial Variability of Precipitation and Temperature Recorded in Peat Deposits

Hongyuan swamp, Hani swamp and Dahu swamp are located in Southwest China, Northeast China and South China respectively (Fig. 3.18), providing an opportunity to study the spatial pattern of environmental change.

The n-alkanes carbon preference index (CPI-ALK), n-alkanol carbon preference index (CPI-ACH), and n-alkane average chain length index (ACL-ALK), are proxies for temperature variations, with lower CPI (both ALK and ACH) and higher ACL indicating higher temperatures. The CPI-ALK in the Dahu peat sequence, CPI-ACH in the Hongyuan peat sequence, and the ACL-ALK in the Hani peat sequence (Fig. 3.21), show that spatial temperature trends are generally in good agreement with one other (Zhou et al. 2010b).

Another index, P_{aq} is defined according to the n-alkane distribution difference between subaerial vascular plants and submerged and floating macrophytes (Ficken et al. 2000; Nichols et al. 2006) and is a good proxy for reconstructing moisture availability in wetlands. Variations of P_{aq} in the above three peat sequences shows temporal asynchronicity between the three regions, especially during the Holocene Optimum (Zhou et al. 2010b). The variation of P_{aq} in the Hani peat, is generally consistent with the variation of C27 n-alkane δD values (Seki et al. 2009). It shows wetter conditions before the Holocene and a relative drier Holocene Optimum, implying that the Hani mire region deviated markedly from the general picture in most of monsoon-influenced Asia (Fig. 3.21). However, an SST record from the Japan Sea suggests that a regional effect may influence the effective moisture in the Hani region by enhancing regional monsoon intensity, a process that depends on the temperature gradient between the land and the Japan Sea.

3.4.4 Mid-Holocene Climatic Transition Recorded in Peat Deposits of the Northeastern Margin of the Tibetan Plateau

The Hongyuan swamp is located on the eastern margin of the TP, where four climate systems converge: 1) southeast monsoon, 2) Indian monsoon, 3) winter monsoon and 4) westerly winds. To understand the climatic dynamics at this site, one must know the relationship between the different climate systems. Combined analyses of pollen concentration (Zhou et al. 2010a), isotope ratios (Hong et al. 2003, 2005; Xu et al. 2002), total organic carbon (TOC), grayscale (Zhou et al. 2002a), humification (Wang et al. 2004c), and biomarkers (Zhou et al. 2010b; Zheng et al. 2011b) of well-dated peat sequences can be used to understand these relationships. However, most of these studies are chiefly concerned with summer monsoon variations and future studies should consider also the influence of winter monsoon and westerly wind variability.

3.4.4.1 Peat Records of Winter Monsoon Intensity Variation during the Holocene

Grain size analyses (Yu et al. 2006b) and surface textures of particles in the peat (Yu and Liu 2010) provide evidence of an eolian origin in the Hongyuan swamp. The dust flux can be used to reconstruct winter monsoon intensity (or westerly wind strength). The proxy shows that the variation of the winter monsoon may be divided into four stages throughout the Holocene (Fig. 3.20e): 1) from 11 to 8.2 cal. ka BP, is characterized by a progressively decreasing winter monsoon; 2) from 8.2 to 5.5 cal. ka BP, is characterized by an increasing winter monsoon with two periods of abrupt strengthening at about 8 cal. ka BP and 6 cal. ka BP; 3) from 5.5 to 4.2 cal. ka BP, winter monsoon intensity reached its lowest level of the Holocene with very low-amplitude fluctuations; and 4) from 4.2 cal. ka BP to present, the winter monsoon intensity increased and exhibited large-scale fluctuations. Seven strong winter monsoon events can be identified in the sequences of dust flux since 11 cal. ka BP, which are generally correlated to the cold events in the North Atlantic Ocean (Bond et al. 1997, 2001; Fig. 3.20f.

3.4.4.2 Changes of Winter and Summer Monsoon Signals in the Mid-Holocene

The summer monsoon proxy and winter monsoon proxy from the same peat core in the Hongyuan swamp (Fig. 3.20d and e) provide an opportunity to investigate the phase relationships between these two monsoon systems. All evidence shows that ~5.5 cal ka can be regarded as the turning point for the Mid-Holocene climatic transition. The general trends of these two proxies on a millennial scale are clearly different before and after 5.5 cal ka BP. Before 5.5 cal. ka BP, two sequences follow a reciprocal pattern where an increasing summer monsoon is observed to correspond to a decreasing winter monsoon (westerly wind), and vice versa. After 5.5 cal. ka BP, the pattern is obviously different with the summer and winter monsoons changing synchronously. Power spectral analyses on these two sequences also shows the difference between the two climatic systems. The dominant periods of the winter monsoon (westerly wind) shifted from a higher frequency before 5.5 cal. ka BP to a lower frequency after 5.5 cal. ka BP. In contrast, the dominant periods of summer monsoon shifted from a lower frequency before 5.5 cal. ka BP to a higher frequency after 5.5 cal. ka BP (Yu et al. 2011).

During the warm epoch, which corresponds to high solar irradiation (Berger 1978), the study area was perched on the southern margin of the winter monsoon regime, and frequent variations in the reconstructed winter monsoon sequence may indicate the advance and retreat of the winter monsoon, which corresponds to solar variations on a centennial timescale. The relatively small swing of the winter monsoon before 5.5 cal. ka BP also indicates that the study area was not close to the center of the winter monsoon regime. During cold epochs, the study area was perched on the northern margin of the summer monsoon regime, so the advance and retreat of the summer monsoon, which corresponds to solar variations at a centennial scale, had a large influence in this area. This results in frequent variations and small swings of the summer monsoon after 5.5 cal. ka BP.

The rate of changes in solar insolation which peaked at about 5.5-6.0 cal ka BP matches well with the Mid-Holocene climatic transition, suggesting that this transition might have resulted from orbital forcing (Yu et al. 2011).

3.4.5 Human Activities Recorded by Peat Deposits

In the period from 5.5 to 4.2 cal. ka BP, Chinese Bronze Age cultures were developing from Neolithic Cultures in Northwest China. Climate in the west of China during this interval was characterized by both attenuated winter and summer monsoons, which produced a stable climate (Yu et al. 2011). This kind of climate in North China might have catalyzed the emergence of Chinese Bronze Age cultures (Yu et al. 2010a). However, at about 4.3 cal ka BP, corresponding to IRD event 3 (Bond et al. 2001), $\delta^{13}C$ records of peat cellulose from the Hongyuan and Hani sites indicate an abrupt increase in mositure for the Indian monsoon region and an abrupt decrease in moisture in Northeast China (Hong et al. 1997, 2003), resulting in floods and severe droughts in the two regions. During this time, many ancient cultures collapsed in West Asia, India, and China, implying that climate conditions directly affected the development of ancient cultures at that time (Wu and Liu 2004).

Ten heavy metallic elements (Pb, Zn, Fe, Mn, Cu, Co, Ni, Ti, V, and Al) in the Hongyuan peat were analyzed to reconstruct wind intensity of the winter monsoon and to extract possible information about atmospheric change relating to human activity in the windward region (Yu and Liu 2010). The results show that variations in the concentration of ten elements are concordant with dust input variations before 6 cal ka BP. However, the abundances of Cu, Pb and Zn are found to be significantly above background levels and follow distinctly different trends compared to other elements, for some stages after 6 cal ka BP, possibly indicating an additional source for those elements. The lithology of the whole core is almost the same, ruling out the possibility of a change of proximal input. It is noteworthy that the elemental differences at the Hongyuan site occurred when the late Neolithic cultures and Bronze Age cultures were prosperous in the windward regions (Shui 2001), for example, the late Neolithic Majiayao culture (6000-4000 a BP), and the Bronze Age Qijia Culture (4200-3800 a BP), Kayue Culture (3600-2800 a BP), Siwa Culture (3300-2500 a BP), Siba Culture (3900-3500 a BP), Shajing Culture (3000-2300 a BP), and Xindian Culture (3600-2800 a BP). Accurate correlation between the elemental anomalies in the Hongyuan peat and cultures in the windward area, based on independent time scales, shows that the peaks of Pb and Zn are responding to the emergence of these cultures.

Four elements (Ti, V, Co and Ni) that have no relation to human activity were used to reconstruct the background level of Pb, according to the method of Shotyk

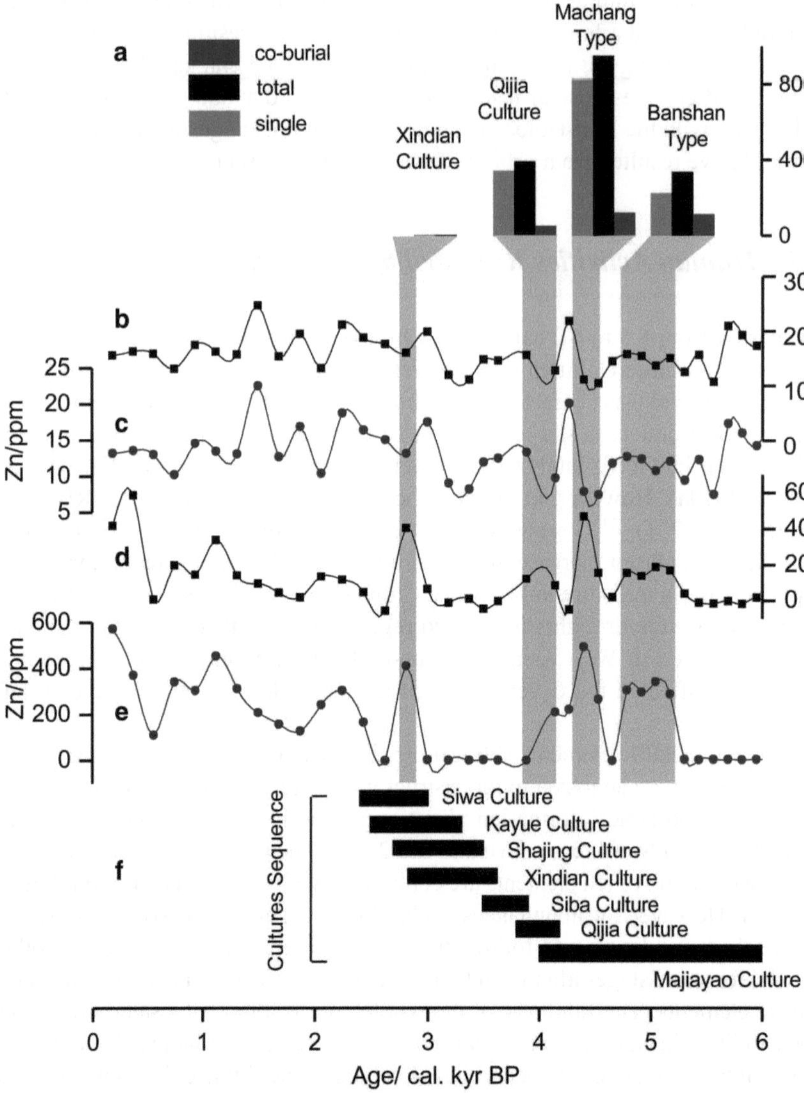

Fig. 3.22 Human-induced elemental content (d for Pb, e for Zn), natural background (b for Pb, c for Zn) as recorded in Hongyuan peat and their comparison with cultural sequences (f, Shui 2001) and population inferred from archaeological excavations at Liuwan Tombs. (a, Shui 2001)

et al. (2001). The reconstructed background level of Pb is around 20 ppm, in agreement with the background in loess (Liu 1985). The human-induced Pb content, which is much higher than its background level at some stages, is well-correlated to inferred population, based on archaeological evidence from Liuwan Tombs (Fig. 3.22). It is interesting that the variations of reconstructed human-induced and

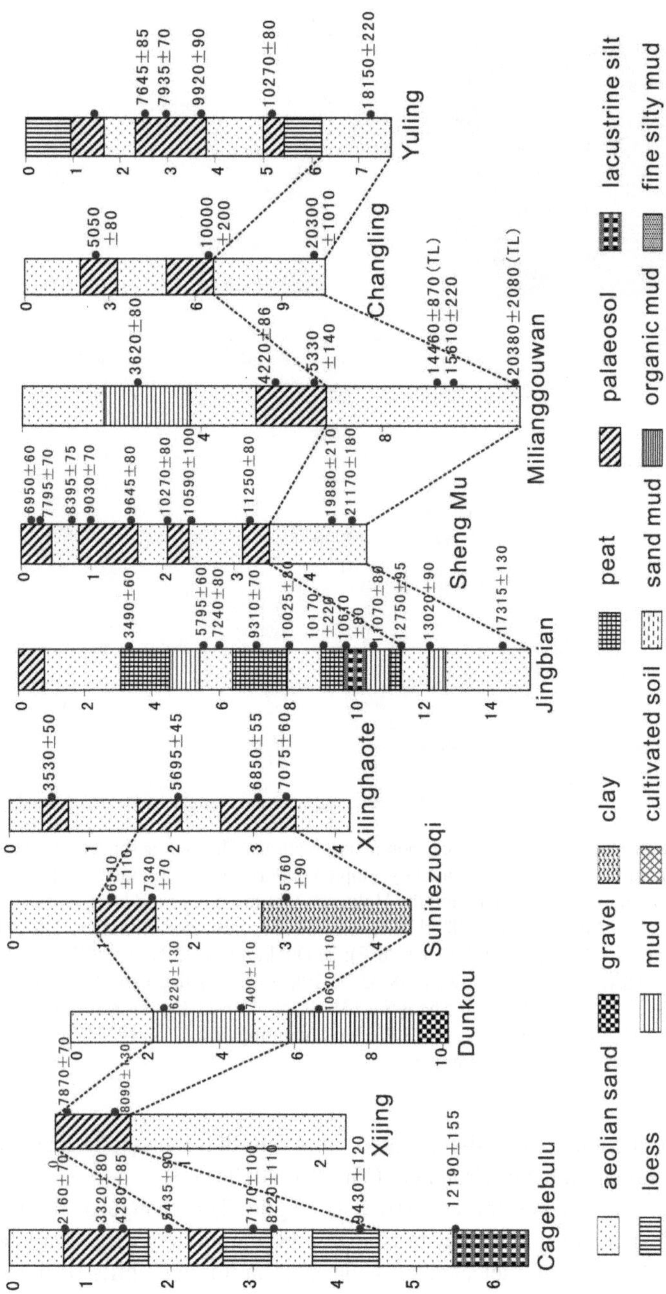

Fig. 3.23 Stratigraphy and radiocarbon dating of 10 sites in the desert-loess transition zone in China

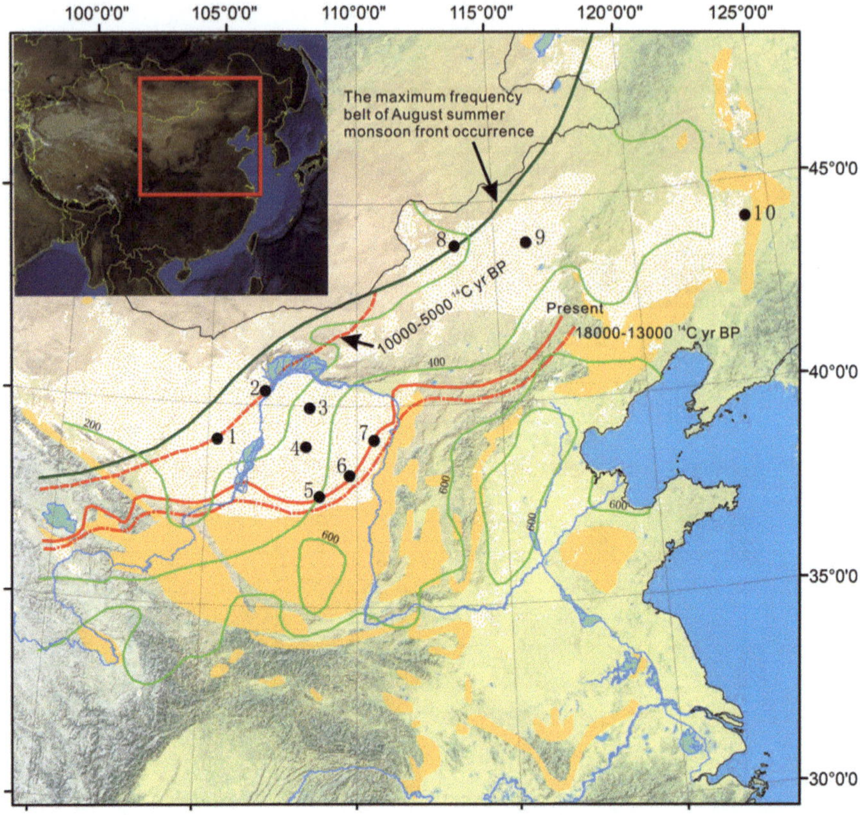

Fig. 3.24 Reconstruction of desert-loess boundary positions. The *yellow* shaded area indicates the Loess Plateau, while deserts are indicated by stippled pattern. Boundary positions (red) during the last glacial maximum (*dashed line*) and the Holocene Optimum (*dashed line*), and their relationship to the present day boundary (38°21`N, *solid line*) are indicated. Numbers 1 to 10 indicate the study locations: 1) Xijing (39°4`N, 105°8`E); 2) Dunkou (40°33`N, 107°0`E); 3) Cagelebulu (39°53`N, 108°18`E); 4) Milanggouwan (38°45`N, 108°35`E); 5) Jingbian (37°39`N, 108°37`E); 6) Yuling (38°21`N, 109°42`E); 7) Shengmu (38°48`N, 110°27`E); 8) Sunitezuoqi (43°14`N, 113°52`E); 9) Xilinghaote (43°56`N, 116°3`E); 10) Changling (44°17`N, 123°59`E). Present mean annual precipitation is contoured in *green*

background elements are anti-correlated, implying that humans at that time were largely dependent on the environment (Yu et al. 2010a).

In the semi-arid zone, the ecosystem is fragile. The change of the margin of the desert responds to both climate change and human activities. Ten sites along the semi-arid zone were selected to reconstruct latitudinal shifts of the southern desert margin over the last 20,000 years (Fig. 3.23). Stratigraphic correlations together with radiocarbon dating show that the southern desert margin during the Last Glacial Maximum (LGM) was very close to its present-day position (Fig. 3.24). The northern boundary of the modern summer monsoon is close to the estimated southern desert margin during the Holocene Optimum (7000-9000 BP), implying that the current southern boundary of the desert should coincide with that of the Holocene Optimum climate, in the absence of external driving mechanisms. But the present-day boundary of the desert is three degrees of latitude north of the Holocene Optimum position (Fig. 3.24). Human activities during the last 3000 years might account for this difference, due to intensified desertification (Zhou et al. 2002b).

3.5 Coral Records

Massive “stony” (scleractinian) corals from modern and fossil reefs of the tropical and subtropical oceans provide an important archive of past climate and ocean variability (Gagan et al. 2000). The aragonite skeletons of reef-building corals carry isotopic and chemical indicators that track water temperature, salinity, sea level, contamination indices, monsoon and, ENSO, and site-specific features including turbidity, runoff, rainfall and, upwelling. As such, coral cores provide a natural archive of tropical paleoclimatic and paleoceanographic conditions on intra-annual-to-centennial timescales. This is especially valuable since multi-century instrumental climate records in the tropics are sparse and century-long sea surface temperature (SST) reconstructions developed from historical records are limited.

3.5.1 Advantages of Coral Studies and Proxy Indices

Coral skeletons serve as excellent natural archives of paleoenvironmental conditions in tropical and subtropical waters. Living colonies can provide continuous environmental records spanning several centuries and, combined with fossil coral, can reveal conditions over recent millennia and even earlier periods.

In the tropical oceans, the isotopic, trace and minor elemental composition of coral skeletons can vary as a result of environmental conditions such as temperature, salinity, cloud cover, river discharge, upwelling and ocean circulation. Massive, symbiotic stony corals act as a tropical proxy record because: 1) annual skeletal banding pattern offers excellent chronological control for living corals, and fossil corals can be precisely dated by the U/Th disequilibrium method (Cobb et al. 2003); 2) modern corals provide continuous climate records, extending several centuries back from the present, and well-preserved fossil corals provide information on climate variability during time-windows throughout the late Quaternary (Klein et al. 1990); 3) skeletons incorporate a variety of isotopic and elemental tracers used for paleoclimatic and paleoceanographic reconstruction, such as SST, sea surface salinity (SSS), precipitation, runoff and other climatic and oceanic features (Gagan et al. 2000); 4) their high growth rate ranging usually from 6 to 20 mm/year allows for sub-seasonal/monthly sampling resolution, 5) they are widely distributed throughout the tropical ocean. Thus, proxy records in corals provide an ad-

Table 3.2 Environmental variables that can be extracted from corals

Proxy	Environmental variable
Skeleton	
Skeletal growth bands	Sea surface temperature, light, stress, water motion, sedimentation
Luminescence	River outflow, ocean productivity
Isotopes	
$\delta^{18}O$	Sea surface temperature, Sea surface salinity
$\delta^{13}C$	Light, cloud cover
$\delta^{14}C$	Ocean ventilation, water mass circulation, upwelling
$\delta^{11}B$	pH,?CO_2
Trace and minor elements	
Sr/Ca, Mg/Ca, U/Ca	Sea surface temperature
Ba/Ca	Upwelling, river outflow, sediment flux
Mn/Ca, Cd/Ca	Upwelling, contamination

vantageous means of obtaining paleoclimate information in the tropics on seasonal/monthly-to-centennial timescales.

Coral-based paleoclimatology has rapidly developed since annual density band couplets that provide time markers for the development of long chronologies were identified by Knutson et al. (1972). Over the past four decades, a series of coral proxies have been developed and used to reconstruct paleorecords. Based on these coral proxy records, scientists commonly reconstruct SST (Charles et al. 1997; Saenger et al. 2009), sea surface salinity (SSS) and (or) precipitation (LeBec et al. 2000; Iijima et al. 2005; Mertz-Kraus 2009), ocean acidification (Pelejero et al. 2005; Wei et al. 2009; Liu et al. 2009a), sea level (Rohling et al. 2008; Yu et al. 2009; Liu et al. 2011a), ocean mixing and upwelling and Suess effect (Druffel 1997; Yu et al. 2010b), and river discharge histories (Isdale 1984; Peng et al. 2002). These paleoclimatic records are important for the assessment of perturbations to natural climate variability by anthropogenic forcing, for climate predictability and for a better understanding of the dominant modes of the global climate system, e.g., the El Niño-Southern Oscillation (ENSO) phenomenon of tropical Pacific origin, the Asian and African monsoon, the Arctic Oscillation (AO)/North Atlantic Oscillation (NAO), the Pacific Decadal Oscillation (PDO), and the mechanisms of decadal climate variability.

Several environmental variables can be reconstructed by growth banding, isotope ratios, trace and minor elemental composition in coral cores (Table 3.2 and following sections). The width, density, fluorescence and geochemical composition of each growth band are generally thought to reflect the average environmental conditions that prevailed during the time over which that portion of the skeleton was calcified. Skeletal structure, isotopes and trace element ratios can be used to reconstruct past environmental and climatic conditions, as described below.

a. Skeleton Commonly, corals deposit high- and low-density band couplets at annual resolution (Knutson et al. 1972). The density variations result from changes of calcification and/or linear extension rates in the skelenton. The annual extension rates of corals can be inferred from the annual density-band pattern (Barnes and Lough 1989). The calcification rate is the product of the linear extension rate and the average skeleton density which can be measured by gamma densitometry (Chalker and Barnes 1990). In general, as light levels decrease due to increased cloud cover, increased sedimentation or due to increasing depth, maximum linear skeletal extension decreases, calcification decreases, and skeletal density increases.

Fluorescence bands appear when coral is exposed to ultra-violet radiation. These bands are due to terrestrial humic and fulvic acids incorporated into the coral skeleton during high river flow events (Isdale 1984). Thus, fluorescence bands can be used to reconstruct past river runoff or precipitation (Klein et al. 1990; Peng et al. 2002).

b. Isotopes The oxygen isotopic ratio ($\delta^{18}O$) of coral aragonite skeleton is a function of both SST and the $\delta^{18}O$ of the ambient seawater ($\delta^{18}O_{sw}$) or SSS (Weber and Woodhead 1972). As temperature increases, the precipitated coral skeletal $\delta^{18}O$ decreases due to temperature-dependent kinetic fractionation effects. Variation in the $\delta^{18}O_{sw}$ can result from evaporation, precipitation, and/or runoff, i.e., such variations reflect the hydrological balance (Juillet-Leclerc et al. 1997); Besides, $\delta^{18}O_{sw}$ and SSS have a strongly linear relationship that is independent of precipitation and evaporation changes (Morimoto et al. 2002). Corals draw on seawater as their primary source of $\delta^{18}O$ for calcification. As $\delta^{18}O_{sw}$ decreases, coral skeletal $\delta^{18}O$ decreases. Hence, in regions where the natural variation in $\delta^{18}O_{sw}$ (or SSS) is small, changes in coral skeletal $\delta^{18}O$ primarily reflect changes in SST (Linsley et al. 2000; Cole et al. 2000). Based on empirical studies, a 1 °C increase in water temperature corresponds to an average decrease of about 0.22 ‰ in coral $\delta^{18}O$ though that slope can vary significantly from 0.15 to 0.24 ‰ based on coral species and local conditions (Wellington et al. 1996). Conversely, in regions with pronounced variability in rainfall and/or river runoff, coral $\delta^{18}O$ values reflect changes in $\delta^{18}O_{sw}$ (and by extension, SSS or rainfall) (Linsley et al. 1994; Urban et al. 2000; Pfeiffer et al. 2004). In sites where the climatic and environmental setting is more complex, coral $\delta^{18}O$ may reflect a composite signal (Quinn et al. 1996).

The carbon isotopic ratio ($\delta^{13}C$) of coral is strongly influenced by photosynthesis (light) and feeding (which directly affects the respired $\delta^{13}C$). For shallow areas, where coral feeding is relatively constant due to non-upwelling activity, changes in $\delta^{13}C$ are predominantly driven by changes in photosynthesis, providing a record of seasonal changes in cloud cover or turbidity (i.e., rainy season vs. dry season) (McConnaughey 1989a, b). In upwelling regions where feeding opportunities can vary dramatically, changes in $\delta^{13}C$ have been used as an indicator of vertical mixing through changes in coral food sources and in dissolved inorganic-$\delta^{13}C$ (Felis et al. 1998). However, coral $\delta^{13}C$ is so complicated that it has been hampered for wide use as a climate proxy.

Radiocarbon (^{14}C) incorporated in coral skeletons during growth reflects the ^{14}C content ($\Delta^{14}C$) of the DIC of the ambient seawater during skeleton precipitation and provides a useful tracer for ocean circulation and upwelling. The $\Delta^{14}C$ of the surface ocean is controlled by the ^{14}C level of the atmosphere and by mixing with waters which have a different $\Delta^{14}C$ value, such as surface waters from other oceanic source regions and deep water (Rodgers et al. 1997). In addition, the nuclear weapon testing in the 1950s and early 1960s increased the $\Delta^{14}C$ of the surface ocean. Consequently, increasing the natural $\Delta^{14}C$ gradient between surface and subsurface waters makes coral $\Delta^{14}C$ records very sensitive to changes in upwelling and to changes in the horizontal advection of water masses which upwelled elsewhere (Guilderson and Schrag 1998; Grottoli et al. 2003; Grumet et al. 2004).

The boron isotope composition ($\delta^{11}B$) in biogenic carbonates is a promising proxy to reconstruct long-term pH records of the ocean, as was first realized in the 1990s (Hemming and Hanson 1992). Afterwards, Pelejero et al. (2005) reconstructed seawater pH since the Industrial Revolution from $\delta^{11}B$ values in modern long-lived coral. Recently, other two pH records have been published, based on corals from Great Barrier Reef and South China Sea, respectively (Wei et al. 2009; Liu et al. 2009a).

c. Trace element ratios Sr/Ca ratios in corals were shown to provide a reliable proxy for SST variability because Sr/Ca is relatively stable in seawater (Beck et al. 1992). As temperature increases, there is a decrease in the Sr/Ca ratio of the coral skeletal aragonite. The average of coral Sr/Ca calibrations published suggests a temperature dependence of 0.06 mmol/mol per 1 °C (Corrége 2006). Although it is a well-established geothermometer, SST-Sr/Ca calibration curves often vary dramatically among species and/or locations, and may be influenced by growth rate (Beck et al. 1992; Marshall and McCulloch 2002). Thus Sr/Ca needs to be calibrated with local or regional temperature records for each genus or species and at every location. A Rarotonga coral Sr/Ca record that dates back to 1726AD indicates that this Sr/Ca record is an excellent proxy for SST, and shows a distinct pattern of decadal variability, with repeated decadal and interdecadal SST regime shifts greater than 0.75 °C (Linsley et al. 2000).

In addition to Sr/Ca ratios, other elemental ratios that are measured for paleoceanographic reconstructions include: uranium to Ca (U/Ca; Min et al. 1995), magnesium to Ca (Mg/Ca; Mitsuguchi et al. 1996) and barium to Ca (Ba/Ca; McCulloch et al. 2003). U/Ca ratios may provide the same level of precision of temperature as reconstruction as Sr/Ca, but this is still being confirmed (Hendy et al. 2002). Mg/Ca ratios are probably not accurate enough to replace other proxies (Schrag, 1999). Ba/Ca ratios in corals can provide records of suspended sediment loads and hence in the nutrients entering reef areas as a result of river floods (McCulloch et al. 2003). Furthermore, combined determinations of coral Sr/Ca and $\delta^{18}O$ should make it possible to reconstruct the $\delta^{18}O_{sw}$, as well as SSTs, through removing the temperature component of the coral $\delta^{18}O$ variations which is derived from the coral Sr/Ca signal. This technique should be capable of yielding information about SST, precipitation, evaporation, and water mass transport, even in oceanic settings where both SST and $\delta^{18}O_{sw}$ change in unexpected ways (Gagan et al. 1998; Ren et al. 2002; Kilbourne et al. 2008).

In general, not only do corals serve as proxies for modern climate, they have also been used to reconstruct conditions in the Holocene (Gagan et al. 1998; Yu et al. 2005), the Last Glacial Maximum (Juillet-Leclerc et al. 1997), and last Interglacial (Rohling et al. 2008). Although calibration issues typically prevent the geochemi-

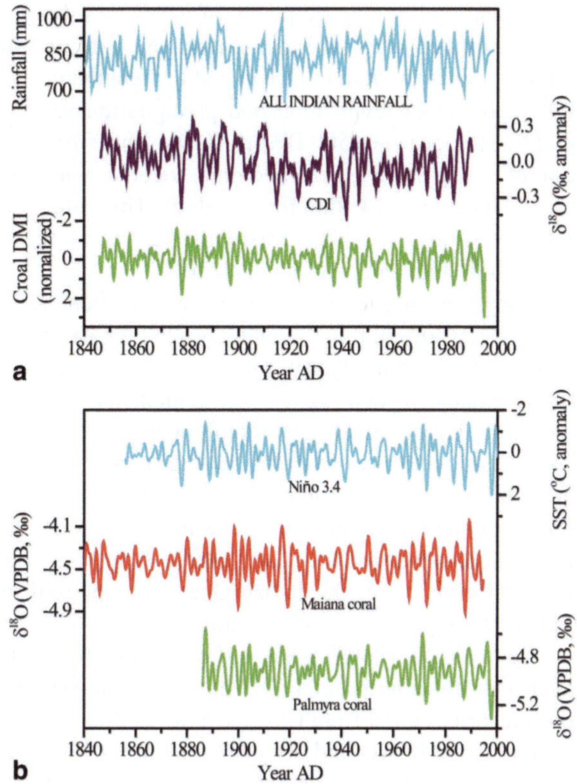

Fig. 3.25 **a** Comparison of all-India monsoon rainfall with coral dipole mode index (DMI; Abram et al. 2008) and coral dipole index (CDI; Charles et al. 2003) of Indian Ocean Dipole (IOD) activity. The data of all-India monsoon rainfall were downloaded from IRI/LDEO Climate Data Library (http://iridl.ldeo.columbia.edu/SOURCES/.Indices/.india/). **b** comparison of coral $\delta^{18}O$ from Maiana (Urban et al. 2000) and Palmyra (Cobb et al. 2003) with Nino 3.4 SST (Kaplan et al. 1998)

cal proxies from being used to directly reconstruct temperatures, these records are primarily used as relative indicators of temperature or other climatic variability.

3.5.2 Coral Records of the Indian Ocean Dipole and ENSO

3.5.2.1 The Indian Ocean Dipole

The Indian Ocean Dipole (IOD) is a coupled ocean-atmosphere phenomenon that involves a reversal of SST gradient and zonal winds over the equatorial Indian Ocean. It is normally characterized by anomalous cooling of SST in the south eastern equatorial Indian Ocean and anomalous warming of SST in the western equatorial Indian Ocean. Associated with these changes, the normal convection situated over the eastern Indian Ocean warm pool shifts to the west and brings heavy rainfall over East Africa and severe droughts and forest fires over the Indonesian region.

Abram et al. (2008) used a suite of coral $\delta^{18}O$ to reconstruct a basin-wide Indian Ocean dipole mode index (DMI) since AD 1846 (Fig. 3.25a). The record indicated an increase in the frequency and strength of IOD events during the 20th century, coinciding with the development of direct, positive IOD-monsoon feedbacks. The authors further suggested that greenhouse warming may lead to redistribution of rainfall across the Indian Ocean and a strengthening interdependence between the IOD and Indian Asian monsoon precipitation variability. Extending back to the middle Holocene (~6500 BP yr), IOD events were characterized by a longer duration of strong surface ocean cooling, together with droughts that peaked later than those expected by El Niño forcing alone. This enhanced cooling and drying was the result of strong cross-equatorial winds driven by the strengthened Asian monsoon of the mid-Holocene (Abram et al. 2007).

Using the long-term coral $\delta^{18}O$ records from Kenya, Nakamura et al. (2009) found the history of rainfall anomalies in East Africa are related to global warming-induced IOD variability. The coral IOD index exhibits an intensification of the IOD and an enhanced coupling with Asian monsoon which drove a decadal mode shift in the 1960s. Charles et al. (2003) built up a coral dipole index (CDI) of IOD based on coral oxygen isotopes from the Seychelles and Bali (Fig. 3.25a). The most prominent feature of CDI is the especially strong periodicity at 12–13 years. Moreover, the ENSO and biennial frequencies also stand out clearly. After much discussion, the authors suggested the 'coral dipole' may be an outcome of strong ENSO-like teleconnections over the Indian Ocean, as opposed to being the result of unique Indian Ocean or monsoonal dynamics.

Pfeiffer et al. (2004) presented the temporal variability in precipitation associated with the Intertropcial Convergence Zone (ITCZ) over 120 years by using a $\delta^{18}O$ record of coral from Chagos in the central Indian Ocean. The record indicated an alternation of wet and dry intervals lasting 15 to 20 years. They then concluded that interdecadal oscillations in coral $\delta^{18}O$ were related to ITCZ precipitation during the prevailing northeast monsoon, and that the ENSO-monsoon relationship is non-stationary and changes through time. Moreover, the boreal summer cooling of the western equatorial Indian Ocean and Arabian Sea resulted from wind-induced mixing and evaporation during the southwest monsoon season, and decadal El Niño-like variability significantly influenced the Indian Ocean (Pfeiffer and Dullo 2006).

3.5.2.2 Historical ENSO Events

ENSO is a global coupled ocean-atmosphere phenomenon and the most prominent known source of inter-annual variability in weather and climate around the world (about 3 to 8 years). The Maiana coral $\delta^{18}O$ record (Fig. 3.25b) is strongly correlated with instrumental indices of ENSO variability (Urban et al. 2000). Reconstructed ENSO extending back to 1840 suggests that ENSO variability in the tropical Pacific is linked to the region's mean climate state. In the mid-late 19th century, cooler and drier climate conditions coincided with prominent decadal variability and weak interannual variability. During a gradual transition towards warmer and wetter conditions in the early 20th century, a periodicity of ~2.9 years intensified. Between 1920 and 1955, periodicities of 2–4 years prevailed. With a shift towards warmer and wetter conditions after 1976, a periodicity of about 4 years became prominent.

Monthly resolved coral $\delta^{18}O$ records, which were generated from nine fossil corals and one modern coral from Palmyra Island, provide information on ENSO and climate variability in the central tropical Pacific for the last 1,100 years (Cobb et al. 2003). The modern coral $\delta^{18}O$ record shares 72 % of its interannual variance with the Nino 3.4 index since AD 1840 (Fig. 3.25b). Although the record is not continuous, it portrays a highly variable ENSO over the last millennium whose amplitude and frequency changed markedly. For example, ENSO activity in the 17th-century sequence is stronger and more frequent than in the late 20th century, whereas ENSO activity was greatly reduced during both the 12th and 14th centuries relative to 20th-century observations. However, the coral records indicate relatively cold and dry mean climate conditions during the 10th century and increasingly warmer and wetter conditions in the 20th century. There is a poor correlation between the behavior of ENSO and these estimates of mean climate, so the Palmyra coral records imply that the majority of ENSO variability over the last millennium may have arisen from dynamics internal to the ENSO system itself.

However, prior to Cobb's work (Cobb et al. 2003), Tudhope et al. (2001) investigated variability in ENSO at intervals through the last glacial-interglacial cycle of Earth history, with a suite of Papua New Guinea fossil corals which individually provide time windows of several decades. The records show that ENSO has existed at least for the past 130,000 years and was active even during glacial times of substantially reduced global temperature and changed solar forcing. The results also suggest that ENSO was weak during the mid-Holocene (~6.5 ka) whereas ENSO during the 20th century has been strong compared with ENSO of previous glacial and interglacial times. Recently, Gergis and Fowler (2009) reconstructed an annual record of ENSO events since AD 1525. The result of their statistical analysis shows that 43 % of extreme ENSO events occurred in the 20th century, and the post-1940 period alone accounts for 30 % of extreme ENSO years observed in the 478-year reconstruction, which suggest that ENSO may operate differently under natural (pre-industrial) and anthropogenic background states. Recent ENSO variability may thus be anomalous in the context of the past five centuries.

3.5.3 Coral Records and the East Asian Monsoon

The East Asian monsoon (EAM) is an integral part of the global climatic system. Monsoon climates, and especially monsoon-associated precipitation, are important to the maintenance of living environments and socially sustainable development in populous regions of East Asia (An 2000).

3.5.3.1 The Holocene

Yu et al. (2005) produced a Sr/Ca-$\delta^{18}O$ record from a coral from the Leizhou Peninsula in the northern South China Sea (SCS). Their record suggested a general decreasing trend in SST in the SCS from about 6800 to 1500 years ago. The 10-year

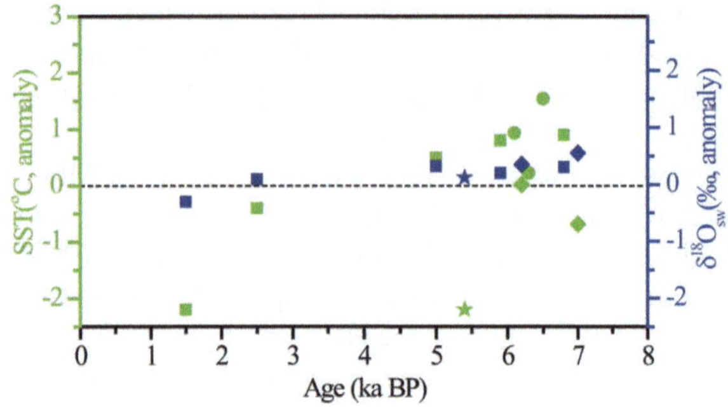

Fig. 3.26 The annual mean SST (*green*) and $\delta^{18}O_{sw}$ (*blue*) records reconstructed from Holocene corals in East Asian monsoon area, relative to present value. The data were cited from papers following: Yu et al. 2005 (*square*); Wei et al. 2007 (*circle*); Morimoto et al. 2007 (diamand); and Su et al. 2010 (*star*)

mean Sr/Ca-SST at 6.8-5.0 ka BP was ~0.9–0.5 °C higher than today, dropped to the present level at 2.5 ka BP, and was 2.2 °C lower at about 1.5 ka BP. The decline in SST was accompanied by a general decrease in seawater $\delta^{18}O$ values (Fig. 3.26), which indicates a decrease in the amount of monsoon moisture transported out of the South China Sea. The results imply that the Asian winter monsoon weakened and the summer monsoon was enhanced during the mid-late Holocene.

Shen et al. (2005b) show that the seasonal anomaly of residual $\delta^{18}O$ was twice that of the modern value in the mid-Holocene, based on Sr/Ca-$\delta^{18}O$ data recorded in a ~6.73 ka BP *Porites* coral from Nanwan Bay, southern Taiwan. By further analysis, the authors speculated that the magnitude of precipitation and the onset of the rainfall season might be influenced by strengthened solar insolation and an enhanced Asian summer monsoon. Wei et al. (2007) investigated three mid-Holocene Sr/Ca-SST records spanning more than 30 years from *Porites* corals from the northern SCS. The results indicate warmer than present climates between about 6100 year BP and 6500 year BP, which appear to correlate to stronger summer monsoon circulation (Fig. 3.26). Coupled records of Sr/Ca and $\delta^{18}O$ of corals, from Kakai Island in the subtropical northwestern Pacific also suggested that the East Asian summer and winter monsoons were more intense in the mid-Holocene, which was likely a function of the mid-Holocene insolation regime (Morimoto et al. 2007; Fig. 3.26).

The coral $\delta^{11}B$ reconstructed paleo-pH records of the SCS show significant variability but a gradual increase from the Holocene thermal optimal and a sharp decrease to modern values. The latter increase in acidity is likely caused by the large amount of anthropogenic CO_2 emissions since the Industrial Revolution, but variations of atmospheric pCO_2 cannot explain the pH change involving less acidity of the SCS before the Industrial Revolution. Therefore, the weaker summer and stronger winter monsoon intensities during the mid-late Holocene may have driven sea surface pH increases from the mid to late Holocene (Liu et al. 2009a). These results are in contrast to those of Su et al. (2010) who analyzed a 40-year $\delta^{18}O$ record of the ~5.4 ka BP *Porites lutea* from the east coast of Hainan Island in the northern SCS. The mean, maximum and minimum of the coral $\delta^{18}O$ were higher than those of modern corals respectively. Thus, the SST for mean annual, winter and summer at ~5.4 ka BP was 2.2, 3.1 and 0.6 °C lower, and accordingly the SSS was 1.5 ‰, 2 ‰ and 1 ‰ higher than today (Fig. 3.26).

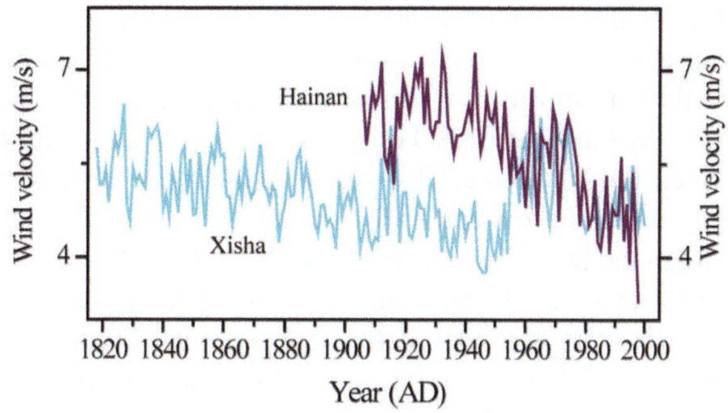

Fig. 3.27 Wind velocity records of East winter monsoon reconstructed from Hianan (Liu et al. 2008b) and Xisha (Song et al. 2012) corals in the SCS

A record of Sr/Ca ratios and oxygen and carbon stable isotopes from a 16 ka fossil coral from the Ryukyus, southwestern Japan, showed that the annual mean SST was approximately 5 °C lower and SSS was locally higher at 16 ka than present conditions, suggesting weak East Asian summer monsoon activity and enhanced winter monsoons during the last glacial (Mishima et al. 2009).

3.5.3.2 The Last 100 Years

Peng et al. (2003) investigated the relation between coral $\delta^{18}O$ and winter monsoon intensity in the SCS and reconstructed a 54-year winter monsoon velocity (WMV) record. The record shows that WMV decreased during the 1940s to the 1960s, increased slightly during the 1970s and then decreased again in the 1980s and 1990s. For the whole period of 1944–1977, WMV had an obvious decrease of 1.29 m/s from 5.92 to 4.63 m/s. Moreover, the averaged WMV value is 5.58 m/s from 1944 to 1976 and this decreased to 4.85 m/s from 1977 to 1998, which is the opposite trend to the observed SST variation. Hence most of the variability of the winter monsoon intensity in the middle of the South China Sea is mainly constrained by changes in the thermal difference between the land and the adjoining sea area, perhaps due to global warming. Subsequently, Liu et al. (2008b) presented a 93-year winter monsoon velocity series based on a coral Sr/Ca record from Hainan Island in the northern SCS (Fig. 3.27). The reconstructed WMV in the 20th century shows significant interannual and decadal variability with a persistent decline in the whole 20th century at a rate of −0.02 (m/s)/yr. The lowest wind velocity occurred during the last two decades of the 20th century. The WMV has decreased significantly by about 30 % from the early to the late 20th century.

Based on the relationship between observed WMV and $\delta^{18}O$ of coral from Xisha Islands in the northern SCS, Song et al. (2012) reconstructed WMV series for years 1818–2000 (Fig. 3.27). The reconstructed WMV sequence shows significant interannual and decadal variability for last 183 years with two distinct declines at the rate of decrease of 0.009 (m/s)/yr for years 1818–1954 and 0.021(m/s)/yr for the years 1955–2000, respectively. Moreover, the WMV has decreased by about 20 % during the late 20th century. Two complete cycles of changes of wind strength have been detected in the coral-based WMV record over past 183 years, and the highest wind velocity occurs in the 1830s, whereas the lowest wind velocity was in the 1940s. During the 20th century, the variability of WMV is very similar to the variability of temperature records from the South China Sea and Chinese landmass, with two significant warm periods for the 1940s and 1980s. Statistics of anomalies of WMV against ENSO events show that most winter monsoons appear weakened when ENSO occurs. This supports the model of interaction between atmospheric circulation in low and mid-latitudes. However, there are distinct discrepancies between the two winter monsoon velocity records from 1906 to 1960. Wind velocities of Hainan were higher than those of Xisha, and the interannual variability of both records were anti-phased during this period. Furthermore, the abrupt shift in WMV in the mid-1950s seen in the Xisha record is not seen in the Hainan record. The factor(s) causing these discrepancies are not clear at present, but further research on many corals may answer those questions.

3.5.4 Summary

We have briefly reviewed various climatic proxies of coral used for paleoclimate research. Based on the geochemical proxy records of corals, we have reviewed the Indian Ocean Dipole, El Niño-Southern Oscillation, and East Asian Monsoon during the Holocene and the last one hundred years. The coral records aimed directly at reconstructing the East Asian Monsoon are limited. Moreover, there are marked discrepancies among several records. This points to the need for more coral research in China, to provide low-latitude evidence for variations in the East Asian Monsoon system.

3.6 Asian Monsoon Variability Recorded by Tree Ring

Tree-rings play a valuable role in understanding spatio-temporal variations in past climate. The indices generally employed in this territory are tree ring width, density and stable isotopes. Tree ring width is a conventional and effective index, by which most tree ring research results are obtained. In warm and moist areas with moderate climate conditions, there is no obvious variation in tree ring width year by year, but there may be variations in wood density. Thus correlation functions between wood density and climate factors can be established and climate change information obtained. In addition, by analyzing the $\delta^{13}C$ in tree rings, the history of atmospheric CO_2 concentration can be reconstructed. Meanwhile $\delta^{18}O$ and δD in tree rings can provide information on isotopic variations in soil moisture and precipitation.

Climatologists require high-resolution indices to extend past instrumental climate records. Tree rings can be used for this purpose because they provide continuous, accurate, high-resolution records that are easy to obtain. Many temperature series (Briffa et al. 1995; Jacoby et al. 1996; Mann et al. 1998, 2002, 2009; Briffa 2000; Esper et al. 2002; Moberg et al. 2005; Liu et al. 2009b), precipitation indices

(Jacoby et al. 2003), and drought indices (D'Arrigo et al. 2006; Cook et al. 2010) from the past 1000 years in the Northern Hemisphere have been reconstructed using tree ring data. From these reconstructions, researchers have gained a better understanding of historical climate change and have become aware of an abnormal climate period: global warming in the 20th century. Compared with other climate proxies, tree-rings are a powerful tool for studying past climate at a high resolution and for understanding historical climate trends.

Tree ring research in the monsoon area of Asia was initiated later than it was initiated elsewhere in the world, but it has made many advances in the reconstruction of Asian monsoon climate. In China, tree ring research work has been continually carried out since the 1930s, but it has developed more extensively in recent years. During the last decade, tree ring research has increased a great deal, especially in China's East Asian monsoon areas. A temperature reconstruction for the TP, 2485 years in length, is the longest reconstruction series calculated for East Asia (Liu et al. 2009b). The history, rate, period, and mechanism of temperature change in the area can now be viewed more clearly, and scientific assessments of developing trends can be made.Tree-ring variations for the past 400 hundred years, associated with the East Asian monsoon, can be used to investigate variations in the monsoon-dry system in China at a centennial timescale (Liu et al. 2007, 2009c, 2010a). This study has provided large datasets for studying the past climate at a high resolution. On the basis of these data, the trend of rainfall at a decadal scale was forecasted (Liu et al. 2004a).

3.6.1 Asian Monsoon Climate Variations

3.6.1.1 Tree-ring Recorded Variations of Temperature

The East Asian monsoon area A lack of long-term climate records for East Asia has limited our understanding of the East Asian monsoon system. Since the 1970s, tree ring climate research has been carried out in Asia. The pioneers of tree ring research, represented by Wu Xiangding, carried out tree ring research in the Qinling Mountains, the Hengduan Mountains, and the middle reaches of the Yellow River, among other areas. Most of the tree ring studies in China have focused on the Qinling Mountains and have been limited because of the area's complex terrain and variable climate (Shao and Wu 1994). However, tree ring research in the Qinling Mountains is considered to be the beginning of tree ring research in China that is in line with international standards. Using modern methods, tree ring research was carried out again in the Qinling Mountains, and significant results were achieved.

The Qinling Mountain range represents an important geographic demarcation line because it is a critical boundary for climate and vegetation distribution in the eastern central mainland of China. In the mid-eastern Qinling Mountains, the temperature reconstruction series includes mean temperature from May to July during 1760–2005 for the northern slope of the Qinling Mountains in the NWT region and

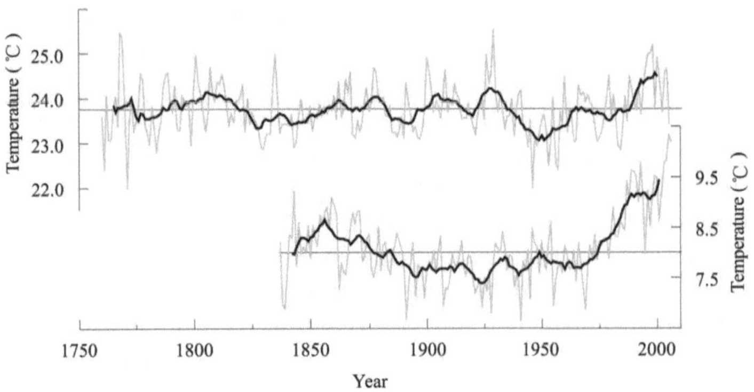

Fig. 3.28 Reconstruction of the May–July mean temperature for the NWT region from 1760 to 2005 (*top*) and the mean temperature from previous September to current April for the MW region from 1837 to 2006 (bottom). Both reconstructions were based on Chinese pine tree ring widths. Smoothed lines represent the 10-year moving average. Horizontal lines represent the long-term mean values. The 246-year mean for the NWT site was 23.71 °C. The 170-year mean for the MW site was 8.11 °C. (from Liu et al. 2009d)

the mean temperature from September to April during 1837–2006 for the southern slope of the Qinling Mountains (Liu et al. 2009d), mean temperature variations from May to June during 1558–2006 for Huashan Mountain (Liu et al. 2009e), mean temperature from January to July during 1814–2003 at the divide sampling site in the mid-Qinling Mountains (Liu et al. 2009f), and average maximum temperature from May to July in the Funiu region during 1874–2007 (Tian et al. 2009).

Cores from 88 Chinese pines (*Pinus tabulaeformis*) collected from the southern (MW site) and northern (NWT site) slopes of the Qinling Mountains were used to reconstruct seasonal temperature variations (Liu et al. 2009d; Fig. 3.28). Prior to the mid-20th century, low September–April temperatures of the southern slope were, in general, followed by high May–July temperatures of the northern slope, which probably reflects variations in the winter and summer monsoons. However, since the mid-20th century, both records show trends of a more pronounced increase in the September–April temperature on the southern slope. These results support the suggestion that the recent warming is unusual in nature, coinciding with the observed instrumental record. These results were comparable to tree-ring-based reconstructions derived for the surrounding regions, which suggests that there are regional signals in the Qinling Mountain reconstructions.

Global warming has led to temperatures increasing at different rates in different regions. In Huashan, at the eastern end of the Qinling Mountains, tree rings were significantly correlated with the average temperature recorded from May to June. As a result, the mean May–June temperature from 1558 to 2006 has been reconstructed for this region (Liu et al. 2009e). There were two important long, warm periods: from 1698 to 1709 and from 1749 to 1777. The long, low temperature period occurred from 1930 to 1950, with the lowest temperature of 12.0 °C occurring in 1933. The reconstructed temperature of Huashan Mountain was significantly correlated with the reconstructed temperature of Nanwutai Mountain, and a synchronous warming trend was found, which is related to global warming.

The mean January–July temperature at the divide sampling site in the mid-Qinling Mountains was reconstructed for 1814–2003 (Liu et al. 2009f). The reconstructed temperature was relatively stable from the early 1930s to 1990 but rapidly increased after 1990. This reconstruction was significantly correlated with the winter temperature reconstruction for Jiuzhaigou region (Song et al. 2007), but they were not synchronized, perhaps due to the different climate patterns. A significant low temperature period was recorded in both areas in 1885. The Galunggung volcano eruption in 1882 and the Krakatau volcano eruption in 1883 may have caused the region to cool in the years following.

The average maximum temperature from May to July was reconstructed for Funiu Mountain in western Henan Province from 1874 to 2007 based on tree ring data (Tian et al. 2009). The most significant warm period during the reconstruction occurred between the 1920s and the 1940s. Afterwards, the temperature decreased and reached its lowest value in the 1950s, and it has increased rapidly since the late 1980s. The Funiu Mountain and Nanwutai Mountain temperature curves were comparable at the 10-year moving average, which showed the same regional trends.

In addition to studying records from the Qinling Mountain region, tree ring researchers have also examined the West Sichuan Plateau, Hengduan Mountain in Yunnan, Changbai Mountain in Northeast China, Yishan Mountain, Mengshan Mountain and the Shandong region, Tianmu Mountain in Zhejiang, and Wuyi Mountain in Fujian, among other locations, to reconstruct temperature variations.

The mean winter temperature (from November to March) lowest temperature since 1750 in the Jiuzhaigou region which is located at the boundary between the TP and the Sichuan Basin, has been reconstructed (Song et al. 2007). Two cold periods (1814–1844 and 1937–1984) and two warm periods (1795–1813 and 1984–2004) were recognized. The cold periods were well-correlated with temperatures recorded in Kathmandu, Nepal and at other sites in west Sichuan. This indicates that the TP affects winter temperatures synchronously in the surrounding regions.

Tree ring samples were used to reconstruct mean winter minimum temperature from 1650 to 1994 for the West Sichuan Plateau (Shao and Fan 1999). In the 345-year reconstruction, 22 cold winters and 26 warm winters were recorded, and there were 5 periods of cooling, 5 periods of warming, and 1 leveling-off stage; 1810–1820 was the warmest decade and had the smallest variability, whereas 1960–1970 was the coldest decade. The 19th century was the warmest century.

The most significant characteristic of the temperature reconstruction in the Wolong National Natural Reserve, western Sichuan (Li et al. 2010), was a significant warming trend after 1940. Before the 1940s, the climate of this region was consistently cold. The low-frequency variation of the reconstruction agreed with previously published proxies (e.g., ice cores of nearby glaciers).

Summer (April–September) temperatures and annual mean temperatures since 1750 for mid-Hengduan Mountain were reconstructed (Fan et al. 2008, 2009a, b). Annually, the central region of Hengduan Mountain experienced some cool periods during the 1810s, 1860s, and 1960–1980. Warm intervals occurred during the

1780s, 1850s, 1940–1960 and during the past two decades. Comparisons showed that the reconstructed droughts were relatively consistent with those found in other regions of Indochina, suggesting similar drought regimes (Fang et al. 2009a).

A temperature reconstruction of February–April for 250 years was completed for the Changbai Mountain area in Northeast China (Zhu et al. 2009). Four cold periods (1784–1815, 1827–1851, 1878–1889, and 1911–1945) and two warm periods (1750–1783 and 1855–1877) were identified. The temperatures were negatively correlated with EAWM intensity records, suggesting that this temperature reconstruction may be a good indicator of EAWM intensity.

Tree rings from Yulin in Shandong Province have been used to reconstruct the mean lowest temperature from October to September for 1616–2007 (Liu et al. 2010a). A significant increasing trend was found for the reconstructed series from 1960 to 1998; the reconstruction suggests an obvious warming occurring in eastern China. After 1998, the reconstructed series displayed a decreasing trend, which was consistent with the observational data. The climate variations observed in Yulin were the result of large-scale ocean, land, and air coupling.

A 207-year chronology has been established using tree ring data for Sha County, Fujian (Chen et al. 2012). The chronology was significantly negatively correlated with the drought season (summer and autumn) temperatures and was correlated with the I_{APO}. This finding suggests that the drought season is the main factor that affects tree growth.

The average temperatures of the previous December to the current March were reconstructed using the regional chronology (RC) which developed from the western Tianmu Mountains and the Xianyu Mountains of southeast China for the period from 1852 to 2006 (Shi et al. 2010a). The reconstruction was significantly correlated with the winter half-year temperature in the eastern Qinling Mountains, suggesting large-scale coherence of winter temperature variability. The reconstruction also corresponds well with an East Asian winter monsoon (EAWM) index in extreme years, reflecting the strong influences of the EAWM in this study region.

Numerous researches showed that tree ring width in the north of the Qinling Mountains reflects the variation of spring and summer temperature, while in the south of the Qinling Mountains (in the western Sichuan area), it mainly represents the winter temperatures. It can be seen that it is completely different for the response of trees to climate north and south of the Qinling Mountains, but they all recorded the increased temperatures (i.e., global warming) during the second half of the 20th century.

Tree ring results from other East Asian countries also reflect the occurrence of the Asian monsoon. For example, warm season temperatures (April–September) were reconstructed for Asahikawa, north of Hokkaido in northern Japan, for 1557–1990 (Davi et al. 2002). The climate in Hokkaido and in the northwestern Pacific region was characterized by a monsoonal reversal of wind direction. In the summer, when the south-to-southeasterly winds from the Hawaiian high pressure system carry warm, moist monsoonal air northward, the spring rainy season and warm temperatures begin.

In Hokkaido, the annual relative humidity (RH) in summer since 1776 was reconstructed using tree ring oxygen isotopes (Hiroyuki et al. 2008). Spectral analysis indicated that the RH included long-term periodicities of 50–100 years. Decadal and centennial variations in the summer RH were negatively correlated with the annual PDO index during all periods and were positively correlated with the reconstructed summer AO index during 1780–1930. Global warming may dramatically alter the causes of variations in the hydrological climate in Hokkaido, from the forcing of the high-latitudes (AO) to those of the mid-latitudes (PDO).

In central Japan, February–April temperatures were reconstructed for the period of 1719–1961 using ring width data (Hitoshi and Dieter 2006). Five cooling events occurred during the Little Ice Age. Cold periods in documentary records from Japan are similar to those of China. A dendroclimatic reconstruction for Korea was also in close agreement, suggesting that the annual activity of the winter monsoon in the western Pacific Rim persisted longer in the 19th century than it currently persists. Records of prolonged famines and unusual climatic phenomena also support the reconstruction of cold periods that occurred in the 1780s, 1810s and 1830s.

In the western arid zone, temperatures from January to August increased slowly and declined rapidly during the last 220 years in the Helan Mountain area (Cai and Liu 2007). Results of tree ring $\delta^{13}C$ studies from Helan Mountain suggest that the variety of summer temperatures recorded since 1890 was significantly correlated with sea surface temperatures recorded in the tropical Pacific from June to August and to the North Atlantic Decadal Oscillation. This series also displayed high-temperatures during the late 1920s to the 1930s (Liu et al. 2002).

Ring-width data collected from Weichang were used to reconstruct the mean May–June temperature from 1884 to 2002 (Liu et al. 2009g). After being smoothed by an 11-year moving average, the comparison between the reconstruction and the mean temperature of May–July in the NWT region suggests large scale similarity. The reconstructed series and the dryness/wetness index, derived from historical documents, were significantly correlated (average of Datong and Beijing), especially after 1900. A synchronous variation was also found for the reconstructed temperature series and the reconstructed Pacific Decadal Oscillation Index (PDO), suggesting a decadal-scale relationship between the climate changes in North China and the PDO.

A regional-scale composite of four tree-ring records from temperature-sensitive alpine timber-line sites in Mongolia was built by D'Arrigo et al. (2000). These trees suggest that recent decades have been some of the warmest in the past 500 years for this region. Although there are obvious differences among the individual chronologies, this result suggests a coherent signal, which is related to temperature. The PCA scores show trends which strongly resemble those seen in recent temperature reconstructions for the NH, very few of which included representation from Eurasia, east of the Ural Mountains. The Mongolia series therefore provides independent corroboration for these reconstructions and their indications of unusual warming during the twentieth century.

A 450-year tree-ring width chronology from the Tarvagatay Mountains in west central Mongolia showed wide annual growth rings for the recent century (Jacoby et al. 1996). Ecological site observations and comparisons with instrumental temperature records indicate that the ring widths of these trees were sensitive to annual temperature variations. Low-frequency variations in the Tarvagatay tree ring record were similar to variations observed in a reconstruction of Arctic annual temperatures, which was based on 20 ring-width series from northern North America, Scandinavia, and western Russia. These results suggest that the recent warming period is unusual relative to the temperature of the past 450 years.

The Tibetan Plateau In the context of global warming, the mean annual surface temperature in China increased significantly over the last 100 years (by 0.5–0.8 °C). This warming primarily occurred during the winter and spring seasons; it did not occur significantly in the summer. In the last 50 years, the mean annual surface temperature in China increased by 1.1 °C (a rate of 0.2 °C per decade), which is significantly greater than the global and the NH average increase over the same time period. The most significant regional warming occurred in northern China and in the TP area (Ding et al. 2006). Thus, as a unique landscape unit of the NH, the TP is in a prominent position in global change research. The thermal effect of the area directly influences the large-scale tropospheric circulation of the NH and influences climate change in the NH by disturbing lower stratospheric processes (Zhou et al. 2009). Because the TP is at a high altitude and because it is a continental area, it is sensitive to climate change and is therefore an ideal site for long-term, high-resolution climate research using tree rings. Chinese tree-ring researchers have completed several studies on the TP.

A 646-year ring-width series was established from tree ring samples collected from the Zaduo area in southern Qinghai Province (Shi et al. 2010b). At the interannual scale, there have been 6 periods with high temperatures (1438–1455, 1572–1612, 1684–1700, 1730–1754, 1812–1829 and 1853–1886) and 5 periods with low temperatures (1547–1571, 1701–1729, 1755–1777, 1830–1852 and 1887–1910) since 1360.

A temperature reconstruction that is based on tree ring data from Animaqing Mountain indicates that the study region experienced three significant cooling periods in the past 830 years. One of the cooling periods occurred during the 1820s–1900s, when global warming was beginning (Gou et al. 2006). In addition, temperature reconstructions of different seasons that are based on tree ring data (Gou et al. 2008a) demonstrate that similar variations in minimum and maximum temperatures occurred in this region during the past 425 years, but minimum temperatures were out of phase and approximately 25 years earlier than the maximum temperatures. If such a pattern continues to occur in the next 30 years, the maximum temperature in this region will significantly increase (Gou et al. 2008b).

Tree-ring-width reconstructions derived from the timberline of southeastern Tibet suggest that the average summer temperature during the last decade was the warmest of the past 250 years (Liang et al. 2009a). However, the mean June–August minimum temperature reconstruction for the source region of the Yangtze River (Liang et al. 2008) indicates that the last decade was not the warmest decade of the past 379 years in that region. The cold and warm periods in the reconstruction coincided with documented glacial advances and retreats on the eastern and southeastern TP. The 1816–1822 cold period coincided with the Tambora eruption that occurred in

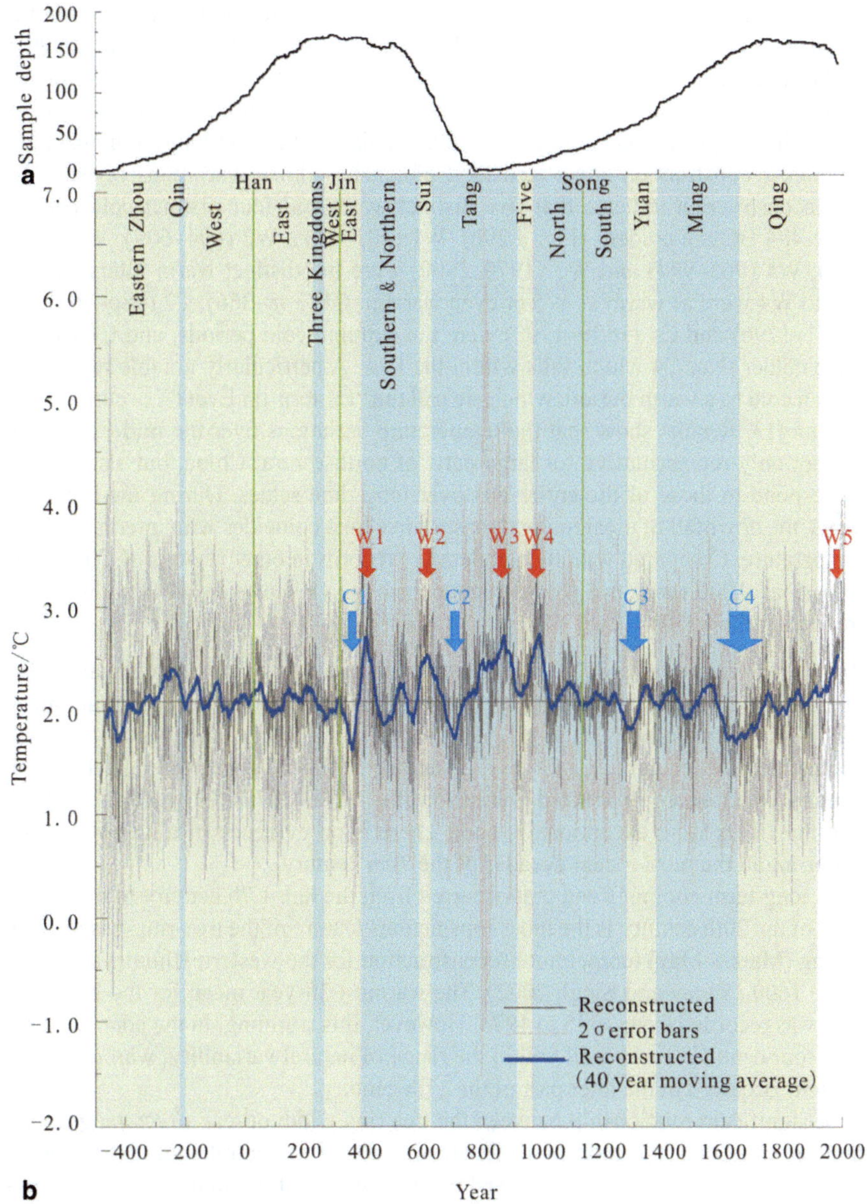

Fig. 3.29 **a** The sample depths **b** The reconstructed temperature over the mid-eastern Tibetan Plateau from 484 BC to 2000 AD (thin lines) with 2σ error bars (dashed lines). The horizontal line is the mean of 484 BC-2000 AD, and the thick line is the 40-year moving average. The colored areas indicate periods covered by different Dynasties. W1 to W5 are the warm periods, and C1 to C4 are the cold intervals. (from Liu et al. 2009b)

Indonesia in 1815 and was detected by the tree ring temperature records from the Jiuzhaigou region and in southeastern Tibet (Song et al. 2007; Liang et al. 2009b).

Mean annual temperatures were reconstructed for the mid-eastern TP for the past 2485 years using tree-ring-width indices from a combination of living trees and archaeological wood (Liu et al. 2009b). This is by far the longest annual temperature reconstruction completed for China. The climate variations revealed by the reconstruction indicate that five distinct warm and four distinct cold periods since 484 BC are evident (Fig. 3.29b). W1 (401–413), W2 (604–609), W3 (864–882), W4 (965–994) and W5 (1970–2000) were the distinct warm intervals, and W1 to W4 were as warm as W5 or even warmer. C1 (348–366), C2 (686–705), C3 (1271–1296) and C4 (1599–1702) were the extreme cold periods, and C1 and C2 were colder than C4 which falls within the LIA. A particularly notable rapid shift from a cold to a warm period, which we call the "Eastern Jin Event", occurred from 348 to 413. Results show that the temperature variations over the mid-eastern TP are not only representative for large parts of north-central China, but also closely correspond to those of the entire NH over long time scales. During the last 2485 years, the downfall of most major dynasties in China coincides with intervals of low temperature. Compared with the temperature records in other regions of China during the last 1000 years, this reconstruction from the TP shows a significant warming trend after the 1950s.

The Indian monsoon area Tree-ring-width chronologies from Himalayan cedar, Himalayan pine, and Himalayan spruce trees have been used to reconstruct mean April–May temperatures back to 1698 (Yadav et al. 1997). The warmest and the coolest decades occurred in 1713–1722 and in 1748–1757, respectively. The 19th century experienced protracted warmth in the 1850s–1870s. However, the 1810s and the 1830s were exceptionally cool. There is no evidence of a warming trend occurring in the most recent decades of the 20th century.

A long-term cooling trend that occurred from the late 17th century to the beginning of the 20th century is the most conspicuous feature of the tree-ring-based mean spring (March–May) temperature reconstruction for the western Himalayan region since 1600 (Yadav and Singh 2002). The warmest 30-year mean for the 20th century was recorded from 1945 to 1974. However, this warming, in the context of the past four centuries, was well within the range of natural variability; warmer spring seasons occurred in the later part of the 17th century.

A significant relationship between the tree ring width of teak (*Tectona grandis*) and pre-monsoon and monsoon climate was found in central India (Borgaonkar et al. 2007). However, tree-ring-based reconstructions of the summer climate (temperature and precipitation) of the Indian Himalayas, Tibet, and the Karakoram region of the Himalayas did not reveal significant increasing or decreasing trends during the past three to four centuries (Borgaonkar et al. 2002; Hughes 2001; Yadav et al. 2004; Brauning and Mantwill 2004; Wu and Shao 1995; Esper et al. 2002). The Little Ice Age (LIA) was observed in these reconstructions over parts of the high latitude areas of the Himalayas. A tree-ring reconstruction of Nepal winter (October–February) temperature (Cook et al. 2003) and Karakoram annual temperatures (Esper et al. 2002) showed that an increasing trend occurred in recent decades. This finding suggests that the Western Himalayas also experienced global warming in the 20th century.

A tree ring chronology network that was suitable for reconstructing temperature-related climate forcing over the past few hundred years was developed for Nepal (Cook et al. 2003). Two temperature seasons were reconstructed: February–June (1546–1991) and October–February (1605–1991). Each reconstruction indicates the occurrence of unusually cold temperatures that occurred in 1815–1822. These periods coincided with the eruption of the Tambora volcano in Indonesia and were confirmed by tree ring research conducted in China (Liang et al. 2009b; Song et al. 2007). The resulting 'adjusted' reconstructions strongly reflect patterns of temperature variability associated with Little Ice Age cooling and warming into the 20th century, with the October–February season exhibiting the strongest temperature increase over the past 400 years. The reconstruction for the October–February season showed evidence of late 20th century warming, whereas the reconstruction for the February–June season showed that temperatures had cooled since 1960.

3.6.1.2 Tree-ring Recorded Variations of Precipitation

In the Asian monsoon area, monsoon-associated precipitation has a major influence on industry, agriculture, animal husbandry and human survival. The spatio-temporal variations in monsoon precipitation have been recorded in tree rings because tree rings are sensitive to these variations. There have been many achievements in China and in other Asian regions regarding precipitation reconstructions for the Asian monsoon using tree ring width, density and stable isotopes.

The East Asia monsoon area July–August (JA) precipitation at Xian (1895–1988) was reconstructed using total ring width (TRW) and maximum latewood density (MXD) from trees in the Qinling Mountains in central China (Garfin et al. 2005). The key feature of the precipitation reconstruction was a prolonged summer drought during the late 1920s and the early 1930s.

The tree ring α-cellulose $\delta^{13}C$ time series was measured by crossdating tree rings from Tianmu Mountain, Zhejiang Province (Qian et al. 2002). The climate change of the last 160 years revealed by the series suggests that there were 57 wet and 49 dry years, which are supported by historical records.

Reconstructions of rainfall patterns derived from tree ring samples of Baekdam-Jang (Park et al. 2001a) and Dae-Seung (Park et al. 2001b), Korea were used to examine the East Asian summer monsoon-related precipitation variation over the past 160 years (Liu et al. 2003a). The reconstructions reveal that the East Asian summer monsoon precipitation abruptly decreased around the mid-1920s and has sustained a reduction since 1960.

The environment-sensitive zone in China is a narrow area stretching from northeast to southwest and is located in an arid and semi-arid transition zone. It is the ecotone of monsoon and non-monsoon areas, wet and dry areas, forest ecosystems and grassland ecosystems, and areas of agriculture and animal husbandry. This zone is a fragile environment. Research regarding climate change in this area is important for understanding global climate change and for promoting pastoral production and

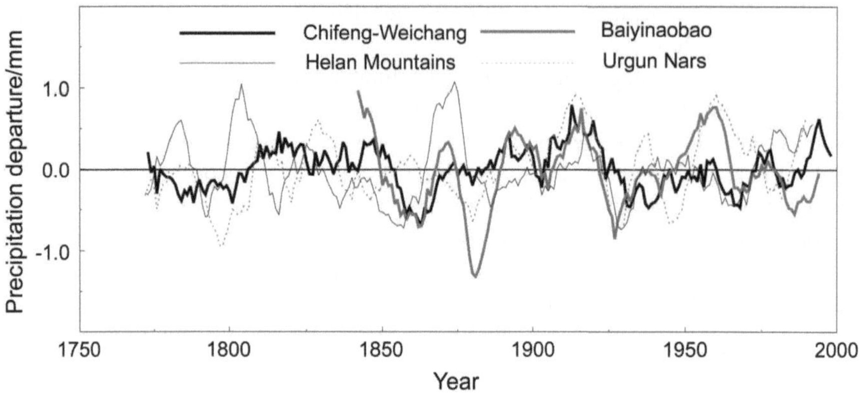

Fig. 3.30 Comparisons of the precipitation departure within Chifeng-Weichang, the Helan Mountains, Baiyinaobao, and Urgun Nars (state of Mongolia). The 4 curves are smoothed by an 11-year moving average. The horizontal line is the zero line (from Liu et al. 2010a)

socio-economically sustainable development. Located in the temperate continental climate zone, the variety of monsoon intensities on the monsoon marginal zone determines the characteristics of the climate. Thus, precipitation is the main limiting factor for tree growth in this area, and tree rings are useful for reconstructing changes in monsoon precipitation.

Precipitation from June to early August from 1742 to 1996 was reconstructed on the basis of tree ring width in Baotou, Inner Mongolia, China (Liu et al. 2001). The East Asian summer monsoon had 8 strong cycles and 8 weak cycles; 1914–1925 was the wettest and 1926–1942 was the driest period. Before and after 1925–1926, the East Asian summer monsoon abruptly weakened in intensity.

Research on precipitation of April–early July in Baiyinaobao, Inner Mongolia reveals three wet periods and four dry periods during the last 160 years (Liu et al. 2003b). Among these, the wet periods of 886–1921 and 1943–1968 and the dry period of 1922–1943 occurred at approximately the same times as the wet and dry periods that occurred in the Qilian Mountain region. The reconstructed precipitation abruptly decreased during the 1920s. It declined after 1950 and increased in the late 1990s.

In the eastern Ortindag Sand Land, a 163-year drought history was reconstructed on the basis of three tree-ring-width chronologies (Liang et al. 2007). The reconstructed May–July PDSI ($PDSI_{5-7}$) agreed with the variation of the average dryness/wetness indices in North China and captured the decline of the East Asian summer monsoon since the mid-1960s. On the basis of the results of the 163-year drought reconstruction, the Ortindag Sand Land experienced its most severe drought during 1966–2004.

The total August to July precipitation was reconstructed for the Chifeng and Weichang regions for the past 236 years (Liu et al. 2010b). The reconstructed series was significantly correlated with the average dryness/wetness index series of Datong and Beijing, as well as with previous results from Baiyinaobao, the Helan

Mountains, and Mongolia (Fig. 3.30), which suggests that the East Asian summer monsoon variations on a decadal scale were coincident with other regions within the environmentally sensitive zone of northern China and that the reconstruction was representative of regional climate patterns.

The precipitation series was reconstructed for the Ma'an Mountain area from July to September since 1942 (Jiang et al. 2010). The precipitation slightly decreased each year. Two dry periods (1942–1957 and 1989–2007) and a rainy period (1957–1989) occurred in the Ma'an Mountain area.

The amount of precipitation from July to June was reconstructed from 1865 to 2003 for Hailar, northeast China (Liu et al. 2009c). Severe drought events (lasting more than 3 years) occurred during 1905–1909, 1926–1929 (a severe drought event in north-central China causing tremendous loss of human lives) and 1968–1970; extreme wet events occurred during 1867–1870, 1932–1934, 1939–1941 and 1955–1957. The reconstructed precipitation was comparable to the Baiyinaobao (Liu et al. 2003b), Ortindag Sand Land (Liang et al. 2007) and northeastern Mongolia (Pederson et al. 2001) rainfall series derived from tree rings. The four curves had variations corresponding with the intensity of the EASM.

The May to July precipitation was reconstructed using latewood ring width for northern Helan Mountain since 1726 (Liu et al. 2004b). On the decadal scale, the rainfall reconstruction for northern Helan Mountain displayed a similar variation pattern to the April to early-July precipitation reconstruction for Baiyinaobao. This pattern may reflect the variety of East Asian summer monsoon front intensities. The 3 dry periods (1831–1863, 1926–1934 and 1954–1971) and the 3 wet periods (1864–1878, 1908–1925 and 1935–1953) coincided with the wet and dry periods observed in Inner Mongolia, China, and Korea.

The precipitation reconstruction from February to July and the PDSI reconstruction from June to August that was based on tree rings for the past 220 years showed that dry periods occurred in the 1750s, 1850s, 1860s and early 1930s. The changing of dryness is slow in drought periods and is fast in wet periods, which reflects the continuity of dry years on Helan Mountain (Liu et al. 2005a, b). The tree ring $\delta^{18}O$ was significantly correlated with the East Asian summer monsoon precipitation. Thus, $\delta^{18}O$ can be an indicator for the variability of East Asian summer monsoon intensities (Liu et al. 2008c).

The annual precipitation from previous August to current July was reconstructed on the basis of tree-ring-width analyses since 1686 for Ningwu (Li et al. 2006a). The reconstructed precipitation series for Ningwu was significantly correlated with precipitation from January to July on Helan Mountain which is also located in an environmentally sensitive zone; precipitation in this zone is controlled by the East Asian monsoon.

Drought reconstruction from 1794 to 2003 in the eastern part of northwestern China was developed by calibrating tree rings of Xinglong Mountain with the PDSI (Fang et al. 2009b). The most significant drought period of the reconstructed series occurred during the 1920s to the 1930s. The drought reconstructions revealed that a decadal-scale climate transition occurred in 1921. This reconstruction is representa-

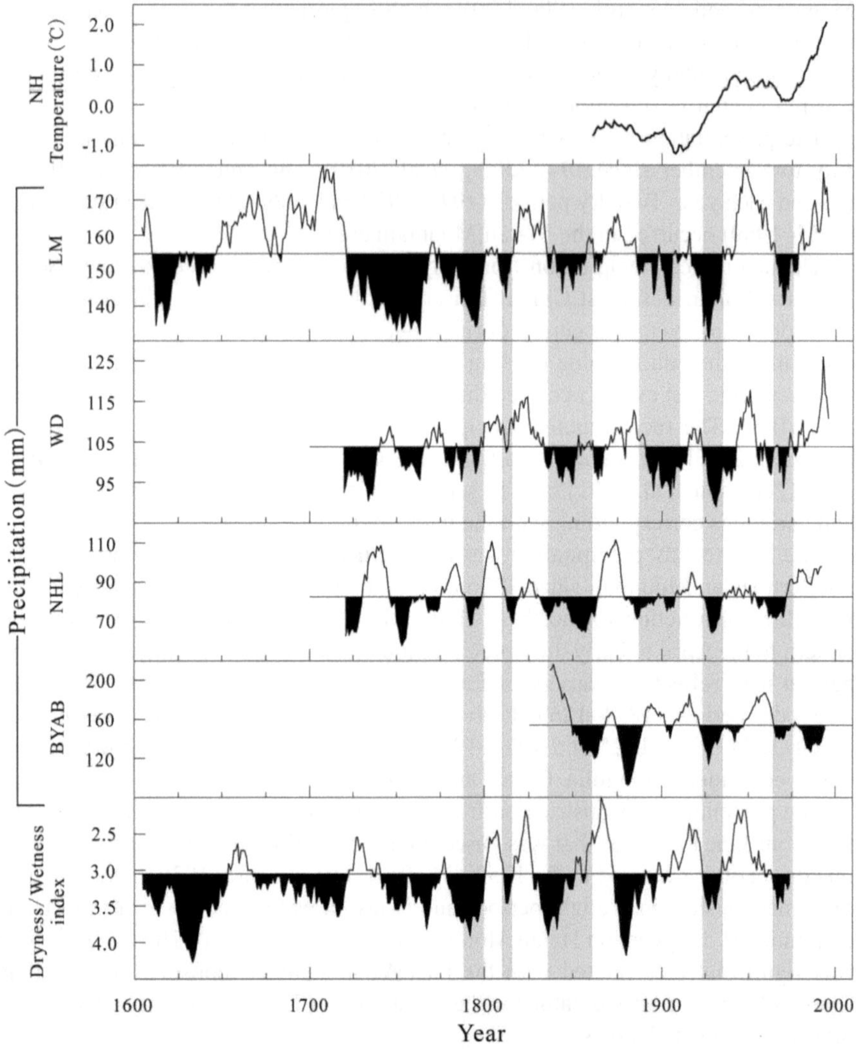

Fig. 3.31 Comparison of dryness/wetness indices and precipitation reconstructions from LM, WD, the northern Helan Mountains, and Baiyinaobao on a decadal to a multi-decadal time scale using a 10-year moving average. The top panel is the Northern Hemisphere temperature curve (Folland, et al. 2001) after a 10-year moving average. Horizontal lines are the means. (from Liu et al. 2007)

tive of regional drought patterns in eastern northwestern China and is different from patterns in western northwestern China.

The PDSI variation was reconstructed for the past 391 years for eastern Guiqing Mountain, located in Gansu (Fang et al. 2009c). The drought in 1997 was focused on the Guiqing Mountain area and appeared to be the most extreme drought to occur during the reconstruction period. The reconstructed droughts were significantly correlated with sea surface temperature in the northern Indian Ocean, the western equatorial Pacific Ocean, the eastern equatorial Pacific Ocean, the eastern China Sea and the Sea of Japan. The linkages to the Indian and Pacific Oceans suggest that regional moisture variations are related to the Asian monsoon patterns.

Precipitation of February—early July for the Wu Dangzhao region and February—mid-July for the La Madong region, in Inner Mongolia were reconstructed (Liu et al. 2007). After applying a 10-year moving average, the precipitation reconstructions and two other tree-ring-based precipitation reconstructions from northern Helan Mountain range and Baiyinaobao were shown to vary synchronously (Fig. 3.31). The most severe drought to occur over a broad area in northern China in the last 375 years occurred in the 1920s. The wettest period occurred in the late 1990s.

Tree-ring-based precipitation reconstruction series from Baotou (Liu et al. 2001), Baiyinaobao (Liu et al. 2003b), Daqing Mountain (Liu et al. 2007), Chifeng-Weichang (Liu et al. 2010b), Ma'an Mountain (Jiang et al. 2010), and Hailar (Liu et al. 2009b) varied in synchrony at the decadal scale, together reflecting the variability of the East Asian summer monsoon during this period. Most noteworthy is that, these precipitation reconstruction series reflect the most serious large-scale drought in northern China (even in the entire East Asia area) in the late 1920s. In addition, precipitation generally contained 7-year, 8-year, and 11-year periods (Liu et al. 2003b, 2004b, 2008c, 2009c, 2010b; Gou et al. 2001), indicating that precipitation in this region is mainly affected by solar activity.

The Tibetan Plateau A 2326-year (326 B.C.–2000) tree-ring-based spring precipitation reconstruction was established for the Dulan area of the northeastern Qinghai-TP (Zhang et al. 2003). The climate on the plateau has undergone oscillations and rapid changes during the last two millennia. The greatest change in spring precipitation during the last two millennia occurred in the second half of the fourth century. The climate proxy data reveal that the North Atlantic MWP and the LIA were accompanied by climate changes on the northeastern Qinghai-TP.

Annual precipitation from July to June in the last 1000 years was reconstructed for Delingha based on tree rings (Shao et al. 2005). Wet periods occurred in 1520–1633 and 1933–2001, whereas dry periods occurred in 1429–1519 and 1634–1741. In addition, the magnitude of precipitation variation was lower before 1430 but increased from 1430 to 1850, and it decreased again after 1850.

The relationship between precipitation variation in Delingha, Qinghai, and solar activities during the last 400 years was statistically examined (Huang and Shao 2005). The main precipitation cycles in Delingha were similar to the solar activity cycles during the last 400 years, and solar activity appears to affect the long-term variation of precipitation at the century scale. The abrupt precipitation changes in Delingha may be linked to the inter-decadal changes of the EASM in the last century. Abrupt changes in Delingha occurred before abrupt changes in eastern China, which may be an important indicator for long-term precipitation forecasts (Huang et al. 2006).

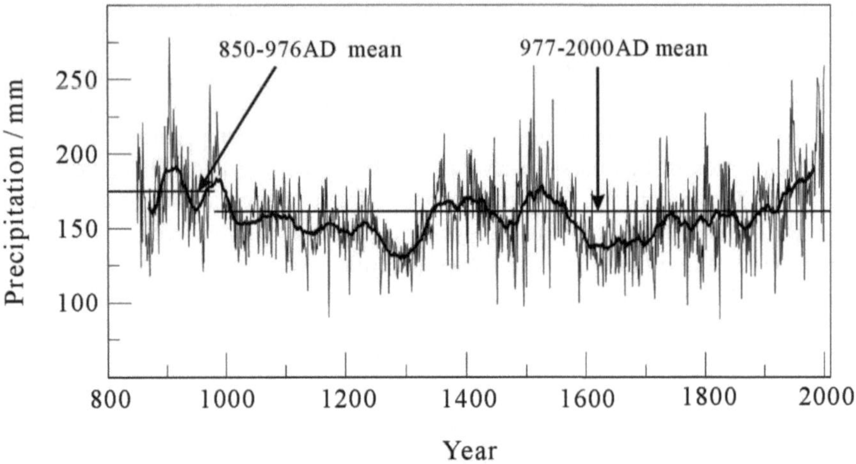

Fig. 3.32 Annual (from July to June) precipitation reconstruction for the Dulan region during 850–2000. The mean precipitation from 977 to 2000 was 157 mm and 1σ=25 mm. The smoothed line represents the 40-year moving average. (from Liu et al. 2006)

Annual precipitation (from previous July to current June) variations in Dulan since 850 were reconstructed (Liu et al. 2006; Fig. 3.32). The highest precipitation over the northeast TP in the last 1000 years occurred from 1900 to 2000; the highest precipitation with highest variability occurred from 1962 to 2000. The variability of annual precipitation was large when there was increased precipitation and small when there was decreased precipitation. During the 20th century, as the temperature increased, precipitation and its variability significantly increased, and drought and flooding occurred more frequently. Treydte et al. (2006) documented that the 20th century was the most humid century to occur in northern Pakistan in the past 1000 years. The series with a 40-year moving average (low-frequency signal) was significantly correlated with seven temperature curves from the NH for different time periods. In general, the observed climatic pattern was a result of the combination of warm-wet and cold-dry periods on the centennial to millennial scale for the Northern TP during the past 1000 years.

Drought variability of the TP is an important component of the Asian monsoon. Long-term information regarding the history of drought is, however, limited because the instrumental records are of short duration. A tree-ring chronology was developed to reconstruct the history of drought for northern Tibet (Wang et al. 2008b). The May–June PDSI reconstruction for the past 500 years showed that the seventeenth century was the driest century, whereas the eighteenth, nineteenth and twentieth centuries were the wettest. Significant abrupt changes were found for 1530 and 1715, indicating the onset of dry and wet conditions during the Little Ice Age. An abrupt change occurred around 1965, which may foreshadow the beginning of prominent regional warming.

Overall, precipitation in the spring displayed by the different curves increased during 929–1031 on the TP, which corresponded to the North Atlantic Medieval Warm Period. In the Little Ice Age, three periods of decreased precipitation occurred during the 1450s–1510s, the 1640s–1720s and the 1790s–1820s. There was more precipitation during the Medieval Warm Period and less precipitation during the Little Ice Age. As a result of the current climate warming, precipitation may again increase.

The Indian monsoon area Great progress has been achieved in East and South Asia with regard to tree-ring-based climatology. The climate of the Indian peninsula is most comparable to the southern Asian monsoon as a result of its geographical location.

Monsoon systems are the dominating feature of the climate throughout southern Asia, and they are characterized not only as being temporally and spatially variable in their precipitation patterns but also as having a reliable pattern of wet and dry seasons. It is important to obtain quantitative information on precipitation and related atmospheric and oceanic parameters prior to the period of recorded meteorological data to understand natural variability, and thus better prepare for projected climate changes indicated by model simulation scenarios.

The tree ring studies of precipitation in Asia suggest that tree ring width, density and stable isotopes are affected by precipitation from April to July which is related to the East Asian monsoon; stronger monsoons result in more precipitation and wider tree-ring widths.

A well-replicated ring-width chronology, derived from a combination of tree-ring samples from two adjacent homogeneous sites, was used to reconstruct precipitation for the non-monsoon months (previous year October to concurrent May) since 1171 (Yadav and Park 2000). The wettest and the driest non-monsoon months during the thirteenth and fourteenth centuries were documented. A precipitation surplus, which has been pronounced since the 1950s, was recorded for the current century.

Spring precipitation since 1560 was reconstructed for the western Himalayan region (Singh et al. 2006). An unprecedented precipitation increase occurred during the late twentieth century. The western Himalayan region experienced unstable springs during the Little Ice Age. Running means for 10-, 20- and 30-year intervals indicate that the wettest period occurred during the twentieth century, while the 10-year mean shows that the driest period occurred in the seventeenth century and the second driest period occurred in the twentieth century.

A 523-year (1481–2003) tree-ring-width index chronology from Kerala, southern India was compiled (Borgaonkar et al. 2010). Since the late 18th century, a higher frequency of low tree growth was observed in drought years (deficient Indian summer monsoon rainfall (ISMR)) associated with El Niño. Prior to that time, many low tree-growth years were detected during known El Niño events, which were probably related to deficient ISMR. The general relationship between ISMR and El Niño is known to be negative, and the spatial correlations between the tree-ring chronology and sea surface temperatures (SSTs) over the Niño regions follow similar patterns.

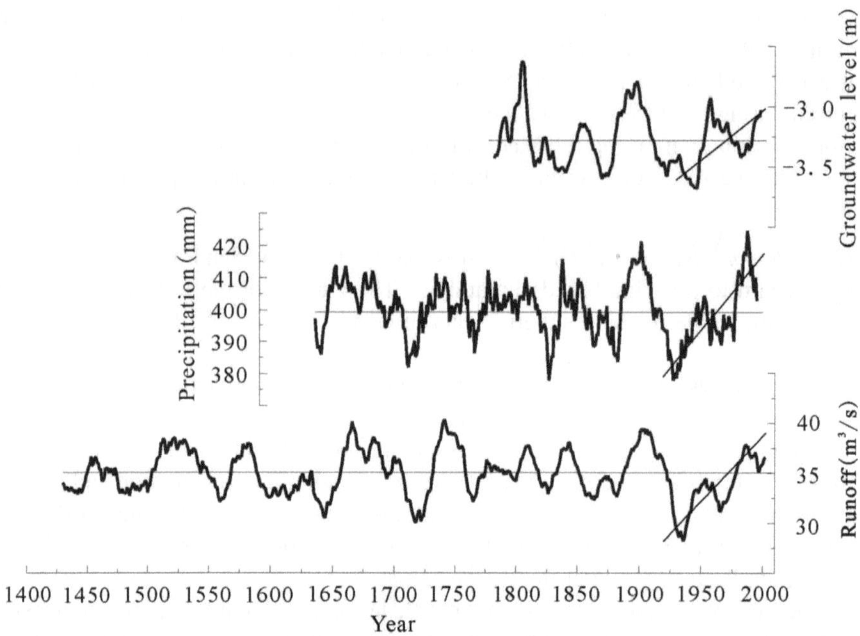

Fig. 3.33 Comparisons of the reconstructed precipitation, runoff, and the groundwater level (Sun et al. 2006) in the watershed of the Heihe River. All curves are smoothed by an 11-year moving average. The oblique lines delineate increasing trends. (from Liu et al. 2010c)

Ring-width chronologies of *Cedrus deodara* and *Pinus gerardiana* from a homogeneous moisture-stressed area in Kinnaur, Himachal Pradesh were developed and March–July precipitation from 1310 to 2004 was reconstructed (Singh et al. 2009). In the past 694 years of the reconstruction, the wettest period occurred during the twentieth century (1963–1992), and the driest period occurred during the eighteenth century (1773–1802). The relationships between the reconstructed precipitation and the Indian summer monsoon on an inter-decadal scale, and the SOI, PDO and NAO indices indicate the potential utility of long-term reconstructions in understanding large scale climate variability.

3.6.2 Human Impact

The expansion of the Industrial Revolution brought prosperity and development to human societies, but human activity has significantly altered the climate and environmental systems. Since the 20th century, the earth has been warming, and extreme climate events have occurred frequently. These changes pose a serious threat to the environment and to ecosystems, and they increasingly impact economic and social development. For example, in recent years, local development in the Heihe

River Basin in western China has been hampered as a result of an increasing population and increasing water usage demands for industrial and agricultural purposes.

Was this situation caused by the deterioration of natural resources or by human activities? In studies that distinguish between natural factors and human impacts, tree ring hydrology has provided a clear answer.

Qilian Mountain is an important mountain system in western China because its surrounding regions depend on it as a source of water (e.g., the Heihe River Basin, the Hexi Corridor, the Ejin Qi). Therefore, the history of the climate, the environment and the water resources of the Qilian Mountains is significant for ecologically, socially and economically sustainable development and for the survival of local residents.

Research conducted on the westernmost forest of the Qilian Mountain demonstrated that juniper (*Juniperus przewalskii* Kom) growth was affected by temperature and precipitation, and that tree ring width was sensitive to the climate variations, including the area's drought history (Tian et al. 2007).

On the basis of tree-ring-width analysis, total precipitation from July to June, from 1634 to 2000, and the average runoff from September to June in the middle section of the Heihe River from 1430 to 2007, were reconstructed (Liu et al. 2010c). The hydrological history of the Heihe River Basin in western China was systematically analyzed by comparing the groundwater level history of Ejin Qi downstream of Heihe (Sun et al. 2006; Fig. 3.33).

The severe drought that occurred in the late 1920s was captured by the precipitation, runoff (1925–1935) and the groundwater level (1925–1940) reconstructions. This event has been found in many tree-ring-based studies that were conducted throughout northern China and in Korea and Mongolia (Gou et al. 2007; Li and Zhang 1992; Xu 1997; Liu and Ma 1999; Liang et al. 2003; Liu et al. 2003a, 2005b, 2007, 2009b, c). These studies reveal that the East Asian summer monsoon circulation was weak and had low precipitation over a large-scale region in East Asia during that time period.

Ninety-one percent of the water resources in the Heihe River watershed were influenced by the precipitation in the recharge region (headwaters), whereas only 9 % of the water resources in the Heihe River watershed were influenced by temperature (Zhang et al. 2004a). The headwaters in the Qilian Mountains are the pace-setter for the spatio-temporal variations in the water resources of the Heihe River watershed. In this region, atmospheric circulation controls precipitation, precipitation controls runoff, and runoff from the middle section controls the groundwater level in the Ejin Qi oasis (Li et al. 2006b). Two long-term periods of low flow occurred (e.g., 1590–1656, 67 years, 33.2 m^3/s and 1923–1978, 56 years, 32.2 m^3/s). The period from 1925 to 1935 was the lowest in the reconstruction, which corresponded to low precipitation in the precipitation reconstruction and severe droughts that occurred throughout northern China.

When precipitation increases in the upper section of the river, the runoff is high in the middle section, and the groundwater level is low (i.e., close to the ground surface level in the downstream section) and *vice versa*. These relationships can be observed from our three tree-ring-based hydrologic reconstructions. We found

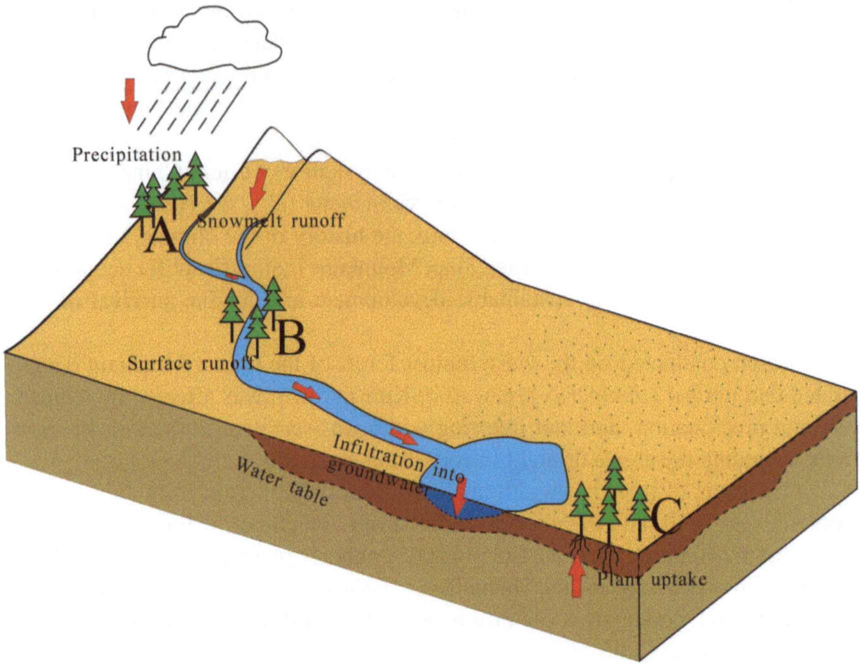

Fig. 3.34 The relationships between precipitation (**a**), runoff (**b**), and groundwater level (**c**) in the Heihe River watershed. (from Liu et al. 2010c)

particularly good correspondence between precipitation and runoff in the natural variations that occurred before 1940 (Fig. 3.34). When precipitation increased in the upper section, the runoff was high in the middle section, and the groundwater level was low in the downstream section and *vice versa*. In addition, precipitation in the upper section was closely related to the downstream groundwater level ($r=0.51$). The three curves displayed synchronized variations before 1940, and they represent the natural variations without any man-made disturbances (Fig. 3.34). The notable low groundwater levels that occurred from 1925 to 1940 corresponded with the low precipitation in the upper section and the low runoff in the middle section. However, the groundwater level had a less direct relationship to the precipitation and the runoff after 1940, which may be the result of increased water consumption in the middle and downstream sections that has occurred since that time.

Human activities, such as the production of dams and reservoirs built for agricultural irrigation, largely disrupted the relationships between precipitation, runoff and groundwater after the 1940s. As recorded, there were 2 reservoirs in the 1940s, and there were 95 reservoirs in 1985 (Wang and Cheng, 1998). As a result, a mass of water has been kept in the middle section. Dams and reservoirs were primarily built between 1954 and 1963 and between 1968 and 1978. During these two time periods, precipitation and runoff were at their maximum levels, but groundwater was at its minimum level (Fig. 3.34).

The runoff data from the Zhengyi Gorge Hydrologic station, which is located at the boundary between the middle and downstream sections, showed that the runoff was reduced from 11.9×10^8 m^3 in the 1950s to 9.42×10^8 m^3 in the 1980s and further to 6.91×10^8 m^3 in the 1990s because water consumption increased in the middle section. However, at the same time, observational data (Wang and Cheng, 1998) and the reconstruction derived in our study show that runoff in the upper section continuously increased during the past 50 years. Since the 1980s, the water output from the middle section to the downstream section has increased because of the management and increases in the groundwater table (Wang and Cheng, 1998), which is what we observed in our results.

Because there was insufficient water flow to the downstream section of the river, the groundwater showed a lag effect (Fig. 3.34), and two big lakes (the western and the eastern Juyan lakes located at the end of the river) disappeared.

Finally, the major periodicities also revealed that three tree-ring-based hydrologic indices had close relationships because they all commonly displayed the most significant cycles of 80 (78–82), 50 (49–58) and 2-year periods.

Some studies have indicated that there have been considerable 70–80-year cycles in the Hetao region (Qian and Lin 2009) (located approximately 500 km east of the Heihe River watershed) that correspond to the oscillation in the global climate system of approximately 60–80-year cycles, which is interpreted as an internal oscillation in the atmospheric-oceanic system (Qian and Lin 2009; Schlesinger and Ramankutty 1994). A 50-year cycle was observed in the upper Yellow River basin (Gou et al. 2007), and it is believed to be associated with drought-flood characteristics in China (Lin et al. 1996). It could be related to an oscillation in the climate system of a 40–50-year cycle, which has been considered to relate to the irregular oscillations of the thermohaline circulation in the North Atlantic Ocean (Qian and Lin 2009; Greatbatch and Zhang 1995). However, a 2-year cycle is similar to the quasi-biennial oscillation (QBO). The effects of the QBO exist at a large scale and may indicate sea-land coupling (Qian et al. 1998). These cycles correspond to large-scale oscillations found in the climate system and are related to atmospheric-oceanic interactions (Qian and Lin 2009).

Our research indicates that the water resources of the Heihe basin have increased since 1950. The precipitation and the middle reaches of the river flow from the source have increased. However, because of a significant human population increase and intensified human activities (e.g., dramatic water consumption by industrial and agricultural entities), the water usage demands in Heihe have increased sharply, and water shortages have occurred throughout the basin. It is clear that human activities have had important negative effects on the environment.

3.6.3 *The Influence of Solar Activity on Climatic Changes in the Last 2500 Years*

The global climatic system is deeply influenced by millennial scale cycles. Numerous geological evidence, from North America to Europe, and from the Middle East

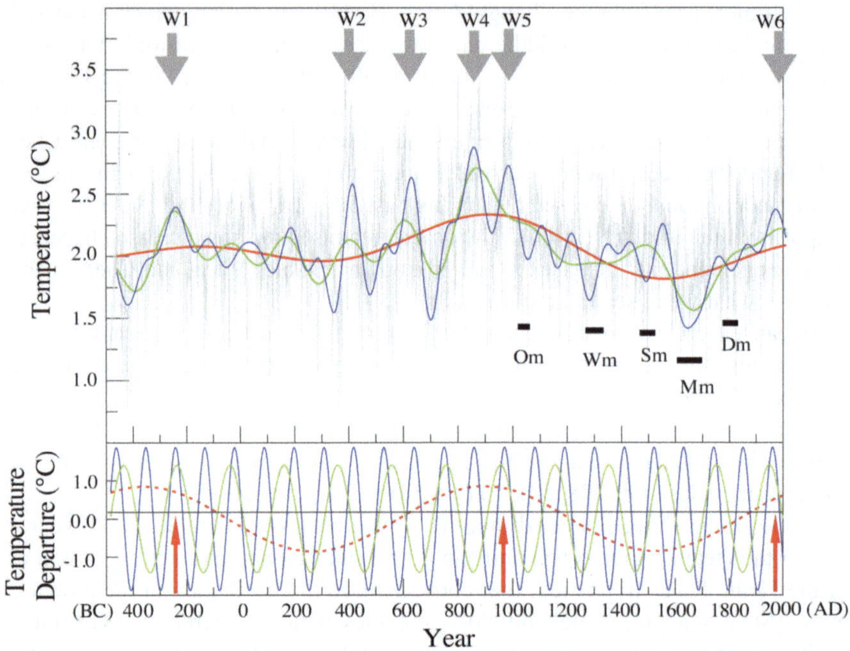

Fig. 3.35 The decomposition of the main cycles of the 2485-year temperature series on Tibetan Plateau and periodic function simulation. Top: Grey line- Original series; red line-1324 cycle; green line-199a cycle; blue line-110a cycle. Bottom: three sine functions with timescales 1324a, red dash line (y=0.848sin (0.005 t+0.23)), 199a, green line (y=1.40sin (0.032 t-0.369)), 110a, blue line (y=1.875sin (0.057 t+2.846)), time t is year from 484BC to 2000AD. (from Liu et al. 2011b)

to the East Asia, demonstrated the existence of them, and the exact value of the millennial scale cycle was modified as 1374±502a by Bond et al. (1997). The inducement mechanism of these cycles may be associated with solar activities (Bond et al. 1997; Xu 1998; Braun et al. 2005), and perhaps, the inherent cycle of solar (Ma 2007; Eddy 1977).

This moderate millennial-scale cycle has lasted for a long time and it has not disappeared during the last 2000 years. However, it has been mentioned rarely in the climate study for the last 2000 years, the possible reason is that the temperature series reported were not long enough.

Cold and warm intervals occurred alternatively in the 2485 years temperature curve, 600 years increase and 600 years decrease. The regularity of climate change did obey this rule in the past, as well as in the future. It could be the turning points from a cold interval to a warm at 343—367AD (the East Jin Event) and 1595—1713AD (LIA). Likewise, MWP is a turning point from a warm to a cold phase under the control of the millennial-scale cycle.

The 1324a cycle in Fig. 3.35 approximates the famous dominant climate change cycles 1374±502a in the Holocene (Bond, et al. 1997) and the 1470a cycle in Dansgaard/Oeschger (D-O) events in the last glacial. All these cycles are called millennial-scale cycles. When we viewed the curve, the 2485-year temperature series could be divided into four stages: stage I (250 BC–369 AD), the temperature gradually decreased for about 600 years; stage II (370–999 AD), the temperature gradually increased for nearly 600 years; stage III (1000–1600 AD) is another 600 years with temperature decrease; and stage IV (1601–2000 AD), temperature increased step by step for about 400 years until now. The entire series from stage I to IV forms sawtooth waveform. Stage I and II showed a complete millennial scale cycle, and stage I to III displayed one and a half millennial scale cycles.

The significant multi-century scale cycles were 199a and 110a in the 2485 years temperature series in the central-eastern TP. They were close to the quasi-200a (Suess cycle or de Vries cycle) (Braun et al. 2005) and quasi-70—100a cycle of solar activity (Gleissberg cycle) (Eddy 1977; Sonett et al. 1990), respectively, which have been proved existing in extensive geological records (Gou et al. 2010; Damon and Sonnett 1991; Wang et al. 1999a; Cook et al. 2000; Hodell et al. 2001; Neff et al. 2001; Agnihotri et al. 2002; Fleitmann et al. 2003; Zhong et al. 2004; Delmonte et al. 2005; Xu et al. 2006; Raspopov et al. 2008; Ogurtsov et al. 2010). Quasi-200a and quasi-100a cycles are extremely notable in global temperature research covering the last millennium.

Multi-century cycles related to solar activities strongly impact the earth climate system. They did leave footprints in various geo- biological carriers. Thus, it is not a surprise to detect these two centurial scale cycles in the Tibetan temperature curve.

It was reported that global warming in the late of 20th century resulted from the overlap of multiple cycles (21a, 64a and 179a) commonly in the uplift phases (Qian and Lu 2010). However, in our opinion, except for the cycles above, the millennium-scale cycles also made a great contribution to the warming, as well as to other warming periods historically.

It is interesting to note that the sunspot minimums corresponded to the cold periods in the 2485 years temperature series very well. For example, the cold period in 1015–1040 AD corresponded to the Oort minimum (Om, 1010–1080 AD), cold 1258–1309 AD (C5) to the Wolf minimum (Wm, 1270–1350 AD), cold 1451–1539AD to the Spörer minimum (Sm, 1430–1520 AD), cold 1620–1715 AD (C6) to the Maunder minimum (Mm, 1620–1710 AD) and cold 1766–1824 AD to the Dalton minimum (Dm, 1787–1843 AD) (Usoskin et al. 2003; Fig. 3.35).

Previous study (Liu et al. 2009b) has stated that the 2485-year temperature curve was quite similar to the solar radiation curve at low frequencies. Further analysis shows that solar activity was a major driving factor for climate variations on the central-eastern TP, perhaps even extending beyond that region. The millennial cycle of solar activity determined long-term trends (>1000a) of temperature variations, meanwhile, centennial cycles determined the amplitude of temperature variations. It was also worth noticing that almost all the sunspot minima correspond to historical low temperature intervals. However, the interannual- to decadal-scale temperature changes may be related to atmospheric circulation. As Zhu (1972) remarked "The big changes of the earth's climate was controlled by solar radiation, but small changes by the atmospheric circulations".

3.7 Variability of Asian Monsoon Recorded by Pollen

3.7.1 Introduction

Pollen grains deposited in sediments reflect vegetation changes that result from climate variability through time. By combining pollen profiles from widely distributed sites across China we can study temporal and spatial changes in the East Asian monsoon. In the Holocene these records tell us that the East Asian monsoon gradually weakened from southeast to northwest in China, followed by humid, semi-humid, semi-arid, arid and extreme arid climate in the northwest. There are three major vegetation regions within the arid-monsoon system: 1) the eastern humid forest zone, 2) the central semi-arid steppe and 3) the western arid desert area (Fig. 1.4). These three regions reflect fully the distribution rule for vegetation zones. Due to the strong influence of frequent cold air invasions from the north, monsoon forest vegetation in China, compared with the same latitudes in wet areas in the Northern Hemisphere, occur much further south. In contrast, desert vegetation in the northwest is displaced far to the north because the TP and adjacent warm temperate deserts serve to increase aridity there.

Fossil pollen is one of the most widely used paleoclimate proxies for their value as indicators of past vegetation and associated climatic conditions. In this chapter, we will first discuss the pollen proxies that reflect the variability of monsoon climate, and then show the sequence of orbital and millennial scale monsoon climate changes recorded in pollen records. We will focus on pollen records from the last interglacial through the Holocene.

3.7.2 Pollen Proxy

Pollen records in monsoon regions can be used to reconstruct the history and variation of the East Asian monsoon by reconstructing vegetation patterns and their responses to climate change (in both temperature and precipitation) over a variety of timescales. Here we introduce the main pollen proxies that are widely used in paleoclimate and paleomonsoon reconstruction.

3.7.2.1 Coniferous and Deciduous Mixed Tree Pollen Percentage

Coniferous and deciduous mixed trees are the typical vegetation in the transition zone from frigid-temperate to temperate climate. They currently grow within a temperate and humid marine climate with annual average precipitation within 550–1100 mm. They are widely distributed within sub-alpine coniferous forests and low-hill deciduous forests that occur in the mountains of the south temperate zone. The relative changes in the proportion of deciduous trees to coniferous trees can provide information on environmental change, with the former increasing relative to the latter under a colder and drier climate (Xia 1988; Ren and Zhang 1998; Jiang et al. 2008).

3.7.2.2 Deciduous Tree Pollen Percentages

Deciduous forests are typical of the vegetation in the east humid part of the warm temperate zone. They are distributed mainly in the southern portion of Northeast China and North China, which have hot-rainy summers and cold-dry winters with annual average precipitation of 500–1000 mm. In addition, these trees can also live at higher elevations, beyond the evergreen forest belt of the sub-tropical and tropical mountains. A thriving deciduous forest indicates a warmer and wetter environment and the strengthening of the summer monsoon, while the increase of coniferous tree pollen suggests a colder and drier climate (Yi et al. 2003).

3.7.2.3 Evergreen and Tropical Tree Pollen Percentages

Evergreen forests are the typical vegetation of subtropical humid regions in China, which are distributed in the Qinling Mountains to the north, along the north of the tropical mountains to the south, in the mountains of Yungui Plateau and South Tibet to the west, and in the mountains of central Taiwan to the east. Tropical forests and rain forests mainly grow on the Hainan Peninsula, Leizhou Peninsula, Taiwan Peninsula and along the southern margin of Yunnan. These areas are subject to subtropical and tropical monsoonal climate, with relatively high temperature and precipitation. The increase of these tree pollen types in the sediments suggests a warm and wet climate, and an increase in summer monsoon intensity (Liu et al. 1992; Zheng et al. 2003; Zhou et al. 2004, 2005; Liew et al. 2006; Xu et al. 2010; Qin et al. 2011).

3.7.2.4 Steppe Pollen Percentage and the Ratio of *Artemisia* to Chenopodiaceae (A/C)

Steppe pollen is mainly composed of Poaceae, *Artemisia*, Leguminosae, Cyperaceae, Compositae, and Chenopodiaceae. At present the steppe can be found in Songliao Plain, Inner Mongolia Plateau, and on the Loess Plateau, which has a typical temperate continental climate with annual average precipitation of 200–450 mm, mostly occurring in the summer. Generally, *Artemisia* and Chenopodiaceae characterize relatively dry environments, while Cyperaceae and Poaceae are more abundant under comparatively moist conditions. Their relative contents can serve as a pollen proxy to indicate relative humidity and aridity (Xu et al. 2005, Fowell et al. 2003).

With respect to ecology, *Artemisia* thrives under relatively humid conditions, and higher percentages of *Artemisia* are consistent with steppe-like environments; while Chenopodiaceae percentages tend to be greater in desert regions. So the A/C ratio was proposed as a measure of relative humidity/aridity, where high A/C ratios represent a wetter climate on the condition that the sum of *Artemisia* and Chenopodiaceae accounts for a high proportion of total pollen (El- Moslimany 1990; Sun et al. 1994; Xu et al. 2007; Zheng et al. 2008). This proxy has been confirmed as an indicator of effective moisture by modern pollen studies in north and northwest China (Huang 1993; Weng et al. 1993; Cour et al. 1999; Liu et al. 1999; Shang et al. 2009; Zhao and Herzschuh 2009), and has been widely applied in palaeoclimate studies (Gasse et al. 1991; Herzschuh et al. 2004, 2006; Chen et al. 2008a; Ren and Zhang 1998; Li et al. 2011b).

3.7.2.5 Desert Pollen Percentages

Dersert vegetation grows mainly in arid areas of Northwest China. The annual average precipitation there is less than 250 mm while evaporation is dozens of times higher. *Ephedra* is a typical desert taxa in northwestern China (Shen 1995; Luo et al. 2009). The pollen percentage of *Ephedra* increases with a decrease of precipitation in surface pollen assemblages from China (Li et al. 2005a). Other high content xerophytic pollen types in desert regions, like *Nitria*, Zygophyllaceae, *Tamarix* and Hippophae, can also serve as characteristic desert indicators (Cour et al. 1999; Jiang and Ding 2008).

3.7.2.6 Arboreal Pollen Percentages and Pollen Concentration

In humid areas strongly influenced by the East Asian monsoon, hygrophyte and subhygrophyte forests grow predominantly along with arboreal trees. The Changbai Mountains can be taken as an example, where arboreal pollen accounts for 70–90 % of the surface sediment pollen. In the arid area of northwestern China at the limit of the summer monsoon, xerophyte herbs account for most of the pollen in the surface sediment, and even in the *Picea* forest of the Kunlun mountains, the arboreal pollen percentage is only 7–29 %, and other types are dominated by xerophytes (An et al. 1991b). Therefore, changes in arboreal pollen percentages and in the arboreal/nonarboreal pollen ratio (AP/NAP) are often used for the palaeoclimatic interpretation of pollen records from forest-steppe and steppe environments, where increases in AP likely correspond to the onset of wetter conditions (Yi et al. 2003; Xiao et al. 2004; Tarasov et al. 2006; Wang et al. 1996; Wu et al. 2007; Zhao et al. 2009; Chen and Wang 2012).

To some extent, pollen concentration may be considered as a basic indicator of environmental change because vegetation coverage is strongly determined by the environment, although there are uncertainties due to the complexity of pollen production, transportation and deposition, and their relationship to vegetation and climate (Luo et al. 2009; Zhao and Sun 2010).

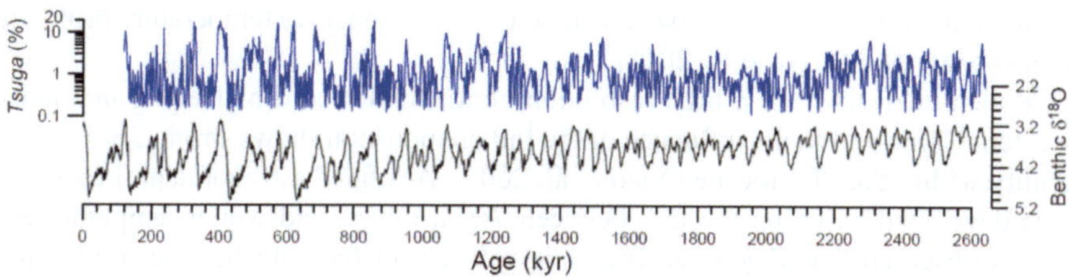

Fig. 3.36 Comparison of Heqing *Tsuga* content with global ice volume (benthic $\delta^{18}O$). (from An et al. 2011)

3.7.2.7 *Picea/Abies* Pollen Percentages

Picea and *Abies* grow mainly on the mountains in northeastern and northern China in frigid-temperate coniferous forests. In Southwest China, Tibet and Taiwan, they live within sub-alpine evergreen coniferous forests or coniferous and deciduous mixed forests with some deciduous trees. The altitude of their distribution decreases from low to high latitudes, from 4300 m in the Himalaya Mountains to less than 2000 m in northeastern China, where their upper limit is also near the tree line in terms of vertical zonality.

These two genera have similar ecological demands and both prefer cold and humid environments. They are in climates with annual average temperatures of 2–8 °C and average temperatures of 10–15 °C for the warmest month. They also require relatively large amounts of water, with precipitation higher than 500 mm and relative humidity over 50 %. A coniferous forest would be replaced by shrubs and steppe vegetation should the annual average precipitation fall below 400–500 mm, even under temperatures suitable to *Picea* and *Abies* (Kong and Du 1980; Xu 1980; Wu 1985; Sun 1989). In the East Asian monsoon area, the expansion of *Picea* and *Abies* indicates decreasing temperatures and increasing regional effective moisture (An et al. 1991c; Ren and Zhang 1998; Yang et al. 1998; Li et al. 2002; Makohonienko et al. 2008; Zhou et al. 2010).

3.7.2.8 *Tsuga* Pollen Percentages

Tsuga flourishes under warm and humid conditions, with moderate temperature ranges at elevations of 2000 to 3500 m in southeastern Asia (Farjon 1990). Researchers found that *Tsuga* is sensitive primarily to winter temperature and annual temperature range (An et al. 1990a; Yang et al. 2009). *Tsuga* is presently found south of the Qinling Mountains (~33°N) under modern warm conditions. During the Holocene Optimum *Tsuga* extended into northern China and even sparsely into the coastal areas of northeastern China (Jia et al. 1989; Kong et al. 1992), when annual temperatures were 2 to 3 °C warmer than today (An et al. 1990a). During the Last Glacial Maximum (LGM), steppe vegetation dominated and *Tsuga* was absent

(Guo et al. 2005). Therefore, *Tsuga* can be used as a sensitive temperature proxy in monsoonal regions (An et al. 2011).

Pollen analysis from a high-resolution continental record of Heqing in southwestern China within the influence of the Indian monsoon shows strong *Tsuga* variability within the Pleistocene (Xiao et al. 2007). A refined astronomical time scale was developed based on the link between *Tsuga* ranges and winter temperatures. It was observed that *Tsuga* percentage is well correlated with benthic $\delta^{18}O$, with high values corresponding to ice volume minima, in response to glacial-interglacial changes. *Tsuga* percentage sequences also show that *Tsuga* variation is controlled by earth orbital parameters. Spectral analyses indicate that proportions of *Tsuga* taxa follow 100-, 41-, 23- and 19-kyr periodicities (Fig. 3.36, An et al. 2011).

3.7.3 East Asian Monsoon Variability Recorded by Pollen

East Asian monsoon variability as recorded in pollen records reflect a broad range of climate cycles and other fluctuations, including millennial-scale events, the Youger Dryas (YD), Heinrich (H*x*) events and Bond cycles.

There are abundant pollen records now available for the most recent 125 ka BP, with good chronological control (Li et al. 1998; 2003) and these are important in the study of orbital-scale monsoon climate variability. The last interglacial (at the time of marine isotope stage MIS-5) was one of the warmest periods in the Quaternary, it was also a time of abundant precipitation in the East Asian monsoon area. During the last interglacial, the landscape in the central portion of northeastern China was covered by a warm temperate woodland to grassland, with dry herb-based vegetation concentrated to the north, associated with scattered birch, pine and oak. Northern vegetation featured broad-leaved forests or mixed forests, containing many sub-tropical species such as *sweetgum, camphor, Albizzia*, belonging to the warm temperate—subtropical range. The entire Ordos Plateau and northern LP from the southeast to the northwest were dominated by forest steppe, shrub steppe or meadow grasslands. Changjiang plain was composed of subtropical evergreens and deciduous broad-leaved forests (Editorial Committee of FOC 2004).

Pollen records from the South China Sea ODP1144 station also reflect East Asian monsoon orbital-scale changes. The sediment from the top 0–225 m contain a rich collection of spores and pollen. In this section, the relative percentage of *Pinus* and herbs rise and fall, and correspond to deep-sea oxygen isotope records for MIS 1–8 (Sun and Luo 2001). *Pinus* dominated the interglacial periods (MIS 7, 5, 3, 1) and herbs were concentrated during glacials (MIS 8, 6, 4, 2), perhaps due to a lowering of sea level which would allow steppe vegetation to develop on the exposed continental shelf. The herb content was lowest during glacial MIS8, but was dominant during the LGM. This change could reflect a response to relatively dry and cold climate conditions during the LGM (Sun and Luo 2001). The pollen zones from sediment in the bottom 225.7–501.3 m correspond to deep-sea oxygen isotope stages MIS 11–29 (Huang and Sun 2002). In the interglacial period, pollen assem-

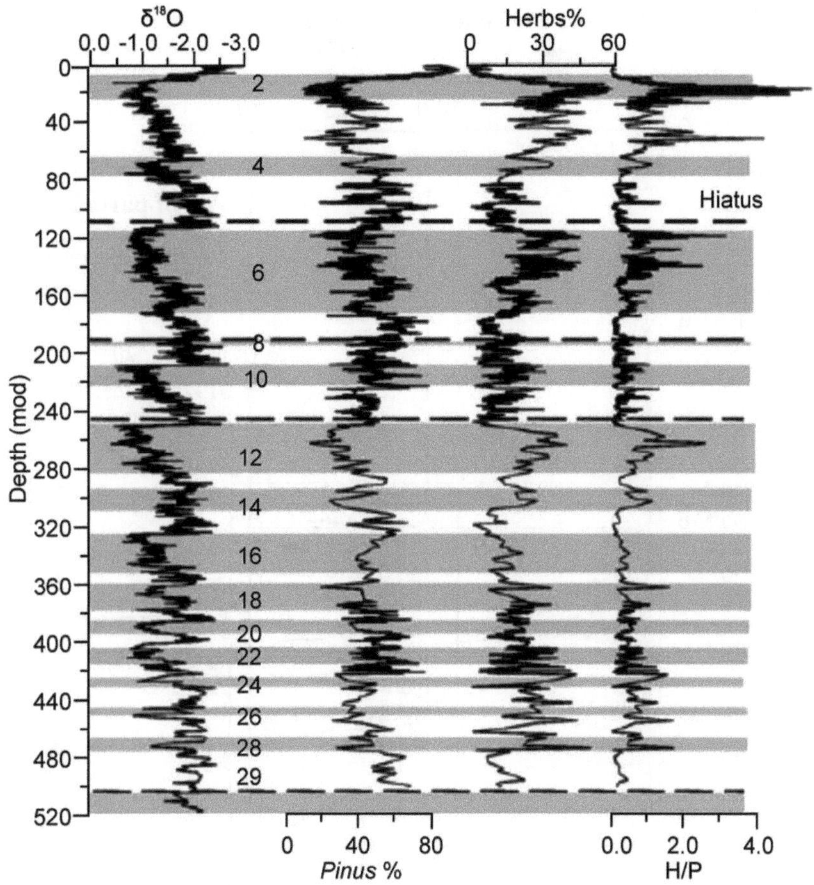

Fig. 3.37 Comparison of pollen percentage curves of *Pinus* and herbs with that of Marine Oxygen Isotope Stages at ODP site 1144, and downcore changes of pollen percentage ratios of herbs vs. *Pinus* (H/P). (from Sun et al. 2003)

blages were characterized by increasing *Pinus* and fern spores and decreasing herb content, with an opposite trend during glacial times. The winter monsoon became strengthened during glacial times, since MIS 16. The summer monsoon was intensified during interglacials, particularly MIS 15 and MIS 5 (Sun et al. 2003). The pollen sequences also show that vegetation changes in the northern South China Sea were controlled by periodic changes in Earth's orbit (Fig. 3.37). Spectral analyses of pollen percentages of pine and herbs from ODP1144 have documented 100-ka, 41-ka and 21-ka Milankovich cyclicities, as well as a tropical-specific 10-ka semi-precession cycle, demonstrating the impact of astronomical forcing on climate (Sun et al. 2003).

Deep-sea records of isotope stage 3 (40–25 ka BP) are at variance with the apparent strengthening of the East Asian summer monsoon evidenced by the appear-

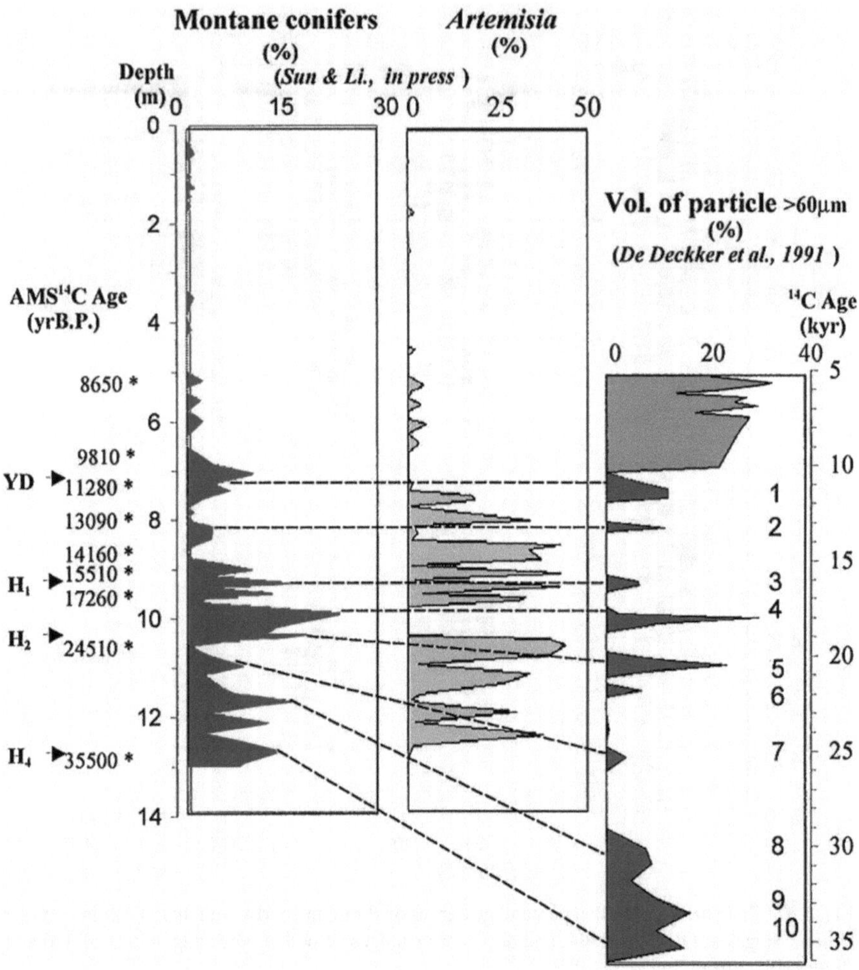

Fig. 3.38 Pollen data from Core 17940 in the South China Sea and comparison with the eolian events preserved in the Carpentaria Core GC-2 (An 2000; De Deckker et al. 1991)

ance of many high continental lake levels in the Asian interior at that time (An 2000) which suggest increased monsoon rainfall. Coniferous forests (dominated with *Picea* and *Abies*) on the LP and eastern plains at that time reflect a colder, wetter interglacial climate (An et al. 1990a). Pollen records from core 17940 in the South China Sea show alternating periods of dominance between mountain conifers (*Tsuga, Picea, Abies*) and *Artemisia* between 36 ~ 10 ka BP. This record shows that alternating periods of relatively cold-wet and warm-dry climate were superimposed on the long term trend of decreasing temperature and humidity during this interval. Dust events recorded in the Gulf of Carpentaria can be compared with the YD, H1, H2 and H4 events recorded in the South China Sea during the latter part of the

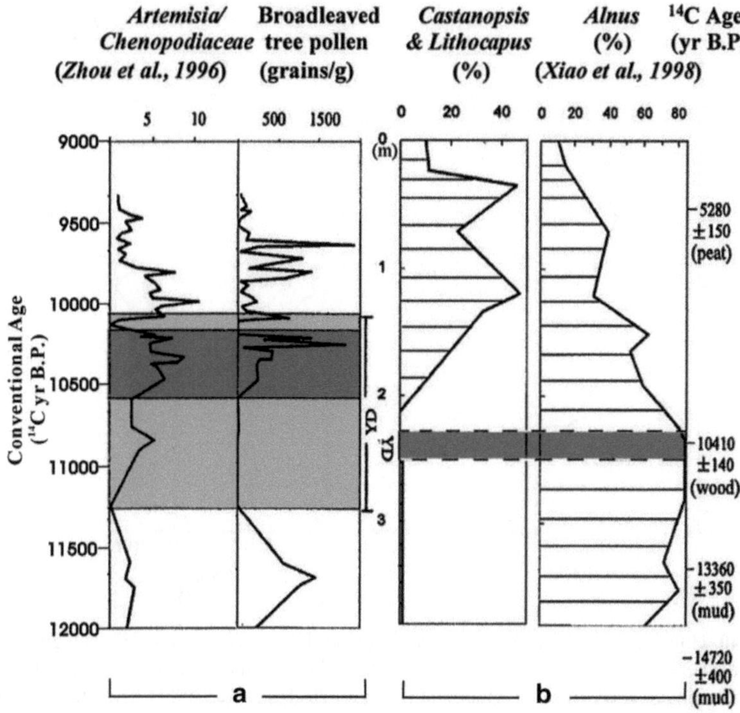

Fig. 3.39 Cold-humid climatic events recorded in **a** the loess-desert transitional zone, Midiwan (Zhou et al. 1996) and **b** Nanling mountain (Xiao et al. 1998) during the last deglaciation. (Modified from An 2000)

last glaciation. This comparison suggests that strengthening of the southeast trade winds in the Southern Hemisphere promoted enhanced cross-equatorial flow, increasing precipitation over the South China Sea (An 2000; De Deckker et al. 1991; Fig. 3.38).

Midiwan peat deposits (Zhou et al. 1996) at the northern margin of the LP show rapid climatic changes during the Younger Dryas, from dry-cold to humid-cold, and to dry-cold again, which are reflected by variations of A/C and broadleaved tree pollen concentration (Fig. 3.39). Changes in the content of evergreen broad-leaf forest and deciduous broadleaf forest tree pollen from a peat deposit in Daping of Nanling (Xiao et al. 2000) also show evidence of a climate event with decreased temperature and increased rainfall at around 10,400 year BP. Such precipitation variability in the YD interval can be ascribed to interactions between the high and low latitude climates, and even to the influences from the Southern Hemisphere (Zhou et al. 1996).

Monsoon precipitation changes in the Holocene inferred from pollen records followed the summer insolation trend of the Northern Hemisphere (Fig. 3.40e). The Holocene optimum is characterized by higher levels of precipitation that mostly

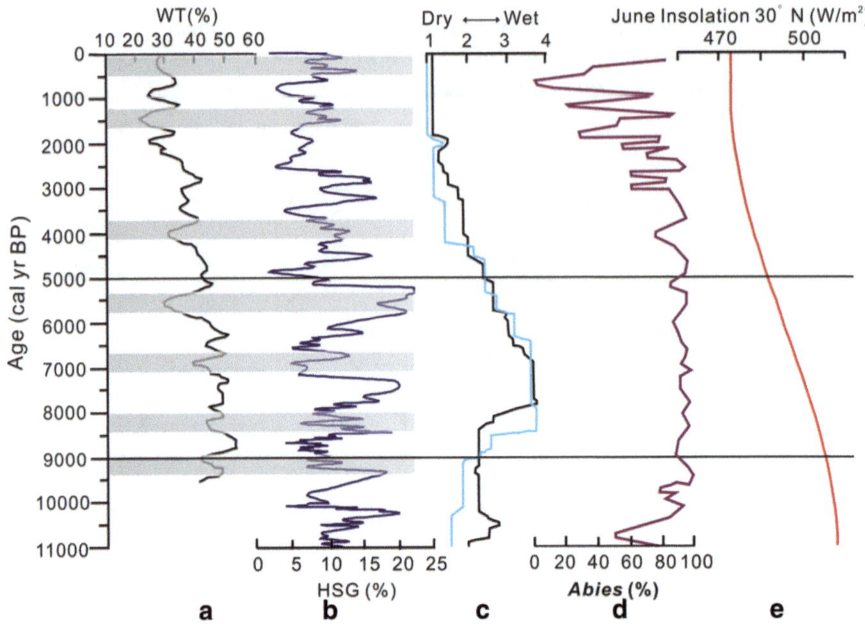

Fig. 3.40 Holocene climate variation inferred from pollen records across various regions in China and related environmental proxies for comparison. **a** Warm-demanding trees (WT) pollen percentages from Jingbo Lake in Inner Mongolia (Li et al. 2011b); **b** Ice-rafted hematite stained grains (HSG, %) in the high latitude North Atlantic ocean (Bond et al. 2001); **c** Synthesized regional time series of relative moisture conditions inferred from fossil pollen data from Eastern Inner Mongolia (black) and Northwestern Loess Plateau (light blue). The moisture conditions were averaged at 100 year intervals and coded by four classes, from 1 (driest) to 4 (wettest) (Zhao et al. 2009); **d** *Abies* pollen percentages from Hongyuan peatland in the southern Zoige Basin on the eastern TP (Zhou et al. 2010); **e** June insolation at 30°N (Berger and Loutre 1991)

occurred during the interval 9000-5000 cal yr BP, as seen in a number of pollen records from northeastern to southwestern China (Fig. 3.40).

The pollen record from Jingbo Lake in northeastern China indicates that the Asian monsoon was strong from 8800-5700 cal yr BP, then weakened after the middle Holocene, consistent with $\delta^{18}O$ Asian monsoon records from Dongge cave. Moreover, a series of monsoon-weakening events on a millennial scale were recorded by pollen, showing low abundances of warm-demanding trees from 9200–9000 cal yr BP, 8400–8100 cal yr BP, 7100–6800 cal yr BP, 5700–5400 cal yr BP, 4100–3800 cal yr BP, 1600–1300 cal yr BP, and 500–200 cal yr BP. These periods correlate with ice rafting events at high latitudes of the North Atlantic Ocean (Fig. 3.40a, Li et al. 2011b). In semi-arid China, synthesized regional time series of relative moisture conditions inferred from fossil pollen data in eastern Inner Mongolia and on the northwestern LP show a maximum moisture period identified by higher tree pollen percentages that occurred from 8500–5500 cal yr BP (Fig. 3.40c, Zhao et al. 2009). On the eastern margin of the TP, a well-dated pollen record from the Hongyuan peatland in the Zoige Basin indicates that monsoon intensity reached a maximum in that region between 8500-4700 cal yr BP when *Abies* pollen abundance peaked (Fig. 3.40d, Zhou et al. 2010). The same trend occurred at other sites from the eastern TP, but their timing exhibited regional temporal and spatial variability, presumably associated with their position relative to the South Asian and East Asian monsoon fronts, and topography. Some dry and cold events reflecting Holocene climate instability are also identified in pollen records (Yan et al. 1999; Joosten et al. 2008; Zhou et al. 2010; Zhao et al. 2011). From comparisons, it seems that the eastern TP saw an earlier moisture increase in the Holocene. All of these regions experienced uniformly dry climate during the late Holocene (~4000 cal yr BP) (Fig. 3.40).

3.7.4 Summary

East Asian monsoon climate inferred from pollen records provide detailed records of temperature and precipitation changes through time. In these records, climate cycles have been identified on orbital time scales, and millennial quasi-periodic fluctuations on sub-orbital time scales. There are still not many pollen records reflecting the whole story of East Asian monsoon variability at millennial time scales, but the existing research has proved that pollen records provide an effective means to identify abrupt climatic fluctuations, like Heinrich events and the Younger Dryas event. These factors all contribute to an understanding of variability in climate in the Asian monsoon-arid environment system during the most recent 125 ka. Pollen records show that this system has responded to high- and low- latitude climate variability in the Northern Hemisphere, as well as climate variability in the Southern Hemisphere. East Asian monsoon climate variation during the Holocene was primarily controlled by changes in NH summer insolation. However, superimposed on these we see evidence from pollen records of abrupt and millennial-scale climate change.

3.8 Climatic and Environmental Records in Ice Cores on the Tibetan Plateau

Ice cores are unique paleoclimate records that preserve continuous, high-resolution, long-term, multi-index information applicable to the study of climate systems and paleoenvironments. Ice core records have greatly improved our understanding about paleoclimatic and environmental change. In Antarctica, ice cores have provided climatic records for the past 8 glacial-interglacial cycles (Lambert et al. 2008). In Greenland, ice core records extend to the beginning of the Last Glacial Stage, and perhaps into the Eemian interglacial period.

Glaciers in alpine low-latitude regions provide an opportunity for ice core paleoclimate research with potential benefits over their polar counterparts. In general, alpine ice records have higher accumulation rates than polar ice cores. In addition, alpine glaciers can offer paleoclimatic and paleoenvironmental information that is directly applicable to the inhabitants of a region, in and around the mountains where the glaciers form. Records from subtropical alpine ice cores serve as a bridge for studies of Antarctic and Greenland ice cores, and as we shall see, certain ice cores contain valuable information specific to the Asian monsoon-arid region.

The TP is sometimes referred to as the "Third Pole" of the Earth. A number of glaciers exist at high altitude in the mountain ranges of the plateau, which spans thousands of kilometers from west to east. Several high and large glaciers, with deep and cold ice have proved to be fruitful sites for ice core research. Ice cores from the TP region have the potential to enhance our understanding about the interaction between different atmospheric circulation patterns, especially the westerly and Indian monsoon. They can also help us to understand the evolution of the Indian monsoon through time.

3.8.1 Climatic and Environmental Proxy Records from Ice Cores

A variety of climatic and environmental proxy data can be acquired from ice cores. Oxygen and hydrogen isotopes are two of the most common, and are used as paleothermometers. Glacial accumulation rate itself is an index that reflects changes in precipitation. Glaciers often contain dust and pollen, which can be useful for study. Even in clear ice, microparticles can be quantified, as well as a variety of chemical species that may be sensitive to climate and the environment.

3.8.1.1 $\delta^{18}O$

Oxygen and hydrogen stable isotopes in ice cores are the primary tools used to reconstruct past climate. Due to the parallel isotopic fractionation in the water cycle, usually only oxygen isotopes are used for ice core research. On a global scale, most atmospheric vapor comes from evaporation of surface water from the tropical oceans. As the vapor moves towards the polar regions, or mountain regions, the heavy isotope will be gradually depleted by isotopic fractionation associated with rainout processes. As a result, $\delta^{18}O$ in polar regions, or at high elevations in mountain regions, will become extremely low. The gradual depletion of heavy isotopes in the vapor condensate is controlled by the cooling of air. With colder air temperatures, $\delta^{18}O$ values of the residual vapor become more negative following precipitation. The "temperature effect" of precipitation $\delta^{18}O$ is found in the spatial relation between precipitation $\delta^{18}O$ and average air temperature at different stations (Dansgaard 1964). It is believed that this relation is maintained with time, and this is the basis for the paleoclimate study from ice core isotope records in the polar regions.

The TP is in the low latitude region, and is affected by strong monsoon precipitation. In monsoon regions a precipitation "amount effect" is generally thought to control $\delta^{18}O$, not the temperature effect. However, a network of precipitation stations to study this effect on the TP was initiated in the early 1990s, and this work shows that in the northern TP it is the temperature effect that controls precipitation $\delta^{18}O$ (Yao et al. 1996a; Tian et al. 2003). In the southern TP precipitation $\delta^{18}O$ is controlled by the amount effect on seasonal and annual timescales. On longer timescales, Thompson et al. (2000) compared the Dasuopu ice core $\delta^{18}O$ record with temperature changes in the Northern Hemisphere, and found that it was most sensitive to temperature. This relationship provides the basis for reconstructing long term temperature variations from ice cores on the TP.

3.8.1.2 Net A Accumulation Rate

Glaciers accumulate primarily from the precipitation of snow. Fresh snow condenses when it recrystallizes as firn, and then to glacier ice at depth due to the weight of subsequent snow on the glacier surface. If snow erosion on the glacier, evaporation, melting and other mass loss or gain processes can be ignored, reconstructed glacial accumulation will reflect real changes in precipitation.

Precise dating of the upper part of an ice core is necessary to measure the annual accumulation depth. However, the measured annual layer depth is not the actual accumulation rate at the surface, from snowfall. The annual snow and ice layer will be thinned and stretched inside the glacier. In alpine glaciers from the TP, annual layers thin rapidly with depth in the upper part of the ice cap, and vertical strain rates in the upper half of the ice cap decrease rapidly. Therefore, the measured annual ice depth in the ice core should be calibrated to obtain the actual glacial accumulation rate (Thompson et al. 1997; Duan et al. 2002). In the Dasuopu glacier, it is believed that under current meteorological conditions the thickness of snow accumulating at the summit each year correctly reflects regional precipitation. Therefore the net accumulation record in the Dasuopu ice core is expected to provide a well-dated, high-resolution precipitation history for the Himalayas. Moreover, because the peaks of $\delta^{18}O$ and ions appear in the dry stage of spring, accumulation recovered in the core represents precipitation from spring to spring (Duan and Yao 2003).

3.8.1.3 Microparticles

Atmospheric dust has important effects on global climate. The Central Asian dust source areas provide an ample supply of dust to the atmosphere. Reconstructing dust records from ice cores on the TP offers a direct means of studying the interaction between Asian dust and climate change. The ice cores on the TP, especially those near the deserts, preserve records of climatic instability in the mid-low latitudes and the environmental evolution of the Central Asian dust source areas.

Microparticles in ice cores are a direct proxy for airborne dust. The total concentration of microparticles can be measured in the laboratory, along with their size distribution. Microparticles have a variety of sources, from volcanic ash to cosmic

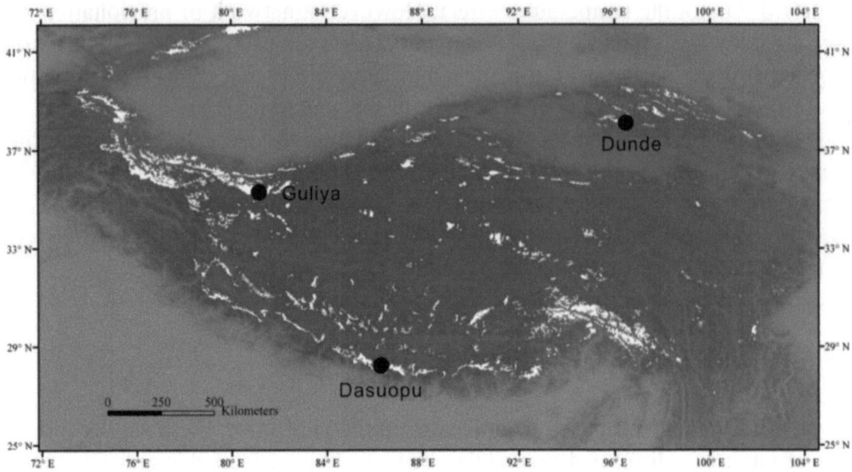

Fig. 3.41 Location of Dunde, Guliya and Dasuopu ice cores on the Tibetan Plateau

dust. There is a high concentration of microparticles contained in ice cores on the TP and the majority are from terrestrial sources. The microparticle record from ice cores tells us the intensity of atmospheric circulation in the past and provides source information.

Large numbers of glaciers in west China are surrounded by deserts, which provide sources for the dust record in ice there. Microparticles in the Guliya and Dunde ice cores are chiefly from desert mountains on the northern plateau. Records from these sites reflect atmospheric circulation and provide dust source information (Yao et al. 2004; Wu et al. 2004b).

3.8.1.4 Chemical Components

Some glacial ice is so clear that we can not see dust layers with the naked eye. However, through laboratory analyses on chemical components from melted ice samples, we can still identify seasonal changes in accumulation, and this approach can also be used to date the ice core. Different ion components can be used to identify different dust sources. The commonly analyzed ions in ice cores include the anions SO_4^{2-}, Cl^-, NO_3^-, $HCOO^-$, $CHCOO^-$, and cations K^+, Na^+, Ca^{2+}, and Mg^{2+}. Ion concentrations in ice core records reflect different dust sources. For instance, Ca^{2+} reflects terrestrial sources, while Na^+ reflects marine sources. NO_3^- has a combined source including both natural and anthropogenic sources. The relationships between different ions in ice core records probably reflect source area changes, so that different dust sources (terrestrial, marine, or cosmic) can be identified. The concentration of Ca^{2+} in the Guliya ice core is roughly correlated with temperature. However, atmospheric circulation changes, land surface conditions and atmospheric humidity might also influence Ca^{2+} concentrations (Yao et al. 2004).

3.8.2 Studies on the Tibetan Plateau Ice Cores

Work on the Dunde ice cap, Qilian Mountains was undertaken from 1984–1987, in a cooperative effort between the Byrd Polar Research Center at Ohio State University, led by Lonnie Thompson, and groups from Lanzhou Institute of Glaciology and Geocryology, Chinese Academy of Sciences, led by Yao Tandong. In 1991–1992, this cooperative work continued to the Guliya ice cap, west Kunlun Mountains, and ice cores were drilled and extracted to a depth of 309 m. In 1996–1997, ice cores were drilled in Dasuopu Glacier, Xixiabangma at the top of the middle Himalayas. The locations of these ice cores are shown in Fig. 3.41.

3.8.2.1 Dunde Ice Core

Dunde ice cap, with an area of about 60 km^2, is located in the Qilian Mountains, northeast of Tibet (38°06′, 96°24.5′). The ice cap has an area of 60 km^2 and a summit elevation of 5325 m. The ice temperature at a depth of 10 m is −7.3 °C. At the bottom (139.8 m below the surface) it is −4.7 °C. The accumulation rate is about 0.42 m water equivalent per year on the glacier. Three cores to bedrock were recovered from the ice cap summit: Core D-1 (139.8 m), Core D-2 (136.6 m) and Core D-3 (138.4 m) (Thompson et al. 1989; Yao et al. 1992). They are the first ice cores to bedrock ever drilled in an Asian low latitude region.

The upper part of the Dunde ice core was dated by the annual and seasonal stratigraphy of microparticles, $\delta^{18}O$ and conductivity. Deeper portions of the ice core were dated by counting visible dust layers. The uncertainties in dating are about ± 10 years between 60 and 105 m, ± 15 years between 105 and 110 m, ± 20 years between 110 and 115 m, and close to ± 40 years at a depth of about 117 m. The dating results in the upper part of the ice core were confirmed by gross beta radioactivity from 1963 atmospheric nuclear tests (Thompson et al. 1989; Yao et al. 1992).

3.8.2.2 Guliya Ice Core

The Guliya ice cap (summit elevation 6710 m above sea level) covers ~200 km^2 and is part of an ice mass extending over 8000 km^2 in the western Kunlun Shan. It is the largest polar-type ice cap in central Asia. The ice cap is surrounded by vertical ice walls 30 to 40 m high and has internal temperatures of − 15.6 °C at 10 m, − 5.9 °C at 200 m, and −2.1 °C at its base. Annual accumulation is quite variable ranging from 140 to 260 mm among the sites (Thomson et al. 1997; Yao et al. 1997).

In the summer of 1992, the cooperative BPRC-LIGG expedition successfully recovered three ice cores (35°179N, 81°299E) 34.5 m, 93.2 m and 308.6 m in length. The nearly 309 m ice core to bedrock is so far the longest ice core, and also the oldest ice ever found on the TP (Thomson et al. 1997; Yao et al. 1997).

In the upper 120 m of this ice core, the ice dating is based on dust layer and $\delta^{18}O$ stratigraphy. In the middle of the ice core, the dating method is based on modeling and a reference method. ^{36}Cl was used to estimate the age at the base of the ice core, and the age of bottom ice is over 700 ka. Also, the apparent correlation between atmospheric CH_4 levels and stadial and interstadial events inferred from $\delta^{18}O$ values in polar cores was also used to establish the Guliya time scale for the past 110,000 years. The age at 268 m of this ice core corresponds to the Last Interglacial.

$\delta^{18}O$ values, dust concentration, and ions in this ice core show large fluctuations. $\delta^{18}O$ value is a proxy for past atmospheric temperature over northern Tibet. Two of the anions (Cl^- and SO_4^{-2}) originate primarily from surface dust, including dust from salt flats and lake beds that dot the western side of the plateau. The probable precursors of NO_3^- in the snowfall are tropospheric nitrogen species originating from sources such as soils, vegetation, and lightning (Thomson et al. 1997).

3.8.2.3 Dasuopu Ice Core

Dasuopu glacier (28°239N, 85°439E) is in Xixiabangma, middle Himalayas. The glacier receives monsoon precipitation. Because of the large annual snowfall, the ice core from this glacier can provide a high-resolution Indian monsoon signal. There is a plateau on the upper part of the glacier, 1.5 km in width and 3 km in length, suitable for ice core drilling. The glacier extends from 7200 m a.s.l. to 5600 m a.s.l. and the snowline is at around 6000 m. The ice temperature at 10 m depth is − 14 °C and − 13 °C at the bedrock according to borehole ice temperature measurements. The glacier flows from 7200 m to 5600 m at the end, with a total length over 10 km. Three ice cores were recovered in 1997 from the Dasuopu glacier in the dry snow zone. The first core (C-1), 159.9 m long, was drilled at 7000 m. The other two cores (C2 and C3), 149.2 m and 167.7 m long, respectively, were drilled to bedrock 100 m apart at 7200 m a.s.l. (Yao 2002; Thompson et al. 2000). Visible stratigraphy showed no hiatus features in any of the cores.

The high annual accumulation allows preservation of distinct seasonal cycles, particularly in $\delta^{18}O$, dust, and NO_3^-, which can be used to count the annual layers of this ice core. Based on the dust and $\delta^{18}O$ stratigraphy, C3 was annually dated to 1440 A.D. From 1439 to 1000 A.D., the ice age was determined by a depth-age model.

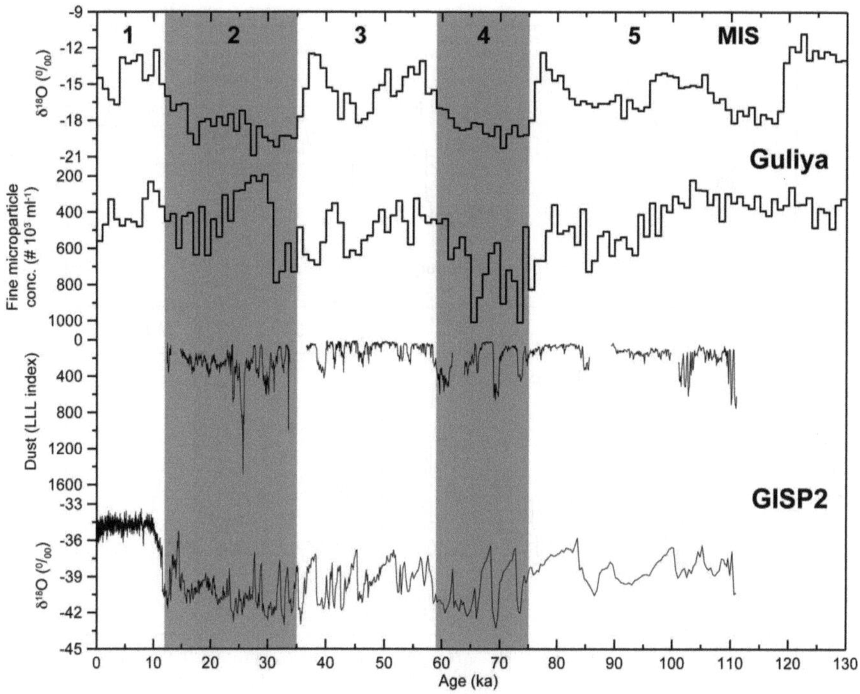

Fig. 3.42 Comparison of $\delta^{18}O$, microparticle concentration in Guliya ice core and $\delta^{18}O$, microparticle concentration in GISP2 ice core. (from Wu et al. 2004b)

3.8.3 Paleoclimatic and Environmental Records from Ice Cores on the Tibetan Plateau

3.8.3.1 Climatic Change Since the Last Interglaciation Recorded in Guliya Ice Core

Dating of the deep Guliya ice core shows that the ice core above 268 m contains ice formed since the Last Interglacial. Glacial-interglacial cycles within the Guliya ice core appear in the $\delta^{18}O$ record as a sequence of stadial and interstadial events from the Last Glacial Maximum. The $\delta^{18}O$ values of the interstadials in Guliya imply that subtropical climate was forced more strongly by precession (23 ka) than by obliquity (41 ka). $\delta^{18}O$ values are the lowest at the Late Glacial Maximum (LGM), and increase from there by about 5.4 ‰ into the Holocene (Thompson et al. 1997).

In the northern TP, precipitation $\delta^{18}O$ is related to temperature change. Guliya ice core $\delta^{18}O$ values reflect temperature changes since the Last Glaciation. For larger fluctuations, we can identify five stages with distinct temperature changes, corresponding to marine oxygen isotope stages (MIS): Stage 1 (deglaciation), Stage 2 (LGM), Stage 3 (interstadial of the Last Glacial Age), Stage 4 (early stadial of the Last Glacial Age) and stage 5 (Last Interglaciation). Stage 5 can be further divided

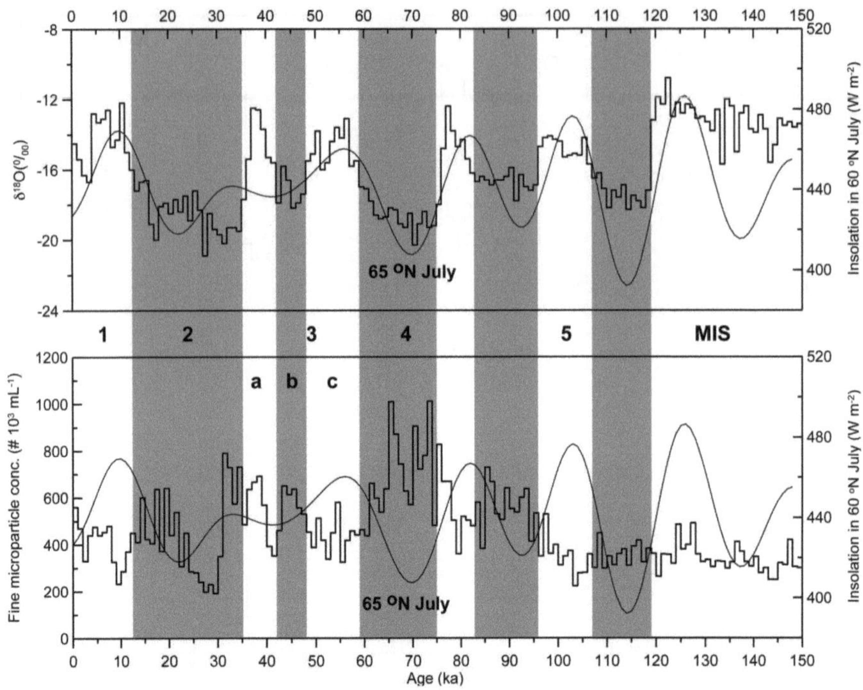

Fig. 3.43 Oxygen isotope and microparticle concentration of the Guliya ice core compared with solar insolation. (from Wu et al. 2004b)

into 5 substages: a, b, c, d and e (Fig. 3.42). Furthermore, the Younger Dryas (YD) is observed in this record (Yao 1999).

The $\delta^{18}O$-derived temperature change from this ice core is quite large. $\delta^{18}O$ in the three warm substages (5a, 5b, 5c) are, respectively, 1.7 ‰, 0.5 ‰ and 3.2 ‰ higher than at present, which translates into temperature increases of about 3, 0.9 and 5 °C. However, the temperatures in substages 5b and 5d are about 3 and 4 °C lower than substages 5e and 5c, respectively. In addition, there also exist temperature variations on a centennial timescale within each substage. The large fluctuations of $\delta^{18}O$-derived temperature changes correspond to solar radiation changes in the Northern Hemisphere (Yao et al. 1997). Between 15 and 33 ka, the ice core record contains approximately 100 $\delta^{18}O$ oscillations with amplitudes from ~2 to 21 ‰ and an average period of 200 years. One possible forcing mechanism is solar variability because there is evidence of a 200-year cycle in sunspot activity (Thompson et al. 1997). The abrupt climate change can be compared with climatic events in GRIP ice core $\delta^{18}O$ record. Seven major warm events can be found in both ice core isotope records. However, the frequency and magnitude are apparently different in the two ice cores. There are more frequent, large magnitude changes in the Guliya ice core, especially during Stage 2.

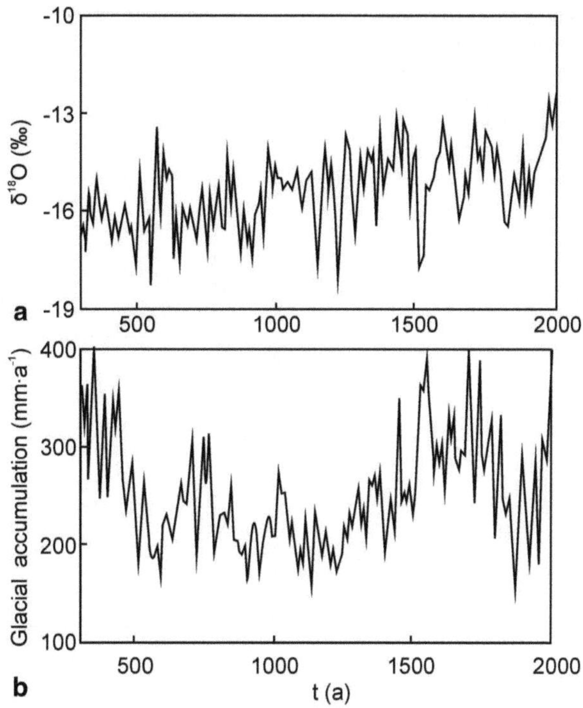

Fig. 3.44 Decadal variations of Guliya ice core $\delta^{18}O$ and glacial accumulation rate in the past 2000 years. (from Yao et al. 1996b)

Dust trapped in the Guliya ice cap is mainly deposited during winter and spring. Ion analyses show the origin of the dust in this ice core is mainly from the Taklamakan Desert and the Tibetan Plateau (Li et al. 1999; Yao et al. 2004). The temporal variation in microparticle concentration in this core is related to $\delta^{18}O$-derived temperature changes. The microparticle concentration is generally lower in the interglacial stage, and higher in the glacial stage (Fig. 3.42). The microparticle concentration increased since 105 ka before present (BP), suggesting a strengthening of dust sources, including the Tibetan Plateau and arid Central Asia. During the MIS 5b cold stage, microparticle concentrations rose to an average of 600×10^3 per milliliter of melted sample, corresponding to cooling in this stage (Wu et al. 2004b).

Figure 3.43 is a comparison between $\delta^{18}O$, microparticle concentration of the Guliya ice core and solar insolation. Microparticle and $\delta^{18}O$ records show that since 110 ka, there is a negative correlation between summer insolation and microparticle concentration. The likely reason for this is that there is less dust in the wet and warm climate of the last interglacial. Increases of atmospheric dust are mainly caused by enhanced aridity and increased wind strength.

A sharp decrease in microparticle concentration occurred around 30 ka. The abnormally low dust concentration during the LGM might be due to the specific environment on the TP. Lacustrine studies shows generally higher lake levels during the period of 30 ~ 20 ka. Because of cool and wet conditions in the region during the LGM, glaciers and snow cover on the plateau were more extensive and persisted

Fig. 3.45 Eleven point running average of Guliya ice core $\delta^{18}O$ and glacial accumulation rate. (from Yao et al. 1996b)

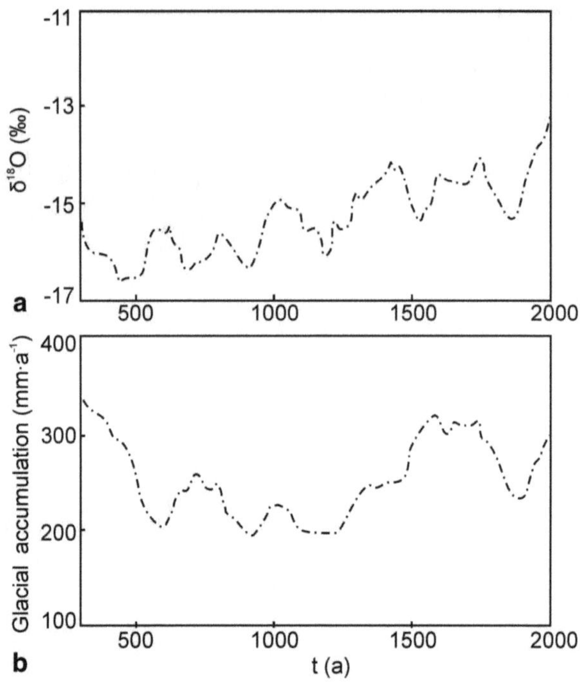

longer. Thus, the dust source area on the TP decreased, reducing the amounts of dust emitted to the air. A shift in atmospheric circulation in the LGM might also have influenced the microparticle record (Wu et al. 2004b).

3.8.3.2 Climatic Change During the Past 2000 Years as Recorded in the Guliya Ice Core

Understanding climate change over the past 2000 years is an essential prerequisite for paleoclimate modeling and global change predictions. The Guliya ice core was continuously deposited over the past 2000 years and the $\delta^{18}O$ record from this time provides a high-resolution picture of recent climate change. The variation of decadally-averaged $\delta^{18}O$, compared with the glacial accumulation record from the past 2000 years is shown in Fig. 3.44. The first century in this period is cold with low precipitation. This is followed by increasing temperatures and precipitation, with higher frequency fluctuations in both parameters. The decadal average of $\delta^{18}O$ and glacial accumulation rate shows that the temperature and precipitation variations follow a common pattern. The smoothed curve of ice core $\delta^{18}O$ shows 7 colder periods and 8 warmer periods. The smoothed accumulation curve shows 4 dry periods and 5 wetter periods. From 500 A.D. to present, all of the $\delta^{18}O$ and glacial accumulation rate data show increasing trends, indicating that climate is becoming warmer and wetter in the northern part of the TP (Fig. 3.45).

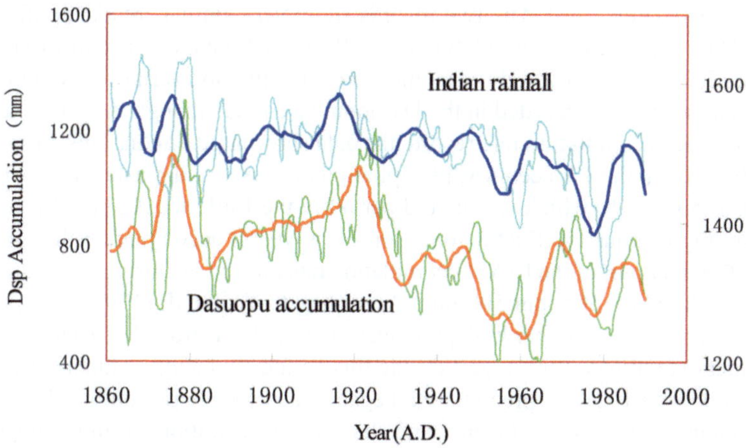

Fig. 3.46 Comparison between Dasuopu accumulation and Northeast Indian Monsoon rainfall. The 5-year unweighted means of both records are also shown. (from Duan et al. 2002)

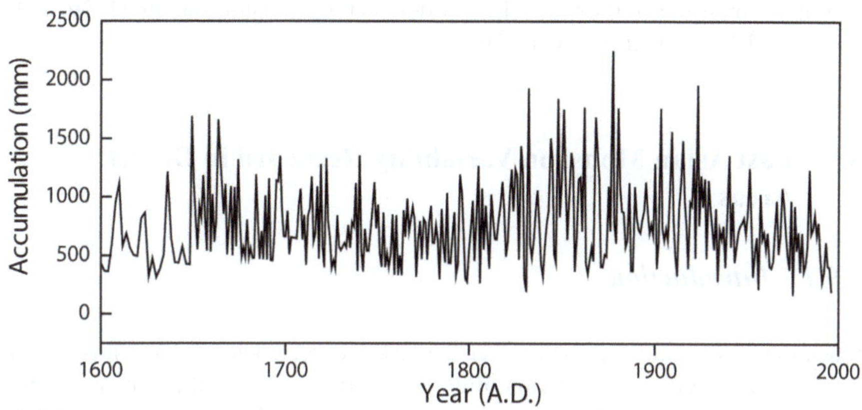

Fig. 3.47 Variability of accumulation recorded in the Dasuopu ice core since 1600 A.D. (from Duan et al. 2002)

3.8.3.3 Monsoon Precipitation Change in the Past 400 Years Recorded from Glacial Accumulation Rate in Dasuopu Ice Core

The high annual accumulation in the Xixiabangma allows preservation of distinct seasonal cycles in $\delta^{18}O$ and ions which make it possible to reconstruct an annual accumulation record for the upper part of the cores. For the Dasuopu ice core, the glacial accumulation rate has been recovered back to 400 years ago (Duan et al. 2002). To validate the recovered accumulation rate from the Dasuopu ice core, the result is compared with rain gauge precipitation data from Northeast India for the past 140 years (Fig. 3.46). The two records show very similar trends, confirming that the Dasuopu glacier accumulation rate reflects past Indian monsoon variations.

Figure 3.47 is the recovered Indian monsoon precipitation variation from the annual precipitation rate recorded in the Dasuopu ice core. Although there is no overall trend in the accumulation rate record, the data show large fluctuations during the past 400 years. In the 17th century, precipitation increased and reached a maximum value in 1650~1670, which was a cold period in the Little Ice Age (LIA). Large, relatively high frequency fluctuations (for a few years) accompanied the gradual rise into the later part of the LIA. Accumulation rates were low during the entire 18th century, with values of about 500 mm. According to Yao et al. (1997), this period was relatively warm. Another high precipitation period was from 1820 and to 1921; inter-annual variations were largest during this period, and temperatures were relatively cold. From 1921 on, precipitation began to decrease to present levels.

The annually averaged $\delta^{18}O$, dust, and chloride concentrations from Dasuopu ice core show several spikes of dust and chloride concentrations, suggesting that these periods had conditions of low precipitation in the source areas. Monsoon failures apparent in the Dasuopu ice core accumulation rate record can be confirmed by known droughts in India and Asia, recorded in historical documents. For instance, two major monsoon failures resulted in devastating Indian droughts (1790–1796 and 1876–1877) (Thompson et al. 2000).

3.9 East Asian Monsoon Variability Recorded in Desert Areas

3.9.1 Introduction

Since the 1990s, and especially over the last decade, scientists have made great progress in studying the evolution of desert environments—in the Sahara and Namibian deserts of Africa (Martin et al. 2006; Bernhard et al. 2006; Arne et al. 2009), the Mojave of North America (Rendell et al. 1994), the Atacama of South America (Mebus et al. 1999), the Strzelecki in Australia (Cohen et al. 2010) and the Arabian Peninsula in Asia (Adrian et al. 2006, 2008). Meanwhile, as scientists identified D/O events since the Last Interglacial in GRIP ice core records (GRIP members 1993), North Atlantic deep-sea sediments (Bond et al. 1999), Chinese loess (An et al. 1990b, 1993, 1994) and Chinese stalagmites (Wang et al. 2001). At the same time, a focused effort has been directed at studying correlations between the desert environment and global climate fluctuations as seen in marine oxygen isotope records (designated as marine isotope stages, MIS). These efforts are aimed at understanding the occurrence, development and variations of Asian deserts (Li et al. 2000, 2007).

Along with global climate variations, the formation and evolution of China's deserts and sandy lands were influenced not only by the intensities of the East Asian

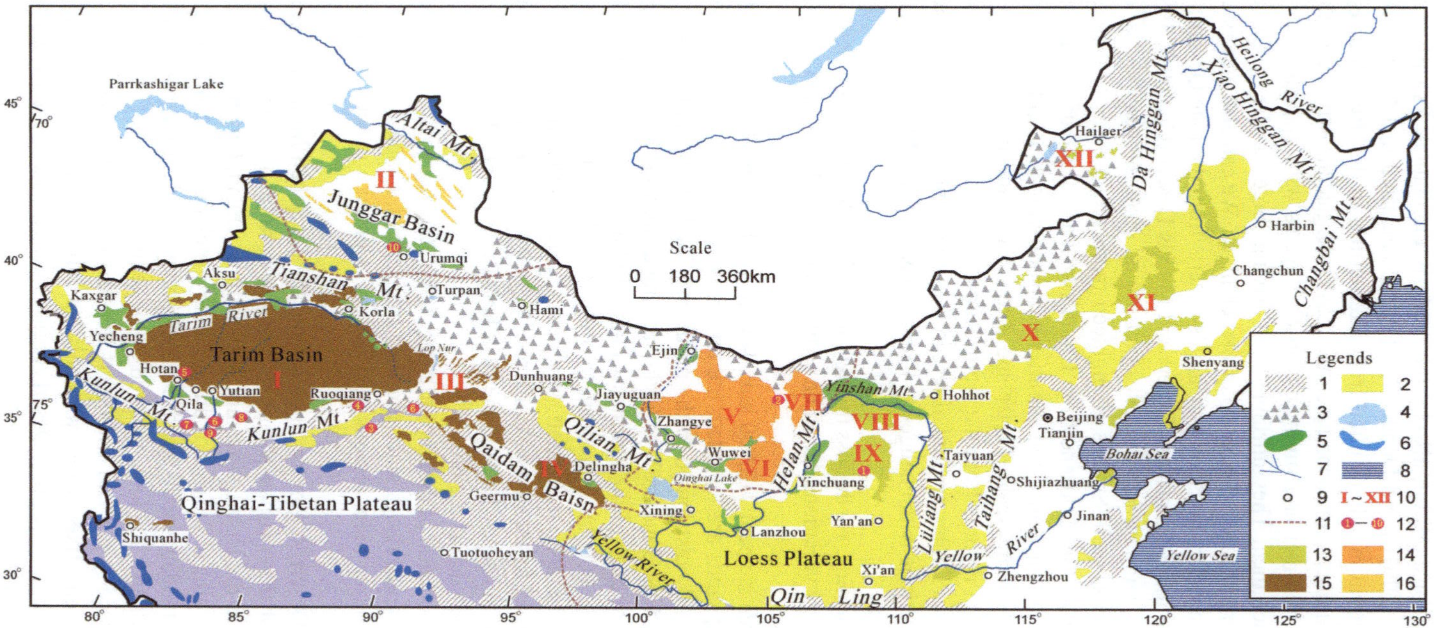

Fig. 3.48 China's deserts and representative section locations since the Last Interglacial. 1–Mountain; 2–loess; 3–Gobi; 4–lake; 5–oasis; 6–glacier; 7–river system; 8–ocean; 9–city;10–desert and sandy land: I–Taklimakan Desert, II–Gurbang Desert, III–Kumtag Desert, IV–Qaidam Desert, V–Badain Jaran Desert, VI–Tengger Desert, VII–Ulan Buh Desert, VIII–Qubqi Desert, IX–Mu Us (Maowusu) Desert, X–Otindag Sandy Land, XI–Korqin Sandy Land, XII–Hulun Buir Sandy Land; 11–boundary between the sandy desert regions; 12–representative stratigraphic profiles since interglacial period and their distribution; 13–Eatern Sandy Desert Region (Monsoon Sandy Desert Region); 14–Central Sandy Desert Region (The region affected by both Monsoon and westerlies); 15–the Extremely Arid Desert Sub-Region in the Western Desert Region (The region affected by westerlies); 16–the Arid Desert Sub-Region in the Western Desert Region (The region affected by westerlies and the air current of the Arctic Ocean current)

winter and summer monsoons but also by the westerly circulation. As early as in the 1990s, An et al. (1991b) and Dong et al. (1998) suggested that China's desert expansion and retreat mainly resulted from alternations in the strength of East Asian winter and summer monsoons since the Last Interglacial, MIS5, and they correlated East Asian winter and summer monsoon cycles with glacials, stadials and interstadials since then. They attributed the driving force for the East Asian winter and summer monsoon variations to orbital changes. We can also understand the impact of the East Asian monsoon on China's deserts from the Late Quaternary geological records in the Badain Jaran Desert and Mu Us Desert (Gao et al. 1996, 2006; Li et al. 2007), and research indicates that orbital changes were the driving force in those regions as well. Studies on eighteen radiocarbon-dated aeolian and paleosol profiles during the Holocene from China's arid and semi-arid areas, also suggest several East Asian winter and summer monsoon climate events occurred, varying on a millennial timescale (Porter and Zhou 2006) and corresponding to similar changes seen in records from the North Atlantic (Bond. et al. 2001). However, there is still only a limited amount of research on Asian monsoon environment variations and correlations with changes in China's deserts since MIS5, especially on the millennial-scale.

To address this question, we discuss the evolution of the desert environment and Asian monsoon climate variations, based on sedimentary records in China's deserts from MIS1 to MIS5.

China's desert area is about 712,900 km^2. If parts of the Gobi region, impacted by the wind-drifting sands are included, the total area amounts to 1,308,000 km^2 (Zhu 1980). It is mainly distributed in the northern inland basins and plateaus of China, from 75°E to 125°E and 35°N to 50°N, ranging as far west as Kaxgar and as far east as the western edge of the Northeast Plain of China, extending ~4500 km from east to west and ~600 km from north to south (Fig. 3.48). It crosses five bioclimate subzones: extremely arid desert, arid desert, arid desert-steppe, semi-arid steppe and semi-humid forest-steppe. Bordered by the Ordos Plateau, it is known as desert in the west and sandy lands to the east; hereafter, we refer to them collectively as China's sandy desert region.

According to the location of deserts and sandy lands, the degree of drought, and their sedimentary characteristics since the Last Interglacial, China's sandy desert can be divided into three main regions, the Eastern, Central and Western Desert Regions (Fig. 3.48).

(1) The Eastern Desert Region is roughly from the Langshan Mountain on the westernmost edge of the Yinshan Mountains, as far north as the Sino-Mongolia border, as far south as the Helan Mountain, Wushaoling, Dulan, Qinghai Lake and Zaling Lake. It is the area to the east of this boundary, including the Mu Us Desert, Kubqi Desert and Otindag Sandy Land, Korqin Sandy Land, Hulun Buir Sandy Land, with an area of 148,300 km^2. Situated in the temperate zone of semi-humid forest-steppe, semi-arid steppe and arid desert steppe, this region is mainly characterized by fixed and semi-fixed dunes, with only a small proportion of mobile-dune sands. It is also known as Monsoon Desert Region because of the great impact of summer monsoons (Li et al. 2001).

(2) The Central Desert Region is located as far west as the Langshan-Helan Mountains, east to the Ruoshui River, north to the Sino-Mongolia border and south to the Qilian Mountains, including Badain Jaran Desert, Tengger Desert, Ulan Buh Desert and some deserts in Hexi Corridor, with an area of 123,300 km^2. This region is located in the arid desert of the temperate zone and consists mainly of mobile dunes, with a certain proportion of fixed and semi-fixed dunes. The dunes in the areas are tall, up to 400 to 500 m high, which are the typical areas of other areas in the world with megadunes. There are also many lakes in them, such as about 140 lakes in the Badain Jaran Desert and more than 400 lakes in the Tengger Desert. North of the Central Desert Region is the Sino-Mongolia Gobi, and the transitional hilly loess to its south. From north to south, the boundaries differentiated by wind force are quite obvious for Gobi, desert and loess (Fig. 3.48). In addition to the wind drifting sand climate impacted by the winter monsoon, the region is also subject to precipitation brought by the summer monsoon's polar front which penetrates into these areas by moving northwestwards. Moreover, after the "Tarim-Hexi Corridor-Lanzhou" high-pressure ridge developed from the "north branch of the westerly jet" in the North Qinghai-Tibet Plateau, drought and an acceleration of desertification in the Central Desert Region further strengthened. Thus, we take this region as under the joint influence of both of monsoon and westerly air masses.

(3) The Western Desert Region lies in the area west of the Ruoshui River, southwestwards to Jiayuguan and Delingha, i.e. the area to the west of 100° E in North China, with an area of 441,300 km^2. This region can be sub-divided by the Tianshan Mountains into two parts, the Extremely Arid Desert Sub-Region in the south and Arid Desert Sub-Region in the north.

The Western Extremely Arid Desert Sub-Region includes the Taklimakan Desert, Qaidam Desert and Kumtag Desert. Of these, the Taklimakan Desert is the largest (365,000 km^2) in China and the second largest mobile desert in the world. The Sub-Region is mainly located in the temperate and warm temperate zone of extremely arid desert. Except for the fixed and semi-fixed dune sands in the region with high-level groundwater along the rivers or lakes, the rest of the vast area is mainly the mobile-dune sands.

In the Western Arid Desert Sub-Region, the Gurbantunggut is the largest, which is the second largest desert in China, with an area of 51,130 km^2. Although located in the temperate zone of the arid desert, the area of fixed and semi-fixed dune sands accounts for more than 95 %, it is China's largest fixed and semi-fixed desert.

In winter, strong Asian winter monsoon wind prevails in the Western Extremely Arid Desert Sub-Region. The westerly jet is located over the TP. Hence the Qaidam Desert is impacted by westerly flow all year-round. Restrained by the high altitude terrain, the north branch of the Westerlies forms a high-pressure ridge over the Tarim Basin, thus strengthening the arid climate in the Western Desert Region. In winter, the westerly jet has played an important part in supplying water vapor for high mountains on the TP, resulting in a mountain snow line that extends to the margin of the desert. When the Siberian-Mongolian cold high pressure moves south, the mountain snow line declines greatly, and the climate is exceptionally cold and arid. In the Tarim Basin, westerly winds from the Pamir mountain pass and the inflowing Siberian-Mongolian high pressure air mass merge with W and NW

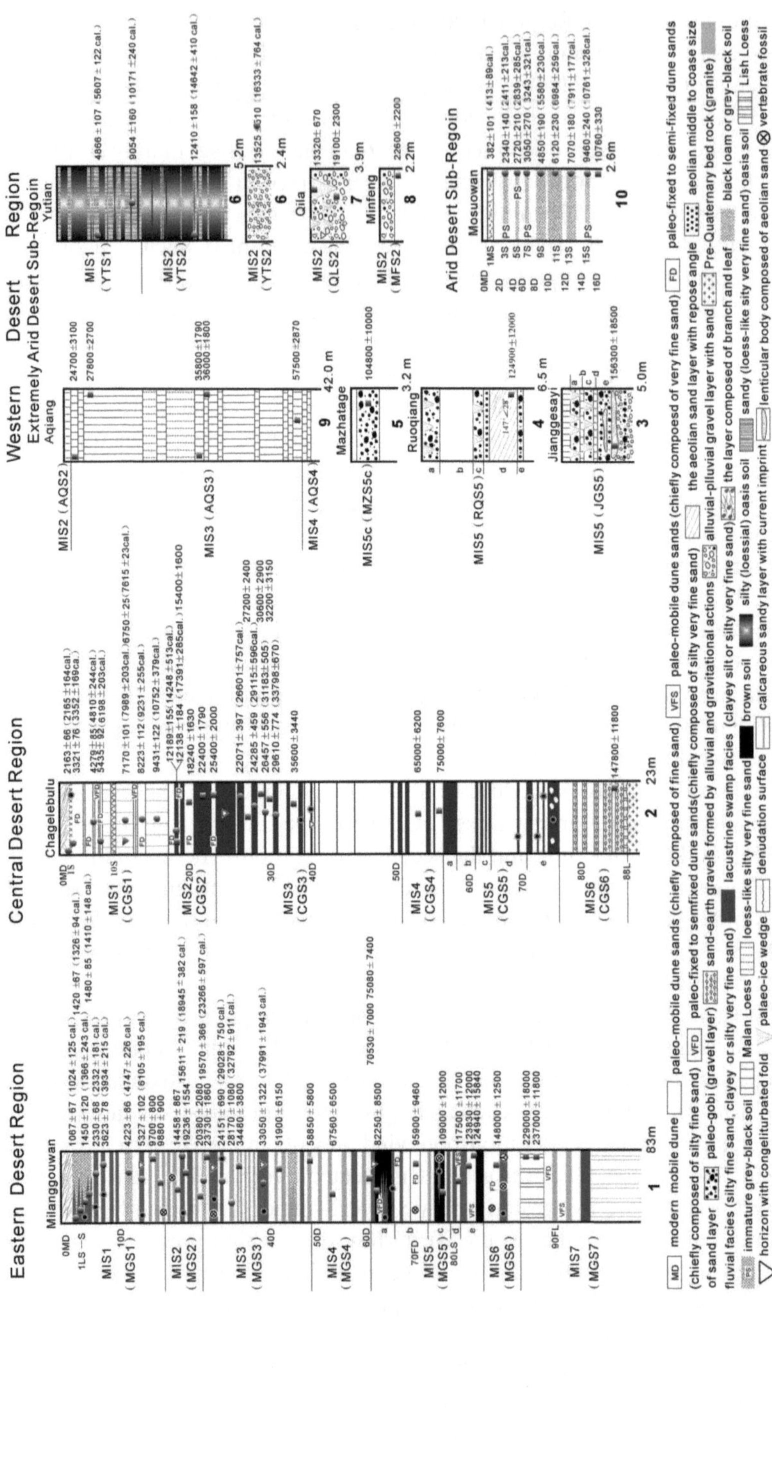

Fig. 3.49 The representative stratigraphical sections in China's deserts since the Last Interglacial. Section 10 "Mosuowan section" was drawn from the research results of Chen et al. 2001b

Fig. 3.50 The palaeo-mobile dune sand layer 58D of MGS4 in the Milangguowan section, Salawusu River valley of the southeastern margin of the Mu Us Desert, in which the palaeo-wind direction indicated by the repose angles (28 ~ 31) are consistent with the modern wind direction at the site

winds, in combination with NE winds, which has resulted in a "dead sea", known as the Taklimakan Desert.

In summer, especially hot summers, the main flow of the Westerlies moves to the north of the Qinghai-Tibet Plateau, but the anti-cyclonic high-pressure ridge of the north branch of the Westerlies still exists, in combination with the hot and dry climate, resulting in scarce precipitation in the Tarim Basin where the Taklikan Desert is located. The water in the Tarim Basin is mainly from melting ice and snow which originated in the stores brought by the westerly jet from the Qinghai-Tibet Plateau, Pamirs and Tianshan Mountains.

The Arid Desert Sub-Region of the Western Desert Region is subject to the impact of the Asian winter monsoon, westerly air-flow and Arctic polar air mass. Precipitation is evenly distributed year round. Since the accumulated snow often appears in winter and spring, some precipitation and snow meltwater from surrounding mountains also appear in summer, the Sub-Region is characterized by fixed and semi-fixed dunes though it is situated in arid desert zone.

Fig. 3.51 The loess and the loess-like silty very fine sand in the segments of AQS1-5 (MIS1-5) in the Aqiang section south of Taklmaken Dersert

3.9.2 *Climate Proxy Indices*

In this section, we use sedimentary facies, Mz (mean particle diameter), σ (standard deviation) of grain-size, primary oxides [SiO_2, Al_2O_3], the ratios Al_2O_3/SiO_2, and (TOFE+Al_2O_3) /SiO_2 and paleontological records as the climate indicators to understand Asian monsoon climate variability and evolution of the desert environment, especially in the deserts of Eastern China.

3.9.2.1 Sedimentary Facies

Dune sands Aeolian dune sands are the most direct and reliable records for the existence of deserts in the geological past. Based on previous studies, dune sands have recorded major climate variations since the Last Interglacial in China's three

Fig. 3.52 The CGS1 (MIS1) loess on both sides of the gully in Chagelebulu of the southeastern margin of Badain Jaran Desert

Fig. 3.53 Fluvial facies 21 FL in the MGS2 (MIS2) in the Milanggouwan section

desert regions (Fig. 3.49, Sect. 1, 2, 4, 5, 8, 10). The dune sands can be divided into paleo-mobile dune sands and paleo-fixed and semi-fixed dune sands, gray-yellow and brown-yellow, and occasional brown-orange red. The former consists mainly of sandy mineral debris, fine sands and very fine sands, loose and evenly sorted; the latter also mainly consists of sandy mineral debris, fine sands and very fine sands, but with a certain proportion of silt and clay, slightly compacted and moderately sorted, with vertical joints developed and no bedding, with visible plant roots.

According to surveys, almost all the wind directions revealed by the repose angles in paleo-dune sands since the Last Interglacial are consistent with those formed by modern winter monsoon *in situ* (Fig. 3.50). Thus, the paleo-dune sands can be taken as the result of the past winter monsoon. Fig. 3.49 shows dune sands in three desert regions and their adjacent areas, related sedimentary sequences, stratigraphical division and dating results, which can account for variations in the desert envi-

Fig. 3.54 Lacustrine facies in CGS5 (MIS5) in the Chagelebulu section

Fig. 3.55 Interbedded palaeo-dune sands, lacustrine facies and swamp facies in MGS1 -2 (MIS1-2) in Sanchahe, Salawusu River valley

ronment since the Last Interglacial. The main stratigraphical division is based on the time-scale of the MIS.

Loess-like sandy sediment The loess-like sandy sediment (silty very fine sand) is a kind of eolian deposit similar to loess (Fig. 3.49, Sect. 3, 9); it is widely spread along the boundary between the desert and loess. There is not a significant difference between the loess and loess-like sandy sediment from its appearance (Fig. 3.51). It is evenly sorted with vertical joints like loess, and with a high content (over 50 %) of finest sands (diameter 50–100 μm). It serves as an important proxy index to indicate the oasis environment changes and border movements between the desert and loess.

Loess The lithology of loess on the edge of the deserts is similar to the sandy loess on the northwestern margin of the Loess Plateau; however, the loess we refer to is often taken as an indicator to show a greatly weakened winter monsoon which was unable to help accumulate desert, and the relatively warm climate appeared with

Fig. 3.56 Fluvial facies in YGS4 (MIS4), bearing plant tubes and iron rust at Yangsigouwan, Salawusu River valley

Fig. 3.57 The CZS3(MIS3) gravels and the "desert varnish" on them in the north piedmont of the Altun mountain

Fig. 3.58 The MGS5 (MIS5) brown drab paleosol 62S in the Milanggouwan section in the Salawusu River Valley

abundant atmospheric dust at that time. According to the survey, the loess on the edge of the Central Desert Region from MIS1 (Fig. 3.52; Fig. 3.49, Sect. 2) and Extremely Arid Desert Sub-Region of Western Desert Region from MIS3 are the most developed (Fig. 3.49, Sect. 9).

Fig. 3.59 The Black loam (MIS1) at Liujiagouwan in the Salawusu River Valley

Fig. 3.60 The oasis in Yutain (on the left bank of Keliya river) on the southern margin of Taklimakan Desert

Fluviolacustrine facies The fluviolacustrine facies often overlaps with dune sands in China's deserts since the Last Interglacial (Fig. 3.49, Sect. 1, 2). The fluvial facies is rusty-yellow and dark-grey-yellow (Fig. 3.53); the lacustrine facies is yellow-green, blue-green, dark-grey (Fig. 3.54), and the swamp facies is gray-dark, dark (Fig. 3.55). Most of the fluviolacustrine facies is of clayey silt-silty very fine sands with some iron rust, calcareous concretions, plant roots, leaf residues and the root tube fossils (Fig. 3.56), and some with fossil mollusks and vertebrates. In addition, the sand-gravel alluvium overlapping with the dune sands or loess-like sandy sediment also appear in the Central and Western Desert Regions (Fig. 3.49, Sect. 2 to

Fig. 3.61 The MIS2 fossil ice wedge and congeliturbated fold (36°37′N, 80°41′E) in the high terrace of alluvial gravel to the south of Qila on the southern margin of Taklimakan Desert

5). The gravel surface has gray-black "desert varnish" on it (Fig. 3.57) and some visible broken plasters are in the gravels. The fluviolacustrine facies in China's deserts means the accumulation of drifting sands was replaced by the water-fluvial deposits, which is interpreted as an "inter-desert period".

Paleosol Paleosol are mainly brown drab soil (Fig. 3.58) and black loam (Fig. 3.59), with significant soil bedding. The upper bed is a sticky layer, the middle bed, a calcium layer of new rich-carbonate matters, and the lower bed is usually eolian dune sands and an occasional matrix of fluviolacustrine deposition. In contrast, the brown soil is stickier than black loam. According to microscope analysis and experimental results, the cement porphyritic micro-structures of both the brown soil and black loam are quite obvious. The matrix of brown drab soil mainly consists of clay, carbonate and iron oxide, and the black loam mainly consists of compounds of clay and some carbonates with gray calcareous pseudohyphae and calcium films. The paleosol in the Eastern Desert Region is the most developed, and appeared in the MIS1, MIS3, and MIS5 periods (Fig. 3.49, Sect. 1), however, it only appeared in the MIS1 stratigraphical sequence in the Tengger Desert (Gao et al. 1993) and the Badain Jaran Desert of the Central Desert Region (Fig. 3.48, Sect. 2) and in the Arid Desert Sub-Region of the Western Desert Region (Fig. 3.49, Sect. 10).

Oasis soil The oasis soil mainly appears on the southern margin of the Taklimakan Desert (Fig. 3.60), with ages ranging from the Last Glacial Maximum to the present. It also developed near Ejin on the northwestern edge of the Badain Jaran Desert, with an age from 2.5 ka BP to the present (Wen et al. 2006). Based on the study of oasis soil in Yutian- Hotan on the southern edge of the Taklimakan Desert, the modern surface oasis soil is brown desert-soil; its underlying stratum is a set of "old oasis soils". This is mainly eolian sand dust accumulation processed by the pedogenesis since 14 ka BP, with a thickness of 5–6 m. It is similar to the modern surface oasis soil, overlying the third alluvial gravel terrace of the Upper-Pleistocene. The soil is light brown-dark brown, slightly compacted with vertical joints developed, many worm holes, plant roots and marks of tree branches, without beddings. We can often find gray calcareous pseudohyphae and many traces of human activities, such as broken pottery and ash. It originally derived mainly from eolian loess-like sandy sediment and loess, and they are known as sandy oasis soil and silty oasis soils.

Fig. 3.62 The calcareous sand layer in the MGS3 (MIS3) at Milanggouwan section (surface with fossil plant roots)

Studies show that the former and the latter represented periods of desertification and oasis development, respectively (Li et al. 2002).

Periglacial configuration There are two types of periglacial configurations, congeliturbated folds and fossil ice wedges (Fig. 3.61). The folds in the Eastern Desert Region mostly developed in lacustrine layers (e.g. the Mu Us Desert), and overlying dune sands; and also in the gravel layers of the Western Desert Region. Usually, scrolls of the folds reach 60 cm, with half-wave length of 30 cm or so, the greater scrolls can reach 1.8 m, with half-wave lengths of 1.0 m or so. The fossil ice wedge is wider at the top and narrower at the bottom, with a depth from 0.5 to 1.2 m (e.g. the Qila site, Fig. 3.61), the maximum reaching 2.1 m. The congeliturbated folds and fossil ice wedges both post-date the ages of the layers in which they formed, and are consistent with the ages of the overlying dune sands of congeliturbated folds, or the sands at the bottom of fossil ice wedges. Based on TL dating results, these two mainly formed in MIS2 and partly in Late MIS3 (Fig. 3.49, Sect. 1, 2, 7; Cui et al. 2004), and fossil ice wedges formed in MIS4 and MIS5 (Fig. 3.49, Sect. 3, 4). This suggests that China's permafrost moved south several times during the Last Glacial, even during the Last Interglacial, with temperatures 8–12° lower than today. Except for the Qinghai-Tibet Plateau, China's present permafrost boundary lies at approximately 46°36′N, thus, the periglacial features in Salawusu (Fig. 3.49, Sect. 1) of the Eastern Desert Region and in Qila and other sites (Fig. 3.49, Sect. 3, 4, 7) of the Western Desert Region indicate that the permafrost moved south at least 8 degrees of latitude several times during the Last Glacial.

Calcareous sand layer within the inter-dune depression The calcareous sand layer is mainly the slightly cemented mixtures of sand and calcium solution which resulted from precipitation in the inter-dune depression (Fig. 3.62). It often appears in the strata of the Salawusu River Valley, grey fine sand-very fine sand, densely compacted and slightly cemented, with a general thickness of only several centimeters. Such deposits are also regarded as having accumulated during the inter-desert periods.

3.9.2.2 Grain-size Distribution

The Wudeng-Wentworth method is used to classify the grain-size, in which 1 mm is the base multiplied or divided by 2, and the formula $\phi=-\log_2 d$, "d" is diameter of mm to be used to calculate the ϕ value (Krumbien 1934). In this paper, we use the formulas $Mz=(\phi16+\phi50+\phi84)/3$, $\sigma=(\phi84-\phi16)/4+(\phi95-\phi5)/6.6$ of Folk and Ward (Folk and Ward 1957) to calculate Mz and σ. Based on the stratigraphical lithology with reference to the grain-size content, we also refer to the values of Mz and σ to determine the strength of wind force when the dune sands accumulated, and how they are sorted. In general, the dune sands and loess-like sandy sediment mostly appear as coarser grain-sized and well-sorted, with low Mz (ϕ) and σ values, whereas the fluviolacustrine, paleosol and loess appear as finer grain-sized and poorly-sorted, with high Mz (ϕ) and σ values because of weaker winds that were not able to accumulate coarser grain-sized dunes, and were even replaced by the processes of water movement and pedogenesis when the sediments developed.

3.9.2.3 Chemical Components [SiO_2, Al_2O_3, Al_2O_3/SiO_2, $(Fe_2O_3+FeO+Al_2O_3)/SiO_2$]

In the Eastern Desert Region, the SiO_2 content in the old and modern dune sands is the highest and nearly the same, ranging from 72.83 to 90.22%. It might correlate with the high content of quartz and silicate minerals, which are mainly feldspar and a little amphibole, pyroxene and epidote, in the old and modern dune sands. In the Central Desert Region, the SiO_2 content in the paleo-dune sands has similar characteristics (Gao et al. 2006). This suggests that elemental Si was enriched when the dune sands accumulated in the Eastern and Central Desert Regions. Although the SiO_2 content in the loess-like sandy sediment and loess is higher in the Western Desert Region, it is lower than that in the sand dunes of the Eastern Desert Regions, ranging from 52.89 to 65.72% (Li et al. 2008a). It is interesting that the SiO_2 content has a positive correlation with the grain-size in the Eastern, Central and Western Desert Regions. However, the Al_2O_3 and TOFE (Fe_2O_3+FeO) contents are negatively correlated with grain-size. Based on the idea that the aggregation of Al_2O_3 and TOFE are usually correlated with a warm-humid climate in the supergene geochemical environment, we thus conclude that the Al_2O_3 and TOFE contents are important proxy indices that reveal climate variations and desert evolution.

3.9.2.4 Animal Fossils

Here we only discuss the fossil vertebrates and terrestrial fossil mollusks. The mollusks are good indicators for paleoecology because of their limited ability to migrate, good preservation, huge numbers and ample species. Based on examination and analyses of mollusk species assemblages, configuration features, embedding conditions, and contents of carbon and oxygen isotopes, and comparison of their living habits with those of their extant species, paleo-sedimentary environments were reconstructed qualitatively or quantitatively. Fossil vertebrates were rarely discovered in China's deserts since the Last Interglacial, but many species appeared in the strata of the Salawusu River Valley, of which some typical species provide important evidence to explain variations in past climate.

3.9.3 Paleoclimatic Variability Recorded in Deserts Since the Last Interglacial

Based on the climate proxies, we attribute primary stages of climate variations to orbital changes since the Last Interglacial in the Western and Central Desert Regions, and recognize millennium-scale climate variations triggered chiefly by the Asian monsoon in the Western, Central and Eastern Desert Regions, which also basically resulted from orbital changes. Climate fluctuations were recorded more completely in the Eastern Desert Region.

3.9.3.1 Desert Region in Western Inland China

Arid Desert Sub-Region Studies of the Holocene desert environment in the Western Arid Desert Sub-Region are mainly from the Mosuowan section (Fig. 3.49, Sect. 10, 86°20′E, 44°40′N) in the southern Gurbantunggut Desert since the Last Interglacial (Chen et al. 2001b). The section contains MIS1 multi-sequence paleosols, which were approximately impacted by orbital changes, and experienced eight cycles of desertification (dune sands) alternating with pedogenesis, indicating millennium-scale climate fluctuations. The foreset laminae inclinations of the dune sands within the section were 100°, ∠22°, indicating that the winds were mainly from the W-NWW. Based on the ages, the dune sands approximately correspond to the Holocene Bond cold events (B0-B7) in the North Atlantic. Studies also show that the >50 μm grain-size content increases in the dune sand but decreases in the paleosol, and the Fe_2O_3 content shows the opposite trends. Chen et al. (2001b) suggest that the accumulation of dune sands in this area mainly resulted from strengthening of Siberia-Mongolia cold high pressure (cold winter monsoon), bringing cold air from the Arctic Ocean, and south movements of the North Hemisphere Westerlies; development of paleosols correlated with the weakened Siberia-Mongolian high pressure and an increase of precipitation brought by the Westerlies.

Extremely Arid Desert Sub-Region The records in the strata of MIS1-MIS3 in this area evidently reflect orbital changes since the Last Interglacial. The MIS1 strata mainly consist of silty oasis soil, the MIS2 strata contain obvious periglacial structures and dune sands, and the MIS3 strata are characterized by dust deposition. Millennium-scale climate variability within these periods also appears clear.

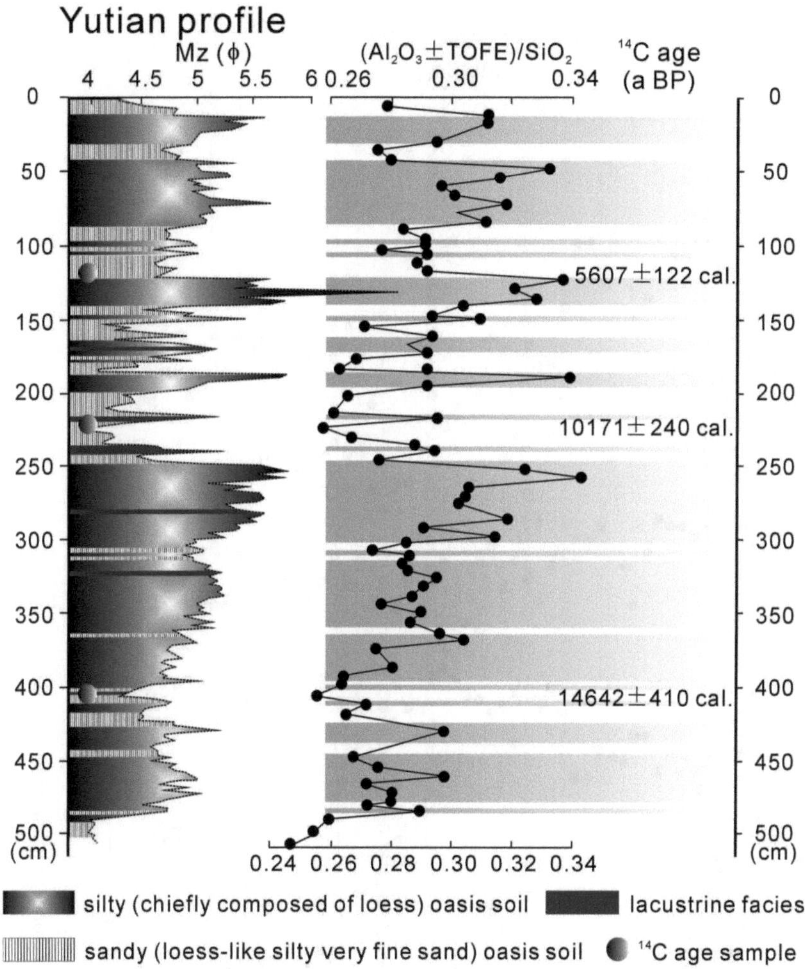

Fig. 3.63 The changes of Mz (ϕ) and (TOFE + Al_2O_3)/SiO_2 in the Yutian oasis section since 15 ka BP

The Yutian oasis section (36°52′N, 81°42′E) is a typical MIS1 stratigraphical sequence. This section contains an oasis soil with a thickness of 500 cm since 15 ka BP, of which the oasis soil with a thickness of 250 cm since 10 ka BP records 12 alternations between the sandy oasis soil and silty oasis soil or lacustrine sediments. The values of Mz (ϕ) and (TOFE + AL_2O_3)/SiO_2 show considerable oscillations between layers (Fig. 3.63), with valleys in the sandy oasis soil and peaks in the silty oasis soil and lacustrine facies, which reveal the repeated alternations of desertification and oasis processes. Five periods of sandy oasis soil in this section since 5.6 ka BP correspond to the Bond cold events, B0 to B4, in the North Atlantic; the period of sandy oasis soil at a depth of 240 cm corresponds to Bond event B7 (Bond et al. 2001).

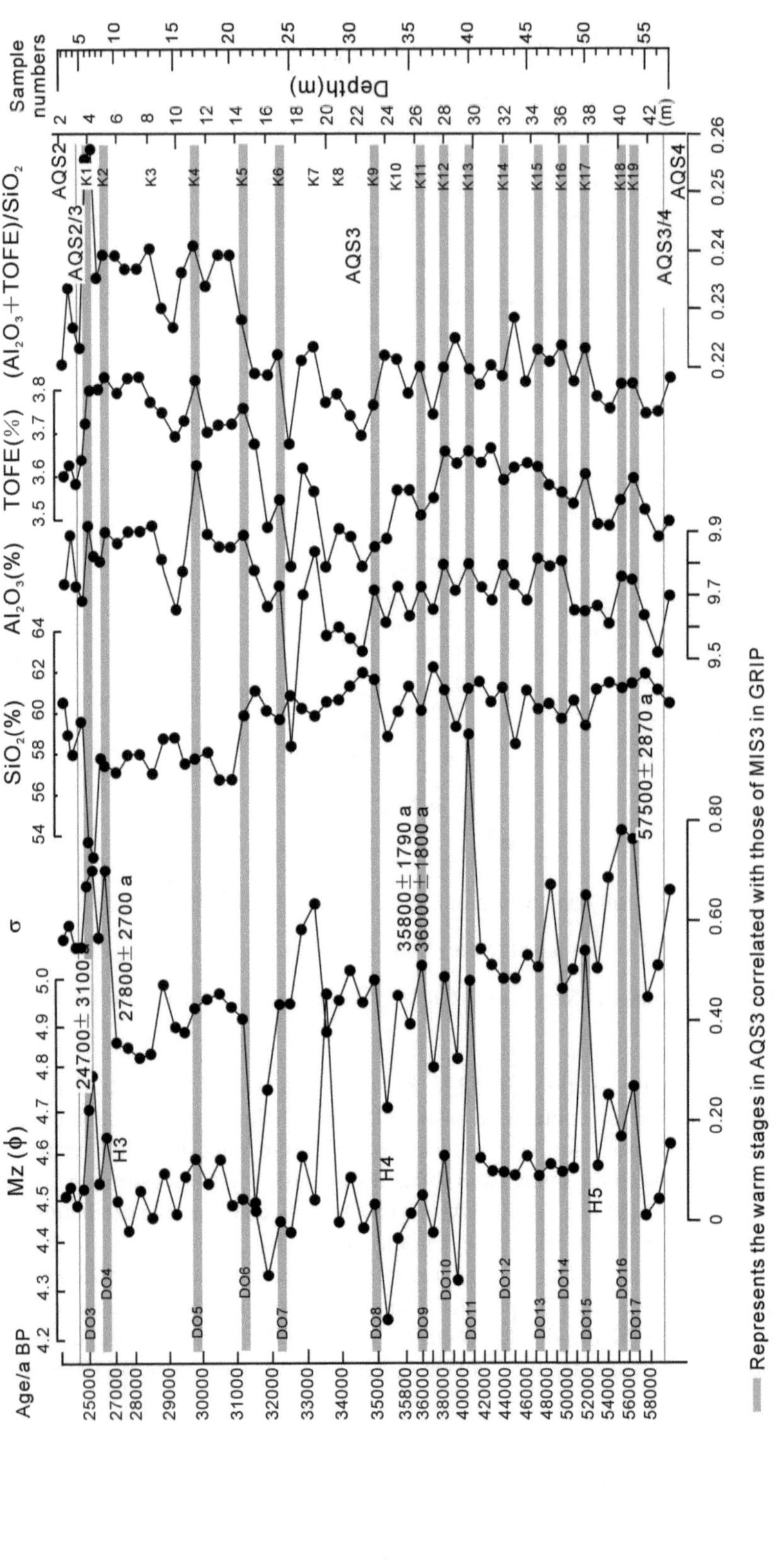

Fig. 3.64 The Mz,σ primary oxides (SiO_2, Al_2O_3, TOFE) and (Al_2O_3+TOFE) /SiO_2 in the MIS3(AQS3) in the Aqiang section

The MIS2 strata are widespread in the Taklimakan Desert, but we still lack a complete section for the whole period. In Fig. 3.49, Sect. 6 to 8 shows the sedimentary characteristics of MIS2 strata-MFS2 (Fig. 3.49, Sect. 8, 37°15′N, 82°58′E), QLS2 (Fig. 3.49, section 7, 36°37′N, 80°41′E) and YTS2 (Fig. 3.49, Sect. 6, 36°52′N, 81°42′E). In addition, there exist dune sands of the Last Glacial Maximum in Shazidaban on the southeast edge of the desert and desert hinterland. The TL age of the former is 17,980±890a BP (Li et al. 1993), and the latter is 17,040±828a BP and 16,700±800a BP (Li et al. 1990). The OSL age of the dune sands in the northwest of Yutian is 14,500±1100a BP (Yang et al. 2006). The MIS2 fossil ice wedges commonly appear on top of the alluvial sandy gravel layers in Minfeng and Yutian. The periglacial structures also often appear on top of the high terraces to the south of Qila (Li et al. 1993), with synchronous congeliturbated folds (Fig. 3.61), TL ages of 19,100±2300a BP and 13,320±670a BP (Fig. 3.49, Sect. 7). In addition, the OSL age of a fossil ice wedge there dated by Yang Xiaoping is 18,300±1300a BP (Yang et al. 2006).

MIS2 is an extremely cold and windy period for desert evolution in Taklimakan. In addition to the development of congeliturbated folds and fossil ice wedges, the gravel cracks and splinters caused by frost weathering also appear in MFS2 and YTS2. Of course, the MIS2 period also experienced warm climate fluctuations in the Taklimakan Desert, such as the extremely thin layers of lacustrine sediments in the loess-like sandy sediment of YTS2 in the Yutian-Hotan Oasis, probably reflecting short warm periods when the winter monsoon weakened (Li et al. 2002).

MIS3 stratigraphical sequences appear in the Aqiang loess section AQS3 (81°56′E, 36°28′N) that outcrops in the south of the Taklimakan Desert (Fig. 3.49, Sect. 9). These are well-preserved, at a depth of 2 to 42 m in the section, with several layers of loess-like sandy sediment (Fig. 3.51). The research results (Li et al. 2008a) show that the AQS3 period generally was warm but with several cold climate fluctuations on a millennium-scale. The Mz, σ, primary oxides (SiO_2, Al_2O_3, TOFE) and (Al_2O_3+TOFE)/SiO_2 ratio show that the AQS3 has experienced 19 cycles of cold-dry and warm-humid climate fluctuations. Of these, 15 cycles of millennium-scale warm climate fluctuations correspond well with D/O cycles in the oxygen isotope record of the GRIP ice core (Fig. 3.64). Probably due to section adjacent to the piedmont glacier at a high altitude, the proxy vaules of grain-size and primary oxides vary within a relatively narrow range.

3.9.3.2 Central Desert Region Influenced by the East Asian Summer Monsoon and the Westerlies

Many researchers have studied environmental changes in the Central Desert Region since the Last Interglacial (Cai et al. 1986; Gao et al. 1993, 1996, 2006; Jäkel 1996; Höffmann 1996; Li et al. 1998, 2005b, c; Yang 2000; Yan et al. 2001; Chen et al. 2001b; Zhang et al. 2004b; Wen et al. 2005; Yang et al. 2008). Nevertheless, a complete high-resolution stratigraphical section is needed to reveal the environmental changes in this desert area. Thus, in 2007, we re-inspected the Chagelebulu

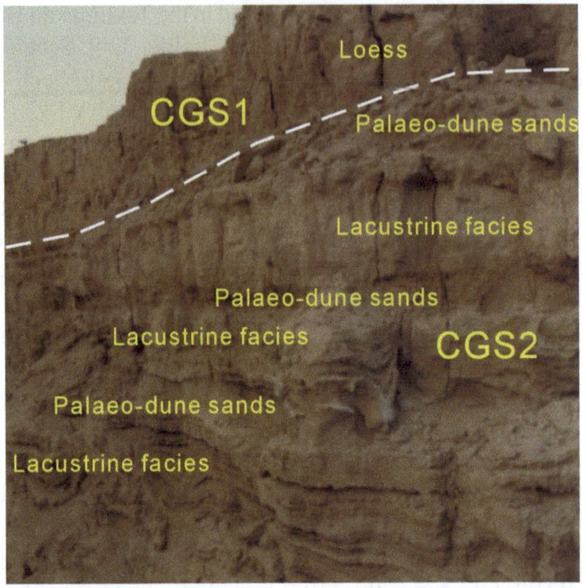

Fig. 3.65 The CGS1 (MIS1) loess and its underlying interbeded dune sands and lacustrine sediments (CGS2) in the Chagelebulu section

section that was found in 1992 (Gao et al. 1996). The section can serve as an ideal geological record, reflecting millennium-scale desert environmental variations in the Central Desert Region since the Last Interglacial.

The Chagelebulu section is situated in the western Gobi desert area, in front of the west piedmont of the Yabulai Mountains, on the southeast edge of the Badain Jaran Desert, which is surrounded by a flat loess surface. Due to fluvial activity, the ground surface was eroded into a valley a depth of 15 to over 20 m, and thus considerable strata are exposed on both sides of the valley. Late Quaternary strata overlie sandy Lishi Loess in the section, and the sedimentary facies are mainly lacustrine and dune sands. The re-examined Chagelebulu section (39°53′50.6″N, 103°18′25.5″E) is in the NW about 50 m away from the original section (Gao et al. 1996), at an altitude of 1748 m. The total thickness of the section is about 23 m and the Late Quaternary sequences contained 88 layers: 1 layer of modern-mobile dune sands, 33 layers of paleo-mobile dune sands, 10 layers of paleo-fixed and semi-fixed dune sands, 6 layers of loess, 2 layers of immature soil, 31 layers of lacustrine and 5 layers of sand-gravel (Fig. 3.49, Sect. 2). In the section, obvious sedimentary sequences consist of the dune sands overlapping with lacustrine (Fig. 3.65) or immature soil. In addition, the Chagelebulu section contained nine fossil layers, of which 8 contained fossil mollusks with shells, except CGS5 bottom which contained plant fossils (*Ranunculaceae*). We have a total of 25 dates for this section.

Based on the MIS ages, we divided the Chagelebulu section into five segments from CGS1 to CGS5 since the Last Interglacial. Of these, the geological boundaries and ages of CGS3/CGS4, CGS5/CGS6 and sub-segments from CGS5a to CGS5e were mainly referred to those of MIS1 to MIS5, particularly the D/O-cycles in the GRIP ice core.

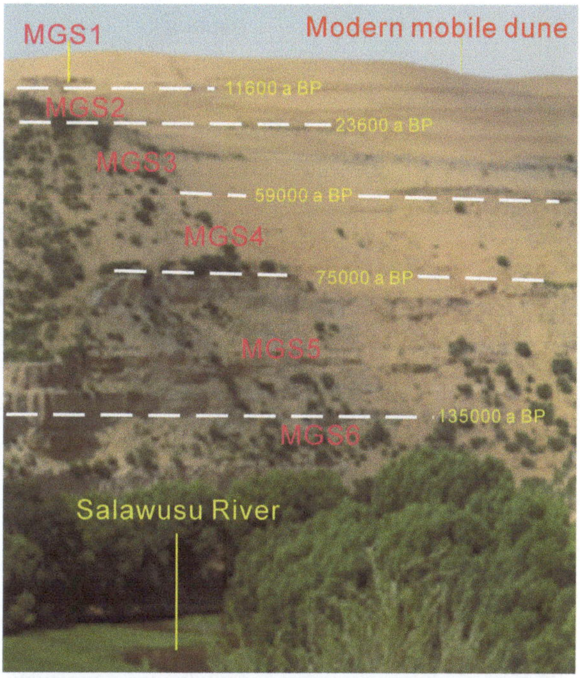

Fig. 3.66 The primary stratigraphic segments and their interface ages of the Milanggouwan section

Fig. 3.67 Calcareous sandy layers with grey-white stripes in the Milanggouwan section

The eolian dune sands overlapping with lacustrine sediments appear in the section from CGS1 to CGS5 and each sub-segment of CGS5. By contrast, coarse-grained dune sands in the section significantly correspond to the MIS2, MIS4,

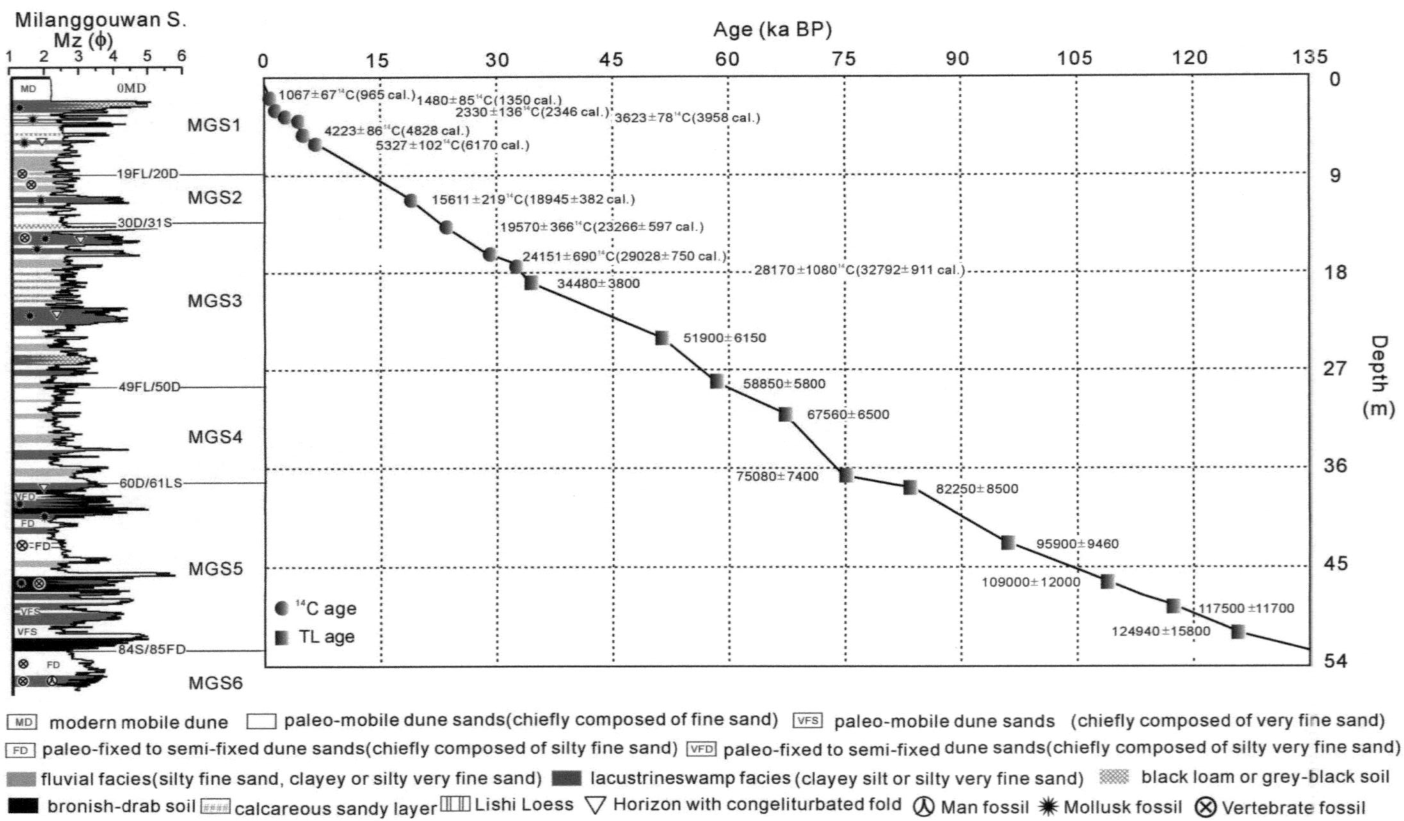

Fig. 3.68 The sedimentary sequences and ages within the Milanggouwan section since the Last Interglacial

Table 3.3 The ^{14}C dating materials, results and calendar ages (a BP) of some horizons in the Milanggouwan section

Layers	Dated ages	Calendar ages	Dating material	Layers	Dated ages	Calendar ages	Dating material
Top of 1S/LS	1067±67	965	Shells of fossil mollusks	Middle of 11S	4223±86	4828	Organic matter
Middle of 1S/LS	1450±120	1330	Silty deposits	Middle of 13LS/FL	5327±204	6170	Shells of fossil mollusks
Lower middle of 1S/LS	1420±67	1307	Silty deposits	Middle of 25LS	15611±219	18945	Silty deposits
Bottom of 1S/LS	1480±85	1350	Silty deposits	Middle of 31S	19570±366	23266	Organic matter
Middle of 3LS	2330±136	2346	Silty deposits	Bottom of 36LS	24151±690	29028	Silty deposits
Middle of 5LS	3623±78	3958	Silty deposits	Middle of 38D	28170±1080	32792	Sandy deposits

MIS5b and MIS5d, whereas fine-grained lacustrine sediments correspond to MIS3, MIS5, MIS5a, MIS5c and MIS5e. The section CGS1 is characterised by loess with two layers of dune sands and immature soil layers, without lacustrine sediments. This shows the overall weakened wind strength which accumulated dune sands there at that time. According to studies on the sedimentary facies at the site by Gao et al. (2006), it probably suggests that the CGS1, CGS3, CGS5 and CGS5a, CGS5c, CGS5e which are mainly characterised by lacustrine sediments or loess, were mainly deposited in warm periods, and CGS2, CGS4 and CGS5b, CGS5d when the dune sands accumulated significantly, were the cold periods. The dune sands and lacustrine sediments in CGS2 to CGS5 corresponded well to D/O-scale cold and warm events in the Greenland ice cores, respectively; and the dune sands, loess and soil layers in CGS1 approximately correlate with millennium-scale climate fluctuations in the North Atlantic (Bond et al. 1999).

3.9.3.3 Central-Eastern Desert Region: East Asian Monsoon Records from the Milanggouwan Locality

Milanggouwan sedimentary sequence since the Last Interglacial Of China's sandy deserts, the Milanggouwan section is the most complete sedimentary facies that records environmental evolution on a millennium-scale (Fig. 3.66, Fig. 3.49, Sect. 1) in the Mu Us Desert of Eastern Desert Region since the Last Interglacial, including its calcareous sandy layers (Fig. 3.67). Thus, we mainly discuss the geological information hereinafter from this section (Fig. 3.68) and attempt to understand Asian monsoon climate variability in China's deserts since that time.

Table 3.4 TL dating results of some horizons and their parameters in the Milanggouwan section

Stratum-Number	depth (m)	U (ppm)	Th (ppm)	K (%)	Annual dose (m Gy)	Total dose (Gy)	TL (ka BP)
38D-PKG588	19.25	2.568	11.633	1.071	110.241±8.82	3.197	34.48±3.80
42D-TGD793	24.23	0.30±0.02	7.19±0.14	1.86±0.015	112.10±8.96	2.16±0.07	51.90±6.15
50D-TGD782	28.42	0.78±0.04	6.72±0.13	1.94±0.016	133.00±9.98	2.26±0.07	58.85±5.80
53FL-TGD783	29.64	0.25±0.01	7.17±0.14	1.95±0.016	152.00±11.40	2.25±0.07	67.56±6.50
60 D-TGD 784	36.95	0.11±0.01	10.06±0.20	2.10±0.02	2.48±0.08	186.20±13.96	75.08±7.40
62 S-TGD 785	37.60	1.07±0.05	12.09±0.24	2.03±0.016	2.31±0.07	190.00±14.25	82.25±8.50
70FD-TGD 786	42.75	0.75±0.04	6.50±0.13	1.99±0.016	2.02±0.06	193.80±14.53	95.90±9.46
76S-TGD 794	46.37	1.49±0.08	12.29±0.25	2.13±0.017	2.65±0.08	288.80±23.10	109.00±12.00
81D-TGD 787	48.47	0.42±0.02	11.04±0.22	1.94±0.016	2.27±0.07	266.76±20.00	117.50±11.70
83 DM- TGD 788	50.29	1.50±0.04	6.10±0.15	1.94±0.018	2.66±0.10	332.27±35.56	124.94±15.84
85 FD-TGD 604	54.18	0.62±0.03	6.94±0.14	2.42±0.019	4.36±0.13	646.80±40.75	148.00±12.50

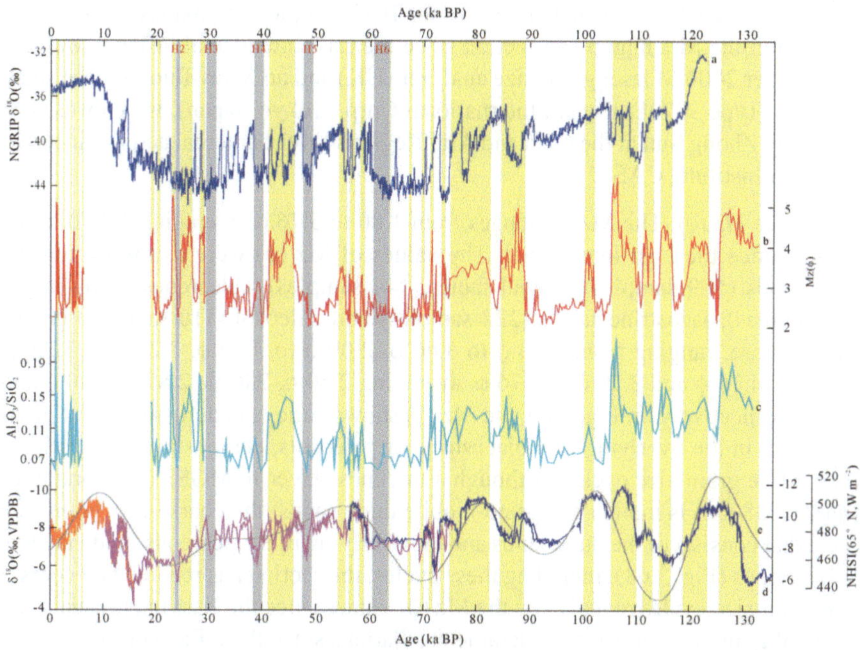

Fig. 3.69 Comparison of Milanggouwan records with other records since 130 ka BP. **a** North GRIP $\delta^{18}O$ (North Greenland Ice Core Project members, 2004); **b** Mean grain size (ϕ) and **c** Al_2O_3/SiO_2 of Milanggouwan section; **d** Dongge (orange), Hulu (purple) and Sanbo (blue) cave speleothem $\delta^{18}O$ records (Dykoski et al. 2005; Wang et al. 2001; Wang et al. 2008a); **e** NHSI (Northern Hemisphere summer insolation, mean July insolation) at 65°N. The yellow bars indicate the lacustrine and paleosols layers and the unshaded parts show the eolian sand layers in Milanggouwan section; Five grey shaded bars indicate Heinrich events 2-6 respectively

The strata in the section were dated by means of ^{14}C and TL. The ^{14}C determination was performed by Zhiyu Hu at the ^{14}C lab of Cold and Arid Regions Environmental and Engineering Research Institute, Chinese Academy of Sciences, China, using a low background value liquid scintillation counter (Table 3.3). The TL age determination (conducted by Liangcai Lu et al. in the TL lab of the Guangzhou Institute of Geochemistry, Chinese Academy of Sciences and the other related institutes) was based on quartz grains smaller than 10 μm. The TL ages and the relevant parameters are listed in Table 3.4.

The primary periods recorded in the Milanggouwan section are approximately correlative with MIS1 to MIS5 since the Last Interglacial. We obtained many dating ages within the section and the dating horizons are approximately evenly distributed. Thus, based on the adjacent ages in the relevant layers and their depths, we used linear regression to estimate the sedimentation rates, and then calculated each stratigraphical sequence age based on different sedimentation rates. The results are shown in Fig. 3.68 with Mz (ϕ) variations.

Climate proxies We analyzed grain-size for 1013 samples and oxides for 351 samples within the Milanggouwan section since the Last Interglacial, using a Malvern Mastersizer 2000 M laser grain-size analyzer (UK) and an X-ray fluorescence spectrometer (Type 3070E, Rigaku International Corp., Tokyo, Japan), which was conducted by Zhong Sun at the Cold and Arid Regions Environmental and Engineering Research Institute, CAS.

Grain-size Mz (ϕ) The Mz (ϕ) ranges from 1.80 to 5.76 ϕ, averaging 2.96 ϕ in the section since the Last Interglacial. The values of Mz (ϕ) were as follows: in the dune sands (509 samples), fluvial facies (144 samples), calcareous sandy layers (28 samples), lacustrine facies (222 samples) and paleosol (110 samples) gradually increase, ranging from 1.80 ϕ to 3.66 ϕ, 2.09 ϕ to 3.89 ϕ, 2.26 ϕ to 3.22 ϕ, 2.07 ϕ to 4.76 ϕ and 2.79 ϕ to 5.76 ϕ, averaging 2.50 ϕ, 2.82 ϕ, 2.87 ϕ, 3.60 ϕ and 4.02 ϕ, respectively. In Fig. 3.69, the Mz (ϕ) shows valleys in dune sands, but obvious peaks in the overlying fluviolacustrine or paleosols, demonstrating significant coarse-fine grain-size cycles. Although a thickness layer in MGS3 (the horizon of 34480 ± 3800a) is mainly the paleo-mobile dune sands, the coarse-fine grain-size cycle that consists of the dune sands and the overlying thin calcareous sandy layer is also obvious (Fig. 3.68). Including these cycles, the section has recorded 44 coarse-fine grain-size cycles revealed by the Mz (ϕ) values since the Last Interglacial. The graph of grain-size shows considerable oscillations since then. Each cycle approximates ~3 ka on average, which represents one cycle of dune sands alternating with fluviolacustrine or paleosols or a calcareous sandy layer.

Al_2O_3, SiO_2 and Al_2O_3/SiO_2 In the primary elements, the SiO_2 content is highest in the section since the Last Interglacial, ranging from 39.10 to 90.50 %, averaging 77.33 %; the Al_2O_3 contents second, ranging from 4.43 to 14.11 %, averaging 7.66 %. In general, the SiO_2 contents in the dune sands (166 samples), calcareous sandy layers (5 samples), fluvial facies (47 samples), lacustrine facies (83 samples) and paleosol (50 samples) gradually decrease, ranging from 73.14 to 90.5 %, 75.99 to 84.70 %, 67.33 to 86.5 %, 49.78 to 83.46 % and 39.1 to 86.84 %, averaging 81.74, 80.85, 78.70, 71.51 and 71.07 %, respectively. On the contrary, the Al_2O_3 contents in these sedimentary facies follow an increasing trend, ranging from 4.43 to 9.28 %, 6.45 to 7.93 %, 5.56 to 9.03 %, 4.75 to 10.74 % and 6.26 to 14.11 %, averaging 6.62, 7.25, 7.56, 8.83 and 9.27 %, respectively.

The Al_2O_3/SiO_2 ratio ranges from 0.05 to 0.36 with an average of 0.10 since the Last Interglacial, and shows synchronous changes with the Al_2O_3 in the sedimentary facies. The values gradually increase in the dune sands, calcareous sandy layers, fluvial facies, lacustrine facies and paleosols in turn, ranging from 0.052 to 0.124, 0.079 to 0.099, 0.067 to 0.134, 0.068 to 0.174 and 0.072 to 0.361, averaging 0.081, 0.090, 0.096, 0.125 and 0.136, respectively.

The Al_2O_3/SiO_2 ratio also shows valleys in dune sands and peaks in the overlying fluviolacustrine or paleosols, with 44 significant Al_2O_3/SiO_2 cycles in the graph. This includes the cycles consisting of dune sands of 38D and the overlying calcareous sandy layers, which are consistent with the sediment and Mz (ϕ) cycles. Obviously, the time and average variability of each Al_2O_3/SiO_2 cycle is the same as their corresponding sediment and Mz (ϕ) cycles.

Because the fluvial facies can reflect a variety of climate signals, we do not consider it here.

Orbital-scale climate variations Records of the main climate stages and some substages from the Milanggouwan section since the Last Interglacial have significant correlations with MIS1 to MIS5 in the Greenland ice cores and deep-sea sediments (Fig. 3.69). It suggests that the Mu Us Desert has experienced analogous environmental variations as the Greenland ice cores and deep-sea sediments, which also corresponds to summer solar radiation changes at 65°N in the NH, and the formation and evolution of China's desert environment is likely to have resulted from variations of Earth's orbit at least from MIS1 to MIS5. Variations of Earth's orbit determined the total solar radiation on its surface, and resulted in the occurrence of Glacial and Interglacial climate changes.

Millennium-scale climate records and its comparison with those from high latitude Previous studies have shown the dune sands in the Salawusu River Valley of the Mu Us Desert, even in the Eastern Desert Region, were the sandy sediments of desert environment as a result of the East Asian winter monsoon. In contrast, the fluvolacustrine and paleosol were strongly correlated with prevalence of the East Asian summer monsoon and its precipitation (Dong et al. 1998). Similar climate variations were found in the Milanggouwan section of the Salawusu River Valley (Li et al. 2005c; Niu et al. 2008, 2011; Wen et al. 2009; Lü et al. 2010; Du et al. 2011, 2012). Therefore, we interpret dozens of sediments, Mz (ϕ) and Al_2O_3/SiO_2 cycles in the stratigraphical sequence to represent climate variations of East Asian winter and summer monsoons since the Last Interglacial. Moreover, we also found many fossils that served as paleo-climatic and paleo-ecological indicators in the Milanggouwan section, to support the interpretation of climate variations during these periods (Li et al. 2007, 2008b):

(1) Fossil egg fragments of Ostrich (*Stuthio* sp.) in the dune sand layers reflected the cold-dry climate then, and both the Mz (ϕ) and Al_2O_3/SiO_2 in them had low values.

(2) The fossil gastropods which indicated the warm climate appear in many lacustrine facies, and the Mz (ϕ) and Al_2O_3/SiO_2 in these layers appear as high peaks. Some fossil gastropods, such as *Gyraulus convexiusculus, Kaliella rupicola* and, *Gryaulus filaris* appear in some lacustrine facies. Their extant species appear widely in the Yangtze River and southern warm tropical and subtropical waters, especially, *Gyraulus convexiusculus*, which occurs not only in Guangdong and Guangxi provinces of China, but even as far south as the Southeast Asia and South Asian.

In addition, a terrestrial snail fossil *Succinea erythrophana* appeared in the lacustrine facies, whose extant species occur widely across China, and live in very warm-humid environments.

(3) The fossil gastropods not only appeared in the Milanggouwan stratigraphycal section, but were widespread in the Salawusu River Valley. Take the MIS3 strata for example. The fossils of *Succinea erythrophana* and *Gyraulus sibiricus* also appeared in the lacustrine sediments of the Yangsigouwan section(37°44′14.5″N, 108°32′24.3″E), upstream of the Milanggouwan section (Li et al. 2008b).

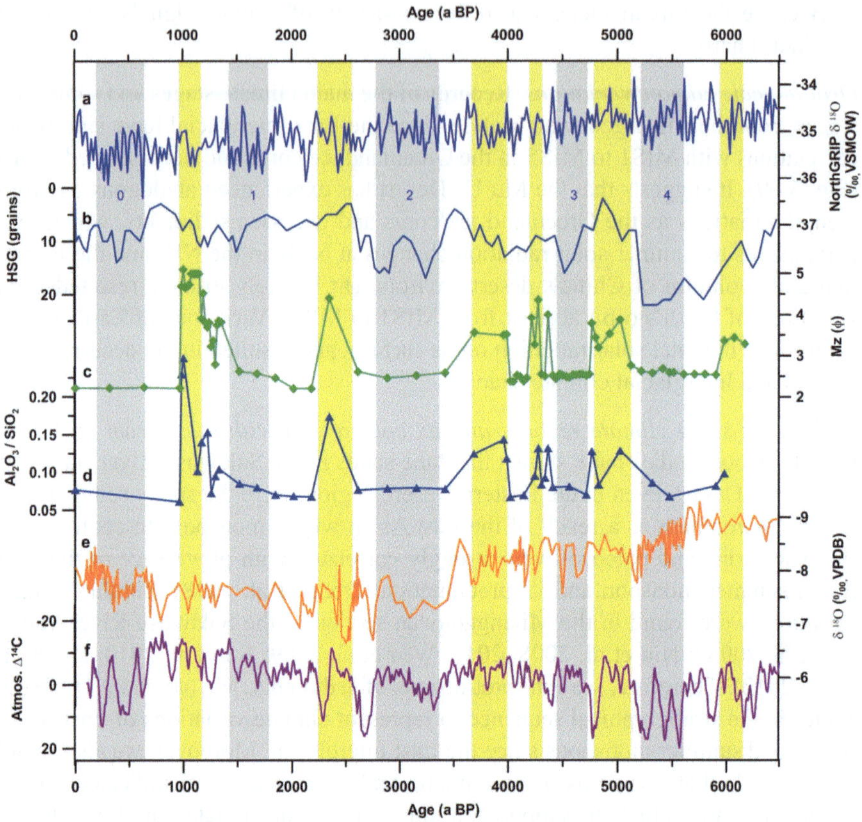

Fig. 3.70 Comparison of Milanggouwan records with other records during the last 6 ka BP. a) NGRIP ^{18}O (Rasmussen 2006); b) North Atlantic ice-rafted hematite stained grains (HSG) (Bond et al. 2001); c) Mean grain size (ϕ) and d) Al_2O_3/SiO_2 of Milanggouwan section; e) Dongge cave speleothem $\delta^{18}O$ records (Dykoski et al. 2005); f) atmosphere $\Delta^{14}C$ (Reimer 2004). The yellow bars indicate the lacustrine and paleosols layers in Milanggouwan section; five grey shaded bars (number 0 to 4) indicate the Bond events respectively

As a result of these data, it is clear that the fossil vertebrates also adapted to warm climate, such as *Bubalus Wansjocki* and *Palaeoloxodon naumanni,* which appear in fluviolacstrine sediments, and, some fossil animals adapted to cold-dry environments, such as *Camelus knoblocki*, *Stuthio* sp. and rodent species, mostly appear in paleo-dune sand deposits.

Millennium-scale climate events in the Holocene The Milanggouwan section contains five layers of dune sands that formed under the winter monsoon since 6000 a BP during the Holocene, and the Mz (ϕ) and Al_2O_3/SiO_2 in the layers appear as low values (Fig. 3.70). These events not only correspond to the cold events "B0 to B4" in the North Atlantic (Bond et al. 2001), but also have a relatively good correlation with cold episodes (low oxygen isotope values) in the North GRIP record (Rasmussen 2006) as well as in stalagmite isotopes (Dykoski et al. 2005) and solar

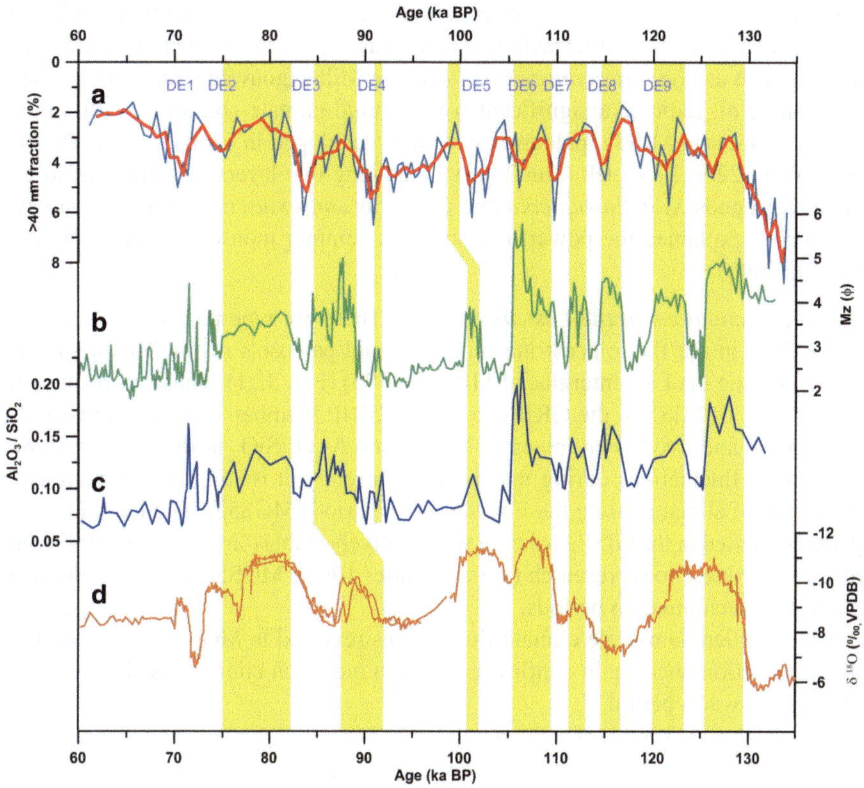

Fig. 3.71 Comparison of Milanggouwan records with other records during the last interglacial. a) Potou section > 40μm quartz size fraction content (An and Porter 1997), red line is the 3-point running average; b) Mean grain size (ϕ) and c) Al_2O_3/SiO_2 of Milanggouwan section; d) Sanbao cave speleothem $\delta^{18}O$ records (Wang et al. 2008a); The yellow bars indicate lacustrine and paleosol layers and the unshaded parts show eolian sand layers in Milanggouwan section; DE1 to DE9 indicates nine dust event intervals, recorded in Potou section, Loess Plateau

activity (Reimer et al. 2004). This shows that the Mu Us Desert also experienced the significant millennium-scale climate fluctuations during the Holocene.

Heinrich events in the Last Glacial Low values of Mz (ϕ) and Al_2O_3/SiO_2 in the Milanggouwan section correspond to H2 to H6 of the Heinrich events in the North Atlantic (Bond et al. 1993; Fig. 3.69). However, climate fluctuations also appear in the Milanggouwan section during the Last Interstadial with many Heinrich-like events, especially in the extremely warm periods of MIS3a and MIS3c, which were incompatible with the increase of global ice volume revealed by deep-sea SPECMAP curve. Sub-stage MIS3c was a warm event during the "anomalous high-temperature precipitation events" period from 30 to 40 ka BP (Shi and Yao 2002). During that period, high lake levels appeared in the Tengger Desert, Qinghai-Tibet Plateau and other places in China (Fang 1991), paleosols generally developed in the

Loess Plateau during the Malan Loess period from about 35 to 42 ka BP (Sun 2005), and a 1.65 m thick lacustrine sediment layer was deposited with warm climate gastropods such as *Gyraulus convexiusculus* in the Milanggouwan stratigraphical section. These all indicate a significant warm-humid climate condition at that time. During Sub-stage MIS3a, paleosols generally developed in China's Loess Plateau from about 22 to 27 ka BP (Sun 2005), containing two layers of warm-humid climate gastropods *Kaliella rupicola* and *Gyraulus convexiusculus*. An and Liu have reasonably explained the powerful East Asian summer monsoon at that time (An and Liu 2000).

Climate fluctuations in the Last Interglacial There are nine peaks in Mz (ϕ) and Al_2O_3/SiO_2 in the fluviolacustrine sediments and paleosols of the Milanggouwan section during the Last Interglacial MIS5 (MGS5) (Fig. 3.71). These correspond to G20 to G25 of MIS5 in the GRIP ice core (NGRIP Members 2004). Significantly, three peaks and two lower values of Mz (ϕ) and Al_2O_3/SiO_2 in the MGS5e shows the climate fluctuations during that period. Currently, it is still controversial as to the nature of climate during the Eemain warm period (MGS5e). The Eemain warm period recorded in the GISP2 ice core was relatively stable (Grootes et al. 1993), but based on Chinese loess research (An and Porter 1997), MGS5e was unstable, with many dust accumulation periods.

Thus, millennium-scale climate fluctuations revealed in MGS5e of the Milanggouwan section once again confirm East Asian monsoon climate instability during the Eemian warm period.

3.9.4 Orbital and Non-orbital Factors Contributing to Monsoonal Desert Environment Change

Orbital factors Based on the summer solar radiation changes in the North Hemisphere since the Last Interglacial, we suggest that orbital changes are vital for desert formation and evolution. Thus, we compare the monsoon climate curves of Mz (ϕ) and Al_2O_3/SiO_2 in the Milanggouwan section with the summer solar radiation changes at 65°N, to further understand the reasons for desert environment evolution and monsoon climate variations.

In Fig. 3.69, the 65° N insolation curve contains, six peaks separated by five lower values since the Last Interglacial (Berger 1978, 1991). The ages of peaks were 10, 31, 55, 81, 103 and 126 ka, and the ages of lower values were 21, 41, 69, 93 and 114 ka. Except for the peak at 10 ka, the warm peaks of Mz (ϕ) and Al_2O_3/SiO_2 correlated well with the other five insolation peaks, and the cold lower values of Mz (ϕ) and Al_2O_3/SiO_2 also corresponded to five lower values of 65°N curve. Although the correlation between the solar radiation peak at 10 ka BP and the warm peaks of Mz (ϕ) and Al_2O_3/SiO_2 (Fig. 3.69) is not clear, the lacustrine deposition of the Dishaogouwan section contained many warm-humid climate gastropods *Gyraulus sibiricus* (according to the investigation by Li et al. recently), with the ages of 9380 ± 30 cal. a BP and 9450 ± 90 cal. a BP (^{14}C AMS 8360 ± 31a BP and 8406 ± 32a BP).

In summary, we believe that orbital changes on the 10 ka-timescale ultimately determined the strength of the East Asian winter and summer monsoons, and triggered the evolution of the desert environment.

Non-orbital factors The millennium-scale warm peaks and cold valleys overlapped with the orbital periods. But most peaks and valleys match D/O cycles recorded by oxygen isotopes in the North Atlantic deep-sea sediments and Greenland ice cores. This suggests that the alternation between desert periods and inter-desert periods that mainly formed under the influence of the winter and summer monsoons, as indicated in the Milanggouwan section, probably resulted from global climate changes during those periods. Because monsoon environment changes were subject to sea ice variations in the sub-polar-North Atlantic, ice coverage on the Eurasian continent and temperature changes associated with the movement of the polar front, they impacted the strength of the Siberian-Mongolian high pressure and associated cold air masses (An et al. 1994). In this view, the monsoon cycles that corresponded to D/O events, as indicated by Mz (ϕ) and Al_2O_3/SiO_2 in the Milanggouwan section since the Last Interglacial, imply that climate variations in the North Atlantic and Greenland are hemispheric in scale. As a result, by comparison between cold lower values in the Milanggouwan and DE1 to DE9 (dust events 1–9) in the Loess Plateau (An et al. 1999), we can easily understand that six millennium-scale dust events from DE1 to DE6 were consistent with six cool events revealed by high oxygen isotope content of foraminifera *Neogloboquadrina pachyderma* (S.) in cores V29 to V191 (Mc Manus 1994) from the North Atlantic.

3.10 Asian Monsoon Variability Recorded in Marine Sediments

3.10.1 Introduction

The Asian monsoon system is a dynamic component of the modern climate system (Zahn 2003). Differential heating of the Asian continent and the Indian Ocean and the western Pacific Ocean causes the reversal of seasonal winds and associated rainfall changes. The Asian monsoon, comprised of the Indian (South Asian) and East Asian Monsoon systems, determines environmental conditions over much of Asia. Monsoonal changes in this convectively active region result in severe droughts or floods over large, densely populated regions (Webster et al. 1998). The reliability of the Asian monsoon has serious implications for the welfare of a large and dense population throughout the region. Given the rapid increases in human activities across Asia in recent years, it is now more urgent than ever to understand past changes in monsoonal climate to assist in developing sound methods for future prediction and social planning.

Over the past few decades, many studies have addressed various aspects of the Asian monsoon evolution and variability, using both terrestrial and marine archives (e.g. An et al. 2001). While some discrepancies remain to be resolved, it is generally accepted that its evolution and variability are linked to the uplift of the TP, ocean gateways, and global boundary conditions (Wang et al. 2005b). This review will largely focus on marine paleomonsoon records that are obtained from the Indian Ocean, South China Sea (SCS) and Indonesian maritime regions. Successful marine paleomonsoon research rests on two conditions: 1) the retrieval of high quality sediments cores and 2) the development of accurate paleomonsoon proxies. Several scientific cruises to the Indian Ocean and the SCS, including the Ocean Drilling Program (ODP) Leg 117 to the Arabian Sea and Oman Margin (Prell et al. 1989), SONNE 90 to the northern Indian Ocean (von Rad et al. 1995), SONNE 94 to the SCS (Sarnthein et al. 1994), IMAGES (International Marine Past Global Change Studies) Leg III to the SCS (Chen et al. 1998a), and ODP Leg 184 to the SCS (Wang et al. 2000), made the former possible. Meanwhile, many lithological, micropaleontological, and geochemical tools have been developed to serve as marine monsoon proxies over the past few decades. However, as most of these proxies may be controlled by more than one factor, assessing the relative importance of monsoonal and non-monsoonal processes on these proxies is still a challenging task (Wang et al. 2005b). The development and validation of marine monsoon proxies, through the deployment of sediment traps and analyses of modern core top sediments, are particularly important in order to further understand Asian paleomonsoon dynamics on various timescales.

An overview of past and current paleomonsoon research on tectonic through interannual timescales, primarily based on marine records, has been compiled by Wang et al. (2005b). We focus here on Asian monsoon variability on millennial and orbital timescales, and within the last 2 Ma, also based on marine records. We compare marine records with land-based monsoonal records (oxygen isotopes from speleothems in particular), and records elsewhere indicating variability in other climate subsystems on these two timescales. This review is organized as follows: in Sect. 3.10.2 marine monsoon proxies are introduced; Asian monsoon variability on millennial and orbital timescales are addressed in Sect. 3.10.3 and 3.10.4, respectively; in Sect. 3.10.5 outstanding issues of Asian monsoon dynamics on orbital timescales are discussed, and a short summary of the review is provided in Sect. 3.10.6.

3.10.2 Proxies

The reversal of seasonal winds and associated rainfall changes can affect many natural processes and leave their imprint in marine sediments. During the summer monsoon, the development of low pressure systems over Asia relative to higher pressure systems over the oceans leads to prevailing landward winds that bring moisture and rainfall. The situation is reversed during the winter monsoon. These fundamentally different weather conditions are reflected by a variety of natural climate proxies that fall into two broad categories, one mainly related to wind direction and strength, and the other primarily responding to rainfall changes (Wang et al. 2005b). Changes in wind direction and strength affect material transport from the land to the ocean floor, and these can be seen for example, in dust and pollen records. Changes in wind direction also affect wind-induced ocean upwelling, and this effect is preserved in a variety of sedimentological and biological proxies. Rainfall changes on the other hand, affect weathering and continental runoff, which is manifest in sedimentological and geochemical proxies, for example. Based on the nature of marine monsoon proxies, we roughly classify them into three groups: lithogenic, micropaleontological and geochemical (elemental and isotopic).

3.10.2.1 Lithogenic Proxies

Lithogenic proxies depend on an understanding of the sedimentological characteristics of the source region and transport pathways to a particular depositional environment. For example, grain sizes of Arabian Sea sediments can be linked to the transport capacity of winds (Clemens and Prell 2003), whereas eolian dust preserved in sediments in the Northwestern Pacific is associated with aridity in the Asian interior (Rea et al. 1998). Clay mineral assemblages have been used to indicate wind directions in the SCS, and hence the relative strength of the summer and winter monsoons (Liu et al. 2003c). Liu et al. (2003c, 2010d) suggested that illite and chlorite in SCS sediments come from Taiwan and the Yangtze River, while smectite and mixed-layer clays originate primarily from Luzon and Indonesia. The ratio of the two sourced clays, smectite and mixed-layer clays over illite and chlorite, was adopted as a proxy for East Asian monsoon variability, with higher ratios indicating a strengthened summer monsoon and weakened winter monsoon, and vice versa.

3.10.2.2 Micropaleontological Proxies—Foram, Radiolarian, Profunda, Pollen

One of the best studied marine monsoon proxies is the abundance of planktonic foraminifera *Globigerina bulloides* (Wang et al. 2005b). Detailed core-top studies in the Arabian Sea (Prell 1984) established a relationship between this subpolar species with monsoon-driven upwelling. Jian et al. (2001) used a similar species of foraminifera, *Neogloboquadrina dutertrei* as an upwelling indicator in the SCS, where *G. bulloides* was not very abundant. *G. falconensis*, indicative of non-upwelling conditions, has been shown to provide an inverse signal for upwelling strength (Schulz et al. 2002). The coccolithophore *Florisphaera profunda*, living deep in the photic zone, has been used as an inverse proxy of primary productivity associated with wind-stress changes in the Arabian and Sulu Seas (Rostek et al. 1997; Beaufort et al. 1997, 2003). Benthic foraminifera (Jian et al. 2001) and radiolaria (Gupta 1999; Chen et al. 2003) have also been used as upwelling indicators. Yu et al. (2006c) defined warm and cold water species factors based on analyses of planktonic foraminifera assemblages and calculated a ratio of these two factors as a hydrographic index. When using these micropaleontological proxies it is important to be aware of potential complications associated with factors such as dissolution and nutrient levels in upwelled waters (Wang et al. 2005b).

Pollen and spores have been analyzed from marine sediments from the North Pacific (Morley and Heusser 1997), SCS (Sun et al. 2003), and Indonesian-Australian regions (Kershaw et al. 2003) to infer terrestrial conditions. Sun et al. (2003) argued that tree pollen influx and fern spore fractions from the SCS can be used as proxies of winter and summer monsoons, respectively. A potential complicating factor in this instance is the effect of changing sea level, which changes the distance of the study area relative to the coastline; however Sun et al. (2003) argued that the pollen ratio of herbs to pine could constrain this factor.

3.10.2.3 Geochemical Proxies

We divide geochemical proxies into two broad categories based on the type of data: 1) proxies based on elemental or molecular analyses, and 2) proxies based on isotopic analyses.

Major, trace elment and biochemical proxies These proxies include major and trace elements, and biogenic constituents such as total organic carbon, carbonate, opal, pigment/chlorine and alkenones. Major and trace elements, often normalized to Al content, in the SCS, are interpreted to reflect chemical weathering in the source region, fluvial input, or monsoon-induced surface biological productivity (Wehausen and Brumsack 2002; Wei et al. 2004). Ba (barium) flux or Ba/Al ratio in the Indian Ocean is linked to upwelling productivity (Wehausen and Brumsack 2002; Clemens and Prell 2003). Applying principal component analyses of down-core records of major and trace elements, Sun et al. (2008) suggested that the first component (e.g. Ca/Al, Ba/Al, Sr/Al) is related to biological productivity; the second (e.g. K/Al, Mg/Al, Rb/Al) is associated with chemical weathering in source regions; and the third component (e.g. Ti/Al, Zr/Al) likely reflects changes in sea level and monsoon-induced fluvial inputs. Major and trace elements used as marine monsoon proxies could potentially be influenced by changes in the source areas of sediment, the relative contributions of biogenic/terrestrial material, chemical weathering in the source areas, and grain-size sorting during the transportation processes (Wehausen and Brumsack 2002; Wei et al. 2004).

Contents of total organic carbon, carbonate, opal, pigment/chlorine and alkenones and their mass accumulation rates indicate biological productivity in coastal and open-ocean upwelling regions and are sometimes used to indicate wind strength (Clemens et al. 1996; Clemens and Prell 2003; Higginson et al. 2003; Wan et al. 2006; Clemens et al. 2008). For these type of proxies, preservation, dissolution, dilution and/or sediment reworking processes are common concerns, although individual proxies may be impacted differently. In addition, the association of biological productivity with wind-driven upwelling is not straightforward. Non-monsoonal processes that could have impacts on water column structure or on nutrient levels could affect biological productivity. Therefore, biological productivity records as monsoon proxies should be used with caution. For instance, Clemens and Prell (2003) rejected organic carbon accumulation rates in the Arabian Sea as a marine monsoon proxy. Budziak et al. (2000) further divided the organic carbon records from the Arabian Sea into three groups and concluded that they may reflect the relative contributions from summer and winter monsoon circulation. Distinguishing forcings of winter and summer monsoon on biological productivity in the SCS is apparently even more complicated (Wei et al. 2003a; Wan et al. 2006).

Isotopic proxies Oxygen isotopic ratios ($\delta^{18}O$) of planktonic foraminifera can potentially be used to indicate hydrological conditions in coastal regions and marginal seas (Wang et al. 1999b; Kudrass et al. 2001), although its application is not always straightforward. Planktonic $\delta^{18}O$ signals are controlled by three factors: temperature, ice volume and surface salinity, and to understand the association of $\delta^{18}O$ with surface hydrological conditions, the effect of each factor needs to be assessed. A number of approaches have been applied to remove the unwanted portion of the planktonic $\delta^{18}O$ signal. For instance, the temperature effect can be accounted for if sea surface temperature (SST) is determined using the alkenone unsaturation index (Wang et al. 1999a) or foraminiferal Mg/Ca ratios from the same core (Dannenmann et al. 2003; Oppo et al. 2009). The planktonic $\delta^{18}O$ gradient between two strategically chosen sites can be used to indicate surface hydrological conditions, as ice volume and temperature effects are effectively removed in the gradient signal (Wei et al. 2003b). SST records from the Indian Ocean (Emeis et al. 1995) and SCS (Zhao et al. 2006) have both been used as indirect indicators of monsoonal changes. SST gradients between two selected sites in the SCS have been used as monsoon indicators as well (Yu et al. 2006c; Tian et al. 2010).

Nitrogen isotopic ratios ($\delta^{15}N$) from the Arabian Sea (Altabet et al. 1995, 1999) are linked to summer monsoon strength via export production and intensity of denitrification within the oxygen minimum zone. Clemens and Prell (2003) suggested that changing waters from the Persian Gulf and Red Sea, linked to sea level changes, could potentially influence $\delta^{15}N$ values. Higginson et al. (2003) suggested that $\delta^{15}N$ values from the SCS can be influenced by equatorial circulation, the intensity of summer and winter monsoons, and relative sea level. Carbon isotopic ratios ($\delta^{13}C$) of near-surface dwelling planktonic foraminifera from the SCS are interpreted as an inverse nutrient signal, which is linked to upwelling intensity and monsoonal changes (Wang et al. 1999b).

Based on the above studies, one might easily conclude that not a single marine monsoon proxy contains a pure monsoon signal. This may be true for all of the proxies that have been used for paleoceanographic and paleoclimatic reconstructions. Therefore the success of paleomonsoon research relies on whether we can accurately assess the extent to which the proxies are influenced by non-monsoonal processes and whether we can successfully remove unwanted signals (Wang et al. 2005b). An effective approach is to use multiple proxies to extract a common signal through stacking or principal component analysis. For instance, Clemens and Prell

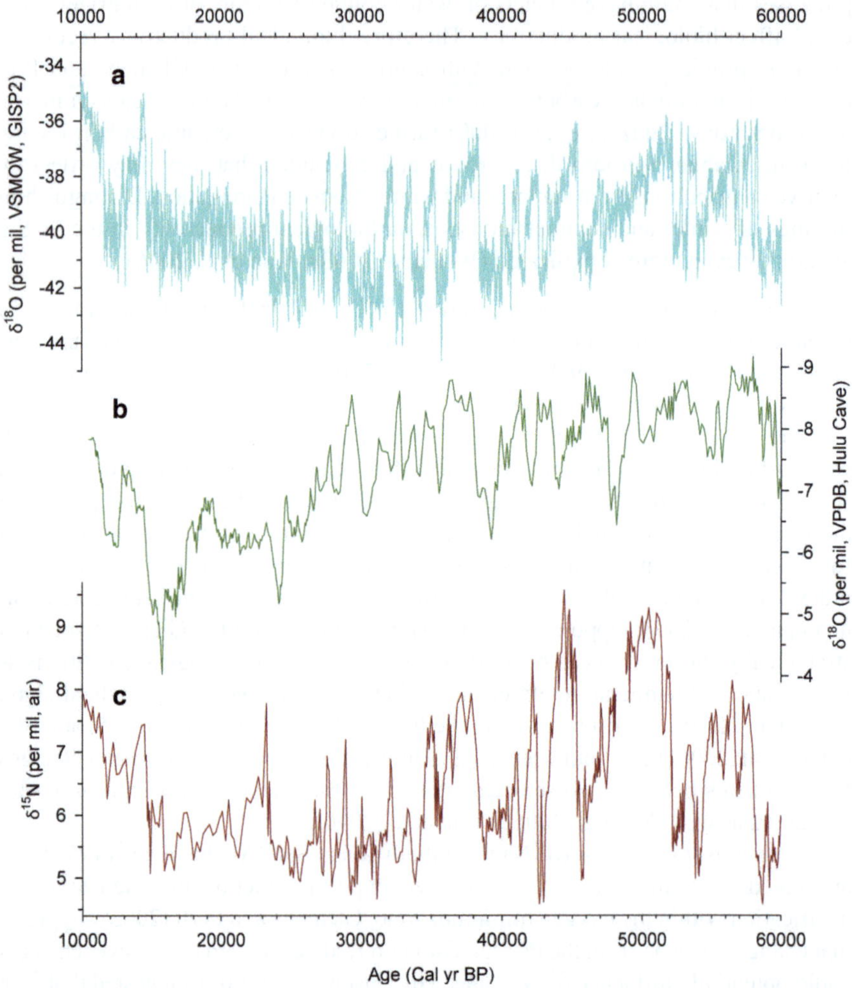

Fig. 3.72 **a** Greenland Ice Sheet Project 2 (GISP2) $\delta^{18}O$ record (Grootes and Stuiver 1997), **b** Hulu Cave $\delta^{18}O$ record (Wang et al. 2001), (3) Oman Margin $\delta^{15}N$ record (Altabet et al. 2002) for the last glacial period (10–60 ka). All records show prominent D/O cycles and they can be correlated with each other within the age uncertainties

(2003) used records of lithogenic grain size, barium mass accumulation rate, $\delta^{15}N$, *G. bulloides* abundance, and opal accumulation rate to derive an Indian summer monsoon stack.

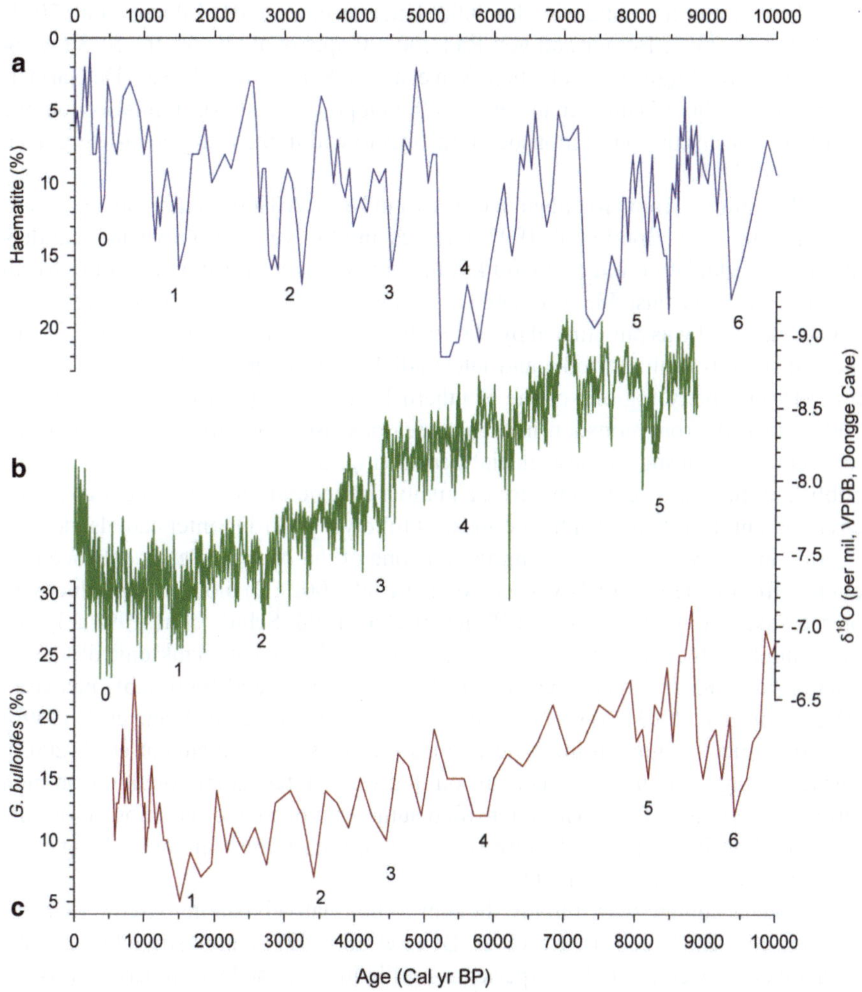

Fig. 3.73 **a** Haematite percentages from the North Atlantic (Bond et al. 2001), **b** Dongge Cave $\delta^{18}O$ record (Wang et al. 2005a), (3) the abundance of *G. bulloides* from the Arabian Sea (Gupta et al. 2003) during the Holocene (0-10 ka). All records show millennial variability and they can be correlated with each other within the age uncertainties

3.10.3 Millennial Variability

Over the past 20 years, the number of high-resolution records from monsoon-influenced regions has increased dramatically, which allows correlation of monsoonal changes with climatic changes elsewhere, particularly at high latitudes of the NH. High resolution records have been obtained from regions with high sedimentation rates, including those from the Arabian Sea and continental margins of Pakistan and West India (Sirocko et al. 1996; Colin et al. 1998; Schulz et al. 1998; Reichart et al. 1998; Leuschner and Sirocko 2000; Kudrass et al. 2001; Lückge et al. 2001; Altabet et al. 2002; Berger and von Rad 2002; Gupta et al. 2003), the SCS (Wang et al. 1999a, b; Higginson et al. 2003; Sun et al. 2008) and the Sulu Sea (De Garidel-Thoron et al. 2001; Dannenmann et al. 2003; Oppo et al. 2009). In this section, we will limit our discussion of monsoonal records to within the last glacial-interglacial cycle, ~0–100 ka.

Earlier studies from Greenland ice cores and North Atlantic sediments (e.g. Dansgaard et al. 1993; Bond et al. 1993; Grootes and Stuiver 1997) revealed that during the last glacial period (~20–60 ka) the climate over Greenland and the North Atlantic was very unstable, oscillated at ~1450-year cycles (Dansgaard-Oeschger, D/O cycles) and was punctuated by Heinrich events. Subsequent studies have documented stadials (cold periods) and interstadials (warm periods) for nearly the entire NH and many regions of the Southern Hemisphere (Leuschner and Sirocko 2000). Records from monsoon-influenced regions also show strong correspondence to D/O cycles during the last glacial period (Fig. 3.72). On the strength of these published studies, we can summarize a number of general findings about the marine environment during interstadial periods, as follows: 1) during interstadials, denitrification intensity in the oxygen minimum zone of the Arabian Sea is enhanced, as evident from increased $\delta^{15}N$ values (Altabet et al. 2002); 2) biological productivity on the Pakistan margin increased (Reichart et al. 1998; Schulz et al. 1998); 3) surface salinity in the northern Bay of Bengal decreased based on planktonic $\delta^{18}O$-SST relationships (Kudrass et al. 2001); 4) $\delta^{15}N$ values from the SCS probably increased (Higginson et al. 2003); 5) surface salinity in the SCS decreased (Wang et al. 1999b) and 6) surface salinity in the Sulu Sea also decreased (Dannemann et al. 2003). These findings lead us to the conclusion that warm interstadial conditions at high latitudes were associated with intensified summer monsoon circulation across the monsoon belt in Asia. This feature is consistent with terrestrial cave (Wang et al. 2001) and loess records (An 2000).

Holocene climate was initially thought to be relatively stable based on Greenland ice core $\delta^{18}O$ records. However, Bond et al. (2001) demonstrated that North Atlantic climate continued to experience oscillating cold and warm periods ("Bond Cycles") during the Holocene. An excellent marine record of the abundance of *G. bulloides* from the Arabian Sea documented Holocene monsoonal changes, which can be correlated with cave $\delta^{18}O$ records in China (Wang et al. 2005a). Both records show a long term weakening trend of monsoonal changes, associated with Earth's orbital changes (Fig. 3.73). In particular, several periods of weakened Asian monsoon activity can be correlated with the "Bond Cycles" in the North Atlantic region, suggesting a persistent link between North Atlantic and Asian monsoon climates.

Studies of late Holocene monsoonal changes are of particular interest. Anderson et al. (2002) showed evidence for a weakened Asian summer monsoon during the relatively cold little ice age (LIA) period, between about 1850 AD and 1400 AD. The inferred association of a weakened monsoon with cold conditions is generally true for the entire Holocene and last glacial periods (see above discussions). However, recent studies from the Sulu Sea (Oppo et al. 2009; Tierney et al. 2010) suggest that wet conditions prevailed in this region during the LIA, opposite to the association identified over the Indian Ocean and Asian continent. Two plausible interpretations for Indonesian hydrological changes during the late Holocene exist: 1) the average position of the intertropical convergence zone and 2) changes in El Nino-Southern Oscillation (ENSO) (Oppo et al. 2009; Tierney et al. 2010). Further studies on this topic will lead to a better understanding of Asian monsoon dynamics and its interaction with ENSO on this timescale.

Although a link between North Atlantic climate and the Asian monsoon appears to be a persistent aspect of global climate for both the Holocene and last glacial periods (e.g. Leuschner and Sirocko 2000; Gupta et al. 2003), our understanding of the teleconnections between climatic subsystems remains largely insufficient. Asian monsoon variability and climate variability appear to be correlated with weak variations in solar irradiance during the Holocene (Bond et al. 2001; Wang et al. 2005a). Amplification mechanisms, or feedbacks within the climate system have to be sought. It remains unclear whether the Asian monsoon responds to high latitude mechanisms or initiates/amplifies low-latitude mechanisms (Wang et al. 2005b).

3.10.4 Orbital Variability

Asian monsoon variability on orbital timescales has been intensively studied. Over the past few decades, numerous orbital-scale monsoonal records have been obtained from the Indian Ocean, SCS and Indonesian maritime regions (e.g. Prell 1984; Morley and Heusser 1997; Rostek et al. 1997; Colin et al. 1998; Reichart et al. 1998; Altabet et al. 1999; Almogi-Labin et al. 2000; Budziak et al. 2000; De Garidel-Thoron et al. 2001; Jian et al. 2001; Wehausen and Brumsack 2002; Beaufort et al. 2003; Clemens and Prell 2003; Kershaw et al. 2003; Liu et al. 2003c; Sun et al. 2003; Wei et al. 2004; Boulay et al. 2005; Clemens et al. 2008; Sun et al. 2008). In general, these records suggest stronger summer monsoons/weaker winter monsoons during interglacial periods and weaker summer monsoons/stronger winter monsoons during glacial periods, consistent with loess records (An et al. 2001). As these records have various lengths, we will discuss monsoonal changes during the late Pleistocene first, followed by changes over the last 2 Ma (from the early Pleistocene to late Pleistocene) based on a few available records.

During the late Pleistocene, major spectral power at three orbital bands, namely ~100-ka, 41-ka, and 23-ka, can be identified from the majority of the records. Substantial spectral power in some non-orbital bands have also been reported, but could be heterodynes associated with the main orbital bands (Clemens and Prell 2003). Based on the relative importance of the spectral power in these three orbital bands, we can roughly divide these records into three groups. The first group of records has very strong spectral power at the ~100-ka band. Representative records include: 1) organic carbon accumulation rates in the Indian Ocean (rejected as a monsoon proxy however, Clemens and Prell 2003), 2) primary production in the Sulu Sea (Beaufort et al. 2003), 3) ratios of smectite and mixed-layer clays over illite and chlorite from the SCS (Liu et al. 2003c; Boulay et al. 2005), and 4) the planktonic $\delta^{18}O$ gradient

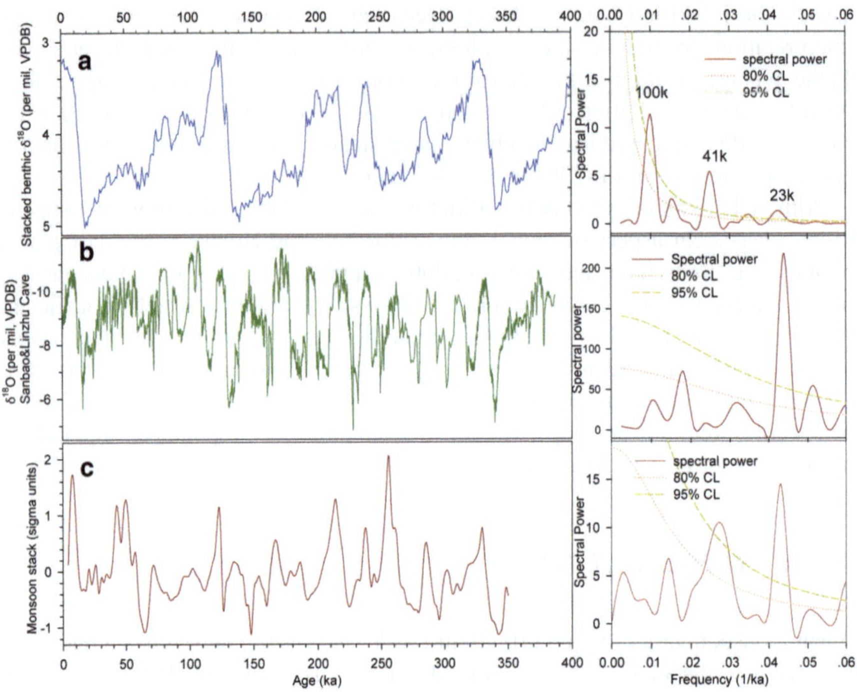

Fig. 3.74 **a** Benthic $\delta^{18}O$ stack (Lisiecki and Raymo 2005), **b** Chinese cave $\delta^{18}O$ records (Cheng et al. 2009), (3) the Indian summer monsoon stack from the Arabian Sea (Clemens and Prell 2003) over the past 400 ka. Both time series (*left*) and spectral results (*right*) of each record are presented. The three records show strikingly different spectral features

in the SCS (Wei et al. 2003b). The second group presents similar spectral power at the 41-ka and 23-ka bands, with little 100-ka spectral power. Representative records are contained in the Indian summer monsoon stack which includes lithogenic grain size, barium mass accumulation rate, $\delta^{15}N$, *G. bulloides* abundance and opal accumulation rate (Clemens and Prell 2003). The third group has very strong spectral power at the 23-ka band, with relatively little 41-ka and 100-ka spectral power. Representative records include: 1) a primary productivity record from the southeastern Arabian Sea (Rostek et al. 1997), and 2) the planktonic foraminifer hydrographic index (Yu et al. 2006c). The last two records should be interpreted with caution as they span less than 2 glacial-interglacial cycles (~200 ka). Clemens and Prell (2003) suggest that only the second group largely represents a true monsoon signal while the other two may be significantly influenced by non-monsoonal processes.

Global climate shifted from predominant cycles lasting 41-ka to cycles lasting ~100-ka, around the mid-Pleistocene (e.g. An et al. 2001; Lisiecki and Raymo 2005). Several marine monsoonal records extending to the mid- and early Pleistocene appear to confirm this shift. Clemens et al. (1996) reported records of lithogenic grain size, the abundance of *G. bulloides*, and opal accumulation rate over the past 2.6 Ma from the Arabian Sea and found very strong 41-ka power, as well as some precessional (23-ka) power, before the mid-Pleistocene. At the same location, the $\delta^{15}N$ record over the past 1 Ma also shows strong 41-ka power in the early portion (Altabet et al. 1999). Clemens et al. (2008) analyzed records of Ba/Al, Ba, Mn, opal, chlorine, alkenones, and lithogenic accumulation rates from the past 5 Ma. These records were interpreted differently, as indicators of summer monsoon circulation and winter circulation, or summer monsoon at the precession band but winter monsoon at the obliquity band. In general, (strong) 41-ka obliquity power is present before the mid-Pleistocene in these records, based on our visual inspection (with no spectral analysis results reported). Liu et al. (2003c) reported a record of the ratio of smectite and mixed-layer clays over illite and chlorite over the past 2 Ma from the SCS, which clearly demonstrates the dominant frequency shift from the 41-ka to 100-ka band around the mid-Pleistocene. African monsoonal changes over this time period, inferred from a dust flux record (deMenocal 1995), shows a similar pattern.

The phase relationship between Asian monsoon and ice volume/orbital insolation (incoming solar radiation changes due to Earth's orbital variations) has also been intensively investigated, especially by Clemens and his colleagues. In a recent paper Clemens et al. (2010) analyzed 19 published proxy records interpreted to reflect aspects of Asian summer monsoon circulation during the late Pleistocene. They found that 18 out of the 19 records clustered at a phase of $-125° \pm 17°$ relative to precession minima, equivalent to a ~8 ka time lag to precessional forcing. Meanwhile at the obliquity band, the Indian summer monsoon stack records appear to be in-phase with obliquity maxima (Clemens and Prell 2003). They propose that insolation forcing, latent heat export from the southern Indian Ocean, and global boundary conditions together drive Asian monsoon variability on orbital timescales.

Monsoonal changes inferred from marine records are remarkably different from those from Chinese cave $\delta^{18}O$ records, which now extend to back ~400 ka (Cheng et al. 2009). The cave records show a nearly zero-phase difference with precession minima, indicating that sensible heating forced by insolation changes is the primary mechanism for orbital monsoon variability (Ruddiman 2006). At the time of this writing there is an ongoing debate on the timing of Asian monsoon changes: whether they are in-phase with or significantly lag the precession minima (Ruddiman 2006; Clemens and Prell 2007; Clemens et al. 2010).

In addition to timing, spectral power distributions in the two records also differ significantly (Clemens and Prell 2007). The cave records show strong dominant precessional power while the monsoon stack shares similar spectral power at both the precession and obliquity bands (Fig. 3.74). Additional late Pleistocene records with a dominant precessional component include African lake levels (Kutzbach and Street-Perrott 1985), Mediterranean sapropel layers (Rossignol-Strick 1983), and $\delta^{18}O_{atm}$ in Antarctic ice (Petit et al. 1999), all of which are thought to be linked to tropical processes (Ruddiman 2006). On the other hand, marine monsoonal records generally show strong obliquity power, in addition to the precession power (e.g. Clemens et al. 2010).

It appears that the problem relies on the interpretation of proxies. Ruddiman (2006) suggests that in addition to summer monsoon circulation, the marine proxies may respond to changes in winter monsoon or changes related to glacial-interglacial cycles, while Clemens et al. (2010) argue that cave $\delta^{18}O$ values may be additionally affected by winter temperature changes. Different proxies can reflect different aspects of the same monsoon system. For instance, Cruz et al. (2005; 2006) reported paired $\delta^{18}O$ and $\delta^{13}C$ changes in cave deposits over the past 116 ka from southern Brazil. $\delta^{18}O$

largely varied with a precessional frequency, and was linked to insolation-driven atmospheric circulation, while $\delta^{13}C$ primarily changed at an obliquity frequency, interpreted to reflect local temperature changes and thus, biological processes in the soil. Another example comes from an early Pleistocene section of Mediterranean sediments. Cleaveland and Herbert (2009) analyzed biomarker alkenones to derive SST and biological productivity changes from this section. They found that during the early Pleistocene, biological productivity, linked to the freshening of surface waters and thus the African monsoon, primarily oscillated at the precession band, while SST largely varied at the obliquity band. Various proxies reflecting different aspects of the monsoon system on orbital timescales would provide a rare opportunity to better understand Asian monsoon dynamics and climatic changes in general, as for example, the universal/similar changes in the monsoon-influenced regions and elsewhere on millennial timescales pose a great challenge to infer driving mechanisms.

3.10.5 Summary

We have reviewed the use of marine climate proxies for paleomonsoon research and discussed their merits and limitations. Asian monsoon variability on millennial and orbital timescales has been summarized in the context of a large number of published studies. Although a link between North Atlantic climate and the Asian monsoon appears to be a persistent aspect of global climate for both the Holocene and last glacial periods, monsoonal variability on orbital timescales inferred from marine reconstructions shows marked differences from cave $\delta^{18}O$ records. It seems clear that the problem rests in the interpretation of monsoon proxies, as proxies may respond differently to common monsoonal influences. We view this discrepancy as an opportunity to further understand Asian monsoon dynamics on a wide range of timescales.

3.11 Monsoon Climate Variability Recorded in Historical Documents

East Asia has a long history of civilization. With the development of civilization came the ability to produce written documents. These historical documents contain a wealth of first-hand climate information, and there are numerous sources to draw on in China, whose contents include both direct and indirect climate data (Chu 1973; Zhang 2004). Examples of direct meteorological data include records of temperature, rainfall, cloud cover, dust storms and winds. Indirect climate records include dates of rivers and lakes freezing, snows and frosts, notable phenological phenomena, crop harvests and famines. Historical records from China extend back for thousands of years. The earliest meteorological information recorded in China can be dated back to about 3000 years BP, as oracle bone inscriptions (Zhang 2004).

Generally speaking, historical climate information comes from three main sources: 1) official dynastic histories, 2) official local chronicles and (3) personal dairies, notes, and collections of writings (Gong et al. 1983). Since the 1970s, the Institute of Geography, Chinese Academy of Sciences, began to collect climate and environmental information from various historical documents. The most recent comprehensive collection of meteorological information gleaned from historical documents includes data from the last 3000 years (Zhang 2004).

In addition to records of climate, geographical and vegetation information is also found in historical documents. For example, Shi (2002) reconstructed the vegetation cover on the Loess Plateau before the 5th century BC. He also determined that the lateral erosion of the Yellow River observed in the Shanxi and Shaanxi provinces occurred after the 3rd century BC (Shi 2002). Hou et al. (2001) and Hou (2007) reconstructed the land use history of the Erdos plateau and studied human adaptation to climate and environmental changes during historical times.

Here we focus on monsoon climate variability during the last 2000 years from historical documents and discuss climate patterns in the characteristic periods such as the "Roman Warm Period" (RWP), "Dark Age Cold Period" (DACP), "Medieval Warm Period" (MWP) and "Little Ice Age" (LIA). The relationship between human activity and environmental change during this time interval is also considered.

3.11.1 Climate Reconstruction from Historical Documents in East Asia

Chu (1925, 1926) was the first to extract climate information from historical documents in China and studied the historical climate conditions since the Southern Song Dynasty (1127–1279 AD). Further study suggested that there were three warm periods during the last 5000 years, which occurred at 3000-1000 BC, ~200 BC-220 AD and ~600–900 AD. Four cold intervals were identified at ~1000BC, ~400AD, ~1200AD and ~1700AD. The temperature history is consistent with snowline variations in Norway, indicating a global climate change connection (Chu 1973). These works together with others, such as Xu (1955) and Wen (1959) laid the groundwork for historical climate studies in China and East Asia.

3.11.1.1 Moisture Records

Every Chinese dynasty knew the value of agriculture and this importance is reflected in the historical records attendant to agricultural concerns. Agriculture was regarded as extremely important and careful attention was paid to climate disasters, especially when the country was stable and flourishing. Hence, drought and flood records in historical documents are good materials to reconstruct local moisture change in the past. Such records have accurate dates and clear climatic information, but they usually consist of qualitative descriptions. To compare them with geologic records, they need to be parameterized. Multiple methods have been applied to accomplish this parameterization (e.g., Zheng et al. 1977; Zhang and Zhang 1979; Central Meteorological Bureau 1981). Among these, the drought/flood (D/F) index method is most common. The 5-level classification method is based mainly on time of occurrence, affected area and degree of drought or flood conditions in spring, summer or autumn (Zhang 1983). This technique has been used to estimate annual D/F levels in the past (Central Meteorological Bureau 1981). Because of the paucity of records prior to 500 years ago, some researchers defined a decadal average D/F index, which depends primarily on records of heavy droughts or floods for a given decade. At the same time, moderate droughts and floods were also considered in the method (Yan et al. 1992; Tan et al. 2008).

Gong and Hameed (1991) studied moisture conditions in China over the last 2000 years, and found distinct regional differences in moisture variations among semi-arid, semi-wet and wet zones. Zhang et al. (1994b) proposed a stage subdivision for climate history in eastern China for the same period, and suggested two abrupt changes in precipitation, at ~280 AD and ~1230 AD, respectively. The abrupt changes, at ~280 AD were also identified in north China by Yan et al. (1992). The spatial and temporal distributions of monsoon precipitation in China seem to be related to the migration of the precipitation band associated with the position of the monsoon front (An and Thompson 1998). When the East Asian summer monsoon is stronger, the monsoonal band stays longer in north China, bringing more rainfall there. In contrast, when the East Asian summer monsoon is weaker, monsoonal bands linger in the mid-lower Yangtze River basin (Ge et al. 2008).

Because more historical documents were kept in the last 1000 years, higher resolution records of dry and wet periods are available for the most recent millennium. For example, Zhang et al. (1997) reconstructed yearly drought/flood series in eastern China during the last 1000 years and studied their periodicities and regional differences. Tan et al. (2008) reconstructed the precipitation variation history of Longxi in the western Loess Plateau since 960 AD, and found the driest period occurred between the late 17th century and the early 18th century. Chinese climatologists have reconstructed yearly drought/flood series in 120 sites over China for the last 500 years, beginning at AD 1470 (Central Meteorological Bureau 1981).

There has been continuous meteorological observation in China since the 18th century, recorded in two documents: 1) *Clear and Rain Records* and 2) *Depths of Rainfall infiltration and Snowfall records* (Zhang 1996). Based on these records, rainfall time series for the mid-lower Yellow River, the mid-lower Yangtze River and north China since the 18th century have been reconstructed (Wang and Zhang 1992; Zhang 1996; Zhang and Liu 2002; Zheng et al. 2005; Ge et al. 2008). The reconstructed rainfall records are consistent with the D/F index time series (Wang and Zhang 1992) and show region-wide synchronous variations in precipitation for the mid-lower Yellow River Basin on a decadal time scale, during the last 300 years (Zheng et al. 2005).

Historical-based climate studies from other parts of East Asia include the study of Kim and Choi (1987), who reconstructed the spring precipitation history in Korea for the last 1000 years. Their results suggest relatively wet spring seasons during the MWP, and dry spring seasons during the LIA. Jung et al. (2001) reconstructed

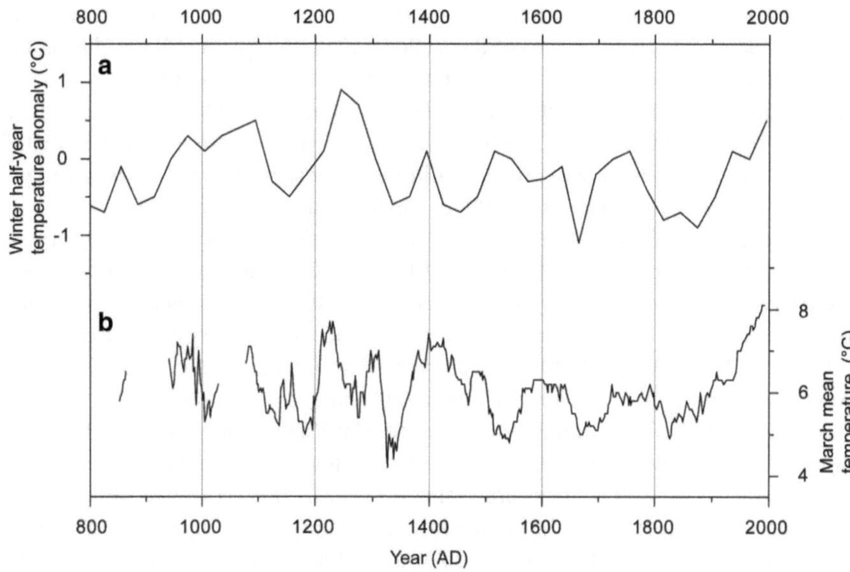

Fig. 3.75 Temperature variations in eastern China (**a**) and Japan (**b**) over the past 1200 years. A – winter temperature reconstruction for the middle and lower reaches of the Yellow River and Yangtze River, China (Ge et al. 2003), B – spring temperature reconstruction from Kyoto, Japan (redrawn from Aono and Kazui 2008)

the annual rainfall in Seoul for the period 1777–1996 AD, using historical meteorological observations. Their study found that annual variations in precipitation were consistent with the reconstructed spring moisture study. For example, the 1900s drought was observed in both records.

3.11.1.2 Temperature Records from Historical Documents

Temperature variations have also been obtained from Chinese historical documents (Zheng et al. 2007 and references therein). For example, Zhang and Gong (1979) reconstructed temperature variations in eastern China since the 16th century based on the recorded frequency of cold winters. A similar method was applied in a winter temperature study in southern China that covered the last 500 years (Zhang 1980). Wang et al. (1998) graded the cold and warm events (such as frost, heavy snow, river and lake freezing, warm winters, etc.) recorded in historical documents, and transformed them into corresponding temperature anomalies by calibration with modern data. Using this method, quantitative temperature variations were reconstructed in eastern China from 800 AD to the present (Wang et al. 1998; Wang and Gong 2000). Zhang (1996) compared the differences between ancient and modern phenological records on plants, hydrometeorology and agriculture to reconstruct annual temperature variations in eastern China for the last 2000 years, with resolutions on the order of several decades to one hundred years. With a similar method, Ge et al. (2003) reconstructed a winter half-year temperature series for every 30-years for the middle and lower reaches of the Yellow River and Yangtze River during the last 2000 years. The temperature reconstructions show high similarity with each other, indicating their reliability (Zhang 1996; Zheng et al. 2007). Several annual winter temperature time series covering the last 200 years were also reconstructed based on observations of snow recorded in the *Depths of Rainfall infiltration and Snowfall records* (e.g., Zhou et al. 1994b; Zheng et al. 2003; Yang 2007).

In Japan, dates of cherry tree flowering were recorded in many diaries and chronicles from the historical period (Aono and Kazui 2008). The inter-annual variation in the flowering times of cherry trees depends closely on the general temperature conditions during the development of the flower buds in February and March (Aono and Omoto 1994; Aono 1998). Hence, springtime temperature changes in Kyoto (Aono and Omoto 1994; Aono and Kazui 2008) and Tokyo (Aono 1998) were reconstructed. They show warm springs in the MWP and cold springs in the LIA. Twentieth century warming was observed both in Kyoto and Tokyo.

Temperature variations in eastern China (Ge et al. 2003) correlate well with Japan (Aono and Kazui 2008) during the last 1200 years on a centennial-scale. Both regions show evidence of the MWP, LIA and 20th century warming (Fig. 3.75). Warm epochs occurred in the 11th, 13th and 20th centuries, while cold epochs occurred in the 16th and 19th centuries. The 12th century was a relatively cold interval within the MWP. The comparison indicates synchronous temperature variations over East Asia on centennial-scales, which may be controlled by the same forcing mechanisms.

On decadal time scales, temperature variations for the two regions show both similarities and discrepancies. For example, there was a relatively cold event that occurred at the end of the warm 13th century in Japan, but it was not observed in eastern China. Another discrepancy occurred in the period 1510–1550 AD. At this time, notably cold conditions were observed in Japan, whereas warm conditions were observed in eastern China. These discrepancies may be caused by the uncertainties in both reconstructions, but it could also be the case that they reflect real temperature differences between eastern China and Japan.

3.11.1.3 Historical Dust Storms and Typhoons

Besides precipitation and temperature, other weather and climate information may also be recorded in historical documents, such as the occurrence of dust storms and typhoons. Sites recording dust storm events during the last 3000 years were distributed across the modern loess area (Zhang 1984). Analysis of the frequency of dust storms from the last 1700 years suggested that high frequency dust storm events occurred during cold, dry periods (Zhang 1984). This was confirmed in related studies (Zhang and Fei 2004; Deng and Jiang 2005). The southern limit of historical dust storms was mainly controlled by moisture conditions in the dust source area, so that dust storm records can be used to infer moisture changes in the north (Zhang and Sun 2001).

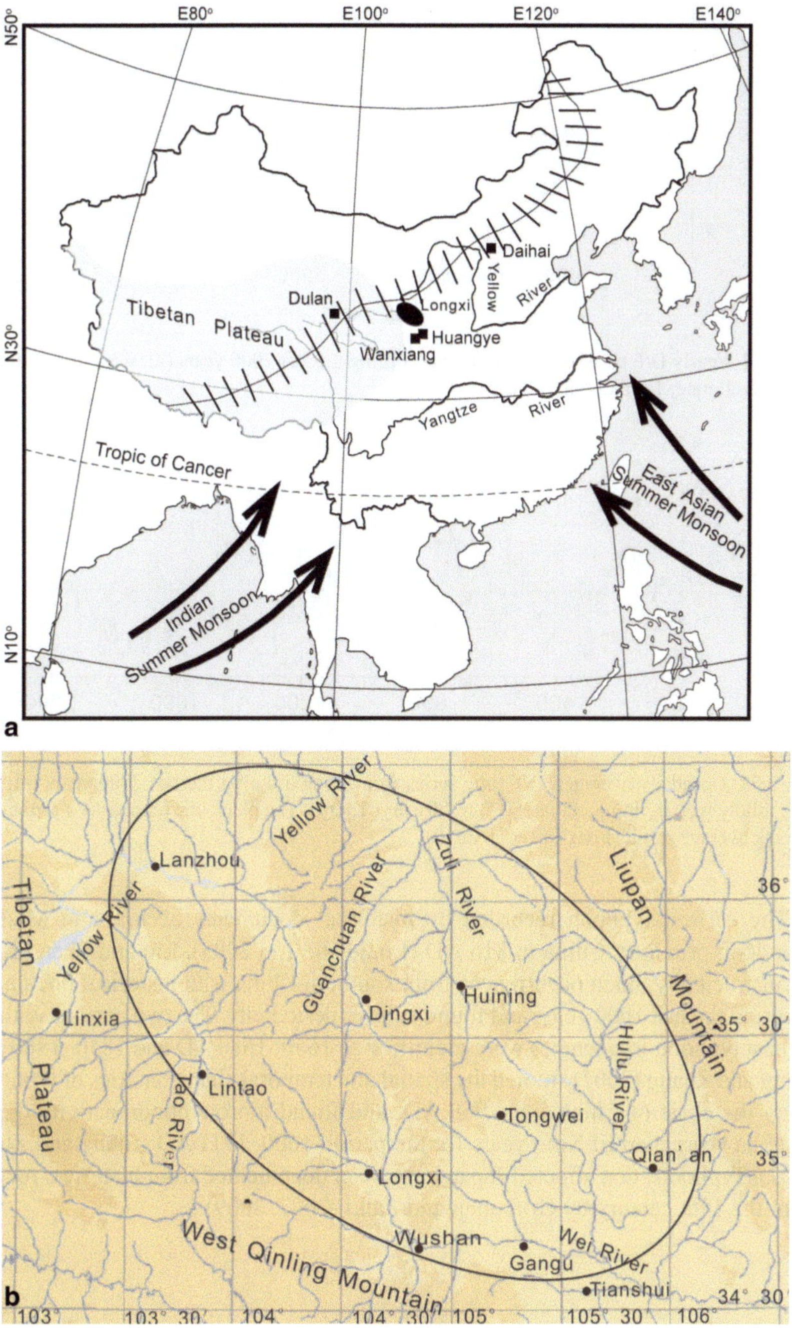

Fig. 3.76 Location of Longxi area (**a**) and the detailed landform (**b**). The ellipses in A and B denote the Longxi area. The cross-hatched area denotes the approximate northern limit of the Asian summer monsoon (modified from Tan et al., 2008)

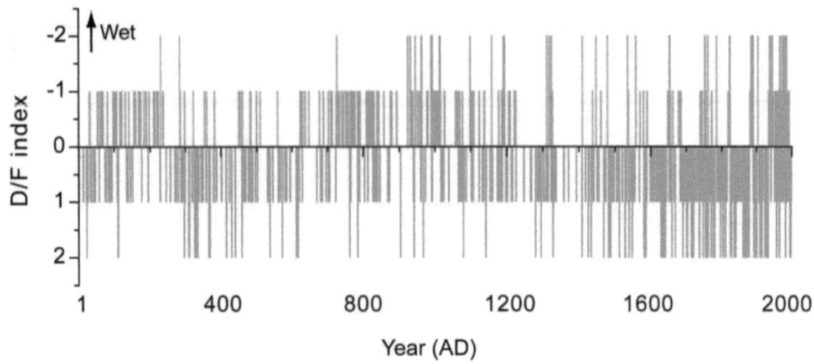

Fig. 3.77 Yearly D/F time series from Longxi during the last 2000 years (-2, wet; -1, mildly wet; 0, fitting climate; 1, mildly dry; 2, dry)

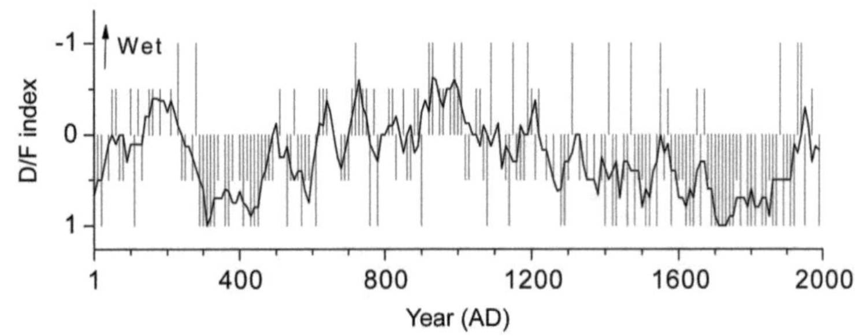

Fig. 3.78 Decadally averaged D/F time series from Longxi during the last 2000 years (-1, wet; -0.5, mildly wet; 0, fitting climate; 0.5, mildly dry; 1, dry). The black line represents a 5-point AA moving average result (from Tan et al. 2010)

The earliest typhoon recorded by historical documents occurred in Mizhou, Shandong province, China in 816 AD (Louie and Liu 2003). Liu and Shen (2001) reconstructed typhoon occurrences from southern China with historical documents covering the last 1000 years and found a 50-yr periodicity. The two periods with the highest typhoon frequencies were identified as 1660–1680 AD and 1850–1880 AD. Liang and Zhang (2007) studied the spatial and temporal characteristics of typhoons along the coast of China since 960 AD, and found more typhoon activity during La Niña years than El Niño years for the period 1600–1911 AD (Liang and Zhang 2007). Typhoons that affected Japan have been documented from historical records from the 19th century by Grossman and Zaiki (2007, 2009).

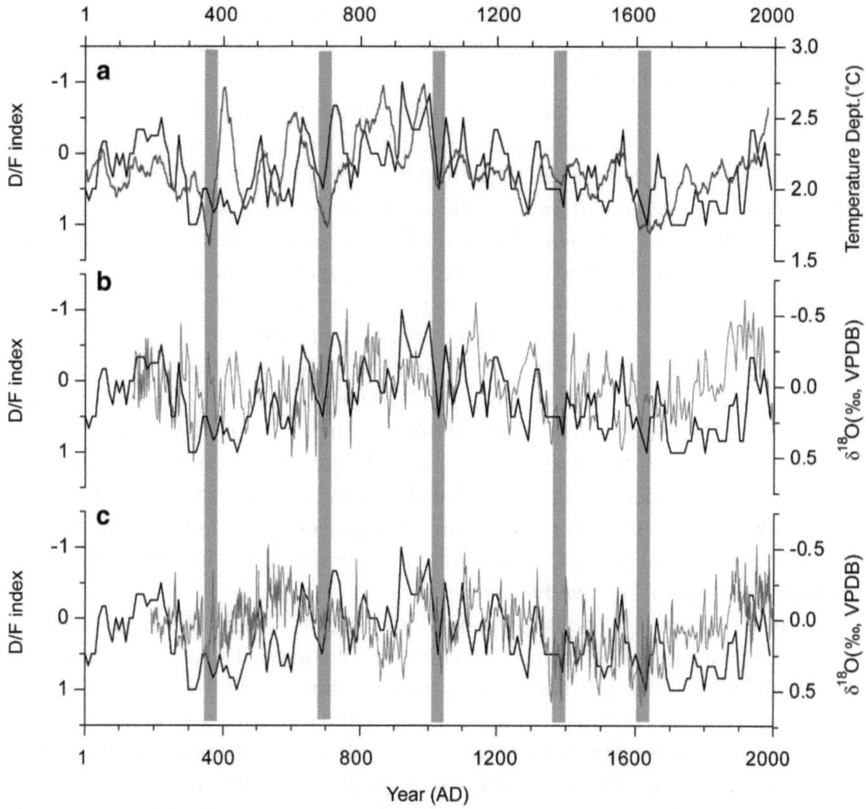

Fig. 3.79 Comparison of records from historical documents, speleothems and tree rings. The grey line in panel A represents a tree ring record from Dulan (after the application of a 40-yr moving average (Liu et al. 2009b). The grey lines in panels B and C are detrended speleothem $\delta^{18}O$ records from Wanxiang (Zhang et al. 2008) and Huangye Caves (Tan et al. 2011), respectively. The black line represents decadally averaged D/F index values from Longxi with 3-point AA moving average

3.11.2 Precipitation Variations in the Western Loess Plateau during the Last 2000 Years

The Longxi area is located on the western Loess Plateau and at the northern edge of the area affected by the Asian summer monsoon. According to historical records and geographical features, we define this area as including what is now Lanzhou, Dingxi, and the adjacent Wushan, Huining, Gangu and Qin'an Counties, as shown in Fig. 3.76. Modern climate here is semi-arid, and annual precipitation is positively correlated with the Asian summer monsoon. When the summer monsoon is strong, the precipitation in this area increases and when the summer monsoon is weak, it decreases (Tang 2006). Meteorological records show that the annual rainfall of this area varied greatly, and depended mainly on long-lasting, heavy rain in the rainy season. These can be used to reconstruct precipitation variations in this area from drought and flood records in historical documents. We have reconstructed the precipitation history of this area since 960 AD (Tan et al. 2008); here we extend this series to the last 2000 years.

The historical climate records came mainly from the official chronicles of the emperors, called *Benji* and the chapters entitled *Wuxingzhi* (*Five Elements Chapter*). In addition, material was gleaned from the *Zizhi Tongjian* (*Dictionary of Economy and Politics*), *Donghua Lu* (*Records of Donghua*), *Donghua Xu Lu* (*Extended Records of Donghua*), *Luzheng Xu Gao* (*Extended Draft of Luzheng*), and local chronicles known as *Zhi*. Documents used in this paper also include: the *Gansu Xin Tong Zhi* (*New Whole Records of Gansu*), compiled during the reign of Emperor Guangxu (AD 1875–1908) of the Qing Dynasty; the *Draft of Whole Records of Gansu* (Liu 1936); the *Abstract History of Gan Qing Ning Area* (Mu 1936); the *Catalogue of Disasters in China* (Chen 1939); the *Brief Natural Disaster Records of each period of Gansu Province* (Zhao 1984); and the *Collection of weather records in China in the last 3000 years* (Zhang 2004). Target sites of the Longxi area have been selected as modern Dingxi City, Lanzhou City and their administrative counties, also Wushan County, Huining County, Gangu County, and Qin'an County. We also refer to historical climate records from neighboring areas, including Pingliang, Tianshui, Longnan, Linxia and Gannan.

Firstly we assign D/F levels to the historical records (Zhang 1983), and then define a decadal average D/F index to calculate the average D/F level for each decade (Yan et al. 1992; Tan et al. 2008). The results are shown in Fig. 3.77 and 3.78. The reconstruction shows that dry climate is usual for the western Loess Plateau. During the last 2000 years, the precipitation variations followed an asymmetric "W" pattern (Fig. 3.78). The precipitation gradually increased from 1 AD to 230 AD, and then decreased. A notable dry period occurred in the period 290–470 AD, after which, precipitation increased. Longxi experienced an extended wet period from 710 AD to 1220 AD, the longest wet period over the past 2000 years. Following this, wet/dry decadal fluctuations followed a more gradual overall decrease in precipitation, with the driest conditions occurring between the end of the 17th century and the 18th centuries. The climate is relatively wet in the 20th century, but has followed a drying trend since the 1950s.

3.11.3 Comparison of Historical, Geological and Biological Proxy Records

The speleothem $\delta^{18}O$ variations from Wanxiang Cave (Zhang et al. 2008) and Huangye Cave (Tan et al. 2011) in the western Loess Plateau are controlled by monsoon precipitation variations. The tree ring width time series from Dulan, along the north margin of the Asian summer monsoon can reflect not only the regional temperature changes, but also the local precipitation changes (Liu et al. 2006, 2009b). Figure 3.79 compares the historical-based D/F indices from Longxi with speleothem $\delta^{18}O$ records from Wanxiang (Zhang et al. 2008) and Huangye Cave (Tan et al. 2011), and the tree ring record from Dulan (Liu et al. 2006, 2009b). As shown in the

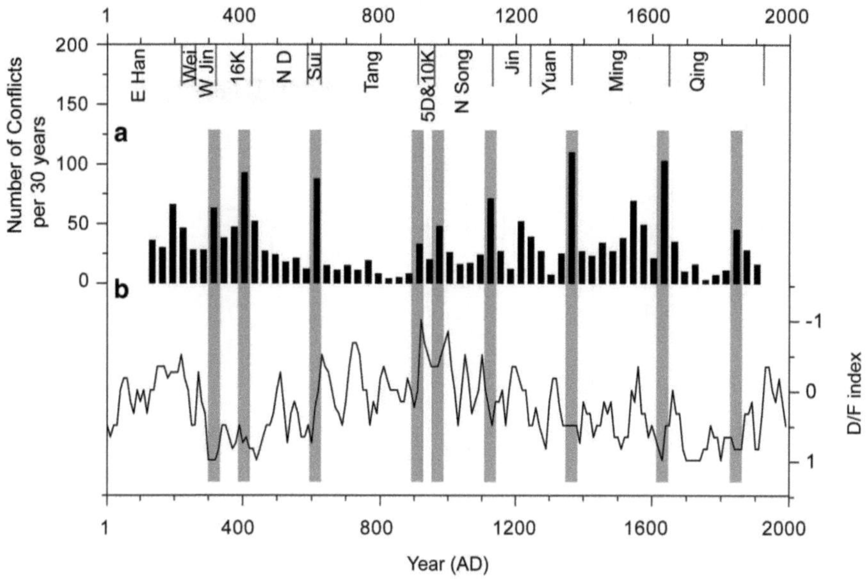

Fig. 3.80 Comparison of climate change and warfare: (**a**) the number of conflicts and wars that occurred in northern China (Editing Committee of China's Military History 1985; Wang 2007); (**b**) the precipitation variations in Longxi, western Loess Plateau revealed by D/F indices

figure, the four precipitation series show broad coherence on centennial- to decadal-scales, although they are very different proxies, and do not have the same temporal resolution or chronology uncertainties. For example, they all indicate wet intervals in the 2nd and 10th centuries, and dry intervals in the 4th, 14th and early 17th centuries. The D/F index record from Longxi, western Loess Plateau also shows remarkable similarities with a low-resolution pollen-based precipitation record from Daihai Lake, Inner Mongolia (Xu et al. 2004). These comparisons prove the reliability of the historical-based precipitation reconstruction. Furthermore, they imply that precipitation variations along the northern limit of the Asian summer monsoon are commonly controlled by variations in summer monsoon intensity. When the summer monsoon strengthens, the rain belt moves northwards and stays longer in the north, bringing more rainfall there. In contrast, precipitation decreases in this region when the summer monsoon weakens.

3.11.4 Connection Between Climate Changes and Social Evolution

Numerous studies have suggested that climate change has had important effects on social evolution (e.g., Cowie 1998; Hsu 1998; deMenocal 2000; Weiss and Bradley 2001; Núñez et al. 2002; Yancheva et al. 2007; Zhang et al. 2008; Liu et al. 2009b). On centennial time scales, the notable dry climate in Longxi, western Loess Plateau from the end of the 3rd century to the 6th century corresponded to the Jin (265–420

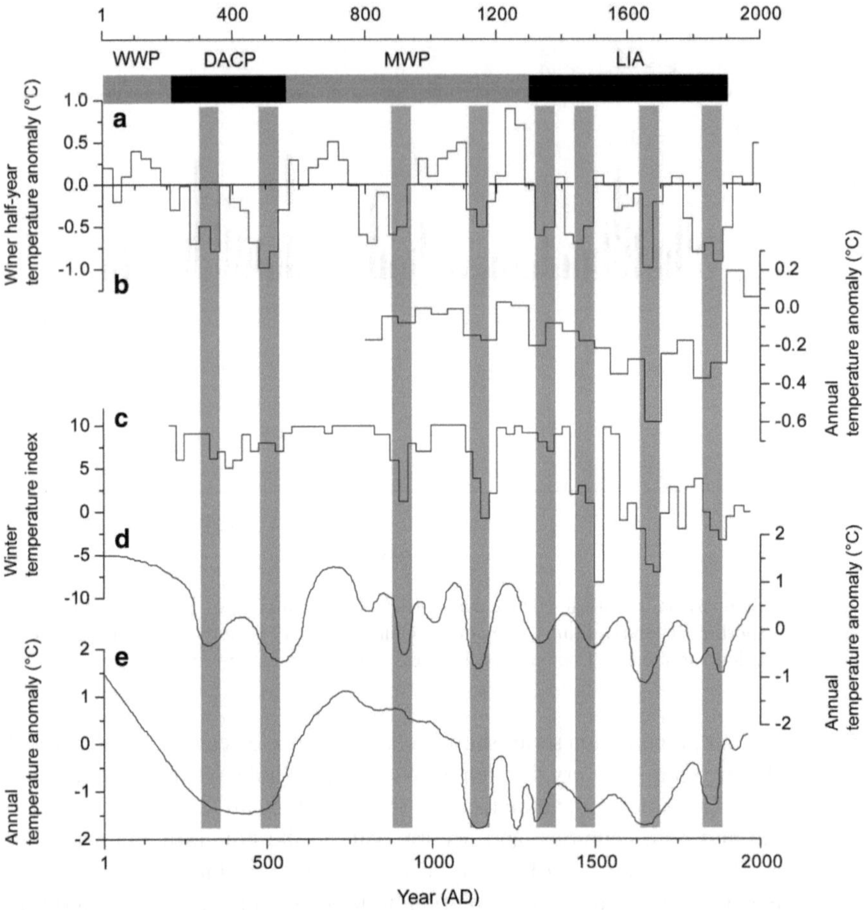

Fig. 3.81 Comparison of five temperature records reconstructed from historical documents in eastern China. (**a**) Ge et al. (2003), (**b**) Wang and Gong (2000), (**c**) Chen et al. (1998b), (**d**) Zhang (1996), (**e**) Chu (1973). Figure modified from Zheng et al. (2007)

AD), Southern and Northern Dynasties (420–589 AD). During this period, wars were frequent and society was quite unstable. The prolonged wet period from the 7th century to the 12th century corresponds with the prosperous Tang (618–907 AD) and Song (960–1279 AD) Dynasties. During this period the economy prospered and society flourished on all fronts, from the arts to science and technology. A comparison between the numbers of conflicts and wars in northern China (Editing Committee of China's Military History 1985; Wang 2007) and inferred precipitation from the Longxi D/F record is shown in Fig. 3.80. Almost all of the peak periods of warfare in dynastic transition times correspond to sharp declines in precipitation. There were as many as 110 conflicts and wars in northern China between 1351 and 1380 AD (the period between the end of the Yuan Dynasty and the beginning of the Ming Dynasty). At that time, a great drought was recorded in the Longxi record. The peak period of warfare between 601 and 630 AD (the period between the end of the Sui Dynasty and the beginning of the Tang Dynasty) also corresponds to a dry epoch. The peak period of warfare at the end of the Ming Dynasty corresponds to a dry climate in the LIA. These comparisons indicate a strong connection between climatic and social change.

3.11.5 Climate Events Recorded in Historical Documents during the Last 2000 Years

The MWP and LIA are the most important two climate events during the last 2000 years, occurring approximately in the periods 900–1300 AD and 1300–1850 AD, respectively in Europe. Although there are discrepancies in the proposed timing and duration of the two events, as reported from different regions, their existence is unambiguous. In China, tree ring records from the TP (Kang et al. 1997; Liu et al. 2009a) and speleothem records from both southern and northern China (Tan et al. 2003; Wang et al. 2005a; Zhang et al. 2008; Tan et al. 2009, 2011) all provide evidence for climate change during the MWP and LIA.

Figure 3.81 shows high coherence of temperature reconstructions from historical documents in China by different studies with different methods (Chu 1973; Zhang 1996; Chen et al. 1998b; Wang et al. 1998; Ge et al. 2003), and all five temperature series support the occurrence of the MWP and LIA in eastern China. Figure 3.81 indicates that the MWP in eastern China occurred between 570 AD and 1300 AD, approximately corresponding to the Sui (581–618 AD), Tang (618–907 AD) and Song Dynasties (960–1279 AD). It seems that the MWP started earlier in eastern China than in Europe, but ended at approximately the same time (Chu 1973). There were three warm intervals (570s–770s, 930s–1110s, 1200s–1290s) and two cool intervals (780s–920s, 1120s–1190s) within the MWP, and the average temperatures for the three warm intervals are higher than in the 20th century (Ge et al. 2003). Fei et al. (2004) also found a cold interval from 794 AD to 844 AD recorded in historical documents. For moisture condition, historical records show a distinct wet climate in the western Loess Plateau (Fig. 3.78) and the Haihe River Basin during the MWP (Yan et al. 1992). According to a comparison of historical, geological and biological records shown in Fig. 3.79, we can conclude that wet and warm climate was the main climatic pattern in monsoon-controlled north China during the MWP (Fig. 3.79). In contrast, it was dry in the lower reaches of the Yangtze River during the MWP (Zheng et al. 2001).

Many studies have reviewed the climate conditions in the LIA in China (e.g., Zhang 1991; Wang et al. 1998; Wang et al. 2003; Chen et al. 2008b). According to the temperature reconstructions from historical documents (Fig. 3.81), the LIA occurred in the period 1300–1900 AD in eastern China, synchronous with that in Europe (Ogilvie and Jónsson 2001). And also, the temperature fluctuated within the LIA. The coldest time appeared in the 17th century, which is consistent with studies in Europe (Grove 2001a, b). Multiple moisture proxy records including different natural archives such as ice-cores, sedimentary records, river terraces, lake-level fluctuations as well as historical documents suggest a relatively humid period in

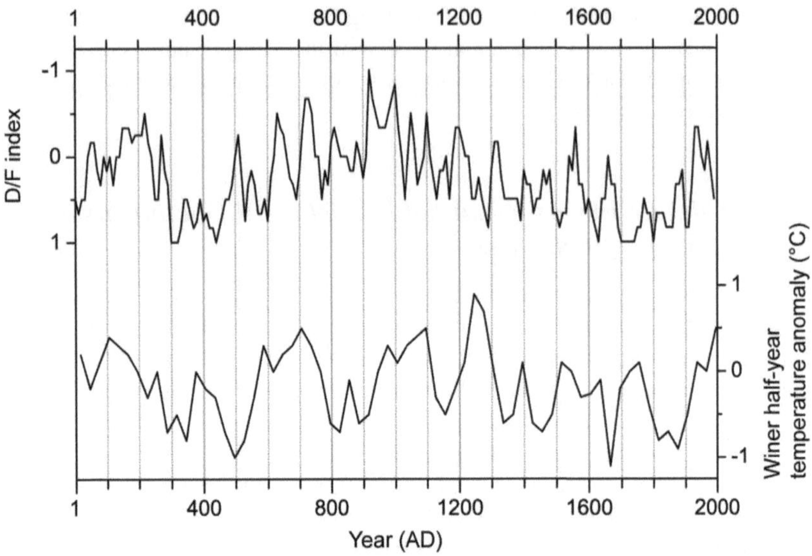

Fig. 3.82 Comparison between the precipitation variations in Longxi, western Loess Plateau and winter half-year temperature variations in eastern China (Ge et al. 2003), which can be considered as a proxy for winter monsoon intensity

arid northwest China during the LIA, contrasting with the most recent 100 years and the MWP (Chen et al. 2008b). Recently, a $\delta^{13}C$ record of plant remains from eolian sand sediments also revealed a relative wet climate in the arid Tarim Basin (Liu et al. 2011c). However, the hydrological changes in monsoonal China during the LIA were quite different from that in arid northwest China. A distinct dry climate was observed for the western Loess Plateau (Fig. 3.79) and the Haihe River Basin (Yan et al. 1992). In contrast, a wet climate with distinct fluctuations was reported in the lower reaches of the Yangtze River during the LIA (Zhang et al. 2004c).

Prior to the MWP and LIA, there were other warm and cold periods in Europe, referred to as the Roman Warm Period (RWP) and Dark Age Cold Period (DACP). The existence of these two climate events is more enigmatic than for the MWP and LIA (Desprat et al. 2003), with disagreements about their global distribution and timing. Pollen records from northwestern Iberia suggest that the RWP occurred between 250 BC and 450 AD, and the DACP occurred between 450 AD and 950 AD (Desprat et al. 2003). This is roughly consistent with another pollen study from south central Iberia (Gil García et al. 2007). Historical records show that the RWP ended at about 210 AD in eastern China, and the DACP occurred between 210 AD and 570 AD. In an analogous fashion to the MWP and LIA, the western Loess Plateau and the Haihe River Basin (Yan et al. 1992) were wet during the RWP, and dry during the DACP (Fig. 3.79).

Modern meteorological studies indicate that the East Asian winter monsoon (EAWM) controls the winter temperature variations in China (Chen et al. 1999;

Gong et al. 2001; Zhu 2008). When the EAWM is strong, a stronger than normal Siberian high and East Asian trough prevails, and the meridional shear of 500 hPa zonal winds between high and low latitudes over East Asia strengthens. Therefore, the southward cold advection becomes strong and leads to a decrease in surface temperature over China. In contrast, weak EAWM will lead to an increase in surface temperature over China (Chen et al. 1999; Zhu et al. 2008). So, here we use the winter half-year temperature series of eastern China (Ge et al. 2003) to reflect winter monsoon intensity. As described above (Sect. 3.11.3), the D/F index series of Longxi can reflect summer monsoon intensity. Figure 3.82 also indicates an antiphase relationship between summer monsoon and EAWM on a centennial scale during the last 2000 years.

3.11.6 Anthropogenic Influences on the Environment in China during the Historical Periods

During the last 100 years, human activities have greatly changed the environment in China, and human impacts can be followed through historical times. Early human activity is thought to have disturbed the vegetation cover on the Loess Plateau, resulting in soil loss and a southeasterly retreat of the southern boundary of the steppe since about 3000 years BP (Wu et al. 1994). The frequency of river disasters in the lower reaches of the Yellow River and elsewhere in the Shaanxi Province has also increased since 600 AD (Wu et al. 1994), which may be a response to anthropogenic impacts (An and Thompson 1998).

Strengthened agricultural activities in historical times have caused major impacts on water bodies in China. The Taihu Lake and surrounding wetlands at the lower reaches of the Yangtze River have been strongly influenced by agricultural production during the historical period. Polders (reclaimed low-lying land) were constructed extensively during the Wu-Yue Kingdom (907–978AD), and the kingdom even set up an army, the "Liaoqian army", to construct and manage the polders. The construction of polders tranformed a large fraction of the wetlands in the Taihu Basin into farm lands. To improve the drainage of the Taihu Lake, the local government started to dredge the original riverways, and also constructed new channels, beginning in the 15th century. These activities have greatly facilitated agricultural development, but resulted in the demise of a large number of wetlands. Furonghu Lake used to be the second largest lake in the Taihu Lake Plain, but was cultivated into polders and had disappeared by the 1550s. Sanmao Lake was also a large lake in the Taihu Lake Plain before the Jin Dynasty (265–420AD). During the Ming Dynasty (1368–1644 AD), polders began to encroach on the lakeshore of Sanmao Lake. The lake shore retreated and the Sanmao Lake became the Maohe River in the middle 19th century. In a similar fashion, lakes in the Songjiang area suffered as the result of the construction of polders at the end of the 12th century (Zhu 1989).

3.11.7 Summary

Past climate variability of monsoonal China is characterized by alternating dry and wet climates (An and Thompson 1998). Temperature variations reconstructed from historical documents show broad similarities throughout East Asia over the past 2000 years, whereas precipitation variations exhibit distinct regional differences. The spatial and temporal distribution of monsoon precipitation in China is likely related to the migration of the precipitation band associated with the monsoonal front (An and Thompson 1998).

Historical documents provide a sound basis for the reconstruction of precipitation at Longxi, western Loess Plateau, during the last 2000 years. The historical-based record is consistent with geological and biological records, including speleothems, lake sediments and tree rings situated along the northern limit of the Asian summer monsoon. During the last 2000 years, precipitation variations through time on the western Loess Plateau followed an asymmetric "W" pattern. Notably dry climate from the end of the 3rd century to the 6th century corresponded to the Jin (265–420 AD), Southern and Northern Dynasties (420–589 AD), with frequent wars and fragmented societies. The prolonged wet period from the 7th century to the 12th century corresponds to the famous Tang (618–907 AD) and Song (960–1279 AD) Dynasties, a time of rapid social and technological development. Further comparisons show that most of the peak periods of warfare in dynastic transition times are marked by periods of sharply declining precipitation, demonstrating a connection between monsoon climate change and social change.

Historical temperature records suggest that the RWP ended around 210 AD in eastern China, and the DACP began at about the same time, lasting from 210 AD until 570 AD. It seems that the MWP started earlier in eastern China (570 AD) than in Europe, but they ended approximately at the same time at about 1300 AD. The LIA occurred during the period 1300–1900 AD in eastern China, with the coldest weather in the 17th century, synchronous with Europe. It was wet in monsoonal north China during the RWP and the MWP, but dry during the DACP and LIA. In contrast, it was dry in the lower reaches of the Yangtze River during the MWP, but wet in arid northwest China and the lower reaches of the Yangtze River during the LIA. As East Asian winter monsoon intensity controls the winter temperature variations in China, and summer monsoon intensity controls the precipitation in monsoonal northern China, the comparison indicates an anti-phase relationship between summer monsoon and EAWM on the centennial timescale during the last 2000 years.

Historical climate records have accurate dates and clear climatic and environmental information, but they also have their shortcomings, including discontinuities and qualitative limitations. Comparisons between historical records and natural records can help to determine the overall sensitivity of climate proxies extracted from historical documents and to place uncertainties on their climate reconstructions. In addition, such an approach can increase our understanding of the climatic interpretation of geological and biological proxies. Human activities during historical times have significantly changed the natural environment, which may be recorded in historical documents. More attention on this aspect will deepen our understanding of the relationship between climate, the environment and human impacts.

References

Abram NJ, Gagan MK, Cole JE et al (2008) Recent intensification of tropical climate variabiliy in the Indian Oecan. Nature Geosci 1(12):849–853

Abram NJ, Gagan MK, Liu ZY et al (2007) Seasonal characteristics of the Indian ocean dipole during the Holocene epoch. Nature 445:299–302

Adrian GP, Andrew SG (2008) Geomorphological and palaeoenvironmental investigations in the southeastern Arabian Gulf region and the implication for the archaeology of the region. Geomorphology 101(3):458–470

Adrian GP, Andrew SG, Stephen S (2006) A record of Holocene climate change from lake geochemical analyses in southeastern Arabia. Quat Res 66(3):465–476

Agnihotri R, Dutta K, Bhushan R et al (2002) Evidence for solar forcing on the Indian monsoon during the last millennium. Earth Planet Sci Lett 198(3–4):521–527

Almogi-Labin A, Schmield G, Hemleben C (2000) The influence of the NE winter monsoon on productivity changes in the Gulf of Aden, NW Arabian Sea, during the last 530 ka as recorded by foraminifera. Mar Micropaleontol 40:295–319

Altabet MA, Francois R, Murray DW et al (1995) Climate-related variations in denitrification in the Arabian Sea from sediment $^{15}N/^{14}N$ ratios. Nature 373:506–509

Altabet MA, Higginson MJ, Murray DW (2002) The effect of millennial-scale changes in Arabian Sea denitrification on atmospheric CO_2. Nature 415:159–162

Altabet MA, Murray DW, Prell WL (1999) Climatically linked oscillations in Arabian Sea denitrification over the last 1 m.y.: implications for marine N cycle. Paleoceanogr 14:732–743

An ZS (2000) The history and variability of the East Asian paleomonsoon climate. Quat Sci Rev 19:171–187

An ZS, Clemens SC, Shen J et al (2011) Glacial-interglacial Indian summer monsoon dynamics. Science 333:719–723

An ZS, Colman SM, Zhou WJ et al (2012) Interplay between the Westerlies and Asian summer monsoon recorded in Lake Qinghai sediments since 32 ka. Scientific Reports 2:619, doi:10.1038/srep00619

An ZS, Kukla GJ, Porter SC (1991a) Magnetic susceptibility evidence of monsoon variation on the Loess Plateau of central China during the last 130,000 years. Quat Res 36:29–36

An ZS, Kutzbach JE, Prell WL et al (2001) Evolution of Asian monsoons and phased uplift of the Himalaya–Tibetan plateau since Late Miocene times. Nature 411:62–66

An ZS, Li L, Lu HY (1999) Instability of the Eastern Asian Winter Monsoon during the last interglacial. In: National Natural Science Foundation research report, Earth Sciences (1997–1998). Beijing: Science Press, pp 18–19

An ZS, Liu TS, Lu YC (1990b) The long-term paleomonsoon variation recorded by the loess-Paleosol sequence in central China. Quat Int (7–8):91–95

An ZS, Liu XD (2000) History and variability of East Asia monsoon climate. Chin Sci Bull 45(3):238–249

An ZS, Porter SC (1997) Millennial-scale climatic oscillations during the last interglaciation in central China. Geology 25(7):603–604

An ZS, Porter SC, Chappell J (1994) Correlation between loess deposit in Luochuan and ice core in Greenland during the past 130,000 years. Chin Sci Bull 39(24):2254–2256

An ZS, Porter SC, Kutzbach JE et al (2000) Asynchronous holocene optimum of the East Asian monsoon. Quat Sci Rev 19:743–762

An ZS, Porter SC, Zhou WJ (1993) Episode of *strengthened summer monsoon climate of younger dryas age on the loess plateau of central china*. Quat Res 39(1):45–54

An ZS, Thompson LG (1998) Paleoclimatic change of monsoonal China linked to global change. In: Galloway JN, Melillo JM (eds.) Asian Change in the Context of Global Climate Change. Cambridge University Press, Cambridge, pp 18–41

An ZS, Wu XH, Lu YC et al (1990a) A preliminary study on the paleoenvironment change of China during the last 20,000 years. In: Liu TS, An ZS (eds.) Loess, Quaternary Geology and Global Change (Part II). Science Press, Beijing, pp 1–26 (in Chinese)

An ZS, Wu XH, Lu YC et al (1992) A preliminary study on the paleoenvironment change of China during the last 20, 000 years. Loess Quat Geol Glob Change 2:1–26 (in Chinese)

An ZS, Wu XH, Wang PX et al (1991b) Paleomonsoons of China over the last 130,000 years – I paleomonsoon variation. Sci China Ser B 34:1016–1024

An ZS, Wu XH, Wang PX et al (1991c) Paleomonsoons of China over the last 130,000 years –II paleomonsoon variation. Sci China Ser B 34:1007–1015

Anderson DM, Overpeck JT, Gupta AK (2002) Increase in the Asian Southwest monsoon during the past four centuries. Science 297:596–599

Aono Y (1998) Climatic change in the March temperature deduced from phenological record for flowering of cherry tree in Tokyo since the late 18th century. Bulletin of Osaka Prefecture University (Ser B) 50:11–19

Aono Y, Kazui K (2008) Phenological data series of cherry tree flowering in Kyoto, Japan, and its application to reconstruction of springtime temperatures since the 9th century. Int J Climatol 28:905–914

Aono Y, Omoto Y (1994) Estimation of temperature at Kyoto since the 11th century using flowering data of cherry trees in old documents. J Agr Meteorol 49:263–272

Arne M, Jussi E, Volker M (2009) The Late Miocene climate response to a modern Sahara desert. Global Planet Change 67(3–4):193–204

Barnes DJ, Lough JM (1989) The nature of skeletal density banding in scleractinian corals: fine banding and second patterns. J Exp Mar Bio Ecol 129:119–134

Beaufort L, de Garidel-Thoron T, Linsley B et al (2003) Biomass burning and oceanic primary production estimates in the Sulu Sea area over the last 380 kyr and the East Asian monsoon dynamics. Mar Geol 201:53–65

Beaufort L, Lancelot Y, Camberlin P et al (1997) Insolation cycles as a major control of equatorial Indian Ocean primary productivity. Science 278:1451–1454

Beck JW, Edwards RL, Ito E et al (1992) Sea-surface temperature from coral skeletal strontium/calcium ratios. Science 257:644–647

Berger AL (1978) Long tern variations of caloric insolation resulting from the Earth's orbital elements. Quat Res 9:139–167

Berger AL, Loutre MF (1991) Insolation values for the climate of the last 10 million years. Quat Sci Rev 10:297–317

Berger WH, von Rad U (2002) Decadal to millennial cyclicity in varves and turbidites from the Arabian Sea: hypothesis of tidal origin. Global Planet Change 34:313–325

Bernhard E, Annette K, Wolf DB (2006) Environmental changes at the eastern Namib Desert margin before and after the Last Glacial Maximum: New evidence from fluvial deposits in the upper Hoanib River catchment, northwestern Namibia. Palaeogeogr Palaeoclim Palaeoecol 234(2–4):201–222

Blunier T, Brook EJ (2001) Timing of *Millennial-Scale Climate Change in Antarctica and Greenland During the Last Glacial Period*. Science 291:109–112

Blunier T, Chappellaz J, Schwander J et al (1998) Asynchrony of Antarctica and Greenland climate during 2 last glacial. Nature 394:739–743

Blyakharchuk TA, Wright HE, Borodavko PS et al (2004) Late-glacial and Holocene vegetational changes on the Ulagan high-mountain plateau, Altai Mountains, southern Siberia. Palaeogeogr Palaeoclim Palaeoecol 209:259–279

Bond GC, Broecker WS, Johnsen SJ et al (1993) Correlations between climate records from North Atlantic sediments and Greenland ice. Nature 365:143–147
Bond GC, Kromer B, Beer J et al (2001) Persistent solar influence on North Atlantic climate during the Holocene. Science 294:2130–2136
Bond GC, Showers W, Cheseby M et al (1997) A pervasive millennial-scale cycle in North Atlantic Holocene and Glacial climates. Science 278:1257–1266
Bond GC, Showers W, Elliot M (1999) The North Atlantic's 1–2 kyr climate rhythm: relation to Heinrich events, Dansgaard/Oeschger cycles and the Little Ice Age. In: Clark P U, Webb R S, Kelgwin L D. Mechanisms of Global Climate Change at Millennial Time Scales. Geophysical Monograhp 112:35–58
Boomer I, Aladin N, Plotnikov I et al (2000) The palaeolimnology of the Aral Sea: a review. Quat Sci Rev 19:1259–1278
Borgaonkar HP, Kumar KR, Pant GB (2002) Tree-ring variations over the western Himalaya: Little evidence of Little Ice Age. PAGES News 10(1):5–6
Borgaonkar HP, Sikder AB, Ram S et al (2007) Dendroclimatological Investigation of High Altitude Himalayan Conifers and Tropical Teak In India. Korean J Quat Res 21(1):15–25
Borgaonkar HP, Sikder AB, Ram S et al (2010) El Niño and related monsoon drought signals in 523-year-long ring width records of teak (Tectona grandis L.F.) trees from south India. Palaeogeogr Palaeoclim Palaeoecol 285(1–2):74–84
Boulay S, Colin C, Trentesaux A et al (2005) Sediment sources and East Asian monsoon intensity over the last 450 ky. Mineralogical and geochemical investigations on South China Sea sediments. Palaeogeogr Palaeoclim Palaeoecol 228(3–4):260–277
Braconnot P, Otto-Bliesner B, Harrison S et al (2007) Results of PMIP2 coupled simulations of the Mid-Holocene and Last Glacial Maximum—Part 1: experiments and large-scale features. Clim Past 3:261–277
Braun H, Christl M, Rahmstorf S et al (2005) Possible solar origin of the 1470-year glacial climate cycle demonstrated in a coupled model. Nature 438(7065):208–211
Brauning A, Mantwill B (2004) Summer temperature and summer monsoon history on the Tibetan plateau during the last 400 years recorded by tree rings. Geophys Res Lett 31(24): Art. No. L24205
Briffa KR (2000) Annual climate variability in the Holocene: interpreting the message of ancient trees. Quat Sci Rev 19:87–105
Briffa KR, Jones PD, Schweinguber FH et al (1995) Unusual twentieth-century summer warmth in a 1000-year temperature record from Siberia. Nature 376:156–159
Broecker WS (1998) Paleocean circulation during the Last Glaciation: A bipolar seesaw? Paleoceanography, 13(2):119–121
Bryson RA, Swain AM (1981) Holocene variations of Holocene rainfall in Rajasthan. Quat Res 16:135–145
Budziak D, Schneider RR, Rostek F et al (2000) Late Quaternary insolation forcing on total organic carbon and C37 alkenone variations in the Arabian Sea. Paleoceanogr 15:307–321
Burns SJ, Fleitmann D, Mudelsee M et al (2002) A 780-year annually resolved record of Indian Ocean monsoon precipitation from a speleothem from south Oman. J Geophys Res 107(D20), doi:10.1029/2001JD001281
Cai HW (1986) The Tentative Probe of Stratigraphical Division of Quaternary Period in Badain Jaran Region. Acta Geologica Gansu 3:142–153
Cai QF, Liu Y (2007) January to August temperature variability since 1776 inferred from tree-ring width of Pinus tabulaeformis in Helan Mountain. J Geograph Sci 17(3):293–303
Cai Y, An Z, Cheng H et al (2006) High-resolution absolute-dated Indian Monsoon record between 53 and 36 ka from Xiaobailong Cave, southwestern China. Geology 34:621–624
Cai Y, Cheng H, An Z et al (2010b) Large variations of oxygen isotope in precipitation over South-Central Tibet during Marine Isotope Stage 5. Geology 38(3): 243–246
Cai Y, Tan L, Cheng H et al (2010a) The variation of summer monsoon precipitation in central China since the last deglaciation. Earth Planet Sci Lett 291:21–31

Cai Y, Zhang M, Peng Z et al (2001) The $\delta^{18}O$ variation of a stalagmite from Qixing Cave, Guizhou Province and indicated climate change during the Holocene: Chin Sci Bull 46(22):1904–1908
Central Meteorological Bureau (1981) Atlas of flood and drought in China in the last 500 years. Map Press, Beijing, China (in Chinese)
Chai X (1990) Peat Geography Beijing: Scientific Press, Beijing (in Chinese)
Chalker BE, Barnes DJ (1990) Gamma densitometry for the measurement of coral skeletal density. Coral Reefs 4:95–100
Chambers FM, Charman DJ (2004) Holocene environmental change: contributions from the peatland archive. The Holocene 14 (1):1–6
Charles CD, Cobb K, Moore MD et al (2003) Monsoon-tropical ocean interaction in a network of coral records spanning the 20th century. Mar Geol 201:207–222
Charles CD, Hunter DE, Fairbanks RG (1997) Interaction between the ENSO and the Asian monsoon in a coral record of tropical climate. Science 277:925–928
Charles CD, Lynch-Stieglitz J, Ninnemann US et al (1996) Climate connections between the hemisphere revealed by deep sea sediment core/ice core correlations. Earth Planet Sci Lett 142:19–27
Charman DJ (2002) Peatland and environmental change. Wiley, Chichester
Chen F, Yuan YJ, Wei WS et al (2011) Correlations between the summer Asian Pacific Oscillation index and the tree-ring width of Pinus massiniana from Sha county, Fujian Province. Quat Sci 31(1):96–103 (in Chinese)
Chen FH, Chen JH, Holmes JA et al (2010b) Moisture changes over the last millennium in the Arid Central Asia: a review, synthesis and comparison with monsoon region. Quat Sci Rev 29:1055–1068
Chen FH, Cheng B, Zhao Y et al (2006) Holocene environmental change inferred from a high-resolution pollen records of Lake Zhuyeze, arid China. The Holocene 16:675–684
Chen FH, Shi Q, Wang JM (1999) Environmental change documented by sedimentation of Lake Yiema in arid China since the last glaciation. J Paleolimn 22:159–169
Chen FH, Yu ZC, Yang ML et al (2008a) Holocene moisture evolution in arid central Asia and its out-of-phase relationship with Asian monsoon history. Quat Sci Rev 27:351–364
Chen FH, Zhu Y, Li JJ et al (2001a) Abrupt Holocene changes of the Asian monsoon at millennial- and centennial-scales: evidence from lake sediment document in Minqin Basin, North China. Chin Sci Bull 46:1942–1947
Chen G (1939) Catalogue of disasters in China. Shanghai Press, Pressed in 1986, Shanghai, China
Chen HF, Song SR, Lee T et al (2010a) A multiproxy lake record from Inner Mongolia displays a late Holocene teleconnection between Central Asian and North Atlantic climates. Quat Int 227:170–182
Chen HZ, Jin J, Dong GR (2001b) Holocene Evolution Processes of Gurbantunggut Desert and Climatic Changes. J Desert Res 21(4):333–339
Chen J, Chen F, Zhang J et al (2008b) Humidity variability in the arid northwest China during LIA derived from different proxy records. Acta Geographica Sinica 63(1):23–33 (in Chinese)
Chen J, Jiang T, Xu P (1998b) Climate changes in Jiangsu Province during the last 2000 years. Scientia Geographica Sinica 18(3):219–226 (in Chinese)
Chen M-H, Wang R, Yang L et al (2003) Development of East Asian summer monsoon environments in the late Miocene, radiolarian evidence from Site 1143 of ODP Leg 184. Mar Geol 201:169–177
Chen M-T, Beaufort L, Shipboard Scientific Party of the IMAGES III/MD106–IPHIS Cruise (1998a) Exploring Quaternary variability of the East Asia Monsoon, Kuroshio Current, and Western Pacific Warm Poll Systems: high resolution investigations of paleoceanography from the IMAGES III (MD106)–IPHIS Cruise. TAO 9:129–142
Chen W, Wang WM (2012) Middle-Late Holocene vegetation history and environment changes revealed by pollen analysis of a core at Qingdao of Shandong Province, East China. Quat Int 254:68–72
Chen Y, Huang PH, Zhu HS (1986) The isotopic paleo-temperature study of Quaternary speleothem from the karst cave in Zhoukoudian, Beijing. Chinese Science Bulletin 20:1576–1578

Cheng H, Edwards RL, Broecker WS et al (2009) Ice Age Terminations. Science 326:248–252
Cheng H, Edwards RL, Wang Y et al (2006) A penultimate glacial monsoon record from Hulu Cave and two-phase glacial terminations. Geology 34(3):217–220
Chiang JCH, Bitz CM (2005) Influence of high latitude ice cover on the marine Intertropical Convergence Zone. Clim Dyn 25:477–496
Chu CC (1926) Climatic pulsations during historical times in China. Geogr Rev 16:274–282
Chu CC (1925) The climate in Southern Song Dynasty. Kexue 10(2):151–164 (in Chinese)
Chu CC (1973) A preliminary study on the climatic fluctuations during the last 5000 years in China. Scientia Sinica (Series A) 16(2):226–256
Cleaveland LC, Herbert TD (2009) Preservation of the alkenone paleotemperature proxy in uplifted marine sequences: a test from the Vrica outcrop, Crotone, Italy. Geology 37:179–182
Clemens S, Murray D, Prell W (1996) Nonstationary phase of the Plio-Pleistocene Asian monsoon. Science 274:943–948
Clemens SC, Prell WL (2003) A 350,000 year summer-monsoon multi-proxy stack from the Owen Ridge, Northern Arabian Sea. Mar Geol 201:35–51
Clemens SC, Prell WL (2007) The timing of orbital-scale Indian monsoon changes. Quat Sci Rev 26:275–278
Clemens SC, Prell WL, Sun Y et al (2008) Southern Hemisphere forcing of Pliocene $\delta^{18}O$ and the evolution of Indo-Asian monsoons. Paleoceanogr 23, PA4210, doi:10.1029/2008PA001638
Clemens SC, Prell WL, Sun YB (2010) Orbital-scale timing and mechanisms driving Late Pleistocene Indo-Asian summer monsoons: Reinterpreting cave speleothem $\delta^{18}O$. Paleoceanogr 25 Pa4207, doi:10.1029/2010pa001926
Cobb KM, Charles CD, Cheng H et al (2003) El Niño/Southern Oscillation and tropical Pacific climate during the last millennium. Nature 424:271–276
Cohen TJ, Nanson GC, Larsen JR (2010) Late Quaternary aeolian and fluvial interactions on the Cooper Creek Fan and the association between linear and source-bordering dunes, Strzelecki Desert, Australia. Quat Sci Rev 29(3–4):455–471
Cole JE, Dunbar RB, McClanahan TR et al (2000) Tropical pacific forcing of decadal SST variability in the western Indian ocean over the past two centuries. Science 287:617–619
Colin C, Kissel C, Blamart D et al (1998) Magnetic peoperties of sediments in the Bay of Bengal and the Andaman Sea: impact of rapid North Atlantic Ocean climatic events on the strength of the Indian monsoon. Earth Planet Sci Lett 160:623–635
Colman SM, Yu S-Y, An Z et al (2007) Late Cenozoic climate changes in China's western interior: a review of research on Lake Qinghai and comparison with other records. Quat Sci Rev 26: 2281–2300
Cook ER, Anchukaitis KJ, Buckley BM et al (2010) Asian Monsoon Failure and Megadrought During the Last Millennium. Science 328(5977):486–489
Cook ER, Buckley BM, D'Arrigo et al (2000) Warm-season temperatures since 1600BC reconstructed from Tasmanian tree rings and their relationship to large-scale sea surface temperature anomalies. Clim Dyn 16:79–91
Cook ER, Krusic PJ, Jones PD (2003) Dendroclimatic signals in long tree-ring chronologies from the Himalayas of Nelpal. Int J Climatol 23:707–732
Corrége T (2006) Sea surface temperature and salinity reconstruction from coral geochemical tracers. Palaeogeogr Palaeoclim Palaeoecol 232:408–428
Cosford J, Qing HR, Eglington B et al (2008) East Asian monsoon variability since the Mid-Holocene recorded in a high-resolution, absolute-dated aragonite speleothem from eastern China. Earth Planet Sci Lett 275:296–307
Cour P, Zheng Z, Duzer D et al (1999) Vegetational and climatic significance of modern pollen rain in northwestern Tibet. Rev Palaeobot Palynol 104:183–204
Cowie J (1998) Climate and human change, disaster or opportunity? Parthenon Publishing, New York
Cruz FW Jr, Burns SJ, Karmann I et al (2006) A stalagmite record of changes in atmospheric circulation and soil processes in the Brazilian subtropics during the Late Pleistocene. Quat Sci Rev 25:2749–2761

Cruz FW, Burns SJ, Karmann I et al (2005) Insolation-driven changes in atmospheric circulation over the past 116,000 years in subtropical Brazil. Nature 434:63–66

Cui ZJ, Yang JQ, Zhao L (2004) Discovery of large ice wedge in the Erdos Area and the southern boundary and environment of permatrost in northern China since 20 ka. Chin Sci Bull 49(13):1304–1310

D'Arrigo R, Wilson R, Palmer J et al (2006) Monsoon droght over Jave, Indonesia, during the past two centuries. Geophys Res Lett 33:L04709, doi:10.1029/2005GL025465

D'Arrigo RD, Jacoby GC, Pederson N et al (2000) Mongolian tree-rings, temperature sensitivity and reconstructions of Northern Hemisphere temperature. The Holocene 10(6):669–672

Damon PE, Sonnett CP (1991) Solar and terrestial components of the atmospheric 14C variation spectrum. In: C P Sonett et al ed. The sun in Time. Tuson: Arizona Univ Press, 360–388

Dannenmann S, Linsley BK, Oppo DW et al (2003) East Asian monsoon forcing of suborbital variability in the Sulu Sea during Marine Isotope Stage 3: Link to Northern Hemisphere climate. Geochem Geophys Geosyst 4(1):1001. doi:10.1029/2002GC000390

Dansgaard W (1964) Stable isotopes in precipitation. Tellus 16(4):436–468

Dansgaard W, Johnsen SJ, Clausen HB et al (1993) Evidence for general instability of past climate from a 250-kyr ice-core record. Nature 364:218–220

Davi N, D'Arrigo RD, Jacoby GC et al (2002) Warm-season annual to decadal temperature variability for Hokkaido, Japan, inferred from maximum latewood density (A.D.1557–1990) and ring width data (A.D.1532–1990). Clim Change 52:201–217

Dayem KE, Molnar P, Battisti DS et al (2010) Lessons learned from oxygen isotopes in modern precipitation applied to interpretation of speleothem records of paleoclimate from eastern Asia. Earth Planet Sci Lett 295:219–230

De Deckker P, Correge T, Head J (1991) Late Pleistocene record of cyclic eolian activity from tropical Australia suggesting the Younger Dryas is not an unusual climatic event. Geology 19:602–605

De Garidel-Thoron T, Beaufort L, Linsley BK et al (2001) Millennial-scale dynamics of the East-Asian winter monsoon during the last 200,000 years. Paleoceanogr 16:491–502

Delmonte B, Petit JR, Krinner G et al (2005) Ice core evidence for secular variability and 2002 year dipolar oscillations in atmospheric circulation over East Antarctica during the Holocene. Clim Dyn 24:641–654

deMenocal P, Ortiz J, Guilderson T et al (2000) Abrupt onset and termination of the African Humid Period: rapid climate responses to gradual insolation forcing. Quat Sci Rev 19:347–361

Deng H, Jiang W (2005) Temporal distribution and characteristics of the sand-dust weathers in North China Plain from 1463 to 1913. Geograph Res 24(3):403–411 (in Chinese)

Desprat S, Goñi MFS, Loutre M-F (2003) Revealing climatic variability of the last three millennia in northwestern Iberia using pollen influx data. Earth Planet Sci Lett 213(1–2):63–78

Ding YH, Ren GY, Shi GY et al (2006) National Assessment Report of Climate Change (I). Climate change in China and its future trend. Adv Clim Change Res 2(1):3–8 (in Chinese)

Dong GR, Wang GY, Li XZ et al (1998) Palaeomonsoon vicissitudes in eastern desert region of China since last interglacial period. Sci Chin (Series D) 41(2):215–224

Druffel ERM (1997) Pulses of rapid ventilation in the North Atlantic surface ocean during the past century. Science 275:1454–1457

Du SH, Li BS, Chen MH et al (2012) Kiloyear-scale climate events and evolution during the Last Interglacial, Mu Us Desert, China. Quat Int 263:63–70

Du SH, Li BS, Niu DF (2011) Age of the MGS5 segment of the Milanggouwan stratigraphical section and evolution of the desert environment on a kiloyear scale during the Last Interglacial in China's Salawusu River Valley: Evidence from Rb and Sr contents and ratios. Chemie Erde-Geochemistry 71(1):87–95

Duan KQ, Wang NL, Pu JC (2002) Events of abrupt change of Indian monsoon recorded in Dasuopu ice core from Himalayas. Chin Sci Bull 47(8):691–696

Duan KQ, Yao TD (2003) Precipitation variability in central Himalayas and its relation to Northern Hemisphere temperature. Chin Sci Bull 48(14):1480–1482

Duplessy JC, Labeyrie J, Lalou C et al (1970) Continental climatic variations between 130,000 and 90,000 years BP. Nature 226:631–633

Dykoski CA, Edwards RL, Cheng H et al (2005) A high-resolution, absolute-dated Holocene and deglacial Asian monsoon record from Dongge Cave, China. Earth Planet Sci Lett 233:71–86

Eddy JA (1977) Climate and the changing sun. Clim Change 1(2):173–190

Editing Committee of China's Military History (ECCMH) (1985) Tabulation of Wars in Ancient China(2). People's Liberation Army Press, Beijing, China

Edwards RL, Chen JH, Wasserburg GJ (1987) ^{238}U-^{234}U-^{230}Th-^{232}Th systematic and the precise measurement of time over the past 500,000 years. Earth Planet Sci Lett 81:175–192

El-Moslimany AP (1990) Ecological significance of common nonarboreal pollen: examples from drylands of the Middle East. Rev Palaeobot Palynol 64:343–350

Emeis K-C, Anderson DM, Doose H et al (1995) Sea-surface temperatures and the history of monsoon upwelling in the northwest Arabian Sea during the last 500,000 years. Quat Res 43:355–361

Esper J, Cook R, Schweingruber FH (2002) Low-frequency signals in long tree-ring chronologies and the reconstruction of past temperature variability. Science 295:2250–2253

Fan ZX, Bräuning A, Cao KF (2008) Annual temperature reconstruction from the central Hengduan Mountains, as deduced from tree rings. Dendrochronologia 26:97–107

Fan ZX, Brauning A, Cao KF et al (2009a) Growth-climate responses of high-elevation conifers in the central Hengduan Mountains, southwestern China. For Ecol Manag 258:306–313

Fan ZX, Brauning A, Yang B et al (2009b) Tree ring density-based summer temperature reconstruction for the central Hengduan Mountains in southern China. Global Planet Change 65:1–11

Fang JQ (1991) Lake evolution during the past 30,000 years in China, and its implications for environmental change. Quat Res 36:37–60

Fang KY, Gou XH, Chen FH et al (2009a) Reconstructed droughts for the southeastern Tibetan Plateau over the past 568 years and its linkages to the Pacific and Atlantic Ocean climate variability. Clim Dyn 35(4):577–585

Fang KY, Gou XH, Chen FH et al (2009b) Drought variation in eastern part of Northwest China over last two centuries: evidence from tree-ring. Clim Res 38:129–135

Fang KY, Gou XH, Chen FH et al (2009c) Tree-ring based drought reconstruction for Guiqing Mountain (China): Linkage to the Indian and Pacific Oceans. Int J Climatol 30(8):1137–1145

Farjon A (1990) Pinaceae. Drawings and descriptions of the genera *Abies, Cedrus, Pseudolarix, Keteleeria, Nothotsuga, Tsuga, Cathaya, Pseudotsuga, Larix* and *Picea*. Koeltz Scientific Books, Königstein, Germany, pp 149–151

Fei J, Zhou J, An ZS (2004) Temperature change in China over 618–959 AD: Based on historical records. Mar Geol Quat Geol 24:109–118

Felis T, Pätzold J, Loya Y et al (1998) Vertical mass mixing and plankton blooms recorded in skeletal stable isotopes of a Red Sea coral. J Geophys Res 103:30731–30739

Feng X, Yan S, Ni J et al (2006) Environmental changes and lake level fluctuation recorded by lakes on the plain in northern Xinjiang during the late Holocene. Chin Sci Bull 51(supp): 60–67

Ficken KJ, Li B, Swain DL, Eglinton G (2000) An n-alkane proxy for the sedimentary inputs of submerged/floating freshwater aquatic macrophytes. Org Geochem 31:745–749

Fleitmann D, Burns SJ, Mangini A et al (2007) Holocene ITCZ and Indian monsoon dynamics recorded in stalagmites from Oman and Yemen (Socotra): Quat Sci Rev 26:170–188

Fleitmann D, Burns SJ, Mudelsee M et al (2003) Holocene Forcing of the Indian Monsoon Recorded in a Stalagmite from Southern Oman. Science 300:1737–1739

Folk PL, Ward WD (1957) Brazos Reviver bar: A study in the significance of grain size parameters. Sedi Petr 27:3–26

Folland CK, Karl TR, Christy JR et al (2001) Observed climate variability and change. In Climate Change. Cambridge University Press, Cambridge, pp 99–181

Fontes JC, Gasse F, Gibert E (1996) Holocene environmental changes in Bangong Co basin (western Tibet), Part 1: Chronology and stable isotopes of carbonates of a Holocene lacustrine core. Palaeogeogr Palaeoclim Palaeoecol 120:25–47

Fowell SJB, Hansen CS, Peck JA et al (2003) Mid to late Holocene climate evolution of the Lake Telmen Basin, North Central Mongolia, based on palynological data. Quat Res 59:353–363

Editorial Committee of FOC (2004) Flora of China, Vol 1. Science Press Beijing, ISIN 7 03-013229-7 (in Chinese)

Gagan MK, Ayliffe LK, Beck JW et al (2000) New views of tropical paleoclimates from corals. Quat Res Rev 19:45–64

Gagan MK, Ayliffe LK, Hopley D et al (1998) Temperature and surface-ocean water balance of the mid-Holocene tropical western pacific. Science 279:1014–1018

Galimov EM, Grinenko VA (1965) The effect of leaching under surface conditions on the isotopic composition of carbon in secondary calcite. Geochem Int 1:79–82

Gao QZ, Dong GR, Zou XY (1996) The Chagelebulu Section: A Strata Record of the Advances and Retreats of the Monsoons in East Asia since Late Pleistocene. J Desert Res 6(2):112–119

Gao QZ, Tao Z, Li BS (2006) Palaeomonsoon variability in the southern fringe of the Badain Jaran Desert, China, since 130 ka BP. Earth Surf Process Landforms 31:265–283

Gao SY, Chen WN, Jin HL (1993) Preliminary research on the desert evolution in northwest of monsoon region of China since Holocene. Sci Chin (Series B) 23(2):202–208

Garfin GM, Hughes MK, Liu Y et al (2005) Exploratory temperature and precipitation reconstructions from the Qinling Mountains, North-central china. Tree-ring Res 61(2):59–72

Gascoyne M (1992) Palaeoclimate determination from cave calcite deposits. Quat Sci Rev 11:609–632

Gasse F, Arnold M, Fontes JC et al (1991) A 13,000-year climate record from western Tibet. Nature 353:742–745

Gasse F, Fontes JC, Van Campo E et al (1996) Holocene environmental changes in Bangong Co basin (western Tibet), Part 4: Discussion and conclusions. Palaeogeogr Palaeoclim Palaeoecol 120:79–92

Gasse F, Van Campo E (1994) Abrupt post-glacial climate events in West Asia and North Africa monsoon domains. Earth Planet Sci Lett 126:435–456

Ge Q, Guo X, Zheng J et al (2008) Meiyu in Middle and Lower Reaches of Yangtze River since 1736. Chin Sci Bull 53:107–114

Ge Q, Zheng J, Fang X et al (2003) Winter half-year temperature reconstruction for the middle and lower reaches of the Yellow River and Yangtze River, China, during the past 2000 years. The Holocene 13(16):933–940

Gergis JL, Fowler AM (2009) A history of ENSO events since A.D. 1525 implications for future climate change. Clim Change 92:343–387

Gil García M, Ruiz Zapata M, Santisteban J et al (2007) Late holocene environments in Las Tablas de Daimiel (south central Iberian peninsula, Spain). Veg Hist Archaeobot 16(4):241–250

Gong D, Wang S, Zhu J (2001) East Asian Winter Monsoon and Arctic Oscillation. Geophys Res Lett 28(10):2073–2076

Gong G, Hameed S (1991) The variation of moisture conditions in China during the last 2000 years. Int J Climatol 11:271–283

Gong G, Zhang P, Wu X et al (1983) Research methods of climate changes in historical time. Science Press, Beijing, China (in Chinese)

Gou XH, Chen FH, Cook ER et al (2007) Streamflow variations of the Yellow River over past 593 years in western China reconstructed from tree rings. Water Res 43:W06434. doi: 10.1029/2006WR005705

Gou XH, Chen FH, Wang YJ et al (2001) Spring precipitation reconstructed in the east of the Qilian Mountain during the last 280 a by tree ring width. J Glaciol Geocryol 23(3):292–296 (in Chinese)

Gou XH, Chen FH, Yang MX et al (2008b) Asymmetric variability between maximum and minimum temperatures in Northeastern Tibetan Plateau: Evidence from tree rings. Sci Chin (D), 51(1):41–55

Gou XH, Deng Y, Chen FH et al (2010) Tree ring based streamflow reconstruction for the Upper Yellow River over the past 1234 years. Chin Sci Bull 55(36):4179–4186

Gou XH, Peng JF, Chen FH et al (2008a) Maximum summer-half-year temperature variations on the Northeastern Tibetan Plateau over past 700 year from tree-rings. Theor Appl Climatol, doi: 10.1007/s00704-007-0336-y

Gou XH, Yang MX, Peng JF et al (2006) Maximum temperature reconstruction for Animaqing mountains over past 830 years based on tree-ring records. Quat Sci 26(6):991–998 (in Chinese)

Greatbatch RJ, Zhang S (1995) An interdecadal oscillation in an idealized ocean-basin forced by constant heat-flux. J Clim 8(1):81–91

GRIP members (1993) Climate instability during the last interglacial period recorded in the GRIP ice core. Nature 371:779–783

Grootes P, Stuiver M (1997) Oxygen 18/16 variability in Greenland snow and ice with 103–105 year time resolution. J Geophys Res 102 C12, 26455–26470

Grootes PM, Stuiver M, White JC (1993) Comparison of oxygen isotope records from GISP2 and GRIP Greenland ice cores. Nature 366:552–554

Grossman M, Zaiki M (2007) Reconstructing typhoon landfalls in Japan using historical documents 1801–1830. In: Harrington LMB, Harrington JA (eds) Papers of the applied geography conference. pp 334–343

Grossman M, Zaiki M (2009) Reconstructing typhoons in Japan in the 1880s from documentary records. Weather 64:315–322

Grottoli AG, Gille ST, Druffel ERM et al (2003) Decadal timescale shift in the ^{14}C record of a central equatorial Pacific coral. Radiocarbon 45:91–99

Grove JM (2001a) The initiation of the "Little Ice Age" in regions around the North Atlantic. Clim Change 48:53–82

Grove JM (2001b) The onset of the Little Ice Age. In: Jones PD, Ogilvie AEJ, Davies TD et al (eds) History and climate—memories of the future? Kluwer Academic Publishers, USA, pp 153–185

Grumet NS, Avram NJ, Beck JW et al (2004) Coral radiocarbon records of Indian Ocean water mass mixing and wind-induced upwelling along the coast of Sumatra, Indonesia. J Geophys Res 109, doi:10.1029/2003JC002087

Gu Z, Liu J, Yuan B et al (1993) The changes in monsoon influence in the Qinghai-Tibetan Plateau during the past 12,000 years: Geochemical evidence from the L. Selin sediments. Chin Sci Bull 38:61–64

Guilderson TP, Schrag DP (1998) Abrupt shift in subsurface temperatures in the tropical pacific associated with changes in El Niño. Science 281:240–243

Guo SQ, Wang SM, Yang LJ (2005) Climatic and Environmental Change in North China Plain during the Last Glacial Maximum. Geol Rev 51(4):423–427

Guo ZT, Peng SZ, Wei LY et al (1999) Weathering Signals of Millennial-scale Oscillations of the East-Asian Summer Monsoon over the Last 220 ka. Chin Sci Bull 44(Suppl.): 20–25

Gupta AK, Anderson DM, Overpeck JT (2003) Abrupt changes in the Asian southwest monsoon during the Holocene and their links to the North Atlantic Ocean. Nature 421:354–357

Gupta SM (1999) Radiolarian monsoonal index Pyloniid group responds to astronomical forcing in the last 500,000 years: evidence from the central Indian Ocean. Man Environ 24:99–107

Harmon RS, Thompson P, Schwarcz HP et al (1978) Late Pleistocene paleoclimates of North America as inferred from stable isotope studies of speleothems. Quat Res 9:54–70

Hayashi T, Tanimura Y, Kuwahara Y et al (2009) Ecological variations in diatom assemblages in the Paleo-Kathmandu Lake linked with global and Indian monsoon climate changes for the last 600,000 years. Quat Res 72: 377–387

Head MJ, Zhou WJ, Zhou MF (1989) Evaluation of ^{14}C ages of organic fractions of paleosol from loess sequence near Xian, China. Radiocarbon 31(3):680–696

Hellstrom J (2003) Rapid and accurate U/Th dating using parallel ion counting multicollector ICP-MS. J Anal At Spectrom 18:1346–1351

Hemming NG, Hanson GN (1992) Boron isotopic composition and concentration in modern marine carbonates. Geochim Cosmochim Acta 56:537–543

Hendy CH (1971) The isotopic geochemistry of speleothems-I: The calculations of the effects of different modes of formation on the isotopic composition of speleothems and their applicability as paleoclimate indicators. Geochim Cosmochim Acta 35:801–824

Hendy CH, Wilson AT (1968) Palaeoclimatic data from speleothems. Nature 216:48–51

Hendy EJ, Gagan MK, Alibert CA et al (2002) Abrupt decrease in tropical Pacific sea surface salinity at end of Little Ice Age. Science 295:1511–1514

Herzschuh U, Tarasov P, Wünnemann B et al (2004) Holocene vegetation and climate of the Alashan Plateau, NW China, reconstructed from pollen data. Palaeogeogr Palaeoclim Palaeoecol 211:1–17

Herzschuh U, Winter K, Wünnemann B et al (2006) A general cooling trend on the central Tibetan Plateau throughout the Holocene recorded by the Lake Zigetang pollen spectra. Quat Int 154–155:113–121

Higginson MJ, Maxwell JR, Altabet MA (2003) Nitrogen isotope and chrolin paleoproductivity records from the northern South China Sea: remote versus local forcing of millennial- and orbital scale variability. Mar Geol 201:223–250

Hiroyuki T, Takeshi N, Koji Y et al (2008) Summer relative humidity in northern Japan inferred from $\delta^{18}O$ values of the tree ring in (1776–2002 A.D.): Influence of the paleoclimate indices of atmospheric circulation. J Geophys Res 113, doi:10.1029/2007JD009080

Hitoshi Y, Dieter E (2006) Reconstruction of early spring temperature for central Japan from the tree-ring widths of Hinoki cypress and its verification by other proxy records. Geophys Res Lett 33 doi:10.1029/2006GL026170

Hodell DA, Brenner M, Curtis JH et al (2001) Solar forcing of drought frequency in the Maya lowlands. Science 292:1367–1370

Hodell DA, Brenner M, Kanfoush SL et al (1999) Paleoclimate of Southwestern China for the past 50,000 year inferred from lake sediment records. Quat Res 52:369–380

Hoffmann G, Heimann M (1997) Water isotope modeling in the Asian monsoon region: Quat Int 37:115–128

Höffmann J (1996) The lakes in the SE part of Badain Jaran Shamo, their Limnology and Geochemistry. Geo-Wissenschaften 14(7/8):275–278

Hong B, Hong YT, Lin QH et al (2010) Anti-phase oscillation of Asian monsoons during the Younger Dryas period: Evidence from peat cellulose $\delta^{13}C$ of Hani, Northeast China. Palaeogeogr Palaeoclim Palaeoecol 297(1):214–222

Hong ES, Peng ZC, Li P (1995) The isotopic geochemical approaches in studying the paleo-temperature with the speleothem. Adv Earth Sci, 10(4):348–352 (in Chinese with English abstract)

Hong YT, Hong B, Lin QH et al (2003) Correlation between Indian Ocean summer monsoon and North Atlantic climate during the Holocene. Earth Planet Sci Lett 211(3–4):371–380

Hong YT, Hong B, Lin QH et al (2009) Synchronous climate anomalies in the western North Pacific and North Atlantic regions during the last 14,000 years. Quat Sci Rev 28 (9–10):840–849

Hong YT, Hong B, Lin QH et al (2005) Inverse phase oscillations between the East Asian and Indian Ocean summer monsoons during the last 12,000 years and paleo-El Nino. Earth Planet Sci Lett 231(3–4):337–346

Hong YT, Jiang H, Tao F et al (1997) The $\delta^{18}O$ temperature record of peat from Jinchuan for recent 5ka. Sci Chin (Series D) 27(6):525–530

Hong YT, Jiang HB, Liu TS et al (2000) Response of climate to solar forcing recorded in a 6000-year $\delta^{18}O$ time-series of Chinese peat cellulose. The Holocene 10(1):1–7

Hong YT, Wang ZG, Jiang Hb et al (2001) A 6000-year record of changes in drought and precipitation in northeastern China based on a $\delta^{13}C$ time series from peat cellulose. Earth Planet Sci Lett 185 (1–2):111–119

Hou Y (2007) The natural environment of the Ordos Plateau and the land use during the Ming and Qing Dynasties. J Chin Hist Geograph 22(4):28–39 (in Chinese)

Hou Y, Zhou J, Wang Y (2001) The natural and human landscape in the Ordos Plateau during the Bei Wei Dynasty (AD 386–534). J Desert Res 21(2):188–194 (in Chinese)

Hsu K (1998) Sun, climate, hunger, and mass migration. Sci Chin (Series D) 41(5):449–472

Hu C, Henderson GM, Huang J et al (2008) Quantification of Holocene Asian monsoon rainfall from spatially separated cave records. Earth Planet Sci Lett 266:221–232

Huang CX (1993) A study on pollen in surface soil from the Western Xizang. Arid Land Geogr 15:75–84 (in Chinese with English abstract)

Huang F, Sun XJ (2002) Pollen record from South China Sea ODP 1144 station during the Pleistocene. Sci Chin (D) 32(11):914–920 (in Chinese)

Huang L, Shao XM (2005) Precipitation variation in Delingha, Qinghai and solar activity over the last 400 years. Quat Sci 25(2):184–192 (in Chinese)
Huang L, Shao XM, Liu HB et al (2006) Abrupt changes of precipitation in Delingha region, Qinghai for the last 1000 years. Acta Geographica Sinica 61(7):713–719 (in Chinese)
Huang XZ, Chen FH, Fan YX et al (2009) Dry late-glacial and early Holocene climate in arid central Asia indicated by lithological and palynological evidence from Bosten Lake, China. Quat Int 194:19–27
Huang, J H (1992) The characteristic of the oxygen and carbon isotopic composition and its paleoclimate significance of the Quaternary speleothem from Shiquan Cave, Chongyang, Hubei. Carsologica Sinica, 11(3):245–249 (in Chinese)
Hughes MK (2001) An improved reconstruction of summer temperature at Srinagar, Kashmir since 1660 A.D. based on tree-ring width and maximum latewood density of Abies pindrow (Royle) Spach. Palaeobotanist 50:13–19
Iijima H, Kayanne H, Morimoto M et al (2005) Inter-annual sea surface salinity changes in the western Pacific from 1954 to 2000 based on coral isotope analysis. Geophys Res Lett 32: doi:10.1029/2004GL022026
Isdale P (1984) Fluorescent bands in massive corals record centuries of coastal rainfall. Nature 310:578–579
Iwamoto N, Inouchi Y (2007) Reconstruction of millennial-scale variations in the East Asian summer monsoon over the past 300 ka based on the total carbon content of sediment from Lake Biwa, Japan. Environ Geol 52:1607–1616
Jacoby G, Pederson N, D'Arrigo R (2003) Temperature and precipitation in Mongolia based on dendroclimatic investigations. Chin Sci Bull 48(14):1474–1479
Jacoby GC, D'Arrigo RD, Davaajamts T (1996) Mongolian Tree Rings and 20th-Century Warming. Science 273(5276):771–773
Jäkel D (1996) The Badain Jaran Desert: its origin and development. Geowissenschafen (7–8): 272–274
Jia CH, Yu L, Du NQ et al (1989) Changes of vegetation and climate in Qian' an county, Jilin province since late Tertiary. Scientia Geographica Sinica 9:274 (in Chinese with English abstract)
Jian Z, Huang B, Kuhnt W et al (2001) Late Quaternary upwelling intensity and East Asian monsoon forcing in the South China Sea. Quat Res 55:363–370
Jiang HC, Ding ZL (2008) A 20 Ma pollen record of East-Asian summer monsoon evolution from Guyuan, Ningxia, China. Palaeogeogr Palaeoclim Palaeoecol 265:30–38
Jiang Q, Shen J, Liu X et al (2007) A high-resolution climatic change since Holocene inferred from multi-proxy of lake sediment in westerly area of China. Chin Sci Bull 52:1042–1049
Jiang SH, Ren LL, Yong B et al (2010) Reconstruction of precipitation series for Ma' an Mountain area from July to September using tree rings. J Hohai Univ (Nat Sci) 38(1):31–35
Jiang WY, Leroy SAG, Ogle N et al (2008) Natural and anthropogenic forest fires recorded in the Holocene pollen record from a Jinchuan peat bog, northeastern China. Palaeogeogr Palaeoclim Palaeoecol 261:47–57
Jin ZD, Bickle MJ, Chapman HJ et al (2009) Early to mid-Pleistocene ostracod $\delta^{18}O$ and $\delta^{13}C$ in the central Tibetan Plateau: Implication for Indian monsoon change. Palaeogeogr Palaeoclim Palaeoecol 280:406–414
Jin ZD, Bickle MJ, Chapman HJ et al (2011) Ostracod Mg/Sr/Ca and $^{87}Sr/^{86}Sr$ geochemistry from Tibetan lake sediments: Implications for early to mid-Pleistocene Indian monsoon and catchment weathering. Boreas 40:320–331
Jin ZD, Cao J, Wu J et al (2006) A Rb/Sr record of catchment weathering response to Holocene climate change in Inner Mongolia. Earth Surface Process Landforms 31:285–291
Jin ZD, Wu JL, Cao JJ (2004) Holocene chemical weathering and climatic oscillations in north China: evidence from lacustrine sediments. Boreas 33:260–266
Jin ZD, Wu YH, Zhang X et al (2005) Role of late glacial to mid-Holocene climate in catchment weathering in the central Tibetan Plateau. Quat Res 63:161–170
Jin ZD, You CF, Wang Y et al (2010) Hydrological and solute budgets of Lake Qinghai, the largest lake on the Tibetan Plateau. Quat Int 218:151–156

Johnson KR, Ingram BL (2004) Spatial and temporal variability in the stable isotope systematics of modern precipitation in China: implications for paleoclimate reconstructions. Earth Planet Sci Lett 220:365–377

Johnson KR, Lynn Ingram B, Sharp WD et al (2006) East Asian summer monsoon variability during Marine Isotope Stage 5 based on speleothem $\delta^{18}O$ records from Wanxiang Cave, central China. Palaeogeogr Palaeoclim Palaeoecol 236:5–19

Joosten H, Clarke D (2002) Wise use the mires and peatlands. International Mire Conservation Group and International Peat Society, p 304

Joosten H, Haberl A, Schumann M (2008) Degradation and restoration of peatlands on the Tibetan Plateau. Peatlands Int 1:31–35

Jouzel J, Masson-Delmotte V, Cattani O et al (2007) Orbital and millennial Antarctic variability over the last 800,000 years. Science 317:793–796

Juillet-Leclerc A, Jouzel J, Labeyrie L et al (1997) Modern and last glacial maximum sea surface $\delta^{18}O$ derived from an atmospheric general circulation model. Earth Planet Sci Lett 146:591–605

Jung HS, Lim GH, Oh JH (2001) Interpretation of the transient variations in the time series of precipitation amounts in Seoul, Korea: Part 1. diurnal variation. J Clim 14(13):2989–3004

Juyal N, Pant RK, Basavaiah N et al (2009) Reconstruction of Last Glacial to early Holocene monsoon variability from relict lake sediments of the Higher Central Himalaya, Uttrakhand, India. J Asian Earth Sci 34:437–449

Kang X, Graumlich LJ, Sheppard PR (1997) Reconstructing a 1835a tree-ring chronology in Dulan area, Qinghai Province and its preliminary analysis. Chin Sci Bull 42(10):1089–1091 (in Chinese)

Kaplan A, Cane MA, Kushnir Y et al (1998) Analyses of global sea surface temperature 1856–1991. Journal of Geophysical Research, 103(C9):18567–18589

Kazanci N, Gulbabazadeh T, Leroy SAG et al (2004) Sedimentary and environmental characteristics of the Gilan-Mazenderan plain, northern Iran: Influence of long- and short-term Caspian water level fluctuations on geomorphology. J Mar Syst 46:145–168

Kelly MJ, Edwards RL, Cheng H et al (2006) High resolution characterization of the Asian Monsoon between 146,000 and 99,000 years BP from Dongge Cave, China and global correlation of events surrounding Termination II. Palaeogeogr Palaeoclim Palaeoecol 236:20–38

Kelts KR, Chen KZ, Lister GS et al (1989) Geological fingerprints of climate history: a cooperative study of Qinghai Lake, China. Eclogae Geologicae Helvetiae 82:167–182

Kershaw AP, vanderKaars S, Moss PT (2003) Late Quaternary Milankovitch-scale climatic change and variability and its impact on monsoonal Australasia. Mar Geol 201:81–95

Kilbourne KH, Quinn TM, Webb R et al (2008) Paleoclimate proxy perspective on Caribbean climate since the year 1751: Evidence of cooler temperatures and multidecadal variability. Paleoceanogr 23, doi:10.1029/2008PA001598

Kim GS, Choi IS (1987) A preliminary study on long-term variation of unusual climate phenomena during the past 1000 years in Korea. In: Ye D, Fu C (eds) The Climate of China and Global Climate. Springer, New York, pp 30–37

Klein R, Loya Y, Gvirtzman G et al (1990) Seasonal rainfall in the Sinai Desert during the late Quaternary inferred from fluorescent bands in fossil corals. Nature 345:145–147

Knutson DW, Buddemeier RW, Smith SV (1972) Smith S.V., Coral chronometers: Seasonal growth bands in reef corals. Science 177:270–272

Kong ZC, Du NQ (1980) Late-glacial floras in eastern China. J Glaciol Geocryol 2(4):29–32 (in Chinese with English abstract)

Kong ZC, Du NQ, Xu QH et al (1992) Palaeoclimatic fluctuations reflected in flora of Holocene megathermal in the northern part of China. In: Shi YF, Kong ZC. The Climates and Environments of Holocene Megathermal in China. China Ocean Press, Beijing, p 48–65

Krumbien WC (1934) Size frequency distribution of sediments. J Sed Petrol 4:65–77

Kudrass HR, Hofmann A, Doose H et al (2001) Modulation and amplification of climatic changes in the Northern Hemisphere by the Indian summer monsoon during the past 80 k.y. Geology 29:63–66

Kutzbach JE, Street-Perrott AF (1985) Milankovitch forcing of fluctuations in the level of tropical lakes from, 18 to 0 kyr BP. Nature 317:130–134
Lambert F, Delmonte B, Petit JR et al (2008) Dust-climate couplings over the past 800,000 years from the EPICA Dome C ice core. Nature 452:616–619
Lappalainen E (1996) Global peat resources. International Peat Society, Finland
Lars GF (1994) Are wetlands the key to the ice-age cycle enigma? Ambio 23:300–308
LeBec N, Juillet-Leclerc A, Corrège T et al (2000) A coral $\delta^{18}O$ record of ENSO driven sea surface salinity variability in Fiji (south-western tropical Pacific). Geophys Res Lett 27(23):3897–3900
Leuschner DC, Sirocko F (2000) The low-latitude monsoon climate during Dansgaard–Oeschger cycles and Heinrich events. Quat Sci Rev 19:243–254
Li BS, Chen DN, Zhang DD et al (2008b) Fossil gastropods from the MGS3 stratigraphic segment in the Salawusu River Valley and their climatic and environmental implications. Sci Chin (Series D) 51(3):339–348
Li BS, David DZ, Zhou XJ (2002) A Preliminary Study of the Sediments in the Yutian-HotanOasis, South Xinjiang, China. Acta Geol Sin 76(2):221–228
Li BS, Dong GR, Ding TH (1990) A few problems about the sand landforms in the eastern of the Takimakan Desert. Chin Sci Bull 35(23):1815–1818
Li BS, Dong GR, Zhu YZ (1993) Sedimentary environment evolution on the desert and loess in Tarim Basin since the last glacial age. Sci Chin (B) 23(6):644–651
Li BS, Gao QZ, Yan MC (2005b) A Recent Study on Sedimentary Sequence of Southeastern Badain Jaran Desert since 150 ka BP. J Desert Res 25(4):457–465
Li BS, Wen XH, Qiu SF (2007) Phases of Environmental Evolution Indicated by Primary Chemical Elements and Paleontological Records in the Upper Pleistocene-Holocene Series for the Salawusu River Valley, China. Acta Geol Sin 81(4):555–565
Li BS, Wen XH, Zhang DD (2008a) Millennial-Scale Climate Fluctuations During the Last Interstadial Recorded in the AQS3 Segment of The AQiang Loess Section in the North Piedmont of the Kunlun Mountains. Quat Sci 28(1):140–149
Li BS, Wu Z, David DZ (2001) Environment and Its Changes in the Monsoon Sandy Region of China during the Late Pleistocene and Holocene. Acta Geol Sin 75(1):125–137
Li BS, Yan MC, Barry BM (1998) Late Pleistocene and Holocene palaeo climate records from the Badain Jaran Desert, China. Current Res Pleistocene, 15:129–131
Li BS, Zhang D D, Jin HL et al (2000) Palaeo-Monsoon activities of Mu Us Desert, China since 150ka–A study of the stratigraphic sequences of the Milanggouwan section, Salawusu River. Palaeogeogr Palaeoclim Palaeoecol 162:1–16
Li BS, Zhang DD, Wen XH (2005c) Multi-cycles of Climatic Fluctuation in the Last Interglacial Period. Acta Geol Sin 79(3):398–404
Li CH, Wu YH, Hou XH (2011b) Holocene vegetation and climate in Northeast China revealed from Jingbo Lake sediment. Quat Int 229:67–73
Li J, Fang X (1999) Uplift of Tibetan Plateau and environmental changes. Chin Sci Bull 44(23):2117–2124
Li KR, Zhang PY (1992) The Variance and Influence for Chinese Climate. China Ocean Press, Beijing, pp 663 (in Chinese)
Li Q, Liu Y, Cai QF et al (2006a) Reconstruction of annual precipitation since 1686A.D. from Ningwu region, Shanxi Province. Quat Sci 26(6):999–1006 (in Chinese)
Li QS, Zhao WZ, Feng Q (2006b) Relationship between the dynamic change of water resources and the oasis development and evolution in the Heihe river basin. Arid Land Geogr 29(1): 21–28 (in Chinese)
Li WY (1998) Chinese Quaternary vegetation and environment. Beijing: Science Press, p 43–48 (in Chinese)
Li XQ, Zhao K, Dodson J et al (2011a) Moisture dynamics in central Asia for the last 15 kyr: new evidence from Yili Valley, Xinjiang, NW China. Quat Sci Rev 30:3457–3466
Li XQ, Zhou WJ, An ZS et al (2003) The vegetation and monsoon variations at the desert-loess transition belt at Midiwan in northern China for the last 13 ka. The Holocene 13(5):825–830

Li Y, Yao T, Sheng W et al (1999) Influence of arid dust on the chemical records of the Guliya ice cap. Mar Geol Quat Geol 19(2):103–108
Li YC, Xu QH, Zhao YK et al (2005a) Pollen indication to source plants in the eastern desert of China. Chin Sci Bull 50:1632–1641
Li YZ, Li BS, Gao QZ (2005) Fluctuations of Main Chemical Elements since 150 ka BP as Indicated in Chagelebulu Stratigraphical Section, Badain Jaran Desert. J Desert Res 25(1):8–14
Li Z, Yao T, Xie Z et al (1994) Modern atmospheric environmental record from Guliya ice core in Tibetan Plateau. Chin Sci Bull 39(23):2200–2201
Li ZS, Liu GH, Zhang QB et al (2010) Tree ring reconstruction of summer temperature variations over the past 159 years in Wolong National Natural Reserve, western Sichuan, China. Chin Plant Ecol 34(6):628–641 (in Chinese)
Liang EY, Eckstein D, Liu HY (2009b) Assessing the recent grassland greening trend in a long-term context based on tree-ring analysis: a case study in North China. Ecol Indic 9:1280–1283
Liang EY, Shao XM, Kong ZC et al (2003) The extreme drought in the 1920s and its effect on tree growth deduced from tree ring analysis: a case study in North China. Ann Forest Sci 60:145–152
Liang EY, Shao XM, Liu HY et al (2007) Tree-ring based PDSI reconstruction since A.D. 1842 in the Ortindag Sand Land, east Inner Mongolia. Chin Sci Bull 52(19):2715–2721
Liang EY, Shao XM, Qin NS (2008) Tree-ring based summer temperature reconstruction for the source region of the Yangtze River on the Tibetan Plateau. Global Planet Change 61:313–320
Liang EY, Shao XM, Xu Y (2009a) Tree-ring evidence of recent abnormal warming on the southeast Tibetan Plateau. Theor Appl Climatol 98:9–18
Liang Y, Zhang D (2007) Landing typhoon in China during the last millennium and its relationship with ENSO. Adv Clim Change Res 3(2):120–121 (in Chinese)
Liew PM, Lee CY, Kuo CM (2006) Holocene thermal optimal and climate variability of East Asian monsoon inferred from forest reconstruction of a subalpine pollen sequence, Taiwan. Earth Planet Sci Lett 250:596–605
Lim J, Matsumoto E, Kitagawa H (2005) Eolian quartz flux variations in Cheju Island, Korea, during the last 6500 year and a possible Sun–monsoon linkage. Quat Res 64:12–20
Lin ZS, Bian WL, You WH (1996) The wavelets and hierarchies of the climate system. Meteorol Atmos Phys 61:19–26
Linsley BK, Dunbar RB, Wellington GM et al (1994) A coral-based reconstruction of Intertropical Convergence Zone variability over Central America since 1707. J Geophys Res 99:9977–9994
Linsley BK, Wellington GM, Schrag DP (2000) Decadal sea surface temperature variability in the subtropical south Pacific from 1726 to 1997 A.D. Science 290:1145–1148
Lisiecki LE, Raymo ME (2005) A Plio-Pleistocene stack of 57 globally distributed benthic $\delta^{18}O$ records. Paleoceanogr 20 PA1003, doi:10.1029/2004PA001071
Lister GS, Kelts K, Chen KZ et al (1991) Lake Qinghai, China: closed-basin lake levels and the oxygen isotope record for ostracoda since the last Pleistocene. Palaeogeogr Palaeoclim Palaeoecol 84:141–162
Liu HY, Cui HT, Pott R et al (1999) The surface pollen of the woodland-steppe ecotone in southeastern Inner Mongolia, China. Rev Palaeobot Palynol 105:237–250
Liu KB, Shen C, S LK (2001) A 1000-year history of typhoon landfalls in Guangdong, southern China, reconstructed from Chinese historical documentary records. Ann Asso Amer Geogr 91(3):453–464
Liu KB, Sun S, Jiang X (1992) Environmental Change in the Yangtze River Delta since 12,000 yr BP. Quat Res 28:32–45
Liu T (1985) Loess and the Environment. China Ocean Press, Beijing (in Chinese)
Liu W, Liu Z, An Z et al (2011c) Wet climate during the 'Little Ice Age' in the arid Tarim Basin, northwestern China. The Holocene 21(3):409–416
Liu XQ, Ulrike H, Shen J et al (2008a) Holocene environmental and climatic changes inferred from Wulungu Lake in northern Xinjiang, China. Quat Res 70:412–425
Liu Y (1936) Draft of whole records of Gansu (1990) Document series of northwest China (27). Lanzhou Historical document Press, Lanzhou, China (in Chinese)

Liu Y, An Z, Linderholm HW et al (2009a) Annual temperatures during the last 2485 years in the Eastern Tibetan Plateau inferred from tree rings. Sci Chin (Series D) 52(3):348–359

Liu Y, An Z, Ma H et al (2006) Precipitation variation in the northeastern Tibetan Plateau recorded by the tree rings since 850 AD and its relevance to the Northern Hemisphere temperature. Sci Chin (Series D) 49(4):408–420

Liu Y, Bao G, Song HM et al (2009c) Precipitation reconstruction from Hailar pine tree rings in the Hailar region, Inner Mongolia, China back to 1865 A.D. Palaegeogr Palaeoclim Palaeoecol 282:81–87

Liu Y, Cai QF, Liu WG et al (2008c) Monsoon precipitation variation recorded by tree-ring $\delta^{18}O$ in arid Northwest China since 1878 A.D. Chem Geol 252(1–2):56–61

Liu Y, Cai QF, Ma LM et al (2001) Tree ring precipitation records from Baotou and the east Asia Summer Monsoon variations for the last 254 years. Earth Sci Frontiers 8(1):91–97 (in Chinese)

Liu Y, Cai QF, Park WK et al (2003b) Tree-ring Precipitation Records from Baiyinaobao, Inner Mongolia since A.D. 1838. Chin Sci Bull 48(11):1140–1145

Liu Y, Cai QF, Shi JF et al (2005a) Seasonal precipitation in the south-central Helan Mountain region, China, reconstructed from tree-ring width for the past 224 years. Can J Fore Res 35(10):2403–2412

Liu Y, Cai QF, Song HM et al (2011b) Amplitudes, rates, periodicities and causes of temperature variations in the past 2485 years and future trends over the central-eastern Tibetan Plateau. Chin Sci Bull 56(28-29):2986–2994

Liu Y, Hans WL, Song HM et al (2009d) Temperature variations recorded in Pinus tabulaeformis tree rings from the southern and northern slopes of the central Qinling Mountains, central China. Boreas 38(2):285–291

Liu Y, Lei Y, Song HM et al (2010a) The annual mean lowest temperature reconstruction based on Pinus Bungeanas (Pinus bungeana Zucc.) ring width in the Yulin region, Shandong, China since A.D.1616. J Earth Environ (1):28–35 (in Chinese)

Liu Y, Liu N, Song HM et al (2009f) Reconstruction mean air temperature form January to July at the divide sampling site in the Mid-Qinling Mountains with tree-ring widths. Adv Clim Change Res 5(5):260–265 (in Chinese)

Liu Y, Liu WG, Peng ZC et al (2009a) Instability of seawater pH in the South China Sea during the mid-late Holocene: Evidence from boron isotopic composition of corals. Geochim Cosmochim Acta 73(5):1264–1272

Liu Y, Ma LM (1999) Seasonal precipitation reconstruction from tree ring widths for the last 376 years in Hohhot region. Chin Sci Bull 44(18):1986–1991

Liu Y, Ma LM, Cai QF et al (2002) Reconstruction of summer temperature (June–August) at Mt. Helan, China, from tree-ring stable carbon isotope values since A.D. 1890. Sci Chin (D), 45(12):1127–1136

Liu Y, Park WK, Cai QF et al (2003a) Monsoonal precipitation variation in the East Asia since A.D. 1840—Tree ring evidences form China and Korea. Sci Chin (D), 46(10):1031–1039

Liu Y, Peng ZC, Chen TG et al (2008b) The decline of winter monsoon velocity in the South Chin Sea through the 20th century: Evidence from the Sr/Ca records in corals. Global Planet Change 63(1):79–85

Liu Y, Peng ZC, Wei GJ et al (2011a) Interannual variation of rare earth element abundances in corals from northern coast of the South China Sea and its relation with sea-level change and human activities. Mar Environ Res 71:62–69

Liu Y, Shi JF, Shishov V et al (2004b) Reconstruction of May–July precipitation in the north Helan Mountain, Inner Mongolia since A.D. 1726 from tree-ring late-wood widths. Chin Sci Bull 49(4):405–409

Liu Y, Shishov V, Shi JF et al (2004a) The forecast of seasonal precipitation trend at north Helan Mountain, Inner Mongolia for the next 20 years. Chin Sci Bull 49(4):410–415

Liu Y, Sun JY, Song HM et al (2010c) Tree-ring hydrologic reconstructions for the Heihe River watershed, western China since 1430 A.D. Water Res doi:10.1016/j.watres. 2010.02.013

Liu Y, Sun JY, Yang YK et al (2007) Tree-ring precipitation records from inner- mongolia, china, and the east asian summer monsoon front variation since 1627 A.D. Tree-ring Res 63(1):3–14

Liu Y, Tian H, Song HM et al (2009g) Tree ring based reconstruction of the May–June mean temperature since A.D. 1884 in Weichang, Hebei Province, China. Quat Sci 29(5):896–904 (in Chinese)

Liu Y, Tian H, Song HM et al (2010b) Tree-ring precipitation reconstruction in the Chifeng-Weichang region, China, and East Asian summer monsoon variation since A.D. 1777. J Geophys Res. doi:10.1029/2009JD012330

Liu Y, Tian QH, Song HM et al (2009e) Tree-ring width based May–June mean temperature re- construction for Hua Mountains since A.D. 1558 and 20th century warming. Quat Sci 29(5):888–895 (in Chinese)

Liu Y, Wang L, Shi JF et al (2005b) Reconstruction of mean drought index from June to August for the last 207 years using tree-ring data in the northern Helan Mountain. Quat Sci 25(5):540–544 (in Chinese)

Liu Z, Trentesaux A, Clemens S et al (2003c) Clay mineral assemblages in the northern South China Sea: implications for East Asian monsoon evolution over the past 2 million years. Mar Geol 201:133–146

Liu ZF, Colin C, Li XJ et al (2010d) Clay mineral distribution in surface sediments of the northeastern South China Sea and surrounding fluvial drainage basins: Source and transport. Mar Geol 277(1–4):48–60

Long H, Lai ZP, Wang NA et al (2010) Holocene climate variations from Zhuyeze terminal lake records in East Asian monsoon margin in arid northern China. Quat Res 74:46–56

Lou JY, Chen CTA (1997) Paleoclimatological and paleoenvironmental records since 4000a BP in sediments of alpine lakes in Taiwan. Sci Chin (Series D) 40:424–431

Lou JY, Chen CTA, Wann JK (1997) Paleoclimatological records of the Great Ghost Lake in Taiwan. Sci Chin (Series D) 40:284–292

Louie K, Liu K (2003) Earliest historical records of typhoons in China. J Hist Geograph 29:299–316

Lowell TV, Heusser CJ, Anderson BG et al (1995) Interhemispheric correlation of late Pleistocene Glacial Events. Science 269:1541–1549

Lü YX, Li BS, Wen XH (2010) Millennial-centennial scales climate changes of Holocene indicated by magnetic susceptibility of high resolution section in Salawusu River valley. J Geograph Sci 20:241–251

Lückge A, Doose-Rolinski H, Khan AA et al (2001) Monsoonal variability in the northeastern Arabian Sea during the past 5000 years: geochemical evidence from laminated sediments. Palaeogeogr Palaeoclim Palaeoecol 167:273–286

Luo C, Peng ZC, Yang D et al (2009) A lacustrine record from Lop Nur, Xinjiang, China: implications for paleoclimate change during Late Pleistocene. J Asian Earth Sci 34:38–45

Ma C, Wang F, Cao Q et al (2008) Climate and environment reconstruction during the Medieval Warm Period in Lop Nur of Xinjiang, China. Chin Sci Bull 53:3016–3027

Ma C, Zhu C, Zheng C et al (2009) Climate changes in East China since the Late-glacial inferred from high-resolution mountain peat humification records. Sci Chin (Series D) 52:118–131

Ma LH (2007) Thousand-year cycle signals in solar activity. Solar Phys 245(2):411–414

Magny M, Bégeot C, Guiot J et al (2003) Contrasting patterns of hydrological changes in Europe in response to Holocene climate cooling phases. Quat Sci Rev 22:1589–1596

Maher BA (2008) Holocene variability of the East Asian summer monsoon from Chinese cave records: a re-assessment: Holocene 18:861–866

Makohonienko M, Kitagawa H, Fujiki T et al (2008) Late Holocene vegetation changes and human impact in the Changbai Mountains area, Northeast China. Quat Int 184:94–108

Mann ME (2002) Little Ice Age. In: Munn T, eds. Encyclopedia of Global Environmental Change, John Wiley and Sons Ltd, London 1:504–509

Mann ME, Bradley RS, Hughes MK (1998) Global-scale temperature patterns and climate forcing over the past six centuries. Nature 392:779–787

Mann ME, Zhang ZH, Rutherford S et al (2009) Global signatures and dynamical origins of the Little Ice Age and Medieval Climate Anomal. Science 326:1256–1260

Marshall JF, McCulloch MT (2002) An assessment of the Sr/Ca ratio in shallow water hermatypic corals as a proxy for sea surface temperature. Geochim Cosmochim Acta 66(18):3263–3280
Martin P, Hamdallah W, Hosny S (2006) Indications for a humid climate in the Western Desert of Egypt 11~10 Myr ago: evidence from Galagidae (Primates, Mammalia). Comptes Rendus Pale 5(8):935–943
Mashiotta TA, Lea DW, Spero HJ (1999) Glacial-interglacial changes in subantarctic sea surface temperature and δ^{18}O-water using foraminiferal Mg. Earth Planet Sci Lett 170:417–432
Maxwell AL (2001) Holocene monsoon changes inferred from lake sediment pollen and carbonate records, northeastern Cambodia. Quat Res 56:390–400
Mc Manus JF, Bond GC, Broecker WS (1994) High Resolution Climate Record from the North Antactic during the Last Interglacial. Nature 371(64–94):326–327
McConnaughey T (1989a) ^{13}C and ^{18}O isotopic disequilibria in biological carbonates: I. Patterns. Geochim Cosmochim Acta 53:151–162
McConnaughey T (1989b) ^{13}C and ^{18}O isotopic disequilibrium in biological carbonates: II. In vitro simulation of kinetic isotope effects. Geochim Cosmochim Acta 53:163–171
McCulloch M, Fallon S, Wyndham T et al (2003) Coral record of increased sediment flux to the inner Great Barrier Reef since European settlement. Nature 421:727–730
Mebus AG, Martin G, Lautaro N (1999) Radiocarbon Reservoir Effect and the Timing of the Late-Glacial/Early Holocene Humid Phase in the Atacama Desert (Northern Chile). Quat Res 52(2):143–153
Meese DA, Gow AJ, Alley RB et al (1997) The Greenland Ice Sheet Project 2 Depth-age Scale:Methods and results: J Geophy Res 102(C12), p 26411–26423
Mertz-Kraus R, Brachert TC, Reuter M et al (2009) Late Miocene sea surface salinity variability and paleoclimate conditions in the Eastern Mediterranean inferred from coral aragonite ^{18}O. Chem Geol 262:202–216
Meyers PA, Lallier-Vergès E (1999) Lacustrine sedimentary organic matter records of late Quaternary paleoclimates. J Paleolimn 21:345–372
Min GR, Edwards RL, Taylor FW et al (1995) Annual cycles of U/Ca in coral skeletons and U/Ca thermometry. Geochim Cosmochim Acta 59(10):2025–2042
Mischke S, Fuchs D, Riedel F et al (2002) Mid to late holocene palaeoenvironment of lake eastern Juyanze (northwestern China) based on ostracods and stable isotopes. Geobios 35:99–110
Mischke S, Wünnemann B (2006) The holocene salinity history of Bosten Lake (Xinjiang, China) inferred from ostracod species assemblages and shell chemistry: Possible palaeoclimatic implications. Quat Int 154:100–112
Mishima M, Kawahata H, Suzuki A et al (2009) Reconstruction of the east china sea palaeoenvironment at 16 ka by comparison of fossil and modern Faviidae corals from the Ryukyus, southwestern Japan. J Quat Sci 24(8):928–936
Mitsuguchi T, Matsumoto E, Abe O et al (1996) Mg/Ca thermometry in coral skeletons. Science 274:961–963
Moberg A, Sonechkin DM, Holmgren K et al (2005) Highly variable northern hemisphere temperatures reconstructed from low- and high-resolution proxy data. Nature 433:613–617
Moore GW, (1952) Speleothem: a new cave term. Natural Speleology Society News, 10(6):2
Morimoto M, Abe O, Kayanne H (2002) Salinity records for the 1997–98 El Niño from western pacific corals. Geophys Res Lett 29(11). doi:10.1029/2001GL013521
Morimoto M, Kayanne H, Abe O (2007) Intensified mid-holocene Asian monsoon recorded in corals from Kikai Island, subtropical northwestern Pacific. Quat Res 67:204–214
Morley JJ, Heusser LE (1997) Role of orbital forcing in east Asian monsoon climates during the last 350 kyr: evidence from terrestrial and marine climate proxies from core RC14–99. Paleoceanogr 12:483–494
Morrill C, Overpeck JT, Cole JE et al (2006) Holocene variations in the Asian monsoon inferred from the geochemistry of lake sediments in central Tibet. Quat Res 65:232–243
Mu S (1936) Abstract history of Gan Qing Ning area. Junhua Press, Lanzhou, China (in Chinese)
Nakamura N, Kayanne H, Iijima H et al (2009) Mode shift in the Indian Ocean climate under global warming stress. Geophys Res Lett 36(23): doi:10.1029/2009GL040590

Neff U, Burns SJ, Mangini A et al (2001) Strong coherence between solar variability and the monsoon in Oman between 9 and 6 kyr ago. Nature 411:290–293
NGRIP Members (2004) High-resolution record of Northern Hemisphere climate extending into the last interglacial period. Nature 431:147–151
Nichols J E, Booth R K, Jackson S T et al (2006) Paleohydrological reconstruction based on nalkane distributions in ombrotrophic peat. Org Geochem 37:1505–1513
Niu DF, Li BS, Du SH (2008) Cold events of Holocene indicated by primary elements distribution of the high-resolution sand dunes in the Salawusu River Valley. J Geogr Sci 18(1):26–36
Niu DF, Li BS, Wen XH (2011) The holocene ka-scale climate variation indicated by trace elements of the MGS1 segment in the salawusu river valley, China. Acta Geol Sinica 85(2):300–308
Núñez L, Grosjean M, Cartajena I (2002) Human occupations and climate change in the Puna de Atacama, Chile. Science 298:821–824
O'Neil JR, Clayton RN, Mayeda TK (1969) Oxygen isotope fractionation in divalent metal carbonates. J Chem Phys 51(12):5547–5558
Ogilvie AEJ, Jónsson T (2001) "Little Ice Age" research: a perspective from Iceland. Clim Change 48(1):9–52
Ogurtsov M, Sonninen E, Hilasvuori E et al (2010) variations in tree ring stable isotope records from northern Finland and their possible connection to solar activity. J Atmos Sol-Terr Phys 73(2–3):383–387
Oppo DW, Rosenthal Y, Linsley BK (2009) 2,000-year-long temperature and hydrology reconstructions from the Indo-Pacific warm pool. Nature 460:1113–1116
Overpeck J, Anderson D, Trumbore S et al (1996) The southwest Indian Monsoon over the last 18000 years. Clim Dyn 12:213–225
Pachur HJ, Wünnemann B, Zhang H (1995) Lake evolution in the Tengger desert, Northwestern China, during the last 40000 years. Quat Res 44:171–180
Park WK, Jeong WS, Liu Y et al (2001a) Reconstruction of April–August precipitation in Mt. Sorak region from tree rings. Korea J Quat Res 15(1):47–52
Park WK, Kim YJ, Lee JH (2001b) Development of tree-ring chronologies of Pinus densiflora from Mt. Sorak and dating the year of construction of the Kyunghoe-ru Pavilion in Seoul. J Korean Phys Soc 39(4):790–795
Paulsen DE, Li HC, Ku TL (2003) Climate variability in central China over the last 1270 years revealed by high-resolution stalagmite records. Quat Sci Rev 22:691–701
Pausata FSR, Battisti DS, Nisancioglu KH et al (2011) Chinese stalagmite ^{18}O controlled by changes in the Indian monsoon during a simulated Heinrich event: Nat Geosci 4:474–480
Pederson N, Jacoby GC, D'Arrigo RD et al (2001) Hydrometeorological reconstructions for Northeastern Mongolia derived from Tree Rings: A.D. 1651–1995. J Clim 14:872–881
Pelejero C, Calvo E, McCulloch MT et al (2005) Preindustrial to modern interdecadal variability in coral reef pH. Science 309:2204–2207
Peng ZC, Chen TG, Nie BF et al (2003) Coral $\delta^{18}O$ records as an indicator of winter monsoon intensity in the South China Sea. Quat Res 59:285–292
Peng ZC, He XX, Zhang JF et al (2002) Correlation of coral fluorescence with nearshore rainfall and runoff in Hainan Island, South China Sea. Prog Nat Sci 12(1):41–44
Petit JR, Jouzel J, Raynaud D et al (1999) Climate and atmospheric history of the past 420,000 years from the Vostok ice core, Antarctica. Nature 399:429–436
Pfeiffer M, Dullo WC (2006) Monsoon-induced cooling of the western equatorial Indian Ocean as recorded in coral oxygen isotope records from the Seychelles covering the period of 1840–1994 AD. Quat Sci Rev 25:993–1009
Pfeiffer M, Dullo WC, Eisenhauer A (2004) Variability of the Intertropical Convergence Zone recorded in coral isotopic records from the central Indian Ocean (Chagos Archipelago). Quat Res 61:245–255
Port SC, Zhou WJ (2006) Synchronism of Holocene East Asia monsoon variation and North Atlantic drift-ice tracers. Quat Res 65:443–449
Porter S C, Zhou W (2006) Synchronism of Holocene East Asian monsoon variations and North Atlantic drift-ice tracers. Quat Res 65(3):443–449

Porter SC, An ZS (1995) Correlation between climate events in the North Atlantic and China during the last glaciation. Nature 375:185–188
Prell WL (1984) Monsoonal climate of the Arabian Sea during the Late Quaternary: a response to changing solar radiation. In: Berger AL, Imbrie J, Hays J, et al (Eds.), Milankovitch and Climate. D. Riedel, Hingham, pp 349–366
Prell WL, Niitsuma N, Shipboard SP (1989) ODP, Init. Reports 117, College Station, TX
Qian JL, Lu J, Tu QP et al (2002) Reconstruction of the climate in the Tianmu Mountain area, Zhejiang Province, in the last 160 years by ^{13}C sequence of tree ring α-cellulose. Sci Chin (Series D) 45(5):409–419
Qian WH, Lin X (2009) An integrated analysis of dry-wet variability in western China for the last 4–5 centuries. Adv Atmos Sci 26(5):951–961
Qian WH, Lu B (2010) Periodic oscillations in millennial global-mean temperature and their causes. Chin Sci Bull 55(35):4052–4057
Qian WH, Zhu YF, Ye Q (1998) Annual and inter-annual sea surface temperature anomalies in the eastern equatorial of Pacific Ocean. Chin Sci Bull 43:1098–1102
Qin JG, Taylor D, Atahan P et al (2011) Neolithic agriculture, freshwater resources and rapid environmental changes on the lower Yangtze, China. Quat Res 75:55–65
Quinn TM, Crowley TJ, Taylor FW (1996) New stable isotope results from a 173-year coral from Espiritu Santo, Vanuatu. Geophys Res Lett 23:3413–3416
Rasmussen SO (2006) A new Greenland ice core chronology for the last glacial termination. J Geophys Res 111 doi:10.1029/2005JD006079
Raspopov OM, Dergachev VA, Esper J et al (2008) The influence of the de Vries (~200-year) solar cycle on climate variations: Results from the central Asian Mountains and their global link. Palaeogeogr Palaeoclim Palaeoecol 259(1):6–16
Rea DK, Snoeckx H, Joseph LH (1998) Late Cenozoic eolian deposition in the North Pacific: Asian drying, Tibetan uplift, and cooling of the northern hemisphere. Paleoceanogr 13:215–224
Reichart GJ, Lourens LJ, Zachariasse WJ (1998) Temporal variability in the northern Arabian Sea oxygen minimum zone (OMZ) during the last 225,000 years. Paleoceanogr 13:607–621
Reimer PJ, Baillie MGL, Bard E et al (2004) IntCal04 terrestrial radiocarbon age calibration, 0–26 Cal kyr BP. Radiocarbon 46:1029–1058
Ren G, Zhang LS (1998) A preliminary mapped summary of Holocene pollen data for Northeast China. Quat Sci Rev 17:669–688
Ren L, Linsley BK, Wellington GM et al (2002) Deconvolving the ^{18}O seawater component from subseasonal coral ^{18}O and Sr/Ca at Rarotonga in the southern subtropical Pacific for the period 1726–1997. Geochim Cosmochim Acta 67(9):1609–1621
Rendell HM, Lancaster N, Tchakerian VP (1994) Luminescence dating of late quaternary aeolian deposits at Dale Lake and Cronese Mountains, Mojave Desert, California. Quat Sci Rev 13(5–7):417–422
Rhodes TE, Gasse F, Lin R et al (1996) A Late Pleistocene–Holocene lacustrine record from Lake Manas, Junggar (northern Xinjiang, western China). Palaeogeogr Palaeoclim Palaeoecol 120:105–121
Rodgers KB, Cane MA, Schrag DP (1997) Seasonal variability of sea surface 14C in the Equatorial Pacific in an ocean circulation model. J Geophy Res 102:18627–18639
Rohling EJ, Grant K, Hemleben C et al (2008) High rates of sea-level rise during the last interglacial period. Nature Geosci 1(1):38–42
Rohling EJ, Liu QS, Roberts AP et al (2009) Controls on the East Asian monsoon during the last glacial cycle, based on comparison between Hulu Cave and polar ice-core records: Quat Sci Rev 28:3291–3302
Rossignol-Strick (1983) African monsoons, an immediate climate response to orbital insolation. Nature 304:46–49
Rostek F, Bard E, Beaufort L et al (1997) Sea surface temperature and productivity records for the past 240 kyr in the Arabian Sea. Deep-Sea Res II 44:1461–1480

Rozanski K, Araguas-Araguas L, Gonfiantini R (1993) Isotopic patterns in modern global precipitation. In: Climate change in continental isotopic records. (Swart P. K., et al Eds) Washington D.C., (Geophysical Monographs) 78:1–36

Rudaya N, Tarasov P, Dorofeyuk N et al (2009) Holocene environments and climate in the Mongolian Altai reconstructed from the Hoton-Nur pollen and diatom records: a step towards better understanding climate dynamics in Central Asia. Quat Sci Rev 28:540–554

Ruddiman WF (2006) What is the timing of orbital-scale monsoon changes? Quat Sci Rev 25:657–658

Saenger C, Cohen AL, Oppo D et al (2009) Surface-temperature trends and variability in the low-latitude North Atlantic since 1552. Nature Geosci 2:492–495

Sarnthein M, Pflaumann U, Wang PX et al (1994) Preliminary Report on SONNE-94 Cruise "Monitor Monsoon" to the South China Sea. Reports, Geol.-Paläont. Inst. Univ. Kiel, No. 68 pages?

Schlesinger ME, Ramankutty N (1994) An oscillation in the global climate system of period 65–70 years. Nature, 367:723–726

Schrag DP (1999) Rapid analysis of high-precision Sr/Ca ratios in corals and other marine carbonates. Paleoceanogr 14(2):97–102

Schrag DP, Hampt G, Murray DW (1996) Pore fluid constraints on the temperature and oxygen isotopic composition of the glacial ocean. Science 272:1930–1932

Schulz H, von Rad U, Erlenkeuser H (1998) Correlation between Arabian Sea and Greenland climate oscillations of the past 110,000 years. Nature 393:54–57

Schulz H, von Rad U, Ittekkot V (2002) Planktonic foraminifera, particle flux and oceanic productivity off Pakistan, NE Arabian Sea: modern analogues and application to the palaeoclimatic record. In: Clift P, Kroon D, Gaedicke C et al (Eds.), Tectonic and Climatic Evolution of the Arabian Sea Region. Geological Society London, Special Publication, 195, London, pp 499–516

Seki O, Meyers P A, Kawamura K et al (2009) Hydrogen isotopic ratios of plant wax nalkanes in a peat bog deposited in northeast China during the last 16 kyr. Org Geochem 40 (6):671–677

Selvaraj K, Chen CTA, Lou JY et al (2011) Holocene weak summer East Asian monsoon intervals in Taiwan and plausible mechanisms. Quat Int 229:57–66

Shang X, Li XQ, An ZS et al (2009) Modern pollen rain in the Lake Qinghai basin, China. Sci Chin (Series D) 52, 1510–1519

Shao XM, Fan JM (1999) Past climate on west Sichuan Plateau as reconstructed from ring-widths of dragon spruce. Quat Sci (1):81–89 (in Chinese)

Shao XM, Huang L, Liu HB et al (2005) Reconstruction of precipitation variation from tree rings in recent 1000 years in Delingha, Qinghai. Sci Chin (Series D) 48(7):939–949

Shao XM, Wu XD (1994) Tree-ring chronologies for Pinus armandi Franch from Huashan, China. Acta Geographica Sinica, 49(2):174–181 (in Chinese)

Shen C-C, Edwards RL, Cheng H et al (2002) Uranium and thorium isotopic and concentration measurements by magnetic sector inductively coupled plasma mass spectrometry. Chem Geol 185:165–178

Shen CC, Lee T, Liu KK et al (2005b) An evaluation of quantitative reconstruction of past precipitation records using coral skeletal Sr/Ca and $\delta^{18}O$ data. Earth Planet Sci Lett 237:370–386

Shen G (1995) Distribution and evolution of the genus Ephedra in China. Acta Botanica Yunnanica. 17(1):15–20 (in Chinese)

Shen J, Liu XQ, Wang SM et al (2005a) Palaeoclimatic changes in the Qinghai Lake area during the last 18,000 years. Quat Int 136:131–140

Shi JF, Cook ER, Lu HY et al (2010a) Tree-ring based winter temperature reconstruction for the lower reaches of the Yangtze River in southeast China. Clim Res 41:169–175

Shi N (2002) Historical geograpghy study in the Loess Plateau. Yellow River Water Conservancy Press, Zhengzhou, China (in Chinese)

Shi XH, Qin NS, Zhu HF et al (2010b) May–June mean maximum temperature change during 1360–2005 as reconstructed by tree rings of Sabina Tibetica in Zaduo, Qinghai Province. Chin Sci Bull 55(26):3023–3029

Shi YF, Yao TD (2002) MIS3b (54-44 ka BP) cold period and glacial advance in middle and low latitudes. J Glaciol Geocryol 24(1):1–9

Shotyk W, Weiss D, Kramers J D et al (2001) Geochemistry of the peat bog at Etang de la Gruere, Jura Mountains, Switzerland, and its record of atmospheric Pb and lithogenic trace metals (Sc, Ti, Y, Zr, and REE) since 12370 14C yr BP. Geochim Cosmochim Acta 65:2337–2360

Shui T (2001) Papers on the Bronze Age Archaeology of Northwest China. Science Press, Beijing (in Chinese)

Sima A, Paul A, Schulz M (2004) The Younger Dryas — an intrinsic feature of late Pleistocene climate change at millennial timescales. Earth Planet Sci Lett 222:741–750

Singh J, Park WK, Yadav RR (2006) Tree-ring-based hydrological records for western Himalaya, India, since AD 1560. Clim Dyn 26:295–303

Singh J, Yadav RR, Wilmking M (2009) A 694-year tree-ring based rainfall reconstruction from Himachal Pradesh, India. Clim Dyn 33:1149–1158

Sirocko F, Garbe-Schönberg D, McIntyre A et al (1996) Teleconnections between the subtropical monsoons and high latitude clim during the last deglaciation. Science 272:526–529

Sirocko F, Sarnthein M, Erienkeuser H et al (1993) Century-scale events in monsoonal climate over the past 24,000 years. Nature 364:322–324

Sonett CP, Finney SA, Berger A (1990) The Spectrum of Radiocarbon. Phil Trans R Soc Lon 330:413–426

Song HM, Liu Y, Ni WM et al (2007) Winter mean lowest temperature derived from tree-ring width jin Jiuzhaigou region, China since 1750 AD. Quat Sci 27(4):486–491 (in Chinese)

Song SH, Peng ZC, Zhou WJ et al (2012) Variation of the winter monsoon in South China Sea over the past 183 years: Evidence from oxygen isotopes in coral. Global and Planetary Change, 98–99:131–138

Spötl C, Mangini A (2002) Stalagmite from the Austrian Alps reveals Dansgaard-Oeschger events during isotope stage 3: Implications for the absolute chronology of Greenland ice cores. Earth Planet Sci Lett 203:507–518

Spötl C, Mangini A, Richards DA (2006) Chronology and paleoenvironment of Marine Isotope Stage 3 from two high-elevation speleothems, Austrian Alps: Quat Sci Rev 25:1127–1136

Strack M (2008) Peatlands and Climate Change. International Peat Society, International Peat Society, Vapaudenkatu 12, 40100 Jyväskylä, Finland

Stuiver M, Grootes PM, Braziunas TF (1995) The GISP2 $\delta^{18}O$ Climate Record of the Past 16,500 Years and the Role of the Sun, Ocean, and Volcanoes. Quat Res 44(3):341–354

Su RX, Sun DH, Chen H et al (2010) Evolution of Asian monsoon variability revealed by oxygen isotopic record of middle Holocene massive coral in the northern South China Sea. Quat Int 213(1–2):56–68

Sun JY, Liu Y, Cai QF et al (2006) Climatic and hydrological changes of Ejin, Inner Mongolia, China during the past 233 years recorded in tree-rings of Populus euphratica. Quat Sci 26: 799–807 (in Chinese)

Sun JZ (2005) Loessology. Hong Kong Archaeology Society, Hong Kong

Sun X, Luo Y, Huang F et al (2003) Deep-sea pollen from the South China Sea: Pleistocene indicators of East Asian monsoon. Mar Geol 201:97–118

Sun XJ (1989) A restudy of the latest Pleistocene paleovegetation at Beizhuangcun, Shaanxi province, Northern China. Quat Res 2, 177–189 (in Chinese)

Sun XJ, Du NQ, Weng CY et al (1994) Plaeovegetation and paleoenvironment of Manas Lake, Xinjiang, Northwestern China during the last 14000 years. Quat Sci 3:239–248

Sun XJ, Li X (1999) A pollen record of the last 37 ka in deep sea core 17940 from the northern slope of the South China Sea. Mar Geol 156:227–244

Sun XJ, Luo YL (2001) Pollen record of the last 280 ka from deep sea sediments of the northern South China Sea. Sci Chin (Series D) 44(10):879–888

Sun Y, Wang X, Liu Q et al (2010) Impacts of post-depositional processes on rapid monsoon signals recorded by the last glacial loess deposits of northern China. Earth Planet Sci Lett 289:171–179

Sun YB, Wu F, Clemens SC et al (2008) Processes controlling the geochemical composition of the South China Sea sediments during the last climatic cycle. Chem Geol 257(3–4):243–249
Swain AM, Kutzbach JE, Hastenrath S (1983) Estimates of Holocene for Rajasthan, Indian, based on pollen and lake-level data. Quat Res 19:1–17
Tan L, Cai Y, An Z (2010) Precipitation variations of Longxi over the last 2000 years and the possible driving forces, J Arid Land Resour Environ 24:109–116
Tan L, Cai Y, An Z et al (2011) Centennial- to decadal-scale monsoon precipitation variability in the semi-humid region, northern China during the last 1860 years: Records from stalagmites in Huangye Cave. The Holocene 21(2):287–296
Tan L, Cai Y, Cheng H et al (2009) Summer monsoon precipitation variations in central China over the past 750 years derived from a high-resolution absolute-dated stalagmite Palaeogeogr Palaeoclim Palaeoecol 280:432–439
Tan L, Cai Y, Yi L et al (2008) Precipitation variations of Longxi, northeast margin of Tibetan Plateau since AD 960 and their relationship with solar activity. Clim Past 4:19–28
Tan M (2009) Circulation effects: Climatic significance of the short term variability of the Oxygen isotopes in stalagmites from monsoonal China–Dialogue between paleoclimate records and modern climate research. Quat Sci 29(5):851–862
Tan M (2011) Trade-wind driven inverse coupling between stalagmite $\delta^{18}O$ from monsoon region of China and large scale temperature–circulation effect on decadal to precessional timescales. Quat Sci 31(6):1086–1097
Tan M, Liu T, Hou J et al (2003) Cyclic rapid warming on centennial-scale revealed by a 2650-year stalagmite record of warm season temperature. Geophys Res Lett 30(12):1617. doi:1610.1029/2003GL017352
Tan M, Liu T, Zhong H et al (1998) Preliminary study on climatic signals of stable isotopes from Holocene speleothems under monsoon condition Chin Sci Bull 43(6):506–509
Tan M, Liu Tungsheng (1995) The stable isotope record of stalagmite from Jiguan Cave, Henan Province. Acta Geologica Sinica, Supp 1:281–284
Tang X (2006) Study on the north edge of summer monsoon over northwest China. Peking University, Beijing, China
Tao SC, An CB, Chen FH et al (2010) Pollen-inferred vegetation and environmental changes since 16.7 ka BP at Balikun Lake, Xinjiang. Chin Sci Bull 55:2449–2457
Tarasov P, Jin G, Wagner M (2006) Mid-Holocene environmental and human dynamics in northeastern China reconstructed from pollen and archaeological data. Palaeogeogr Palaeoclim Palaeoecol 241:284–300
Thompson LG, Mosley-Thompson E, Davis ME et al (1989) Holocene-Pleistocene climate ice core record from Qinghai-Tibetan. Science, 246:474–477
Thompson LG, Mosley-Thompson E, Wu X et al (1988) Wisconsin/Wurm Glacial Stage Ice in the Subtropical Dunde Ice cap, China. Geo J 17(4):517–523
Thompson LG, Yao T, Davis ME et al (1997) Tropical climate instability: the Last Glacial Cycle from a Qinghai-Tibetan ice core. Science 276:1821–1825
Thompson LG, Yao T, Mosley-Thompson E et al (2000) A high-resolution millennial record of the South Asian Monsoon from Himalayan ice cores. Science 289:1916–1919
Thompson P, Schwarcz HP, Ford DC (1976) Stable isotope geochemistry, geothermometry and geochronology of speleothems from West Virginia. Bull Geol Soc Amer 87:1730–1738
Tian J, Huang EQ, Pak DK (2010) East Asian winter monsoon variability over the last glacial cycle: Insights from a latitudinal sea-surface temperature gradient across the South China Sea. Palaeogeogr Palaeoclim Palaeoecol 292(1–2):319–324
Tian LD, Yao TD, Schuster PF et al (2003) Oxygen-18 concentrations in recent precipitation and ice cores on the Tibetan Plateau. J Gepphys Res,108(D9), 4293. doi:10.1029/ 2002JD00217
Tian QH, Gou XH, Zhang Y et al (2007) Tree ring based drought reconstruction (A.D.1855–2001) for the Qilian mountains, northwestern China. Tree-Ring Res 63(1):27–36
Tian QH, Liu Y, Cai QF et al (2009) The Maximum Temperature of May–July inferred from treeing in Funiu Mountain since 1874A.D. Acta Geographica Sinica, 64(7):879–887 (in Chinese)

Tierney JE, Oppo DW, Rosenthal Y et al (2010) Coordinated hydrological regimes in the Indo-Pacific region during the past two millennia. Paleoceanogr 25, PA1102. doi:10.1029/2009PA001871
Treydte KS, Schleser GH, Helle G et al (2006) The twentieth century was the wettest period in northern Pakistan over the past millennium. Nature 440:1179–1182
Tudhope AW, Chilcott CP, McCulloch MT et al (2001) Variability in the El Niño-Southern Oscillation through a glacial-interglacial cycle. Science 291:1511–1517
Urban FE, Cole JE, Overpeck JT (2000) Influence of mean climate change on climate variability from a 155-year tropical pacific coral record. Nature 407:989–993
Usoskin IG, Solanki SK, Schüssler M et al (2003) Millennium-scale sunspot number reconstruction: evidence for an unusually active sun since the 1940s. Phys Rev Lett 91, doi:10.1103/Phys Rev Lett. 91.211101
van Campo E, Cour P, Hang S (1996) Holocene environmental changes in Bangong Co basin (western Tibet), Part 2: The pollen record. Palaeogeogr Palaeoclim Palaeoecol 120:49–63
van Campo E, Gasse F (1993) Pollen- and diatom-inferred climatic and hydrological changes in Sumxi Co Basin (western Tibet) since 13000 year BP. Quat Res 39:300–313
von Rad U, Schulz H, SONNE 90 Scientific Party (1995) Sampling the oxygen minimum zone off Pakistan: glacial–interglacial variations of anoxia and productivity (preliminary results, SONNE90 cruise). Mar Geol 125:7–19
Wan SM, Li AC, Clift PD et al (2006) Development of the East Asian summer monsoon: Evidence from the sediment record in the South China Sea since 8.5 Ma. Palaeogeogr Palaeoclim Palaeoecol 241(1):139–159
Wang GX, Cheng GD (1998) Changes of hydrology and ecological environment during late 50 years in Heihe river basin. J Desert Res 18 (3):233–238 (in Chinese)
Wang H, Hong Y, Zhu Y et al (2004c) Humification degrees of peat in Qinghai-Xizang Plateau and palaeoclimate change. Chin Sci Bull 49(5):514–519
Wang J (2007) Research of the relationship between climatic changes and wars in China. Zhejiang Normal University, Hanzhou, China
Wang L, Sarnthein M, Erlenkeuser H et al (1999a) Holocene variations in Asian monsoon moisture: a bidecadal sediment record from the South China Sea. Geophys Res Lett 26:2889–2892
Wang L, Sarnthein M, Erlenkeuser H et al (1999b) East Asian monsoon climate during the Late Pleisocene: high-resolution sediment records from the South China Sea. Mar Geol 156:245–284
Wang P, Prell W, Blum P et al (2000) Proceedings of Ocean Drilling Program. Initial Reports 184, College Station, TX
Wang PK, Zhang D (1992) Reconstruction of the 18th century summer precipitation of Nanjing, Suzhou and Hangzhou, China, based on the Clear and Rain Records. In: Bradley RS, Jones PD (eds) Climate Since AD 1500. Routledge, London, UK, p 184–209
Wang PX (1999) Response of Western Pacific marginal seas to glacial cycles: paleoceanographic and sedimentological features. Mar Geol 156:5–39
Wang PX, Clemens S, Beaufort L et al (2005b) Evolution and variability of the Asian monsoon system: state of the art and outstanding issues. Quat Sci Rev 24(5–6):595–629
Wang RL, Scarpitta SC, Zhang SC et al (2002) Later Pleistocene/Holocene climate conditions of Qinghai-Xizang Plateau (Tibet) based on carbon and oxygen stable isotopes of Zabuye Lake sediments. Earth Planet Sci Lett 203:461–477
Wang S, Gong D (2000) Climate in China during the four special periods in Holocene. Progress in Natural Progress 10(5):379–386 (in Chinese)
Wang S, Liu J, Zhou J (2003) Climate and environment in LIA maximum in China. J Lake Sci 15 (4):369–376 (in Chinese)
Wang S, Ye J, Gong D (1998) Climate during the Little Ice Age in China. Quat Sci 1:54–62 (in Chinese)
Wang SM, Yang XD, Ma Y et al (1996) Environmental change of Gucheng lake of Jiangsu in the past 15 ka and its relation to palaeomonsoon. Sci Chin (Series D) 39(2):144–151
Wang WG, Feng ZD, Li XQ et al (2004b) Holocene abrupt climate shifts recorded in Gun Nuur lake core, northern Mongolia. Chin Sci Bull 49:520–526

Wang X, Auler AS, Edwards RL et al (2004a) Wet periods in northeastern Brazil over the past 210 kyr linked to distant climate anomalies. Nature 432:740–743

Wang XC, Zhang QB, Ma KP et al (2008b) A tree-ring record of 500-year dry-wet changes in northern Tibet, China. The Holocene 18(4):579–588

Wang XF, Auler AS, Edwards RL et al (2006) Interhemispheric anti-phasing of rainfall during the last glacial period: Quat Sci Rev 25:3391–3403

Wang XF, Auler AS, Edwards RLet al (2007) Mil- lennial-scale precipitation changes in southern Brazil over the past 90,000 years. Geophys Res Lett 34:L23701. doi:10.1029/2007GL031149

Wang XY (1985) The radioactive age and characteristic of oxygen and carbon isotope composition of the speleothem from the Maomaotou Cave, Guilin. Chin Sci Bull 30(1-2):149–159 (in Chinese)

Wang YJ, Cheng H, Edwards RL et al (2001) A high-resolution absolute-dated Late Pleistocene monsoon record from Hulu Cave, China. Science 294:2345–2348

Wang YJ, Cheng H, Edwards RL et al (2005a) The Holocene Asian Monsoon: Links to solar changes and North Atlantic climate. Science 308:854–857

Wang YJ, Cheng H, Edwards RL et al (2008a) Millennial- and orbital-scale changes in the East Asian monsoon over the past 224,000 years. Nature 451:1090–1093

Wang Z, Mysak LA (2006) Glacial abrupt climate changes and Dansgaard-Oeschger oscillations in a coupled climate model. Paleoceanogr 21:PA2001. doi:10.1029/2005PA001238

Weber JN, Woodhead PMJ (1972) Temperature dependence of Oxygen-18 concentration in reef coral carbonates. J Geophy Res 77(3):463–473

Webster PJ, Magana VO, Palmer TN et al (1998) Monsoons: processes, predictability, and the prospects for prediction. J Geophy Res 103(C7):14451–14510

Wehausen R, Brumsack HJ (2002) Astronomical forcing of the East Asian monsoon mirrored by the composition of Pliocene South China Sea sediments. Earth Planet Sci Lett 201:621–636

Wei GJ, Deng WF, Yu KF et al (2007) Sea surface temperature records in the northern South China Sea from mid-Holocene coral Sr/Ca ratios. Paleoceanogr 22:PA3206. doi: 10.1029/2006PA001270

Wei GJ, Liu Y, Li XH et al (2003b) High resolution elemental records from the South China Sea and their paleoproductivity implications. Paleoceanogr 18:1054–1065

Wei GJ, Liu Y, Li XH et al (2004) Major and trace element variations of the sediments at ODP Site 1144, South China Sea, during the last 230 ka and their paleoclimate implications. Palaeogeogr Palaeoclim Palaeoecol 212:331–342

Wei GJ, McCulloch MT, Mortimer G et al (2009) Evidence for ocean acidification in the Great Barrier Reef of Australia. Geochim Cosmochim Acta 73:2332–2346

Wei K, Gasse F (1999) Oxygen isotopes in lacustrine carbonates of West China revisited: implications for post glacial changes in summer monsoon circulation. Quat Sci Rev 18:1315–1334

Wei KY, Chiu TC, Chen YG (2003a) Toward establishing a maritime proxy record of the East Asian monsoon. Mar Geol 201:67–79

Weiss H, Bradley RS (2001) What drives societal collapse? Science 291(5504):609–610

Wellington GM, Dunbar RB, Merlen G (1996) Calibration of stable oxygen isotope signatures in Galapagos corals. Paleoceanogr 11(4):467–480

Wen H (1959) Climate in the mid-low Yellowe River Basin during the Qin and Han Dynasties. Shangwu Press, Shanghai, China (in Chinese)

Wen RL, Xiao JL, Chang ZG et al (2010a) Holocene precipitation and temperature variations in the East Asian monsoonal margin from pollen data from Hulun Lake in northeastern Inner Mongolia, China. Boreas 39:262–272

Wen RL, Xiao JL, Chang ZG et al (2010b) Holocene climate changes in the mid-high latitude monsoon margin reflected by the pollen record from Hulun Lake, northeastern Inner Mongolia. Quat Res 73:293–303

Wen XH, Li BS, Li S (2005) Grain-size Characteristics of Sand Dunes in the Ejin Oasis since 2.5 ka BP and Related Sedimentary Process. Acta Geol Sin 79(5):710–718

Wen XH, Li BS, Wang W (2006) Deposition of sandstorms in a vegetation covered sand dune in Ejin Oasis and its characreristics. J Geograph Sci 16(4):502–508
Wen XH, Li BS, Zheng YM (2009) Climate variability in the Salawusu River valley of the Ordos Plateau (Inner Mongolia, China) during Marine Isotope Stage 3. J Quat Sci 24(1):61–74
Weng CY, Sun XJ, Chen YS (1993) Numerrical characteristics of pollen assemblages of summer samples from the west Kunlun mountains. Acta Botanica Sinica 35: 69–79 (in Chinese with English abstact)
Wu FL, Fang XM, Ma YZ et al (2007) PlioQuaternary stepwise drying of Asia: Evidence from a 3-Ma pollen record from the Chinese Loess Plateau. Earth Planet Sci Lett 257:160–169
Wu G, Yao T (2005) A comparison of the records of dust originated from Central Asia since the Last Interglacial. J Glaciol Geocryol 27(1):80–87
Wu GG, Yao TD, Thompson LG et al (2004b) Microparticle record in the guliya ice core and its comparison with polar records since the last interglacial. Chin Sci Bull 49(6):607–611
Wu JL, Liu JJ, Wang SM (2004a) Climatic change record from stable isotopes in lake aibi, xinjiang during the past 1 500 years. Quat Sci 24:585–590
Wu JL, Shen J, Wang SM et al (2005) Characteristic of an early Holocene climate and environment from lake sediments in Ebinur region, NW China. Sci China (D) 48:258–265
Wu WX, Liu TS (2004) Variations in East Asia monsoon around 4 000a BP and the collapse of Neolithic cultures around central plain. Quat Sci 24(3):278–284 (in Chinese with English Abstract)
Wu X, Niu Z, Wang S (1994) Historical changes of environment and water and sediment in the Yellow River basin. In: Ye Q (ed) Researches on environment changes of the Yellow River basin and law of water and sediment transportation. Shandong Science and Technology Press, Jinan, China, pp 30–57 (in Chinese)
Wu XD, Shao XM (1995) Status and prospects of dendrochronological study in Tibetan Plateau. Dendrochronologia 13:89–98
Wu XH (1985) A study of palaeotemperatures research recorded by the Pleistocene Picea-Abies floras in east and southwest China. Bull Institute of Geomechanics CAGS 6:155–166 (in Chinese with English abstract)
Wünnemann B, Chen FH, Riedel F et al (2003) Holocene lake deposits of Bosten Lake, southern Xinjiang, China. Chin Sci Bull 48(14):1429–1432
Xia YM (1988) A preliminary study of the vegetation development and climatic changes in the Sanjiang Plain since 12,000 year BP. Scienta Geographica Sinica 8:240–249 (in Chinese)
Xiao JL, Chang ZG, Si B et al (2009) Partitioning of the grain-size components of Dali Lake core sediments: evidence for lake-level changes during the Holocene. J Paleolimn 42:249–260
Xiao JL, Inouchi Y, Kumai H et al (1997) Eolian quartz flux to Lake Biwa, central Japan, over the past 145,000 years. Quat Res 48:48–57
Xiao JL, Porter SC, An ZS et al (1995) Grain Size of Quartz as an indicator of Winter Monsoon Strength on the Loess Plateau of Central China during the Last 130,000Yr: Quat Res 43:22–29
Xiao JL, Si B, Zhai DY et al (2008) Hydrology of Dali Lake in central-eastern Inner Mongolia and Holocene East Asian monsoon variability. J Paleolimn 40:519–528
Xiao JL, Wu JT, Si B et al (2006) Holocene climate changes in the monsoon/arid transition reflected by carbon concentration in Daihai Lake of Inner Mongolia. The Holocene 16:551–560
Xiao JL, Xu QH, Nakamura T et al (2004) Holocene vegetation variation in the Daihai Lake region of north-central China: a direct indication of the Asian monsoon climatic history. Quat Sci Rev 23:1669–1679
Xiao JY, Wang J, An ZS et al (2000) Evidence for the Younger Dryas event in the eastern part of Nanling region. Acta Micropalaeontologica Sinica 17(2):171–177
Xiao XY, Shen J, Wang SM et al (2007) Palynological evidence for vegetational and climatic changes from the HQ deep drilling core in Yunnan Province, China. Sci Chin (Series D) (8):1189–1201
Xie S, Nott CJ, Avsejs LA et al (2000) Palaeoclimate records in compound-specific δD values of a lipid biomarker in ombrotrophic peat. Org Geochem 31 (10):1053–1057

Xu DK, Lu HY, Wu NQ et al (2010) 30000-Year vegetation and climate change around the East China Sea shelf inferred from a high-resolution pollen record. Quat Int 227(1):53–60

Xu J (1955) Study on climatic record during historic period of Huang-Huai River plain. Acta Geographica Sinica 21(2):181–190 (in Chinese)

Xu GC (1997) Climate Change in Arid and Semiarid Regions of China. China Meteorological Press, Beijing, pp 101 (in Chinese)

Xu H, Hong YT, Lin QH et al (2002) Temperature variations in the past 6000 years inferred from delta O-18 of peat cellulose from Hongyuan, China. Chin Sci Bull 47(18):1578–1584

Xu H, Hong YT, Lin QH, et al (2006) Temperature responses to quasi-100-yr solar variability during the past 6000 years based on $\delta^{18}O$ of peat cellulose in Hongyuan, eastern Qinghai–Tibet plateau, China. Palaeogeogr Palaeoclim Palaeoecol 230(1–2):155–164

Xu JH (1998) Sun, climate, famine and folk nation transfer (in Chinese). Sci Chin (Series D) 28 (4):366–384

Xu Q, Xiao J, Toshio N et al (2004) Climate changes of Daihai Basin during the past 1500 from a pollen record. Quat Sci 24(3):341–347 (in Chinese)

Xu QH, Li YC, Yang XL et al (2007) Quantitative relationship between pollen and vegetation in northern China. Sci Chin (Series D) 50:582–599

Xu QY, Yang X, Zheng Z (2005) Study on surface pollen of major steppe communities in northern China. Geogr Res 24(3):394–402 (in Chinese)

Xu R (1980) The Picea-Abies Flora Complexes in the Pleistocene of China and their Significance toward Quaternary Research. Quat Sci 5(1) (in Chinse), pages

Yadav RR, Park WK, Bhattacharyya A (1997) Dendroclimatic reconstruction of April–May temperature fluctuations in the Western Himalaya of India Since AD 1698. Quat Res 48:187–191

Yadav RR, Park WK, Singh J et al (2004) Do the western Himalayas defy global warming. Geophys Res Lett 31: doi:10.1029/2004GL02020

Yadav R, Park W-K (2000) Precipitation reconstruction using ring-width chronology of Himalayan cedar from western Himalaya: Preliminary results. Journal of Earth System Science 109:339–345

Yadav RR, Singh J (2002) Tree-Ring-Based spring temperature patterns over the past four centuries in Western Himalaya. Quat Res 57:299–305

Yamada K (2004) Last 40 ka climate changes as deduced from the lacustrine sediments of Lake Biwa, central Japan. Quat Int 123–125:43–50

Yamamoto S, Kawamura K, Seki O et al (2010a) Environmental influences over the last 16 ka on compound-specific $\delta^{13}C$ variations of leaf wax n-alkanes in the Hani peat deposit from northeast China. Chem Geol 277(3–4):261–268

Yamamoto S, Kawamura K, Seki O et al (2010b) Paleoenvironmental significance of compound-specific $\delta^{13}C$ variations in n-alkanes in the Hongyuan peat sequence from southwest China over the last 13 ka. Org Geochem 41(5):491–497

Yan G, Wang FB, Shi GR et al (1999) Palynological and stable isotopic study of palaeoenvironmental changes on the northeastern Tibetan Plateau in the last 30,000 years. Palaeogeogr Palaeoclim Palaeoecol 153:147–159

Yan MC, Wang GQ, Li BS (2001) Formation and growth of high megadunes in Badain Jaran Deser. Acta Geographica Sinica 56(1):83–91

Yan Z, Ye D, Wang C (1992) Climatic jumps in the flood/drought historical chronology of Central China. Clim Dyn 6:153–160

Yancheva G, Nowaczyk NR, Mingram J et al (2007) Influence of the intertropical convergence zone on the East Asian monsoon. Nature 445:74–77

Yang QS, Xing YW, Zhou ZK (2009) Modern geographical distribution of Tsuga and its climatic condition in the Asian monsoon region. Acta Botanica Yunnanica 31:389–398 (in Chinese with English abstract)

Yang XD, Wang SM, Tong GB et al (1998) The Late Pleistocene pollen record in the lake sediments from ancient Heqing lake and its significance for palaeoclimate. Quat Sci 4: 335–343 (in Chinese with English abstract)

Yang XP (2000) Landscape types and its formation mechanism in the Badain Jaran Desert and its surrounding areas. J Desert Res 20(2):166–170

Yang XP, Frank P, Ulrich R (2006) Late Quaternary environmental changes in the Taklamakan Desert, western China, inferred from OSL-dated lacustrine and aeolian deposits. Quat Sci Rev 25:923–932

Yang Y (2007) The preliminary study in reconstructing the mean winter temperature in Kunming during 1721–1900 AD. J Chin Hist Geograph 22(1):17–31(in Chinese)

Yang Y, Li BS, Qiu SF (2008) Climatic changes indicated by trace elements in the Chagelebulu stratigraphic section, Badain Jaran Desert, China, since 150kyr BP. Geochem Int 46(1):96–103

Yao T (1999) Abrupt climatic changes on the Tibetan Plateau during the Last Ice Age--Comparative study of the Guliya ice core with the Greenland GRIP ice core. Sci Chin (Series D) 42(4):358–368

Yao TD, Duan KQ, Xu BQ et al (2002) Temperature and methane changes over the past 1000 years recorded in Dasuopu glacier (central Himalaya) ice core. Ann Glaciol 35:379–383

Yao TD, Jiao K, Zhang X et al (1992) Glaciologic studies on Guliya Ice Cap. J Glaciol Geocryol 14(3):233–241

Yao TD, Thompson LG, Shi YF et al (1997) Climate variation since the Last Interglaciation recorded in the Guliya ice core. Sci Chin (Series D) 40(6):662–668

Yao TD, Thompson LG, Thompson EM et al (1996a) Climatological significance of delta O-18 in north Tibetan ice cores. J Geophys Res 101(D23):29531–29537

Yao TD, Thompson LG, Qin DH et al (1996b) Variations in temperature and precipitation in the past 2000 years on the Xizang (Tibet) Plateau–Guliya ice core record. Sci Chin (Series D) 39(4):425–433

Yao TD, Wu GJ, Pu JC et al (2004) Relationship between calcium and atmospheric dust recorded in Guliya ice core. Chin Sci Bull 49(7):706–710

Yi S, Saito Y, Oshima H et al (2003) Holocene environmental history inferred from pollen assemblages in the Huanghe (Yellow River) delta, China: climatic change and human impact. Quat Sci Rev 23:609–628

Yin S (1999) Peat resources of China. Earth Sci Frontiers 6(supp):116-124 (in Chinese with English abstract)

Yu KF, Hua Q, Zhao JX et al (2010b) Holocene marine ^{14}C reservoir age variability: Evidence from ^{230}Th- dated corals in the South China Sea. Paleoceanogr 25:PA3205, doi: 10.1029/2009PA001831

Yu KF, Zhao JX, Done T et al (2009) Microatoll record for large century-scale sea-level fluctuations in the mid-Holocene. Quat Res 71:354–360

Yu KF, Zhao JX, Wei GJ et al (2005) Mid-late Holocene monsoon climate retrieved from seasonal Sr/Ca and $\delta^{18}O$ records of Porites lutea corals at Leizhou Peninsula, northern coast of South China Sea. Global Planet Change 47:301–316

Yu PS, Huang CC, Chin Y et al (2006c) Late Quaternary East Asian Monsoon variability in the South China Sea: evidence from planktonic foraminiferal faunal and hydrographic gradient records. Palaeogeogr Palaeoclim Palaeoecol 236:74–90

Yu XF, Liu Z (2010) Surface texture and mineralogical characters of the peat in Hongyuan Swamp. J Earth Environ 1(2):122–125 (in Chinese with English abstract)

Yu XF, Zhou W, Liu Z et al (2011) Different patterns of changes in the Asian summer and winter monsoons on the eastern Tibetan Plateau during the Holocene. The Holocene 21(7):1–6

Yu XF, Zhou WJ, Franzen LG et al (2006b) High-resolution peat records for Holocene monsoon history in the eastern Tibetan Plateau. Sci Chin (Series D) 49(6):615–621

Yu XF, Zhou WJ, Liu XQ et al (2010a) Peat records of human impacts on the atmosphere in Northwest China during the late Neolithic and Bronze Ages. Palaeogeogr Palaeoclim Palaeoecol 286 (1–2):17–22

Yu Y, Yang TB, Li JJ et al (2006a) Millennial-scale Holocene climate variability in the NW China drylands and links to the tropical Pacific and the North Atlantic. Palaeogeogr Palaeoclim Plalaeoecol 233:149–162

Yuan DX (1995) The sensitivity and records of karstification to the environmental changes. Chin Sci Bull 40(13):1210–1213

Yuan D, Cheng H, Edwards RL et al (2004) Timing, Duration, and Transitions of the Last Interglacial Asian Monsoon. Science 304:575–578

Yuan DX, Qin JM, Lin YS (1999) High-resolution paleoenvironmental reconstruction up to 20 ka BP with speleothem from Guilin. Guilin: Guangxi Normal University Press, 32–56 (in Chinese)

Zahn R (2003) Global change: Monsoon linkages. Nature 421:324–325

Zhang D (1980) A preliminary study on winter temperature changes in southern China during the last 500 years. Chin Sci Bull 6:270–272 (in Chinese)

Zhang D (1983) The method for reconstruction of climate series for the last 500 years and its reliability In: Bureau CM (ed) Collected Papers of Meteorological Science and Technology (4) Meteorological Press (in Chinese), Beijing, China, pp 17–26 (in Chinese)

Zhang D (1984) Synoptic-climatic studies of dust fall in China since the historic times.Sci Chin (Series B) 24(3):278–288 (in Chinese)

Zhang D (1991) Climate in LIA in China and its relationship with global changes. Quat Sci 2:104–111 (in Chinese)

Zhang D (2004) Collection of weather records in China in the last 3000 years. Jiangsu Education Press, Nanjing, China (in Chinese)

Zhang D, Liu C, Jiang J (1997) Reconstruction of dry/wet series in six regional of Eastern China and the abrupt changes analysis during the last 1000 years. Quat Sci 17(1):1–11 (in Chinese)

Zhang D, Liu Y (2002) A new approach to the reconstruction of temporal rainfall sequences from 1724–1904 Qing-Dynasty weather records for Beijing. Quat Sci 22(3):199–208 (in Chinese)

Zhang D, Sun X (2001) Variations of the southern limits of dust fall records and inferred historical climate drought in Northern China. Quat Sci 21(1):1–8 (in Chinese)

Zhang GH, Liu SY, Zhang CY et al (2004a) Evolution of groundwater circulation in the Heihe River drainage area. Geol Chin 31(3):289–293 (in Chinese)

Zhang HC, Pachur HJ, Wunnemann B (2004b) Late Quaternary development and levels of palaeo lakes in Tengger desert, NW China. Palaeogeor Palaeoclim Palaeoecol 211:41–58

Zhang J, Zhang X (1979) Climatic fluctuations in the last 500-years in China and their interdependence. Acta Meteorologica Sinica 37(2):50–57 (in Chinese)

Zhang P (1996) Historical climate variations in China. Shandong Science and Technology Press, Jinan, China (in Chinese)

Zhang P, Gong G (1979) Climate in China since 16th century. Acta Geographica Sinica 34 (3):238–247 (in Chinese)

Zhang P, Wang Z, Liu X et al (1994b) The modes of climate change for the past 2000 years in China. Sci Chin (Series B) 24:998–1008 (in Chinese)

Zhang PX, Zhang PZ, Qian GM et al (1994a) The study of paleoclimatic parameter of Qinghai Lake since Holocene. Quat Sci 3:225–228

Zhang PZ, Cheng H, Edwards RL et al (2008) A test of climate, sun, and culture relationships from an 1810-Year Chinese cave record. Science 322:940–942

Zhang Q, Fei J (2004) Statistics and analysis on the dust weathers in the Loess-distributed regions in northwest and north China during the Ming and Qing Dynasties. Arid Zone Res 21 (2):108–121 (in Chinese)

Zhang Q, Jiang T, Liu C et al (2004c) A preliminary study on climate changes and flood events in Hai'an area during the last 1000 years. J Natural Disaster 13(3):127–132 (in Chinese)

Zhang QB, Cheng GD, Yao TD et al (2003) A 2326-year tree-ring record of climate variability of the northeastern Qinghai-Tibetan Plateau. Geophys Res Lett 30(14):1739–1742

Zhang R, Delworth TL (2005) Simulated Tropical Response to a Substantial Weakening of the Atlantic Thermohaline Circulation. J Clim 18:1853–1860

Zhang SY, Zhao SS, He YB (1985) ^{230}Th - ^{234}U ages of speleothems in eastern mainland of China and paleoenvironment studies. Earth Sci 10(1):65–72 (in Chinese)

Zhao M, Huang CY, Wang CC et al (2006) A millennial-scale U_{37}^{K} sea-surface temperature record from the South China Sea (8°N) over the last 150 kyr: Monsoon and sea-level influence. Paleogeogr Paleoclim Paleoecol 236(1–2):39–55

Zhao SS, Liu ML, Qiao GS (1989) The U-series ages of speleothem of karst caves in east China. Carsologica Sinica 8(1):53–61 (in Chinese with English abstract)

Zhao SY (1984) Brief nature disaster records in each period of Gansu province. Gansu People's Press, Lanzhou, China (in Chinese)

Zhao Y, Herzschuh U (2009) Modern pollen representation of source vegetation in the Qaidam Basin and surrounding mountains, north-eastern Tibetan Plateau. Vegetation History and Archaeobotany 18:245–260

Zhao Y, Sun QF (2010) Reliability of pollen concentration as the indicator of effective moisture in arid and semi-arid regions of China. J Arid Environ 74:423–427

Zhao Y, Yu ZC, Chen FH (2009) Spatial and temporal patterns of Holocene vegetation and climate changes in arid and semi-arid China. Quat Int 194(1–2):6–18

Zhao Y, Yu ZC, Zhao WW (2011) Holocene vegetation and climate histories in the eastern Tibetan Plateau: controls by insolation-driven temperature or monsoon-derived precipitation changes? Quat Sci Rev 30:1173–1184

Zheng J, Ge Q, Fang X et al (2007) Comparison on temperature series reconstruction from historical documents in China for the last 2000 years. Acta Meteorologica Sinica 65(3):428–439 (in Chinese)

Zheng J, Ge Q, Hao Z (2003) Reconstruction of winter mean temperature series from 1736 to 1999 in Xi'an and Hanzhong. Geograph Res 22(3):343–348 (in Chinese)

Zheng J, Hao Z, Ge Q (2005) Variation of precipitation for the last 300 years over the middle and lower reaches of the Yellow River. Sci Chin (Series D) 48(12):2182–2193

Zheng J, Zhang P, Ge Q et al (2001) Centennial changes of drought/ flood spatial pattern in Eastern China for the last 2 000 years. Prog Nat Sci 11(1):65–70 (in Chinese)

Zheng S, Zhang F, Gong G (1977) The change of the moist state of climate in southeastern China during recent 2000 years. In: Central Meteorological Bureau (ed) Proceedings of symposium on climate change and prediction for a special long time. Science Press Beijing, China, pp 50–57 (in Chinese)

Zheng Y, Xie S, Liu X et al (2009a) n-alkanol ratios as proxies of paleovegetation and paleoclimate in a peat-lacustrine core in southern China since the last deglaciation. Frontiers of Earth Science in China 3(4):445–451

Zheng Y, Zhou W, Liu X et al (2011a) n-Alkan-2-one distributions in a northeastern China peat core spanning the last 16 kyr. Org Geochem 42(1):25–30

Zheng Y, Zhou W, Liu Z et al (2011b) The n-alkanol paleoclimate records in two peat deposits: a comparative study of the northeastern margin of the Tibetan Plateau and Northeast China. Environ Earth Sci 63:135–143

Zheng Y, Zhou W, Xie S et al (2009b) A comparative study of n-alkane biomarker and pollen rcords: an example from southern China. Chin Sci Bull 54(6):1065–1072

Zheng Z, Huang KY, Xu QH et al (2008) Comparison of climatic threshold of geographical distribution between dominant plants and surface pollen in China. Sci Chin (Series D) 51:1107–1120

Zheng Z, Wang JH, Wang B et al (2003) Holocene high-resolution environmental record from Shuangchi Maar Lake in Hainan Island. Chin Sci Bull 48:282–286

Zhong W, Wang LG, Tyip T et al (2004) Possible solar forcing of climate variability in the past 4000 years inferred from a proxy record at the southern margin of Tarim Basin. Chin Sci Bull 49(11):1190–1195

Zhong W, Xue JB, Li XD et al (2010) A Holocene climatic record denoted by geochemical indicators from Barkol Lake in the northeastern Xinjiang, NW China. Geochem Int 48:792–800

Zhong W, Xue JB, Zheng YM et al (2010a) Variations of monsoonal precipitation over the last 16,000 years in the eastern Nanling Mountains, South China. J Paleolimnol 44(1):177–188

Zhong W, Xue JB, Zheng YM et al (2010b) Climatic changes since the last deglaciation inferred from a lacustrine sedimentary sequence in the eastern Nanling Mountains, south China. J Quat Sci 25(6):975–984

Zhou Q, Zhang P, Wang Z (1994b) Reconstruction of yearly winter temperature in Hefei area during 1736–1991 AD. Acta Geographica Sinica 49(4):332–337 (in Chinese)

Zhou WJ, An ZS (1996) Discussion on the Reliability of ^{14}C Dating of pollen concentrates from loess-paleosol sequences by accelerator mass spectrometry. Radiocarbon 38(1):132

Zhou WJ, An ZS (1991a) ^{14}C chronology of loess plateau in China, in Quaternary Geology and Environment in China. in Liu TS, Editor. Science Press pp 192–200

Zhou WJ, An ZS, Head MJ (1994a) Stratigraphic division of Holocene loess in China. Radiocarbon 36(1):37–45

Zhou WJ, An ZS, John H et al (1991b) ^{14}C dating and measurements of climatic proxy indices of loess sequence to record paleomonsoon variation on the loess plateau of China during the last 16,000 years. Radiocarbon 33(2):259

Zhou WJ, An ZS, Porter SC et al (1997) Correlation of climatic events between East Asia and Norwegian Sea during last deglaciation. Sci Chin (Series D) 40(5):496–501

Zhou WJ, An ZS, Lin BH (1992) Chronology of the Baxie loess profile and the history of monsoon climates in China between 17,000 and 6000 years BP. Radiocarbon 34(3):818–825

Zhou WJ, Dodson J, Head MJ et al (2002b) Environmental variability within the Chinese desert-loess transition zone over the last 20,000 years. The Holocene 12(1):107–112

Zhou WJ, Head MJ, An ZS et al (2001) Terrestrial evidence for a spatial structure of tropical-polar interconnections during the Younger Dryas episode. Earth Planet Sci Lett 191:231–239

Zhou WJ, Head MJ, Lu XF et al (1999) Teleconnection of climatic events between East Asia and polar, high latitude areas during the last deglaciation. Palaeogeogr Palaeoclim Palaeoecol 152:163–172

Zhou WJ, Li XQ, Dung GR et al (1996) The high resolution peat records from desert loess transition zone during the Younger Dryas: The eveidence of eastern Asian monsoon vibration. Sci Chin (Series D) 26(2):118–124 (in Chinese)

Zhou WJ, Lu XF, Wu ZK et al (2002a) Peat record reflecting Holocene climatic change in the Zoigê Plateau and AMS radiocarbon dating. Chin Sci Bull 47(1):66–70

Zhou WJ, Song SH, Burr GS et al (2007) Is there the time-transgressive Holocene Optimum in East Asian monsoon area? Radiocarbon 49(2):865–875

Zhou WJ, Xie SC, Meyers PA et al (2005) Reconstruction of Late-Glacial and Holocene Climate Evolution in Southern China from Geolipids and Pollen in the Dingnan Peat Sequence. Org Geochem 36(9):1272–1284

Zhou WJ, Yu SY, Georges B et al (2010a) Postglacial changes in the Asian summer monsoon system: a pollen record from the eastern margin of the Tibetan Plateau. Boreas 39:528–539

Zhou WJ, Yu XF, Timothy Jull AJ et al (2004) Highresolution evidence from southern China of an early Holocene optimum and a mid-Holocene dry event during the past 18,000 years. Quat Res 62:39–48

Zhou WJ, Zheng YH, Meyers PA et al (2010b) Postglacial climate-change record in biomarker lipid compositions of the Hani peat sequence, Northeastern China. Earth Planet Sci Lett 294 (1-2):37–46

Zhou XJ, Zhao P, Liu G (2009) Asian-Pacific Oscillation index and variation of East Asian summer monsoon of the past millennium. Chin Sci Bull 54:3768–3771

Zhu C, Ma C, Yu S et al (2010) A detailed pollen record of vegetation and climate changes in Central China during the past 16000 years. Boreas 39(1):69–76

Zhu HF, Fang XQ, Shao XM et al (2009) Tree ring-based February–April temperature reconstruction for Changbai Mountain in Northeast China and its implication for East Asian winter monsoon. Clim past 5:661–666

Zhu HS, Zhang Y (1992) The speleothem recorded paleo-temperature changes in Beijing during the past 440 ka BP. Chin Sci Bull 20:1880–1883

Zhu KZ (1972) A preliminary study on climate changes since the last 5000 years in China. Acta Archaeologica Sinica, (1):168–189

Zhu P (1989) Historical geography of Shanghai. Xuelin Press, Shanghai, China (in Chinese)

Zhu K (1973) Climatic changes in the past 5000 years in China. Sci Chin 3:1–5

Zhu Y (2008) An index of East Asian winter monsoon applied to description the Chinese mainland winter temperature changes. Acta Meteorologica Sinica 66(5):781–788

Zhu ZD, Wu Z, Liu S (1980) Introduction of Desert Research in China (Revision). Beijing: Sci Press, 68–76

Chapter 4
Asian Dust, Eolian Iron and Black Carbon—Connections to Climate Changes

Xiaoye Zhang, Yongming Han, Youbin Sun, Junji Cao and Zhisheng An

Abstract This chapter focuses on the aridification of the Asian interior during the Late Cenozoic, with an emphasis on modern dust and black carbon, and examines global iron connections which originate from dust derived from the arid interior of Asia. The modern formation of Asian dust, the connection of black carbon, paleofires and climate change, and relationships among dust, global iron cycles, marine productivity and climate change were discussed. The aridification history of the Asian interior were reconstructed from Chinese loess record.

Keywords Dust sources · Black carbon · Soot · Iron loading · Asian interior · Eolian deposits

4.1 Introduction

The sediments deposited on the LP are eolian in origin. Modern observations of windblown dust can help us understand how the LP formed, why its characteristics vary, and how these topics relate to changes in our climate. Key factors and

X. Zhang (✉)
Chinese Academy of Meteorological Sciences, 100081, Beijing, China
e-mail: xiaoye@cams.cma.gov.cn

Y. Han · Y. Sun · J. Cao · Z. An
State Key Laboratory of Loess and Quaternary Geology, Institute of Earth Environment, Chinese Academy of Sciences, No. 10 Fenghui S. Rd, High-Tech Zone, Xi'an,
PO Box 17, 710075, Xi'an, China

Y. Han
email: yongming@ieecas.cn

Y. Sun
email: sunyb@ieecas.cn

J. Cao
e-mail: cao@loess.llqg.ac.cn

Z. An
e-mail: anzs@loess.llqg.ac.cn

Z. An (ed.), *Late Cenozoic Climate Change in Asia,*
Developments in Paleoenvironmental Research 16,
DOI 10.1007/978-94-007-7817-7_4,

processes discussed below include the origins and sources for dust, transport patterns, deposition, and post-depositional alterations, etc. The transport and deposition of dust into the oceans links the land with the atmosphere and oceans through the dusts' effects on bio-geochemical cycles and primary production; these effects also have feedbacks to climate. The environmental conditions that cause aridification not only increase the emissions of mineral dust from arid lands but also increase the numbers of fires and hence the composition of the air; these effects can lead to changes in radiative fluxes and thus the earth's heat balance. In addition to trace gases, wildfires produce large quantities of black carbon (BC) aerosol via incomplete combustion, and analyses of loess samples can be used to reconstruct aerosol BC emissions from paleo-fires. Variations in environmental conditions affect the occurrence of both dust storms and forest fires, and therefore records of dust and BC preserved in Chinese loess provide powerful tools for studying paleoclimate.

In this chapter, we will: 1) review the processes involved in the formation of Asian dust, giving special attention to how these processes relate to climate change; 2) discuss climate connections involving black carbon and natural fires; 3) highlight relationships among desert dust, global iron cycles, marine productivity and climate change; and 4) reconstruct the aridification history of the Asian interior as recorded by Chinese loess.

4.2 Formation of Asian Dust and its Impact on Climate

4.2.1 Dust Sources

Dust, also called dust aerosol, atmospheric dust, mineral aerosol, sand-dust aerosol, soil aerosol and so forth, is a major component of the global tropospheric aerosol (Zhang et al. 2012). Dust particles play an important role in weather and climate, not only through their direct effects on the Earth's radiative balance, but also as a result of their indirect effects, especially their ability to act as cloud condensation nuclei, which can be enhanced as a result of heterogeneous chemical reactions with trace gases.

Asian dust naturally refers to the mineral aerosol that originates from Asia, mainly in the Northern Hemisphere mid-latitudes. Before the 1990's, the locations of the sources of Asian dust and the loess material were most often described in general terms. The main source regions were said to lie in the vast arid and semi-arid regions of Asia (Liu et al. 1985; Prospero et al. 1989; Uematsu et al. 1983). In the mid-1990's, elemental tracer systems have provided more detailed information on two major source regions for Asian dust, the first being the western Chinese deserts and the deserts north of it with Taklimakan as its centre, and the second the northern high- and low-dust deserts as well as the desert areas north of it, which have the Badain Juran Desert as their main body (Zhang et al. 1996a, 1996b,). The elemental tracer studies made use of samples from nine Chinese desert sites, and they provided a way of evaluating the sources of contemporary dust.

Further studies using the same elemental tracers system showed that the Tibetan Plateau and potential desert source areas to the west of China are not major sources for Asian dust, and they also were not major sources for the loess deposited during the last glacial cycles (Zhang et al. 1996a). Studies of dust loadings (Chen et al. 1999a) and fluxes (Zhang et al. 1998) also support the importance of the two main dust sources mentioned above. Similarly, the locations of major Asian dust sources (the Taklimakan, the Central Gobi and the deserts on the Alxa Plateau) have been deduced using calculations of dust emissions based on a formula developed by the U.S. Environmental Protection Agency (Xuan et al. 2000; Xuan and Sokolik 2002). Studies leading to similar conclusions have been based on analyses of dust-storm frequency reports made by meteorological observers (Sun et al. 2001; Zhou 2001).

The visibility-based approach is one of another commonly used methods for identifying dust sources (Littmann 1991; Middleton et al. 1986; Natsagdorj et al. 2003), but this approach is subject to uncertainties. These include human observer bias which can lead to discrepancies in the reported intensities for the same storm. As an example of the latter, the strength of 20 dust storm reports for one area will not necessarily be higher than 10 dust storm reports for the same storm. More important, the dust storm frequencies and the strengths of the dust sources are not necessarily directly related because dust transported into a region, as well as newly raised dust, can reduce visibility.

Numerical models provide yet another approach for identifying and evaluating dust source regions (Claquin et al. 1999; Ginoux et al. 2001; Shao 2001). Numerical methods are crucially dependent on parameterizations of dust deflation, and as such they require accurate information on the geographical distributions of deserts, their surface roughness elements, grain size, soil moisture, vegetative cover, etc. (Marticorena and Bergametti 1995). Asian dust emissions over the past 43 years have been estimated, and sources identified, based on the particle size-dependent soil dust emission and transport NARCM model (Northern Aerosol Regional Climate Model) along with supporting data from a network of surface stations (Zhang et al. 2003b; Fig. 4.1). The deserts in Mongolia and in western and northern China (mainly the Taklimakan and Badain Juran, respectively) were found to contribute ~70 % of the total dust emitted. Hence, these three areas were concluded to be the major sources for Asian dust. Non-Chinese sources account for ~40 % of this. The desert in the Qaidam Basin and the Kumutage Desert, Mu Us Desert, Onqin Daga sandy land, and the Horqin sandy land each contribute a few percent of the dust produced. The regions south of Tibet and the Gurbantunggut Desert are minor sources, combining for less than 3 % of the dust production.

Satellite retrievals are among the most effective tools for characterizing the overall magnitude and spatial/temporal patterns of atmospheric dust plumes. Studies based on satellite technology have made use of data from various sensors, including the Advanced Very High Resolution Radiometer (Husar et al. 1997), the ozone mapping spectrometer (Herman et al. 1997; Prospero et al. 2002), the coastal zone colour scanner (Stegmann and Tindale 1999) and at a regional scale, METEOSAT visible-channel images (Moulin et al. 1998). There are difficulties in quantifying dust optical thickness and dust loadings from satellites, however, and these are due to uncertainties in assumptions concerning particle size, refractive index and particle shape (King et al. 1999) as well as interference from cloud cover. The limited temporal coverage of the satellites presents other problems.

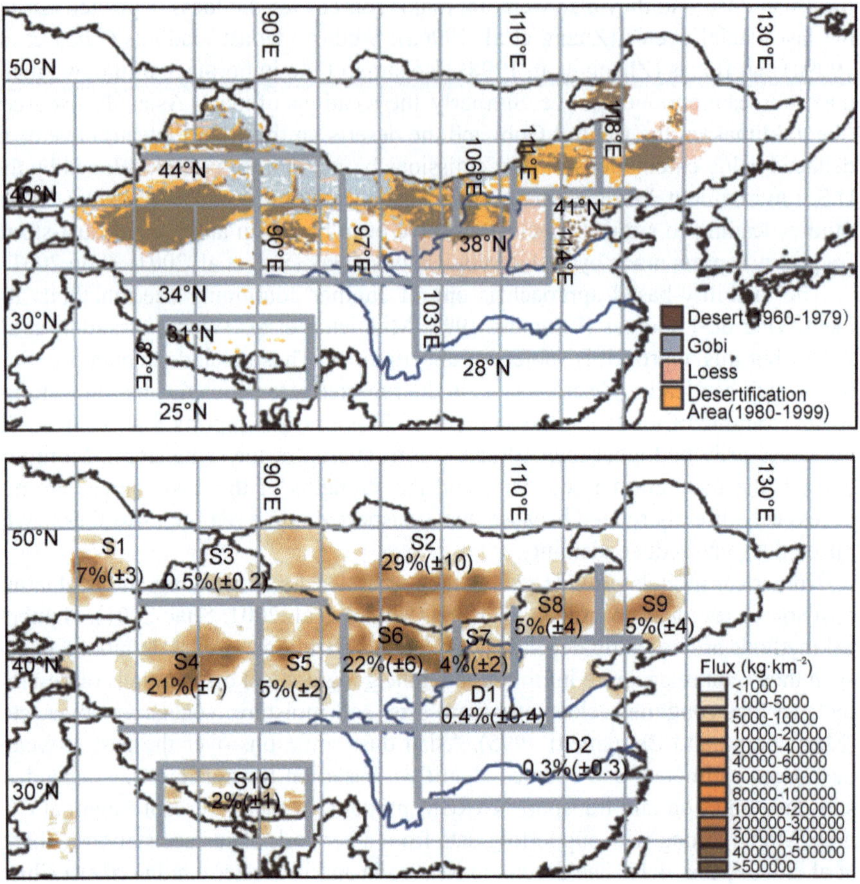

Fig. 4.1 Top diagram: Chinese desert distributions during the 1960s–1970s and deserts plus desertified areas during the 1980s and 1990s. Bottom diagram: Sources (S1–S10) and Depositional areas (D1 and D2) for Asian dust and the average springtime dust emissions (kg km^{-2} 3 months^{-1}) between 1960 and 2002. The percentages with standard deviations in parentheses denote the average contribution to the total dust production from each source and depositional area over the last 43 years. (Zhang et al. 2003b)

4.2.2 Dust Production

The Loess Plateau (LP) in China was initially thought to be a potentially important source of dust (Gao et al. 1992; Iwasaka et al. 1983), but more recent studies have shown that it is a major depositional area for Asian dust, not a significant source (Zhang et al. 2003b). For instance, an assessment of Asian dust emissions over a 43-year period led to the conclusion that the deserts and sandy lands in China, Mongolia and other parts of Asia that have persisted over geologic time are the main sources for Asian dust (Zhang et al. 2003b). This conclusion was drawn following an evaluation of data from a network of stations operated as part of the China

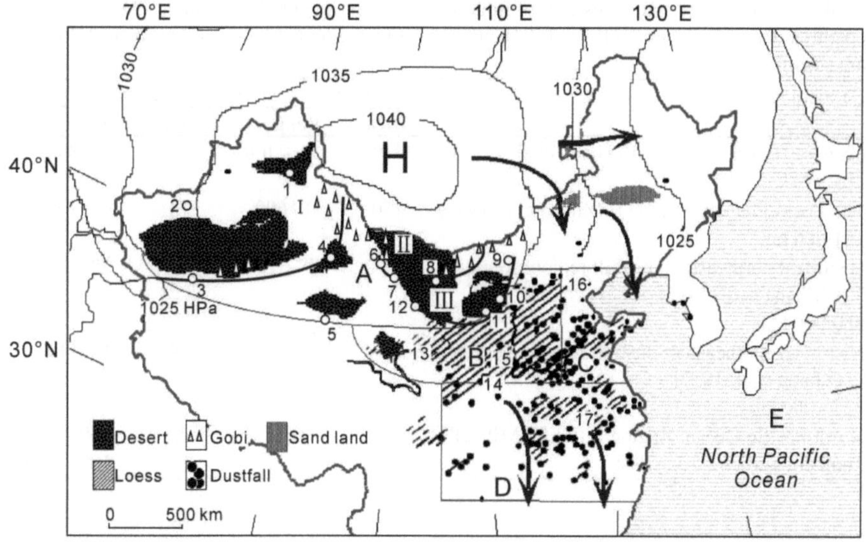

Fig. 4.2 Map showing the winter monsoon regimes of eastern Asia (Zhang and Lin 1992), aerosol sampling locations (1–12) in Chinese deserts, the sites (open squares with numbers) for the data cited (13–17), source regions (Sources I, II, III) and depositional regions (Regions A, B, C, D and E) for Asian dust. The prevailing northwesterly winds (*arrows*) associated with the Siberian High (HPa) and westerly winds from central Asia entrain the bulk of the Chinese desert dust delivered to the depositional areas. 1. Fukang (44°17', 88°7'), 2. Aksu (41°22', 80°43'), 3. Qira (37°6', 82°34'), 4. Dunhuang (40°16', 94°10'), 5. Golmud (36°52', 95°54'), 6. Jiayuguan (40°38', 98°31'), 7. Heiquan (40°26', 100°16'), 8. Jartai (40°34', 106°34'), 9. Dalad Qi (40°53', 110°5'), 10. Yulin (38°37', 109°46'), 11. Dingbian (37°37', 107°34'), 12. Minqin (39°17', 103°10'), 13. Lanzhou, 14. Xi'an, 15. Luochuan, 16. Beijing, 17. Hefei. Chinese deserts with representative sites in parentheses: Gurbantunggut Desert (1), Taklimakan Desert (2, 3), Kumutage Desert (4), Desert in the Tsaidam Basin (5), Badain Juran Desert (6, 7), Ulan Buh Desert (8), Hobq Desert (9), Mu Us Desert (10, 11), Tengger Desert (12). Source Regions: I, western deserts (2, 3, 4); II, northern high-dust deserts (6, 7, 8); III, northern low-dust deserts (9-12). Depositional Regions: A, Chinese desert regions (1–12), excluding three sandy lands in northeastern China (Zhang 1984); B, Chinese Loess Plateau (13, 14) (Liu et al. 1985); C, Historical northeastern dustfall region, 34.3–41°N, to the east of 114°E (16) (Zhang 1984); D, Historical southeastern dustfall region, 27.3–34.3°N, to the east of 104.7°E (17) (Zhang 1984); E, North Pacific Ocean. (GESAMP 1989; Zhang et al. 1997)

Dust Storm Research Project and also though tests against visibility reports from the China Meteorological Administration (Gong et al. 2003; Zhang et al. 2003a, b; Zhao et al. 2003).

Much of the dust raised in Asia is re-deposited onto the deserts (Region A, see Fig. 4.2), but a substantial amount is transported to, and then deposited onto, the LP, especially along the valley of the Yellow River, i. e., Region B in Fig. 4.2 (Liu et al. 1985). Historically, "dust showers" have been recorded over extensive areas of northeastern and southeastern China, i. e., Regions C and D in Fig. 4.1 (Zhang 1984). Westerly winds also carry the dust thousands of kilometers out over the Pacific Ocean, i. e., Region E in Fig. 4.2 (Duce et al. 1980; GESAMP 1989; Merrill et al. 1989; Prospero et al. 1989; Shaw 1980).

Table 4.1 Total flux and deposition of mineral dust onto the Qinghai-Xizhang Plateau and five potential depositional areas (1991–1994), and the estimated total emission of Asian dust. Regions are shown in Fig. 4.2

Depositional Area	Total Flux ($g \cdot m^{-2} \cdot a^{-1}$)		Deposition ($Tg \cdot a^{-1}$)	
	Mean	Range	Mean	Range
Chinese Deserts: Region A (Zhang et al. 1997) 585,970 km^2	410	14–2100	240	50–840
The Loess Plateau: Region B (Zhang et al.1997) 291,600 km^2	250	10–600	73	3.0–180
Historical northeastern dustfall region: Region C(Zhang et al. 1997) 360,000 km^2	95	78–120	34	28–43
Historical southeastern dustfall region: Region D (Zhang et al. 1997) 1250,000 km^2	27	14–89	34	18–110
North Pacific Ocean: Region E (GESAMP 1989)			420	400–500
Total emission of Asian dust			800	500–1100

The total quantity of Asian dust produced has been estimated by assuming that all of the Chinese desert dust produced is deposited in Regions A, B, C, D and E in Fig. 4.2; that is, deposition was assumed to equal production (Zhang et al. 1997). For this, the amount of dust deposited was calculated as the sum of the dry plus the wet deposition averaged over each of the depositional regions. The dry deposition of dust (F_d) was parameterized as the product of a dry deposition velocity (V_d) and the concentration of Al in air (C_{Al}), which for a global average is about 8% of the mineral dust source material by weight (Taylor and McLennan 1995)[1]. Values of V_d used for the estimates were calculated by fitting a log-normal distribution to the mass-particle size distribution of Al obtained with cascade impactors, and calculating deposition rates for 100 particle-size intervals using a two-layer deposition model (Slinn and Slinn 1981; Arimoto et al. 1985; Dulac et al. 1989; Zhang et al. 1993). Not included in the flux calculations was the wet deposition (F_w) for Region A; this was because of the extremely low precipitation in that hyperarid region (Liu et al. 1985). Methods for estimating the wet deposition fluxes for Regions B, C, D and E have been described in Zhang et al. (1993); Gao (1994); Hashimoto et al. (1994); GESAMP (1989), respectively.

The total dust production from Chinese deserts, obtained by summing the amounts of dust deposited on the five study regions, is ~800 Tg yr^{-1} with an estimated range of 500–1100 Tg yr^{-1} (Table 4.1; (Zhang et al. 1997). This amounts to about half of the global production of dust, which has been variously estimated to be ~1500 Tg yr^{-1} (Andreae 1995), or 1000–2000 Tg yr^{-1} (Duce et al. 1991), and this highlights the important global contribution of Asian dust. For Region A, the deposition (240 Tg yr^{-1}) is ~30% of the total emissions from the Chinese deserts (Table 4.1). The estimated dust deposition per unit area in some desert regions, e.g.,

[1] As more chemical data have become available, Asian dust is now known to actually contain about 7% Al (and 4% Fe) by weight (Zhang et al. 2003a)

Table 4.2 Comparison between mode values of particle fractions, aluminium content and dust flux for present-day dust aerosol and loess/paleosol at the center of the Loess Plateau

	MSD %			Al Conc.	Dust Flux
	>20 μm	2~20 μm	<1 μm	$\mu g \cdot m^{-3}$	$g \cdot cm^{-2} \cdot 10^3 a^{-1}$
Desert dust in non-dust storm period (n^a=111)	6.4	58	33	11	0.49[c]
Luochuan S_1 Paleosol (n=21)	7.0	59	34	55000[b]	7.8
Desert dust in dust storm period (n=9)	22	55	23	40	8.3[c]
Louchuan L_1 Loess (no L_1SS_1 samples) (n=46)	31	46	24	49000[b]	20
Luochuan L_1LL_1 Loess (n=19)	32	47	22	48000[b]	18
Luochuan L_1LL_2 Loess (n=27)	27	45	25	62000[b]	21

[a] Number of samples
[b] Unit is 10^{-6} $\mu g \cdot m^{-3}$
[c] Unit is $g \cdot m^{-2} \cdot d^{-1}$

the Gurbantunggut and the desert in the Tsaidam Basin, actually were lower than what was estimated for Lanzhou, which borders the western margin of the LP. The LP itself (Region B) receives far more desert dust per unit area (250 g $m^{-2}yr^{-1}$) than continental Regions C and D (Table 4.2). The total dust input to continental China (the sum of Regions B, C and D) is equivalent to ~20 % of the total desert production. About half of the dust produced is transported to Region E, the remote North Pacific (Zhang et al. 1997).

In addition to the natural dust production in Asian deserts, human activities have created another source for dust, especially along the margins of deserts and sandy lands (Sokolik and Toon 1996; Tegen et al. 1996). The perturbations caused by the so-called "disturbed" or "anthropogenic" dust sources are thus related to the expansion of deserts caused by human activities, a process known as desertification. In one study, ~50 % of the total global dust emissions was attributed to "disturbed source" dust (Tegen et al. 1996). Indeed, some experts have argued that dust primarily originates from degraded grasslands and open farmlands, not deserts.

Several areas in China have in fact experienced significant recent expansion of their deserts and sandy areas. Examples from the past 20 years include the Onqin Daga Sands, Horqin Sands, and Mu Us Desert, and all of these areas have seen increased dust emissions. Desertification has exposed about 8 million square kilometers (Fig. 4.1; Zhang et al. 2003b), but this is a small fraction (6–7 %) of the total pre-existing desert and sandy land areas, which together total about 1.67 million km^2 (Wang 1995). This modest expansion of deserts cannot account for the observed Asian dust storm occurrences in modern times because these areas are not large producers of Asian dust. Further studies have shown that meteorological and climatic conditions in the key source areas have a more important role in determining dust emissions and dust storm occurrences rather than desertification (Zhang et al. 2003b).

Zhang et al. (2003b) simulated Asian dust emissions for the period from 1960 to 2002 to further examine the effects of desertification across Asia. The simulations were performed using the NARCM model driven by re-analyzed meteorology from the National Centers for Environmental Prediction. The model was run on a polar stereographic projection with a horizontal resolution of 100 × 100 km and 22 vertical levels on a Gal-Chen terrain-following coordinate system from the ground to 30 km. The model was restricted to the suspension of particles <40 μm because particles of that size are most readily transported by turbulent eddies and advection. Twelve dust particle-size bins with diameters from 0.01 to 41 μm were used to represent the size distribution of soil dust, and dust fluxes and concentrations were calculated for each size bin. The model domain covered East Asia, the North Pacific and western North America. A soil texture map for China was generated to drive the particle-size specific soil dust emission scheme (Alfaro et al. 1997; Marticorena and Bergametti 1995). Surface observations of the size distribution of dust aerosols during dust storms at nine Chinese desert regions (Zhang et al. 2003a) were used to constrain the size distribution of the vertical dust flux. Information taken from Chinese desert maps from the 1960's to 70's was used to generate the distribution of deserts, Gobi, sandy lands, loess and other land types prior to 1980. That information was mainly obtained from aviation images and surveys conducted from the 1950's to the 70's (Wang 1995). In contrast, the maps of desert and expanded deserts and sand areas for the 1980's to 1990's were mainly based on surveys and satellite images (CAS 1998; Fig. 4.1).

The simulated 43-yr time series of total dust emissions from the various source areas differ when stratified for the months of March, April and May, but the trend for April (~40 % of the springtime emissions) is similar to that for springtime emissions and yearly dust storm frequencies as recorded at the Chinese Meteorological Administration stations (Fig. 4.3). This agreement indicates that Chinese sources produce most of the Asian dust and attests to the overall reliability of the model results. We note that some inconsistencies between the quantitative model simulations versus semi-quantitative dust storm reports are to be expected. For instance, the simulations were made for spring whereas the dust storm frequencies were reported on an annual basis. In addition, there were differences in the areas considered, that is, the dust reports were restricted to China while the model simulations included regions outside of China.

After a trend of increasing Asian dust emissions from 1961 to 1963, followed by lower activity from 1964 to 1967, the model simulations indicate that the amount of dust produced generally decreased after 1969. Interestingly, the simulated dust fluxes often switched from high- to low-emissions after intervals of 3–4 year (Fig. 4.3a). It is also noteworthy that no abrupt change in the pattern of emissions was observed from 1979 to 1981 which is when the data inputs to the model changed. The most recent simulations showed dust emissions began a two-year rise from 2000 to 2001, reaching the highest value of the 43-yrs in 2001 then dropping again in 2002. The average springtime dust emissions for particles <40 μm diameter from 1960 to 1979 (~ 1.2×10^{11} kg) was ~24 % higher than during 1980–1999 (~0.97×10^{11} kg).

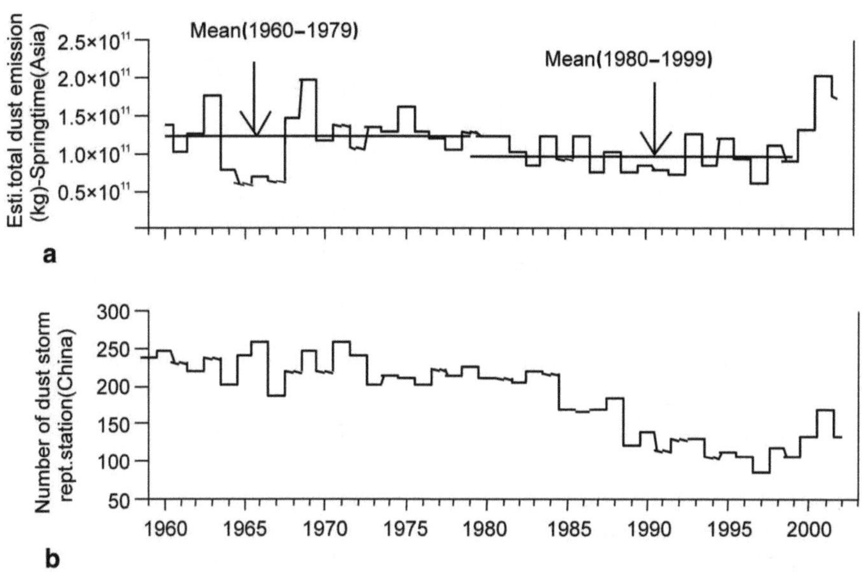

Fig. 4.3 **a** Time series of the estimated total dust emission (kg) in springtime for the sum of all investigated areas over the last 43-years and **b** the stations reporting dust storms in China during the same period. (Zhang et al. 2003b)

The observed trends and relatively low dust levels from 1980 to 1999 suggest that climate change has had a stronger influence on dust emissions than the expansion of deserts. Over the past 43 years, the area of China covered by deserts increased from ~2% (Zhong 1999) to ~7% (Zhu and Zhu 1999) on average (Fig. 4.1), and if desertification were the predominant influence on dust production, the emissions and dust storm reports should have increased proportionately. By extension, the emissions from 1980 to 1999 would be expected to be larger than those from 1960 to 1979, but this is not the case. In addition, the large increases in dust- and sandstorms simulated for the last three year period of the model (Fig. 4.3) can be more logically explained by changes in weather and climate than desertification. This is because the land area affected by desertification changes relatively little over a few years time (Zhong 1999). Therefore, two lines of evidence support the idea that meteorological factors and climate have been more important than desertification with respect to the occurrence of the Asian dust storms.

The growth of China's deserts has occurred mainly at the rim of the Onqin Daga sandy land, the southwestern and northwestern fringes of the Horqin sandy land, the central and northern parts of Mu Us Desert, and to some degree in the desert in Qaidam Basin, the Kumutage Desert, and the eastern part of Badain Jaran Desert (Fig. 4.1). There also were minor expansions in the sandy lands at the margins of the Taklimakan and Gurbantunggut Deserts and other areas of China. Model results indicate that there was an increase in the 5-yr average dust emissions relative to the 40-year average over the past 20 years (Fig. 4.4) in the Onqin Daga sandy

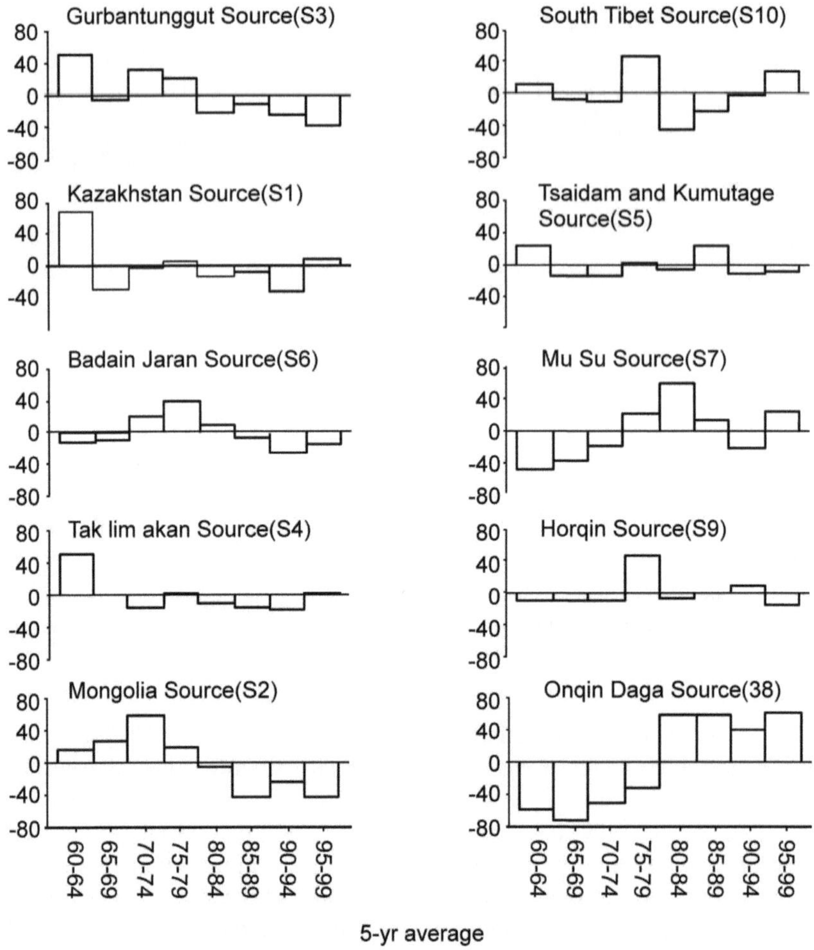

Fig. 4.4 Variations in 5-year averaged total dust emissions relative to the 40-year mean between 1960 and 1999 at each source region. (Zhang et al. 2003b)

land (S8), Horqin sandy land (S9), and to some degree in the Mu Us Desert (S7) and Kumutage Desert (S5). These increases in dust production can be attributed to desert expansion. On the other hand, the amounts of dust ejected from the three major sources, Mongolia (S2), Taklimakan (S4) and Badain Jaran Deserts (S6), and from sources in the Gurbantunggut (S3) and Kazakhstan (S1), exhibited a decreasing trend over the last 20-yrs (Fig. 4.4). It was the changes in dust emissions from these latter five regions that were largely responsible for the overall trends shown in Fig. 4.3.

To further assess the contributions of the Asian deserts, Gobi, degraded grasslands, open farmlands and other potential source areas to Asian dust production, threshold friction velocities (TFVs) have been calculated and compared for different types of surfaces (Mei et al. 2004). These studies made use of data for soil texture, mineral content, land-use/cover, desertification maps etc. for sand dunes, sandy farmlands, gravelly farmlands, sandy-gravel plains, and gravel Gobi collected in or near the Taklamakan Desert, Badain Jaran Desert, Tsaidam Basin Desert, Kumutage Desert, Mu Us Desert, Hobq Desert, Onqin Daga sandy land and Horqin sandy land. The TFV values were ~20 cm s^{-1} for sand dunes and sandy farmlands, ~58 cm s^{-1} for gravelly farmlands, and 60–80 cm s^{-1} for sandy-gravel plains and gravel Gobi.

Dust deflation occurs when the surface wind speed exceeds the TFV, and therefore smaller TFVs are characteristic of surface materials that are more readily eroded. The low TFVs for dunes and farmlands show that these surfaces are the most favourable for dust production on a per unit area basis. Nonetheless, the final result from this analysis is that deserts are the largest source for windblown dust in Asia, and this is because sand dunes cover a much larger area than sandy farmlands.

The simulations of emissions and surface concentrations for Asian dust presented above, along with validations from a network of observations, indicate that the deserts in Mongolia and in western and northern China (mainly the Taklimakan and Badain Juran, respectively) contribute ~70 % of the total Asian dust emissions and non-Chinese sources account for ~40 % of that (Zhang et al. 2003b). Desertified lands within China account for only about 6 % to the total dust emissions. This latter conclusion is supported by sensitivity tests for reductions in dust concentrations run with and without desertified areas (Gong et al. 2004). These findings suggest that efforts to mitigate desertification may have little impact on the quantity of Asian dust produced because the relatively small desertified areas are not major dust sources. Meteorology and climate have a far greater influence on Asian dust storms and associated dust emissions than desertification. Globally, estimates of the contribution of "disturbed source" dust to the total dust emissions have been revised downward from ~50 % (Tegen et al. 1996) to less than 10 % (IPCC 2007; Tegen et al. 2004)

4.2.3 Dust Transport

4.2.3.1 Asian Dust Transport Under Different Climatic and Meteorological Conditions

Asian dust is a major, sometimes dominant, component of the tropospheric aerosol, especially in the mid-latitudes of the Northern Hemisphere (Duce et al. 1980; Husar et al. 2001; Merrill et al. 1989; Prospero et al. 1989; Zhang et al. 1991). Early studies showed that westerly winds are largely responsible for loess transport (Liu et al. 1985), and subsequent studies have highlighted the importance of the East Asian winter monsoon for the transport of mineral particles deposited on the LP over geologic time (An et al. 1991b; An et al. 1995). Dust storms have also been shown to be crucial for eolian loess accumulation (Liu et al. 1985; Zhang 1984), and the seasonal transport and deposition of dust into the remote North Pacific Ocean (NPO) has been linked to Asian dust storm frequencies and intensities as well (Duce et al. 1980; Merrill et al. 1994; Parrington et al. 1983; Prospero et al. 1989; Uematsu et al. 1983).

The relative inputs of mineral dust from dust storms versus days without dust storms have been evaluated by using the elemental composition of dust (Zhang et al. 1993). Two major findings of the study were: 1) on an annual basis, the dry deposition input of dust to the LP is much higher than that from wet deposition; the latter accounts for only about 7 % of the total; and 2) the annual dry deposition during non-dust storm days accounts for the bulk of the dust input to the LP. Averaged over the course of a year, the "normal" input, that is, deposition on non-dust storm days, was 6–39 times that from dust storm events. The second finding was unexpected because it was at odds with traditional views.

Measurements of the particle-size distributions of mineral dust and sedimentary materials provide a means for studying the dynamics of dust transport and deposition. Present-day dust was found to be composed of three modes, each characterized by a log-normal size distribution, with size ranges of about 20–200, 2–20, and 0.04–1 μm in diameter (Patterson and Gillette 1977). Zhang et al. (1993) found that the mass-particle size distributions (MSDs) of dust-derived elements (Al, Ca, Fe, K, Si, Ti) in aerosol samples from the LP also were approximately log-normal. Three modes also were found for loess samples from the LP (Zhang et al. 1994), with the >20 μm particle fraction (i. e., the heavy mode) mainly associated with heavy dust-transport events. In contrast, the 2–20 μm particles were representative of the ubiquitous dust, deposited under normal conditions (normal or central mode), and the <1 μm particles mainly reflected the background deposition of dust plus weathering and reworking products (background mode).

Further comparative studies of the relative influences of dust storms and non-dust storm processes on Asian dust transport and deposition have been made using data for dust fluxes, MSDs, and the aluminium content in mineral dust, loess and paleosols. For these studies, aerosol samples were collected during dust storms and on non-dust storm days from desert sites in China and loess samples from the centre of the LP (Table 4.2). The MSD of present-day mineral dust was determined from cascade impactor Al data (Zhang et al. 1993), and the grain-size distributions of the loess were measured with a Sedigraph 500 ET device.

Differences between the MSDs for dust storm versus non-dust storm samples were quite large, particularly for the coarse particles greater than 20 μm (Table 4.2). For the three particle-size ranges, the MSDs of non-dust storm dust particles were all similar to paleosol from stage S1, indicating that during interglacial climate conditions, the bulk of the Asian dust was transported to the LP during non-dust storm conditions. This is in line with observations of present-day dust inputs to the LP (Zhang et al. 1993). The MSDs for modern dust during dust storms, in contrast, are similar to the size distributions of Malan loess in all three particle modes (Table 4.2). These results and the observations above led to the conclusion that during the relatively warm and humid interglacials, the dust storm contributions to the loess sediments are less than the non-dust storm component (Zhang et al. 1999).

The LP experienced the highest monthly dust deposition rates in China during a study conducted during the high-dust season (Zhang et al. 1998). To the southeast of the LP, the Hua Bei Plain and western area of northeast China had the second highest depositional fluxes. The lowest monthly dust depositional fluxes were in the South China Sea and associated coastal cities. The dust transported to all of these regions follows pathways driven by surface winds (Zhang et al. 1999). The direction of these transport pathways also corresponds with the northwest to southeast decrease in particle sizes seen in Malan loess (Liu et al. 1985). Variations in the Si concentrations in dust from the surface atmosphere over the LP have been linked to changes in the strength of the East Asian winter monsoon (Zhang et al. 1995). These results support the notion that the regional-scale transport of Asian dust during interglacial periods is mainly attributable to non-dust storm processes and northwesterly surface winds drive the transport. Conversely, the accumulation of loess during glacial stages is more strongly linked to dust storms (Zhang et al. 1999).

During interglacials, including the present, dust storms inject desert mineral particles into the mid-troposphere where strong westerly winds can lead to hemispheric or near global-scale transport. This explains why seasonal peaks in dust concentrations over the NPO are closely related to the frequencies of dust storms in Asia (Uematsu et al. 1983). On the other hand, the regional transport of desert dust is more strongly affected by near-surface conditions, especially non-dust storm processes, and the dust flux to the southeast is strongly influenced by the Asian winter monsoon, as noted above (Zhang et al. 1999).

During glacial stages, dust storms undoubtedly do lift some mineral particles to the altitudes where long-range transport can occur, but at the same time, dust storms exert a stronger influence on regional-scale dust transport than during interglacials (Zhang et al. 1999). This conclusion is based on observed similarities between the particle-size distributions of the dust collected during dust storms and the grain-size distributions of glacial-age loess (Table 4.2). In glacial times, a large proportion of the dust deposited on the LP is due to dust storms, and more coarse particles (>20 μm) are carried by those storms than under non-dust storm conditions (Zhang et al. 1999).

4.2.3.2 Transport Paths of Present-day Asian Dust

Studies of present-day dust transport, which is arguably representative of interglacial climate conditions, have made extensive use of air mass trajectories calculated backwards in time. Results produced using Environment Canada's global trajectory model have shown that low-level northerly winds at 850, 950, and 1000 h Pa frequently pass across the Northern High Dust Desert, which has the Badan Juran Desert at its centre and is a major source for Asian dust and (we present only the 850 hPa data in Fig. 4.5). Interestingly, fewer trajectories have been found to pass over the other major dust source region, the Western High Dust Desert, which includes the Taklimakan (Fig. 4.5).

Five general groups of surface trajectories that reached Beijing during spring 2001 have been identified, and they are differentiated by color in Fig. 4.5d. The first type of pathway, labelled "Northeast Sand Land Path" (black line), accounted

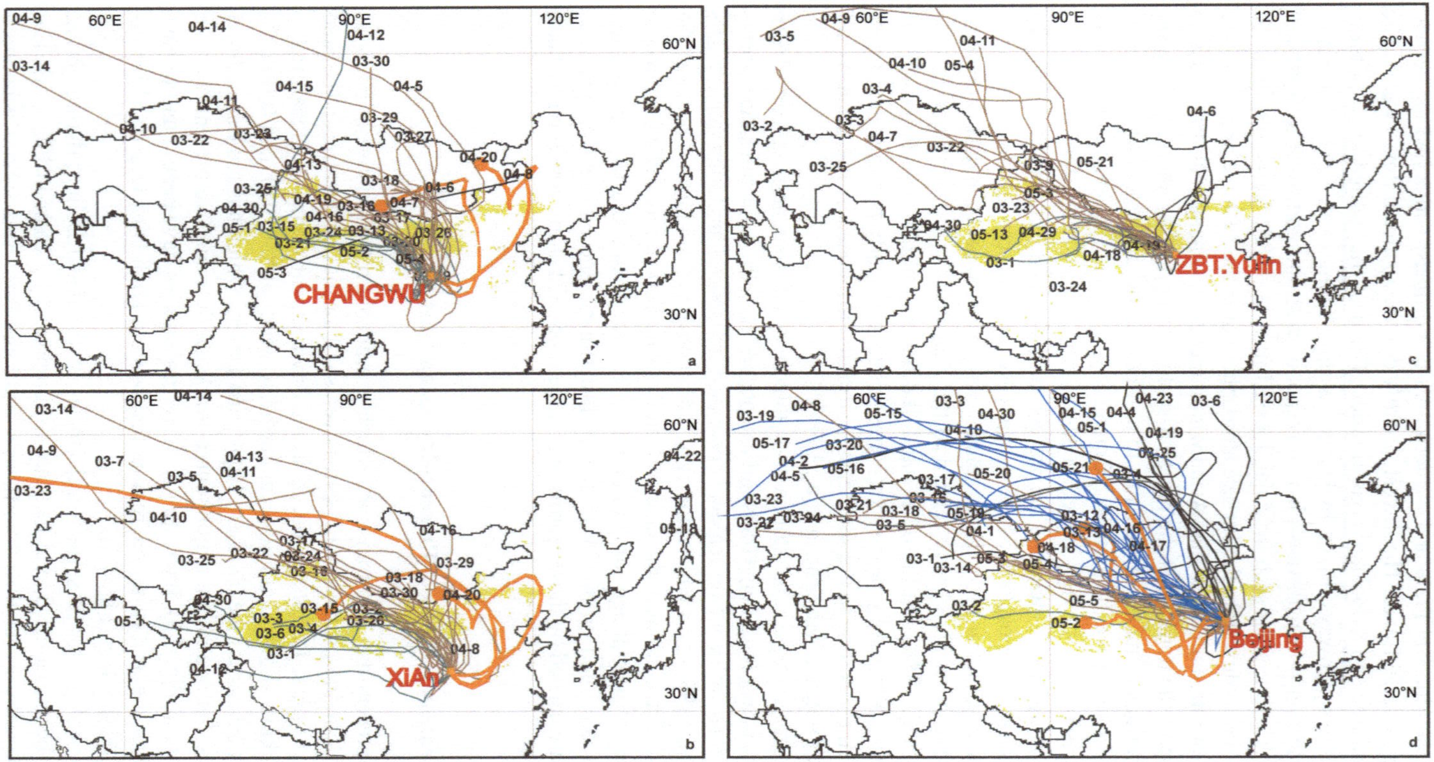

Fig. 4.5 A 120-h air mass backward-trajectory analyses at 850 mbar, 0600 UTC for **a** Changwu, **b** Xi'an, **c** Zhenbeitai, and **d** Beijing. *Black* and *blue* lines denote the Northeast Sand Land Path and Northerly Mongolia Path for the dust transport, respectively. Northern Desert Path and Western Desert Path are shown by *mauve* and *green* lines, respectively. The bold *red* lines represent the Turning Path. Six major mountain ridges are marked approximately as black lines on Figures b and d for the Yin, Taihan, Qingling, Liupan, Helan, and Qilian Mountains. (Zhang et al. 2003a)

for ~22 % of the 43 total trajectories. Under this type of flow, desert dust was entrained from source regions in the sandy lands of northeastern China, especially the Onqin Daga sandy land, and then surface northeasterly winds brought the dusty air to Beijing. Pathway 2 ("Northerly Mongolia Path") was characterized by air that passed over Mongolia and the northern slopes of Ying Mountain and then to Beijing (blue line); this type pathway was quite common, accounting for ~40 % of the trajectories. The third type of trajectory was mainly from the Ala Mountain Pass and surrounding areas, and the pathway continued through the Northern High Dust Desert along the south slope of Yin Mountain. This type of trajectory passed between the Helan and Yin Mountains and then to Beijing. Labelled the "Northern Desert Path" (mauve line), these trajectories accounted for ~25 % of the total. A fourth trajectory occurred only once in 2001, and it originated from the Western High Dust Desert and moved easterly to Beijing ("Western Desert Path," green line). The final pathway (bold red line "Turning Path") occurred five times (~11 % of the total) in 2001 and featured a change in flow from south to north, often passing through the southern and northern mountain passes of Taihang Mountain and then turning north to Beijing.

The transport pathways that bring soil dust to Changwu (Fig. 4.5a) and Xi'an (Fig. 4.5b) are initially similar to those leading to Beijing; that is, they start out resembling two of the major pathways just discussed, the Northern Desert Path and Western Desert Path. Another interesting pathway that brings dust to Xi'an and Changwu (bold red line in Fig. 4.5a and Fig. 4.5b) passes over the sandy lands of northeastern China, moves southward along the western part of Taihang Mountain, travels through the Taihang-Qingling mountain pass, and finally northwestward to Xi'an and Changwu. At Zhenbeitai, Yulin (Fig. 4.5c), the Northeast Sand Land Path (back line), Northern Desert Path (mauve line), and Western Desert Path (green line) were most common during the spring of 2001.

The long-range transport of Asian desert dust contrasts sharply with regional-scale transport. Air-mass back-trajectories have shown that Chinese desert dust is carried out over the NPO by mid-latitude westerly winds (Merrill et al. 1994; Merrill et al. 1989). During late winter and early spring, desert dust is lifted from source regions in western China and carried by these winds to the NPO ~5–8 km above sea level at potential temperatures of ~305–315 K and pressures of ~350–600 hPa. After the dust-laden air masses pass to the east of subtropical highs, the plumes typically move southward, gradually descending into the surface northeasterly trade winds, and finally into the marine boundary layer.

4.2.3.3 Transport Height of Present-day Asian Dust

Simulated time-series showing the vertical distributions of dust during spring 2001 at selected locations in China and across the Pacific Ocean are shown in Fig. 4.6 (Gong et al. 2003). The contours in that figure depict the vertical distributions of soil dust over selected sites and demonstrate the layered structures of soil dust transport. Peak dust concentrations generally occur below 1000 m for sites close

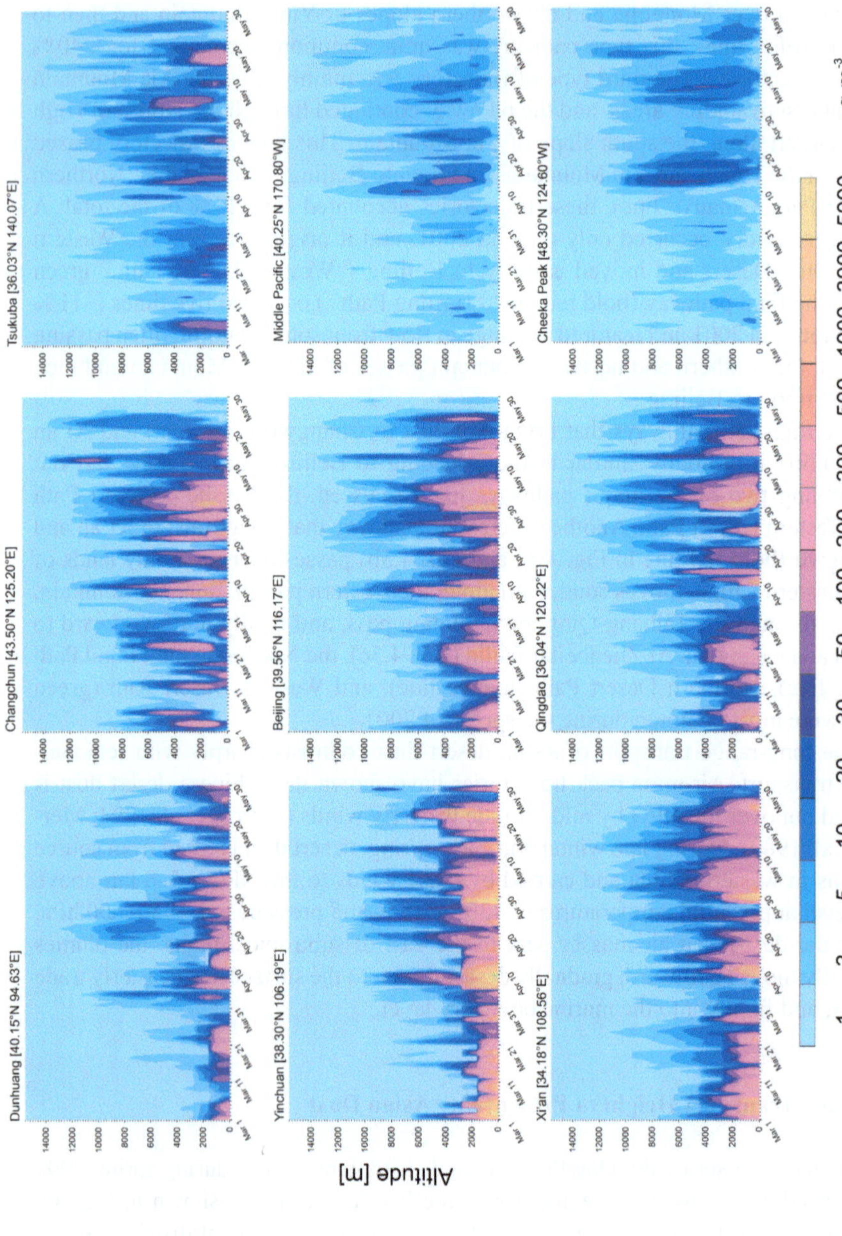

Fig. 4.6 Time vertical profiles of soil dust concentrations for selected locations for spring 2001 ACE-Asia. Local time is used for the plots. (Gong et al. 2003)

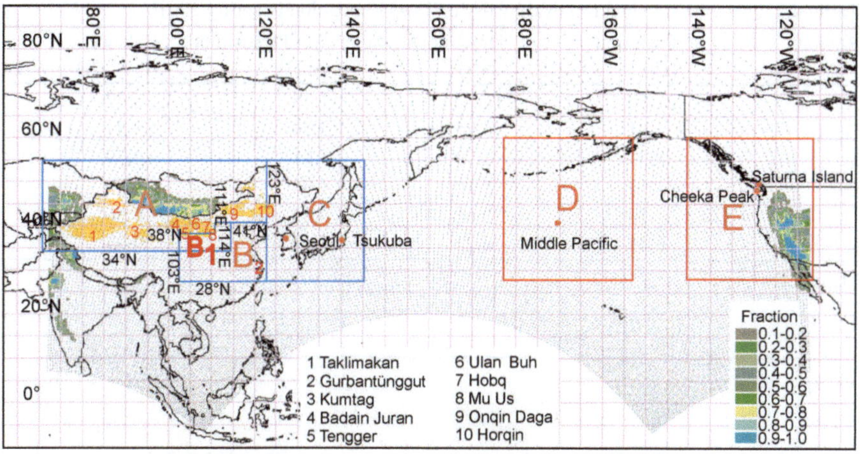

Fig. 4.7 Model simulation domain (*dotted area*) with desert distributions in China (*gray scale*) and outside China expressed as fractions of deserts in a grid and shown as contours (Gong et al. 2003). Regions A-E are regions used to simulate emission and deposition in the Asian Pacific Region Aerosol Characterization Experiment

to the source regions (e.g., Dunhuang and Yinchuan). The variations in dust concentrations with altitude are largely controlled by vertical mixing processes, such as convection and turbulent diffusion, and for downwind locations in China, such as Xi'an, Beijing, Qingdao, and Changchun, they increase to 1000–3000 m. The peak altitudes continue to increase downstream of China, reaching 2000–4000 m in Tsukuba, 4000–5000 m in the middle Pacific, and 5000–7000 m at Cheeka Peak on the western coast of North America (Fig. 4.1). As the dust moves across the Pacific, concentrations decrease dramatically as a result of dry and wet deposition. It is worth mentioning that depending on the strength and location of the dust source, the heights at which dust passes over a downstream location can vary substantially from one episode to another.

4.2.4 Dust Deposition and Post-Depositional Disturbance

A simulation of Asian dust was performed as part of the Asian Pacific Region Aerosol Characterization Experiment (ACE-Asia) to evaluate the balance between the emission and deposition of Asian dust in each of five major regions (A, B1, B2, C, D, E; Fig. 4.7). Only Region A, identified as the Asian desert area, was also found to be an unambiguous dust source region (Table 4.3; Zhao et al. 2003). Coincidentally, this is the same region where the largest quantities of dust are both produced and deposited. The results of this simulation are thus consistent with previous findings discussed above (Zhang et al. 1996b, 2003b).

Table 4.3 dust emission and removal by dry and wet deposition in spring 2001[a]

Region	Emission[a] (E)	Dry Deposition (DD)	Wet Deposition (WD)	Balance (E − (DD + WD))
A	21.50	11.72	1.35	8.42
B_1	2.81	5.75	2.90	−5.84
B_2	0.31	1.79	1.81	−3.29
C	1.17	1.10	1.96	−1.89
D	1.13	0.66	0.53	−0.06
E	0.45	0.27	0.11	0.07

[a] Units: ton · km^{-2}

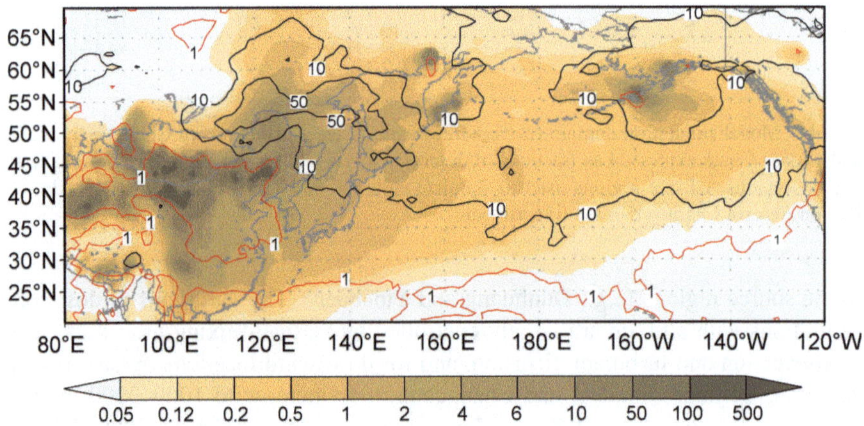

Fig. 4.8 Total deposition (t km^{-2}) and ratio of wet and dry depositions for spring 2001. The contour lines are for ratio between wet and dry deposition (Zhao et al. 2003). *Red* lines encompass areas where dry deposition exceeds wet

In a study of present-day dust deposition (Zhang et al. 1994), calculated dry deposition separately for dust storm and non-dust storm conditions. Wet deposition was estimated from scavenging ratios, and that amounted to ~7 % of the total annual deposition, with a range between 4.3% and 10 % (Zhang et al. 1993). Therefore, dry deposition apparently dominates the total atmospheric deposition of mineral dust to the LP (Zhang et al. 1993). An analysis of the spatial distribution of dust deposition (Zhao et al. 2003; Fig. 4.8) provides further evidence to suggest that dry deposition has dominated the deposition of Asian dust in China's inland regions. This is in contrast to the situation over the remote Pacific Ocean, where precipitation is more abundant, and wet deposition consequently plays a more important role than dry deposition for transferring particles from the atmosphere to the ocean (Zhang et al. 1993, 1998).

An investigation based on the MSDs of loess sediments indicated that more than 90 % of the loess-paleosol material in the LP from the late Quaternary was originally deposited as Asian dust, and less than 10 % of this material was weathered or reworking products (Zhang et al. 1994). These results are significant because they imply that rate-based chronologies for loess and paleosol sequences based on dust deposition, as well as those for bulk loess samples, are suitable for paleoclimatic reconstructions (Zhang et al. 1994).

4.2.5 Shifts in Sources of Asian Dust and Loess Material During Different Climate Conditions

The major sources for Asian dust lie in deserts of northern and northwestern China, southern Mongolia and other desert areas where high winds inject hundreds of teragrams of mineral matter into the atmosphere each year (Zhang et al. 2003b; 1996a; 1996b). From a study based on 43 years of dust and weather reanalysis data (Zhang et al. 2003b), it has become apparent that the dust output from different major sources of Asian dust change to some extent from year to year. Similarly, there is compelling evidence for high-frequency variability in the contributions of the western desert sources vs. northern high-dust and low-dust desert sources to the eolian deposits from the center of the LP. Samples from loess-paleosol sequences, long-term records of Asian dust transport and deposition, and elemental tracer systems based on dust-derived elements (Al, Fe, Mg and Sc) all provide evidence of this variability. Comparisons of the patterns with climate signals recorded in sediments from the remote North Atlantic from the last glaciation show that shifts in source areas of Asian dust were synchronous with large-scale variations in atmospheric circulation (Zhang et al. 1997).

Since the main structures of Chinese deserts formed before the Late Pleistocene (Dong et al. 1991), one can assume that the source signatures (elemental tracers) have changed little since the last glaciation (Zhang et al. 1997). Accordingly, the dust contributions to the well-known Luochuan loess/paleosol sequence (An et al. 1991b; Liu et al. 1985) can be apportioned between Source I and Sources II + III using a chemical element balance model. All of the dust tracer elements showed similar variations, and therefore only the curves for Al showing the source contributions vs. age are presented in Fig. 4.9.

In Fig. 4.9, the contributions from three source areas to loess from the Luochuan loess/paleosol sequence are presented along with data for several climate proxies. The loess core was sampled at 10 cm intervals and the elemental composition of the material was determined Zhang et al. 1994); the core was dated using the sedimentation model of (Porter and An (1995). The contributions of three sources of desert dust, labelled as I denoting Source 4 (S4), S5 and the sources north of them in Fig. 4.1, II, and III representing S6, S7, S8, S9 and the other sources north of them in Fig. 4.1, were then calculated (Zhang et al. 1997).

The contribution of Source I exhibits high-frequency variability synchronous with paleoclimatic proxy records from the North Atlantic that date back to the last glaciation (Fig. 4.9). Maxima in Source I contributions are generally coherent with: 1) minima in $\delta^{18}O$ values from the Greenland ice core project cores (GRIP), i. e., Dansgaard-Oeschger cycles (Dansgaard et al. 1993); 2) maxima in the percentages

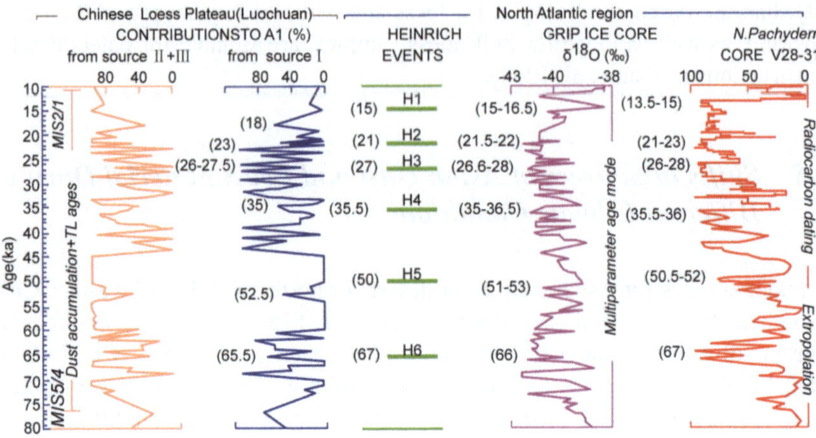

Fig. 4.9 Time series comparing variations in the contribution of desert sources to Al concentrations in the Luochuan loess/paleosol succession with Greenland ice core $\delta^{18}O$ variations (Dansgaard 1993), North Atlantic Heinrich events (Heinrich 1988) and *N. pachyderma* (s) variations (Bond et al. 1993) during the last glaciation (80–10 kyr). Ages of Heinrich events H1–H3 were based on accelerator mass spectrometry (AMS) dates bracketing layers of ice-rafted detritus (Bond et al. 1992); the age of H4 was based on one AMS age (Bond et al. 1993), whereas ages of H5 and H6 were obtained by extrapolation (Bond et al. 1993). The upper part of the *N. pachyderma* (s) record was controlled by AMS ages, but ages of the lower part were based on extrapolation of ^{14}C-based sedimentation rates (Bond et al. 1992). The chronology of the upper 8.6 kyr of the GRIP core was controlled by annual layer counting and volcanic-acid reference horizons, whereas the lower part was based on a multiparameter method (Johnsen et al. 1992); estimated error limits ranged from 0.05 kyr at 8.6–2 kyr at 40 kyr. Ages of peaks (in parentheses) are rounded to the nearest 0.5 kyr in all time series. (Zhang et al. 1997)

of *Neogloboquadrina pachyderma* (s) in marine records, i. e., Bond cycles (Bond et al. 1993); and 3) Heinrich layers (H1–H6) from a core in the North Atlantic (Heinrich 1988). Therefore, increased quantities of dust exported from the western desert sources evidently coincided with cooling of Atlantic surface waters (Bond et al. 1993; Heinrich 1988) and with atmospheric cooling over Greenland (Dansgaard et al. 1993). These cooling episodes are thought to have affected sites outside the North Atlantic region (Broecker 1994), including regions directly downwind in Europe (Rind et al. 1986) and China (Porter and An 1995). Furthermore, it is believed that the cooling episodes in the Northern Hemisphere were caused by a strengthening and meandering of the westerly jet and associated changes in atmospheric pressure systems (e.g. Siberian-Mongolian High, Porter and An 1995; XD. Liu, Institute of Plateau Atmospheric Physics Research, Chinese Academy of Sciences, Lanzhou, People's Republic of China, personal communication 1997).

Historically, surges of cold air associated with the monsoonal circulation increased in strength and frequency as the westerlies intensified and latitudinal circulation developed (Zhang and Wang 1995). Similarly, the increased dust inputs from Source I observed in Chinese loess can be attributed to shifts in circulation that favored the western fraction of the cold surge. Conversely, during periods when the dust from Sources II + II was increasingly important, the transport of cold air from the north was associated with a weakening of the westerlies. These shifts in Asian dust sources apparently were synchronous with climatic events in the North Atlantic region, a strong indication that changes in the dust fluxes from the Asian source regions were linked to large-scale variations in atmospheric circulation.

4.2.6 Dust Impacts and Feedbacks to Weather and Climate

The long-range transport and air/sea exchange of dust links the biogeochemical cycles of land, atmosphere and oceans (Bergametti and Dulac 1998; Martin and Gordon 1988), possibly of a magnitude large enough to affect the global carbon cycle (Ridgwell 2002). Mineral aerosol particles also scatter and absorb solar radiation, and in this way plumes of dust can significantly affect regional radiative balances (Kinne and Pueschel 2001; Sokolik and Toon 1996; Sokolik et al. 2001). Dust concentrations are clearly sensitive to variations in climate, but it is also possible that the dust loadings themselves have feedbacks to production: these result from differences in the dusts' radiative effects over land versus water surfaces (Zhang et al. 2002).

Consistent with reconstructions from marine sediments (Hovan et al. 1991; Rea 1994) and ice cores (Petit et al. 1999), estimates of dust inputs to the Chinese LP indicate that the dust fluxes during periods of extensive glaciation were higher than those during interglacial times (An et al. 1991a; Zhang et al. 1994). Prior studies (Porter and An 1995; Zhang et al. 1999) have uncovered strong evidence for millennial-scale oscillations of Asian dust during times of extensive glaciation. These events have raised questions about the climatic and meteorological factors that govern the generation and transport of dust, especially those operating during the winter monsoon.

Valuable insights into the relationships between climate and Asian dust have been obtained through investigations of historical dust loads reconstructed from LP samples. For example, dust concentrations over the last two glacial cycles have been reconstructed from two central LP sections (Zhang et al. 2002). Results of that study indicated that the loadings were higher and more variable during glacial periods compared with interglacials and also that high-dust loadings during glacials frequently preceded decreases in winter sea-surface temperatures in the NPO (Fig. 4.10). As part of that study, Zhang et al. also proposed a conceptual model to explain the results. That is, high-dust loadings could lead to enhanced temperature gradients between Eurasia and surrounding oceans as a result of the dusts' radiative effects; this is because high dust loads tend to warm the land but cool water surfaces. The enhanced thermal gradients could either trigger or modulate millennial-scale variations in winter monsoon climate, and in this way feedbacks between dust and climate may help explain the rapid transitions in glacial dust loadings.

Liu et al. (2004) analyzed the spring dust storm frequencies in northern China during 1982–2001, and these authors found that the time series of the spring dust-

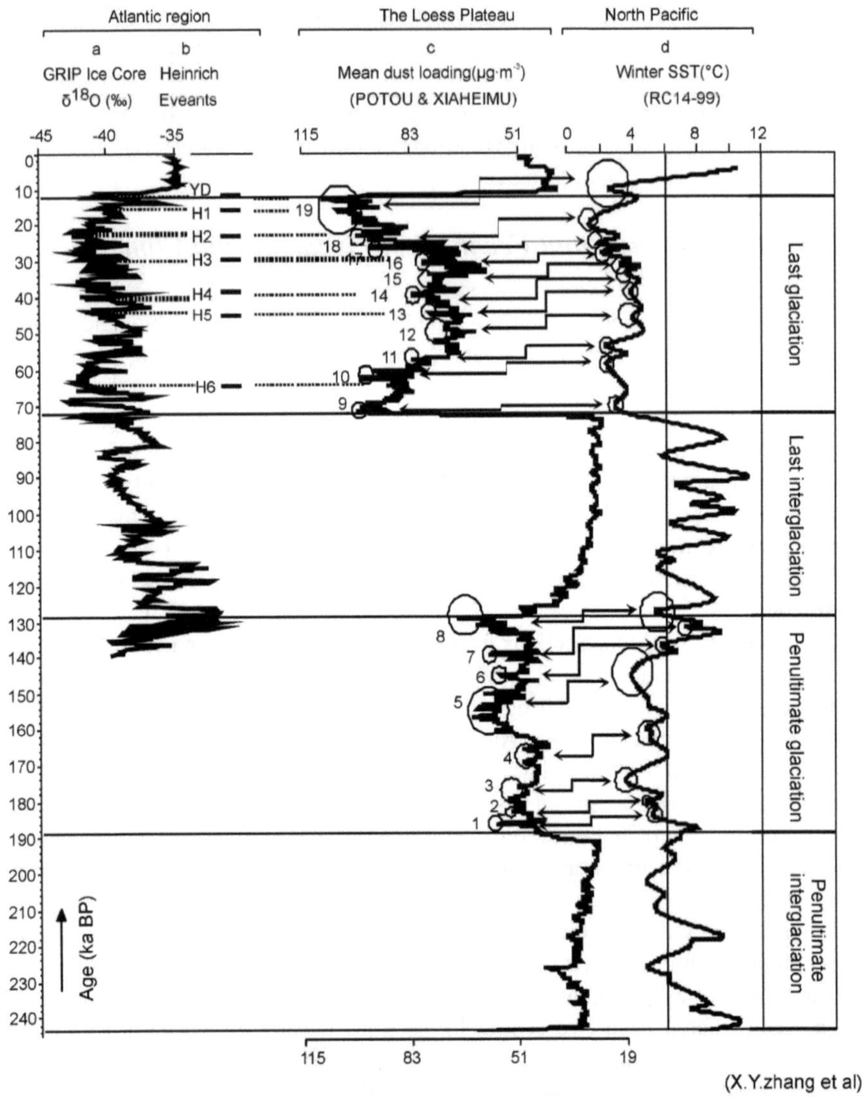

Fig. 4.10 Timeseries comparing **a** $\delta^{18}O$ variations in Greenland ice core, **b** North Atlantic Heinrich events (Heinrich 1988), **c** mean dust loading in the Luochuan loess/paleosol succession, and **d** winter sea surface temperature (WSST) of the North Pacific Ocean (Morley and Heusser 1997). The mean dust loading was averaged from the Potou and Xiaheimu profiles. We assumed that 7 % and 2 % (Zhang et al. 1993) of the total mass fluxes were attributable to wet deposition in the interglacial and glacial stages, respectively. The ages of the Yonger-Dryas (YD) cold phase (23) and six Heinrich events (H1–H6) are after Bond et al. (1997). Open circles mark nineteen high-dust episodes and corresponding low WSST periods

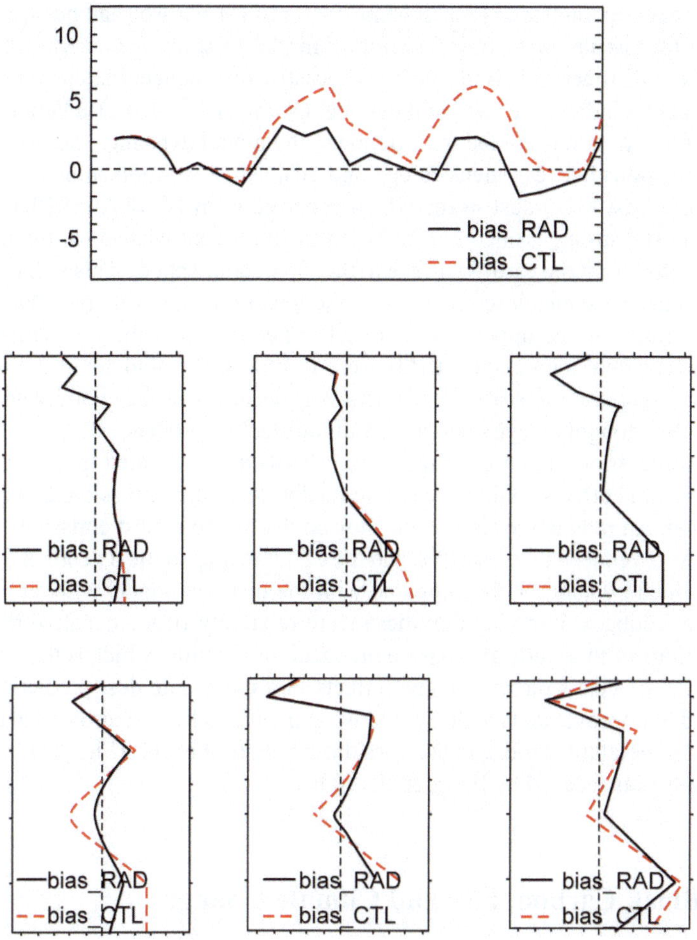

Fig. 4.11 **a** The seven station-averaged bias of surface temperature within 72 h modeled by CTL [for the experiment period without DS feedback, the model produced a 72-h DS forecast named control test (CTL)] and RAD experiments [then, with the radiative effects online, the model repeated the same forecast (RAD) for the same DS case]; **b** Vertical profiles of temperature bias modeled by RAD and CTL experiment from surface to 100 hpa.

storm frequencies are strongly positively correlated with wind velocities upwind. However, the dust storm frequencies were strongly negatively correlated with the precipitation during the same year (June of the prior year to May) as well as that during the preceding summer (June–August). The frequencies of the dust storms also were negatively correlated with soil moisture anomalies. These relationships held across a broad region running northeast-southwest from the northeast China and China-Mongolia border to the Taklimakan Desert, and the results strongly imply that soil moisture plays an important role in determining whether dust storms are likely to occur.

A radiative parameterization scheme for the dust aerosol has been developed within a mesoscale dust-storm forecasting model to study the direct radiative effects of the dust aerosol. Both online forecasted dust concentrations and updated dust reflective indices are used in the model (Wang et al. 2010). Radiation-induced temperature variations are fed back online to the model dynamics, resulting in two-way interactions between meteorology and dust aerosols. Application of the model to a typical East Asian dust storm, which occurred from 16–18 April 2006, showed that the dusts' strong extinction effects led to significant changes in the longwave radiative flux from the earth's surface to the top of atmosphere. This radiative effect tended to decrease the air temperature in the lower dust-aerosol layers but increase the temperature in the upper dust layers. The net effects of these variations in the three-dimensional temperature fields were to reduce the cold air in the upper atmosphere, increase the sea-level air pressure, decrease surface wind velocity, and weaken the Mongolian cyclones as a result of blocking effects.

The forecasting model of Wang et al. (2010) also indicated that dust-induced meteorological effects could have strong feedbacks to dust emission, transport, and deposition. An interactive simulation showed that the total dust emissions from the ground were reduced by over 50 %, and the 72-hr average optical depth of dust aerosols declined by about 33 % compared with the one-way model without the dust's radiative feedback, implying that these were relatively strong negative feedbacks. The findings of this study also have a practical implication, which is that the online calculation of dust's direct radiative effects in a mesoscale dust prediction model may lead to improvements in the predictive capabilities for meteorological elements such as temperature, wind, and pressure during the dust events (Fig. 4.11). A similar conclusion was reached by Pérez et al. (2006).

4.3 Black Carbon, Fire and Climate Change

4.3.1 Introduction

Black carbon (BC), sometimes called elemental carbon (EC), is produced by the incomplete combustion of fossil fuels and biomass. This substance is important in the context of loess research because it can serve as a tracer of the Earth's fire history (Goldberg 1985). The earliest fires can be traced back to ~360 million years ago, that is, during the Late Devonian when atmospheric oxygen concentrations first became sufficient to support fires (Cope and Chaloner 1980). Large fires can emit sizeable quantities of both particles and gases, and thus they have widespread effects on the composition of the atmosphere (Andreae and Merlet 2001). Human evolution and the accompanying reliance on fire have led to greater impacts from fires since the origin of our species on Earth. Around the time of the Industrial Revolution, anthropogenic emissions first become a major concern to society owing to their harmful effects on human health. More recently concerns have been raised about anthropogenic substances in the context of global warming. Greenhouse gases produced by fires, especially carbon dioxide (CO_2), were the initial concern for climate scientists, but more recently BC in the form of soot has become a matter of increased interest.

Various species in the atmosphere have specific optical properties that impact the Earth's climate system. Combustion aerosols can perturb the radiation budget by virtue of their light-scattering and light-absorbing effects and by their influence on cloud microphysical processes (Crutzen and Andreae 1990). Early estimates of pyrogenic emissions suggested that biomass burning could rival fossil fuel use as a source for some atmospheric pollutants (Crutzen and Andreae 1990; Seiler and Crutzen 1980). The contribution of BC to global warming may be second only to that of CO_2 (Jacobson 2001b); however, these two substances affect climate in slightly different ways. First, greenhouse gases warm the atmosphere by absorbing infrared (outgoing) terrestrial radiation, whereas BC warms the air by absorbing both incoming sunlight[2] and reflected longwave radiation. Second, BC exists as aerosol particles which have shorter lifetimes (several days to tens of days) than CO_2, (~50–300 years), and therefore the spatial distribution of BC is much more patchy than CO_2. This may help explain the lack of coherence in global long-term (millennial-scale) historical reconstructions of atmospheric BC, especially in form of soot.

A history of the carbonaceous aerosol would be invaluable for investigating the role of BC in past climate change. Unfortunately, the historical records of atmospheric BC are limited in number, relatively short in duration, and from only a few locations (Lavanchy et al. 1999; McConnell et al. 2007; Xu et al. 2009a). Records of estimated BC emissions from fuel consumption (Ito and Penner 2005; Novakov et al. 2003) are also quite limited. In this section, we focus on the relationships between BC, paleo-fires, and climate change. Differences among various types of BC are discussed along with the methodologies used to quantify them. The reconstructed history of BC and the relationship between BC and climate change are then explored.

4.3.2 Definitions of BC and its Various Forms

Although it has long been known that the atmosphere contains carbonaceous particles that are black in color, a systematic discussion of BC can only be traced back a few decades, beginning with the publication of Edward Goldberg's book, *Black carbon in the environment* (Goldberg 1985). The characteristics of carbonaceous particles are typically defined on the basis of their optical properties, not their chemical properties (Hansen and Novakov 1989; Novakov 1984). Two categories of carbonaceous aerosols have been suggested in previous studies, organic carbon (OC) and BC. Of these, BC was defined as the carbonaceous aerosol that strongly

[2] Mineral dust also absorbs sunlight but due to its strong reflective characteristics, it is generally thought to be an agent of atmospheric cooling.

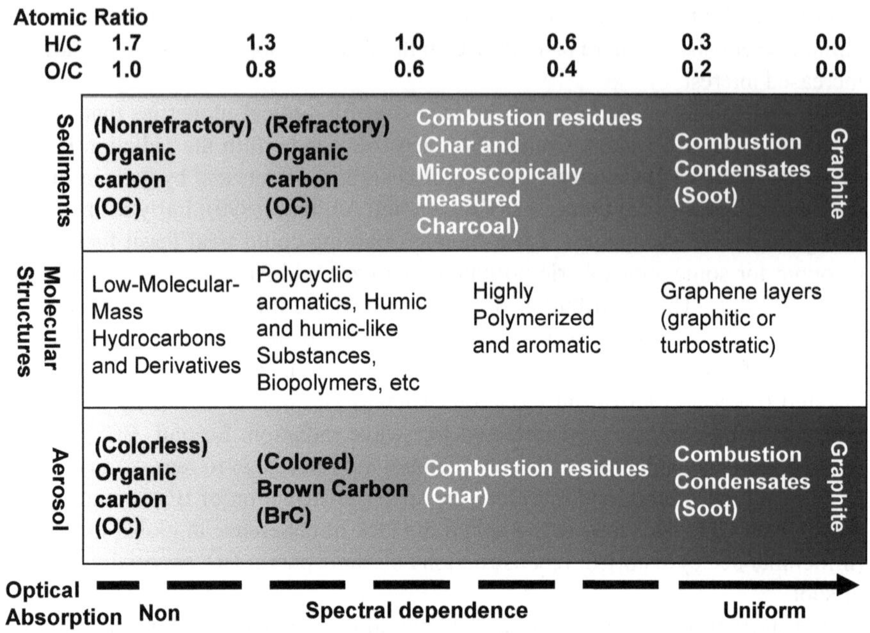

Fig. 4.12 Optical and thermochemical classification and molecular structures of different terms used to describe sediments and aerosols, including soot, char, brown carbon (BrC), and organic carbon (Modified from Hammes et al. 2007; Hedges et al. 2000; Masiello 2004; Pöschl 2005). The sum of char and soot is defined as black carbon (or termed elemental carbon). The term brown carbon is generally only used in the atmospheric science. In sediment and soil science, microscopic combustion residues are termed charcoal

absorbed visible radiation, and the remaining carbonaceous aerosols were taken to be OC. However, some OC aerosols also absorb visible light, and thus another type of carbonaceous aerosol was proposed: the light-absorbing OC termed "brown carbon" (BrC) (Andreae and Gelencser 2006). Studies suggest that BrC absorbs visible light relatively weakly, with increasing absorption towards ultraviolet wavelengths (Andreae and Gelencser 2006).

It is not surprising that the operational definition of BC does not correspond to a single compound or a group of compounds with well-defined characteristics. Indeed, BC is a mixture of compounds occurring along a "combustion continuum" (Elmquist et al. 2006; Goldberg 1985; Masiello 2004; Schmidt and Noack 2000), that is, the substances that compose BC range from slightly charred biodegradable material to highly condensed, refractory soot-EC. Figure 4.12 summarizes the different forms of carbon considered by sedimentary geologists and atmospheric scientists, and it illustrates the lack of a distinct division between OC and BC.

There are two fundamentally different pathways leading to the formation of the two types of BC (Goldberg 1985). In the first case, atmospheric volatiles condense to form highly graphitized soot-BC; and in the second, solid residues form char-BC directly. Definitions for "char" and "soot" have been proposed by various researchers (Elmquist et al. 2006; Goldberg 1985; Jones et al. 1997; Kuhlbusch 1998; Masiello 2004; Schmidt and Noack 2000). Slight differences exist in these definitions, and some terms have been used interchangeably, such as "char" and "charcoal"; but the basic ideas behind the nomenclature are the same. That is, they all follow Goldberg's (1985) definitions, "charcoals from biomass burning and soot from fossil fuel and wood combustion", although some char comes from fossil fuel combustion and motor vehicle emissions (Han et al. 2009a). Here we follow the main idea of Goldberg (1985) and adhere to his definitions of char and soot, "Char is used for all black-colored material derived from solid or liquid organic matter that has had its chemical composition and ultrastructure significantly altered as a result of heating and retains the recognizable anatomic structure of the parent organic matter, even if only in fragmentary form"; and "Soot summarized all carbonaceous particles that are emitted by the combustion process and formed via gas-to-particle conversion". From these definitions we can see that char and soot are clearly distinguishable because they have different morphological, chemical and physical characteristics (Goldberg 1985; Masiello 2004; Han et al. 2010).

Char and soot also have different optical properties, and therefore they can have somewhat different effects on climate even though they both absorb visible light. Soot exhibits strong light absorption but with little spectral dependence (Kirchstetter et al. 2004; Schnaiter et al. 2003) while in contrast, light-absorption of char is generally weak with a tendency to absorb most strongly in the ultraviolet (Bond et al. 2001, 2002; Kirchstetter et al. 2004; Lewis et al. 2008). The same is true for BrC, leading some scientists to suspect that char is closely related to BrC, a humic-like material. From our perspective, the two main types of BC, that is, char and soot, are distinct (Goldberg 1985), and their origins and physical and chemical properties differ from BrC. For example, although BrC has been suggested to have combustion origins (Bonasoni et al. 2010; Chakrabarty et al. 2010; Lack and Cappa 2010; Lukács et al. 2007), it also can originate from naturally-occurring, colored, humic substances (Andreae and Gelencser 2006). In contrast, char is only produced via combustion. Furthermore, because BrC has been proposed mainly in relation to its light-absorbing characteristics, the term is used solely with reference to aerosol studies and is not meaningful for sediment or soil research. We favor retaining the term "char" and using well defined measurement criteria to differentiate between char and BrC in aerosols (Graber and Rudich 2006).

For sediment and soil studies, the term "charcoal," is sometimes used interchangeably with "char." Charcoal has long been used to describe the material in sediments that has been used to reconstruct the histories of paleo-fires. More technically, "charcoal" is defined as combustion residue that can be recognized under the microscope, and this is the definition we adopt here. Thus, charcoal can be seen as a special kind of large char particle, but these two materials are identified using different techniques. Again, charcoal particles, which often have some associated organic matter, can be recognized with a microscope. Char, on the other hand, is identified by other means, typically by chemical or thermal methods.

4.3.3 Measurement of BC and its Different Forms

A variety of methods have been used for BC quantification. Summaries of the advantages and disadvantages of these methods have been made by various researchers (Elmquist et al. 2006; Hammes et al. 2007; Han et al. 2011; Masiello 2004; Nguyen et al. 2004; Schmidt and Noack 2000; Watson et al. 2005). For aerosol studies, optical, thermal and mixed thermal methods are the most popular while for sediment and soil samples, thermal, chemical, chemo-thermal and biomarker methods are most often used. As BC is produced as a result of a "combustion continuum," its composition can be quite variable (Masiello 2004), and the practical problem for chemists is that the various analytical methods in use can measure different parts of the BC continuum. In addition, analytical artifacts exist: some methods can incorrectly identify non-BC carbon as BC while others oxidize BC during pretreatment. Another technical problem is that the detection efficiencies of different methods can vary across the BC continuum (Schmid et al. 2001).

These confounding factors contribute to the large variability in concentrations measured using different techniques; indeed, the concentration of a given material can vary by up to a factor of 500 measured with different methods (Schmidt et al. 2001). In the atmospheric sciences, differences between measurement techniques are usually smaller, up to a factor of 7 for BC concentrations using different methods, with factor of 2 differences being more typical (Watson et al. 2005). Differences between methods are not consistent, however, with some methods showing higher BC for one set of samples and lower values for others. The causes for the differences remain unclear (Watson et al. 2005). We suggest that the different forms of BC material may vary at least quasi-independently, and therefore it is essential to separate the different forms across the BC continuum. This also will facilitate quantitative comparisons among the different methods.

Recently, (Han et al. 2007b) applied the thermal optical reflectance (TOR) method, which is the most popular method used for the analysis of carbonaceous aerosols, to measure the BC concentrations in sediments and soils. In the TOR method (Chow et al. 1993), OC is evolved in a pure He environment as the sample is heated stepwise to 120, 250, 450 and 550°C; this produces four OC fractions (OC1, OC2, OC3, and OC4). The system is then switched to a He/O_2 mixture, and the oven temperature is raised from 550 to 700°C and then 800°C, producing three EC fractions (EC1, EC2, and EC3). As O_2 enters the oven, the pyrolyzed/charred OC (POC) that was produced in the pure He environment is rapidly oxidized, and a concurrent increase in filter reflectance/transmittance occurs. Correction for the charring effect is accomplished by measuring the time it takes for the filter reflectance/transmittance to return to its initial value. The Interagency Monitoring of Protected Visual Environments (IMPROVE) protocol defines OC as OC1 + OC2 + OC3 + OC4 + POC and EC as EC1 + EC2 + EC3 − POC (Chow et al. 1993). The use of the same analytical methods for both sedimentary and aerosol BC studies is advantageous because that approach will facilitate direct comparisons of the short atmospheric EC time series with the much longer sedimentary EC records.

In addition, (Han et al. 2007a) suggested that the TOR method could be used to separate soot from char by employing a stepwise oxidation of the various carbon fractions under different temperatures and ambient atmospheres. They found that pure char reference materials always evolved at temperatures lower than 550°C in an oxidizing atmosphere while pure soot reference materials always evolved at temperatures >550°C in an oxidizing environment. Furthermore, BC measured by the chemothermal oxidation (CTO) method, which is frequently referred to as soot-EC (Gustafsson et al. 2001; Gelinas et al. 2001; Elmquist et al. 2006), always occurs as EC2 and EC3 in the TOR method. Hence the TOR method can be used to operationally distinguish between char and soot by defining char as EC1 − POC and soot-EC as EC2 + EC3 (Han et al. 2007a). This procedure offers a new way of interpreting the data for thermal carbon fractions commonly obtained and reported using the IMPROVE-TOR protocol. Indeed, this method has been used in aerosol studies to explore differences in the characteristics of char and soot in the atmosphere (Han et al. 2010, 2008, 2009b).

Significant uncertainties exist in BC quantification, especially with regard to analyses of BC in complex matrices (Hammes et al. 2007). Some methods may be optimal for detecting a certain fraction of the total BC pool (e.g., soot) while others may incorrectly identify non-BC material, or in some cases fail to detect BC at all. Thus multi-method inter-comparison studies involving different laboratories are needed. Recently, an inter-comparison was conducted for 12 recommended BC reference materials by 17 laboratories from different disciplines, using 7 different methods. While that comparison provided useful insights into some of the major the problems with current BC methodologies, no consensus could be reached with respect to the best method for BC quantification.

Early comparisons of EC methods only tested potentially interfering materials, laboratory-produced BC-rich materials, BC-containing environmental matrices, and some soils (Hammes et al. 2007; Schmidt et al. 2001). More recently, Han et al. (2011) compared three BC methods (TOR, CTO, and thermal optical transmittance, TOT), on a dated lake-sediment core from Lake Chaohu to explore possible associations between industrial pollution and the data produced using the different analytical methods. Their study showed that the profiles of TOR soot-EC were similar to the profiles of high-temperature TOT EC (that is EC3 + EC4 + EC5 by the TOT method) and CTO EC (EC2 + EC3 measured using the TOR method after the CTO-375 pretreatment) in the sediment cores. Yet, more revealing, these profiles were consistent with the industrialization history of China, suggesting that high-temperature thermal protocols are suitable for differentiating between soot and other carbon fractions. This study also confirms the usefulness of differentiating between different BC forms.

Charcoal is large-particle BC (generally >10 μm), and it is determined microscopically. There are no standard techniques for charcoal measurement in sediments, and a variety of procedures are used for sample preparation, identification, quantification and calibration (Conedera et al. 2009). Information on the particle-size distributions of charcoal in sedimentary materials can provide insights into paleo-fire histories, including some information on their distance from the sampling site. Thin section and sieving techniques have proven to be useful for reconstructing the histories of local fires (Tinner et al. 1998), but there are drawbacks to both techniques. Working with thin sections is expensive and time consuming, whereas sieving is more convenient but lacks standardization. Therefore, most analysts measure particle areas or count the numbers of particles > 100–200 μm. Using areas instead of number count has the advantage that larger particles (which have a higher likelihood of originating from local fires) are more heavily weighted.

Microcharcoal series, which are more indicative of regional fire histories compared with larger particles, are usually obtained by analyzing slides prepared using the same methods as for pollen studies (Faegri et al. 1989). Very small particles < 10 μm can be rapidly analyzed via quantification of the surface area of microcharcoals in these types of slides. This is typically accomplished using an automated light microscope coupled with a digital camera sensor (Conedera et al. 2009).

4.3.4 Concentrations, Spatial and Temporal Distributions of BC in Eastern Asia

As aerosol particles, atmospheric BC loadings vary markedly from place to place, primarily as a result of human activities. Although some aerosol monitoring networks, such as the well-known IMPROVE and STN networks in the USA (Malm et al. 2004; Peterson and Richards 2002) have provided information on regional or national BC distributions, information on the near surface aerosol BC loadings at the global scale is still lacking. At least in a general sense, emissions figures can provide some basic information on the spatial distributions of OC and EC in the atmosphere. For example, global and regional emission inventories indicate that Asia, especially east China and India have the world's largest BC emissions (Bond et al. 2004; Streets et al. 2003), and near surface observations confirm that the highest BC aerosol concentrations occur in these areas.

Currently, no systematic BC monitoring networks have been established in Asia. In China, most of the data for near-surface BC concentrations are scattered around the country, brief in duration, and concentrated in urban areas. Seldom studies have focused on BC concentrations in rural or remote areas. A summary and comparison of these reports (Table 4.4) suggests that aerosol BC concentrations in urban areas vary from 2.0 to more than 10 ug m^{-3}; in rural areas the BC concentrations vary from 0.6 to 4.0 ug m^{-3}; while in some remote areas such as in the Tibetan Plateau the concentrations are generally one or two orders-of-magnitude lower than in urban areas (Cao et al. 2009b). Cao et al. (2007) conducted synchronous monitoring in 14 Chinese cities, and that study showed that both OC and EC varied substantially with season and also varied significantly among cities (Fig. 4.13).

The two different forms of BC, char and soot (Han et al. 2007a), have recently been studied in aerosols ($PM_{2.5}$) in China. In that study, a re-analysis of aerosol monitoring data collected from 14 cities by Cao et al. (2007) showed that char is a major component of $PM_{2.5}$ BC in urban areas, accounting for about 87.3 % and

Table 4.4 Comparison of black carbon concentrations in the atmosphere in Asia

Locations	Type of site	Sampling period	Sample type	Method[a]	BC Concentration ($\mu g\ m^{-3}$)	Reference
Nam Co, central Tibet	Remote	2006.7–2007.1	TSP	IMPROVE	0.082±0.07	(Ming et al. 2010)
Akdala, Shangri-La, Zhuzhang	Remote				0.35±0.01	
Nine rural sites, China	Rural	2006	PM_{10}	IMPROVE	3.6±0.93	(Zhang et al. 2008)
Seven large cities, China	Urban				11.2±2.0	
Zhuzhang, China	Remote	2004.8–2005.2	PM_{10}	IMPROVE	0.34	(Qu et al. 2008)
Akdala, Xinjiang, China					0.35	
Tengchong Country, Tibet	Remote	2004.4–5	$PM_{2.5}$	IMPROVE	1.10±0.80	(Engling et al. 2011)
			PM_{10}		1.50±1.0	
Beijing, China	Urban	2003.8	TSP	Aethalo-meter	0.42±0.244	(Chan et al. 2005)
			$PM_{2.5}$	IMPROVE	6.7	
			PM_{10}		8.7	
	Suburban		$PM_{2.5}$		9.1	
			PM_{10}		13	
Lake Daihai, Inner Mongolia, China	Remote	Four seasons 2005–2007	TSP	IMPROVE	3.1	(Han et al. 2008)
			$PM_{2.5}$		1.85	
Pearl River Delta Region, China	Urban	2002.1–2	$PM_{2.5}$	IMPROVE	6.1±4.0	(Cao et al. 2003)
			PM_{10}		7.8±5.0	
Hangzhou, China	Urban	Four seasons 2001.9–2002.8	PM_{10}	IMPROVE	4.06±4.13	(Cao et al. 2009a)
14 Cities, China	Urban	2003 Winter	$PM_{2.5}$	IMPROVE	9.9	Cao et al. 2007)
		2003 Summer			3.6	
Beijing, China	Urban	2002.7.23–8.2	$PM_{2.5}$	TMO	5.4±2.6	(He et al. 2004)
Gwangju, South Korea	Sub-urban	2002.7.23–8.2			0.3±0.1	
Kyoto, Japan	Near coast	2002.7.23–8.1			1.3±0.8	
Ulan-Bator, Mongolia	No industry	2002.7.24–7.30			0.4±0.2	

Table 4.4 (continued)

Locations	Type of site	Sampling period		Sample type	Method[a]	BC Concentration ($\mu g\ m^{-3}$)	Reference
Kwun Tong, Hongkong Urban		2000.11–2001.2		$PM_{2.5}$	TMO	4.33 ± 0.57	(Ho et al. 2006)
				PM_{10}		4.33 ± 0.54	
Hok Tsui, Hongkong Back-ground		2001.6–2001.8		$PM_{2.5}$		0.42 ± 0.07	
				PM_{10}		0.42 ± 0.07	
Nanjing, China	Urban	2001.2		$PM_{2.5}$	NIOSH	0.61–6.27	(Yang et al. 2005)
Zijin Mountain, China	Sub-urban					1.19–5.69	
Taiyuan, China	Urban	2005.12–2006.2		$PM_{2.5}$	NIOSH	4.8 ± 2.2	(Meng et al. 2007)
Guangzhou, China	Urban	Summer 2004		$PM_{2.5}$	NIOSH	5.7	(Duan et al. 2007)
		Winter 2005				4.4	
Kangwha, Korea	Back-ground	1996–1999		$PM_{2.5}$	TMO	0.26–2.45	(Kim et al. 2000)
Kosan, Korea						0.09–0.43	
Monora Peak, India	Back-ground	2004.12		TSP	NIOSH	0.89 ± 0.33 (0.34–1.40)	(Rengarajan et al. 2007)
Hisar, India	Urban					3.80 ± 1.40 (1.90–8.60)	
Monora Peak, India	Background	2005.2–2006.11		TSP	NIOSH	1.1 ± 0.7 (0.14–2.7)	(Ram et al. 2008)
Mt. Abu, India		2005.5–2006.2				0.5 ± 0.5 (0.05–2.3)	
Nepal Climate Observatory Pyramid, Nepal	Remote	2006–2008	Pre-monsoon	TSP	Multi-angle absorption photo-meter	0.317 ± 0.343	(Bonasoni et al. 2010)
			Monsoon			0.050 ± 0.061	
			Post-monsoon Winter			0.118 ± 0.081	
			ABC hotspots			1.97 ± 0.896	
Sapporo, Janpan	Near coast	2005.5–7		TSP	IMPROVE-NIOSH	1.0 ± 0.68 (2.0–16)	(Aggarwal and Kawamura 2009)

[a] *IMPROVE* thermal optical reflectance method; *Aethalometer* a kind of optical method

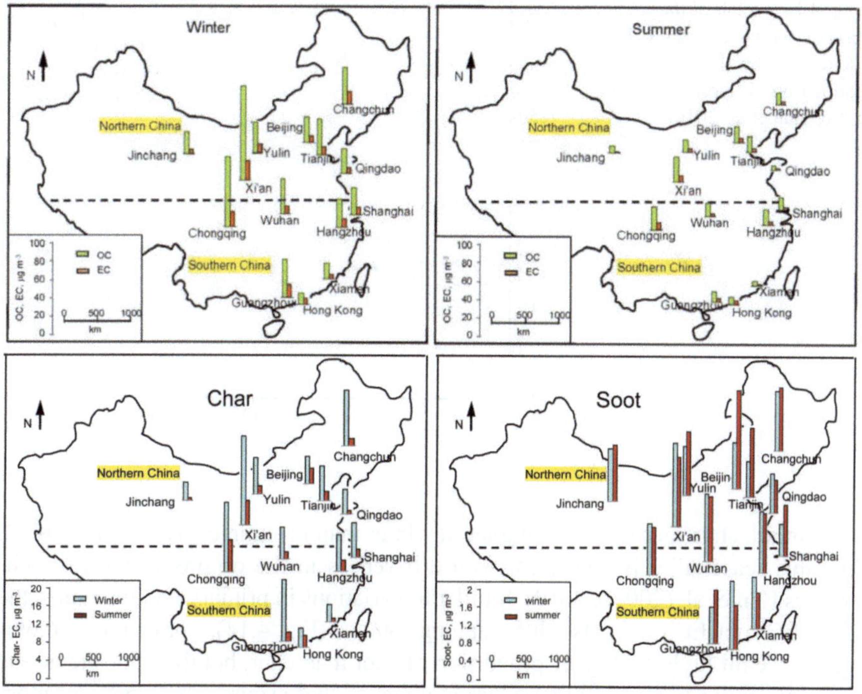

Fig. 4.13 Spatial distribution of organic carbon (OC), elemental carbon (EC), char and soot concentrations in winter and summer in $PM_{2.5}$ from 14 different cities in China. OC and EC are from Cao et al. (2007), and char and soot are from Han et al. (2009b)

66.6% of total BC in winter and summer, respectively (Han et al. 2009b). Char concentrations show similar variations with total BC, that is, large differences both between summer and winter and among the different cities (Fig. 4.13). The average concentrations of char in the 14 cities were 8.67 and 2.41 μg m^{-3} in winter and summer, respectively. In comparison, soot concentrations showed smaller seasonal and regional variations among the 14 cities. The average soot concentrations are 1.26 and 1.21 μg m^{-3} in winter and summer, respectively.

Annual monitoring data from Xi'an (Han et al. 2010) similarly showed that soot concentrations vary less with season than those of char, and the highest concentrations of soot occurred in spring. These results can be explained by the fact that char, which is associated with large BC particles, is more strongly impacted by local sources than soot. The particle-size distribution of soot, on the other hand is skewed toward fine particles, and therefore this component of BC can be regionally or even globally dispersed. In rural areas, both char and soot have relatively low concentrations compared with urban areas. For example, in Daihai, a mountain village region, the average char and soot concentrations during four different seasons were 1.15 and 0.69 μg m^{-3}, respectively (Han et al. 2008). The proportion of soot in total BC in rural areas appears to be higher than in urban areas, and this is in agreement with the different impacts of distant versus nearby sources for char and soot.

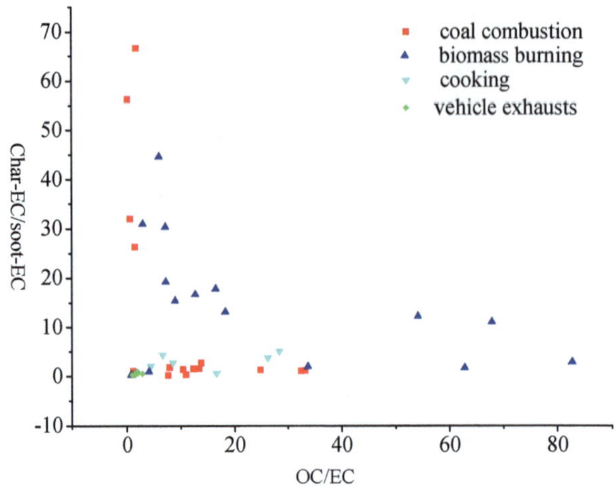

Fig. 4.14 Graph of the primary emissions of OC/EC ratios and char-EC/soot-EC ratios from different sources. (after Han et al. 2010)

Overall, char/soot ratios are higher in winter than in summer (Han et al. 2009b), and this seasonality is consistent with differences in the origins of the two substances Han et al. (2009b, 2010). Similar to variations in primary OC/EC ratios, the char- EC/soot-EC ratios also differ among sources (Fig. 4.14). For example, motor vehicles emit higher a higher proportion of soot than char, but the opposite is true for coal combustion and biomass burning; that is, these sources both produce more char than soot (Han et al. 2010). Thus the higher char/soot ratios observed during the winter can be attributed to seasonal increases in coal and biomass burning, which are widely used for heating in China.

4.3.5 Geological History of BC and the Relationship to Climate Change

Combustion products preserved in sediments can be used to reconstruct past fire histories, and Conedera et al. (2009) presented a comprehensive review of the various materials used for this purpose. In addition to analyzing sediments for BC and charcoal, there are other tools and techniques used to reconstruct the histories of fires; these include dendrochronology (tree rings), molecular markers of combustion (i. e. laevoglucose) and magnetic parameters, among others. However, the most commonly used proxies for long-term fire reconstructions are still charcoal and BC.

Charcoal is the combustion residue that has been most widely used for reconstructing fire histories since the pioneering work of Iversen (Iversen 1941). However, the lack of standardized methods for charcoal analysis makes it difficult to compare charcoal data, and thus there are difficulties in determining fire strength, intensity, scope and extent. Local, regional, and even semi-hemispheric scale fire histories have been reconstructed based on the distribution of different sized charcoal particles (Clark 1988). Macrocharcoal, that is, particles > 100–200 μm, has

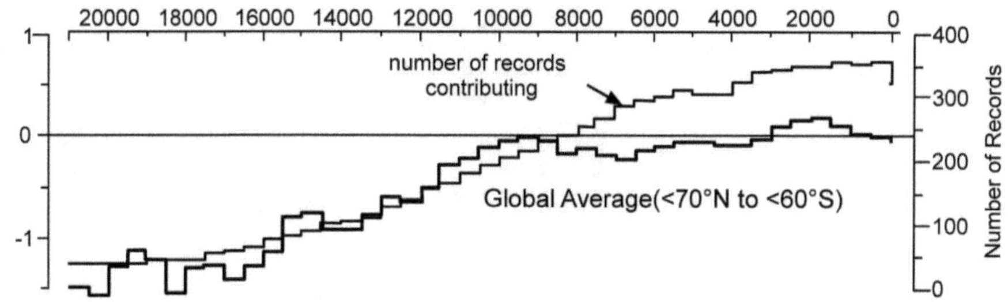

Fig. 4.15 Global summaries of average anomaly Z scores of charcoal values (Z score is a standardized measurement of charcoal accumulation rates, please refer to Power et al. (2008) for calculation details). The number of sites (*gray line*) contributing to each regional summary are compared to the regional average anomaly Z scores (*black line*), revealing the potential influence of increasing sites for each regionally averaged time series. (Power et al. 2008)

been suggested to be a useful indicator of local fires; and microcharcoal, with particles down to about 10 µm, is a better indicator of regional fires. In the last decade the measurement of very small particles < 10 µm also has been used to study fire histories (Daniau et al. 2009, 2010a; Thevenon et al. 2004).

A large-scale (continental to hemispheric) history of fire is needed to understand how the different scales of biomass burning relate to global climate and climate change (Carcaillet et al. 2002; Marlon et al. 2008; Power et al. 2008). Recently, Power and co-workers (Marlon et al. 2008; Power et al. 2008) compiled charcoal records from different regions around the earth to reconstruct a global fire history. A statistical summary of global charcoal data sets has also been used to reconstruct global and regional biomass fires and to study the relationship of those fires to climate events (Marlon et al. 2009). Since information on the global distribution of charcoal older than ~20 kyr is sparse, the available reconstructed global fire history only covers the period from the Last Glacial Maximum (LGM) to the present (Fig. 4.15; Power et al. 2008). The global record reveals spatial heterogeneity in fire activity; however, regional coherence in the patterns of changing fire activities is evident throughout the post-glacial period. From the LGM to the Holocene, the global frequency of biomass fires increased (Power et al. 2008), and this may be the result of climatic controls on biomass burning, e.g., the increase in greenhouse gases and Northern Hemisphere insolation, and the decline in global ice volume (Power et al. 2008) and their consequent effects on temperature (Marlon et al. 2008).

Biomass burning influences the cycling of carbon both in the biosphere and atmosphere, and these effects can contribute to global climate change. For instance, it has been suggested that biomass fires contributed to the increases in atmospheric CO_2 mixing ratios over the past 7 kyr (Carcaillet et al. 2002). Marlon et al. (2008) and Power et al. (2008) reconstructed fire histories at different spatial scales for the past 2 ka, and their studies showed that increases in global biomass burning occurred during a period of global warming; and conversely that a decline in biomass burning occurred at times of global cooling. It is also evident that human activities promote global fires.

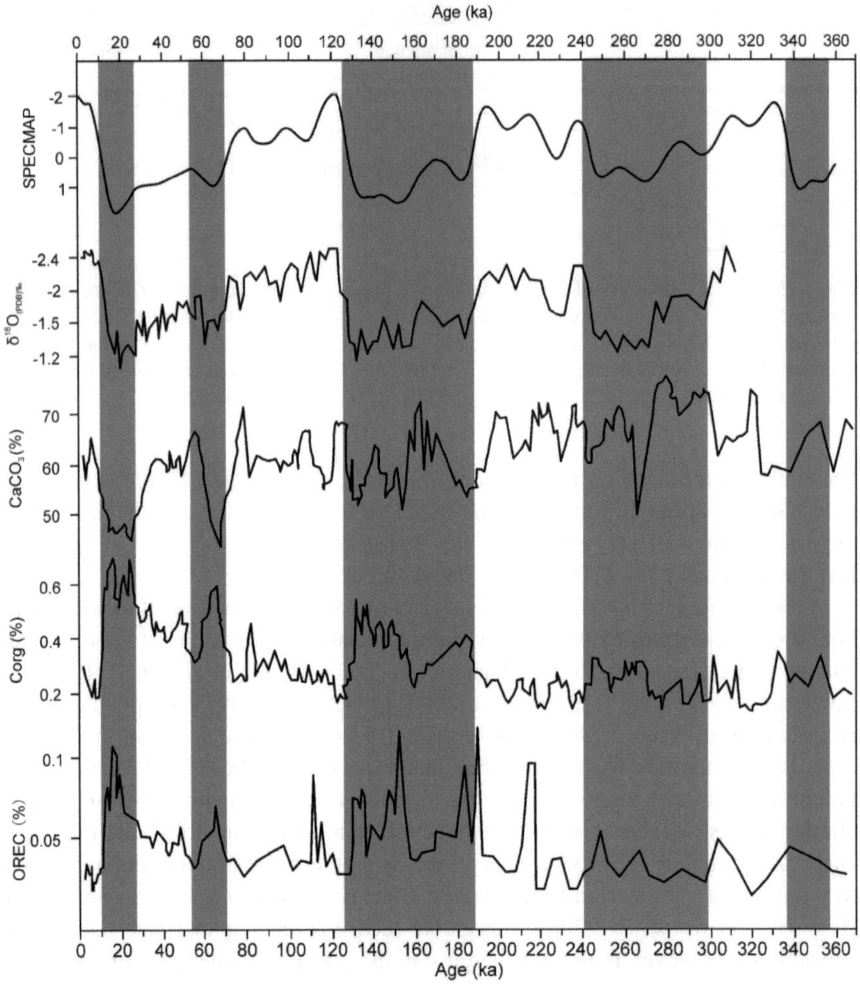

Fig. 4.16 $\delta^{18}O$, $CaCO_3$, organic carbon, and black carbon contents versus age (ky) for marine sediment core MD97-2140, compared to SPECMAP stacked $\delta^{18}O$ record. *Shaded areas* indicate the glacial periods, *unshaded* areas indicate interglacial periods. (after Thevenon et al. 2004)

Records of past biomass fires reconstructed from marine sediments appear to have different BC patterns when compared with those from lacustrine sediments (Fig. 4.16). In particular, the marine records show that intense burning of vegetation occurred during interglacial to glacial transitions (Bird and Cali 1998; Thevenon et al. 2004; Verardo and Ruddiman 1996). Reasons for these discrepancies in the marine versus lacustrine records are as yet unresolved. We suggest that this may be associated with the atmospheric transport and subsequent atmospheric deposition of a much higher proportion of relatively small BC particles to the ocean. The discussion above concerning dust transport (Sect. 4.1.3.1) may be relevant in this regard;

that is, during cold periods the winds are strong and more particles can make their way to the ocean. Further studies are needed to know the real pattern of BC between glacial and inter-glacial intervals

The earliest use of BC as a marine indicator of biomass burning was in the 1970s (Smith et al. 1973). Sediment BC is inert and can persist intact for tens of millions of years (Cope and Chaloner 1980), and this material has provided the basis for a variety of studies of paleo-biomass fires extending far back in time. For example, it has been proposed that global biomass burning occurred during the famous global extinction of the Late Cretaceous (Wolbach et al. 1988).

BC in nearshore marine sediments is transported from terrestrial source regions by both riverine transport, which is relatively slow, and by atmospheric transport, which is much faster, and thus the records in nearshore sediments are not related to isolated fire events in a straightforward manner. Deep-sea sediments from the open ocean, on the other hand, contain BC that is mainly transported through the atmosphere, and therefore these sediments can be used as an indicator of hemispheric or global biomass burning. Indeed, studies of the spatial distributions of BC suggest that BC fluxes to the remote ocean are several orders of magnitude lower than in nearshore areas (Suman et al. 1997). Long-term BC records from marine sediments also suggest that extensive burning of vegetation generally occurred during periods of abrupt global climate change, especially during transitions from interglacial to glacial conditions (Bird and Cali 1998, 2002; Daniau et al. 2010b).

Long-term loess-sediment records provide clear evidence for glacial to interglacial cycles of BC variability (Wang et al. 2005; Zhou et al. 2007). However, published records from loess are contradictory even for sections taken from the same locality. For example, BC concentrations in the Lingtai section of the Chinese LP measured by (Wang et al. 2005) and (Zhou et al. 2007) have differences of an order-of-magnitude despite the fact that the same chemical method was used in both studies. Zhou et al.'s BC ranged quite widely, from 0.05 % to 0.43 %, and the average of 0.22 % from that study was far different from the average of 1.78 % reported by Wang et al. (2005). More important, the interpretations regarding the BC cycles between glacial and interglacial periods were completely different. (Zhou et al. 2007) concluded that "the EC abundance is generally higher in the paleosols than in the loess layers", and this suggests significant increases in biomass fires during interglacial periods. In contrast, Wang et al. (2005) stated that "black carbon mass sedimentation rates in glacial periods are 2–3 times higher than in interglacial periods." From these conflicting results we can see that even though BC abundance vary cyclically between glacial and interglacial times, further efforts are needed to determine why there are such profound differences between samples.

Several comparisons of the variability in BC and charcoal abundances from long-term records also have been conducted. One would expect that the reconstructed charcoal and BC records would exhibit some coherence, but apart from some similarities for specific time periods, significant differences are evident in the long-term records (Thevenon et al. 2010; Wang et al. 1999). Compelling explanations for the observed dissimilarities have not been put forth, but we can offer two possible reasons. One is associated with the analytical methods. That is, charcoal concentrations are determined microscopically, and some organic materials are measured along with the charcoal. BC, on the other hand, is measured using thermal, chemical and optical methods, and these methods generally exclude the organic fraction. Differences related to the presence or absence organic materials could easily confound the results. Another reason for the lack of coherence is that the two substances may be transported in different ways. Both charcoal and BC are transported to their ultimate repository via rivers and through the atmosphere. However, if the relative proportions of charcoal and BC delivered by the riverine versus atmospheric pathways should differ or change with time, then this could lead to the observed divergences in the historical patterns for the two substances.

4.3.6 Atmospheric Soot History

The interrelations between charcoal, BC, paleo-fires and climate change are generally connected to changes in vegetation susceptible to burning and land surface albedo, both of which can influence the Earth's radiative balance. However, BC, especially in the form of soot, is itself an important constituent of the atmosphere in the context of climate: soot is a dark aerosol and strongly absorbs solar radiation (Hansen and Nazarenko 2004; Hansen et al. 2005; Jacobson 2001b; Ramanathan and Carmichael 2008; Ramanathan et al. 2001; Han et al. 2010). The fact that soot concentrations in the atmosphere have been linked to global temperature raises several questions: 1) How did the regional and global atmospheric soot loadings vary over time? 2) What are the relationships between historical variations of atmospheric soot and regional and global temperature? and 3) How much do variations in soot concentrations contribute to global and regional climate change? The answers to these questions hinge on our ability to understand the historical variations in atmospheric BC and soot loadings, and this is the subject of the following discussion.

To date, long-term (millennial) historical variations of atmospheric BC or soot are still poorly documented. Previous attempts to reconstruct historical atmospheric BC variations have been conducted in three ways: 1) from modern observations (no more than 50 years), 2) from the statistics of regional and global energy usage relative to BC production, and 3) from ice cores. Modern observations and statistical analyses of BC production can extend the atmospheric soot history from the last several decades back about 150 years into the past (Bond et al. 2007; Novakov et al. 2003). Ice core records can be used to reconstruct atmospheric soot variations in polar areas and high-altitude snow-covered mountainous areas over even longer times, i.e. to several millennia (Chylek et al. 1992). In eastern Asia there is a famous so-called "third polar region" (Menon et al. 2010), the Himalayas; these mountains are covered by thick ice layers which provide an ideal medium for the preservation of atmospheric BC. Recent studies in this region (Ming et al. 2008; Xu et al. 2009a) showed that in the past fifty years atmospheric BC concentrations have increased about 2.5 times (Fig. 4.17), and this has been attributed to increasing anthropogenic emissions on the Indian subcontinent (Xu et al. 2009a).

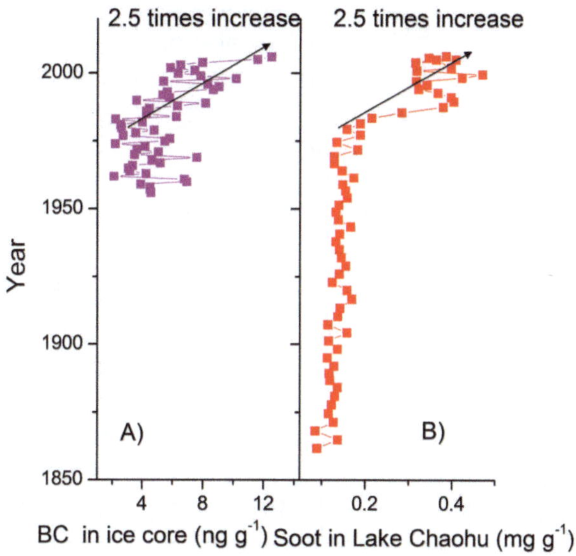

Fig. 4.17 Comparison of BC concentrations in the Zuoqi-upu ice core from the Tibetan Plateau (Xu et al. 2009a) with soot concentrations measured with the TOR method at Lake Chaohu (Han et al. 2011). Similar increases in BC and soot concentrations can be seen for the past three decades

Atmospheric BC concentrations are normally higher over inland areas, especially urban regions, than over the sea. Even though reconstructing historical inland BC has proven challenging, long-term histories of atmospheric BC concentrations are critical to our understanding of relationships between atmospheric BC and climate change. Lake sediments taken from inland environments could provide BC records to address this need. However, as discussed above, the techniques used in the analysis of sedimentary BC measurements differ from those used atmospheric BC samples, and therefore it can be difficult to directly compare and interpret the results for different types of environmental samples.

The TOR method is capable of bridging this gap because it can be used to determine the concentrations of BC in sediments and soils as well as aerosol particles. This method not only provides useful information on the differences among sediments but also can be used to derive information on global BC cycling, given its similarities with aerosol BC measurements. Furthermore, by differentiating char from soot in both types of samples, valid comparisons between sediment and aerosol samples can be made directly.

Although char and soot share combustion sources, their main modes of environmental transport differ because of differences in their particle-size distributions, and this has important implications for the ways in which they are spread around the globe (Fig. 4.18). Soot is more likely to be regionally to globally dispersed than char because the fine (submicron) soot particles can remain suspended in the atmosphere for months (Cofer et al. 1997; Kuhlbusch 1998; Masiello 2004; Ogren and Charlson 1983). Char, is comparably larger particle and therefore has a shorter atmospheric residence time than soot and less easily transported. As a result, the soot concentration in sediments can integrate atmospheric background concentrations over regional or even global scales but this is not the case for char.

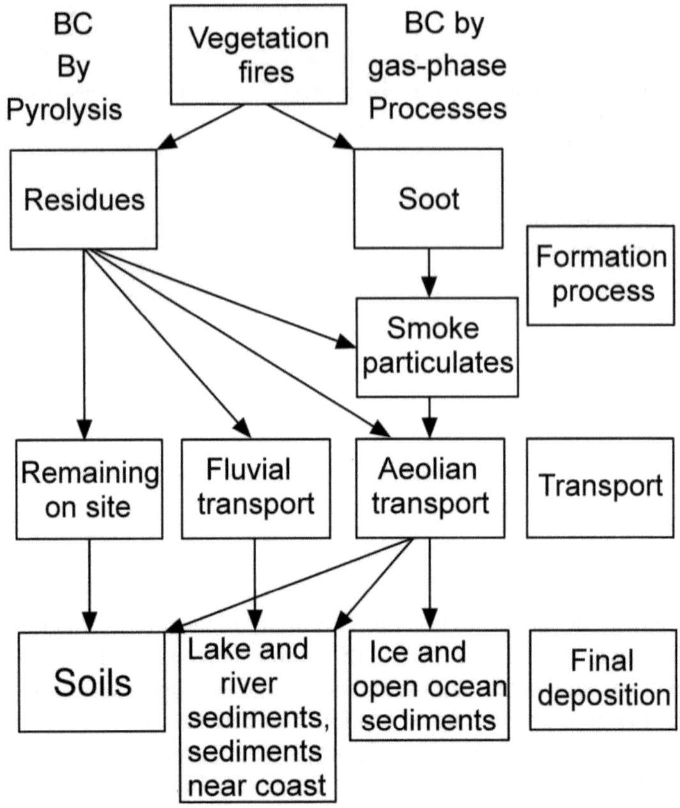

Fig. 4.18 Different transport modes of char and soot in the environment. (after Kuhlbusch 1998)

The ability to distinguish between char and soot in sediments (Han et al. 2007a) provides an opportunity to characterize the historical atmospheric soot variations on a global scale, not only for the pre-industrial period but also over geologic time. Recently, Han et al. (2012) analyzed soot concentrations in sediments from Lake Daihai in Inner Mongolia, China, which is located in a transitional zone between semi-arid and semi-humid conditions. Results of the study showed that sedimentary soot concentrations varied synchronously with several paleoclimate indicators; that is, low soot concentrations occurred during cold periods, including ice-rafted debris events (Bond et al. 1997), the 8,200 cold event and the Little Ice Age. High concentrations of soot occurred during warm periods (Fig. 4.19).

The reconstruction of atmospheric char concentrations seems more complicated than soot because char is on larger particles and therefore tends to be deposited over local to regional scales. The consequence of this is that in contrast to soot, local biomass fires dominate the production and deposition of char, and the impacts of local sources may obscure the signals from long-range transport. Indeed, as in the case of charcoal, reconstructions of regional or larger scale atmospheric char histories will likely require statistical analyses of large datasets for numerous inland sediment cores (Marlon et al. 2008).

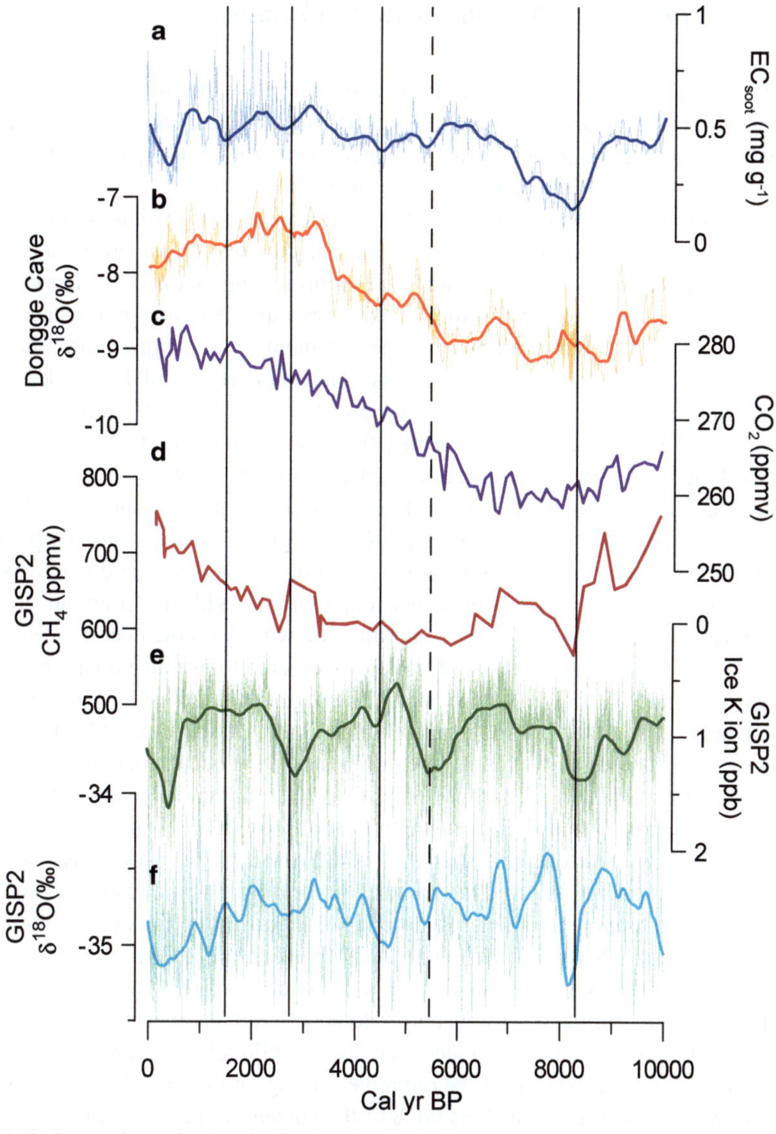

Fig. 4.19 Comparison of **a** Standardized and smoothed ECsoot from Lake Daihai; **b** original and smoothed oxygen isotope data, a proxy for changes in regional moisture availability from Dongge Cave (Dykoski et al 2005); **c** CO_2 from Antarctica Taylor Dome C (Monnin et al 2004); **d** GISP2 CH4 (Blunier et al 1995) from Dome C; **e** GISP2 K (Mayewski et al 1997) and **f** oxygen stable isotope (Stuiver et al 1995). Vertical black lines indicate the ice-rafted debris (IRD) events (Bond et al. 1997), with the exception of the 5,900 yr BP IRD event, which is shown as a dashed line at 5,300 yr BP here. Two conspicuous cold periods in the Holocene—the 8,200 yr BP and the Little Ice Age are also evident in the ECsoot record. All records were smoothed with a 500-year lowess smoothing window with one robustness iteration (Power et al 2010)

4.3.7 Influences of BC and Fires on Climate

The relationships between climate and fire are complex. On the one hand, human activities and climate are major controls on fire occurrences. On the other, fires directly influence the environment that spawned them, with consequences that touch on climate dynamics, the carbon cycle, atmospheric chemistry, terrestrial ecosystems and biodiversity. Nevertheless, long-term records of large biomass fires are as yet unavailable, and therefore, the full impact of climate change on biomass fires largely remains a matter of speculation. One complicating factor is that reconstructed records of local to regional scale biomass fires are generally not synchronous (Power et al. 2008). For example, marine sediment cores suggest that intensive biomass fires occur during glacial periods (Thevenon et al. 2004; van der Kaars et al. 2000; Verardo and Ruddiman 1996), but Power et al. (2008) found that from the LGM to the Holocene, global biomass fires increased markedly.

Two things are normally needed to start and sustain a fire: one is an ignition source and the other is sufficient available fuel (Pyne and Goldammer 1997). It has been estimated that a change of just 1 K in the global average land temperature would result in a change in lightning activity of about 40 % (Reeve and Toumi 1999), and this may help to explain why higher temperatures evidently promote fires. In biomes with sufficient fuel, sources of ignition are what limit the occurrence of fire, so areas with dry climates may be reasonably expected to produce a larger number of fires compared with more humid areas (Flannigan et al. 2000; Fry and Stephens 2006; Huerta et al. 2009; Margolis and Balmat 2009; Olsson et al. 2010). In some arid or semi-arid areas, however, the factor that limits the occurrence of fire is available fuel. For example, black carbon analyses from the LP in China have shown that paleo-fires increase when the Asian summer monsoon, which carries warm moist air to north China, is strong (Zhou et al. 2007). From a global perspective, warmer temperatures would be expected to increase evaporation, and the resultant drying of fuels could lead to more and perhaps larger and longer-lasting fires (Carcaillet et al. 2002; Daniau et al. 2010b; Marlon et al. 2008; Power et al. 2008).

Apart from preserving information on paleo-biomass fires, BC also acts as an agent for climate change, that is, its ability to absorb radiant energy can directly affect the earth's radiative balance (Jacobson 2001b). Globally averaged atmospheric BC is thought to exert a net positive radiative forcing[3] at the top of the atmosphere, and this may amount to as much as 60 % of the current radiative forcing due to the CO_2 greenhouse effect (Jacobson 2001a; Ramanathan et al. 2001). In contrast to greenhouse gases, which warm the atmosphere by absorbing infrared (terrestrial) radiation, BC absorbs incoming solar radiation and also reflects some of it back to space. Both of these effects of BC lead to a reduction of solar radiation at the surface, a process sometimes referred to as "dimming" (Ramanathan et al. 2001). This stabilizes the vertical thermal structure of the atmosphere, and in so doing reduces

[3] Radiative forcing is the change in stratospherically adjusted radiative flux at the tropopause, compared to 1750AD as defined by the Intergovernmental Panel on Climate Change (Forster et al. 2007).

evaporation and rainfall. It should be noted that dimming at the surface due to the absorption of solar energy by BC can be factor of 3 larger than dimming due to reflection, and therefore the warming caused by BC generally exceeds its cooling effects (Ramanathan et al. 2001). In addition to the absorption of solar radiation, BC also absorbs outgoing (longwave) radiation reflected both by clouds and the Earth's surface back towards space: this is the main contributor to the positive radiative forcing by BC (Ramanathan et al. 2001).

Soot not only alters the Earth's radiation balance by absorbing and scattering radiant energy (the direct effects) but also by acting as cloud condensation nuclei (the indirect effect); the combination of these direct and indirect effects can influence weather in the short term and climate over the longer term (Hansen and Nazarenko 2004; Jacobson 2001b; Ramanathan and Carmichael 2008). Recognizing that biomass burning produces copious amounts of char (Han et al. 2010), the potential radiative forcing of char also needs to be taken into account to fully understand the interactions between BC and climate. Compounding this, char can affect climate indirectly by altering the composition of ecosystems, especially vegetation, over local to regional scales (Clark 1988).

Char exhibits much weaker light absorption than soot (Han et al. 2010), and a practical consequence of the difference in light-absorbing properties between char and soot (Han et al. 2010) is that the effects of Chinese BC on the Earth's radiation budget may have been overestimated. As noted above, the BC from China is comparatively rich in the less absorbing char. The existing priorities for controlling pollution emissions have been based on total BC, without taking into account the differences in light absorption by soot and char, but this clearly oversimplifies the science.

In developed countries, BC originates primarily from motor vehicles, which emit a high proportion of soot. From this perspective, efforts to reduce motor vehicle emissions may be more effective in mitigating current warming trends than reductions in the sources that produce more char. The ultimate objective, of course, is to control emissions of all sources for BC, and there are other considerations involved besides climate forcing, but the efforts to mitigate the effects of BC should take into account the differences in physico-chemical properties of soot and char.

The fundamental driver for evaporation is the absorption of solar radiation at the Earth's surface, particularly by the ocean, and here too, BC can disrupt the natural balance. Due to the dimming effects of BC, the evaporation of water both inland and over the open ocean may be suppressed, thus potentially affecting the genesis of the Asian monsoon. In addition, precipitation over land is driven by two major source terms: 1) evaporation from the land surface and 2) the long-range transport of moisture from the oceans and its subsequent convergence over inland regions. Solar heating caused by BC has also been linked with the evaporation of low-level clouds, further perturbing the hydrological cycle. Thus, the recent increases in atmospheric BC emissions and concentrations over Asia may be contributing to the observed weakening of the monsoon system. Considering that BC emissions will likely persist at or near the existing levels for the foreseeable future, their effects on the Asian monsoon could lead to aridification over large inland areas in Asia, and in this way the cycles of BC and dust may become even more directly and tightly coupled.

After BC is deposited on land surfaces it can continue to have an analogous influence on climate. In particular, when BC is deposited on sea ice and snow, it increases the absorption of solar radiation, thus warming the surface and causing melting. Such an effect could lead to an order-of-magnitude increase in the melting of sea ice and snow (Hansen and Nazarenko 2004). In this case the Himalayan Mountains, which are covered by a thick ice layer, such melting would provide more water to surrounding areas. The evaporation of this meltwater could then enhance the regional hydrological cycle. Indeed, the observed increase in precipitation in northwest China over the past three decades may be associated with melting of the ice covering the Himalayas, and there may be other, as yet unknown, consequences of the BC induced melting of ice for regional climate.

There remain many unknowns concerning how BC and fires interact with climate, but the research conducted to date has convincingly demonstrated that the relationships are complex and likely to affect humans and life in general in significant ways. For a truly comprehensive evaluation of the contributions of char and soot to climate change, additional sampling, methods development, and modeling efforts are needed. Eventually, information obtained on long-term variations of both char and soot in Chinese loess and other sedimentary materials will provide a stronger foundation for climate assessments and also a means for validating models that deal with the workings of climate.

4.4 Global Iron Connections for Desert Dust, Marine Productivity and Climate Change

4.4.1 Introduction

Dust emissions from central Asia have been ranked as the second largest source of dust on a global scale (Duce 1995). The transport of eolian dust from central Asia began at least 22 Ma ago (Guo et al. 2002), and around 6–8 Ma before the present (BP), large quantities of Asian dust began to be produced, transported and deposited. These processes led to the formation of the expansive loess deposits in central China (e.g. Sun et al. 1998b; Qiang et al. 2001; An et al. 2001; Xu et al. 2009b). Mineralogical, geochemical and sedimentological analyses have established that arid lands in central Asia also are the main source for deep-sea sediments of the NPO (Rea 1994; Petit et al. 1999).

Atmospheric dust is important for the dynamics of modern climate (IPCC 2001), and variations in dust fluxes in the past have been linked to changes in paleoclimate (An et al. 1991b; Zhang et al. 2002). In addition to direct effects (Arimoto 2001), the influence of dust on global biogeochemical cycles has far-reaching implications for climatic evolution (Martin 1990). Indeed, dust can be viewed a biogeochemical link connecting the land, air and oceans (Ridgwell 2002). Most notably, the emission of dust and the terrestrial nutrients it carries (particularly iron) when deposited in the open ocean can promote primary production; the ensuing growth in marine phytoplankton populations in turn can increase the drawdown of atmospheric CO_2, potentially cooling the planet. (Martin et al. 1994; Kumar et al. 1995; Boyd et al. 2000, 2004; Watson et al. 2000; Johnson et al. 2003; Röthlisberger et al. 2004). The most recent models estimate that atmospheric CO_2 removed by global airborne dust can amount to 30 ppm (Bopp et al. 2003). This is equivalent to approximately one-third of the total change in CO_2 during glacial-interglacial periods (Petit et al. 1999).

The "iron hypothesis" for glacial-interglacial control of atmospheric CO_2 was first proposed by Martin (1990) who caused a major rethinking of what limits primary productivity in the ocean. The main premise behind this hypothesis is that biological productivity in the modern Southern Ocean is limited by an insufficient supply of the micro-nutrient iron, not the "classical" nutrients nitrate and phosphate. Despite the availability of NO_3- in some parts of the oceans, the standing stocks of phytoplankton there are relatively low, leading to the designation of these areas as 'high nitrate low chlorophyll' (HNLC) regions. Chemical and biological oceanographers have shown that the modern supply of iron to biota in these HNLC regions is insufficient to drive the so-called biological pump at its maximum efficiency (i. e., complete NO_3- uptake). Therefore, any reduction in the dust/iron supply would be expected to intensify the Fe limitation where it already exists and possibly reduce productivity where it was previously not limiting. The net result of this would be a reduction in the rate of CO_2 uptake by the ocean.

Recent studies have begun to focus on the underlying mechanisms that control the global dust/iron budget. For instance, Jickells et al. (2005) developed a framework for the connections between iron in desert dust, ocean biogeochemistry, and climate. Figure 4.20 offers a schematic view of the linkages between iron and dust and climate. Highlighted are four essential components of the system (shown clockwise from top): 1) the state of the land surface and dust availability; 2) atmospheric aerosol loading; 3) marine productivity; and 4) a measure of the climatic state (such as mean global surface temperature). The sign of the connections linking these components varies; where the correlation is positive (for example, increased atmospheric aerosol loading varies directly with increased marine productivity), the line connecting the two components is terminated with a solid arrowhead. Where the correlation is negative (for example, increased marine productivity varies inversely with CO_2 and colder climate), the termination is an open circle. Connections with an uncertain sign are terminated with an open arrowhead. Mechanisms that influence a particular link are displayed in italics (for example, *radiative forcing* in ocean CO_2 sequestration). Finally, the "water tap" symbols represent a secondary mechanism modulating the effect of a primary mechanism; for instance, a change in global precipitation strength and distribution will alter the efficiency with which entrained dust is transported to the open ocean.

If pathways of successive connections can be traced from any given component back to itself, a closed or feedback loop is formed. An even number (including

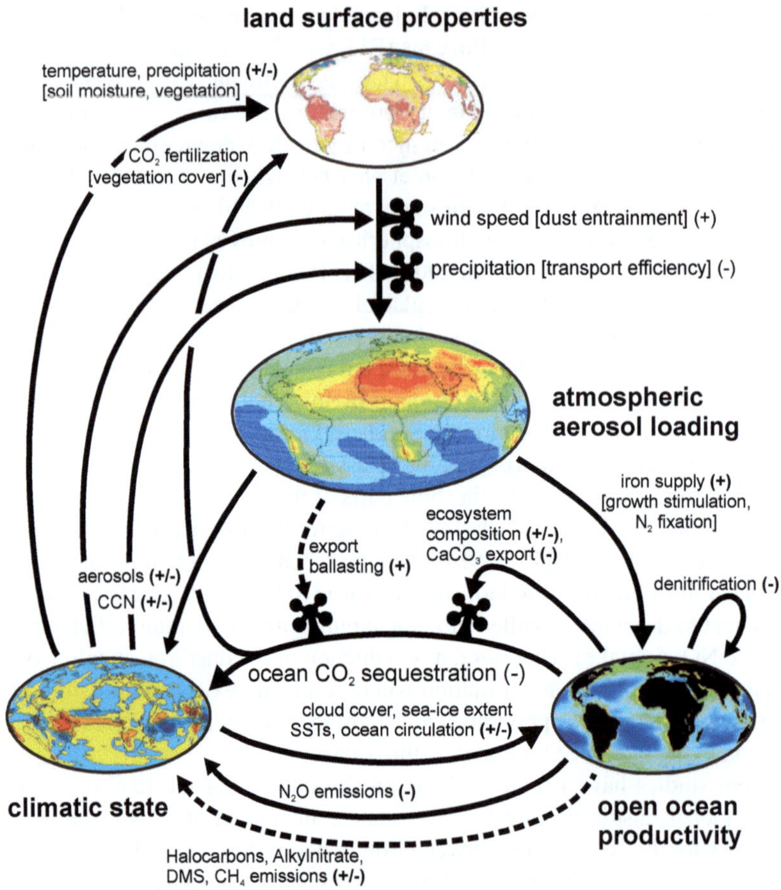

Fig. 4.20 Schematic view of global iron and dust connections. (Jickells et al. 2005)

zero) of negatively correlated connections counted around the loop gives a positive feedback, which will act to amplify a perturbation and tend to destabilize the system. Conversely, an odd number of negative correlations gives a negative feedback, dampening any perturbation and thus stabilizing the system. For instance the pathway, atmospheric aerosol loading → marine productivity → climatic state →dust availability → atmospheric aerosol loading contains two negative and two positive correlations and is considered to be positive overall. In contrast, marine productivity looping back onto itself contains a single negative correlation and thus represents a negative feedback.

Because of the profound climatic implications inherent in global iron connections (GICs), a variety of geologic paleoclimate archives have been examined to study the relationships between dust fluxes and climate change (Martin et al. 1994; Kumar et al. 1995; Boyd et al. 2000, 2004; Watson et al. 2000; Johnson et al. 2003). Records of dust in continental sediments, marine sediments and ice cores have been used for this purpose because they reflect GICs under conditions different from those today. Examination of these connections provides information on historical patterns that can serve to establish boundary conditions for modeling studies. A core of over 3 km in length recovered from Vostok (in Antarctica) has been found to record over 420 thousand years (ka) of Earth climatic history (Petit et al. 1999). Recent analyses of the CO_2 content of air bubbles contained in this core revealed a pattern similar to that of temperature, but CO_2 was found to be anti-correlated with the dust record. Relatively high atmospheric pCO_2 (ca. 280 ppm) occurred during interglacials, and low atmospheric pCO_2 (ca. 190 ppm) during the most intense glacial periods. This anti-correlation between dust and pCO_2 may be explained by GICs operating in the past. As the relationships between dust supply and atmospheric CO_2 are not direct, there are two main considerations to keep in mind when examining these results for possible dust impacts on pCO_2.

First, the dust records in the Vostok core reflect emissions from the Patagonian desert in Argentina (Petit et al. 1999). The issue here is that the dust fluxes there may differ from the global picture because the emission and transport of dust do not necessarily follow the same trends in the Northern and Southern Hemispheres (Duce et al. 1991). Second, one would expect that increased dust input into the ocean during full glacials would enhance the "biological pump" and promote the uptake of atmospheric CO_2. But paleoproductivity variations in the surface ocean are not well documented, especially in HNLC regions. Therefore, the observed relationship between dust and CO_2 in the Vostok ice core provides only provisional support to for the GIC mechanism. To consider the relationships in greater detail, the following discussion covers a wide range of topics concerned with dust, productivity, and CO_2 levels in the past. Taken together, this information makes a strong case that GICs play a significant role in climate and climate change over different spatial and temporal scales.

4.4.2 Contemporary Iron-Productivity Connections in the Asia-Pacific Region

The high-latitude North Pacific is one of the three largest HNLC regions in the world, and it represents a sizeable reservoir available for carbon sequestration. It is also a major depositional area for Asian dust. The annual flux of Asian dust into this region is estimated to be 420 Tg (GESAMP 1989), which is about half of the total dust production from Chinese deserts (Zhang et al. 1997). Asian dust contributes 90% of the clay minerals to the NPO, and it provides about two-thirds of the biologically available iron. Marine productivity is normally limited by the concentrations of biologically available nutrients, and the air/sea exchange of Asian dust can provide micronutrients to marine phytoplankton. More specifically, it is the input of eolian iron that promotes the growth of marine organisms in HNLC waters, and this in turn can stimulate biological nitrogen fixation, which is a key driver for biogeochemical cycles in oligotrophic waters. Silicon is another element in the deposited dust that has biological significance: it is essential for diatoms because their cell walls, called frustules or tests, are composed of biogenic silica (hydrated silicon dioxide). Diatoms are the main primary producers in the NPO, and the air/sea exchange of Asian dust is thus linked to primary production through these micronutrients.

Even a single dust storm can affect marine ecosystems. For example, a marked increase in the blue-green alga, *Trichodesmium thiebautii* in the East China Sea was observed after a large dust storm there (Asaoka et al. 1975). Young et al. (1991) observed an analogous increase in primary productivity after a dust storm reached a sub-Arctic region of the NPO, where Fe concentrations were low but other nutrients relatively abundant. Bishop et al. (2002) deployed two autonomous robotic floats to measure particulate organic carbon in the NPO sub-Arctic region in 2001. Those sensors showed a doubling of biomass in the mixed-layer that coincided with the passage of a large dust storm. Tsuda et al. (2003) conducted a mesoscale iron-seeding experiment in the NPO for which 350 kg of iron were added to a "patch" of the ocean. The researchers observed an increase in phytoplankton standing stocks and decreases in macronutrients and dissolved carbon dioxide. Yuan and Zhang (2006) used an array of land-based meteorological stations and sediment traps to document dust events between December 1997 and April 2000. They found a correlation between dust events on land and the fluxes of opaline and pennate diatoms in the sea. Iwamoto et al. (2011) found that Asian dust aerosols were scavenged by sea fog, a process that enhanced their deposition in the ocean. These authors attributed the increases in productivity they observed at several sites in the NPO to dust deposition, and they calculated that the increase in bio-available iron was comparable to the amount used in the Tsuda et al. (2003) seeding experiments.

4.4.3 Biogeochemical Records of Past Global Iron Connections

Here we summarize the past changes of dust, paleoproductivity and atmospheric CO_2 as recorded in continental sediments, marine sediments and ice cores; these provide evidence for GICs on geological time scales.

4.4.3.1 Global Distribution of Dust in the Past

The natural sources of mineral dust today mainly lie in the arid and semi-arid tropics and subtropics (Fig. 4.21), with the most important source regions being the Sahara–Sahel in northern Africa and the Gobi–Taklimakan in central Asia (Middleton et al. 1986; Middleton 1991; Bergametti et al. 1992). The amount of dust trans-

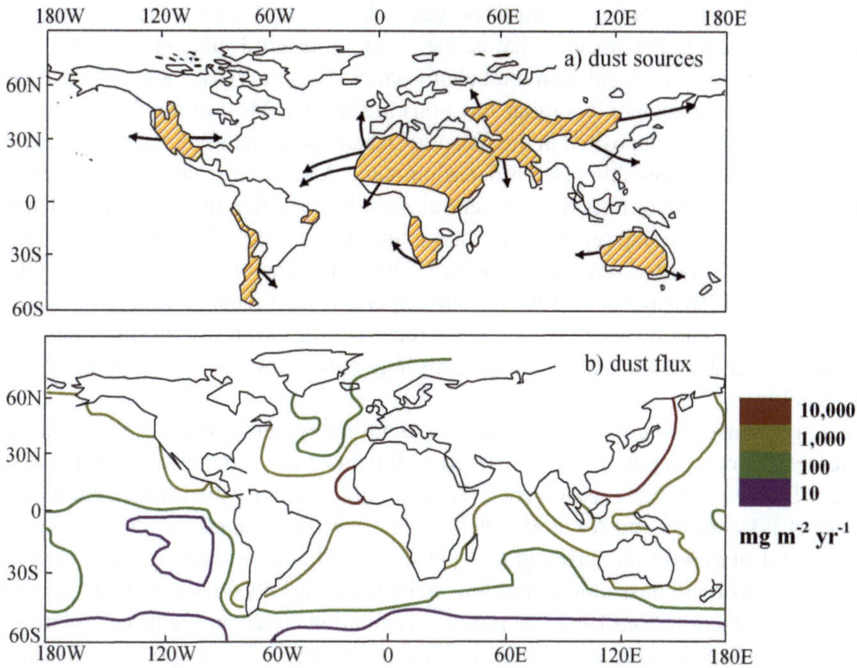

Fig. 4.21 The locations of modern dust sources, transport pathways and deposition zones: **a** modern dust source regions and wind trajectories reconstructed from observations of dust storms after Livingstone and Warren (1996); **b** global fluxes of mineral aerosols to the ocean (mg m^2 yr^{-1}), after Duce et al. (1991)

ported from the Sahara has been estimated at 0.6–0.7 Pg yr^{-1} (d'Almeida 1989), of which ~0.22 Pg is deposited in the North Atlantic (Duce et al. 1991). The total emission of Asian dust is around 0.8 Pg yr^{-1} (Zhang et al. 1997) of which ~0.3–0.4 Pg yr^{-1} is deposited into NPO (Prospero et al. 1989; Zhang et al. 1997). Other, relatively minor source areas for dust include the Middle East (Middleton et al. 1986; Middleton 1986; Pease et al. 1998), the northern part of the Indian subcontinent (Middleton 1986), southern South Africa (Prospero 1981), the interior basins of the southwestern USA and the southern High Plains (Orgill and Sehmel 1976; Lee and Tchakerian 1995; Bach et al. 1996), southern South America (Ares 1994; Buschiazzo et al. 1999) and central Australia (Middleton 1984; McTainsh and Pitblado 1987; Shao and Leslie 1997). Dust deflation can also occur under periglacial conditions at high latitudes (Pye 1995; Landvik 1998), but in today's climate, the areas affected are of limited extent and do not contribute significantly to the total global dust budget. Due to the larger source regions for dust north of the equator, ~8 times more atmospheric Fe is deposited in the Northern Hemisphere than in the Southern Hemisphere (Duce et al. 1991).

The global distribution of dust sources today is similar with that of the past, especially the late Pleistocene; that is, the major Northern Hemisphere dust source regions are presently located in the Sahara-Sahel region of Africa and central Asia, and they were in generally similar locations in the past. Most of the dust produced in Africa is rapidly deposited in the Atlantic Ocean and generally not well preserved in continental basins although on occasion some African dust does reach Europe. In comparison, dust emitted from central Asia has a different fate: much of it is deposited in downwind regions proximal to the sources, especially the LP, with the remainder transported onward to the NPO. Indeed, some of the dust travels as far as Greenland (Biscaye et al. 1997). Complementary dust records from land, ocean, and ice cores can provide an integrated record of Asian dust concentrations and dynamics, and the array of available sampling sites allows for the study of spatial variability.

Large-scale changes in dust flux on glacial/interglacial cycles are well documented in loess deposits, ice cores and sediments from around the world. These records show that dust fluxes were 2–20 times higher during the last glaciation (see Sect. 4.1.3.1 above and An et al. 1991a; Rea 1994; Delmonte et al. 2004). The increased dust deposition during glacial periods could be due to a number of factors: 1) an expansion of dust source areas due to increased aridity could result an increase in dust production (Rea 1994), 2) reduced vegetation cover, and reduced soil moisture could expose more erodible soils (Petit et al 1981), 3) stronger glacial winds (Cohmap members 1988) could entrain more dust and carry it to remote areas, and (4) reduced intensity of the hydrological cycle could increase the residence time for dust in the atmosphere allowing it to be transported farther from its source(s) (Yung et al. 1996).

Dust flux records preserved in continental sediments, marine sediments and ice cores have been scrutinized by scientists as they have sought to understand the environmental cycling of dust and how that cycling relates to climate. In the past several decades, loess deposits have been used to recover the spatio-temporal distribution of Asian dust over different time scales (An et al. 1991a; Zhang et al. 1994). During interglacial periods, the dust fluxes ranged from 8–12 g cm^{-2} kyr^{-1}, but during glacial periods, the dust deposition increased by 2–3 fold, varying between 30 and 45 g cm^{-2} kyr^{-1}. The increased dust flux during the glacial period observed in four continental profiles from China has been attributed to increased aridity in the dust source regions and strengthening of Asian winter monsoon.

Large regions of the ocean floor, especially the deep basins, are covered almost entirely with clay minerals derived from the atmosphere. These regions thus appear to be ideal for reconstructing dust records. Rea (1994) estimated that only 5% of all the terrigenous material entering the oceans is transported there through the atmosphere, but these authors also noted that most of the riverine sediments are deposited before they reach the deep ocean. In hemipelagic regions, for example, terrigenous fluxes are usually of the order of 18–30 g m^{-2} yr^{-1}, 1–2 orders of magnitude higher than typical eolian fluxes found in open-ocean sediments. In general, dust records confirm the conclusion from regional studies, viz the earth was considerably dustier during the LGM than today, but the dust was also heterogeneously distributed. Records of high-latitude changes in dust deposition contained in polar ice cores suggest a glacial-interglacial difference of 2–20-fold in dustiness. Marine sediment records from low- to mid-latitude regions suggest a smaller glacial-interglacial difference of up to 5 times, and some equatorial regions actually may have experienced reduced dust fluxes during the LGM.

The amounts and characteristics of dust particles found in ice cores vary greatly in response to changes in climate. Under colder conditions, higher concentrations and smaller mineral particles are generally found in ice cores far from the source areas. Dust fluxes to the polar ice caps have changed dramatically over time, too, often in concert with major climate events, such as glacial-interglacial transitions and Dansgaard-Oeschger (D-O) events. Analyses of Greenland ice cores suggest that the magnitude of dust fluxes there changed abruptly. For the last glacial cycle, dust concentrations at NGRIP (North Greenland Ice Core Project) fall into four general categories: 1) C_N (insoluble particle number concentration) of $\sim 0.01 \times 10^6\,\mathrm{ml}^{-1}$ during the early Holocene, 2) $\sim 0.05 \times 10^6\,\mathrm{ml}^{-1}$ during the warm interstadials, 3) $\sim 0.5 \times 10^6\,\mathrm{ml}^{-1}$ during cold stadials and 4) $\sim 1 \times 10^6\,\mathrm{ml}^{-1}$ for the LGM. Further, the dust concentrations in the Greenland cores have been found to follow the climatic changes depicted by δD (the ratio of $^2H/^1H$ in a sample relative to that in standard seawater). During the last glacial period, the dust concentrations show an inverse correlation with δD values. The highest dust values are found during the cold marine oxygen-isotope stages (MIS) 2, 4, and 6, and increases in dust flux for these periods are estimated to be ~ 25, ~ 20 and ~ 12 respectively, as compared with the Holocene (Delmonte et al. 2004).

Antarctic ice cores have lower dust accumulation rates than Greenland cores, but they reach further back in time. For instance, the Vostok (Petit et al. 1999) ice core spans 420 kyr, the Dome Fuji record extends back 340 kyr (Watanabe et al. 1999), and the new EPICA Dome C record is projected to extend backwards in time 800 kyr. The calcium (indicator of dust) record from Taylor Dome (Steig et al. 2000) is not shown here, but for the glacial period it is quite similar to the Dome C calcium record (Röthlisberger et al. 2002). Analyses of Ca^{2+}concentrations and dust size distributions have been performed on the Dome Fuji ice cores (Watanabe et al. 1999; Fujii et al. 2003). The dust concentrations in Dome Fuji and Vostok ice cores are similar over the past three glacial cycles, and this suggests a similar climate history at these two sites.

4.4.3.2 Paleoproductivity Variations in the Past

During glacial times, an enhanced dust supply to the ocean, particularly in the main HNLC areas of the open ocean (primarily the Pacific subarctic, equatorial Pacific, and Southern Ocean) could have 'fertilized' the marine biota and driven atmospheric CO_2 concentrations downward (Martin 1990; Berger and Wefer 1991). Over the last decade, considerable progress has been made in investigating how the strength of the biological pump changed over time in these HNLC areas.

North Pacific Ocean

Palaeodata show glacial increases in paleoproductivity in the mid-latitude northwest Pacific (Kawahata 1999; Kawahata et al. 2000) but little change in the Gulf of Alaska (McDonald et al. 1999). Primary productivities estimated from cores S2612 and LH3166 in the mid-latitudes of the western Pacific (~32°N and 35°S, respectively) during the late Pleistocene demonstrate that maxima were observed at late (MIS) Stage 2, late Stage 4 (middle Stage 4 for L3187) and late Stage 6 while minimum values were observed at Stage 5 (Kawahata 1999). Mass accumulation rates of organic carbon and biogenic opal show similar profiles in these cores. These results and paleontological evidence show that the transition zone between subtropical and subarctic waters migrated nearly synchronously along a latitudinal gradient during the last 150 kyr.

A gravity core (H3571) recovered from the eastern corner of Hess Rise at 34°54.25'N and 179°42.18'E was used to study paleoproductivity fluctuations in the mid-latitude NPO (Kawahata et al. 2000). The mass accumulation rate of organic carbon ($MAR_{organic,}$ mg cm^{-2} kyr^{-1}) exhibited two prominent maxima in MIS 2 and 4 and relatively high values in middle and late MIS 6. Enhanced primary production is likely responsible for the high burial rate of organic carbon in those sediments. Downcore geochemical data from ODP Site 887 in the Gulf of Alaska indicate exceptional episodic increases in primary productivity, marked by rapid settling of biogenic detritus and deposition of diatomacous oozes throughout the Brunhes Chron (McDonald et al. 1999). The diatom-rich intervals are characterized by an abundance of large, rapidly accumulating, diatom frustules. This deduction is supported by ^{230}Th data that indicate a doubling to a quintupling of the sedimentation rate in the two large diatom bands occurring in MIS 5.2 and the 5/4 transition. We speculate that episodic inputs of meltwater- or dust-borne iron from Asian or Alaskan sources may have promoted the extraordinary diatom production events recorded in these sediments.

Equatorial Pacific Ocean

High productivity during glacial periods has been reported for the central equatorial Pacific (Murray et al. 2000), the western equatorial Pacific (Kawahata et al. 1998), and the eastern equatorial Pacific (Weber and Pisias 1999). In one study, temporal and spatial variability in several chemical proxies of export production were studied in a set of sediment cores collected along a cross-equator transect (Murray et al. 2000). These included sites at 5°S, 2°S, 0°, 2°N, and 4°N, and the proxies included Ba, Al, and P. During (interglacial) MIS 11, a lower particle flux and export production was inferred from the chemical data, relative to (glacial) MIS 12. While the spatial geometry of the transect suggested that maximum productivity did occur during both glacial and interglacial conditions, the absolute magnitude of export production, integrated from 5°S to 4°N during MIS 11 was 25–50 % less than during MIS 12 and also 25–50 % less than today. In sediments from the western equatorial Pacific in the West Caroline Basin (site C4402), production increased during (glacial) MIS 2 and 3, late MIS 6, the boundary MIS 6/7 and MIS 8 (Kawahata et al. 1998). Similar observations of enhanced productivity in low-latitude regions have been noted elsewhere. High-resolution records of biogenic carbonate from 15 sites in the eastern equatorial Pacific (sampling resolution ~1 kyr) demonstrate that accumulation rates are uniformly higher during glacials and glacial-interglacial transitions compared with interglacials (Weber and Pisias 1999).

Southern Ocean

Data from the South Pacific suggest little change or even slight decreases in export production during glacial times (Berger and Wefer 1991; Chase et al. 2003). On the other hand, a study of N isotopes in the modern Southern Ocean supports the view that sediment $\delta^{15}N$ records reflect increased surface nitrate utilization during glacial periods, hence increased productivity (Altabet and Francois 2001). Data from the South Atlantic and Indian Oceans indicate lower export production south of the modern-day polar front but higher export production between approximately 53.5°S and 40°S (Kumar et al. 1995; Anderson et al. 1998; Nees et al. 1999; Moore et al. 2000). A variety of complementary radionuclide proxies in sediments of the southernmost Atlantic Ocean over the past 140 kyr show that glacial periods were characterized by a greatly increased biogenic detrital flux from surface waters (Kumar et al. 1995). This increase in export production, which may have contributed to lower concentrations of CO_2 in the glacial atmosphere, was coincident with a greater than five-fold increase in the accumulation of lithogenic iron transported by winds from the Patagonian deserts.

The ratio of $^{10}Be/^{230}Th$ is sensitive to both particle flux and accumulation rate, and it serves as a proxy for the flux of organic carbon to the sea floor. This ratio showed dramatic increases, compared to the present, in the export production of the Subantarctic zone (approximately the region between the present-day positions of the Subtropical Convergence and the Antarctic Polar Front) during glacial periods (Anderson et al. 1998). The last two glacial-interglacial cycles represented in core MD88-779 in the southeastern Indian Ocean over the South Tasman Rise (47°50.69'S, 146°32.75'E; 2260 m water depth) have also been studied (Nees et al. 1999). Benthic foraminifera data suggest that ocean surface productivity at that site increased significantly during glacial periods, particularly during MIS 2, late 3 and 6. Moore et al. (2000) estimated that export production in the Southern Ocean as a whole was by 2.9–3.6 Gt C yr^{-1} higher during the LGM compared with the modern era.

In summary, the sediment data for nine cores taken from major HNLC regions support the view that paleoproductivity increased significantly during glacial peri-

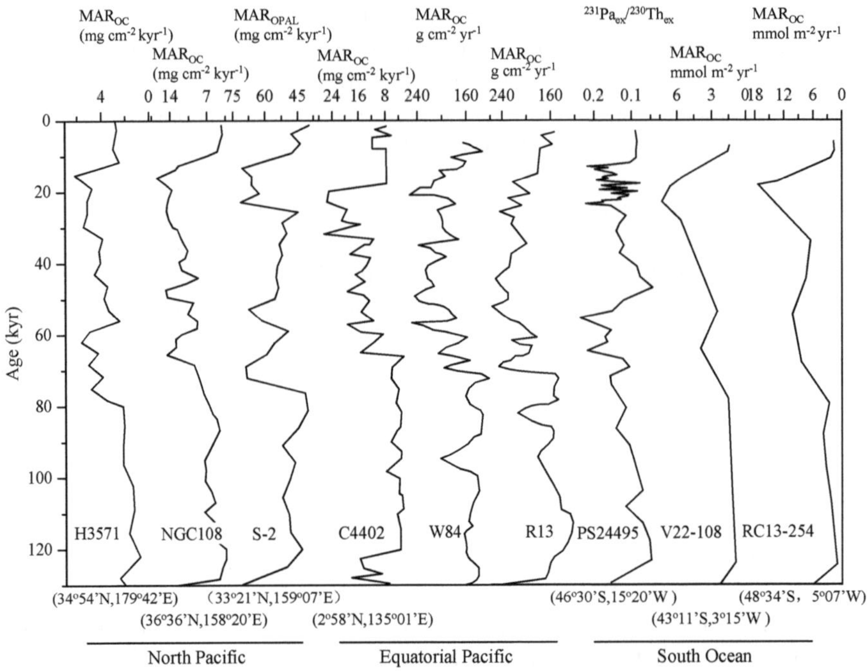

Fig. 4.22 Variations of paleoproductivity recorded in nine marine cores from major high nutrient low chlorophyll (HNLC) regions. H3571 from Kawahata et al. (2000); NGC108 from Maeda et al. (2002); S-2 from Yamane (2003); C4402 from Kawahata (1999); W83 and R13 from Beaufort et al. (2001), PS2449-5 from Asmus et al. (1999); V22-108 and RC13-254 from Anderson et al. (1998)

ods (Fig. 4.22). Though the increases in magnitude are not the same for all regions, it is evident that paleoproductivity during glacial periods is enhanced by as much as double when compared with interglacial periods.

4.4.3.3 Impact of Dust on Atmospheric CO_2

Air preserved in Antarctic ice records atmospheric pCO_2 levels during peak glacial times that are roughly 80–100 ppm lower than corresponding interglacials, with a typical value of 280 ppm (Barnola et al. 1987). These records also show a consistent correlation between atmospheric pCO_2 and temperature for the last four glacial/interglacial cycles (Petit et al. 1999). Since these findings became known, numerical models of global ocean biogeochemistry were developed to quantify the impact of high dust-deposition rates on marine biota and atmospheric pCO_2. For a present-day to LGM increase in dust deposition (Mahowald et al. 1999), current models predict a CO_2 drawdown in the range of roughly 5–45 ppm. A study by Watson et al. (2000) showed a drawdown of atmospheric CO_2 of 40 ppm while on the other hand,

Archer et al. (2000), using an ocean general circulation model with a more primitive biogeochemical cycle, showed that the impact of iron fertilization resulted in only an 8 ppm drawdown of atmospheric CO_2. A state-of-the-art ocean biogeochemistry model with more realistic representations of the dusts' effects showed that the maximum impact of high dust-deposition on atmospheric CO_2 must be <30 ppm (Bopp et al. 2003).

There are substantial uncertainties in our current understanding of the linkages between the ocean and atmospheric carbon and iron cycles, especially how their interactions relate to climate change. The GICs contain positive feedbacks to climate, that is, higher fluxes of eolian iron can perturb the climate system in ways that can in turn increase the production of dust. Therefore, this element and the dust that delivers it to the oceans are keys to understanding potential instabilities inherent in the climate system. Improving the representation of iron biogeochemistry in ocean carbon-cycle models is a prerequisite to gaining a fuller understanding of late Quaternary atmospheric CO_2 and climatic variability.

4.4.4 Relationships Between Asian Dust Concentrations and Ocean Productivity and Atmospheric CO_2 Over Different Time Scales

Although it is well known that arid and semi-arid regions in central Asia are a major source of dust and that the NPO is a primary repository, the connections between Asian dust, ocean productivity, and global climate remain unclear. A synthesis of geologic records that reflect the coupling of dust/iron to paleoproductivity and CO_2 provides a means of clarifying and validating possible dust/iron connections as manifested in global biogeochemical pathways.

4.4.4.1 Iron Loading and Paleoclimate from 6.0 to 2.0 Ma Before the Present

Variations in atmospheric iron loadings from 6.0 to 2.5 Ma ago as reconstructed from the Bajiazui profile in central China are illustrated in Fig. 4.23a. Prior to 4.2 Ma BP, the iron loadings fluctuated from approximately 0.1–0.4 μg m^{-3} (Table 4.5), and then from 4.2 to 3.6 Ma BP, eolian Fe increased from ~0.2 to 0.8 μg m^{-3}. From 3.6 to 2.7 Ma BP, the Fe loadings rose quickly, with the highest values reaching 1.2 μg m^{-3}. After 2.7 Ma BP, atmospheric Fe decreased from ~0.9 μg m^{-3} to 0.2 μg m^{-3}. For most of the record, the dust flux follows a pattern similar to the iron loadings, but exhibits a different trend after 2.7 Ma BP when the dust flux increased and atmospheric Fe decreased (Fig. 4.23a and Fig. 4.23b). This effect is apparently caused by differences in the elemental composition of the red clay matrix of the core, and it is also a function of particle size; that is, smaller particles appear redder because they have a higher Fe content.

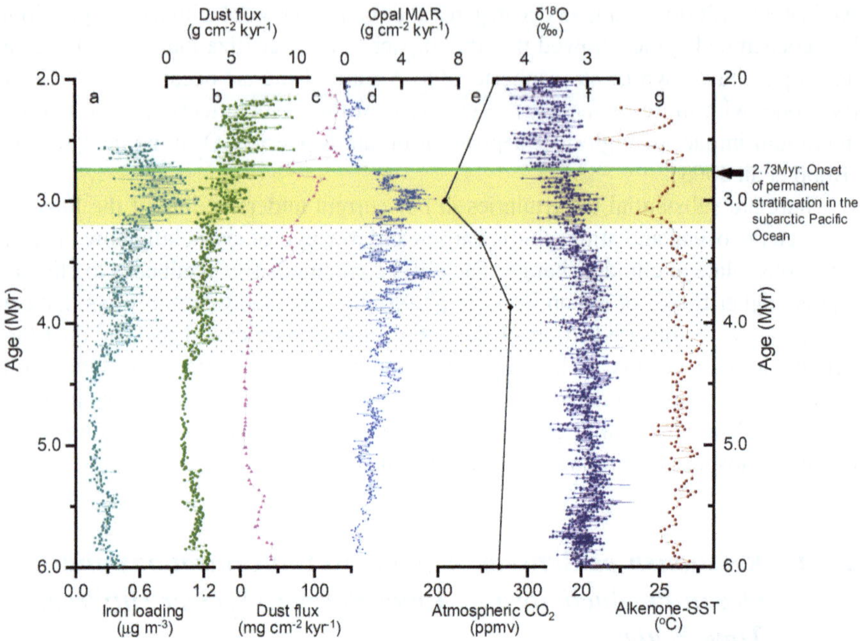

Fig. 4.23 Time series plots of atmospheric Fe loading, dust flux, biogenic opal MAR, atmospheric CO_2, ice volume and sea surface temperatures (SST) from 6.0 to 2.0 Ma BP. Shading highlights the period from 4.2 to 2.73 Ma and *yellow shading* highlights 3.2 to 2.73 Ma. *Green line* at 2.73 Ma marks the onset of permanent stratification in the subarctic Pacific Ocean. **a** Atmospheric Fe loading, reconstructed from the Bajiazui profile (note: Fe loading could not be determined for the period, 2–2.5 Ma), **b** dust flux recorded in the Bajiazui profile, **c** dust flux at OPD sites 885/886 (Rea et al. 1998), **d** biogenic opal MAR at OPD site 882 (Haug et al. 1999), **e** atmospheric CO_2 concentrations (Pearson and Palmer 2000), **f** global ice volume indicated by $\delta^{18}O$ from ODP 846 (Shackleton 1995), **g** alkenone-based sea surface temperature (SST) recorded at ODP site 1084. (Marlow et al. 2000)

The climate was warm and wet ~5 Ma ago (Pettke et al. 2000), and this is consistent with the relatively low dust fluxes and iron loadings as observed from 5.2 to 4.5 Ma BP (Table 4.5). The dust flux at OPD sites 885/886 was found to increase around 4.2 Ma BP (Fig. 4.23c), and according to Rea et al. (1998) that was when the Asian dust source regions became sufficiently arid to produce significant quantities of dust. At ~3.6 Ma BP, the uplift of the Tibetan Plateau accelerated, and this caused further increases in dust production and fluxes (An et al. 2001). From 3.6 to 2.7 Ma BP, the dust source regions continued to dry, and the East Asian winter monsoon strengthened; these two factors led to further increases in dust production (An et al. 2001; Fig. 4.23a and b).

As shown in Fig. 4.23, the reconstructed dust flux from Bajiazui and the sedimentation record at ODP site 885/886 show generally similar patterns from 6.0 to 2.0 Ma BP. The eolian depositional fluxes to both of these environments were relatively low before the drying period which began around 4.2 Ma BP. At that time the

Table 4.5 Iron loadings, dust loadings, and dust fluxes in the Bajiazui profile compared with other studies

Variable	Site	Baijiazui profile age (My BP)				Value and reference
		6.0–4.2	4.2–3.6	3.6–2.7	2.7–2.5	
Iron loading ($\mu g\ m^{-3}$)	Bajiazui	0.22[a] (0.1–0.37)[b]	0.38 (0.17–0.68)	0.59 (0.26–1.01)	0.51 (0.2–0.81)	
	Central China					7–28 (Zhang et al. 2003b)
Dust loading ($\mu g\ m^{-3}$)	Bajiazui	5.1 (2.8–7.9)	7.7 (4.0–12.1)	12.0 (5.3–22.8)	12.9 (7.4–18.3)	
	Luochuan					41–171 (Zhang et al. 2002)
Dust Flux ($g\ m^{-2}\ yr^{-1}$)	Bajiazui	2.0 (1.0–3.3)	2.8 1.5–4.8	4.2 (1.9–8.0)	6.1 (3.7–9.9)	
	Luochuan					9–41 (Zhang et al. 2002)

[a] Mean
[b] Range in parentheses

dust fluxes at both sites apparently increased, with sharper increases after 3.6 Ma BP. The dust flux after 4.2 Ma BP increased by a factor of ~2 for the continental sediments (Fig. 4.23b) and by a factor of ~ 10 for the marine sediments (Fig. 4.23c).

To evaluate temporal changes in dust fluxes and iron concentrations, arithmetic mean values were calculated for different time intervals and compared with one another. The results summarized in Table 4.5 show that the average iron concentration increased from 0.25 $\mu g\ m^{-3}$ from 6.0 to 4.2 Ma BP to 0.43 $\mu g\ m^{-3}$ from 4.2 to 3.6 Ma BP. The highest Fe loading (0.67 $\mu g\ m^{-3}$) occurred between 3.6 and 2.7 Ma BP and then decreased to 0.58 $\mu g\ m^{-3}$ from 2.7 to 2.5 Ma BP (Table 4.5). The reconstructed iron loading from 6.0 to 2.5 Ma BP was less than one-tenth of that observed in the contemporary atmosphere of central China (7–28 $\mu g\ m^{-3}$; Zhang et al. 2003b).

The highest average atmospheric dust load (12.9 $\mu g\ m^{-3}$) at Bajiazui occurred during the interval between 2.7 and 2.5 Ma BP; and the concentration there is roughly 10–40 % of that estimated for the glacial-interglacial atmosphere of Luochuan (35°45'N, 109°25'E; Zhang et al. 2002; Fig. 4.23). Average values for dust fluxes at Bajiazui follow the same trends as the dust loadings there; these are 2–5 times lower than those calculated for the Luochuan section (Zhang et al. 2002; Tab. 4.5). These comparisons indicate that the atmosphere over central Asia from 6.0 to 2.0 Ma BP was much less dusty compared with glacial-interglacial or present conditions.

Haug et al. (1999) produced a high-resolution paleoproductivity MAR record, based on biogenic opal (Fig. 4.23d), at ODP site 882 in the subarctic Pacific. This record provides a useful comparison for the reconstructed Fe time-series history at Bajiazui. The iron loadings and opal MAR patterns covary, especially from 3.6 to 2.7 Ma BP (Fig. 4.23), that is, both sets of values increased from 3.6 to 3.0 Ma BP, and then both decreased at about the same time. The records of opal MAR and iron loadings also show increasing trends prior to 2.7 Ma BP. Opal MAR values increased from ~1.5 g cm^{-2} kyr^{-1} at 5.8 Ma BP to 3.5 g cm^{-2} kyr^{-1} at 4.5 Ma BP; then decreased to 2 g cm^{-2} kyr^{-1} at 4.2 Ma BP before increasing again to 5.5 g cm^{-2} kyr^{-1} around 3.0 Ma BP. After 3.0 Ma BP, opal MAR levels decreased to ~2.0 g cm^{-2} kyr^{-1} at 2.7 Ma BP and then dropped sharply to 0.3 g cm^{-2} kyr^{-1}, ostensibly due to the onset of permanent stratification in the subarctic Pacific Ocean (Haug et al. 1999).

Correlation coefficients (r) between iron loadings and opal MARs were 0.29 for 2.53–2.73 Ma BP, 0.42 for 2.73–3.6 Ma BP, and 0.32 for 3.6–6.0 Ma BP. The highest correlation coefficient ($r = 0.42$) was statistically significant, with a probability for chance occurrence of $p < 5\,\%$, and this is indicative of stronger connections between iron inputs and biological productivity during that period. Biogenic silica data from sites 885/886 also indicate a peak in paleoproductivity in the subarctic Pacific at ~3.0 Ma BP (Dickens and Owen 1996). The diatom MAR record at Deep Sea Drilling Project (DSDP) site 438, which is to the northeast of Japan, showed that paleoproductivity reached its highest value (4 g cm^{-2} kyr^{-1}) from 3.1 to 2.7 Ma BP (Barron 1998). Taken together these widely dispersed records attest to relatively high levels of paleoproductivity over a broad expanse of the NPO at ~3 Ma BP.

Increased quantities of dust in the atmosphere imply greater inputs of iron into the ocean, and for iron-limited phytoplankton communities, nitrogen fixation would likely increase (Orcutt et al. 2001). Under that scenario, more of the available major nutrients would be consumed. Previously, Berger and Wefer (1991) speculated that dust from the Asian deserts may have stimulated the productivity of coccolithophores in the northern Pacific in the late Miocene. Indeed, there are strong similarities between modern oceanic conditions and those of the subarctic Pacific after the late Miocene, about 6 Ma BP (Barron 1998; Haug et al. 1999). Furthermore, a connection between Asian dust/iron inputs and paleoproductivity has been invoked to explain the patterns in carbon isotope records from the late Miocene (~ 7.6–6.6 Ma BP) and marine biological productivity at ODP site 846 in the Eastern Pacific (Fig. 4.23; Diester-Haass et al. 2006).

As noted above, the biogeochemical response of the oceans to enhanced iron inputs can be envisioned as a priming of the biological pump: this causes phytoplankton populations to grow, and the increase in biomass results in the drawdown of atmospheric CO_2. The stimulation of biological activity in surface seawater by eolian iron also could conceivably affect the air/sea exchange of radiatively-active trace gases and aerosol particles (Charlson et al. 1987; Jickells et al. 2005). Since current models only predict a maximum interglacial to glacial reduction of atmospheric pCO_2 of less than 30 ppm due to increased iron loadings (Bopp et al. 2003), there must be other forces at work to account for the additional 50–70 ppm change in atmospheric pCO_2 found in ice cores. These are considered below.

Atmospheric CO_2 exhibited a minimum at around 3.0 Ma BP (Fig. 4.23e), and one explanation for this is that an increase in phytoplankton abundances led to the drawdown of CO_2. Sigman et al. (2004) suggested that increased dust/iron inputs affected the nitrate uptake in the subarctic Pacific around 2.7 Ma BP, and as noted above, Haug et al. (1999) linked opal MAR to stratification of the polar ocean.

Therefore, an increase in the supply of eolian iron and changes in oceanic circulation may have worked together to increase the opal MAR and decrease CO_2. Yet another hypothesis concerning the reduction of CO_2 at 3.0 Ma BP is that it was consumed by chemical weathering of rocks on the Tibetan Plateau (Raymo 1997).

Temporal changes in atmospheric Fe loadings and dust fluxes to the LP, dust fluxes at ODP site 885, and biogenic opal MARs at ODP site 882 all showed similar increases from 6.0 to 2.7 Ma BP. Coincident increases also were observed for dust production in Asia (Fig. 4.23a, Fig. 4.23b) and ocean paleoproductivity in the Pacific (Fig. 4.23d) starting around 4.2 Ma BP. These changes occurred at the same time that CO_2, global ice volume (Fig. 4.23f), and sea surface temperature (SST) all decreased (Fig. 4.23g). Reconstructed atmospheric Fe loadings, dust fluxes at ODP 885 and biogenic opal MARs continued to covary from 3.2 to 2.7 Ma BP. In contrast, variations in atmospheric CO_2 and SST diverged at around 3.0 Ma BP, with CO_2 falling to a minimum value and SST dropping to about 5°C. After 2.73 Ma BP, oceanic productivity appears to be insensitive to the increased dust supply, and this may be a result of the aforementioned subarctic Pacific stratification (Haug et al. 1999).

The coupled evolution of dust, paleoproductivity and CO_2 implies dynamic links amongst them, and these can be summarized as follows. Beginning around 6 Ma BP, dust emissions from central Asia began to increase due to drying on the continent, and as a result larger quantities of dust were transported to the NPO as the westerly winds coincidently increased in strength (Rea et al. 1998). From 3.6 to 2.6 Ma BP, dust concentrations increased and dust transport pathways were altered by the uplift of the Tibetan Plateau (Rea et al. 1998; An et al. 2001). The ensuing increase in dust/iron deposition eased the iron limitation in the surface waters of the NPO, and this promoted the growth of plankton (Jickells et al. 2005) and stimulated N_2 fixation (Orcutt et al. 2001; Falkowski 1997). The operation of the biological pump increased with the greater iron flux, and as more atmospheric CO_2 was sequestrated by the ocean, the climate cooled.

A flowchart of these relationships can be envisioned as follows: increased atmospheric iron loading and flux → increased marine productivity → lower CO_2 → colder climate (Jickells et al. 2005). Feedbacks may have arisen because large-scale cooling, as indicated by the increase in global ice volume (Fig. 4.23f) at ~2.7 My BP, could have led to an expansion of arid lands and an increase of mineral aerosol loadings over Asia (Fig. 4.23a, b, c). The deposition of this fine dust is what formed the red clay deposits in China (e.g. Lingtai section in Sun et al. 1998; Jiaxia section in Qiang et al. 2001; Zhaojiachuan section in An et al. 2001; Qin'an section in Guo et al. 2002; Shilou section in Xu et al. 2009b; Fig. 4.23).

Positive feedbacks between atmospheric dust loads, CO_2 and large-scale cooling (Ridgwell 2002) may thus have contributed to the development or the intensification of the Northern Hemisphere glaciation (NHG; An et al. 2001; Mudelsee and Raymo 2005). Tectonic forcing, especially the closing of the Panama Seaway and the uplift of the Tibetan plateau, has been implicated as the likely root cause of NHG (Mudelsee and Raymo 2005), but several other factors, such as Milankovitch orbital motions, changes in oceanic circulation, and changes in sea ice, also may have been involved (Rea et al. 1998; Hay et al. 2002; Mudelsee and Raymo 2005; Reynolds et al. 2008). Our studies suggest that the increased dust/iron loadings and fluxes that accompanied the Tibetan uplift may have further contributed to global cooling due to both short-term radiative effects (Rea et al. 1998; Arimoto 2001) and longer-term biogeochemical feedbacks (Jickells et al. 2005).

4.4.4.2 Iron Loadings and Paleoclimate During the Last 130 ka

The wind-blown loess-palesol and red clay sequences on the LP compose the most complete record of Asian dust loadings, spanning at least 6–8 Ma BP (An et al. 2001). Loess is deposited rapidly and changes little during pedogenesis, and these properties make it an ideal material for reconstructing high-resolution records of dust loadings and chemistry.

The Lingtai profile (35°04′N, 107°39°E) is located on the western side of the LP, and it is considered a classical loess profile. Dust deposition at the site can be traced back 7 Ma BP (Sun et al. 1998b). The upper portion of the section preserves a continuous dust record for the past 130 kyr, and it has been characterized in terms of magnetic susceptibility, particle-size distribution, and the concentrations of 10 major elements (Si, Al, Fe, Ca, K, P, Mn, Mg, Na, Ti). Noting a correlation between particle-sizes in the section and oxygen isotopes from the GRIP ice core, (SS09 time scale), we can identify three rapid changes in particle-size distributions; these occurred at 1.08 m, 8.24 m and 10.68 m in the profile and correspond to the Holocene-Younger Dryas transition (11.48 kyr), MIS4/MIS5 boundary (79 kyr) and MIS5/MIS6 boundary (129.3 kyr; Johnsen et al. 2001; Yuan et al. 2004). These markers provide the age control needed to apply the grain-size vs. age model of Porter and An (1995); the assumption for this analysis is that the >43 μm coarse particle fraction is linearly related to deposition rate. Elemental fluxes (EFx) are calculated using the equation:

$$EFx = Cx\rho LDR$$

where Cx is the concentration of element x, ρ is the bulk density of dust particles ($1.6\ g \cdot cm^{-3}$), and LDR is the linear deposition rate of loess. The dust flux (F_{dust}) can be calculated using an aluminum concentration in Asian dust of 6 % by weight or by using the same approach for other crustal elements.

Figure 4.24 shows the dust fluxes during the past 130 kyr as reconstructed from the Lingtai loess profile. The dust flux varied from 7.1 g cm^{-2} kyr^{-1} to 27.6 g cm^{-2} kyr^{-1}, with higher fluxes during glacial stages and lower fluxes during interglacials. During the last interglacial (MIS5), the dust flux decreased to its minimum, with an average value of 10.3 g cm^{-2} kyr^{-1}. Dust fluxes increased during the last glacial, with an average value of 20.1 g cm^{-2} kyr^{-1}, or about twice those during the last interglacial. During the Holocene, the dust flux decreased markedly, averaging 19.0 g cm^{-2} kyr^{-1}, but that flux is still higher than during the last interglacial.

We can compare the loess records with those for marine sediments from the mid-latitude NPO (core H3571) to investigate the relationships between dust deposition and marine productivity. During the past 130 kyr, the organic mass accumulation rate at H3571 (MAR_{OC}) fluctuated on glacial-interglacial timescales, with high values during glacial times and relatively low values during interglacials (Kawahata et al. 2000). C/N atomic ratios of organic matter from the site showed that the organic C there is mainly marine in origin, and therefore the MAR_{OC} can be used as a proxy for primary productivity. Hence, paleoproductivity in the NPO evidently increased during glacial times and decreased during interglacials. The subarctic water mass in the NPO shows high productivity, and this region is usually thought to be responsible for the increased productivity during glacial periods, but a shift in the subarctic water mass at H3571 during the past 130 kyr has never been documented.

It can be seen in Fig. 4.24 that the dust/iron fluxes in Lingtai loess and MAR_{OC} values for the NPO (site H3571) followed similar patterns over the past 130 kyr. That is, high marine productivity in the NPO occurred during glacial times when dust fluxes in Asia were high, and conversely, low productivity and low-dust fluxes occurred during interglacials. This demonstrates that glacial/interglacial variations in dust fluxes from Asia are linked to changes in productivity in the NPO. Moreover, a connection between global dust fluxes and CO_2 concentrations in the atmosphere is plausible, if not likely, and this is because biological activity has been identified as a key modulator of atmospheric pCO_2 over glacial-interglacial time scales (Broecker and Henderson 1998). Indeed, productivity in the NPO and CO_2 concentrations in the Vostok ice core are inversely correlated (Fig. 4.24), and this suggests that primary production in the NPO impacted atmospheric CO_2 levels over the past 130 kyr. Variations in dust fluxes from Africa could presumably impact marine productivity in downwind HNLC waters in an analogous manner, and in this way atmospheric CO_2 levels could be influenced by GICs.

4.4.4.3 GICs During the Past 1800 Years

More recent geological and historical records also demonstrate possible dust/iron connections in the Northern Hemisphere. For instance, Zhang (1984) reconstructed the frequency of dust events from ancient Chinese literature, and that study provides a history of dust storm frequencies over the last 1800 years (Fig. 4.25a). Before 1200 AD, records of dust storm activity were relatively low, averaging only 1.5 events per year. From 1200 AD to 1700 AD, the number of dust events increased to 3 per year, on average. For comparison, Fig. 4.25b presents a record of paleoproductivity over the past 1800 years based on $\delta^{15}N$ measurements in lake sediments from Kodiak Island, Alaska. The Karluk lake $\delta^{15}N$ profile (‰) serves as a proxy for sockeye salmon abundances in Alaska, and it can be considered a measure of productivity in the northeastern Pacific Ocean.

The paleoproductivity time-series is plotted as the difference of each $\delta^{15}N$ value from the series mean calculated over the time period presented (Fig. 4.25b); that is, positive $\delta^{15}N$ values indicate increases in fisheries productivity, and vice versa.

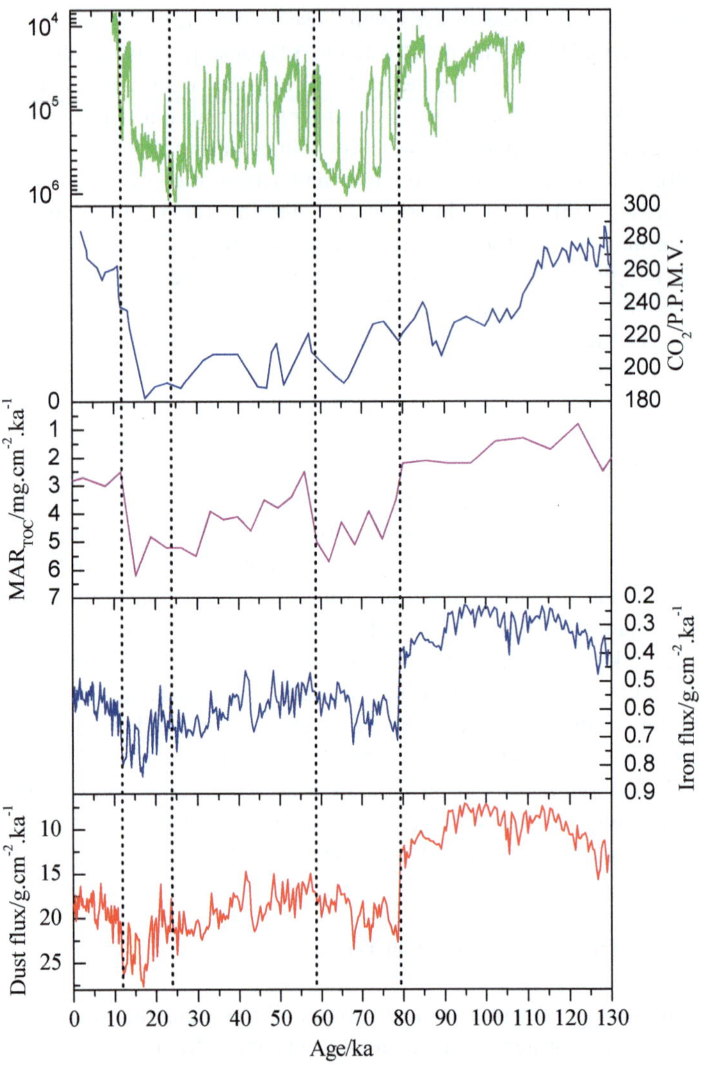

Fig. 4.24 Time series of Asian dust and iron fluxes at Lingtai, paleoproductivity at H3571, and CO_2 in Vostok ice over the last 130 kyr

Before ~AD 1200, the $\delta^{15}N$ was much lighter (~−0.5 %) than average, but then it reached ~2.0 % from AD 1200–1700, indicating that paleoproductivity increased in the NPO during that 500 year interval (Finney et al. 2002). This period of elevated $\delta^{15}N$ corresponds to more frequent dust storms in Asia (Fig. 4.25a.), suggesting once again the existence of a teleconnection between Asian dust and productivity in the NPO. During the same period, atmospheric pCO_2 dropped by about 10 ppm based on ice core records from Low Dome in Antarctica. Although the possible

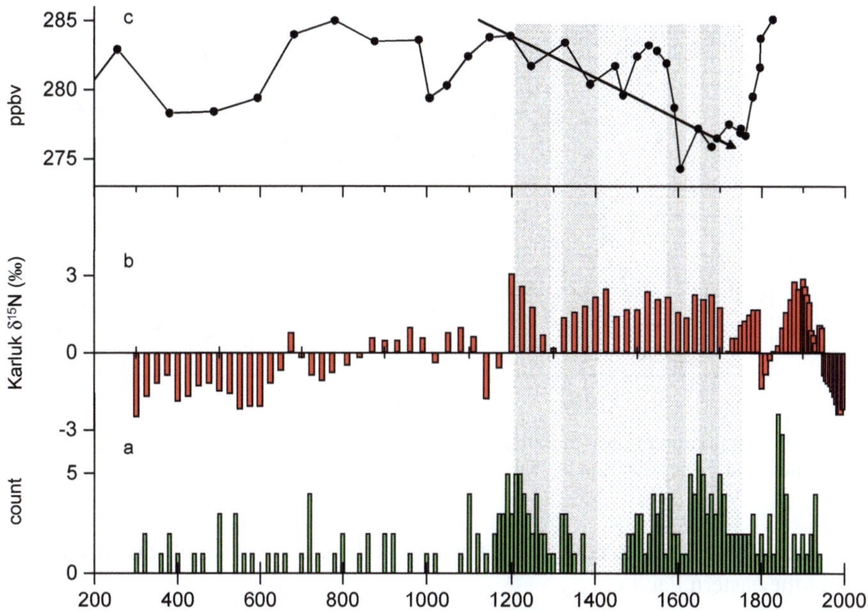

Fig. 4.25 Correlations of dust storm events in China and fishery productivity in the northeastern Pacific Ocean, plotted with CO_2 during the last 1800 years. **a** Dust storm frequencies in China (Zhang 1984), **b** fisheries productivity in the northeastern Pacific Ocean (Finney et al. 2002), **c** atmospheric CO_2 concentrations (after 1000 AD, data from Etheridge et al. 1996; before 1000 AD, data from Indermühle et al 1999; after the industrial Revolution, atmospheric CO_2 concentrations were perturbed by anthropogenic activities and therefore those data are not shown)

explanations for the decline in CO_2 are numerous, dust/iron connections must be included in them.

4.4.4.4 GICs During the Past 50 years

The influence of Asian dust on ocean productivity has been detected in modern times as well (Johnson et al. 2003). There are, however, only a few studies that have addressed the changes that have occurred over the last several decades. Here we consider the results from the available literature in light of the GIC framework. Figure 4.26a presents the Asian dust emissions from 1960 to 2002 as simulated by the Northern Aerosol Regional Climate Model (NARCM; Zhang et al. 2003b). The emissions varied from 0.55×10^{10} kg springtime^{-1} to 2.0×10^{11} kg springtime^{-1}, with a minimum in 1965 and maximum in 1969; from the late 1960s–1990, the total dust emissions decreased. Figure 4.26b shows $\delta^{15}N$ variations of Pacific salmon, which reflect primary productivity changes in the eastern Pacific (Satterfield and Finney 2002). Even with the uncertainties in Asian dust emission estimates, seven matching peaks can be found in the two time series from 1968 to 1988, with a 2–3 year lag in the productivity records. The correspondence between the two records points to a

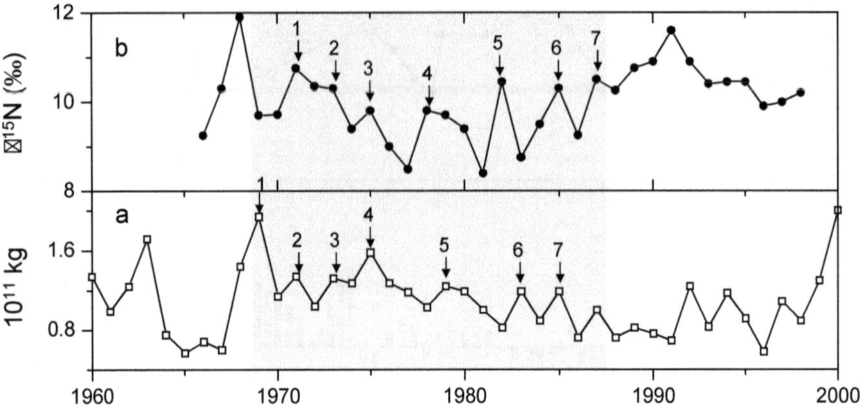

Fig. 4.26 Comparison of dust emissions in Asia (Zhang et al. 2003b) with fishery productivity (Satterfield and Finney 2002) in the Pacific Ocean since 1960

possible high-frequency connection between Asian dust/iron supply and productivity in the eastern Pacific.

4.4.5 Further Considerations Regarding Global Iron Connections

Even as dust-borne iron has attracted increased attention as a possible limiting agent in oceanic productivity in HNLC regions, the roles of other components in dust also have been considered in recent years. For example, some studies have emphasized that complex interactions govern biological activities in the ocean and that limitation by eolian iron alone is likely an oversimplification of the system. Indeed, there is evidence that a multitude of factors can limit primary production in some areas; these include iron, light, macronutrients (Mills et al. 2004), and trace nutrients (such as Co and Zn) (Morel and Price 2003; Schulz et al. 2004). Furthermore, the atmospheric deposition of dust provides more than just iron to marine biota, most notably other micronutrients and carbonate, which under some circumstances may be important for sustaining life in the oceans (Baker et al. 2003; Oba and Pedersen 1999).

At the same time, other chemical componenets of dust may actually have the opposite effect of iron, that is inhibit primary production. For instance, recent studies demonstrate that elemental copper carried by the dust can be toxic to phytoplankton, and so an increase in the deposition of eolian copper could exert a negative impact on marine productivity (Paytan et al. 2009). In a somewhat related vein, a series of studies of African dust emphasized the role of fungal “hitchhikers,” that is, biological contaminants transported along with the dust that can be harmful to corals (Shinn et al. 2000).

As research continues to unravel the ways in which the air/sea exchange of dust and other substances affect marine phytoplankton, it will be important to determine exactly how each component in the puzzle participates in the biogeochemical cycles and how the pieces interconnect. Studies of Chinese loess can provide insights into the concentrations of dust and trace elements in the past, and that information is likely to be useful for these types of assessments.

4.5 Aridification History of Asian Inland as Recorded by Eolian Deposits

4.5.1 Introduction

Since the 1950s numerous studies have been conducted to understand the geologic processes that led to the formation and evolution of the Chinese LP. The first step in this process was to map the broad physiographic zones that compose central Asia. These efforts produced the now familiar picture of rocky deserts to the west, LP to the east, Gobi and sandy deserts to the north, and coarse silt to fine clay bands extending in a thinning sequence outwards from the LP. The establishment of this spatial pattern was complemented by the recognition of stratigraphic boundaries of the typical loess (L1, L9, L15, etc.) and paleosol (S1, S5, etc) layers, based on sedimentological and paleomagnetic characteristics. A northwest to southeast thinning in the last glacial Malan loess also has been observed, e.g., loess deposits at Jingyuan and Lanzhou of the northwestern LP are notably thicker than those found at Luochuan or Weinan of the central or southern LP. The orientation of this thinning pattern suggests that the northwesterly near-surface wind (winter monsoon) is the major vehicle for transporting eolian dust.

Indeed, the westerly jet also carries Asian dust from inland sources over the Asian continent (Liu et al. 1985; Pye 1987) to the NPO (Rea et al. 1985; Rea 1994) and even as far as Greenland (Biscaye et al. 1997; Bory et al. 2002). The LP, adjacent to the major Asian dust sources, is the first sink for the dust particles, and much of the dust is deposited there. Loess deposits on the LP cover an area of about 500,000 km^2 with a thickness of 150–450 m. Loess and red clay sequences on the LP have been studied intensively to reconstruct the history and variability of the East Asian paleomonsoon climate (e.g. Liu et al. 1985; An 2000; An et al. 2001). Aridity in central Asia has been dynamically linked to the evolution of the East Asian paleomonsoon (An 2000; An et al. 2001; Guo et al. 2002) and to the phased uplifting of the Himalaya-Tibetan Plateau (Rea et al. 1998; An et al. 2001).

The distribution of loess from different periods shows a clear spatial pattern. Wucheng loess of the Early Pleistocene is concentrated north of 35°N and west of 110°E while Lishi loess of the Middle Pleistocene extends further to the southeast, with a larger areal extent than the Wucheng loess. The Lishi deposit is also complemented by its southern contemporary, the Xiashu loess, which was deposited in the

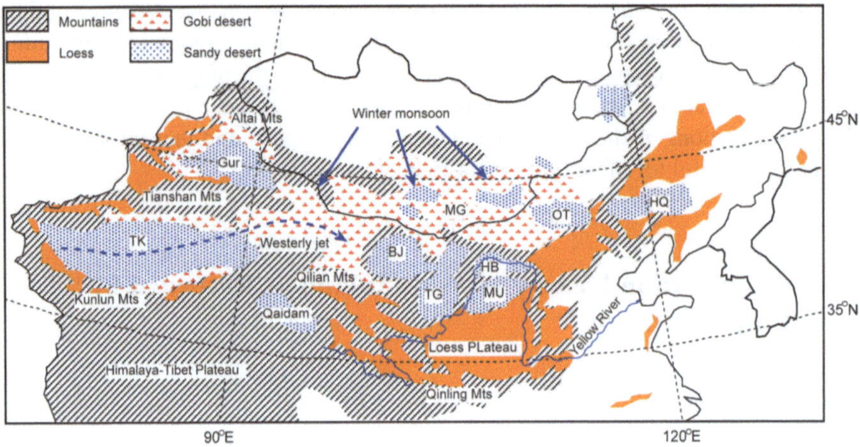

Fig. 4.27 Map showing the distribution of loess deposits, Gobi Desert and sandy deserts in North China (Sun et al. 2008)

middle and lower reaches of Yangtze River. The Late Pleistocene Malan loess has an even larger depositional area, extending to the Miaodao Islands to the east, and to the middle and lower reaches of Yangtze River to the south. The continuously growing loess coverage since the Pleistocene suggests that aridity gradually has increased during the Quaternary. The gradual thinning and fining of those typical loess layers southeastward confirm the eolian origin of the loess on the LP, which therefore can be regarded as a reliable recorder of the history of aridity in inland Asia.

4.5.2 Provenance of Chinese Loess-Red Clay Deposits

Vast desert areas lie to the west and north of the LP including the Gobi (gravel) and the sandy deserts in northern China and southern Mongolia. These deserts, which are strongly influenced by a high-altitude westerly jet and the near-surface winter monsoon circulation, are all possible sources for eolian sediments on the LP (Fig. 4.27). Source apportionments for the sediments have most often been conducted through direct comparisons of mineral composition, elemental characteristics, and Sr-Nd isotopic ratios of loess samples and desert surface soils. The results of such studies have, however, generated disparate opinions on the relative importance of the Gobi Desert to the north, or the western deserts especially the Taklimakan and the Qaidam Basin, as primary dust sources for the LP.

For example, Liu et al. (1985) argued that the Gobi and sandy deserts adjacent to the northern LP and northwestern inland basins were the primary sources for Chinese loess. After a detailed comparison of Sr-Nd isotopic characteristics of sediment samples from the Tarim Basin, North Tianshan, and LP, Liu et al. (1994) later concluded that the geochemical character of loess sediments on the LP was most

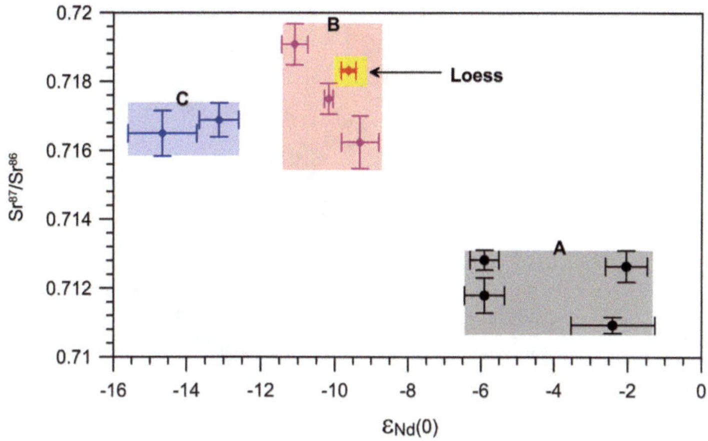

Fig. 4.28 $^{87}Sr/^{86}Sr$ versus $^{\varepsilon}Nd(0)$ for desert surface sediments and loess deposits. A, B, and C are the three isotopically distinct regions. (Chen et al. 2007)

consistent with a Tarim Basin source. Based on the elemental composition of dust samples from the deserts and LP, Zhang et al. (1996) identified the Gobi desert in southern Mongolia, the Taklimakan and surrounding desert, and the Badain Jaran and surrounding desert as the main sources of loess on the LP. More recent Sr-Nd isotopic data suggest that the Tarim Basin, deserts in the mid-western Inner Mongolia, and the northern Tibetan Plateau are the primary sources of dust (Chen et al. 2007; Fig. 4.28).

Based on a study of contemporary meteorology combined with isotopic, elemental, and mineralogical data for the <20 μm size fraction of loess collected from the LP and associated inland basins, Sun and An (2002) concluded that the main sources of dust lie in midwestern Mongolia, the Gobi Desert, and adjacent Chinese sandy deserts immediately north of the LP. A related study based on Sr-Nd data indicated that the Taklimakan Desert is not a primary source of dust but deserts on the Alashan and Erdos Plateaus were good candidates (Yokoo et al. 2004). Recently, Sun et al (2008) analyzed the electron spin resonance (ESR) signal intensity and crystallinity index of fine quartz from two loess sequences. Comparisons with samples from presumptive dust sources indicate that the fine dust deposited on the LP since the last interglacial mainly originated from the deserts in northern China and the Gobi Desert in southern Mongolia (Fig. 4.29).

There are also other areas that have been suggested to be the primary sources for loess. For example, Bowler et al. (1987) asserted, based on geochemical constraints, that the Qaidam basin was the main source for loess. Derbyshire et al. (1998) suggested that large alluvial fans at the northern margin of the Qilian Mountains were the main source for loess on the western LP. That conclusion was reached by comparing monthly dust fluxes from alluvial fans, deserts and sand dunes, within loess and paleosols from Gansu Province. Fang et al. (2004) analyzed forty years of remote sensing data that were used to track storms, and suggested that the Tibetan Plateau itself is a primary source of dust.

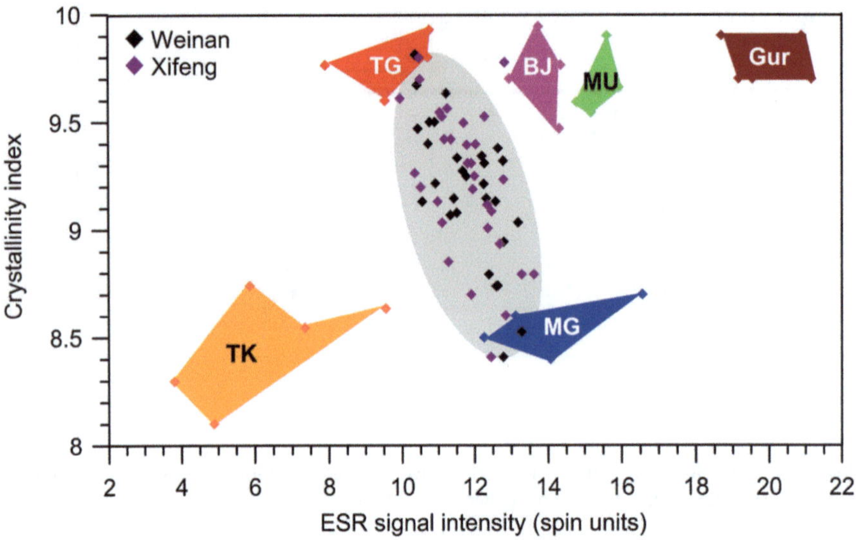

Fig. 4.29 Comparison of electron spin resonance (ESR) signal intensities and crystallinity indices of fine-grained quartz contained in desert surface samples and two loess-paleosol sequences. (Sun et al. 2008)

Continuous red clay and loess-paleosol sequences have developed widely in the central LP, for example, the Lingtai, Xifeng, Chaona, and Jingchuan profiles. Geochemical studies conducted at those sites suggest that the source for dust has not been stable over the last 7 Ma. For instance, Sun (2005) studied changes in the Sr-Nd isotopic ratios in fine loess particles (<20 μm) from the Jingchuan profile, suggesting that regional tectonic activities and climatic change were possible causes of the observed variability in dust sources. This finding is especially meaningful for understanding the loess provenance fluctuations because tectonic and glacial activity in the area was vigorous during the late Cenozoic (Fig. 4.30).

Guo et al. (1999) examined changes in Fe concentrations in late Quaternary loess/paleosol sequences from Changwu and Yichuan, and discovered that eolian dust from different times varied in chemical composition, possibly due to climatic influences. Variations of Fe/Al ratios from the Lingtai profile suggest that aside from the intervals from 4.5 to 3.2 and 2.6–0.5 Ma BP, the dust source regions were variable (Gu et al. 1999). Other studies have led to different conclusions. For example, Hou et al. (2003) determined the oxygen isotopic compositions of fine (4–16 μm) quartz particles from Lingtai while Wang et al. (2007) looked at Nd isotopes from the same site. Those studies showed that late Tertiary red clay deposits and Quaternary loess and paleosols were isotopically indistinguishable, which implied consistency in sources. This conclusion is also consistent with studies of Sm/Nd ratios and the Nd isotopes in eolian dust extracted from ODP site 885/886 in the NPO (Pettke et al. 2000). Taken together, these latter results, in contrast to those above, suggest that over the last 12 Ma, eolian dust in the NPO originated from a

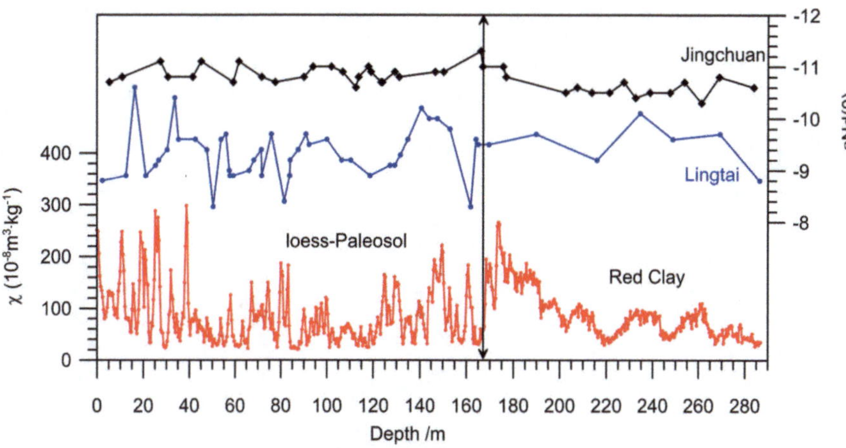

Fig. 4.30 Variations of Nd isotopic composition of two red clay- loess-paleosol sequences in the Jinchuan (Sun 2005) and Lingtai (Wang et al. 2007) sections

stable, relatively homogeneous source area that was not greatly affected by regional tectonics.

4.5.3 Long-Term Drying of Inland Asia Since the Late Oligocene

The growth of ice sheets in high-latitude regions of both hemispheres since the Oligocene has been linked to an expansion of the world's deserts. This period of drying has imposed hardships on those living in arid and semi-arid regions, and it has consequences that reach far beyond Asia through dust-driven teleconnections (Jickells et al. 2005). It has been challenging to reconstruct the history of this desert expansion from the sedimentary sequences in the Asian deserts. However, eolian deposits preserved on land and in the ocean can be used for this purpose. For example, the late Cenozoic eolian deposits that accumulated on the LP and pelagic sediments in the NPO have been extensively used to infer the history and variability of Asian inland drying (Rea et al. 1985, 1998; Rea 1994; An et al. 1991b; Sun and An 2002; Guo et al. 2002). The well-developed loess and red clay sequences on the LP contain high-resolution information on dust production, transport and deposition that spans many millions of years (Liu et al. 1985; An 2000). From these records, we can document the history of desert expansion (Ding et al. 2005), and obtain information on dust fluxes, particle size, etc., that is useful for developing and refining dynamic models of paleoclimate (Kohfeld and Harrison 2001; Harrison et al. 2001; Sun and An 2002, 2005).

An et al. (1991a) first addressed the historical variability in dust input to the LP using reconstructed flux records, and their studies showed that the maximum dust flux to the central LP occurred during the last glacial maximum. Moreover, the deposition rate then was more than three times higher than during the last interglacial or earlier glacial times (Fig. 4.31). Subsequently, the variability in the aridity of dust

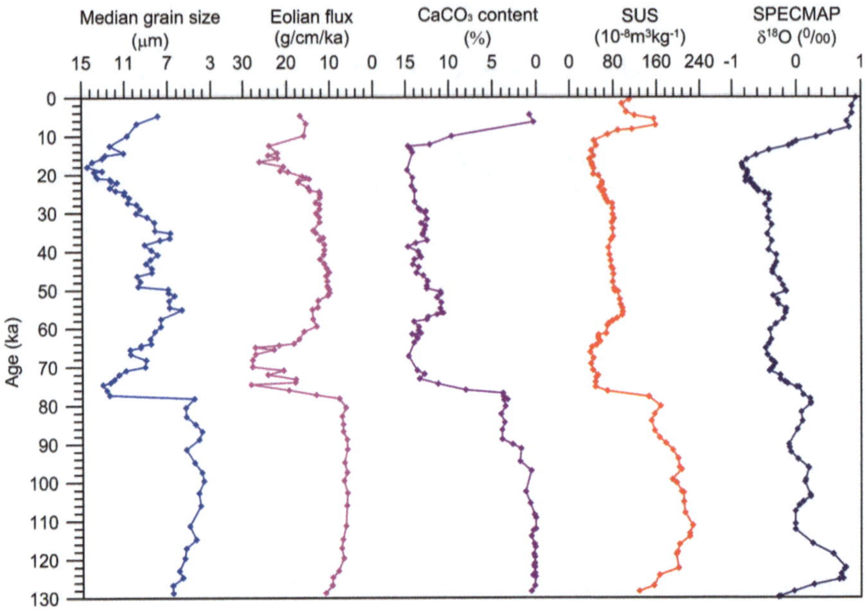

Fig. 4.31 Comparison of median grain size, eolian flux, carbonate content and magnetic susceptibility of Luochuan loess section with the SPECMAP $\delta^{18}O$ curve (An et al. 1991a)

source regions has been inferred from changes in sedimentation rates (Sun et al. 1998a, b; Guo et al. 2002), Al flux variations (Zhang et al. 1999, 2002; An et al. 2001) and dust flux records (Kohfeld and Harrison 2001, 2003; Sun and An 2002, 2005). In addition, dust fluxes to the NPO also have been used as an indicator of aridity in the Asian interior (Rea et al. 1985, 1998).

Analyses of eolian deposits from the Qin'an section on the LP indicate that large source areas for desert dust might have existed in the Asian interior beginning in the early Miocene at least (Guo et al. 2002). This conclusion is consistent with the timing of the formation of gypsum and salt deposits in northwestern China (Liu et al. 1998), and the essential implication of these studies is that desertification in Asia evidently began at least 22 Ma ago. Dust fluxes to the NPO also increased around the late Oligocene and subsequently increased in a stepwise fashion (Rea et al. 1985). Guo et al. (2002) found a peak in the dust accumulation rate at Qin'an around 16–14 Ma BP, and suggested that large-scale drying occurred in the Asian interior at that time (Fig. 4.32). Red clay deposits dated from 11 to 7 Ma BP are widely distributed in the central LP (e.g. Sun et al. 1998a; Ding et al. 1999; Qiang et al. 2001; Xu et al. 2009b), and together with several remarkable peaks in the eolian flux to the NPO (Fig. 4.32; Rea et al. 1998), imply that another significant drying phase occurred in Asia then. Over the last 5 Ma, eolian flux records from the land and ocean show a significant wet-dry variability superimposed on a gradual drying trend in the Asian dust source regions (Sun and An 2002, 2005). The observed increases in eolian fluxes indicate that drying of the dust source region intensified from 3.6 to 2.6 Ma BP (Fig. 4.32).

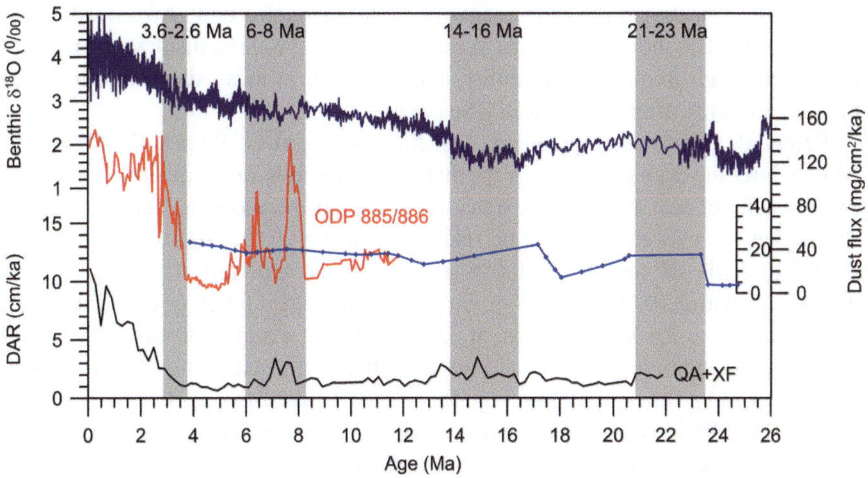

Fig. 4.32 Asian inland drying over the last 22 Ma as indicated by dust fluxes at ODP 885/886 (Rea et al. 1998), dust accumulation rates DAR at Qin'an and Xifeng (QA+XF, Guo et al. 2002) and changes in global ice volume as reflected in benthic $\delta^{18}O$. (Zachos et al. 2001)

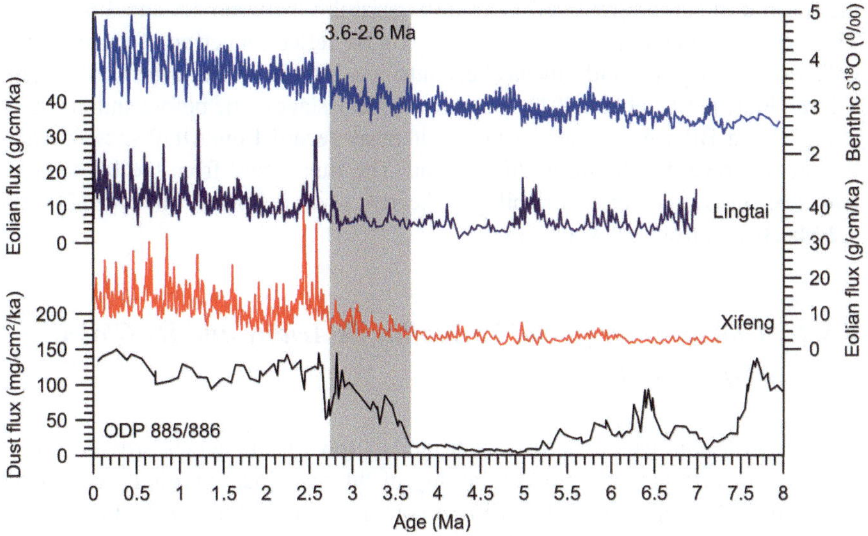

Fig. 4.33 Comparison of dust flux on the Chinese Loess Plateau (Sun and An 2002, 2005) and the North Pacific (Rea et al. 1998) with the benthic $\delta^{18}O$ record (Zachos et al. 2001) over the last 7 Ma

Variations in the eolian flux reconstructed from the Lingtai and Zhaojiachuan profiles of the central LP, show four evolutionary stages (Fig. 4.33). Flux variations during the period 7–4.5 Ma BP mark the beginning of increasing aridity of the dust source regions. The dust flux at two sections averaged ~5.30 $g \cdot cm^{-2} \cdot kyr^{-1}$ from 4.5 to 3.6 Ma BP, and then gradually increased to 7.11 $g \cdot cm^{-2} \cdot kyr^{-1}$ from 3.6 to

2.6 Ma BP, indicating an enhancement of aridity during this time. Afterwards, from 2.6 to 1.2 Ma BP, the eolian fluxes at both Lingtai and Xifeng displayed large-amplitude and high-frequency variability, and the fluxes at those sites exhibited even higher variability after ~1.2 Ma BP (Sun and An 2002).

Rea et al. (1998) used the eolian dust flux records at ODP sites 885/886 in the NPO to infer the drying history for the dust source regions and to determine the relative intensity of past atmospheric transport in the Northern Hemisphere (Fig. 4.33). The eolian fluxes were found to be relatively constant (0.011–0.093 $g \cdot cm^{-2}\ kyr^{-1}$ from 7 to 5 Ma BP (average = 0.036 $g \cdot cm^{-2}\ kyr^{-1}$). A period of remarkably low dust accumulation (0.0003–0.017 $g \cdot cm^{-2}\ kyr^{-1}$; average = 0.009 $g \cdot cm^{-2}\ kyr^{-1}$) followed, from 5 to 3.6 Ma BP. Then, at 3.6–2.6 Ma BP, a major change in the North Pacific sedimentation occurred, and the dust fluxes increased by an order-of-magnitude to 0.144 $g \cdot cm^{-2}\ kyr^{-1}$ (average flux = 0.079 $g\ cm^{-2}\ kyr^{-1}$). The eolian fluxes at ODP 885/996 have fluctuated between 0.093 and 0.15 $g \cdot cm^{-2}\ kyr^{-1}$ (average = 0.122 $g \cdot cm^{-2}\ kyr^{-1}$) over the last 2.6 Ma (Fig. 4.33).

The eolian fluxes reconstructed from LP and NPO sediments show similar variations and evolutionary stages over the past 7 Ma (Fig. 4.33). A dust flux peak that began at ~7 Ma BP at both ODP sites 885/886 correlates well with a drying event in the Asian dust source areas as seen in the Lingtai profile. Subsequent to that event, the Asian dust sources returned to modest variability between 4.5 and 3.6 Ma BP. The increase in eolian flux from 3.6 to 2.6 Ma BP reflects a period of drying in the eolian source regions, and this has been attributed to the uplift of the northeastern Tibetan Plateau (An et al. 2001). Two evolutionary stages, one before and the other after 1.2 Ma BP, can be seen in the Quaternary record from ODP sites 885/886 despite the lower resolution in this interval. The flux record from the LP displays two clear stages of distinct variability, indicative of drying phases, between 2.6 and 1.2 Ma BP and from 1.2 Ma BP on.

4.5.4 Glacial-Interglacial Variability in Aridity and Its Global Implications

Dust records from ice cores that cover the last few glacial cycles provide another source of information about the aridity of the dust source regions (e.g. GRIP members 1993; Petit et al. 1999; EPICA community members 2004). Figure 4.34 compares ice core dust records from Greenland and Antarctica with those from sediments from the northwest Pacific (Hovan et al. 1989) and loess from the central LP over the past 800 ka. Also included for comparison are the stacked mean grain sizes of quartz particles (MGSQ; Sun 2005) and benthic $\delta^{18}O$ curve (Lisiecki and Raymo 2005). Although the time references for the ice cores, deep-sea sediment core V21–146, and Chinese loess were generated by different methods, it is clear that the MARs, and by implication dust concentrations, were higher in glacial times than in interglacials. Furthermore, the MARs changed synchronously with the stacked MGSQs and benthic $\delta^{18}O$ records; that is, enhanced aridity in

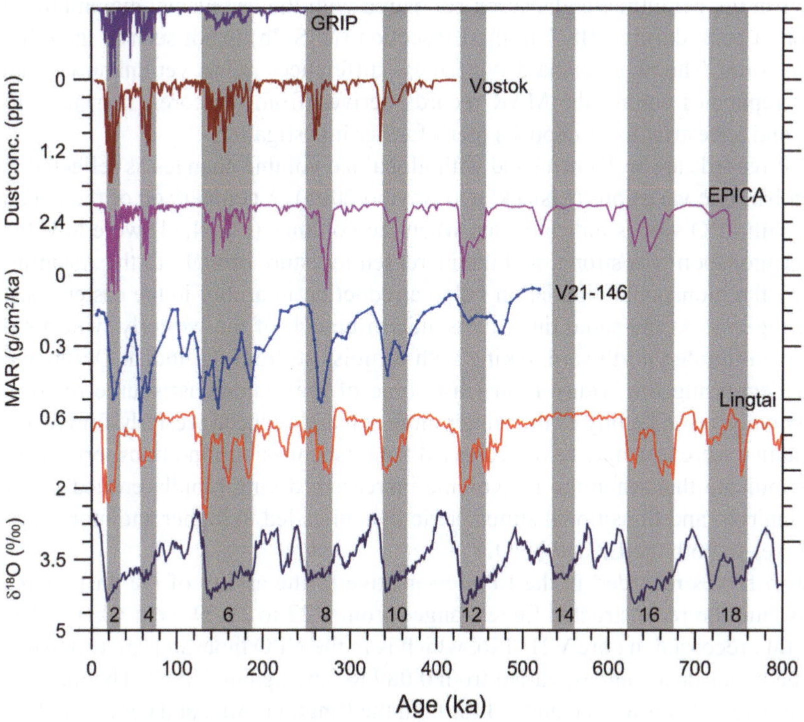

Fig. 4.34 Comparison of dust records from terrestrial loess (Sun and An 2005), marine (Hovan et al. 1989) and ice-core sediments (GRIP members 1993; Petit et al. 1999; EPICA community 2004). $\delta^{18}O$ results for ODP site 677 from Lisiecki and Raymo 2005. Shaded areas mark the marine isotope stages

the source regions apparently was synchronous with stronger paleowinds and increased ice volume during glacial times. Furthermore, correlations among the various climate proxies from ice cores, land, and ocean provide further evidence that in periods of extensive glaciations, changes in source-area aridity were related to the strengths of paleowinds (Clemens and Prell 1990; An et al. 1991a; Zhang et al. 1999, 2002).

The eolian flux variability during the last 800 kyr BP was examined by comparisons of the inferred flux record from the Lingtai profile with marine core V21–146 (37°41′N, 163°02′E, 3,968 m depth) in the northwestern Pacific (Hovan et al. 1989, Fig. 4.34). It is clear that flux maxima began around 430, 350, 250, 140 and 18 kyr BP for both continental and marine sediments. Unlike the dust records from ice cores and marine sediments, the MAR record from the LP displays prominent peaks within marine isotope stages (MISs) 6, 12 and 16 (Fig. 4.34). The periodicity of these MAR peaks is ~21 ka, that is, approximately at the same frequency as the coarsening of the MGSQ; this implies a strong precessional forcing on the aridity of central Asia and paleowind intensity during glacial periods. The MAR and stacked MGSQ records indicate that the Asian interior was more arid and paleowinds were stronger in the penultimate glaciation compared with the last glacial maximum. In addition, a peak during MIS 7 in the LP section (MIS 7b) is not seen in the other MAR records. The phenomena driving these differences are as yet unknown, and the discrepancies among the MAR records derived from ice cores, marine sediments, and terrestrial loess deposits merit further investigation.

MAR records are well correlated with global ice volume changes as reflected in oxygen isotopic variations (Lisiecki and Raymo 2005). A comparison of the eolian fluxes with $\delta^{18}O$ values indicates that when ice volumes (Fig. 4.34) were low, the summer monsoon was strong, and the increased moisture brought to the Asian interior by the monsoonal circulation led to a reduction in aridity in the eastern dust source regions. At the same time, the southern branch of the westerly wind field may have migrated northward, taking with it moisture from the Indian Ocean and further decreasing the aridity in at least some of the major dust source regions. Alternatively, a weakening of the winter monsoon and reduced westerly flow when ice volumes were low may have decreased the dust emissions and transport. These results indicate that when the ice volume increased, drying rapidly ensued in the Asian interior, and transitional atmospheric dynamics led to higher and more variable fluxes to both the LP and NPO.

Eolian fluxes recorded in the LP are sensitive to the aridity of the dust source regions, and the reconstructed fluxes ranged from 5.22 to 29.09 g cm^{-2} kyr^{-1}. The eolian flux recorded in core V21–146, which is in the NPO thousands of kilometers from the major dust sources, varied from 0.089 to 0.678 g cm^{-2} kyr^{-1}. The ratios of highest/lowest fluxes are 5.6 and 7.4 for both the Lingtai profile and core V21–146. Average fluxes show that the dust input to the LP is 50–80 times higher than that to the NPO during glacial maxima and about 40–60 times higher during interglacials. This can be explained by differences in dust production, presumably due at least in part to changes in the arid lands and differences in transport dynamics between glacial and interglacial times. The increased eolian fluxes seen in both continental and marine sediments over the last ~3 Ma provide evidence that the Asian interior has become increasingly arid, and this trend has been superimposed on the cyclical flux maxima and minima observed during glacial and interglacial periods.

4.5.5 Millennial-Scale Oscillations Linked to Variations in High-Latitude Climate

The regional-scale transport of dust from northern China is dominated by northwesterly near-surface winter monsoon winds (An et al. 1991a). Surges of cold air associated with this circulation historically increased in strength and frequency as the westerlies intensified. Similarly, variations in the dust inputs from western China can be attributed to shifts in the western fraction of the cold surge, or the westerlies. In contrast, dust from the sources in northern China evidently became increasingly important when cold air from the north dominated and the westerlies weakened. These shifts in Asian dust source regions were apparently synchronous with climate events in the North Atlantic, a strong indication that millennial-scale changes in dust flux from Asian source regions were linked to global- or near global-scale variability in atmospheric circulation.

A proposed link between increased dust loads and an attenuation of the land/sea temperature gradient (Zhang et al. 2002) is also consistent with the observed decreases in sea surface temperature following glacial high-dust events. These links connecting dust loadings and climate are the foundation for a conceptual model that may be applied to the interpretation of millennial-scale variations of the Asian winter monsoon climate, especially those that occurred during glacial stages. This model also provides the basis for better understanding climate changes beyond Asia. Although changes in atmospheric dust concentrations may trigger large-scale climate responses, as suggested above, the feedback between dust and climate also can be viewed as a mechanism by which the monsoon climate is modulated by the dusts' radiative effects, including any attendant effects on the hydrological cycle (Ramanathan et al. 2001).

On tectonic timescales, the development of aridity in the interior of Asian is primarily related to two factors: 1) the tectonic uplift of the Tibetan Plateau (Rea et al. 1998; An et al. 2001); and 2) the expansion of ice sheets in the Northern Hemisphere (Zachos et al. 2001). The uplift of the Tibetan Plateau blocked the flow of Indian Ocean moisture to western China (Rea et al. 1998), and thus reduced the summer precipitation in inland Asia (An et al. 2001). Evaporation simultaneously increased owing to the updraft current on the top of the plateau and the sinking airflow on the northern plateau in summer (Chen et al. 1999b). Dust flux records reconstructed from marine sediments and ice-cores and MARs from the LP suggest that the gradual drying of the Asian interior was associated with significant aridity-humidity fluctuations on millennial and orbital timescales. These in turn are closely linked to high-latitude climate changes in the Northern Hemisphere, especially ice volume and North Atlantic cooling events.

Indeed, synchronous increases in eolian fluxes seen in the LP and NPO sediments from 3.6 to 2.6 Ma BP indicate that the sudden drying of dust source regions also may be related to an increase in global ice volume (Ruddiman 1997). Positive correlations among glacial climate (benthic $\delta^{18}O$), paleowind intensity (MGSQ) and atmospheric dust loadings (MAR) suggest that the variability of the East Asian monsoon and Asian interior aridity have been dynamically linked to the development of Northern Hemisphere ice sheets. Even though the details concerning the forcing/response relationships among tectonic uplift, global ice-volume changes, Asian interior aridity and the East Asia monsoonal circulation are not yet well understood, analyses of terrestrial (MAR and MGSQ) and marine (MAR and $\delta^{18}O$) records attest to the existence of strong climatically-mediated teleconnections involving eolian dust.

4.6 Summary

This chapter has highlighted historical connections involving aridity in Asia; the production, transport, and deposition of desert dust; the formation of the Loess Plateau, eolian iron's impacts on oceanic primary production and atmospheric CO_2; and black carbon and paleo-fires. These relationships have been viewed in the context of climate, or more precisely, the changes in climate that have occurred over millions of years. Keys to understanding these complex relationships may be found in the historical reconstructions made possible by studies of Chinese loess.

The total dust production from Chinese deserts is estimated to be ~800±300 Tg yr^{-1} (Zhang et al. 1997); this may amount to as much as half of the global dust production although most studies indicate that the amount of desert dust exported from Africa is greater than that from Asia (Andreae 1995; Duce 1995). While the major sources for modern dust are reasonably well known, the relative contributions to the LP sediments in the past from the western deserts, especially the Taklimakan and the Qaidam Basin, and the Gobi and Badain Jaran to the north remain a matter of some dispute. Moreover, there is evidence that the major sources of Asian dust have changed over time, and very recently, from the 1950s to the 1990s, the area of China covered by deserts has increased from ~2% (Zhong 1999) to 7% (Zhu and Zhu 1999). To this point, climate change apparently has been a more important determinant of dust emissions than the expansion of desert lands due to human activities.

From March to May each year, when the Asian dust storm activity peaks, mineral dust is often a major component of the tropospheric aerosol in the Northern Hemisphere mid-latitudes. The regional-scale transport of this mineral matter is dominated by surface-level winds of the winter monsoon (An et al. 1991b; Zhang et al. 1996a). Air mass back-trajectory analyses show that Chinese desert dust is carried far out over the NPO by upper-level, mid-latitude westerlies (Merrill et al. 1994). During interglacial periods, including the present, dust carried from Asia by these winds can be transported great distances, that is, over hemispheric to global-scales. This explains in large measure why seasonal peaks in dust concentration over the NPO are closely related to dust storm activities in Asia (Uematsu et al. 1983).

The deposition of Asian desert dust over geologic time formed the Loess Plateau, and the loess sediments have been used to reconstruct the spatio-temporal distribution of Asian dust over different time scales. The dust fluxes during interglacial periods were found to be ~2–3 times lower than those in glacial periods. In glacial times, an enhanced dust supply to the ocean supplied micronutrients to marine biota thus potentially 'fertilizing' surface waters and leading to a drawdown of atmospheric CO_2. Variations in atmospheric iron loadings from 6.0 to 2.5 Ma BP, as reconstructed from the Bajiazui profile in central China, display patterns similar to the reconstructed fluxes from ODP site 885/886; that is, both show an increase in dust fluxes after 4.2 Ma BP. A high-resolution record of the biogenic opal MAR recovered from ODP site 882 (Fig. 4.23) parallels the patterns seen in the reconstructed Fe data. Increased dust inputs to the ocean imply an increased supply of iron and other micronutrients; the importance of this is that greater supplies of windborne nutrients may have stimulated the growth of phytoplankton communities. If primary production was increased, nitrogen fixation would likely also have increased, and a greater proportion of the available major nutrients would have been consumed.

The relationships among Asian dust, ocean productivity and atmospheric CO_2 over historical time scales suggest that dust-borne iron played an important role in global biogeochemical cycles in the past. This scenario is supported by the results of modern iron-seeding experiments. The global iron connection has several significant implications for our understanding of climate change: 1) it suggests that fertilization of HNLC waters could at least to some degree offset the effects of the anthropogenic CO_2 released into the environment; 2) it helps explain why analyses of ice cores have shown that atmospheric CO_2 concentrations are always relatively low during glacial periods; and 3) it provides a plausible geologic mechanism for global controls on marine primary productivity in some large areas of the world's oceans over geologic time scales.

Studies of eolian deposits from the Qin'an section, covering the interval from 22 to 6.2 Ma BP, indicate that large source areas of eolian dust must have existed in the Asian interior by the early Miocene (Guo et al. 2002). The development of aridity in the Asian interior is doubtlessly related to the tectonic uplift of the Tibetan Plateau, but other possible causes for the increased aridity observed around 8–7 Ma BP include growth in the Northern Hemisphere or global ice volume, increased evaporation on land (Chen et al. 1999b), and decreased evaporation from the ocean (Sun and An 2002). The uplift of the Tibetan Plateau led to drying in the Asian interior, an expansion of the deserts, and changes in the seasonality of the westerlies in the Northern Hemisphere. The westerlies are divided into two branches by the Tibetan Plateau in the winter, but in spring, the southern branch migrates northward, and the northern branch becomes unstable. In this way, variations in the strength of the winter monsoon are linked to dynamic conditions that control dust emissions and transport.

Late Cenozoic eolian deposits on the Loess Plateau changed from red clays to loess-paleosols around 2.6 Ma BP, coincident with the onset of a major Northern Hemisphere glaciation (Sun and An 2002). Since that time, alternating glacial/interglacial periods reflect an increased variability in inland Asian aridity and eolian dust fluxes. The resulting pattern of frequent dry-wet fluctuations is superimposed on a background of gradual but persistent long-term drying. A change in eolian flux variability since 1.2 Ma BP, together with a distinct shift in magnetic susceptibility and grain size characteristics from the Luochuan profile (Xiao and An 1999) may be the result of more recent uplift of the Tibetan Plateau and a related change in East Asian monsoon circulation.

The uplift of the Himalaya-Tibetan Plateau partially blocked the flow of moisture from the Indian Ocean to the interior of continental Asia (Rea et al. 1998), and this caused a decrease in the precipitation associated with the East Asian monsoon (An et al. 2001). Meanwhile, correlations among glacial climate (benthic $\delta^{18}O$), paleowind intensity (MGSQ) and atmospheric dust loadings (MAR) suggest that since the late Pliocene, the history and variability of the East Asian monsoon and

Asian interior aridity have been dynamically linked to glaciation in the Northern Hemisphere.

Black carbon is produced during the incomplete combustion of biomass and fossil fuels, and the cycles of BC and dust are linked because variations in both are connected to the relative wetness or dryness of the climate. BC concentrations and fluxes in loess and other sediments have been used as an indicator of paleo-fires, and as shown in this chapter, the concentrations and fluxes of BC as well as dust are affected by some of the same environmental factors, most notably precipitation and wind strength. That is, at least to a first order, fires will be more frequent and dust production higher when the environment is dry and winds are strong.

BC forms in two ways: 1) as solid residues that form char by pyrolysis, or 2) as volatiles that recondense into highly graphitized soot via gas-to-particle conversion. Char and soot have distinctly different physical and chemical characteristics. In particular, the optical properties of char and soot are quite different, and these differences have implications for climate dynamics. More specifically, although both char and soot absorb visible light, soot features strong light absorption with little spectral dependence while char, in contrast, is a relatively weak absorber with a strong spectral dependence. Thus, char and soot should be considered separately when assessing the climatic impacts of BC pollution.

Aerosol BC concentrations vary sharply from place to place, with the highest concentrations occurring close to sources. Fluvial inputs confound the reconstruction of inland atmospheric BC records from lacustrine sediments, but deep-sea sediments are not materially affected by these inputs. Although both char and soot originated mainly from biomass burning during pre-industrial times, they have different modes of formation and transport. Char particles, which are larger than soot, are more likely to be incorporated into soils and sediments close to their sources, and therefore char is often considered a better indicator of regional fires than soot. Soot forms much smaller (micrometer-sized) particles than char, and therefore, soot particles can remain airborne longer and thus be transported greater distances than char. Because of this, soot becomes a major component of the atmospheric background aerosol, and it is a better indicator of hemispheric-scale fire activities. Combined studies of soot and char may prove to be an effective new tool for reconstructing atmospheric BC concentrations. For example, a soot record from Lake Daihai (Han et al. 2012) correlates with the Northern Hemisphere temperature records of Mann and Jones (2003) and Moberg et al. (2005). This result suggests that soot can be dispersed over regional- to hemispheric-scales and also that atmospheric soot may have a direct feedback to climate (temperature). The exact nature of this feedback and the observed correlation between char and local-to-regional temperature (Mann and Jones 2003) remain subjects for future study.

Liu et al. (1985) noted that loess is a product of atmospheric activity and also that together with studies of deep-sea sediments, loess research has provided meaningful insights into the causes for long-term changes in regional and global climate. Continued studies of Chinese loess will surely shed more light on how the relationships discussed in this chapter were manifest in the past and how they might affect climate in the future. Finally, the development of new analytical procedures and advances in technology will provide new opportunities for addressing old issues and learning more about what may happen to our environment in the future.

References

Aggarwal SG, Kawamura K (2009) Carbonaceous and inorganic composition in long-range transported aerosols over northern Japan: Implication for aging of water-soluble organic fraction. Atmos Environ 43:2532–2540

Alfaro SC, Gaudichet A, Gomes L, Maille M (1997) Modeling the size distribution of a soil aerosol produced by sandblasting. J Geophys Res 102:11239–11249

Altabet MA, Francois R (2001) Nitrogen isotope biogeochemistry of the Antarctic Polar Frontal Zone at 170 W. Deep Sea Res Part II: Top Stud in Oceanogr 48:4247–4273

Andreae MO (1995) Climate effects of changing atmospheric aerosol levels. World survey of climatology, future climates of the world. Elsevier, Amsterdam, pp 341–392

Andreae MO, Gelencser A (2006) Black carbon or brown carbon? The nature of light-absorbing carbonaceous aerosols. Atmos Chem Phys 6:3131–3148

Andreae MO, Merlet P (2001) Emission of trace gases and aerosols from biomass burning. Global Biogeochem Cycles 15:955–966

Anderson RF, Kumar N, Mortlock RA, Froelich PN, Kubik P, Dittrich-Hannen B, Suter M (1998) Late-quaternary changes in productivity of the Southern Ocean. J Mar Syst 17:497–514

An ZS, Kukla G, Porter SC, Xiao JL (1991a) Late Quaternary dust flow on the Chinese Loess plateau. Catena 18:125–132

An ZS, Kukla GJ, Porter SC, Xiao J (1991b) Magnetic susceptibility evidence of monsoon variation on the Loess Plateau of central China during the last 130,000 years. Quat Res 36:29–36

An ZS, Kutzbach JE, Prell WL, Porter SC (2001) Evolution of Asian monsoons and phased uplift of the Himalaya-Tibetan plateau since Late Miocene times. Nature 411:62–66

An ZS, Sun DH, Zhang XY, Zhou WJ, Porter SC, Shaw J, Zhang DZ (1995) Accumulation sequence of Chinese loess and climatic records of Greenland ice core during the last 130 ka. Chin Sci Bull 40:1272–1276

An ZS (2000) The history and variability of the East Asian paleomonsoon climate. Quat Sci Rev 19:171–187

Archer D, Winguth A, Lea D, Mahowald N (2000) What caused the glacial/interglacial atmospheric pCO2 cycles? Rev Geophys 38:159–189

Ares J (1994) Long range long term transport and decay of PAHs in a semiarid coastal area of Argentina. Toxicol Environ Chem 41:109–123

Arimoto R (2001) Eolian dust and climate: relationships to sources, tropospheric chemistry, transport and deposition. Earth-Sci Rev 54:29–42

Arimoto R, Duce RA, Ray BJ, Unni CK (1985) Atmospheric trace elements at Enewetak Atoll: 2. Transport to the ocean by wet and dry deposition. J Geophys Res 90:2391–2408

Asaoka O (1975) Phytoplankton distribution with reference to hydrographic condition in the East China Sea. Mar Sci 7:38–45 (in Japanese)

Asmus T, Frank M, Koschmieder C, Frank N, Gersonde R, Kuhn G, Mangini A (1999) Variations of biogenic particle flux in the southern Atlantic section of the Subantarctic Zone during the late Quaternary: Evidence from sedimentary 231Paex and 230Thex. Mar Geol 159:63–78

Bach AJ, Brazel AJ, Lancaster N (1996) Temporal and spatial aspects of blowing dust in the Mojave and Colorado deserts of southern California, 1973–1994. Phys Geogr 17:329–353

Baker AR, Kelly SD, Biswas KF, Witt M, Jickells TD (2003) Atmospheric deposition of nutrients to the Atlantic Ocean. Geophys Res Lett 30:2296. doi:10.1029/2003GL018518

Barnola JM, Raynaud D, Korotkevich YS, Lorius C (1987) Vostok ice core provides 160,000-Year record of atmospheric CO2. Nature 329:408–414

Barron JA (1998) Late Neogene changes in diatom sedimentation in the North Pacific. J Asian Earth Sci 16:85–95

Beaufort L, de Garidel-Thoron T, Mix AC, Pisias NG (2001) ENSO-like forcing on oceanic primary production during the late Pleistocene. Science 293:2440–2444

Bergametti G (1992) Atmospheric cycle of desert dust. Encycl Earth Syst Sci 1:171–182

Bergametti G, Dulac F (1998) Mineral aerosols: renewed interest for climate forcing and tropospheric chemistry studies. IGACtivities Newsletter 1:13–17

Berger WH, Wefer G (1991) Productivity of the Glacial Ocean: Discussion of the Iron hypothesis. Limnol Oceanogr 36:1899–1918

Bird MI, Cali JA (1998) A million-year record of fire in sub-Saharan Africa. Nature 394:767–769

Bird MI, Cali JA (2002) A revised high-resolution oxygen-isotope chronology for ODP-668B: Implications for Quaternary biomass burning in Africa. Global Planet Change 33:73–76

Biscaye PE, Grousset FE, Revel M, VanderGaast S, Zielinski GA, Vaars A, Kukla G (1997) Asian provenance of glacial dust (stage 2) in the Greenland Ice Sheet Project 2 Ice Core, Summit, Greenland. J Geophys Res-Oceans 102:26765–26781

Bishop JK, Davis RE, Sherman JT (2002) Robotic observations of dust storm enhancement of carbon biomass in the North Pacific. Science 298:817–821

Blunier T, Chappellaz J, Schwander J, Stauffer B, Raynaud D (1995) Variations in atmospheric methane concentration during the holocene epoch. Nature 374:46–49

Bonasoni P, Laj P, Marinoni A, Sprenger M, Angelini F, Arduini J, Bonafe U, Calzolari F, Colombo T, Decesari S, Di Biagio C, di Sarra AG, Evangelisti F, Duchi R, Facchini MC, Fuzzi S, Gobbi GP, Maione M, Panday A, Roccato F, Sellegri K, Venzac H, Verza GP, Villani P, Vuillermoz E, Cristofanelli P (2010) Atmospheric Brown Clouds in the Himalayas: first two years of continuous observations at the Nepal Climate Observatory-Pyramid (5079 m). Atmos Chem Phys 10:7515–7531

Bond G, Broecker W, Johnsen S et al (1993) Correlations between climate records from North Atlantic sediments and Greenland ice. Nature 365:143–147

Bond G, Heinrich H, Broecke W, Labeyrie L, McManus J, Andrews J, Huon S, Jantschik R, Clasen S, Simet C, Tedesco K, Klas M, Bonani G, Ivy S (1992) Evidence for massive discharges of icebergs into the North Atlantic ocean during the last glacial periods. Nature 360:245–249

Bond G, Kromer B, Beer J, Muscheler R, Evans MN, Showers W, Hoffmann S, Lotti-Bond R, Hajdas I, Bonani G (2001) Persistent solar influence on North Atlantic climate during the Holocene. Science 294:2130–2136

Bond G, Showers W, Cheseby M, Lotti R, Almasi P, deMenocal P, Priore P, Cullen H, Hajdas I, Bonani G (1997) A pervasive millennial-scale cycle in North Atlantic Holocene and glacial climates. Science 278:1257–1266

Bond TC, Bhardwaj E, Dong R, Jogani R, Jung SK, Roden C, Streets DG, Trautmann NM (2007) Historical emissions of black and organic carbon aerosol from energy-related combustion. Global Biogeochem Cycles 21:1850–2000

Bond TC, Covert DS, Kramlich JC, Larson TV, Charlson RJ (2002) Primary particle emissions from residential coal burning: Optical properties and size distributions. J Geophys Res 107:ICC 9-1-ICC 9–14

Bond TC, Streets DG, Yarber KF, Nelson SM, Woo JH, Klimont Z (2004) A technology-based global inventory of black and organic carbon emissions from combustion. J Geophys Res 109 D14203

Bopp L, Kohfeld KE, Le QC, Aumont O (2003) Dust impact on marine biota and atmospheric CO_2 during glacial periods. Paleoceanography 18:1046. doi:10.1029/2002PA000810

Bory AJM, Biscaye PE, Svensson A, Grousset FE (2002) Seasonal variability in the origin of recent atmospheric mineral dust at NorthGRIP, Greenland. Earth Planet Sci Lett 196:123–134

Bowler JM, Chen KZ, Yuan BY (1987) Systematic variations in loess source areas: evidence from Qaidam and Qinghai basins, western China. In: Liu TS (ed) Aspects of loess research. China Ocean Press, Beijing

Boyd PW, Law CS, Wong CS, Nojiri Y, Tsuda A et al. (2004) The decline and fate of an iron-induced subarctic phytoplankton bloom. Nature 428:549–553

Boyd PW, Watson AJ, Law CS, Abraham ER, Trull T, Murdoch R, Bakker DCE, Bowie AR, Buesseler KO, Chang H, Charette M, Croot P, Downing K, Frew R, Gall M, Hadfield M, Hall J, Harvey M, Jameson G, LaRoche J, Liddicoat M, Ling R, Maldonado MT, McKay RM, Nodder S, Pickmere S, Pridmore R, Rintoul S, Safi K, Sutton P, Strzepek R, Tanneberger K, Turner S, Waite A, Zeldis J (2000) A mesoscale phytoplankton bloom in the polar Southern Ocean stimulated by iron fertilization. Nature 407:695–702

Broecker WS, Henderson GM (1998) The sequence of events surrounding termination II and their implications for the cause of glacial-interglacial CO2 changes. Paleoceanography 13:352–364

Broecker WS (1994) Massive iceberg discharges as triggers for global climate change. Nature 372:421–424

Buschiazzo DE, Zobeck TM, Aimar SB (1999) Wind erosion in loess soils of the semiarid Argentinian Pampas. Soil Sci 164:133–138

Cao J, Shen Z, Chow JC, Qi G, Watson JG (2009a) Seasonal variations and sources of mass and chemical composition for PM_{10} aerosol in Hangzhou, China. Particuology 7:161–168

Cao JJ, Lee SC, Chow JC, Watson JG, Ho KF, Zhang RJ, Jin ZD, Shen ZX, Chen GC, Kang YM, Zou SC, Zhang LZ, Qi SH, Dai MH, Cheng Y, Hu K (2007) Spatial and seasonal distributions of carbonaceous aerosols over China. J Geophys Res 112:D22S11

Cao JJ, Lee SC, Ho KF, Zhang XY, Zou SC, Fung K, Chow JC, Watson JG (2003) Characteristics of carbonaceous aerosol in Pearl River Delta Region, China during 2001 winter period. Atmos Environ 37:1451–1460

Cao JJ, Xu BQ, He JQ, Liu XQ, Han YM et al (2009b) Concentrations, seasonal variations, and transport of carbonaceous aerosols at a remote Mountainous region in western China. Atmos Environ 43:4444–4452

Carcaillet C, Almquist H, Asnong H, Bradshaw RHW, Carrion JS, Gaillard MJ, Gajewski K, Haas JN, Haberle SG, Hadorn P, Muller SD, Richard PJH, Richoz I, Rosch M, Goni MFS, von Stedingk H, Stevenson AC, Talon B, Tardy C, Tinner W, Tryterud E, Wick L, Willis KJ (2002) Holocene biomass burning and global dynamics of the carbon cycle. Chemosphere 49:845–863

CAS (1998) Chinese desertification map, resource and environment database. Chinese Academy of Sciences, Beijing

Chakrabarty RK, Moosmuller H, Chen LWA, Lewis K, Arnott WP, Mazzoleni C, Dubey MK, Wold CE, Hao WM, Kreidenweis SM (2010) Brown carbon in tar balls from smoldering biomass combustion. Atmos Chem Phys 10:6363–6370

Charlson RJ, Lovelock JE, Andreae MO, Warren SG (1987) Oceanic phytoplankton, atmospheric sulphur, cloud albedo and climate. Nature 326:655–661

Chase Z, Anderson RF, Fleisher MQ, Kubik PW (2003) Accumulation of biogenic and lithogenic material in the Pacific sector of the Southern Ocean during the past 40,000 years. Deep Sea Res Part II: Top Stud in. Oceanogr 50:799–832

Chan CY, Xu XD, Li YS, Wong KH, Ding GA, Chan LY, Cheng XH (2005) Characteristics of vertical profiles and sources of $PM_{2.5}$, PM_{10} and carbonaceous species in Beijing. Atmos Environ 39:5113–5124

Chen J, Li GJ, Yang JD et al. (2007) Nd and Sr isotopic characteristics of Chinese deserts: implications for the provenances of Asian dust. Geochim Cosmochim Acta 71:3904–3914

Chen LX, Liu JP, Zhou XJ et al. (1999b) Impact of uplift of Qinghai-Xizang Plateau and change of land-ocean distribution on climate over Asia. Quat Sci 4:314–329 (in Chinese)

Chen W, Fryrear DW, Yang Z (1999a) Dust fall in the Takla Makan desert of China. Phys Geogr 20:189–224

Chow JC, Watson JG, Pritchett LC, Pierson WR, Frazier CA, Purcell RG (1993) The DRI thermal/optical reflectance carbon analysis system: description, evaluation and applications in US air quality studies. Atmos Environ 27:1185–1201

Chylek P, Johnson B, Wu H (1992) Black carbon concentration in Byrd Station ice core—from 13,000 to 700 years before present. Ann Geophys 10:625–629

Claquin T, Schulz M, Balkanski YJ (1999) Modeling the mineralogy of atmospheric dust sources. J Geophys Res 104:22243–22256

Clark JS (1988) Particle motion and the theory of charcoal analysis—source area, transport, deposition, and sampling. Quaternary Res 30:67–80

Clemens SC, Prell WL (1990) Late Pleistocene variability of Arabian Sea summer monsoon winds and continental aridity: Eolian records from the lithogenic component of deep–sea sediments. Paleoceanography 5:109–145

Cofer III WR, Koutzenogii KP, Kokorin A, Ezcurra A (1997) Biomass burning emissions and the atmosphere. Springer, Berlin Heidelberg 51:189–206

COHMAP (1988) Climatic changes of the last 18,000 years: observations and model simulations. Science 241:1043–1052

Conedera M, Tinner W, Neff C, Meurer M, Dickens AF, Krebs P (2009) Reconstructing past fire regimes: methods, applications, and relevance to fire management and conservation. Quaternary Sci Rev 28:555–576

Cope MJ, Chaloner WG (1980) Fossil charcoal as evidence of past atmospheric composition. Nature 283:647–649

Crutzen PJ, Andreae MO (1990) Biomass burning in the tropics: Impact on atmospheric chemistry and biogeochemical cycles. Science 250:1669–1678

d'Almeida GA (1989) Desert aerosol: characteristics and effects on climate, Paleoclimatology and paleometeorology: modern and past patterns of global atmospheric transport. Springer, Netherlands, pp 311–338

Daniau AL, d'Errico F, Goñi, MFS (2010a) Testing the hypothesis of fire use for ecosystem management by Neanderthal and Upper Palaeolithic modern human populations. PLoS ONE 5 e9157

Daniau AL, Goñi MFS, Duprat J (2009) Last glacial fire regime variability in western France inferred from microcharcoal preserved in core MD04–2845, Bay of Biscay. Quaternary Res 71:385–396

Daniau AL, Harrison SP, Bartlein PJ (2010b) Fire regimes during the last glacial. Quaternary Sci Rev 29:2918–2930

Dansgaard W, Johnsen SJ, Clansen HB et al. (1993) Evidence for general instability of past climate from a 250-kyr ice-core record. Nature 364:218–220

Delmonte B, Basile-Doelsch I, Petit J-R, Maggi V, Revel-Rolland M, Michard A, Jagoutz E, Grousset F (2004) Comparing the Epica and Vostok dust records during the last 220,000 years: stratigraphical correlation and provenance in glacial periods. Earth-Sci Rev 66:63–87

Derbyshire E, Meng X, Kemp RA (1998) Provenance, transport and characteristics of modern aeolian dust in western Gansu Province, China, and the interpretation of the quaternary loess record. J Arid Environ 39:497–516

Dickens GR, Owen RM (1996) Sediment geochemical evidence for an early-middle Gilbert (early Pliocene) productivity peak in the North Pacific Red Clay Province. Mar Micropaleontol 27:107–120

Diester-Haass L, Billups K, Emeis KC (2006) Late Miocene carbon isotope records and marine biological productivity: was there a (dusty) link? Paleoceanography 21:PA4216. doi:10.1029/2006PA001267

Ding ZL, Derbyshire E, Yang SL, Sun JM, Liu TS (2005) Stepwise expansion of desert environment across northern China in the past 3.5 Ma and implications for monsoon evolution. Earth Planet Sci Lett 237:45–55

Ding ZL, Xiong SF, Sun JM, Yang SL, Gu ZY, Liu TS (1999) Pedostratigraphy and paleomagnetism of a ~ 7.0 Ma eolian loess-red clay sequence at Lingtai, Loess Plateau, north-central China and the implications for paleomonsoon evolution. Palaeogeogr Palaeoclimatol Palaeoecol 152:49–66

Dong GR, Li S, Li B et al. (1991) The preliminary research on the formation and evaluation of Chinese desert. Chinese Desert 4:26–31

Duan J, Tan J, Cheng D, Bi X, Deng W, Sheng G, Fu J, Wong MH (2007) Sources and characteristics of carbonaceous aerosol in two largest cities in Pearl River Delta Region, China. Atmos Environ 41:2895–2903

Duce RA, Liss PS, Merrill JT et al. (1991) The atmospheric input of trace species to the world ocean. Global Biogeochem Cycles 5:193–259

Duce RA (1995) Sources, distributions, and fluxes of mineral aerosols and their relationship to climate. In: Charlson RJ, Heintzenberg J (eds) Aerosol forcing of climate. Wiley, New York, pp 43–72

Duce RA, Unni CK, Ray BJ, Prospero JM, Merrill JT (1980) Long-range atmospheric transport of soil dust from Asia to the tropical North Pacific: temporal variability. Science 209:1522–1524

Dulac F, Buat-Menard P, Ezat U, Melki S, Bergametti G (1989) Atmospheric input of trace metals to the western Mediterranean: uncertainties in modelling dry deposition from cascade impactor data. Tellus B 41:362–378

Dykoski C, Edwards R, Cheng H, Yuan D, Cai Y et al. (2005) A high-resolution, absolute-dated Holocene and deglacial Asian monsoon record from Dongge Cave, China. Earth Planet Sci Lett 233:71–86

Elmquist M, Cornelissen G, Kukulska Z, Gustafsson Ö (2006) Distinct oxidative stabilities of char versus soot black carbon: Implications for quantification and environmental recalcitrance. Global Biogeochem Cycles 20:GB2009. doi:2010.1029/2005GB002629

Engling G, Zhang Y-N, Chan C-Y, Sang X-F, Lin M et al. (2011) Characterization and sources of aerosol particles over the southeastern Tibetan Plateau during the Southeast Asia biomass-burning season. Tellus B 63:117–128

EPICA community members (2004) Eight glacial cycles from an Antarctic ice core. Nature 429:623–628

Etheridge DM, Steele LP, Langenfelds RL, Francey RJ, Barnola JM, Morgan VI (1996) Natural and anthropogenic changes in atmospheric CO2 over the last 1000 years from air in Antarctic ice and firn. J Geophys Res 101:4115–4128

Faegri K, Kaland PE, Krzywinski K (1989) Textbook of pollen analysis, 4th edn. John Wiley & Sons Ltd., Chichester

Falkowski PG (1997) Evolution of the nitrogen cycle and its influence on the biological sequestration of CO2 in the ocean. Nature 387:272–275

Fang XM, Han YX, Ma JH, Song LC, Yang SL, Zhang XY (2004) Dust storms and loess accumulation on the Tibetan Plateau: a case study of dust event on 4 March 2003 in Lhasa. Chin Sci Bull 49:953–960

Finney BP, Gregory-Eaves I, Douglas MS, Smol JP (2002) Fisheries productivity in the northeastern Pacific Ocean over the past 2,200 years. Nature 416:729–733

Flannigan MD, Stocks BJ, Wotton BM (2000) Climate change and forest fires. Sci Total Environ 262:221–229

Forster P, Ramaswamy V, Artaxo P, Berntsen T, Betts R, Fahey DW, Haywood J, Lean J, Lowe DC, Myhre G, Nganga J, Prinn R, Raga G, Schulz M, Dorland, R.V (2007) Changes in atmospheric constituents and in radiative forcing. Cambridge University Press, Cambridge

Fry DL, Stephens SL (2006) Influence of humans and climate on the fire history of a ponderosa pine-mixed conifer forest in the southeastern Klamath Mountains, California. Forest Ecol Manag 223:428–438

Fujii Y, Kohno M, Matoba S, Motoyama H, Watanabe O (2003) A 320 k-year record of microparticles in the Dome Fuji, Antarctica ice core measured by laser-light scattering (scientific paper). Mem Natl Inst Polar Res, Spec Issue 57:46–62

Gao Y, Arimoto R, Zhou MY et al. (1992) Relationships between the dust concentrations over eastern Asia and the remote North Pacific. J Geophy Res 97:9867–9872

Gao Y (1994) Characterization of atmospheric aerosols over East Asia and the western North Pacific. University of Rhode Island, Narragansett, pp 178–190

Gelinas Y, Prentice KM, Baldock JA, Hedges JI (2001) An improved thermal oxidation method for the quantification of soot/graphitic black carbon in sediments and soils. Environ Sci Technol 35:3519–3525

GESAMP (1989) GESAMP Working Group 14. In: GESAMP group of experts on scientific aspects of marine pollution. World Meteorological Organization, Geneva

Ginoux P, Chin M, Tegen I, Prospero JM, Holben B, Dubovik O, Lin SJ (2001) Sources and distributions of dust aerosols simulated with the GOCART model. J Geophys Res 106:20255–20273

Goldberg ED (1985) Black carbon in the environment: properties and distribution. Wiley, New York

Gong SL, Zhang XY, Zhao TL, Barrie LA (2004) Sensitivity of Asian dust storm to natural and anthropogenic factors. Geophy Res Lett 31:L07210

Gong SL, Zhang XY, Zhao TL, McKendry IG, Jaffe DA, Lu NM (2003) Characterization of soil dust aerosol in China and its transport/distribution during 2001 ACE-Asia. 2. Model simulation and validation. J Geophys Res:108. doi:10.1029/2002JD002633

Graber ER, Rudich Y (2006) Atmospheric HULIS: How humic-like are they? A comprehensive and critical review. Atmos Chem Phys 6:729–753

GRIP members (1993) Climate instability during the last interglacial period recorded in the GRIP ice core. Nature 364:203–207

Gu ZY, Ding ZL, Xiong SF, Liu TS (1999) A seven million geochemical record from Chinese red-clay and loess-paleosol sequence: weathering and erosion in northwestern China. Quat Sci 4:357–365 (in Chinese)

Guo ZT, Peng SZ, Hao QZ, Chen XH, Liu TS (1999) Late Tertiary development of ariditfication in north western China: link with the arctic ice-sheet formation and Tibetan uplifts. Quat Sci 6:556–567 (in Chinese)

Guo ZT, Ruddiman WF, Hao QZ, Wu HB, Qiao YS, Zhu RX, Peng SZ, Wei JJ, Yuan BY, Liu TS (2002) Onset of Asian desertification by 22 Myr ago inferred from loess deposit in China. Nature 416:159–163

Gustafsson Ö, Bucheli TD, Kukulska Z, Andersson M, Largeau C, Rouzaud JN, Reddy CM, Eglinton TI (2001) Evaluation of a protocol for the quantification of black carbon in sediments. Glob Biogeochem Cycles 15:881–890. doi:10.1029/2000GB001380

Han YM, Marlon J, Cao JJ, Jin ZD, An ZS (2012) Holocene linkages between char, soot, biomass burning and climate from Lake Daihai, China. Glob Biogeochem Cycles 26:GB4017. doi:4010.1029/2012GB004413

Han YM, Cao JJ, Chow JC, Watson JG, An ZS, Jin ZD, Fung KC, Liu SX (2007a) Evaluation of the thermal/optical reflectance method for discrimination between char- and soot-EC. Chemosphere 69:569–574

Han YM, Cao JJ, Chow JC, Watson JG, An ZS, Liu SX (2009a) Elemental carbon in urban soils and road dusts in Xi'an, China and its implication for air pollution. Atmos Environ 43:2464–2470

Han YM, Cao JJ, Lee SC, Ho KF, An ZS (2010) Different characteristics of char and soot in the atmosphere and their ratio as an indicator for source identification in Xi'an, China. Atmos Chem Phys 10:595–607

Han YM, Cao JJ, Yan BZ, Kenna TC, Jin ZD, Cheng Y, An ZS (2011) Comparison of elemental carbon in lake sediments measured by three different methods and 150-year pollution history in Eastern China. Environ Sci Technol 45:5287–5293

Han YM, Han ZW, Cao JJ, Chow JC, Watson JG, An ZS, Liu SX, Zhang RJ (2008) Distribution and origin of carbonaceous aerosol over a rural high-mountain lake area, Northern China and its transport significance. Atmos Environ 42:2405–2414

Han YM, Lee SC, Cao JJ, Ho KF, An ZS (2009b) Spatial distribution and seasonal variation of char-EC and soot-EC in the atmosphere over China. Atmos Environ 43:6066–6073

Han YN, Cao JJ, An ZS, Chow JC, Watson JG, Jin Z, Fung K, Liu SX (2007b) Evaluation of the thermal/optical reflectance method for quantification of elemental carbon in sediments. Chemosphere 69:526–533

Hansen ADA, Novakov T (1989) Aerosol black carbon measurements in the Arctic haze during AGASP-II. J Atmos Chem 9:347–361

Hansen J, Nazarenko L (2004) Soot climate forcing via snow and ice albedos. Proc Nat Acad Sci 101:423–428

Hansen J, Sato MKI, Ruedy R, Nazarenko L, Lacis A, Schmidt GA, Russell G, Aleinov I, Bauer M, Bauer S, Bell N, Cairns B, Canuto V, Chandler M, Cheng Y, Del Genio A, Faluvegi G, Fleming E, Friend A, Hall T, Jackman C, Kelley M, Kiang N, Koch D, Lean J, Lerner J, Lo K, Menon S, Miller R, Minnis P, Novakov T, Oinas V, Perlwitz J, Perlwitz J, Rind D, Romanou

A, Shindell D, Stone P, Sun S, Tausnev N, Thresher D, Wielicki B, Wong T, Yao M, Zhang S (2005) Efficacy of climate forcings. J Geophys Res 110:D18104

Hammes K, Schmidt MWI, Smernik RJ, Currie LA, Ball WP, Nguyen TH, Louchouarn P, Houel S, Gustafsson O, Elmquist M, Cornelissen G, Skjemstad JO, Masiello CA, Song J, Peng P, Mitra S, Dunn JC, Hatcher PG, Hockaday WC, Smith DM, Hartkopf-Froeder C, Boehmer A, Luer B, Huebert BJ, Amelung W, Brodowski S, Huang L, Zhang W, Gschwend PM, Flores-Cervantes DX, Largeau C, Rouzaud JN, Rumpel C, Guggenberger G, Kaiser K, Rodionov A, Gonzalez-Vila FJ, Gonzalez-Perez JA, de la Rosa JM, Manning DAC, Lopez-Capel E, Ding L (2007) Comparison of quantification methods to measure fire-derived (black/elemental) carbon in soils and sediments using reference materials from soil, water, sediment and the atmosphere. Global Biogeochem Cycles 21:GB3016. doi:3010.1029/2006GB002914

Harrison SP, Kohfeld KE, Roelandt C, Claquin T (2001) The role of dust in climate changes today, at the last glacial maximum and in the future. Earth-Sci Rev 54:43–80

Hashimoto Y, Kim HK, Chen ZL, Yang ZM (1994) Atmospheric fingerprints of east Asia, 1986–1991: an urgent record of aerosol analysis by the JACK Network. Atmos Environ 28:1437–1445

Haug GH, Sigman DM, Tiedemann R, Pedersen TF, Sarnthein M (1999) Onset of permanent stratification in the subarctic Pacific Ocean. Nature 401:779–782

Hay WW, Soeding E, DeConto RM, Wold CN (2002) The Late Cenozoic uplift-climate change paradox. Int J Earth Sci (Geol Rundsch) 91:746–774

He Z, Kim YJ, Ogunjobi KO, Kim JE, Ryu SY (2004) Carbonaceous aerosol characteristics of $PM_{2.5}$ particles in Northeastern Asia in summer 2002. Atmos Environ 38:1795–1800

Hedges JI, Eglinton G, Hatcher PG, Kirchman DL, Arnosti C, Derenne S, Evershed RP, Kogel-Knabner I, de Leeuw JW, Littke R, Michaelis W, Rullkotter J (2000) The molecularly-uncharacterized component of nonliving organic matter in natural environments. Org Geochem 31:945–958

Heinrich H (1988) Origin and consequences of cyclic ice rafting in the northeast Atlantic Ocean during the past 130,000 years. Quaternary Research 29:142–152

Herman JR, Bhartia PK, Torres O, Hsu C, Seftor C, Celarier E (1997) Global distribution of UV-absorbing aerosols from Nimbus 7/TOMS data. J Geophys Res 102:16911–16922

Ho KF, Lee SC, Cao JJ, Li YS, Chow JC, Watson JG, Fung K (2006) Variability of organic and elemental carbon, water soluble organic carbon, and isotopes in Hong Kong. Atmos Chem Phys 6:4569–4576

Hou SS, Yang SL, Sun JM, Ding ZL (2003) Oxygen isotope compositions of quartz grains (4–16 μm) from Chinese eolian deposits and their implications for provenance. Sci China Ser D-Earth Sci 46:1003–1011

Hovan SA, Rea DK, Pisias NG (1991) Late Pleistocene continental climate and oceanic variability recorded in northwest Pacific sediments. Paleoceanography 6:349–370

Hovan SA, Rea DK, Pisias NG, Shackleton NJ (1989) A direct link between the China loess and marine δ18O records: aolian flux to the north Pacific. Nature 340:296–298

Huerta MA, Whitlock C, Yale J (2009) Holocene vegetation-fire-climate linkages in northern Yellowstone National Park, USA. Palaeogeogr Palaeoclimat Palaeoecol 271:170–181

Husar RB, Prospero JM, Stowe LL (1997) Characterization of tropospheric aerosols over the oceans with the NOAA advanced very high resolution radiometer optical thickness operational product. J Geophys Res 102:16889–16909

Husar RB, Tratt DM, Schichtel BA, Falke SR, Li F et al. (2001) Asian dust events of April 1998. J Geophys Res—Atmos 106:18317–18330

Indermuhle A, Stocker TF, Joos F, Fischer H, Smith HJ et al. (1999) Holocene carbon-cycle dynamics based on CO_2 trapped in ice at Taylor Dome, Antarctica. Nature 398(6723):121–126.

IPCC (2001) Climate change 2001: the scientific basis. Contribution of working group I to the Third Assessment Report of the Intergovernmental Panel on Climate Change. Cambridge University Press, Cambridge

IPCC (2007) Couplings between changes in the climate system and biogeochemistry, fourth assessment report: climate chang 2007: The scientific basis. Cambridge University Press, New York, pp 555–556

Ito A, Penner JE (2005) Historical emissions of carbonaceous aerosols from biomass and fossil fuel burning for the period 1870–2000. Global Biogeochem Cycles:19:GB2028

Iversen J (1941) Land occupation in Denmark's Stone Age. Danmarks Geologiske Undersogelse II 66:9

Iwamoto Y, Yumimoto K, Toratani M, Tsuda A, Miura K, Uno I, Uematsu M (2011) Biogeochemical implications of increased mineral particle concentrations in surface waters of the northwestern North Pacific during an Asian dust event. Geophys Res Lett 38: L01604. doi:10.1029/2010GL045906

Iwasaka Y, Minoura H, Nagaya K (1983) The transport and spatial scale of Asian dust-storm clouds: a case study of the duststorm event of April 1979. Tellus B 35:189–196

Jacobson MZ (2001a) Global direct radiative forcing due to multicomponent anthropogenic and natural aerosols. J Geophys Res 106:1551–1568

Jacobson MZ (2001b) Strong radiative heating due to the mixing state of black carbon in atmospheric aerosols. Nature 409:695–697

Jickells TD, An ZS, Andersen KK, Baker AR, Bergametti G et al. (2005) Global iron connections between desert dust, ocean biogeochemistry, and climate. Science 308:67–71

Johnsen SJ, Clausen HB, Dansgaard W, Fuhrer K, Gundestrup N, Hammer CV, Lversen P, Jouzel J, Stauffer B, Steffensen JP (1992) Irregular glacial interstadials recorded in a new Greenland ice core. Nature 359:311–313

Johnsen SJ, DahlJensen D, Gundestrup N, Steffensen JP, Clausen HB et al. (2001) Oxygen isotope and palaeotemperature records from six Greenland ice-core stations: Camp Century, Dye-3, GRIP, GISP2, Renland and NorthGRIP. J Quat Sci 16:299–307

Johnson KS, Elrod VA, Fitzwater SE, Plant JN, Chavez FP, Tanner SJ, Gordon RM, Westphal DL, Perry KD, Wu JF, Karl DM (2003) Surface ocean-lower atmosphere interactions in the northeast Pacific Ocean gyre: Aerosols, iron, and the ecosystem response. Glob Biogeochem Cycles 17:1063 doi:10.1029/2002GB002004

Jones TP, Chaloner WG, Kuhlbusch TAJ (1997) Proposed bio-geological and chemical based terminology for fire-altered plant matter. Springer-Verlag, New York

Kawahata H, Suzuki A, Ahagon N (1998) Biogenic sediments in the West Caroline Basin, the western equatorial Pacific during the last 330,000 years. Mar Geol 149:155–176

Kawahata H (1999) Fluctuations in the ocean environment within the western Pacific warm pool during late Pleistocene. Paleoceanography 14:639–652

Kawahata H, Okamoto T, Matsumoto E, Ujiie H (2000) Fluctuations of eolian flux and ocean productivity in the mid-latitude North Pacific during the last 200 kyr. Quat Sci Rev 19:1279–1291

Kim YP, Moon KC, Lee JH (2000) Organic and elemental carbon in fine particles at Kosan, Korea. Atmos Environ 34:3309–3317

King MD, Kaufman YJ, Tanré, D, Nakajima T (1999) Remote sensing of tropospheric aerosols from space: past, present, and future. Bull Am Meteorol Soc 80:2229–2259

Kinne S, Pueschel R (2001) Aerosol radiative forcing for Asian continental outflow. Atmos Environ 35:5019–5028

Kirchstetter TW, Novakov T, Hobbs PV (2004) Evidence that the spectral dependence of light absorption by aerosols is affected by organic carbon. J Geophys Res 109:D21208

Kohfeld KE, Harrison SP (2001) DIRTMAP: The geologicalrecord of dust. Earth Sci Rev 54:81–114

Kohfeld KE, Harrison SP (2003) Glacial-interglacial changes in dust deposition on the Chinese Loess Plateau. Quat Sci Rev 22:1859–1878

Kuhlbusch TAJ (1998) Black carbon in soils, sediments, and ice cores. John Wiley, New York

Kumar N, Anderson R, Mortlock R, Froelich P, Kubik P, Dittrich-Hannen B, Suter M (1995) Increased biological productivity and export production in the glacial Southern Ocean. Nature 378:675–680

Lack DA, Cappa CD (2010) Impact of brown and clear carbon on light absorption enhancement, single scatter albedo and absorption wavelength dependence of black carbon. Atmos Chem Phys 10:4207–4220
Landvik J (1998) Wind erosion and transport in an arctic environment. An analogue to past processes in mid latitude regions. In: Busacca A (ed) Dust aerosols, loess soils & global change. Conference Proceedings, pp 19–20
Lavanchy VMH, Gäggeler HW, Schotterer U, Schwikowski M, Baltensperger U (1999) Historical record of carbonaceous particle concentrations from a European high-alpine glacier (Colle Gnifetti, Switzerland). J Geophys Res 104:21227–21236
Lee JA, Tchakerian VP (1995) Magnitude and frequency of blowing dust on the Southern High Plains of the United States, 1947–1989. Ann Assoc Am Geogr 85:684–693
Lewis K, Arnott WP, Moosmüller H, Wold CE (2008) Strong spectral variation of biomass smoke light absorption and single scattering albedo observed with a novel dual-wavelength photoacoustic instrument. J Geophys Res 113:D16203
Lisiecki LE, Raymo ME (2005) A pliocene-pleistocene stack of 57 globally distributed benthic δ18O records. Paleoceanography 20:PA1003. doi:10.1029/2004PA001071
Littmann T (1991) Dust storm frequency in Asia: climatic control and variability. Int J Climatol 11:393–412
Liu CQ, Masuda A, Okada A, Yabuki S, Fan ZL (1994) Isotope geochemistry of Quaternary deposits from the arid lands in northern China. Earth Planet Sci Lett 127:25–38
Liu TS et al. (1985) Loess and the Environment. Ocean Press, Beijing
Liu XD, Yin ZY, Zhang XY, Yang XC (2004) Analyses of the spring dust storm frequency of northern China in relation to antecedent and concurrent wind, precipitation, vegetation, and soil moisture conditions. J Geophys Res 109:D16210
Liu TS, Zheng MP, Guo ZT (1998) Initiation and evolution of the Asian monsoon system timely coupled with the ice-sheet growth and the tectonic movements in Asia. Quat Sci 18:194–204 (in Chinese)
Livingstone I, Warren A (1996) Aeolian geomorphology: an introduction. Addison Wesley Longman Ltd, UK
Lukács H, Gelencsér A, Hammer S, Puxbaum H, Pio C, Legrand M, Kasper-Giebl A, Handler M, Limbeck A, Simpson D, Preunkert S (2007) Seasonal trends and possible sources of brown carbon based on 2-year aerosol measurements at six sites in Europe. J Geophys Res 112:D23S18
Maeda L, Kawahata H, Nohara M (2002) Fluctuation of biogenic and abiogenic sedimentation on the Shatsky Rise in the western North Pacific during the late Quaternary. Mar Geol 189:197–214
Mahowald N, Kohfeld K, Hansson M, Balkanski Y, Harrison SP et al (1999) Dust sources and deposition during the last glacial maximum and current climate: A comparison of model results with paleodata from ice cores and marine sediments. J Geophys Res 104:15895–15916
Malm WC, Schichtel BA, Pitchford ML, Ashbaugh LL, Eldred RA (2004). Spatial and monthly trends in speciated fine particle concentration in the United States. J Geophys Res 109:D03306
Mann ME, Jones PD (2003) Global surface temperatures over the past two millennia. Geophys Res Lett 30:1820. doi:10.1029/2003GL017814
Margolis EQ, Balmat J (2009) Fire history and fire-climate relationships along a fire regime gradient in the Santa Fe Municipal Watershed, NM, USA. Forest Ecol Manag 258:2416–2430
Marlon JR, Bartlein PJ, Carcaillet C, Gavin DG, Harrison SP, Higuera PE, Joos F, Power MJ, Prentice IC (2008) Climate and human influences on global biomass burning over the past two millennia. Nat Geosci 1:697–702
Marlon JR, Bartlein PJ, Walsh MK, Harrison SP, Brown KJ, Edwards ME, Higuera PE, Power MJ, Anderson RS, Briles C, Brunelle A, Carcaillet C, Daniels M, Hu FS, Lavoie M, Long C, Minckley T, Richard PJH, Scott AC, Shafer DS, Tinner W, Umbanhowar CE, Whitlock C (2009) Wildfire responses to abrupt climate change in North America. Proc Nat Acad Sci 106:2519–2524
Marlow JR, Lange CB, Wefer G, Rosell-Mele A (2000) Upwelling intensification as part of the Pliocene-Pleistocene climate transition. Science 290:2288–2291

Marticorena B, Bergametti G (1995) Modeling the atmospheric dust cycle, 1. Design of a soil-derived dust emission scheme. J Geophys Res 100:16415–16430
Martin JH (1990) Glacial-interglacial CO_2 change: the iron hypothesis. Paleoceanography 5:1–13
Martin JH, Coale KH, Johnson KS, Fitzwater SE, Gordon RM et al. (1994) Testing the iron hypothesis in ecosystems of the equatorial Pacific-Ocean. Nature 371:123–129
Martin JH, Gordon RM (1988) Northeast Pacific iron distributions in relation to phytoplankton productivity. Deep Sea Res. 35:177–196
Masiello CA (2004) New directions in black carbon organic geochemistry. Mar Chem 92:201–213
Mayewski PA, Meeker LD, Twickler MS, Whitlow S, Yang Q, Lyons WB, Prentice M (1997) Major features and forcing of high-latitude northern hemisphere atmospheric circulation using a 110,000-year-long glaciochemical series. J Geophys Res 102:26345–26366
McConnell JR, Edwards R, Kok GL, Flanner MG, Zender CS, Saltzman ES, Banta JR, Pasteris DR, Carter MM, Kahl JDW (2007) 20th-century industrial black carbon emissions altered arctic climate forcing. Science 317:1381–1384
McDonald D, Pedersen T, Crusius J (1999) Multiple late quaternary episodes of exceptional diatom production in the Gulf of Alaska. Deep Sea Res Part II: Top Stud in. Oceanogr 46:2993–3017
McTainsh G, Pitblado J (1987) Dust storms and related phenomena measured from meteorological records in Australia. Earth Surf Proc Land 12:415–424
Mei FM, Zhang XY, Lu HY, Shen ZX, Wang YQ (2004) Characterization of MASDs of surface soils in north China and its influence on estimating dust emission. Chin Sci Bull 49:2169–2176
Meng ZY, Jiang XM, Yan P, Lin WL, Zhang HD, Wang Y (2007) Characteristics and sources of $PM_{2.5}$ and carbonaceous species during winter in Taiyuan, China. Atmos Environ 41:6901–6908
Menon S, Koch D, Beig G, Sahu S, Fasullo J, Orlikowski D (2010) Black carbon aerosols and the third polar ice cap. Atmos Chem Phys 10:4559–4571
Merrill J, Arnold E, Leinen M, et al. (1994) Mineralogy of aeolian dust reaching the North Pacific Ocean: 2. Relationship of mineral assemblages to atmospheric transport patterns. J Geophys Res 99:21025–21032
Merrill JT, Uematsu M, Bleck R (1989) Meteorological analysis of long range transport of mineral aerosol over the North Pacific. J Geophys Res 94:8584–8598
Middleton N (1984) Dust storms in Australia: frequency, distribution and seasonality. Search 15:46–47
Middleton N (1986) A geography of dust storms in South-West Asia. J Climatol 6:183–196
Middleton N, Goudie A, Wells G (1986) The frequency and source areas of dust storms. In: Nickling WG (ed) Aeolian geomorphology. Allen & Unwin, New York, pp 237–259
Middleton N (1991) Dust storms in the Mongolian people's republic. J Arid Environ 20:287–297
Mills MM, Ridame C, Davey M, La Roche J, Geider RJ (2004) Iron and phosphorus co-limit nitrogen fixation in the eastern tropical North Atlantic. Nature 429:292–294
Ming J, Cachier H, Xiao C, Qin D, Kang S, Hou S, Xu J (2008) Black carbon record based on a shallow Himalayan ice core and its climatic implications. Atmos Chem Phys 8:1343–1352
Ming J, Xiao C, Sun J, Kang S, Bonasoni P (2010) Carbonaceous particles in the atmosphere and precipitation of the Nam Co region, central Tibet. J Environ Sci-China 22:1748–1756
Moberg A, Sonechkin DM, Holmgren K, Datsenko NM, Karlen W (2005) Highly variable Northern Hemisphere temperatures reconstructed from low- and high-resolution proxy data. Nature 433:613–617
Monnin E, Steig EJ, Siegenthaler U, Kawamura K, Schwander J et al (2004) Evidence for substantial accumulation rate variability in Antarctica during the Holocene, through synchronization of CO2 in the Taylor Dome, Dome C and DML ice cores. Earth Planet Sci Lett 224:45–54
Moore JK, Abbott MR, Richman JG, Nelson DM (2000) The Southern Ocean at the last glacial maximum: a strong sink for atmospheric carbon dioxide. Glob Biogeochem Cycles 14:455–475
Morley JJ, Heusser LE (1997) Role of orbital forcing in East Asian monsoon climates during the last 350 kyr: Evidence from terrestrial and marine climate proxies from core RC14–99. Paleoceanography 12:483–493

Morel F, Price N (2003) The biogeochemical cycles of trace metals in the oceans. Science 300:944–947

Moulin C, Lambert CE, Dayan U, Masson V, Ramonet M, Bousquet P, Legrand M, Balkanski YJ, Guelle W, Marticorena B, Bergametti G, Dulac F (1998) Satellite climatology of African dust transport in the Mediterranean atmosphere. J Geophys Res 103:13137–13144

Mudelsee M, Raymo ME (2005) Slow dynamics of the Northern Hemisphere glaciation. Paleoceanography 20:PA4022. doi:10.1029/2005PA001153

Murray R, Knowlton C, Leinen M, Mix A, Polsky C (2000) Export production and terrigenous matter in the central equatorial Pacific Ocean during interglacial oxygen isotope Stage 11. Glob Planet Chang 24:59–78

Natsagdorj L, Jugder D, Chung YS (2003) Analysis of dust storms observed in Mongolia during 1937–1999. Atmos Environ 37:1401–1411

Nees S, Armand L, De Deckker P, Labracherie M, Passlow V (1999) A diatom and benthic foraminiferal record from the South Tasman Rise (southeastern Indian Ocean): implications for palaeoceanographic changes for the last 200,000 years. Mar Micropaleontol 38:69–89

Nguyen TH, Brown RA, Ball WP (2004) An evaluation of thermal resistance as a measure of black carbon content in diesel soot, wood char, and sediment. Org Geochem 35:217–234

Novakov T (1984) The role of soot and primary oxidants in atmospheric chemistry. Sci Total Environ 36:1–10

Novakov T, Ramanathan V, Hansen JE, Kirchstetter TW, Sato M, Sinton JE, Sathaye JA (2003) Large historical changes of fossil-fuel black carbon aerosols. Geophys Res Lett 30: 1324, DOI: 10.1029/2002GL016345

Oba T, Pedersen TF (1999) Paleoclimatic significance of eolian carbonates supplied to the Japan Sea during the last glacial maximum. Paleoceanography 14:34–41

Ogren JA, Charlson RJ (1983) Elemental carbon in the atmosphere—cycle and lifetime. Tellus B 35:241–254

Olsson F, Gaillard MJ, Lemdahl G, Greisman A, Lanos P, Marguerie D, Marcoux N, Skoglund P, Wäglind J (2010) A continuous record of fire covering the last 10,500 calendar years from southern Sweden—The role of climate and human activities. Palaeogeogr Palaeoclimat Palaeoecol 291:128–141

Orcutt KM, Lipschultz F, Gundersen K, Arimoto R, Michaels AF et al (2001) A seasonal study of the significance of N_2 fixation by *Trichodesmium* spp at the Bermuda Atlantic Time-series Study (BATS) site. Deep Sea Res Part II-Top Stud in. Oceanogr 48:1583–1608

Orgill M, Sehmel G (1976) Frequency and diurnal variation of dust storms in the contiguous USA. Atmos Environ 10:813–825 (1967)

Pérez C, Nickovic S, Pejanovic G, Baldasano JM, Özsoy E (2006) Interactive dust-radiation modeling: A step to improve weather forecasts. J Geophysical Research 111:D16206

Parrington JR, Zoller WH, Aras NK (1983) Asian dust: Seasonal transport to the Hawaiian Islands. Science 220:195–197

Patterson EM, Gillette DA (1977) Commonalities in measured size distributions for aerosols having a soil-derived component. J Geophy Res 82:2074–2082

Paytan A, Mackey KR, Chen Y, Lima ID, Doney SC, Mahowald N, Labiosa R, Post AF (2009) Toxicity of atmospheric aerosols on marine phytoplankton. Proc Natl Acad Sci 106:4601–4605

Pearson PN, Palmer MR (2000) Atmospheric carbon dioxide concentrations over the past 60 million years. Nature 406:695–699

Pease PP, Tchakerian VP, Tindale NW (1998) Aerosols over the Arabian Sea: geochemistry and source areas for aeolian desert dust. J Arid Environ 39:477–496

Peterson MR, Richards MH (2002) Thermal-optical-transmittance analysis for organic, elemental, carbonate, total carbon, and OCX2 in $PM_{2.5}$ by the EPA/NIOSH method. In: Tropp RJ (ed) AWMA air quality measurement symposium. Air and Waste Management Association, Pittsburgh

Petit JR, Briat M, Royer A (1981) Ice age aerosol content from East Antarctic ice core samples and past wind strength. Nature 293:391–394

Petit JR Jouzel J, Raynaud D, et al. (1999) Climate and atmospheric history of the past 420,000 years from the Vostok ice core, Antarctica. Nature 399:429–436

Pettke T, Halliday AN, Hall CM, Rea DK (2000) Dust production and deposition in Asia and the north Pacific Ocean over the past 12 Myr. Earth Planet Sci Lett 178:397–413

Porter SC, An ZS (1995) Correlation between climate events in the North Atlantic and China during the last glaciation. Nature 375:305–308

Pöschl U (2005) Atmospheric aerosols: Composition, transformation, climate and health effects. Angewandte Chemie-International Edition 44:7520–7540

Power MJ, Marlon J, Ortiz N, Bartlein PJ, Harrison SP, Mayle FE, Ballouche A, Bradshaw RHW, Carcaillet C, Cordova C, Mooney S, Moreno PI, Prentice IC, Thonicke K, Tinner W, Whitlock C, Zhang Y, Zhao Y, Ali AA, Anderson RS, Beer R, Behling H, Briles C, Brown KJ, Brunelle A, Bush M, Camill P, Chu GQ, Clark J, Colombaroli D, Connor S, Daniau AL, Daniels M, Dodson J, Doughty E, Edwards ME, Finsinger W, Foster D, Frechette J, Gaillard MJ, Gavin DG, Gobet E, Haberle S, Hallett DJ, Higuera P, Hope G, Horn S, Inoue J, Kaltenrieder P, Kennedy L, Kong ZC, Larsen C, Long CJ, Lynch J, Lynch EA, McGlone M, Meeks S, Mensing S, Meyer G, Minckley T, Mohr J, Nelson DM, New J, Newnham R, Noti R, Oswald W, Pierce J, Richard PJH, Rowe C, Goni MFS, Shuman BN, Takahara H, Toney J, Turney C, Urrego-Sanchez DH, Umbanhowar C, Vandergoes M, Vanniere B, Vescovi E, Walsh M, Wang X, Williams N, Wilmshurst J, Zhang JH (2008) Changes in fire regimes since the Last Glacial Maximum: an assessment based on a global synthesis and analysis of charcoal data. Clim Dynam 30:887–907

Power MJ, Marlon JR, Bartlein PJ, Harrison SP (2010) Fire history and the global charcoal database: a new tool for hypothesis testing and data exploration. Palaeogeogr Palaeoclimatol Palaeoecol 291:52–59

Prospero JM (1981) Eolian transport to the World Ocean. In: Emiliani C (ed) The oceanic lithosphere. The Sea, vol 7. Wiley Interscience, New York, pp 801–874

Prospero JM, Ginoux P, Torres O, Nicholson SE, Gill TE (2002) Environmental characterization of global sources of atmospheric soil dust identified with the NIMBUS-7 TOMS absorbing aerosol product. Rev Geophys 40:1002. doi:10.1029/2000RG000095

Prospero JM, Uematsu M, Savoie DL (1989) Mineral aerosol transport to the Pacific ocean. In: Riley JP, Chester R, Duce R A (eds) Chemical oceanography. Academic Press, San Diego, pp 188–218

Pye K (1987) Aeolian dust and dust deposits. Academic Press, London

Pye K (1995) The nature, origin, and accumulation of loess. Quat Sci Rev 14:653–667

Pyne SJ, Goldammer JG (1997) The culture of fire: an introduction to anthropogenic fire history. Springer, Berlin

Qiang XK, Li ZX, Powell CM, Zheng H (2001) Magnetostratigraphic record of the Late Miocene onset of the East Asian monsoon, and Pliocene uplift of northern Tibet. Earth Planet Sci Lett 187:83–93

Qu WJ, Zhang XY, Arimoto R, Wang D, Wang YQ, Yan LW, Li Y (2008) Chemical composition of the background aerosol at two sites in southwestern and northwestern China: potential influences of regional transport. Tellus B 60:657–673

Ram K, Sarin MM, Hegde P (2008) Atmospheric abundances of primary and secondary carbonaceous species at two high-altitude sites in India: Sources and temporal variability. Atmos Environ 42:6785–6796

Ramanathan V, Carmichael G (2008) Global and regional climate changes due to black carbon. Nat Geosci 1:221–227

Ramanathan V, Crutzen PJ, Kiehl JT, Rosenfeld D (2001) Aerosols, climate, and the hydrological cycle. Science 294:2119–2124

Raymo M (1997) Carbon cycle models: how strong are the constraints? In: Ruddiman WF (ed) Tectonic uplift and climate change. Plenum Press, New York, pp 367–381

Rea DK (1994) The paleoclimatic record provided by eolian deposition in the deep sea: the geologic history of wind. Rev Geophys 32:159–195

Rea DK, Leinen M, Janecek TR (1985) Geologic approach to the long-term history of atmospheric circulation. Science 227:721–725

Rea DK, Snoeckx H, Joseph LH (1998) Late Cenozoic eolian deposition in the North Pacific: Asian drying, Tibetan uplift, and cooling of the northern hemisphere. Paleoceanography 13:215–224

Reeve N, Toumi R (1999) Lightning activity as an indicator of climate change. Q. J. Roy. Meteor. Soc 125:893–903

Rengarajan R, Sarin MM, Sudheer AK (2007) Carbonaceous and inorganic species in atmospheric aerosols during wintertime over urban and high-altitude sites in North India. J Geophys Res 112: D21307

Reynolds B, Frank M, Halliday A (2008) Evidence for a major change in silicon cycling in the subarctic North Pacific at 2.73 Ma. Paleoceanography 23:PA4219. doi:10.1029/2007PA001563

Ridgwell AJ (2002) Dust in the Earth system: the biogeochemical linking of land, air and sea. Phil Trans R Soc Lond 360:2905–2924

Rind D, Peteet D, Broecker W, et al. (1986) The impact of cold North Atlantic sea-surface temperatures on climate: Implications for the Younger Dryas cooling (10–11k). Climate Dynamics 1:3–33

Röthlisberger R, Mulvaney R, Wolff EW, Hutterli MA, Bigler M, Sommer S, Jouzel J (2002) Dust and sea salt variability in central East Antarctica (Dome C) over the last 45 kyrs and its implications for southern high-latitude climate. Geophys Res Lett 29:2421–2424

Röthlisberger R, Bigler M, Wolff E, Joos F, Monnin E, Hutterli M (2004) Ice core evidence for the extent of past atmospheric CO_2 change due to iron fertilisation. Geophys Res Lett 31:L16207. doi:10.1029/2004GL020338

Ruddiman WF (ed) (1997) Tectonic uplift and climate changes. Plenum Press, New York

Satterfield FR, Finney BP (2002) Stable isotope analysis of Pacific salmon: insight into trophic status and oceanographic conditions over the last 30 years. Prog Oceanogr 53:231–246

Schmid H, Laskus L, Jurgen Abraham H, Baltensperger U, Lavanchy V, Bizjak M, Burba P, Cachier H, Crow D, Chow J, Gnauk T, Even A, ten Brink HM, Giesen K-P, Hitzenberger R, Hueglin C, Maenhaut W, Pio C, Carvalho A, Putaud J-P, Toom-Sauntry D, Puxbaum H (2001) Results of the "carbon conference" international aerosol carbon round robin test stage I. Atmos Environ 35:2111–2121

Schmidt MWI, Noack AG (2000) Black carbon in soils and sediments: Analysis, distribution, implications, and current challenges. Global Biogeochem Cycles 14:777–793

Schmidt MWI, Skjemstad JO, Czimczik CI, Glaser B, Prentice KM, Gelinas Y, Kuhlbusch TAJ (2001) Comparative analysis of black carbon in soils. Global Biogeochem Cycles 15:163–167

Schnaiter M, Horvath H, Möhler O, Naumann KH, Saathoff H, Schöck O W (2003) UV-VIS-NIR spectral optical properties of soot and soot-containing aerosols. J Aerosol Sci 34:1421–1444

Schulz K, Zondervan I, Gerringa L, Timmermans K, Veldhuis M, Riebesell U (2004) Effect of trace metal availability on coccolithophorid calcification. Nature 430:673–676

Seiler W, Crutzen PJ (1980) Estimates of gross and net fluxes of carbon between the biosphere and the atmosphere from biomass burning. Climatic Change 2:207–247

Shackleton N (1995) New data on the evolution of Pliocene climatic variability. Paleoclimate and evolution, with emphasis on human origins. Yale University Press, New Haven, pp 242–248

Shao Y (2001) A model for mineral dust emission. J Geophys Res 106:20239–20254

Shao YP, Leslie LM (1997) Wind erosion prediction over the Australian continent. J Geophys Res 102:30091–30105

Shaw GE (1980) Transport of Asian desert aerosol to the Hawaiian Islands. J Appl Meteorol 19:1254–1259

Shinn EA, Smith GW, Prospero JM, Betzer P, Hayes ML, Garrison V, Barber RT (2000) African dust and the demise of Caribbean coral reefs. Geophys Res Lett 27:3029–3032

Sigman DM, Jaccard SL, Haug GH (2004) Polar ocean stratification in a cold climate. Nature 428:59–63

Slinn SA, Slinn WGN (1981) Modeling of atmospheric particulate deposition to natural water. In: Eisenreich SJ (ed) Atmospheric Pollutants in Natural Waters. Science Publishers, Inc. Michigan, Ann Arbor, pp 23–53

Smith DM, Griffin JJ, Goldberg ED (1973) Elemental carbon in marine sediments— abaseline for burning. Nature 241:268–270

Sokolik IN, Toon OB (1996) Direct radiative forcing by anthropogenic airborne mineral aerosols. Nature 381:681–683

Sokolik IN, Winker DM, Bergametti G, Gillette DA, Carmichael G, Kaufman Y, Gomes L, Schuetz L, Penner JE (2001) Introduction to special section: outstanding problems in quantifying the radiative impacts of mineral dust. J Geophys Res-Atmos 106:18015–18027

Stegmann PM, Tindale NW (1999) Global distribution of aerosols over the open ocean as derived from the coastal zone color scanner. Global Biogeochem Cycles 13:383–397

Steig EJ, Morse DL, Waddington ED, Stuiver M, Grootes PM, Mayewski PA, Twickler MS, Whitlow SI (2000) Wisconsinan and Holocene climate history from an ice core at Taylor Dome, western ross embayment, Antarctica. Geografiska annaler: Series A. Phys Geogr 82:213–235

Streets DG, Bond TC, Carmichael GR, Fernandes SD, Fu Q, He D, Klimont Z, Nelson SM, Tsai NY, Wang MQ, Woo JH, Yarber KF (2003) An inventory of gaseous and primary aerosol emissions in Asia in the year 2000. J Geophys Res 108: 8809, DOI: 10.1029/2002JD003093

Stuiver M, Grootes PM, Braziunas TF (1995) The GISP2 δ18O climate record of the past 16,500 years and the role of the Sun, Ocean and Volcanoes. Quat Res 44:341–354

Suman DO, Kuhlbusch TAJ, Lim B (1997) Marine sediments: a reservoir for black carbon and their use as spatial and temporal records of combustion. In: JS Clark, Cachier H, Goldammer JG, Stocks B (eds) Sediment records of biomass burning and global change. Springer, Berlin

Sun DH, An ZS, Shaw J, Bloemendal J, Sun YB (1998a) Magnetostratigraphy and paleoclimatic significance of late tertiary sequences in the Chinese Loess Plateau. Geophys J Int 134:207–212

Sun DH, Shaw J, An ZS, Chen MY, Yue LP (1998b) Magnetostratigraphy and paleoclimatic interpretation of a continuous 7.2 Ma late Cenozoic eolian sediments from the Chinese Loess Plateau. Geophys Res Lett 25:85–88

Sun JM (2005) Nd and Sr isotopic variations in Chinese eolian deposits during the past 8 Ma: implications for provenance change. Earth Planet Sci Lett 240:454–466

Sun J, Zhang M, Liu T (2001) Spatial and temporal characteristics of dust storms in China and its surrounding regions, 1960–1999: Relations to source area and climate. J Geophys Res 106:10325–10333

Sun YB, An ZS (2002) History and variability of the Asian interior aridity recorded by eolian flux in the Chinese Loess Plateau during the past 7 Ma. Sci China (Ser D) 45:420–429

Sun YB, An ZS (2005) Late Pliocene-Pleistocene changes in mass accumulation rates of eolian deposits on the central Chinese Loess Plateau. J Geophys Res -Atmos 110:D23101

Sun YB, Tada R, Chen J, Liu Q, Toyoda S, Tani A, Ji JF, Isozaki Y (2008) Tracing the provenance of fine-grained dust deposited on the central Chinese Loess Plateau. Geophys Res Lett 35:L01804. doi10.1029/2007GL031672

Taylor SR, McLennan SM (1995) The geochemical evolution of the continental crust. Rev Geophy 33:241–265

Tegen I, Lacis AA, Fung I (1996) The influence on climate forcing of mineral aerosols from disturbed soils. Nature 380:419–422

Tegen I, Werner M, Harrison SP, Kohfeld KE (2004) Relative importance of climate and land use in determining present and future global soil dust eimssion. Geophys Res Lett 31:L05105. doi:510.1029/2003GL019216

Thevenon F, Bard E, Williamson D, Beaufort L (2004) A biomass burning record from the West Equatorial Pacific over the last 360 ky: methodological, climatic and anthropic implications. Palaeogeogr Palaeoclimat Palaeoecol 213:83–99

Thevenon F, Williamson D, Bard E, Anselmetti FS, Beaufort L, Cachier H (2010) Combining charcoal and elemental black carbon analysis in sedimentary archives: Implications for past fire regimes, the pyrogenic carbon cycle, and the human-climate interactions. Global Planet Change 72:381–389

Tinner W, Conedera M, Ammann B, Gaggeler HW, Gedye S, Jones R, Sagesser B (1998) Pollen and charcoal in lake sediments compared with historically documented forest fires in southern Switzerland since AD 1920. Holocene 8:31–42

Tsuda A, Takeda S, Saito H, Nishioka J, Nojiri Y, Kudo I et al. (2003) A mesoscale iron enrichment in the Western Subarctic Pacific induces a large centric diatom bloom. Science 300:958–961

Uematsu M, Duce RA, Prospero JM et al. (1983) Transport of mineral aerosol from Asia over the North Pacific Ocean. J Geophy Res 88:5343–5352
van der Kaars S, Wang X, Kershaw P, Guichard F, Setiabudi DA (2000) A Late Quaternary palaeoecological record from the Banda Sea, Indonesia: patterns of vegetation, climate and biomass burning in Indonesia and northern Australia. Palaeogeogr Palaeoclimat Palaeoecol 155:135–153
Verardo DJ, Ruddiman WF (1996) Late Pleistocene charcoal in tropical Atlantic deep-sea sediments: Climatic and geochemical significance. Geology 24:855–857
Wang H, Zhang XY, Gong SL, Chen Y, Shi GY, Li W (2010) Radiative feedback of dust aerosols on the East Asian dust storms. J Geophys Res 115:D23214. doi:23210.21029/22009JD013430
Wang J (1995) Distributions of nature disaster in China. Chinese Science and Technology, Beijing
Wang X, Peng PA, Ding ZL (2005) Black carbon records in Chinese Loess Plateau over the last two glacial cycles and implications for paleofires. Palaeogeogr Palaeoclimat Palaeoecol 223:9–19
Wang X, van der Kaars S, Kershaw P, Bird M, Jansen F (1999) A record of fire, vegetation and climate through the last three glacial cycles from Lombok Ridge core G6–4, eastern Indian Ocean, Indonesia. Palaeogeogr Palaeoclimat Palaeoecol 147:241–256
Wang YX, Yang JD, Chen J, Zhang KJ, Rao WB (2007) The Sr and Nd isotopic variations of the Chinese Loess Plateau during the past 7 Ma: Implications for the East Asian winter monsoon and source areas of loess. Palaeogeogr Palaeoclimatol Palaeoecol 249:351–361
Watanabe O, Fujii Y, Kamiyama K, Motoyama H, Furukawa T, Igarashi M, Kohno M, Kanamori S, Kanamori N, Ageta Y (1999) Basic analysis of Dome Fuji deep ice core part 1: stable oxygen and hydrogen isotope ratios, major chemical compositions and dust concentration. Polar Meteorol Glaciol 13:83–89
Watson A, Bakker D, Ridgwell A, Boyd P, Law C (2000) Effect of iron supply on Southern Ocean CO2 uptake and implications for glacial atmospheric CO2. Nature 407:730–733
Watson JG, Chow JC, Chen L-WA (2005) Summary of organic and elemental carbon/black carbon analysis methods and intercomparisons. Aero Air Qual Res 5:65–102
Weber M, Pisias N (1999) Spatial and temporal distribution of biogenic carbonate and opal in deep-sea sediments from the eastern equatorial Pacific: implications for ocean history since 1.3 Ma. Earth Planet Sci Lett 174:59–73
Wolbach WS, Gilmour I, Anders E, Orth CJ, Brooks RR (1988) Global fire at the Cretaceous-Tertiary boundary. Nature 334:665–669
Xiao JL, An ZS (1999) Three large shifts in East Asian monsoon circulation indicated by loess-paleosol sequences in China and late Cenozoic deposits in Japan. Palaeogeogr Palaeoclimatol Palaeoecol 154:179–189
Xu BQ, Cao JJ, Hansen J, Yao TD, Joswia DR, Wang NL, Wu GJ, Wang M, Zhao HB, Yang W, Liu XQ, He JQ (2009a) Black soot and the survival of Tibetan glaciers. Proc Nat Acad Sci 106:22114–22118
Xu Y, Yue L, Li J, Sun L, Sun B, Zhang J, Ma J, Wang J (2009b) An 11-Ma-old red clay sequence on the Eastern Chinese Loess Plateau. Palaeogeogr Palaeoclimatol Palaeoecol 284:383–391
Xuan J, Liu G, Du K (2000) Dust emission inventory in Northern China. Atmos Environ 34:4565–4570
Xuan J, Sokolik IN (2002) Characterization of sources and emission rates of mineral dust in Northern China. Atmos Environ 36:4863–4876
Yamane M (2003) Late Quaternary variations in water mass in the Shatsky Rise area, northwest Pacific Ocean. Mar Micropaleontol 48:205–223
Yang H, Yu JZ, Ho SSH, Xu JH, Wu WS, Wan CH, Wang XD, Wang XR, Wang LS (2005) The chemical composition of inorganic and carbonaceous materials in PM(2.5) in Nanjing, China. Atmos Environ 39:3735–3749
Yokoo Y, Nakano T, Nishikawa M, Quan H (2004) Mineralogical variation of Sr-Nd isotopic and elemental compositions in loess and desert sand from the central Loess Plateau in China as a provenance tracer of wet and dry deposition in the northwestern Pacific. Chem Geol 204:45–62

Young R, Carder K, Betzer P, Costello D, Duce R, DiTullio G, Tindale N, Laws E, Uematsu M, Merrill J, Feely R (1991) Atmospheric iron inputs and primary productivity: Phytoplankton responses in the North Pacific. Glob Biogeochem Cycles 5:119–134

Yuan D, Cheng H, Edwards RL, Dykoski CA, Kelly MJ, Zhang M, Qing J, Lin Y, Wang Y, Wu J (2004) Timing, duration, and transitions of the last interglacial Asian monsoon. Science 304:575–578

Yuan W, Zhang J (2006) High correlations between Asian dust events and biological productivity in the western North Pacific. Geophys Res Lett 33: L07603 doi:10.1029/2005GL025174

Yung YL, Lee T, Wang C-H, Shieh Y-T (1996) Dust: a diagnostic of the hydrologic cycle during the last glacial maximum. Science 271:962–963

Zachos J, Pagani M, Sloan L, Thomas E, Billups K (2001) Trends, rhythms, and aberrations in global climate 65 Ma to present. Science 292:686–693

Zhang DE (1984) Synoptic-climatic studies of dust fall in China since historic times. Scientia Sinica 27:825–836 (in Chinese)

Zhang DE, Wang JH (1995) Preliminary studies on atmospheric circulation and synoptic-dynamic conditions for dustfall processes, m Loess. In: Liu TS (ed) Quaternary geology, global change (in Chinese). Science, Beijing, pp 151–157

Zhang JC, Lin ZG (1992) Climate of China. Wiley, New York

Zhang XY, Arimoto R, An Z, et al. (1993) Atmospheric trace elements over source regions for Chinese dust: Concentrations, sources and atmospheric deposition on the Loess Plateau. Atmos Environ 27:2051–2067

Zhang XY, An ZS, Chen T, Zhang GY, Arimoto R, Ray BJ (1994) Late quaternary recrods of the atmospheric input of eolian dust to the Center of the Chinese Loess Plateau. Quaternary Research 41:35–43

Zhang XY, An Z, Liu T, Arimoto R, Chen T, Zhang G, Zhu G, Wang X (1991) Study on three dust storms in northern and northwestern China: concentration and source characterization of trace elements in mineral aerosol particles. In: Liu T (ed) Loess environment and global change. Science Press, Beijing, pp 199–212

Zhang XY, Arimoto R, An ZS (1997) Dust emission from Chinese desert sources linked to variations in atmospheric circulation. J Geophy Res 102:28041–28047

Zhang XY, Arimoto R, An ZS (1999) Glacial and interglacial patterns for Asian dust transport. Quaternary Science Review 18:811–819

Zhang XY, Arimoto R, Zhang D, et al. (1995) Atmospheric transport of wind-blown mineral particles in dust storm over northwestern China. Science Geology Sinica Sup Iss (1):135–151

Zhang XY, Arimoto R, Zhu GH, Chen T, Zhang GY (1998) Concentration, size-distribution and deposition of mineral aerosol over Chinese desert regions. Tellus B 50:317–330

Zhang XY, Gong SL, Shen ZX, Mei FM, Xi XX, Liu LC, Zhou ZJ, Wang D, Wang YQ, Cheng Y (2003a) Characterization of soil dust aerosol in China and its transport/distribution during 2001 ACE-Asia. 1. Network Observations. J. Geophys. Res:108. doi:10.1029/1004 2002JD002632

Zhang XY, Gong SL, Zhao TL, Arimoto R, Wang YQ, Zhou ZJ (2003b). Sources of Asian dust and role of climate change versus desertification in Asian dust emission. Geophysical Research Letters 30:2272 doi:10.1029/2003GL018206

Zhang XY, Lu HY, Arimoto R, Gong SL (2002) Atmospheric dust loadings and their relationship to rapid oscillations of the Asian winter monsoon climate: two 250-kyr loess records. Earth Planet Sci Lett 202:637–643

Zhang XY, Shen Z, Zhang G, Chen T, Liu H (1996a) Remote mineral aerosol in westerlies and their contributions to the Chinese loess. Science in China (Series D) 39:134–143

Zhang XY, Wang YQ, Niu T, Zhang XC, Gong SL et al. (2012) Atmospheric aerosol compositions in China: spatial/temporal variability, chemical signature, regional haze distribution and comparisons with global aerosols. Atmos Chem Phys 12(2):779–799

Zhang XY, Wang YQ, Zhang XC, Guo W, Gong SL (2008) Carbonaceous aerosol composition over various regions of China during 2006. J Geophys Res 113:D14111

Zhang XY, Zhang GY, Zhu GH, Zhang DE, An ZS, Chen T, Huang XP (1996b) Elemental tracers for Chinese source dust. Science in China (Series D) 39:512–521

Zhao TL, Gong SL, Zhang XY, McKendry G (2003) Characterization of soil dust aerosol in China and its transport/distribution during 2001 ACE-Asia. 3: Size-segregated wet and dry deposition budgets. J Geophys Res: 108. 8665 doi:8610.1029/2002JD003363

Zhong DC (1999) The dynamic changes and trends of modern desert in China. Adv Earth Sci 14:229–234 (in Chinese)

Zhou B, Shen C, Sun W, Zheng H, Yang Y, Sun Y, An Z (2007) Elemental carbon record of paleo-fire history on the Chinese Loess Plateau during the last 420 ka and its response to environmental and climate changes. Palaeogeogr Palaeoclimat Palaeoecol 252:617–625

Zhou ZJ (2001) Blowing-sand and sandstorm in China in recent 45 years. Quaternary Sciences 21:9–17 (in Chinese)

Zhu JF, Zhu ZD (1999) Combating desertification in China. Chinese Forest, Beijing

Chapter 5
Mammalian Evolution in Asia Linked to Climate Changes

Yongxiang Li, Yunxiang Zhang, Xinzhi Wu, Hong Ao, Li Li, and Zhisheng An

Abstract The environmental significances of climate change as reflected in the evolution of mammalian fauna, including humans, in China during the Cenozoic are reviewed in this chapter. The mammal distributions and evolution of mammalian zoogeographic provinces in China during the Cenozoic are suggested to be significantly affected by the Asian monsoon development and the uplift of the Tibetan Plateau. The evolutionary history of these mammals is linked to different habitat scenarios in central China related to the development of the Asian monsoon-arid environment system. Some mammals are sensitive to the Asian summer and winter monsoon variations. Early humans are quite adapted to the wide environmental changes in East Asia during the Quaternary.

Keywords Cenozoic · Mammalian faunas · Fossil · Early human · Environmental and climatic changes

---

Y. Li (✉) · Y. Zhang
Institute of Cenozoic Geology and Environment, State Key Laboratory of Continental Dynamics, Department of Geology, Northwest University, Xi'an 710069, China
e-mail: mzlyx11@163.com

Y. Zhang
e-mail: yxzhang@nwu.edu.cn

X. Wu
Institute of Vertebrate Paleontology and Paleoanthropology, Chinese Academy of Sciences, 100044 Beijing, China
e-mail: wuxinzhi@ivpp.ac.cn

H. Ao · L. Li · Z. An
State Key Laboratory of Loess and Quaternary Geology, Institute of Earth Environment, Chinese Academy of Sciences, No. 10 Fenghui S. Rd, High-Tech Zone, Xi'an PO Box 17, 710075 Xi'an, China
e-mail: aohong@ieecas.cn

L. Li
e-mail: lili@ieecas.cn

Z. An
e-mail: anzs@loess.llqg.ac.cn

Z. An (ed.), *Late Cenozoic Climate Change in Asia*, Developments in Paleoenvironmental Research 16,
DOI 10.1007/978-94-007-7817-7_5,

The study of mammalian fossils is central to our understanding of the Cenozoic biostratigraphy in China, and has a history of more than a century. Up to now, abundant mammal fossils have been recovered from the widespread Cenozoic continental sediments in China. Investigations of fossils indicate that the taxonomic composition and faunal properties of the mammals are sensitive to changes in climate and environment. Mammalian faunas from the Asian monsoon-arid zone have changed substantially in the Cenozoic. This evolutionary history is important for our understanding of past changes in paleoenvironments and paleozoogeographic provinces. Of particular interest is the change from mammals that lived before the onset of Asian monsoon and aridification, to the mammals that developed later. This study reports on the land mammal evolution in Asia and its linkages with environmental and climatic changes, based on previously established mammalian fossil data. Our discussions include four parts.

In the first part, we deal with the general evolution of mammals from the Paleocene to the Pleistocene across China, and explore its environmental significance, especially the implications for the onset and development of the Asian monsoon system, aridification in the Asian interior and uplift history of the Tibetan Plateau. Evolutionary tendencies and distribution of mammals show that the initiation of the Asian monsoon may have occurred as early as the Eocene. A relatively high Tibetan Plateau, strong monsoon, and high contrast in climate and environments between western and eastern China developed since the Oligocene.

In the second part, we investigate the relationship between mammalian composition on the Chinese Loess Plateau (LP) and nearby semi-arid areas, and the variability of the monsoon and environment in these areas. The mammalian fossil records include a comprehensive mammalian profile on the LP, including: 1) Late Miocene-Pleistocene micromammals from the Lingtai red clay-loess sequence on the central LP; 2) Nihewan Fauna, Zhoukoudian Fauna and Salawusu Fauna on the margins of the LP; and 3) Linxia faunas (Late Oligocene *Dzungariotherium* fauna, Middle Miocene *Platybelodon* fauna, Late Miocene and Early Pliocene *Hipparion* fauna, and the Early Pleistocene *Equus* fauna) on the western margin of the LP. The evolutionary history of these mammals indicates different habitat scenarios in central China related to the development of the Asian monsoon-arid environment system.

In the third part, the environmental significances of several representative mammals are reported, including climatic events reflected by the evolutionary trend of Concave-Occiput siphneids on the LP, distribution of microtine in mid-Pleistocene Qinling Mountains indicative of rapid oscillations of cold and warm climate, expansion of *Equus przewalskii*, *Mammuthus-Coelodonta* fauna and *Elaphurus davidianus* in North China linked to the East Asian winter monsoon, mammal composition of the Guanzhong area on the LP, indicative of extreme dry and cold conditions during the Last Glacial Maximum (LGM), and the implication of strong seasonal Indian summer monsoon implicit in the antler morphology of cervids from the Yuanmou Basin in South China.

In the last part, we discuss early human evolution from *H. erectus* to archaic and modern *H. sapiens* in China and their adaptability to climate changes.

5.1 Cenozoic Evolution of Chinese Mammalian Faunas

The general distribution of Chinese mammal fossil localities, the contours of zoogeographic provinces and the geomorphic outlook in different epochs of the Cenozoic are shown in Fig. 5.1. Table 5.1 presents the Chinese representative mammalian taxa listed by family and epoch. At present, the low elevation in South and Southwest China, where the climate is affected by the summer monsoon, is dominated by mammals preferring warm and wet environments. The arid or sub-arid Mongolia-Xinjiang uplands are characterized by mammals adapted to a dry environment. Because of the high elevation, thin air, and low temperature, the Tibetan Plateau is generally unsuitable for many mammals and is colonized by only a small number of mammals that are adapted to rigorous climates on the plateau. Distribution of extant land vertebrates indicates two distinct zoogeographic provinces in China. Roughly, a line following Huaihe River, Qinling Mts. and Hengduan Mts. separates the Palearctic Province in the north, where mammals live in frigid to temperate climates, from the Oriental Province in the south, where mammals live in tropical or subtropical zones. During the Cenozoic, Asian monsoon development and the uplift of the Tibetan Plateau had profound impacts on mammal distributions and evolution of mammalian zoogeographic provinces in Asia, and these are evident in the fossil record. We show in this chapter how mammalian fossils can be used to understand paleoenvironmental and paleozoogeographic changes.

5.1.1 Cenozoic Mammalian Faunal Distributions

Paleocene mammals The Paleocene is the earliest epoch of the Cenozoic. Paleocene mammal localities are mainly found in central and southern China (Fig. 5.1a). All Paleocene mammalian fossil-bearing beds are red terrestrial deposits, which lay directly on Cretaceous strata. Mammals recovered from the Paleocene generally have small body sizes and very primitive characteristics. Most of the mammals belong to extinct orders, such as Multituberculata, Pantodonta, Dinocerata, Tillodonta, Condylarthra, Creodonta and Glires. There are more than 10 families of archaic mammals during the Early Paleocene, including Bemalambdidae, Anagalidae, Viverravidae, Astigalidae and Pseudictopidae. During the Late Paleocene, Bemalambdidae, which is a common family in the Early Paleocene, became extinct and was replaced by more advanced families such as Phenacolophidae, Arctostylopidae, Prodinoceratidae, Eurymylidae and Mimotonidae. All Paleocene faunas found in central and southern China are quite similar in composition at the family and genus level. All the families and the overwhelming majority of the orders from the Paleocene faunas became extinct, thus it is impossible to make confident inferences about the environments in which these mammals lived, because of a lack of living relatives. However, similarities in faunal compositions from Xinjiang and the intermountain basins of Nanling, Qinling and Dabie Mts. seem to indicate no obvious topographic differences between eastern and western China. The vast area

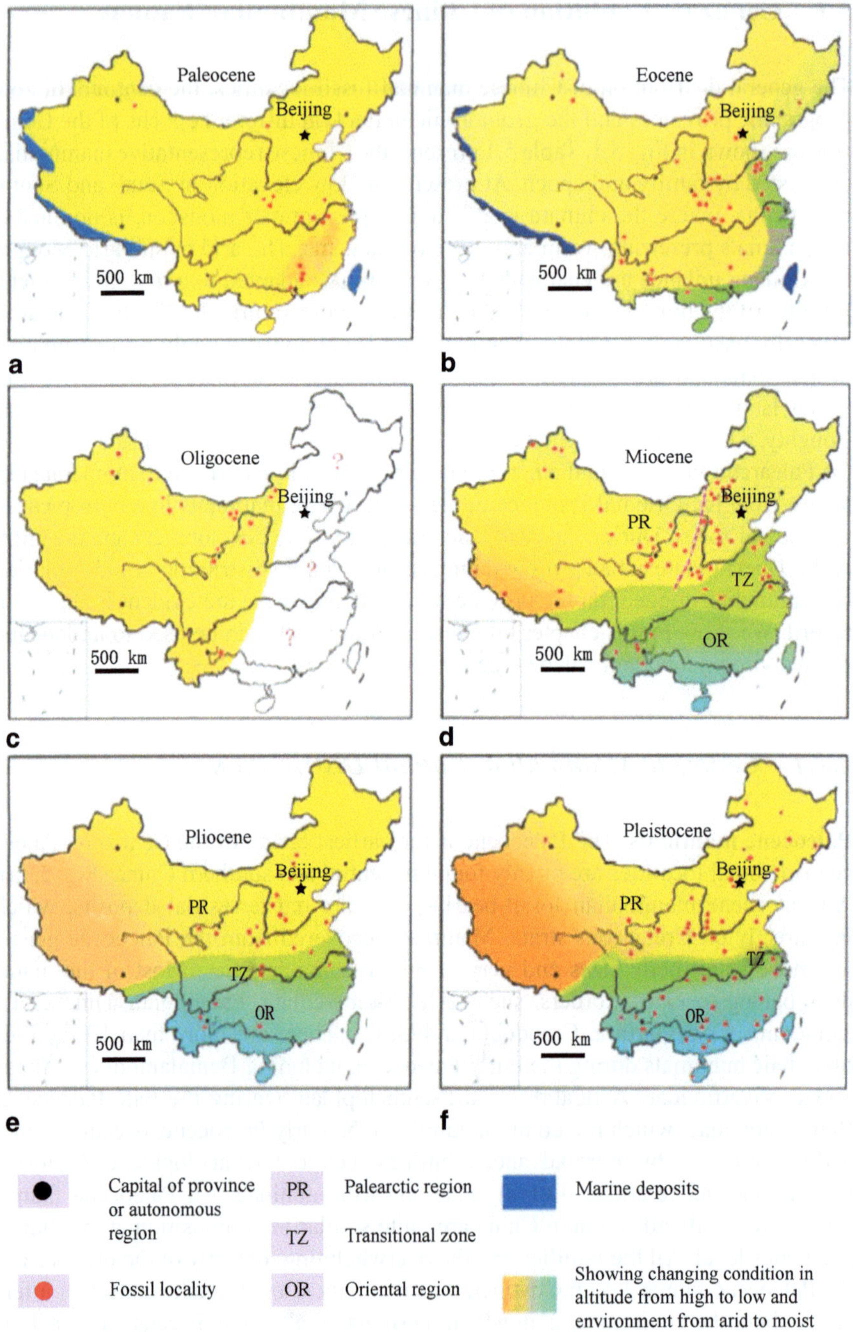

Fig. 5.1 Map showing Cenozoic localities, contours of paleozoogeographic provinces and geomorphic outlook in different ages of China (Qiu and Li 2005)

Table 5.1 Ranges of Chinese representative Cenozoic mammals at family level (Qiu and Li 2005)

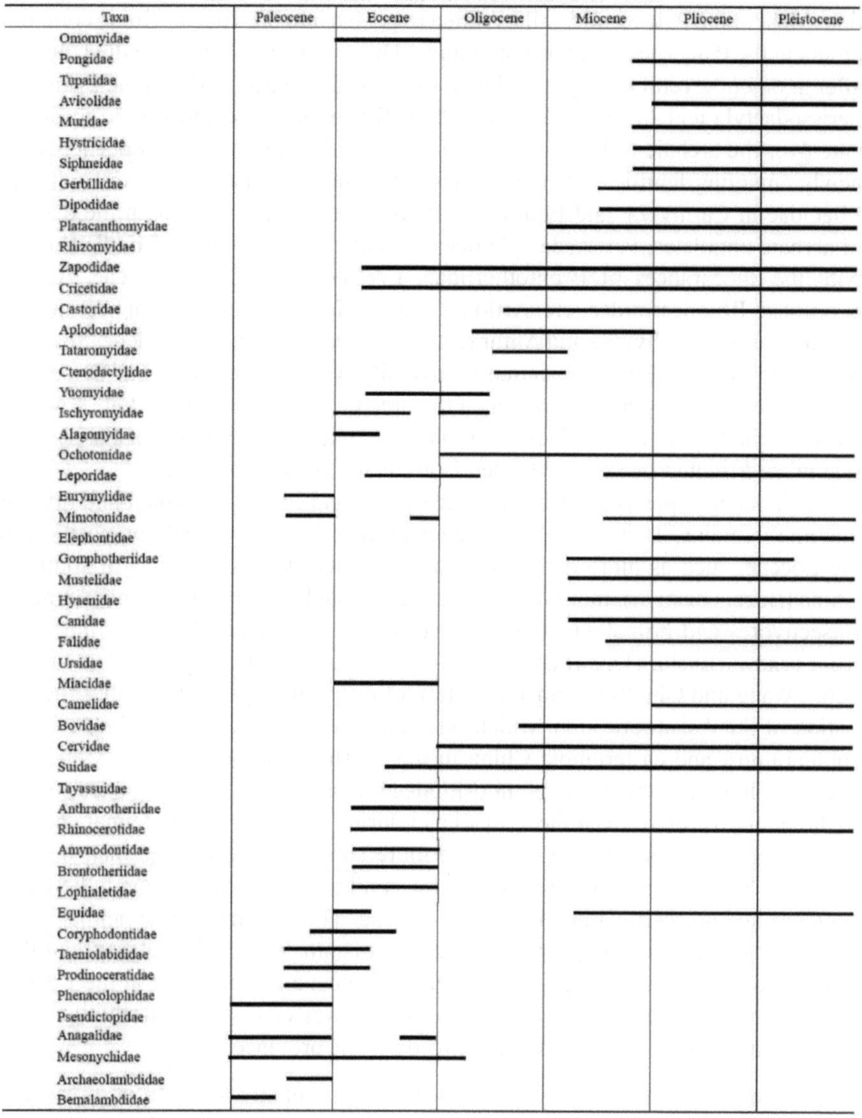

of central and southern China was probably dominated by an east-west planetary wind system at that time.

Eocene mammals Eocene mammal sites are mainly found in eastern China (Fig. 5.1b). The fossil-bearing beds are brownish red continental deposits, which consist mainly of lacustrine and swamp mudstones and marls, with frequent occurrence of gypsum in western China, and lignite and oil shale in eastern China. The grains of the sediments are fine and the color is usually lighter than observed in Paleocene sediments. The Eocene marked the initiation of mammalian modernization. Most of the Paleocene families failed to persist into the Eocene. Some common orders in the Paleocene, such as Pantodonta, Dinocerata and Anagalida died out one after another. Several widely distributed extant orders, such as Rodentia, Primates, Perissodactyla and Artiodactyla, made their first appearance and started to dominate over the archaic orders in their communities. During the Early Eocene, some newly emerging families, such as Cocomyidae and Tamquammyidae in Rodentia, Miacidae in Carnivora, and Equidae in Perissodactyla appeared. With the decline of archaic ungulates, Perissodactyla began a major radiation in the Middle Eocene with the appearances of Brontotheriidae, Lophialetidae, Deperetellidae, Amynodontidae, Rhinocerotidae, etc. Artiodactyla at this time made their appearance in the families of Helohyidae and Anthracotheriidae and showed the tendency towards an initial radiation. Rodentia started to diversify with the appearance of Cricetidae and Zapodidae and gradually replaced the archaic glires. During the Late Eocene, Artiodactyla had an initial diversification and broke the dominance of Perissodactyla in the Middle Eocene, and Rodentia gradually flourished as well.

The Eocene deposits in China are widespread and contain abundant and diverse mammal remains. There were some regional differences in faunal composition in the Eocene, such as more diversity of anthracotheriids in the faunas of southern China (Bose, Guangxi), more taxa preferring forested, warm and wet environments (insectivores and primates) from the east (Wutu, Shandong; Huadian, Jilin) than from central (Lushi, Henan) and northwestern (Hami, Xijiang) China (Qiu and Li 2005; Wang and Qiu 2003). Such a pattern of distribution possibly implies the initiation of the Asian monsoon, which resulted in a relatively humid climate in the southernmost and easternmost China in the coastal region (Fig. 5.1b). Generally speaking, however, the faunal compositions in South and North China were similar at the family level, or even the genus level during the Eocene. The faunal compositions between these two regions were more similar in the Eocene than in later intervals. Such a similarity might suggest similar natural environments between southern and northern China during the Eocene. There seemed to be neither great topographic barriers to impede the spread of mammals, nor distinct regional differentiation in faunal elements. It is likely that most areas of China in the Eocene belonged to one zoogeographic province, which was consistent with the limits of the arid paratropical region of the phytogeographic province.

Oligocene mammals Oligocene mammal sites are mainly found in northwestern China, and are rare in eastern and southwestern China, including the Tibetan Plateau (Fig. 5.1c). The fossiliferous beds consist mainly of red, yellowish brown, grayish yellow and green fluvio-lacustrine deposits, usually with some intercalated conglomerate beds, even gravel conglomerates. The Oligocene deposits have lighter color and coarser grains than those of the Eocene. In some Paleogene basins on the northern edge of the Tibetan Plateau, such as the Lanzhou Basin (Gansu Province), sediments are rather fine in the Paleocene and Eocene, but are significantly coarser and contain frequently thickly-bedded poorly-rounded conglomerates starting from the Middle Oligocene.

The Oligocene was a turning point in the history of mammalian modernization. The family and genus level of the mammals changed significantly. All archaic ungulates in the Eocene are nearly absent. Perissodactyla, which flourished in the Eocene, began to decline gradually. For example, the quite prosperous Lophialetidae, Deperetellidae, Eomoropidae from the Eocene failed to persist into the Oligocene. Artiodactyla began a major radiation, and Rodentia started a new developmental stage. In the Early Oligocene, a remarkable change (relative to the Eocene faunas) is the appearance of Ochotonidae in the Mongol-Xinjiang Region, which are presently adapted to temperate grasslands. A great turnover of taxa in mammalian faunas took place at the end of the Early Oligocene. All of the families of Paleocene origin such as Mesonychidae and Mimotonidae, and about half of the families present in the Eocene, such as Brontotheriidae, Miacidae and Ischyromyidae, became extinct. Some primitive rodents such as Tsaganomyidae, Aplodontidae and Tataromyidae appeared. A so-called "*Indricotherium* Fauna", which consisted of indricotheriines and brontotheriids in the large mammals and tsaganomyids, tataromyids and ochotonids in the small mammals, formed during this period. In Late Oligocene, two new steppe-adapted groups, Tachyoryctoididae and Distylomyidae, joined this fauna.

Two obvious phenomena occurred in the evolution of Oligocene mammals, i.e., disappearance of a great quantity of archaic taxa and appearance of higher-crowned and steppe-adapted small mammals (Tsaganomyids and Ochotonids) in the northwestern regions adjacent to the Tibetan Plateau (Qiu and Li 2005; Wang and Qiu 2003). The turnover of mammals and changes in sedimentation took place almost simultaneously. This coincidence implies a change of ecological and geographic environment, e.g., significant environmental changes may have taken place during the Middle Oligocene. These changes include the topographic trend from a relatively flat Eocene landscape to a west-tilting topography, and a cooler, more arid climate with steppe biota in Northwest China. These changes are possibly related to the result of initial uplift of the Tibetan Plateau in the Oligocene. The uplift of the Tibetan Plateau might have changed not only the sedimentary environment and depositional fabric, but also the climatic pattern, such as the intensification of Asian monsoon with a decrease of the east-west planetary wind system. Because of the changes in environment, the composition and distribution of floras and faunas changed accordingly. However, judging from the recovered indricotheriines of Late Oligocene in Xinjiang and Yunnan of China and in the Indian subcontinent, the altitude of the Plateau did not hinder the migration and spread of some mammals between central and southern Asia.

Miocene mammals Miocene mammal sites are mainly found in northern and northwestern China, with a few in Yunnan and Xizang (Fig. 5.1d). The Miocene was an important period of mammalian modernization. All the mammals from the Miocene are of extant orders, and most of the modern mammalian families appeared during this period. During the Early Miocene, the dominance of Protrogomorpha was replaced by Sciuromorpha and Myomorpha in Rodentia. The archaic carnivores were completely replaced by modern ones. Proboscidea immigrated into Asia from Africa. The archaic Perissodactyla became extinct and the Artiodactyla diversified prominently. During the Middle Miocene, proboscideans and artiodactyls increased greatly. Some taxa of rodents that were common in northern China during the Late Oligocene and Early Miocene, such as Ctenodactylidae and Tachyoryctoidae, were replaced by drought-adapted newcomers, such as Gerbillidae and Dipodidae. A so-called "*Platybelodon* Fauna", consisting mainly of cricetodont rodents, *Anchitherim* horses, rhinoceros, shovel-tusked elephants, *Kubanochoerus* and *Stephanocemas*, was formed in this interval. During the Late Miocene, mammals became more modernized. Myomorphs were dominant in rodents. Cricetodonts were replaced by cricetids. Rodent families like Muridae, Siphneidae, Rhizomyidae and Hystricidae rose abruptly. Mustelidae, Hyaenidae and Felidae in Carnivora became greatly diversified. Proboscidea further diversified. *Hipparion* and *Chilotherium* were dominant for Perissodactyla. Artiodactyla flourished to an unprecedented degree. Mammals in this interval formed the well-known "*Hipparion* Fauna", which features the modern mammal pattern in community structure at higher taxonomic ranks. The Late Miocene was probably a time when Chinese mammals thrived during the Cenozoic, and the diversity and abundance of mammals in northern and northwestern China was comparable to those in the African savanna today.

During the Miocene, mammals, especially the small mammals, differentiated distinctly in distribution between North and South China. The micromammals in the Xiejia fauna (Early Miocene), the Tunggur Fauna (Middle Miocene) and the Ertemte Fauna (Late Miocene) of northern China, such as Castoridae, Aplodontidae, Gliridae, Zapodidae, Dipodidae, Gerbillidae and Ochotonidae, are restricted to or mainly distributed over the Palearctic Region, but none is found in the present Oriental Region. The Yuanmou and Lufeng faunas in South China, however, contain extant families which live in the current Oriental Region or in tropical-subtropical zones, such as Tupaiidae, Echinosoricinae, Pteropidae, Platacanthomyidae, Rhizomyidae and Hystricidae. The faunas in North China reflect a temperate, relatively arid steppe environment, similar to the present Palearctic Region; while the faunas in South China indicate a tropical or subtropical forest environment with characteristics of the Oriental Region. This indicates that the initiation of the Oriental Region occurred in the Miocene. In addition, there was a transitional zone of zoogeographic region between North and South China, examplified by the Early Miocene Sihong Fauna, which contains taxa from the modern Palearctic Region, such as Castoridae, Gliridae and Ochotonidae, and members restricted to the Oriental Region, such as Echinosoricinae, Rhizomyidae and Platacantomyidae. The taxonomic composition of Chinese Miocene mammals showed not only a latitudinal ecological variation, but also a longitudinal variation. Take the earlier and middle Miocene faunas for example, the Sihong Fauna and the Shanwang Fauna consist mainly of mammals preferring warm and moist habitats, whereas the Xiejia Fauna of Qinghai and the Quantougou Fauna of Gansu are composed of taxa adapted to dry and cold climates (Qiu and Li 2005; Wang and Qiu 2003). Between them, in central China, such as in Gansu and Ningxia provinces, there existed transitional faunas which were characterized by not only steppe rodents but also abundant proboscidians, pigs and deer. The occurrence of both latitudinal and longitudinal variations in mammals during the Miocene is consistent with the presence of an uplifted Tibetan Plateau, with concomitant monsoon circulation in place.

Pliocene mammals The Pliocene is a short period (5.3–2.6 Ma). Pliocene fossil sites are mainly found in North China (Fig. 5.1e), and much reduced in number as compared with the Miocene. Characteristics of the fossiliferous sediments are similar to those of the Late Miocene, mainly consisting of fluvio-lacustrine and reddish silty deposits, with the frequent occurrence of thick conglomerates along the northern edge of the Tibetan Plateau. For example, the thick "Jishi Conglomerates" and the widespread "Yumen Conglomerates" in Gansu.

During the Pliocene, the faunas show a continued development of the Late Miocene faunas, with faunal variations only at the family and genus level. Obvious changes in the large mammals include the replacement by new carnivores, the decline of some perissodactyls such as *Hipparion* and *Chilotherium*, the continued diversity of the artiodactyls and the appearance of Caninae, Elephantidae and Camelidae. Among the small mammals, protrogomorphs became completely extinct, and the Late Miocene survivors such as Aplodontidae and Eomyidae disappeared at this time. Sciuromorphs declined greatly at this time; and myomorphs became highly diversified, along with the families that originated and flourished in the Late Miocene. Dipodidae, Gerbillidae, Siphneidae and Muridae, and a new family with high crowned cheek teeth, Arvicolidae, appeared at this time. All of the families of Chinese extant mammals were present until the Late Pliocene, with some genera extending to the present day.

Pliocene mammal fossils from southern China are scarce, but collections from Yuanmou (Yunnan province) and Huainan (Anhui province) indicate clear differences in composition between southern and northern China. Compared with the Late Miocene, during the Pliocene rodents not only increased in diversity and abundance, some also strengthened their ability to adapt to hard-vegetation through heightening of crowns. Such a phenomenon might imply an environment of drier and more grassland-dominant conditions during the Pliocene than the Miocene in northern and northwestern China (Qiu and Li 2005; Wang and Qiu 2003). The composition of faunas and the thick conglomerates widely distributed along the northern edges of the Tibetan Plateau suggest that the elevation of the plateau increased after the Late Miocene, and achieved a rather high altitude in the Pliocene. It is likely that the Chinese modern environmental pattern, as well as the community structure of modern mammals, was initiated at the end of the Late Pliocene.

Hipparion Fauna spread across most China during the Pliocene, with representatives from Xinjiang, Jiangsu, Yunnan and Inner Mongolia. They generally consisted of *Hipparion*, *Chilotherium*, *Stegodon* and *Gazella*. In southeastern China (e.g., Jiangsu, Shandong, Hubei and Yunnan), however, they were usually represented by a large number of species adapted to humid environments, such as tragulidae, *Muntiacus lavustris*, pigs, rhizomyids, insectivores, chiropterans and primates. In northwestern China (e.g., Gansu, Qinghai and Xinjiang), they were usually represented by grassland mammals such as rhinos, gazelles and pikas. This differentiation of mammalian distribution illustrates the strong contrast between northwestern and eastern China at that time, associated with the influence of a monsoon climate in the relatively humid east.

Pleistocene mammals Pleistocene mammal sites are widely spread across China, yielding abundant fossils (Fig. 5.1f). The Pleistocene deposits include loose, silty, eolian, fluvio-lacustrine and cave deposits. During the Pleistocene, Proboscidea and Perissodactyla further declined, Cervidae and Bovidae in Artiodactyla further flourished, and Myomorpha in Rodentia further diversified. The evolution of mammals is characterized by the faunal turnover at the genus and species level. Differentiation in the mammalian composition between North and South China became more distinct than ever before. A so-called "*Ailuropoda-Stegodon* Fauna," including *Tapirus*, *Rhinoceros*, *Rhizomys* and *Hystrix*, appeared in South China. These mammals were generally adapted to warm and humid environments. In North China, however, mammals are dominated by forms accustomed to temperate and dry conditions, such as Bovidae, Cricetidae, Siphneidae, Avicolidae, Dipodidae and Ochotonidae. Almost all the Late Pleistocene taxa can be grouped into extant genera, and the distribution of the Late Pleistocene mammals at the family level is basically consistent with that of the present day. The composition and distribution of Pleistocene mammals suggest that the zoogeographic divide between the Palearctic region and the Oriental realm was quite distinct by this time. This zoogeographic boundary followed a narrow east-west trending zone from the Huaihe River through the Qinling Mountains to the Hengduan Mountains, representing the traditional climate divide between temperate and sub-arid North China and subtropical and humid South China.

During the Pleistocene when the humid monsoonal zone in eastern China, arid zone in western China and semi-arid zone between eastern and western China, became more obvious than during previous periods, and a large number of mixed faunas occurred in the semi-arid LP. On the eastern and northern margin of the LP, desert-grassland mammals were dominant, such as Dipodidae, *Ochotona* and camelidae. On the southern margin, however, a large number of humid-adaptated mammals (e.g., water buffalo, sika deer, *Apodemus*) occurred and even some mammals of the Oriental Realm, such as *Cercopithecidae*, *Ailuropoda* and *Stegodon*. Under the influence of the oscillations of the Asian summer monsoon, the distribution of these mammals showed not only zonal but also temporal variations. Along with the southeastward retreat of the Asian summer monsoon during the glacial periods, some mammals adapted to cold and dry climate in northwestern China spread towards southeastern China. For example, *Mammuthus* in Siberia could expand to North China during glacial periods. *Equus przewalskii* in the Gobi desert in northeastern China also expanded significantly towards the southeast during glacial periods (Deng 1999).

5.1.2 Fossil Mammalian Evidence for Cenozoic Environmental Change in Asia

The composition and distribution of the Chinese Cenozoic mammalian faunas (Qiu and Li 2005; Qiu and Qiu 1995; Yuan and Du 1984) show an essentially united zoogeographic province and similar environment between South and North China during the Paleocene, and demonstrate the appearance of a great quantity of taxa adapted to temperate steppe in southern and eastern China during the Eocene. There was a progressive differentiation of mammalian faunas since the Miocene, and a gradual increase of mammals adapted to abrasive diets in North China and mammals adapted to tropical and sub-tropical forests in South China during the Pliocene and Pleistocene. The mammal evolution and environmental changes are consistent with a gradual uplift of the Tibetan Plateau since the Oligocene. This is because the uplift changed the environment that resulted in the differentiation in animals and plants distribution and in the changes of adaptation of the organisms living within. The boundary of the Palearctic Region and the Oriental Region in China compared to that boundary between the Palearctic Region and the Ethiopian Region in Africa also provides evidence for an elevated plateau. The uplift resulted in the initiation and strengthening of the Asian monsoon. The strengthened summer monsoon caused a northwestward expansion of the mammals from the Oriental Province. As a result, the latitude of the boundary between Palearctic Region and the Oriental Region increased.

The elevated plateau formed an effective barrier to the penetration of moist air from the Indian Ocean and the Pacific Ocean. As a result, aridity increased in central Asia, including the northwestern part of China, and warmth and seasonal moisture characterized the southeast and southwest. It is likely that the Tibetan uplift, and associated global climatic changes resulted in the environmental differentiation in China, including: high altitude and cold temperature conditions in the Qinghai and Xizang areas; gradual aridity in Inner Mongolia and the Xinjiang regions; and warm and moist conditions in the South and Southwest China. With these environmental changes the distribution of mammals and plants was progressively differentiated and biogeographic provinces were formed. With the continual rise of the plateau, the two zoogeographic provinces in China became increasingly distinct, as the modern pattern of the Asian monsoon-arid climate system was gradually established. The elevation of the Tibetan Plateau in the Late Cenozoic significantly affected the evolution of mammals in southeastern Asia. The strengthened summer monsoon from South Asia and winter monsoon from North Asia, made the climate steadily drier in the northwestern part, and colder in the Qinghai and Xizang areas. The aridity brought about grassland conditions in Northwest China until the Miocene, with a harsh ecological environment on the Tibetan Plateau. The high altitude and cold climate made the Plateau an impassable barrier for many mammals since the Pliocene.

5.2 Environmental Significance of Mammalian Faunal Evolution on the Loess Plateau

5.2.1 Comprehensive Mammalian Profile on the Loess Plateau

Thick loess deposits of the Quaternary were accumulated on the LP of North-central China, which covers an area of about 440,000 km^2 (Fig. 5.2). Loess deposits are commonly underlain by reddish silt-clay sediments of Late Neogene age, tradi-

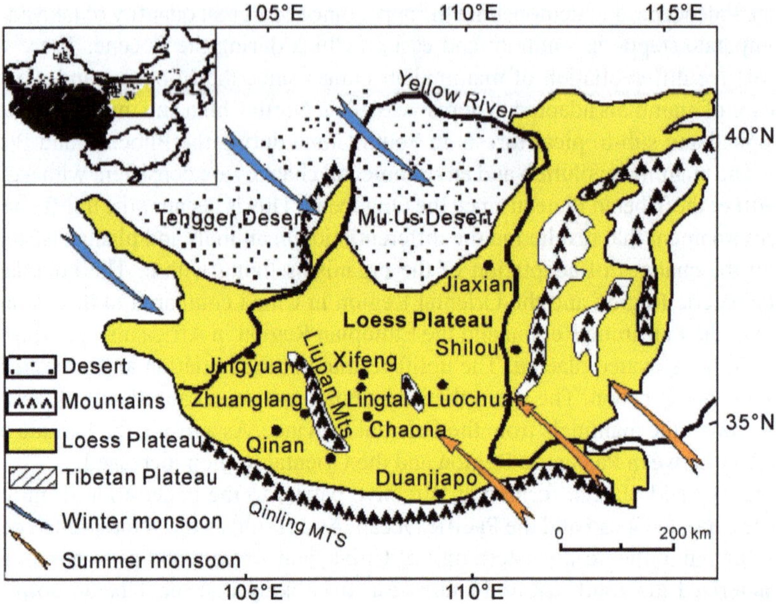

Fig. 5.2 Map showing the typical loess-red clay sections

tionally known as the *Hipparion* red clay. Eolian red clays and loess layers have drawn much attention of the world for their continuous and lengthy records of paleoclimate information. Red clay-loess sections on the LP yielded a large number of mammalian fossils. Paleomagnetic studies of these sections provide important chronologic control for the interpretation of the fossils. However, the quantity and kinds of fossils are quite uneven. Some groups are very abundant, while others are much less so. Tens of thousands of kilograms of fossils were unearthed from the red clay deposits of Laogaochuan, Fugu County, Shaanxi Province, Longjiagou, Wudu County, Gansu Province and Baode County of Shanxi Province and from the loess deposits of Longdan (Linxia Authonomous Region, Gansu Province). However, the mammal fossils are less abundant and scattered in other areas, and mainly consist of isolated teeth, partial jaws, or a few bone fragments. Such a situation is not very useful for reconstructing the conditions of the fossil-bearing bed. Therefore, Xue et al. (2006) established a comprehensive mammalian profile by combining a series of magnetostratigraphically-dated red clay-loess sections that contain mammalian fossils in order to investigate the mammal evolution on the LP and its paleoenvironmental significances (Fig. 5.3). Having analyzed the characteristics of mammalian assemblages, we can infer the paleoenvironmental changes during the past 8 Ma as follows.

Lamagou Fauna The Lamagou Fauna was discovered from the lower part of Laogaochuan red clay profile in Fugu County, Shaanxi Province. It is a typical Late Miocene fauna on the LP, with a paleomagnetic chronology of ~7.8 Ma (Xue

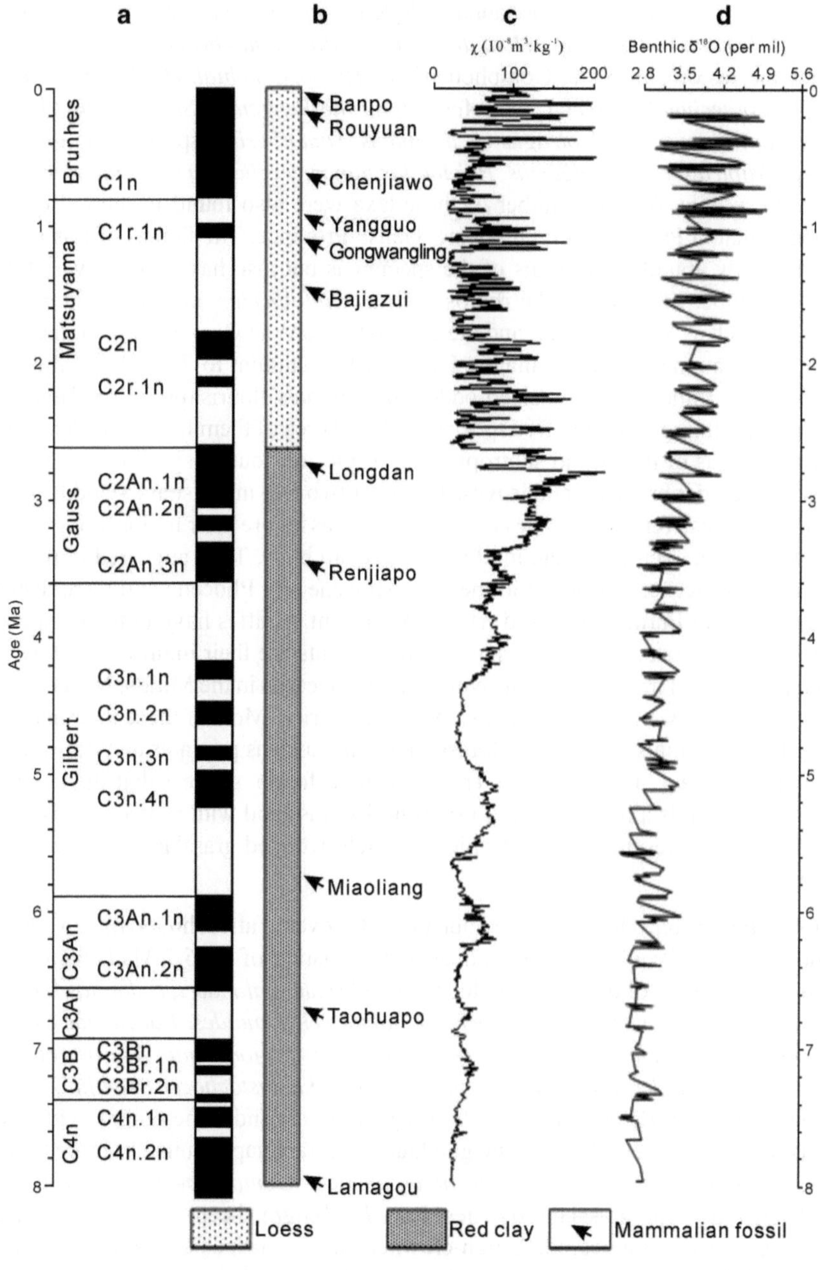

Fig. 5.3 A composite diagram showing the magnetostratigraphically-dated faunas on the Chinese Loess Plateau within the framework of the geomagnetic polarity timescale and temporal variability of regional and global climate during the past 8 Ma. (a) geomagnetic polarity timescale (GPTS) (Cande and Kent 1995), (b) mammal faunas and (c) magnetic susceptibility of the red clay-loess sequence (Nie et al. 2008; Xue et al. 2006). (d) Global deep-sea oxygen isotope record based on data compiled from more than 40 Deep Sea Drilling Project (DSDP) and Ocean Drilling Program (ODP) sites (Zachos et al. 2001)

et al. 1995). The fossils of this fauna include *Prosiphneus* sp., *Dinocrocuta gigantea, Ictitherium wongi, I.* sp., *Homotherium* sp., *Plesiogulo brachygnathus*, several other species of Carnivora, Gomphotheridae, *Hipparion chiai, H.* cf. *forstenae, H.* sp., *Chilotherium habereri, C.* sp., *Acerorhinus hezhengensis, Sinotherium lagrelii. Chalicotherium* sp., *Paleotragus* cf. *decipiens, Samotherium* sp., *Gazella gaudryi, G.* sp., *Miotragocerus gregarius, Plesiaddax minor, Hezhengia* sp., *Tragoceros* sp., *Testudo* sp., etc. A large number of these taxa were also found in the red clay at Baode, Shanxi Province and Qingyang, Gansu Province. The faunas of this period are not only abundant in terms of the specimens but also have a large variability of the number of species. The rhinos such as *Chilotherium* and *Acerorhinus*, the giraffes such as *Samotherium*, and the gigantic hyeana as *Dinocrocuta* are the most common mammals in the Lamagou Fauna, which amount to 70–80% of the total fossils. The rhinos are large-sized odd-toed mammals flourishing in the Neogene. Today only four genera and five species are left. Three of them live in the dense jungle grasslands of the humid subtropics and tropics of South Asia. They eat mainly fresh and tender branches and leaves. The other two live in the sparse-woody grasslands of Africa. Tender branches of shrubs and grasses are their major food. Giraffe is a kind of an even-toed mammal with a very tall body. They appeared initially in the Early Miocene, flourished in the Late Miocene and Pliocene and were distributed widely in Europe, Asia and Africa. At present, giraffes have only one species surviving in Africa. Fresh leaves and tender sprouts are their main food. *Dinocrocuta* is a hyaenid in the order Carnivora, which appeared in the Middle Miocene and was dispersed extensively in Europe, Asia, and Africa. Most of these mammals are grassland mammals. The typical forest mammals, such as primates and deer are not abundant in the Lamagou fauna. These mammal fossils suggest that the environment of the LP is a hot-semiarid to semi-moist grassland with sparse woods at ca. 8 Ma, possibly resembling the tropical sparsely wooded grasslands and savannas of Africa.

Taohuapo Fauna The Taohuapo Fauna was discovered at Taohuapo hill, in Lingtai County, Gansu Province, with a paleomagnetic dating of ca. 6.7 Ma (Zhang et al. 2002). The fossils of this fauna include *Homotherium palanderi, Indarctos lagrelii, Adcrocuta eximia variabilis, Ictitherium wongi, I. hyaenoides, Parataxidea sinensis, Gomphotherium* sp., *Hipparion* (*Cremohipparion*) *forstenae, Acerorhinus* sp., *Chilotherium anderssoni, Sinotherium lagrelii, Chleuastochoerus stehlini, Samotherium* sp., *Gazella dorcadoides, Palaeoryx sinensis*, and some other bovid mammals. In comparison with the Lamagou fauna, the Taohuapo fauna has more typical forest mammals such as sabertooth, *Indarctos* bear, *Gomphotherium* and *Chleuastochoerus* pig. The grassland dwellers (e.g. *Hezhengia* sheep) and the huge sized *Dinocrocuta* all disappeared. A high-crowned gazelle such as *Gazella dorcadoides* appeared. Despite these differences between the Lamagou and Taohuapo faunas, the general characteristics of genera and species are quite similar. They probably reflect a climate that was becoming moister, but the principal environment of the LP had not changed. It is also possible that, owing to stable conditions, the mammalian assemblage was maintained for a relatively long time.

Miaoliang Fauna The Miaoliang Fauna was collected from the upper fossil bed of the latest of the Late Miocene red clay layers at Laogaochuan Valley (Shaanxi Province) (Xue et al. 1995; Zhang et al. 1995, 2002). It has a paleomagnetic age of ca. 5.7 Ma, 40 m higher than the stratum of the Lamagou fauna in the same profile. Its main members include *Eostylocerus blainvillei*, *Muntiacus lacustris*, *Cervavitus novorossiae*, *C. demissus*, *Procapreolus latifrons*, *Hipparion platyodus*, *Acerorhinus*, *Gomphotherium* sp., *Stegodon zdanskyi*, *Elasmotherium*, *Honanotherium*, *Adcrocuta eximia variabilis*, *Chleuastochoerus stehlini*, and Rhizomyidae. The same or similar faunas are found at Longjiagou (Wudu County, Gansu Province) and at Locality 49 in Baode County (Shanxi Province). Wudu fauna includes *Dryopithecus wuduensis*, *Eomellivora* sp., *Pliohyaena* sp., *Machairodus* sp., *Paramachairodus* sp., Hyracoidea, Gomphotheridae, *Hipparion platyodus*, *H. parvum*, *H. dermatorhinus*, *Chilotherium* spp., *Acerorhinus* sp., *Chalicotherium wuduensis*, *Chleuastochoerus stehlini*, *Honanotherium* sp., *Samotherium* sp., *Eostylocerus* sp., *Cervavitus demissus*, *Metacervulus bidens*, *M. brevis*, *Muntiacus* aff. *lacustris*, *M.* sp., *Miotragocerus* sp., *Gazella gaudyi*, several kinds of Bovinae, and *Testudo* sp.. These faunas have a higher proportion of forest mammals than the Lamagou and Taohuapo faunas. Forest ape, *Dryopithecus*, the larger sable-toothed tiger, the clawed-toed *Chalicotherium*, and Gomphotheriidae elephants were the typical forest animals. Muntjacs are a small to medium-sized members of the deer family. They have simply forked antlers and lower crowned teeth. This kind of deer flourished in the Neogene. Now only one genus and four species survive in China and southern Asia. They live in the sparse-wooded, jungle-forest and near water bodies in hilly lands and plains. The tooth crown of the giraffe *Honanotherium* were lower than that of the earlier *Samotherium*, a probable feature brought-up by consuming more fresh and tender leaves and branches. Different from other species of *Hipparion*, the metacarpal and metatarsal bones of *H. platyodus* were all shorter. This might be related to their different living conditions. The comprehensive characters of mammalian fossils indicate that the environment of the LP was warmer and much moister at around 6–5.3 Ma than before, and characterized by forests or densely wooded grasslands. These environmental changes are consistent with the intensification of the East Asian monsoon (An et al. 2001).

Renjiapo Fauna The Renjiapo Fauna was discovered in the Renjiapo red clay-loess profile, in Lingtai County, Gansu Province (Zhang et al. 1999). It has a paleomagnetic age of ca. 3.4 Ma. Its main members include *Chardinomys* sp., *Nyctereutes sinensis*, Gomphotheriidae, *Hipparion houfenense*, Rhinocerotidae, *Paracamelus* sp., *Gazella blacki*, *G.* sp., *Antilospira licenti*. The mammals favoring bushwood or wood elements among the previous assemblages, such as *Eostylocerus* deer, *Chleuastochoerus* pig, *Homotherium* sabertooth and *Acerorhinus* rhino either decreased or disappeared in this period. Some new members such as the ancestor of the modern camel, *Paracamelus* and the raccoon-dog first appeared. Artiodactyls were the chief members of this fauna, most of them with high crowned teeth (hypsodont). All of these show clearly that the environment of this period changed distinctly. It became drier and cooler, the forests shrank and the grasslands expanded. The environment on the LP seems to have been temperate grasslands and temperate desert during this period.

Longdan Fauna The Longdan Fauna was discovered in the lower part of the loess profile of Longdan, Linxia Autonomous Region, Gansu Province, with a paleomagnetic age between 2.55 and 2.16 Ma (Qiu et al. 2004). It consists of 29 genera and 31 species, including *Aepyosciurus orientalis*, *Marmota parva*, *Mimomys* cf. *M. gansunicus*, *Bahomys* sp., *Sericolagus brachypus*, *Macaca* cf. *M. anderssoni*, *Paradolichopithecus gansuensis*, *Vulpes chikushanensis*, *Canis teilhardi*, *C. longdanensis*, *C. brevicephalus*, *Sinicuon* cf. *dubius*, *Eirictis robutsta*, *Meles teilhardi*, *Chasmaporthetes progressus*, *Pachycrocuta licenti*, *Crocuta honanensis*, *Homotherium crenatidens*, *Megantereon nihowanensis*, *Sivapanthera linxiaensis*, *Panthera palaeosinensis*, *Felis teilhardi*, *Lynx shansius*, *Hipparion* (*Proboscidipparion*) *sinense*, *Equus eisenmannae*, *Coelodonta nihowanensis*, *Hesperotherium* sp., *Nipponicervus longdanensis*, *Gazella* cf. *G. blacki*, *Leptobos brevicornis*, *Hemibos gracilis*. Among these animals, *Chalicotherium* took the tree leaves for their food. *Aepyosciurus*, *Bahomys*, *Macaca* and *Nipponicerus* liked to eat the leaves and berries of the trees and shrubs. Horses and even-toed animals lived separately in grasslands or at their margins. *Felis* and *Lynx* dwelled mainly in the bushes. Animals adapted to deserts and forests were absent in this fauna. The characteristics of the mammal fossils suggest that the Longdan area had plenty of water, with grass and brushes and thin forests during the Early Pleistocene. Its climate was not very cold-dry but warm-cool, a semiarid or semi-moist condition.

Bajiazui Fauna The Bajiazui Fauna was found at the lower part of the Bajiazui loess profile, in Xifeng County, Gansu Province (Wang et al. 1985). It was one of the best representatives of the Early Pleistocene faunas, with an age between 1.8 and 1.2 Ma. It consists of 11 genera and 22 species, including *Myospalax* sp., *Canis chihliensis*, *Cuon* cf. *dubius*, *Vulpes* cf. *corsac*, *Meles* cf. *chia*, *Hyaena* cf. *licenti*, *Megantereon* sp., *Homotherium* cf. *crenatidens*, *Felis* sp., *Coelodonta antiquitatis*, *Hipparion* (*Proboscidipparion*) sp., *Equus wangi*, *E. teihlardi*, *E. qingyangensis*, *Sus lydekkeri*, *S.* sp., *Cervus* (cf. *Euctenoceros*) sp., *Gazella sinensis*, *Bison paleosinensis*, *Leptobos brevicornis*. The main characteristics of Bajiazui fauna are the dearth of micromammals, like rodents and rabbits. It contains more carnivores but no *Nyctereutes* and *Ursus*. It has no typical forest dwellers such as elephants, *Chalicotherium*, *Macaca*, etc. but more inhabitants of grasslands such as horses, gazelles, etc. The former two characteristics are similar to that of Longdan faunas, but the latter one is quite different from that at Longdan. The differences probably reflect the disparity in duration and variation of the environment between the Longdan period and that of Bajiazui. The Bajiazui fauna is younger than that at Longdan, and the environment of Bajiazui period was much drier and cooler than that of Longdan.

Gongwangling Fauna The Gongwangling Fauna was found in loess layer L_{15} of Gongwangling loess profile in Lantian County, Shaanxi Province. Its paleomagnetic age is 1.15 Ma (An and Ho 1989). In addition to the hominid *Homo erectus* *lantianensis*, 41 species of fossil mammals are found, including *Scaptochirus moschatus*, *Rhinopithecus lantianenss*, *Arvicola terraerubrae*, *Microtus epiratticeps*, *Myospalax tingi*, *M. fontanieri*, *M.* sp., *Cricetulus* cf. *griseus*, *C. varians*, *Bahomys hyposodonta*, *Gerbillus* sp., *Petaurista* sp., *Apodemus* cf. *sylvaticus*, *Hystrix* cf. *subcristata*, *Ochotonoides complicidens*, *Ochotona* cf. *thibetana*, *Canis variabilis*, *Ursus* cf. *etruscus*, *Ailuropoda melanoleuca fovealis*, *Meles* cf. *leucurus*, *Hyaena sinensis*, *Panthera* cf. *tigris*, *P. pardus*, *Sivapanthera pleistocaenicus*, *Megantereon lantianensis*, *Stegodon orientalis*, *Equus sanmemiensis*, *Dicerorhinus lantianensis*, *D.* cf. *merckii*, *Megatapirus augustus*, *Tapirus sinensis*, *Nestoritherium sinensis*, *Sus lydekkeri*, *Elaphodus cephalaphus*, *Pseudaxis grayi*, *Sinomegaceros konwanlinensis*, *Leptobos brevicornis*, *L.* sp., *Capricornis sumatraensis qinlingensis*. Among these fossils, the *Ailuropoda*, *Stegodon*, *Megatapirus*, *Tapirus*, *Nestoritherium*, *Elaphodus*, *Capricornis* and *Rhinopithecus* are not only southern genera, but also forest dwellers. In addition to the location of Gongwangling fauna at the northern piedmont of Qinling Mountains, it is very close to the natural boundary of northern and southern China. The conditions at that time were much warmer and moister. In addition, during that time the Qinling Mountains were possibly not uplifted high enough to represent a geographic barrier. As a result, some of the animals living in South China could have wandered northward and even crossed the mountain ridge to the southern margin of the LP when the climate was suitable. If we do not consider the southern members, the Gongwangling fauna has only a couple of forest animals such as *Ursus* cf. *etruscus*, or *Sus lydekker*; some of the dwellers of forest margins like *Pseudaxis grayi*, *Megaloceros konwanlinensis*, etc. and many grassland inhabitants, such as *Equus sanmeniensis*, *Leptobos* and many kinds of rodents. According to these typical northern elements, we could imagine that the climatic environment of the Lantian area was cool-warm and semi-humid. In short, in addition to the forest on the mountain there was a shrub-grassland in this area.

Yangguo Fauna The Yangguo Fauna was found in the S_{10}–S_{11} layers of Yangguo loess section, in Weinan County, Shaanxi Province (Ji 1975; Yue and Xue 1996), slightly later than the Gongwangling fauna. It has 21 genera and species of mammals, including *Allocricetus teilhardi*, *Cricetulus farabensis griseus*, *Myospalax chaoyatseni*, *M. tingi*, *M. arvicolinus*, *Hystrix subcristata*, *Ursus kokeni*, *Hyaena licenti*, *Lynx shanxius*, *Panthera palaeosinensis*, *Proboscidipparion sinensis*, *Hipparion* sp., *Equus sanmeniensis*, *Axis* cf. *shanxius*, *A.* cf. *rugosus*, *Cervus* (*R.*) *elegans*, *Elaphurus chinnaniensis*, *E. lantianensis*, *Gazella sinensis*, *Bison* sp., *Leptobus brevicornis*, *L. amplifrontalis*, *L. laochihensis*. Among them, the *Lynx*, *Felis* and *Ursus* were forest animals. *Elaphurus* and *Rusa* lived near water bodies. *Myospalax*, *Proboscidippsrion*, *Equus*, *Gazella* and *Bison* were grassland dwellers. Grassland animals dominate this fauna. These mammal fossils indicate a warm-humid climate with grasslands and some forests in the southeastern area of the LP at ca. 1 Ma. This sort of environment favored the formation of thick paleosols.

Chenjiawo Fauna Chenjiawo Fauna was found in the S_5–S_6 layers of the loess profile at Chenjiawo village, Lantian County, Shaanxi Province (Zhou and Li 1965). It has an age of ca. 0.65 Ma (An and Ho 1989). In addition to a fossil of *Homo erectus lantianensis*, 14 genera and species of animals were unearthed, including *Myospalax tingi*, *M.* cf. *fontanieri*, *Bahemys hypsodonta*, *Hystrix* sp., *Lepus wongi*, *Ochotonoides complicidens*, *Cuon alpinus*, *Meles* cf. *leucurus*, *Panthera tigris*, *Elephas* sp., *Sus* cf. *lydekkeri*, *Megaloceros* sp., *Cervus* (*P.*) *grayi*. The species of smaller mammals like rodents and rabbits amount to half of the mammalian fauna, and the specimens of *Pseudaxis grayi* are dominant. Except for *Elephas*, *Sus lydekkeri* and *Apodemus sylvaticus*, which are adapted to forests, the other species had wide ranges and mostly lived in grasslands or in bush-grasslands. These fossils indicate that the environment around Chenjiawo was a broad grassland with sparse woods and shrubs about 0.65 Ma ago (An and Ho 1989). Some woods may have developed under the warm semi-humid or semiarid conditions. The thickest fifth palaeosol S_5 developed about 0.5 Ma ago. It is situated in the middle part of this unit, composed of three thinner, about one meter thick, red palaeosols. It has been regarded as an index layer seen everywhere in the LP.

Rouyuan Fauna The Rouyuan Fauna was found in the Malan loess of the Rouyuan loess profile, in Huachi County, Gansu Province (Xie and Xu 1988). It includes *Panthera tigris*, *Crocuta ultima*, *Equus hemionus*, *E. przewalskii*, *Coelodonta antiquitatis*, *Cervus* (*E.*) *canadenensis*, *Gazella przewalskii* and *Bos primigenius*. From the Rouyuan fauna we may imagine that during this period the Huachi area was a warm and humid wooded grassland. This reconstruction probably only refers to a limited area rather than to the whole LP.

Banpo Fauna Banpo Fauna was found in an ancient village of Neolithic relics at the eastern suburb of Xi'an city (Li and Han 1959). It has an age of ca 7 ka. *Myospslax*, *Rhizomys*, *Lepus*, *Ochotona*, *Canis*, *Vulpes*, *Equus*, *Sus*, *Hydropotes*, *Pseudaxis*, *Ovis*, *Gazella*, *Bos*, and some bones of birds and fish were found. All of these mammals are modern genera and species, which indicate that the central part of Shaanxi Province was a grassland with forests and bamboo forests under the warm and moist climate at that time.

In summary, the mammal evolution on the LP can be subdivided into three stages: 1) The early stage (Late Miocene) was characterized by an abundance of species, larger differentiation of genera and species, more frequently large-sized animals, and a huge number of fossils. Most mammals lived in sparse-shrub grasslands. In addition, there were forest dwellers in near-water or in-water environments. Fossil assemblages reflect hot-semiarid and warmer semi-moist to warm-moister conditions on the LP. Grasslands with sparse woods were the principle environments. Over this long period the climate fluctuated, but the changes were not very pronounced. Probably these were the principal conditions stabilizing the mammal assemblages. 2) The middle stage (Pliocene), only fossils of the Late Pliocene were found. Their most obvious characteristics are a reduced number of species. The cold- and dry-resistant, even desert-adjusted animals appeared. This demonstrated that the climate was changing towards colder and drier conditions. Wooded areas were reduced, while grasslands expanded. Dry grasslands and small areas of desert appeared. 3) The late stage (Quaternary), many species of Neogene mammals died out, and the proportion of forest animals decreased sharply. The mammals adjusted to dry and cold climates and especially, the grassland dwellers became more frequent. Large animals became fewer, and the medium and small sized ones became more frequent. Rodents developed quickly and formed a major part of the fauna. Rabbits, horses, antelopes and buffalos became the common mammals of that time. These reflected the change of climate and environments of the LP during the Quaternary, which became very different from those of the Neogene. It became increasingly dry and cold with large orbital-timescale climate fluctuations. Such tremendous alternation forced a rapid replacement of species and large changes in mammalian assemblages. The 8-Ma evolution of mammalian communities on the LP reflects the shift from a hot-semiarid and warmer semi-moist, warm moister climate of the Late Miocene, to a slightly warm moist, cool-dry, slightly warmer and moister climate of the Pliocene, to cold-dry and frequently alternating mild semi-moist climate of the Quaternary. The vegetation shifted from grasslands with sparse woods and shrubs to grasslands with more woods and shrubs, and then to grasslands with sparse woods, to wooded shrub grasslands, and further to subtropical forest-grasslands during the Late Miocene. It was subsequently characterized by temperate sparse grasslands, cold-dry sparse grasslands during the Pliocene, and by sparse grasslands, dry grasslands and tundra during the Quaternary.

The palaeoclimate and the environment of the late Cenozoic on the LP reflected by the evolution of fossil mammals suggest a small fluctuation of an average warm-moist climate during the Late Neogene but larger fluctuations during the Quaternary. The large fluctuations during the Quaternary were probably linked to the formation and expansion of the ice cover in the Arctic (Fig. 5.3). The late Cenozoic environmental changes on the LP indicated by the mammalian assemblages are roughly consistent with those indicated by other climate proxy studies.

The main features of the climate and environment of the LP between the Late Neogene and recent were not the same and coeval everywhere. It appears that the eastern and southeastern part of the LP and some other special areas such as Longdan, Gongwangling and Rouyuan were warmer and moister than the bulk of the LP.

5.2.2 Late Miocene-Pleistocene Micromammals from the Lingtai Red Clay-loess Sequence on the Loess Plateau

The Lingtai section is located in the central LP (Fig. 5.2). The Lingtai red clay-loess deposit has a large number of micromammal fossils (Zheng and Zhang 2000; 2001). The succesion can be subdivided into 5 biostratigraphic units based on the fossil contents. Unit 1 (ca. 8–6 Ma) has dominant murids (41 % of the total species in the assemblage) and endemic taxa prevously found in North China. Some species such as *Ochotona lagrelli*, *Prosiphneus* sp., *Micromys tedfordi*, *Huaxianmys* n. sp., *H. primitivus*, and *Pseudomeriones abbreviatus* etc, can be compared directly with those from the Mahui Formation and the lower Gaozhuang Formation in the Yushe Basin (Flynn et al. 1991, 1995, 1997; Teford et al. 1991). Unit 2 (ca. 6–5 Ma) is a conglomerate layer, in which no fossils were found. This conglomerate layer covers a large area in Lingtai County. Its source material may be from far away from the south Qinling Mountains as suggested by geological survey. Unit 3 (ca. 5–3.6 Ma) has several first-appearances of mammals such as *Chardina sinensis* and *Micromy* cf. *M. Chalceus*, in addition to some taxa of unit 1, such as *Pseudomeriones abbreviatus*、*Micromys tedfordi*、*Huaxiamys primitivus* and *Chardinomys yusheensis*. The murids are still dominant. With higher crowned molars, *Chardina sinensis* is more derived than *Prosiphneus* from the Late Miocene (Zheng 1997). Unit 4 (ca. 3.6–2.6 Ma) has the most diverse (43 species) and abundant fossils in the Lingtai succesion. The component of murids decreased significantly in the number of species, while the siphneids, cricetids, arvicolids, scuirids, ochotonids and others flourished. Unit 5 (ca. 2.6–1.6 Ma) has a significantly decreased diversity of species. However, the Early Pleistocene elements from North China, such as *Yangia* cf. *Y. trassaerti* and *Allophaiomys terrae-rubrea* appeared then.

The micromammal fossils from the Lingtai area indicate several clear bioevents. Many turnover events took place in siphneids between 3.6 and 3 Ma. At about 3.5 Ma, the most derived close-rooted concave-occipital siphneids, *Mesosiphneus intermedius*, evolved from *M. praetingi*, and slightly later, itself evolved into the open-rooted *Yangia* n. sp. and *Y. omegodon*. Meanwhile, the most advanced close-rooted convex-occipital *Pliosiphneus* n. sp. 2 evolved into the most primitive rootless *Eospalax* n. sp., and another rootless *Allosiphneus teilhardi* at ca. 3.3 Ma. The cricetids, *Bahomys* might be evolved from the large-sized *Cricetinus mesolophidus* at ca. 3.4 Ma with great changes of the check tooth morphology. *Pseudomeriones complicidens* evolved from *P. abbreviatus* by increasing m1 complexity at ca. 3.5 Ma. These morphologic changes in bioevents indicate that the habitat of these animals became toward drier and colder grassland. This is consistent with the magnetic study on the red clay on the LP (An et al. 1999). At ca. 2.4 Ma, *Allophaiomys terraerubrae*, the most primitive species in *Allophaiomys*, first appeared, which might have evolved from the *Cromeromys gansunicus*. Meanwhile, *Yangia trassaerti* evolved from *Yangia* n. sp. At ca. 5.8 Ma, *Micromys tedfordi* made its first appearance. These two major bioevents are generally synchronous with the onset of Northern Hemisphere glaciation during the Late Pliocene (Raymo 1994) and associated decrease in global sea levels during the Late Miocene (Zachos et al. 2001).

During the past 7 Ma, almost all the taxa from the Lingtai section are endemic elements of North China. In order to adapt to increasing variations of environment during the late Cenozoic, all micromammals (whether endemic or immigrants) have to constantly change their morphologic adaptation. Some successors survived, whereas others died out during ecological changes. Meanwhile, some massive bioevents happened, such as the bioevents of micromammal evolution at about 5.8, 3.6, 3 and 2.4 Ma revealed by the fossils from the Lingtai red clay-loess sediments.

During the late Cenozoic, global climate changes and the growth of the Tibetan Plateau produced signficant effects on the climates of North China. From the base of the Lingtai section, *Pseudomeriones*, a dry grassland dweller, migrated to China from West Asia. The typical grassland animals, *Ochotona* and *Prosiphneus* also flourished. However, the existence of a large number of murid rodents also indicates a warm, humid forest or forest-grassland ecological environment on the LP.

During the Late Pliocene, signifiant bioevents took place with tooth morphology changes. The loss of tooth roots in siphneids that makes teeth grow throughout its life span, which significantly enhanced masticatory ability and leads to the flourishing of *Yangia* and *Eospalax* in the Pleistocene (Zheng 1997). Their high crown, complicated enamel folds, and much more crests of *Bahomys*, enable this genus to survive until the Middle Pleistocene (Zhou and Li 1965). By complication of the tooth enamel, *Pseudomeriones complicidens* evolved from *P. abbreviatus*. The body size of *Chardinomys louisi* minimized, the root numbers increased. *Cromeromys* and *Borsodia* developed high crowned teeth. All these functional morphology changes occurred at almost the same time, which indicates that the habitat of these animals became a much drier and colder grassland. During the Pleistocene, the uplift of the Tibetan Plateau to the present level (Harris 2006), sustained major ice-sheet buildup in the Northern Hemisphere (Raymo 1994), and the intensification of the East Asian monsoon circulation (An et al. 2001) may have significantly impacted mammals living on the LP.

5.2.3 Typical Nihewan Fauna, Zhoukoudian Fauna and Salawusu Fauna on the Margin of the Loess Plateau

5.2.3.1 Nihewan Fauna

The Nihewan Basin is located in the transitional zone between the North China Plain and the Inner Mongolian Plateau, on the northeastern margin of the LP (Fig. 5.4). The basin is filled with late Cenozoic fluvio-lacustrine deposits (Deng et al. 2008; Zhou et al. 1991), which have been named the Nihewan Formation (or Nihewan Bed) (Barbour 1925). The Nihewan Formation is rich in mammalian fossils and Paleolithic sites (see review by Xie 2006; Xie et al. 2006; Zhou et al. 1991; Zhu et al. 2007). These mammalian fossils are known as the Nihewan Fauna, which traditionally correspond to the Villafranchian fauna in Europe (Barbour 1925; Qiu 2000; Teilhard de Chardin and Piveteau 1930). The Nihewan Fauna (sensu lato) comprises a series of mammalian faunas in the Nihewan Basin, such as the Hongya (Huang and Tang 1974), Huabaogou-I (Wang 1982; Zhou et al. 1991), Huabaogou-II (Wang 1982), Majuangou-III (Zhu et al. 2004), Xiaochangliang (Tang et al. 1995; You et al. 1980), Banshan (Wei 1994), Donggutuo (Wei 1991; Wei et al. 1985) and Maliang faunas (Wei 1991). The ages of these mammal faunas have been estimated using magnetostratigraphy (see review of Deng et al. 2008). Combining these estimated ages, we established a magnetochronological framework for the Nihewan faunas (Fig. 5.5). The components of these mammalian faunas in the eastern Nihewan Basin are also listed in Table 5.2.

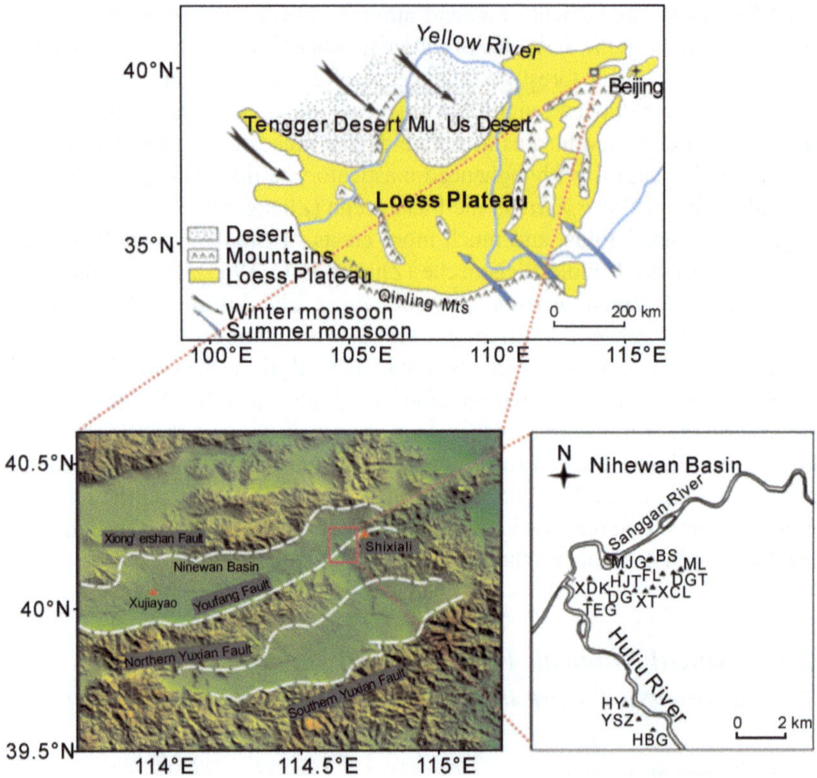

Fig. 5.4 Schematic map showing the eastern Nihewan Basin and the mentioned sections

The Hongya Fauna (see its location at Fig. 5.4), located at the Hongya village, is currently the oldest of the Nihewan Fauna. It mainly includes *Hipparion* sp. and *Chilotherium* sp. (Huang and Tang 1974) (Table 5.2). Magnetostratigraphy suggested that this fauna was located within the Matuyama chron bracketed by the late Gauss chron and the Olduvai subchron, yielding an estimated age of ca. 2.3 Ma (Deng et al. 2008) (Fig. 5.5). The Xiashagou Fauna was also named the Nihewan Fauna (sensu stricto), which yields a rich faunal list (Teilhard de Chardin and Piveteau 1930) (Table 5.2). The faunal elements of *Vulpes* sp., *Canis chihliensis, C. chihliensis palmidens, C. chihliensis minor, Ursus* cf. *U. etruscus, Pachycrocuta licenti, Chasmaporthetes* cf. *ossifragus, Megantereon nihowanensis, Homotherium* cf. *H. crenatidens, Sus* sp. and *Eucladoceros boulei* show close morphologic similarity with the corresponding forms of the Olivola fauna in the early Villafranchian age. The *Nyctereutes sinensis, Lynx shansius, Pseudodama elagans, Megalovis piveteaui* can be compared with the Senèze fauna in the late Villafranchian age. Consistent with their similarities to the Villafranchian Fauna in Europe, recent magnetostratigraphic investigation indicates that the Xiashagou Fauna is located between the pre-Reunion and post-Olduvai Matuyama chron, yielding an estimated

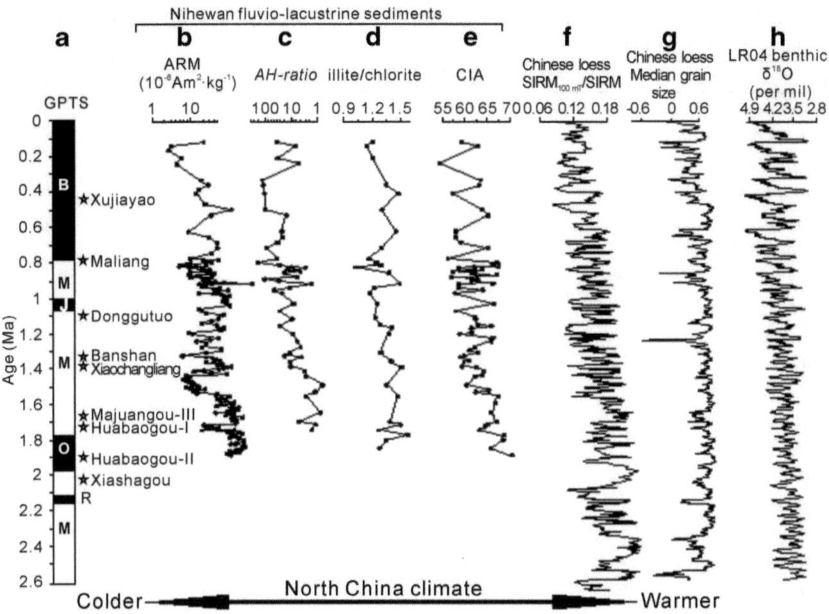

Fig. 5.5 A composite diagram with all magnetostratigraphically dated Nihewan faunas within the framework of the geomagnetic polarity timescale and temporal variability of regional and global climate during the Quaternary. (a) GPTS (Cande and Kent 1995). (b) Well-dated Nihewan faunas using magnetostratigraphy. Magnetoclimatological proxies (c) Anhysteretic remanent magnetization (ARM) (Ao 2010) and (d) *AH*-ratio (Ao et al. 2009) (*AH*-ratio characterizes the relative amount of newly formed fine-grained ferromagnetic minerals during heating, which is linked to the relative concentration of chlorite in the samples. For more detailed description see the cited reference), (e) clay mineralogical proxy illite/chlorite (Ao et al. 2009) and (f) Chemical Index of Alteration (CIA) (Ao et al. 2010a) from the Nihewan Formation. (g) Changes in the $SIRM_{100mT}$/SIRM ratio from the Jingbian sequence (SIRM is the saturation isothermal remanent magnetization, and $SIRM_{100mT}$ represents the residual SIRM after 100-mT alternating field demagnetization) (Deng et al. 2006a). (h) Grain size record of the Chinese loess (Ding et al. 2002). (i) LR04 benthic $\delta^{18}O$ stack (Lisiecki and Raymo 2005)

age of ca. 2.2–1.7 Ma (Liu et al. 2012). The Huabaogou Fauna is located at about 2 km southeast of the Hongya Fauna (Fig. 5.4). There are two layers preserving rich mammalian faunas. The fauna at Huabaogou-I includes *Postschizotherium* sp., *Hipparion houfenense, Viverra* sp., *Gazella blacki, Antilospiroides hobeiensis, Sinoryx* sp. and *Gazella* sp. (Wang 1982; Zhou et al. 1991) (Table 5.2). Huabaogou-II Fauna includes Leporidae gen. et sp. indet., *Canis multicuspus, C.* sp., *Nyctereutes sinensis, Hipparion* cf. *hippidiodus, H.* sp., *Palaeotragus* sp., *Gazella blacki* and *G.* sp. (Wang 1982) (Table 5.2). Magnetostratigraphy suggested that Huabaogou-II Fauna was located at the lower Olduvai subcrhon, with an estimated age of ca. 1.9 Ma, while Huabaogou-I is fairly close to the upper boundary of the Olduvai subcrhon, with an estimated age of ca. 1.7 Ma (Deng et al. 2008). The Majuangou-III Fauna is located at the eastern margin of the Nihewan Basin (Fig. 5.4), which contains a large number of vertebrate fossil remains. Most of them are attributable to *Elephas* sp.,

Table 5.2 List of mammalian faunas in the Nihewan Basin directly dated by magnetostratigraphy (Deng et al. 2008; Løvlie et al. 2001)

Taxa	DGT	ML	XCL	MJG-III	BS	HY	HBG-I	HBG-II	XSG	XJY
Allophaiomys cf. *A. pliocaenicus*			+							
Antilospiroides hobeiensis							+			
Bison sp.	+									
Bovinae indet.			+							
Canis sp.	+			+				+		
Canis multicuspus								+		
Carnivora gen. et sp. indet.				+						
Cervus sp.			+	+						
Chilotherium sp.						+				
Coelodonta antiquitatis	+	+	+	+	+				+	+
Elephas sp.				+					+	
Equus sanmeniensis	+		+	+					+	
Equus sp.		+			+					
Gazella sp.	+		+	+			+	+		+
Gazella blacki							+	+		
Hipparion sp.			+			+		+		
Hipparion houfenense							+			
Hipparion cf. *hippidiodus*								+		
Pachycrocuta licenti			+							
Pachycrocuta sp.				+						
Leporidae gen. et sp. indet								+		
Martes sp.			+							
Mimomys chinensis			+							
Myospalax cf. *fontanieri*	+									+
Nyctereutes sinensis								+	+	
Palaeoloxodon sp.	+	+	+							
Palaeotragus sp.								+		
Postschizotherium sp.							+			
Proboscidipparion sinensis			+						+	
Sinoryx sp.							+			
Struthio sp.				+						+
Viverra sp.							+			

and other taxa include hyena *Pachycrocuta* sp., horse *Equus sanmeniensis*, rhinoceros *Coelodonta antiquitatis*, deer *Cervus* sp. and bovid *Gazella* (Zhu et al. 2004) (Table 5.2). This fauna is also close to the upper boundary of the Olduvai subcrhon, with an estimated age of ca. 1.66 Ma (Zhu et al. 2004). The Xiaochangliang Fauna includes *Allophaiomys* cf. *A. pliocaenicus, Mimomys chinensis, Hyaena* (*Pachy-*

crocuta) *licenti*, *Martes* sp., *Hipparion* sp., *Proboscidipparian sinensis*, *Equus sanmeniensis*, *Coelodonta antiquitatis*, *Palaeoloxodon* sp., *Cervus* sp. and *Gazella* sp. (Tang et al. 1995; You et al. 1980) (Table 5.2). This fauna is located between the termination of the Olduvai subchron and the onset of the Jaramillo subchron within the Matuyama chron, with an estimated age of ca. 1.36 Ma (Zhu et al. 2001). The Donggutuo Fauna includes *Myospalax* cf. *fontanieri*, *Canis* sp., *Palaeoloxodon* sp., *Equus sanmeniensis*, *Coelodonta antiquitatis*, *Gazella* sp. and *Bison* sp. (Wei 1991) (Table 5.2). It is located just below the lower boundary of the Jaramillo subchron, with an estimated age of ca. 1.1 Ma (Wang et al. 2005). The Maliang Fauna includes *Palaeoloxodon* sp., *Equus* sp. and *Coelodonta antiquitatis* (Wei 1991) (Table 5.2). It is located just below the Matuyama-Brunhes boundary, yielding an estimated age of ca. 0.8 Ma (Wang et al. 2005). The Xujiayao Fauna is located at the western Nihewan Basin (Fig. 5.4). It contains more than 20 taxa, such as *Ochotona* sp., *Myospalax fontanieri*, *Microtus brandtioides*, *Canis lupus*, *Panthera* cf. *tigris*, *Sus* sp., *Palaeoloxodon* cf. *naumanni*, *Equus przewalskii*, *E. hemionus*, *Coelodonta antiquitatis*, *Cervus elaphus*, *C. nippon grayi*, *Megaloceros ordosianus*, *Spirocerus hsuchiayaocus*, *S. peii*, *Procapra picticaudata przewalskii*, *Gazella subgutturosa*, *G.* sp., *Bos primigenius* and *Struthio* sp. (Jia et al. 1979; Jia and Wei 1976) (Table 5.2). In addition to a small percentage of forest-dwelling taxa (e.g. *Cervus* (deer) and *Sus* (pig)), the majority of these forms represent grassland-adapted large mammal associations. Magnetostratigraphy suggested that the Xujiayao is located at the middle Brunhes chron, with an estimated age of ca. 0.4–0.5 Ma (Løvlie et al. 2001).

In addition to these precisely dated faunas, there are also some faunas without precisely estimated ages, such as Dongyaozitou (Tang 1980; Tang and Ji 1983), Danangou (Li 1984) and Laowogou faunas (Li et al. 2008; Zhang et al. 2003) (Table 5.3). However, almost all of the mammals from these undated faunas have occurred in the dated Nihewan faunas, which indicate that the ages of these faunas should be within the aforementioned Pleistocene chronological framework (a time range of ca. 2.6–0.3 Ma).

The mammal fossils indicate that the Nihewan faunas contain typical Pleistocene mammals in North China. The composition of the faunas also indicates different mammal habitation during the Pleistocene. *Hipparion*, *Gazella* and hamster are adapted to the grassland or forest-grassland conditions. *Axis* sp., *Apodemus* and *Sminthoides* sp. are adapted to forest conditions. Dipodidae and *Ochotona* lived in arid grasslands or desert steppe. Thus the characteristics of the faunas indicate that Nihewan Basin may have a mixed condition of forests and grasslands. From Early to Late Pleistocene, the grasses and herbs increase gradually. There is a long-term decrease in East Asian monsoon precipitation and humidity in the Nihewan Basin during the Pleistocene suggested by the mineral magnetic and geochemical studies (Ao 2010; Ao et al. 2009, 2010a). At present, the basin shows a moderate temperate semiarid climate. The regional climate (mean annual temperature of 6.4–7.3 °C; annual rainfall of 359–418 mm) is characterized by cold and dry winters under the influence of the East Asian winter monsoon. In contrast, summers are relatively warm and humid under the influence of the East Asian summer monsoon (the rainfall during summer amounts to 80 % of the total annual rainfall). Therefore, the

Table 5.3 List of mammalian fauna from the Laowogou (LWG), Danangou (DNG) and Dongyaozitou (DYZT) in the eastern Nihewan Basin (Zhou et al. 1991)

LWG	DNG	DYZT
Sorex sp.	*Soriculus* sp.	*Lynx variabilis*
Paenelimnoecus chinensis	*Blarinella* sp.	*Dipoides* sp.
Lunanosorex cf. *L.ii*	*Erinaceus* sp.	*Nyctereutes* cf. *sinensis*
Quyania sp.	*Ochotona* sp.	*Zygolophodon* sp.
Ochotona sp.	*Phodopus* sp.	*Coelodonta antiquitatis*
Ochotona minor	*Sinocricetus progressus*	*Proboscidipparion sinense*
Trischizolagus sp.	*Nannocricetus mongolicus*	*Hipparion* cf. *houfenense*
Cricetinae indet.	*Mimomys* sp.	*Paracamelus* sp.
Kowalskia similis	*M.*sp.2	*Palaeotragus progressvus*
Sinocricetus progressus	*Borsodia chinensis*	*Gazella sinensis*
Nannocricetus mongolicus	*Ungaromys* spp.	*Axis* sp.
Ungaromys spp.	*Mesosiphneus paratingi*	*Antilospira yuxianensis*
Mimamys sp.	*Paralactaga anderssoni*	
Mimamys sp. 2	*Dipus fraudator*	
Chardina truncatus	?*Dipoides* sp.	
Mesosiphneus praetingi	*Micromys tedfordi*	
Pliosiphneus sp. 2	*Apodemus zhangwagouensis*	
Pseudomeriones complicidens	*Chardinomys nihowanicus*	
Dipus fraudator	*Lynx variabilis*	
Paralactaga anderssoni	*Paracamelus* sp.	
Castor anderssoni	*Antilospira yuxianensis*	
Micromys tedfordi	*Palaeotragus progressus*	
? *Huaxiamys* sp.		
Huaxiamys downs		
Apodemus zhangwagouensis		
Chardinomys yusheensis		
Chardino nihowanicus		
? *Karnimata* sp.		
Saidomys sp.		

Nihewan Basin is very sensitive to monsoon changes. During the strong summer monsoon period with increased forests, the mammals adapted to forests would increase. During the strong winter monsoon period with increased grasslands, the grassland mammals would increase.

5.2.3.2 Zhoukoudian Fauna

The cave site of Zhoukoudian Locality 1 has an inventory of 6 fairly complete hominin crania and other bones representing at least 40 individuals, 98 species of non-hominin mammalian fossils and tens of thousands of stone artefacts (Li and Ji 1981; Wu et al. 1985; Xu and Ou 1982) (Table 5.4). It is the largest single source of *Homo erectus* and one of the most important Palaeolithic sites in the world. The site is a sedimentary infill within a vertical karstic fissure. Its 40-m-thick depositional sequence can be divided into 17 layers. Mammalian fossils were mainly found in layers 1–13 (Tables 5.4 and 5.5). The mammal fossils in the layer 13 mainly include

Table 5.4 List of mammalian fauna from the Zhoukoudian site (Li and Ji 1981; Wu et al. 1985; Xu and Ou 1982)

Layers	Mammal fossils
1–3	*Homo erectus pekinensis, Macaca robustus, Neomys sinensis, Neomys bohlinni, Crocidura* sp., *Erinaceus olgae, Ia io, Miniopterus schreiberii, Cricetinus varians, Cricetulus* cf. *grisen, Cricetulus* cf. *obscurus, Micromys* cf. *minutus, Rhizomys* sp., *Canis lupus, C. l. variabilis, Nyctereutes sinensis, Vulpes* cf. *corsac, Ursus thibertanus kokeni, Ursus arctos, Meles* cf. *leucurus, Hyaena sinensis, Crocuta ultima, Panthera* cf. *tigris, Dicerorhinus choukotienensis, Coelodonta antiquitatis yenshensis, Equus sanmeniensis, Sus lydekkeri, Pseudaxis grayi, Megaceros pachyosteus, Cervus* sp., *Gazella* sp.
4	*Homo erectus pekinensis, Macaca robustus, Scaptochirus primitivus, Neomys sinensis, Neomys bohlinni, crocidura* sp., *Erinaceus olgai, Rhinolophus pleistocaenicus, Miniopterus schreiberii, Ia io, Ochotona koslowi, Cricetinus varians, Cricetulus* cf. *griseus, Cricetulus* cf. *obscurus, Micromys* cf. *minutus, Rattus ratttus, Gerbillus roborowskii, Microtus brandtioides, Microtus oeconomus (epiratticeps), Canis lupus variabilis, Ursus thibetanus kokeni, Ursus arctos, Palaeoloxodon* cf. *namadicus, Dicerorhinus choukoutienensis, Coelodonta antiquitatis yenshensis*(?), *Eguussan meniensis, Moschus moschiferus pekinensis, Megaceros pachyosteus, Ovis* sp., *Bison* sp.
5	*Homo erectus pekinensis, Macaca robustus, Scaptochirus primitivus, Erinaceus olgai, Trogontherium cuvieri, Castor* sp. (?), *Cricetinus varians, Cricetulus obscurus, Micromys* cf.*minutus, Rattus ratttus, Microtus brandtioides, Microtus oeconomus (epiratticeps), Canis lupus, Canis lupus vriabilis, Nyctereutes sinensis, Vulpes* cf. *corsac, Ursus thibetanus kokeni, Ursus arctos,? Ailuropoda* sp., *Meles* cf. *leucurus, Mustela* cf. *sibirica, Hyaena brevirostris sinensis, Machairodus inexpectatus, Panthera* cf. *tigris, Panthera* cf. *pardus, Felis teilhaedi, Felis* sp. A, *Felis* sp. B, *Felis* cf. *microtis, Acinonyx* sp., *Cynailurus* sp., *Moschus moschiferus pekinensis, Pseudaxis grayi, Megaceros pachyosteus, Gazella* sp., *Ovis* cf. *ammon*
6	*Homo erectus pekinensis, Macaca robustus, Erinaceus olgai, Castor* sp.?, *Trogontherium cuvieri, Cricetinus varians, Cricetinus* cf. *obscurus, Micromys* cf. *minutus, Canis lupus vriabilis, Nyctereutes sinensis, Vulpes* cf. *corsac, Ursus thibetanus kokeni, Ursus arctos, Meles* cf. *leucurus, Hyaena brevirostris sinensis, Panthera* cf. *tigris, Dicerorhinus choukoutienensis, Equus sanmeniensis, Moschus moschiferus pekinensis, Pseudaxis grayi, Megaceros pachyosteus, Gazella* sp., *Ovis* cf. *ammon*
7	*Macaca robustus, Marmota complicidens, Trogontherium cuvieri, Cricetinus* cf. *griseus, Cricetinus* cf. *obscurus, Micromys* cf. *minutus, Canis lupus, Canis lupus variabilis, Nyctereutes sinensis, Lutra melina, Hyaena brevirostris sinensis, Panthera* cf. *tigris, Felis* sp. B, *Dicerorhinus choukoutienensis, Equus sanmeniensis, Sus lydekkeri, Moschus moschiferus pekinensis, Pseudaxis grayi, Megaceros pachyosteus, Ovis* cf. *ammon, Bubalus teihardi*
8–9	*Homo erectus pekinensis, Macaca robustus, Rhinolophus pleistocaenicus, Myotis* sp. A, *Myotis* sp. B, *Ia io, Miniopterus schreiberii, Marmota* sp., *Trogontherium cuvieri, Hystrix* cf. *subcristata, Canis lupus, Canis lupus vriabilis, Canis cyonids, Nyctereutes sinensis, Cuon antiquus, Vulpes* cf. *vulgaris, Vulpes* cf. *corsac,* Canidae gen. et sp. indet., *Ursus thibetanus kokeni, Ursus arctos, Ursus* cf. *spelaeus, Meles* cf. *Leucurus, Gulo* sp.? *Hyaena brevirostris sinensis, Machairodus inexpectatus, Panthera* cf. *tigris, Panthera* cf. *pardus, Felis teilhaedi, Felis* sp. A, *Felis* cf. *microtis, Palaeoloxodon* cf. *namadicus, Dicerorhinus choukoutienensis, Coelodonta antiquitatis yenshensis, Equus sanmeniensis, Pseudaxis grayi, Megaceros pachyosteus, Spirocerus peii, Ovis* cf. *ammon, Bubalus teilhardi*

Table 5.4 (continued)

Layers	Mammal fossils
10–11	*Homo erectus pekinensis, Macaca robustus, Scaptochirus primitives, Neomys sinensis, Nenmys bohlini, Ochotona koslowi, Marmota bobak, Marmota complicidens, Cricetuus* cf. *obscurus, Cricetulus* cf. *griseus, Apodemus sylvaticus, Mus sylvaticus, Mus musculus bieni, Gerbillus roborowskii, Microtus oeconomus (epiratticeps), Canis lupus variabilis, Nyctereutes sinensis, Vulpes* cf. *vulpes, Ursus arctos, Ursus* cf. *spelaeus, Mustela* sp., *Hyaena brevirostris sinensis, Machairodus inexpectatus, Panthera* cf. *pardus, Felis teilhardi, Dicerorhinus choukoutienensis, Equus sanmeniensis, Sus lydekkeri, Paracamelus gigas*(?), *Pseudaxis grayi, Megacceros flabellatus, Megacceros pachyosteus,* Bovidae gen.et sp. indet

fallow deer, Hyaenidae, *Equus sanmeniensis* and *Myospalax*, which indicate large grasslands around this area. *Dicerorhinus choukoutiensis* is found in the layer 12, which also indicates a grassland environment. There are 28–29 kinds of mammals in the layer 11. In addition to the mammals adapted to grassland environment such as fallow deer and *Equus sanmeniensis*, mammals adapted to forests such as *Sus lydekkeri* and *Pseudaxis grayi* also existed. These mammal components indicate increased grasslands during this period. The composition of the mammals in layer 10 is similar to that in layer 11, which shows slightly increased grassland mammals. Layers 8–9 have 37–38 kinds of mammals. Forest mammals increased significantly, while the grassland mammals decreased. This indicates a warmer climate. Layers 7–6 still have more forest mammals than grassland mammals, with increased mammals preferring waters. This indicates a warm and humid climate. Although the amount of mammals preferring warm climate still existed in layers 5–4, the amount of mammals preferring cold climate increased. In layers 3–1, the amount of mammals preferring cold climate decreased, while the amount of mammals preferring warm climate such as *Rhizomys*, *Ia io* and *Miniopterus* appeared again. This indicates expanded steppe with climate change from cold and dry to warm and wet.

Based on the mammal components, the Zhoukoudian Fauna indicates three major environmental units (Table 5.5). In layers 11–10, the mammals preferring warm forest have comparable number with the mammals preferring cold/dry grassland. In layers 9–5, the mammals preferring warm forest have larger number than the mammals preferring cold/dry grassland. In layers 4–1, the mammals preferring warm forest have smaller number than the mammals preferring cold/dry grassland.

5.2.3.3 Salawusu Fauna

The Sjara-Osso-Gol area is located between the southern Mu Us Desert and the northern LP (Fig. 5.6). The fluvio-lacustrine strata of this area represent the type section of Late Pleistocene in North China, which contains a large number of Late Pleistocene mammlian fossils, known as the Salawusu fauna (Xie et al. 1995; Yuan 1978). It is also one of the sites yielding the first evidence of the presence of Late Pleistocene *Homo sapiens* in Asia (Teilhard de Chardin and Licent 1924). In ad-

Table 5.5 Depositional cycles and mammal environmental implications of the Zhoukoudian site (Huang 1995; Li and Ji 1981; Zheng 1983)

Layer	depth (m)	Age (ka)	Number of mammals preferring cold and warm climate	The proportion of forest-mammals to grassland-mammals	Climatic implication
1	1.5				Larger propertion of grasslands Dominated by relatively cold/dry climate
2	1.7	200 250			
3	3.6	300			
4	6.92				Large propretion of forests Dominated by relatively warm and climate
5	0.45	340			
6	7.12	380			
7	1.53	420			
8-9	6.15				Large propretion of grasslands Dominated by relatively cold/dry climate
10	0.60	460 530			
11	0.80	580			
12-13					

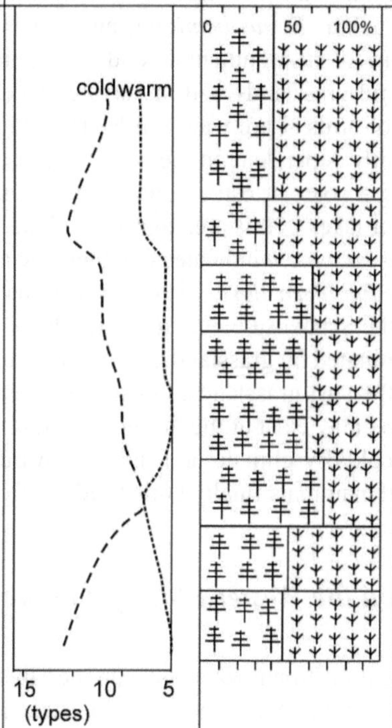

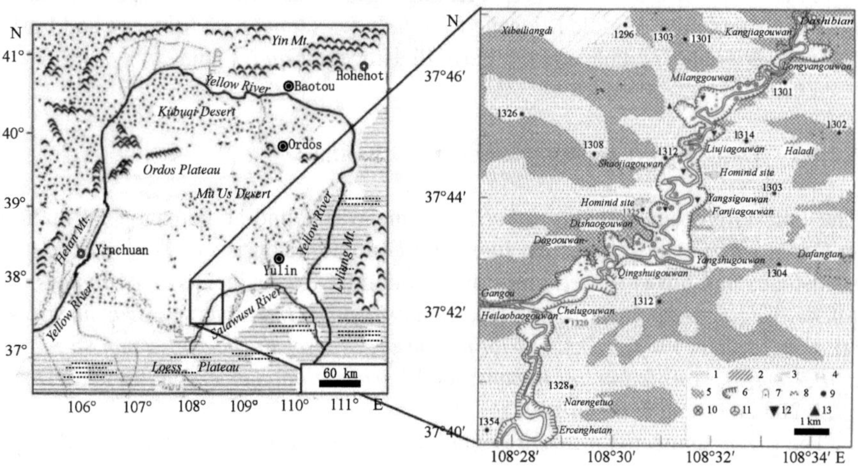

Fig. 5.6 Schematic map showing the Sjara-Osso-Gol area

dition, Late Pleistocene stone tools were also found in this fluvio-lacustrine strata (Huang et al. 2004; Teilhard de Chardin and Licent 1924).

Table 5.6 shows the mammalian faunas of the Salawusu fauna (Tong et al. 2008; Xie et al. 1995). The Salawusu fauna contains not only animals that prefer hot and humid conditions such as *Bubalus wansjocki* and *Bos primigenius*, but also animals that prefer the cold climate such as *Equus hemionus* and *Equus* cf. *E przewalskii.* This is consistent with the current environmental condition, which is a transitional zone between the warm/humid grasslands with associated forests and dry desert steppe. Among the 11 kinds of birds, 4 are water birds. The life and oviposition of these water birds are signifiantly dependent on the water. For example, the *Tadorna tadorna* generally lives along a coast or estuary, and sometimes in alkaline lakes. Thus it needed the presence of lakes in this area during the Late Pleistocene. However, the presence of *Camelus knoblochi* indicates that there are also desert steppes. In addition, the presence of *Panthera tigris* also indicates that there exist some forests nearby. In summary, the Salawusu Fauna indicates that the Late Pleistocene Sjara-Osso-Gol area was dominated by grasslands, coupled with forests and deserts nearby. Lakes may exist on the vast grasslands. Open forests and bushwoods may exist on the highlands. Deserts may exist at the northern Sjara-Osso-Gol area. Possibly due to the increasing cold/dry climate and the southward expansion of the deserts, the early humans and mammals finally emigrated out from this area.

5.2.4 Linxia Faunas on the Western Margin of the Loess Plateau

Linxia Basin is situated at the transitional zone between the Tibetan Plateau and LP (Fig. 5.7). Thick deposition from the Oligocene to the Pleistocene is developed and exposed in the basin, which contains a large number of mammal fossils (Fig. 5.8). The Linxia Basin has the richest Late Cenozoic mammal fossils in China as well as in the whole of Eurasia. The representative faunas are the Late Oligocene *Dzungariotherium* fauna, Middle Miocene *Platybelodon* fauna, Late Miocene and Early Pliocene *Hipparion* fauna, and the Early Pleistocene *Equus* fauna.

5.2.4.1 Late Oligocene *Dzungariotherium* Fauna

The Late Oligocene *Dzungariotherium* fauna was found at Jiaozigou Formation in Dongxiang County (Fig. 5.8). Jiaozigou Formation consists of brownish yellow sandstones and brownish red mudstones. Mammal fossils include *Tsaganomys altaicus*, *Megalopterodon* sp., *Schizotherium ordosium*, *Triplopus* sp., *Ardynia* sp., *A.* sp. nov., *Allacerops* sp., *Dzungariotherium orgosense*, *Paraceratherium* sp., *Ronzotherium* sp., *Aprotodon lanzhouensis* and *Paraentelodon macrognathus* etc (Deng et al. 2004a, b; Qiu and Qiu 1990).

The giant rhinoceros is a representative mammal in Asia, and infrequently discovered in Eastern Europe. The giant rhinoceros was diversified in the Middle Oli-

Table 5.6 Faunal list of the Salawusu Fauna in North China (Tong et al. 2008; Xie et al. 1995)

Taxa
Erinaceus sp.
Scaptochirus moschatus
Chiroptera gen et sp indet
Ochotona sp.
Lepus sp.
Spermophilus dauricus
Citellus mongolicus
Dipus sagitta
Allactaga sibirica
Meriones meridianus
Eospalax fontanier
Eolagurus luteus
Prometheomys schaposchnikowi
Microtus oeconomus
Lasiopodomys brandti
Cricetulus barabensis
Canis lupus
Crocuta ultima
Panthera tigris
Meles meles
Palaeoloxodon naumanni
Coelodonta antiquitatis
Equus hemionus
Equus cf. *przewalskii*
Sus scrofa
Camelus knoblochi
Cervus elaphus
Sinamegaloceros ordosianus
Procapra (Gazella) przewalskii
Gazella subgutturosa
Spirocerus kiakhtensis
Ovis ammon
Bubalus wansjocki
Bos primigenius
Buteo cf. *ferox*
Vultur monaclans
Passereus
Perdix cf. *perdix*
Coturunis sp.
Syrrhaptes paradoxus
Echassier
Podioeps auritus
Anas boschas
Tadorna tadorna
Struthio sp.

gocene, and it became very advanced in the Late Oligocene. *Dzungariotherium orgosense* was firstly found from the Junggar Basin in Xinjiang, with large rudimental lower incisors, well-developed antecrochets and wide foot bones. In Xinjiang, *D.*

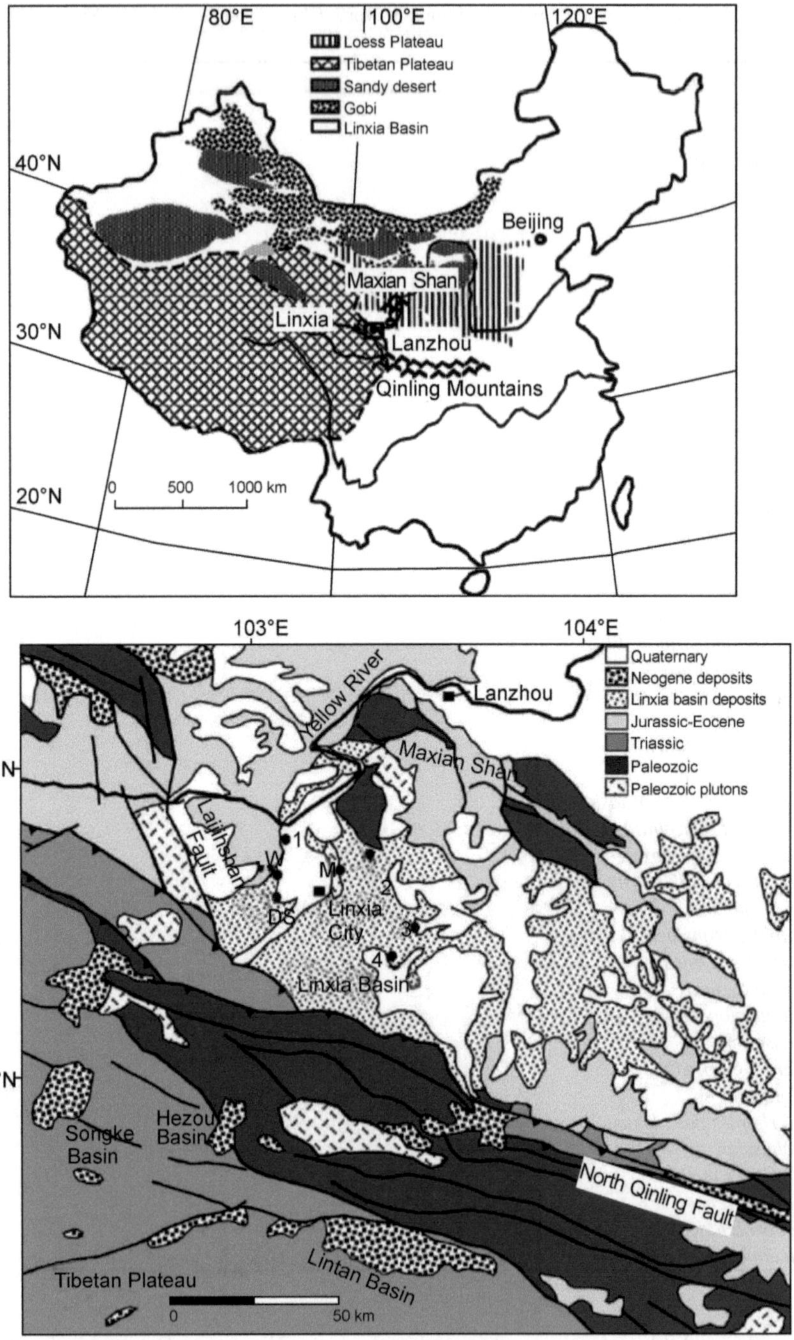

Fig. 5.7 Map of the Linxia Basin on the northeastern margin of the Tibetan Plateau (Fan et al. 2007; Fang et al. 2003)

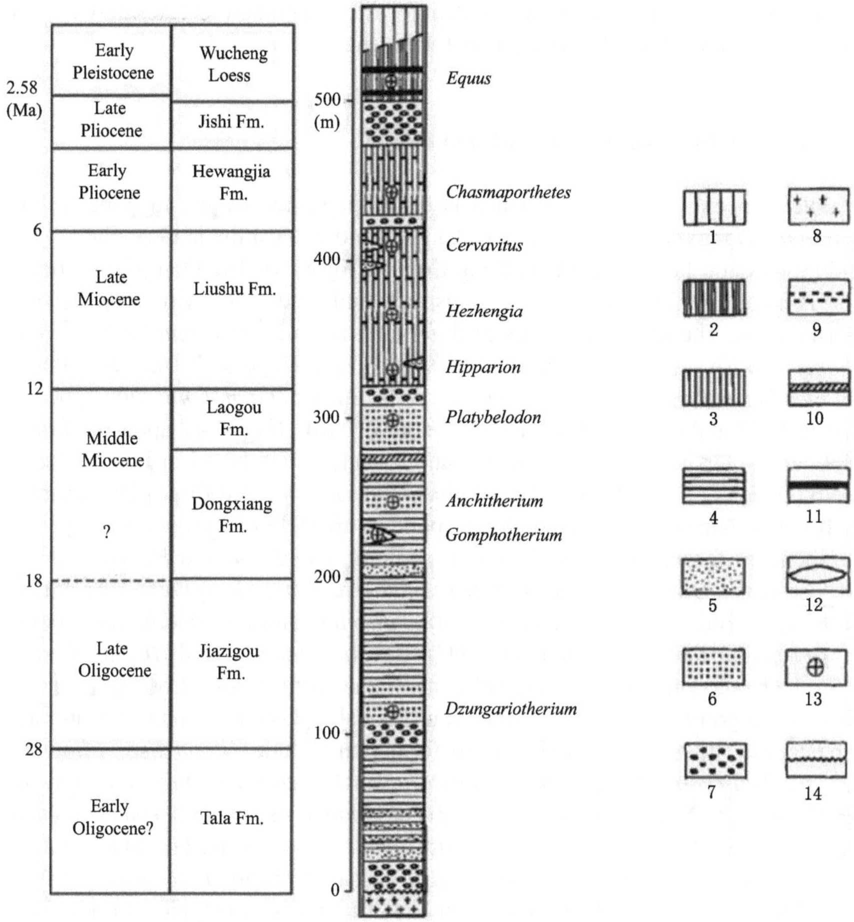

Fig. 5.8 Cenozoic stratigraphic column in the Linxia Basin (modified from Deng et al. 2004b). 1. Malan loess; 2. Wucheng loess; 3. Red clay; 4. Mudstone; 5. Siltstone; 6. Sandstone; 7. Conglomerate ; 8. Granite ; 9. Carbonate concretions; 10. Marlite; 11. Paleosol; 12. Lenticular body; 13. Fossil bed; 14. Unconformity

orgosense coexists with *Lophiomeryx*, the last record of the latter is from the middle late Stampian Age in Europe (Qiu 1973; Qiu and Qiu 1990). *Allacerops* was a rhinocerotid living in the Oligocene of Asia, and it was found from the Oligocene in the Lanzhou Basin adjacent to the Linxia Basin (Qiu and Wang 1999). *Schizotherium* was also a characteristic Oligocene form in China, and it was found from the Oligocene Nanpoping Fauna in the Lanzhou Basin (Qiu et al. 1998). *Aprotodon* was previously found only from Pakistan, Kazakstan, and the Lanzhou Basin, which coexisted with the giant rhinoceros in these three regions (Qiu and Xie 1997) like in the Jiaozigou Formation of the Linxia Basin. The occurrence of giant rhinoceros in the Linxia Basin indicates that abundant forests and numerous lakes may exist in the northwestern China during the Late Oligocene.

5.2.4.2 Middle Miocene *Platybelodon* Fauna

Middle Miocene *Platybelodon* fauna is best represented by the mammal fossils within the Dongxiang Formation at Dalanggou and Wangshijie in Guanghe County and the Laogou Formation in Hezheng County (Fig. 5.8). The Dongxiang Formation is composed of brownish red sandstones and mudstones with greyish white marlite beds. The mammal fossils of this formation include *Hemicyon* sp., *Gomphotherium* sp., *Anchitherium* sp., *Alicornops* sp., *Hispanotherium matritense, Chalicotherium* sp. and *Kubanochoerus* sp. In Europe, *Gomphotherium* appeared first in MN3 at Artenay in France (Deng et al. 2004b). *Alicornops* appeared first in MN3 at Wintershof-West in Germany, and is widely distributed in Europe during MN6 (Deng et al. 2004b). In China, *Kubanochoerus* appeared first at Dingjiaergou in Tongxin, Ningxia with the age corresponding to MN6 of Europe, accompanied by abundant fossils of *Platybelodon.* In addition, *Hemicyon* and *Hispanotherium* coexist in the Dingjiaergou Fauna; *Hispanotherium* and *Gomphotherium* coexist in the Lengshuigou Fauna; *Hemicyon, Anchitherium* and *Hispanotherium* coexist in theTunggur Fauna (Qiu and Qiu 1995). But *Anchitherium* and *Hispanotherium* from the Dongxiang Formation are obviously smaller than those from Tunggur.

This Laogou Formation is composed of grayish yellow fine conglomerates and sandstones. The mammal fossils of this formation include *Alloptox* sp., *Pliopithecus* sp., *Hemicyon teilhardi, Amphicyon tairumensis, Percrocuta tungurensis, Gomphotherium* sp., *Platybelodon grangeri, Zygolophodon* sp., *Anchitherium gobiensis, Alicornops* sp., *Hispanotherium matritense, Chalicotherium* sp., *Listriodon mongoliensis, Kubanochoerus gigas, Palaeotragus tungurensis* and *Turcocerus* sp. (Deng et al. 2004a, b). The fossils from the Laogou Formation are the typical Middle Miocene forms, represented by Tunggurian *Platybelodon* and *Anchitherium* (Qiu and Qiu 1995). As mentioned above, *Platybelodon* has not appeared in the Dongxiang Formation, and *Anchitherium* from the Laogou Formation is obviously larger than its ally from the Dongxiang Formation. The shared genera between the Laogou Formation and the Tunggur Fauna include *Alloptox, Hemicyon, Amphicyon, Platybelodon, Zygolophodon, Anchitherium, Hispanotherium, Kubanochoerus, Listriodon, Palaeotragus* and *Turcocerus*. *Alloptox* was widely found from the Middle Miocene faunas in China, such as Tunggur in Nei Mongol (Young 1932), Lengshuigou in Lantian, Shaanxi (Li 1978), Qijia in Minhe, Qinghai (Qiu et al. 1981), and Dingjiaergou in Tongxin of Ningxia Hui Autonomous Region (Wu et al. 1991). *Pliopithecus* existed during MN5–9 in Europe, and it was found from the Dingjiaergou and Halamagai (Junggar, Xinjiang) Faunas of the Middle Miocene in China (Qiu and Guan 1986; Wu et al. 2003). In Eurasia, *Hispanotherium matritense* was found in Spain, Portugal and France of Europe, and Turkey, Pakistan, Mongolia and China of Asia. In China, *H. matritense* is distributed widely, found from the Middle Miocene Dingjiaergou, Lengshuigou and Erlanggang (Fangxian, Hubei) Faunas (Deng et al.

2004b). *Alicornops* was widespread in Europe during MN6–10 (Cerdeño and Sánchez 2000), and was found from the Middle Miocene strata in Turkey (Deng et al. 2004b). *Kubanochoerus* is a typical Middle Miocene form, which appeared in the strata of MN6 in Caucasia, and it was found at Dingjiaergou, Tunggur and Koujiacun (Lantian, Shaanxi) (Deng et al. 2004b).

5.2.4.3 Late Miocene *Hipparion* Fauna

The Late Miocene *Hipparion* fauna was found at the Liushu Formation (12–6 Ma), which is composed of red clay. Possibly due to differential erosion, the thickness of the Liushu Formation is variable in different sections. At Laogou, it has a thickness of only ca. 17 m and is directly capped by the Jishi conglomerate. At some localities near the southern margin of the Linxia Basin, however, its thickness is close to 200 m. The lower, middle and upper part of the Liushu Formation all contain a large number of mammal fossils.

The mammal fossils of the lower part of the Liushu Formation are found at Guonigou and Wangji in Dongxiang County (Fig. 5.8), including *Dinocrocuta gigantea, Machairodus* sp., *Tetralophodon* sp., *Hipparion dongxiangense, Parelasmotherium simplum, P. linxiaense,* and *Shaanxispira* sp. (Deng et al. 2004b). These fossils are characteristic of early Late Miocene age. *Dinocrocuta igantea* is a representative of the early *Hipparion* fauna in Asia, and it is discovered mainly from the early Late Miocene, corresponding to the European Vallesian Age or MN9–10 (Howell and Petter 1985). The materials discovered in China show the same trend. The fossils of *Dinocrocuta* were found from Bahe in Lantian and Fugu (Shaanxi Province), Songshan in Tianzhu, (Gansu Province), and Bulong in Biru (Tibet), which are interpreted to be equivalent to the European Turolian and Vallesian age (Liu et al. 1978; Qiu and Qiu, 1995; Xue et al. 1995; Zheng 1982).

The mammal fossils of the middle part of the Liushu Formation are found at Dashengou in Hezheng County, Houshan, Shanzhuang and Sigou in Guanghe County, and Shuanggongbei in Dongxiang County (Fig. 5.8), and include: *Pararhizomys hipparionum, Promephitis* sp., *P. hootoni, Melodon majori, Sinictis* sp., *Ictitherium* sp., *Hyaenictitherium wongii, H. hyaenoides, Dinocrocuta gigantea, Machairodus palanderi, Felis* sp., *Tetralophodon exoletus, Hipparion* sp., *H. chiai, H. weihoensis, Acerorhinus hezhengensis, Chilotherium wimani, Iranotherium morgani, Chleuastochoerus stehlini, Dicrocerus* sp., *Samotherium* sp., *Honanotherium schlosseri, Gazella* sp., *Hezhengia bohlini,* and *Miotragocerus* sp.. This fauna is still composed of early Late Miocene taxa, such as *Dinocrocuta gigantea.* Other important components of the Bahe Fauna are relatively richly represented in the middle part of the Liushu Formation, such as *Hipparion weihoense* and *H. chiai.* These two species of *Hipparion* have a large size, deep preorbital fossa far from the orbit, and narrow and long protocones. These characters show that both of them apparently belong to the *H. primigenium* group, and the hipparionines of this group in Europe and Africa are predominantly Vallesian in age (Qiu et al. 1988). The shared species in the Bahe Fauna and the middle part of the Liushu Formation also include

Tetralophodon exoletus and *Chleuastochoerus stehlini* (Liu et al. 1978). Besides the Bahe Fauna, the Lamagou Fauna in Fugu, Shaanxi (Xue et al. 1995) may be contemporaneous with the fauna of the middle part of the Liushu Formation. There are many common taxa between the Lamagou Fauna and the middle part of the Liushu Formation, such as *Dinocrocuta gigantea, Hyaenictitherium wongii, Hipparion chiai, Chilotherium wimani, Samotherium* sp. and *Miotragocerus* sp.

The mammal fossils of the upper part of the Liushu Fomration are found at Yangjiashan, Heilinding, and Shanjiawan in Hezheng County, Songshugou, Shilei, Huaigou, Yangwapuzi, Nanmiangou and Longjiawan in Guanghe County, including *Hystrix gansuensis, Pararhizomys hipparionum, Simocyon* sp., *Promephitis* sp., *P. hootoni, Parataxidea sinensis, Pleisiogulo* sp., *Ictitherium* sp., *Hyaenictitherium wongii, H. hyaenoides, Adcrocuta variabilis, Machairodus* sp., *M. palanderi, Metailurus* sp., *M. minor, Felis* sp., *Hipparion* sp., *H. coelophyes, H. dermatorhinum, Acerorhinus hezhengensis, Chilotherium wimani, Dicerorhinus ringstromi, Ancylotherium* sp., *Chleuastochoerus stehlini, Microstonyx major, Metacervulus* sp., *Cervavitus novorossiae, Honanotherium schlosseri, Palaeotragus microdon, Miotragocerus* sp., *Sinotragus* sp., *Protoryx* sp. and *Gazella* sp. (Deng et al. 2004b). In this fauna, the early Late Miocene representatives, such as *Dinocrocuta gigantea* and *Hezhengia bohlini* have disappeared. Similar to the Baode Fauna, this fauna is the typical fauna of the Late Miocene in China. Like the Baode Fauna, rhinocerotids are absolutely dominant in the upper part of the Liushu Formation, although both of them are named as the *Hipparion* faunas. On the other hand, the more primitive *Chilotherium wimani* in the upper part of the Liushu Formation is replaced by the more derived *C. anderssoni* in the Baode Fauna. The primitive characters of *C. wimani* include the low position of orbit, well-developed supraorbital tubercle, weak postorbital process, concave dorsal skull profile, little separated parietal crests, narrow braincase, and strong paracone rib on premolars, while the derived characters of *C. anderssoni* include the high position of orbit, absence of supraorbital tubercle, well-developed postorbital process, flat dorsal skull profile, broadly separated parietal crests, rounded braincase, and weak or absent paracone rib on premolars.

5.2.4.4 Early Pliocene *Hipparion* Fauna

The Early Pliocene *Hipparion* fauna was found at the Hewangjia Formation (Deng et al. 2004b), which is composed of red clay and a basal conglomerate (Fig. 5.8). The fauna of the Hewangjia Formation comes from the red clay at Shilidun in Guanghe County and Yinchuan in Jishishan County (Fig. 5.7), including *Hystrix gansuensis, Promephitis* sp., *Chasmaporthetes* sp., *Hyaenictitherium wongii, Hipparion* sp., *Shansirhinus ringstromi, Cervavitus novorossiae, Palaeotragus* sp., *Sinotragus* sp. and *Gazella* sp. Among them, the *Chasmaporthetes* sp. and *Shansirhinus ringstromi* newly appeared in the Hewangjia Fm., while the dominant *Chilotherium* in the Late Miocene *Hipparion* fauna disappeared in this fauna. Other members of the Late Miocene *Hipparion* fauna, such as *Adcrocuta, Ictitherium* and *Chleuastochoerus* are absent in the Hewangjia Formation. *Chasmaporthetes* has a widespread distribution in the world, and its fossils were discovered from Europe, Asia, Africa, and North America; it first appeared in the Early Pliocene, corresponding to the Ruscinian Age of Europe. In China, *Chasmaporthetes* was discovered from Yushe and Shouyang in Shanxi, Nihewan in Hebei, and Mianchi in Henan. In the Yushe Basin, *Chasmaporthetes* appeared firstly in the Gaozhuang Fauna with a paleomagnetic age of 5. 2~ 3. 4 Ma, when *Chilotherium* had disappeared (Qiu and Qiu 1995). The Gaozhuang Fauna should be correlated to MN14–15 of the European mammal ages (Qiu and Qiu 1995).

5.2.4.5 Early Pleistocene *Equus* Fauna

Early Pleistocene *Equus* fauna in the Linxia Basin is mainly found in the yellowish brown massive siltstones of the Dongshan Formation or the contemporaneous Wucheng Loess (Fig. 5.8). The abundant mammal fossils include *Aepyosciurus orientalis*, *Marmota parva*, *Castor* sp., *Mimomys* cf. *gansunicus*, *Bahomys* sp., *Sericolagus brachypus*, *Macaca* cf. *anderssoni*, *Paradolichopithecus gansuensis*, *Vulpes chikushanensis*, *Canis teilhardi*, *C. longdanensis*, *C. brevicephalus*, *Sinicuon* cf. *dubius*, *Eirictis robusta*, *Meles teilhardi*, *Chasmaporthetes progressus*, *Pachycrocuta licenti*, *Crocuta honanensis*, *Homotherium crenatidens*, *Megantereon nihowanensis*, *Sivapanthera linxiaensis*, *Panthera palaeosinensis*, *Felis teilhardi*, *Lynx shansius*, *Hipparion sinense*, *Equus eisenmannae*, *Coelodonta nihowanensis*, *Hesperotherium* sp., *Sus* sp., *Nipponicervus longdanensis*, *Gazella* cf. *blacki*, *Leptobos brevicornis* and *Hemibos gracilis* (Deng and Downs 2002; Qiu et al. 2004).

5.2.4.6 Mammalian Implication for the Central Asian Environmental Changes

The ecological features of the distinct faunas indicate great climatic and environmental changes in the Linxia Basin from the Late Oligocene to Pleistocene. There are warm and humid forests mixed with some open lands in the Late Oligocene, denser forests with richer water bodies in the Middle Miocene, tropical semiarid savannas with strong seasonal variation in the Late Miocene, and cold and arid climate with high elevation in the Early Pleistocene. Especially, the Linxia Basin is situated at the transitional zone between the Tibetan Plateau and LP. During the important period of the uplift of the Tibetan Plateau in the Late Cenozoic, the Linxia Basin has the mammal fossil assemblages, which can determine geological ages and reflect the uplift process of the Tibetan Plateau and its influence on the climate and environment. The Linxia Basin is comparable to the classical Siwaliks on the south margin of the Tibetan Plateau. The uplift of the Tibetan Plateau was not enough to baffle the dispersal of large mammals between north and south of this plateau in the Late Oligocene, but it became an obvious barrier for the mammal migration in the Middle Miocene and more obvious in the Late Miocene. The Linxia Basin reached a relatively high elevation in the Early Pleistocene, so that a plateau or alpine mammalian fauna appeared in this area.

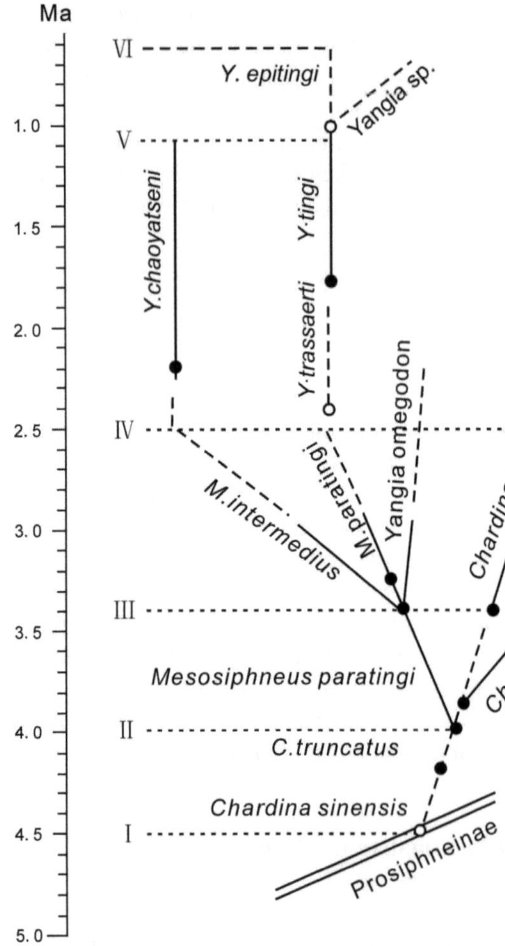

Fig. 5.9 Diagram showing the evolutionary trend of Concave-Occiput siphneids (Zheng 1997).

5.3 Environmental Significances of Representative Mammal Fossils

5.3.1 Evolution of Mesosiphneinae and its Implication for Environmental Changes

The Concave-Occiput siphneids has become an important fossil micromammal group in China since the establishment of *Yangia tingi* in the 1920s. Combining the skull patterns and dental morphlology, especially the last upper molar (M3) and the first lower molar (ml), Zheng (1997) revised the known materials from Asia. The evolutionary trend of Concave-Occiput siphneids is shown in Fig. 5.9. Thirteen species belonging to three genera are suggested (Zheng 1997). *Chardina, Mesosiphneus* and *Yangia* are recognized in this subfamily. The distribtion of these species are limited to a small range of Norhtern China from 34–41°N and from 107–118°E geographically.

Chardina sinensis was the earliest representative of this subfamily. It appeared at ca. 4.5 Ma and might have derived from the convex-occipital Prosiphneinae because its hypsodonty molar has never been found in any other species of the Siphneids before it appeared. *Yangia epitingi* was the latest species, which firstly appeared at 3.4 Ma and disappeared at ca. 0.65 Ma. Thus the existing time span of this subfamily is less than 3 Ma. In such a short time interval, their molar underwent a series of changes. The main evolutionary trend is from brachyodont to hypsodont and from large to small in rooted molar stage, but from small to large in rootless molar stage.

Both first and last-occurrence events are evident in the evolutionary course of the mesosiphneinae (Zheng 1997). The short-lived species (less than 0.4 Ma) always appeared in both originating and declining stages, while the long-lived species (more than 1 Ma) generally appeared in flourishing period of the subfamily. There are six siphneid events during its evolution: 1) first appearance of *Chardina* at about 4.5 Ma, 2) first appearance of *Mesosiphneus* at ca. 4 Ma, 3) first appearance of *Yangia* associated with *Chardina* sp. 2, *M. paratingi* and *M. intermedius* at ca. 3.4 Ma, 4) disappearance of all rooted molar Mesosiphneine and appearance of *Y. Chaoyatseni* and *Y. trassaerti* at ca. 2.5 Ma, 5) disappearance of both *Y. chaoyatseni* and *Y. tingi* at ca. 1.1 Ma, 6) Final disappearance of the Mesosiphneine at 0.65 Ma. These events were shown by fauna turnover or terrestrial events in Lingtai, Lantian, Luochuan and Yushe sections. Each event indicates a climate change from warm to cold. Both the 3.4 and 2.5 Ma events clearly correspond to the global climate changes.

5.3.2 The Microtine Distribution and its Environmental Significance Since the Middle Pleistocene in Zhangping Caves of the Qinling Mountains

At the northern bank of Luohe River, behind Zhangping Village, in the Qinling Mountains, southern Shaanxi, there exists a cave with six layers of deposition, of which five contain vertebrate fossils. Fossil comparison, terrace analysis and optically stimulated luminescence (OSL) dating suggest that the six layers of cave depositon were formed in different stages since the Middle Pleistocene, i.e., during the last 493 ka (Li and Xue 2008). Of the fossil-bearing layers, four are rich in microtine fossil voles. A statistical analysis of the vole fossils suggests different fossil compositions in different layers of the cave (Fig. 5.10). Since different voles have their unique habitats, these data may indicate environmental changes of this area and its vicinity during the past ca. 500 ka.

The highest layer in the cave has an age of ca. 493 ka. *Caryomys eva*, which was adapted to warm and humid climate, is predominant, at about 38.1 %. *Microtus oeconomus*, which was adapted to cold and wet climate, amounts to 28.6 %. *Caryomys inez*, which was adapted to warm and dry climate, amounts to 33.3 %. The fourth layer has an age of about 259 ka ago. The composition of microtine

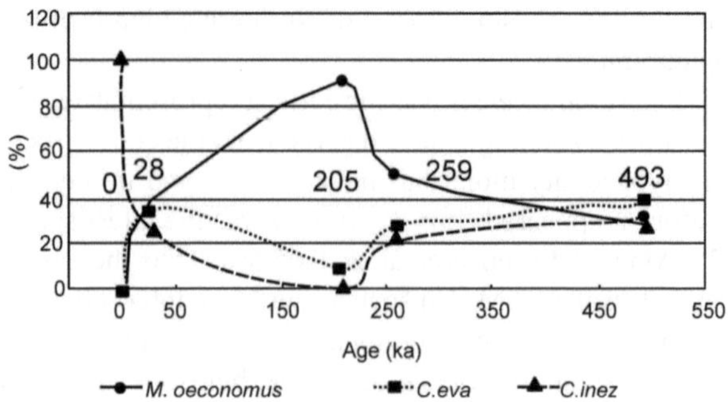

Fig. 5.10 Distribution trends of relative numbers of microtines in the Qinling Mountains (Li and Xue 2008)

fauna changed greatly, with *C. inez* only amouting to about 22 %, and *C. eva* about 28 %, but *M. oeconomus* surging to 50 %. This probably reflects gradual cooling in the area, and a relatively cold and humid environment. The third layer has an age of about 205 ka. The complete absence of *C. inez*, overwhelming of *M. oeconomus* (accounting for 90 %), and rareness in *C. eva* (10 %) all suggest a very cold and wet period. The lowest layer has an age of ca. 28 ka, within the transitional period from the interglacial stage to the LGM. Both *C. eva* and *M. oeconomus* amount to 37.5 %, while *C. inez* occupies only 25 %. This indicates a cool moist environment.

At present, *M. oeconomus*, which dominated in the Middle and Late Pleistocene, not only disappeared within the studied area, i.e. Luonan, but also vanished in other areas of the eastern Qinling Mountains. Those almost extinct *C. inez* in the Middle and late Middle Pleistocene have reappeared in the Luonan area and its vicinity in modern times, however, *C. eva* no longer appears in the Luonan area. Generally, the distribution of the three voles during the past ca. 500 kyr suggests a warm and humid environment at 493 ka, cool and wet at 259 ka, cold and damp at 205 ka, cool and wet at 28 ka, and warm and dry at present.

5.3.3 Expansion of Equus przewalskii in North China Linked to the East Asian Winter Monsoon

The *Equus przewalskii* is highly adapted to a dry and cold climate. In other words, the warm and humid climate is less helpful for its colonization. At present, it lives mainly in the northern Xinjiang of China and the Kobdo Basin of Mongolia, which are generally characterized by cold and dry climate. The earliest *E. przewalskii* occurred in the Dingcun Fauna (Shanxi Province, North China), which has an age of ca. 130 ka, corresponding to a glacial period (You and Xu 1981). During the last interglacial period when a strong summer monsoon resulted in a relatively warm and humid North China, however, no *E. przewalskii* ever appeared there. For example, during the warm Holocene, there are more than a hundred mammal faunas in North China, such as Yinxu Fauna (Henan Province), Banpo Fauna (Shaanxi Province) and Cishan Fauna (Hebei), but none of them shows the presence of *E. przewalskii*. During the last glacial period when the East Asian winter monsoon led to a relatively cold and dry North China, most of the faunas in North China contain *E. przewalskii* fossils, such as the Hutouliang Fauna (Nihewan Basin, Hebei Province), Wurji Fauna (Nei Mongol), Qingshantou Fauna (Jilin Province), Xueguan Fauna (Puxian County, Shanxi Province), Shanchengzi (Benxi County, Liaoning Province) and the Angangxi Fauna (Heilongjiang Province). These facts indicate that the historical distribution of *E. przewalskii* is closely related to the temporal and spatial variation of the Asian monsoon system (Deng 1999).

5.3.4 The Evolution of Mammuthus-Coelodonta Fauna and Elaphurus Davidianus

5.3.4.1 The Evolution of *Mammuthus-Coelodonta* Fauna

Like the *Equus przewalskii*, the *Mammuthus-Coelodonta* fauna is adapted to cold climate. It mainly occurred in China during two time intervals of the last glacial period, 40–20 ka and 15–11 ka. These two intervals were characterized by very cold climate. During these two cold periods, *Mammuthus* mainly lived in the high latitude areas of North China (>41°N). In comparison with the *Mammuthus*, *Coelodonta* lived in the >35°N latitude areas of North China. This indicates that the *Mammuthus* is more dependent on a cold climate, as compared to *Coelodonta*, although both of them prefer cold climate. After 11 ka when climate became warmer, however, both *Mammuthus* and *Coelodonta* disappeared in North China, which indicates a great change in environment, i.e, the winter monsoon intensity decreased but the summer monsoon intensity increased significnalty. This may have made the environment unsuitable for *Mammuthus* and *Coelodonta.*

5.3.4.2 *Elaphurus davidianus*

Generally, *Elaphurus davidianus* is more adapted to warm climate than to cold climate. It has flourished during the Holocene, espeically between 7.5 and 5 ka. At 4–3.5 ka when climate became somewhat colder, its abundance decreased. At 3.3 ka when climate became warmer, its abundacne increased significantly. After 2 ka when the climate became colder, its abundance decreased again. Thus the changes in abundance of *Elaphurus davidianus* may be linked to climate changes associated with Asian monsoon fluctuations and human activities.

5.3.5 The Mammal Composition of the Guanzhong Area in the Loess Plateau During the Last Glacial Maximum

The Xianyang Fauna (20.8 ka), Pucheng Fauna (20.4 ka) and Dali Fauna (17.3 ka) are three important mammal faunas of the Late Pleistocene in the Guanzhong area of the southern LP (Li et al. 2012). Their combined changes in composition have

Table 5.7 Distribution changes of the three faunas about 20 ka BP ago in Guanzhong area, Shaanxi province (Li et al. 2012)

Species	Xianyang fauna 20.796 ka BP	Pucheng fauna 20.400 ka BP	Dali fauna 17.330 ka BP
Canis lupus		----------	
Panthera tigris		----------	
Palaeoloxodon naumanni	----------	----------	
Palaeoloxodon sp.			----------
Mammuthus sp.			----------
Dicerorhinus cf. *mercki*		----------	
Coelodonta antiquitatis	----------	----------	
Cervus nippon	----------		
Equus przewalskii		----------	----------
Equus hemionus	----------	----------	----------
Sus cf. *scrofa*		----------	
Cervus cf. *hortulorum*		----------	
Procapra przewalskii		----------	----------
Ovis sp.	----------		
Bubalus youngi	----------		
Bos primigenius	----------		

been interpreted to indicate the effects of the East Asian monsoon on environmental and climatic changes during the LGM (Li et al. 2012). The Xianyang Fauna was collected from gravel layers within a pit on the second terrace of the Weihe River, about 10 km west of Sanqiao (between Xi'an and Xianyang cities) (Xue et al. 2000). The Pucheng Fauna was collected from the second terrace of a tributary of the Weihe River in Pucheng County (The Institute of Archaeology of Shaanxi Province and the CPAM of Dali County 1996). Specimens were collected from the lower alluvial layer of the terrace in three localities (PN6, PN16 and PB17) on the bank of the Luohe River. The Dali fauna, represented by diverse fossils, was also collected from a second terrace (DY11) of a tributary of the Weihe River in Dali County (The Institute of Archaeology of Shaanxi Province and the CPAM of Dali County 1996). The distance between localities of the three animal faunas is less than 140 km, thus it is reasonable to assume a common living area for these mammals. Table 5.7 shows the composition changes of the three faunas (Li et al. 2012). Overall, the mammalian fossils from the three faunas were all common in northern China in the Late Pleistocene. Because of the presence of *Palaeoloxodon*, none of the three faunas belongs to the typical *Mammothus-Coelodonta* fauna.

The earliest Xianyang fauna is characterized by water buffalo and sika deer, which usually inhabit warm moist forests and forest edge meadows. At present in Shaanxi, buffalo is adapted to warm and moist environments, and lives strictly in areas south of the Qinling Mountains, such as Hanzhong, but not in areas to the north, such as the Guanzhong area (Xue et al. 2000). Thus the composition of the Xianyang Fauna suggests that the Guanzhong area was a warm and humid environment with open water and forests at ca. 20.8 ka, a relatively moderate humid environmental condition.

The Pucheng Fauna contains more than ten mammal species (Table 5.7). The buffalo and sika deer disappeared, but there are wild horses, wild boar, Beijing spotted deer, antelopes, wolves and tigers, which is a group of animals living in the forest edge or Gobi grassland.

The Dali Fauna consists of the fewest species, mainly including *Equus przewalskii*, which is now mainly distributed in the Gobi in northwest China. *E. przewalskii* is a desert animal strictly adapted to dry and cold climates, and lives in areas affected by the winter monsoon. In addition to *E. przewalskii*, wild donkey (*E. hemionus*) and *Procapra* (*Gazella*) *przewalskii* are also found in the Dali Fauna, which are mainly distributed in the cold dry steppe, desert and semi-desert areas of northwest China (Zhang 1997). Except for grassland or Gobi desert steppe animals, such as wild horses, wild donkey and wild gazelle, few animals living in the forest edge grassland were found in this fauna.

Overall, the earlier Xianyang fauna indicates a warm-cool humid environment. The slightly younger Pucheng fauna indicates a climate changing from cool humid to dry and cold. The latest Dali fauna indicates a steppe environment. Evolutionary changes of species composition in the three faunas reveal the processes of climate and environmental evolution from warm-cool humid to extreme dry and cold conditions during the LGM. The rapid changes in the composition of faunas at 20.79–20.4 ka provide important mammalian evidence for the rapid fluctuations of the monsoonal climate on the LP.

5.3.6 The Late Miocene Antlers of the Cervids from the Yuanmou Basin and its Climate Implications

The *Lufengpithecus hudienensis* Fauna was found in the Yuanmou Basin (Yunnan Province), South China. It is an important Late Miocene fauna, which contains more than a hundred taxa (Qi and Dong 2006). The cervids from the *L. hudinenensis* Fauna are brachyodont, which implies that they feed on juicy and tender leaves of dicotyledon. Such vegetation may be found in a southern subtropical evergreen forest, which is generally developed in a humid and temperate climate. The burr on the antlers of the cervids is well developed (Fig. 5.11), which implies that the Late Miocene climate of the Yuanmou Basin had distinct seasonal variations. The seasonal replacement of antlers cause the burr to become well-developed (Fig. 5.11). Based on the study of extant cervids, the replacement of antlers is generally modulated by the seasonal cycles of the cervid endocrine. This cycling is significantly influenced by an annual difference in sunlight. The Yuanmou Basin is located within 25–26°N latitudes, which has small annual differences of sunlight irradiation. However, the humid climate in the Yuanmou Basin is significantly controlled by Indian monsoon circulation. The Indian summer monsoon brings humidity from the Indian Ocean that would increase cloud coverage, which would in turn decrease sunlight during

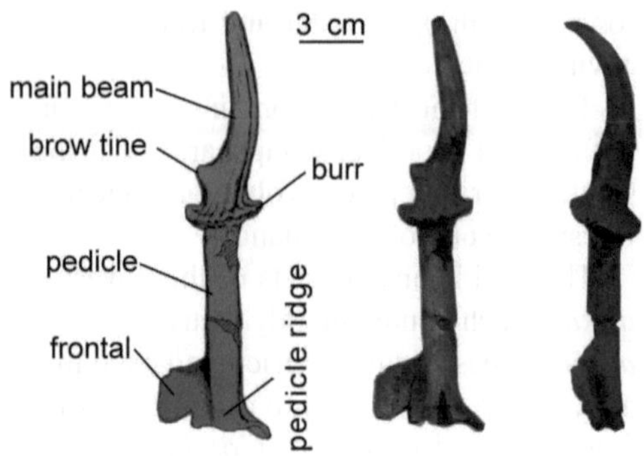

Fig. 5.11 Antler of *Muntiacus leilaoensis* (Qi and Dong 2006)

cloudy seasons. On the contrary, the retreat of the Indian summer monsoon would decrease the cloud coverage and subsequently increase sunlight during cloudy seasons. Thus the seasonal replacement of antlers indicates that the seasonal variability of the Indian summer monsoon may have appeared since the Late Miocene (Qi and Dong 2006).

5.4 Early Human Evolution in China

An increasing body of evidence suggests that changes or shifts in environmental scenarios have played a pivotal role on the origin and evolution of early humans (deMenocal 1995; Potts 1996, 1998). For example, the savanna hypothesis claims that the rise of African dry savanna during the Pleistocene is a critical event behind the development of terrestrial bipedality, stone tool making, encephalized brains and other traits. In addition, the variability selection hypothesis states that large disparities in environmental conditions were responsible for important episodes of adaptive evolution. The resulting adaptations enhanced behavioral versatility and ultimately ecological diversity in the human lineage. The earliest humans are suggested to originate from ape in Africa at 6–7 Ma, based on the discovery of *Sahelanthropus tchadensis*, a fossil hominid which exhibits both primitive and derived features and is thought to be close to the purported 'last common ancestor' (Brunet et al. 2002). These earliest humans gradually evolved into *Homo erectus* in Africa at 2–1.8 Ma. With an increasing population of early humans in Africa, competition for food and water were enhanced. This may have promoted the expansion of *H. erectus* to other areas, out of Africa. *H. erectus* are suggested to have expanded to western Eurasia and East Asia as early as 1.85–1.7 Ma (Ferring et al. 2011; Gabunia et al. 2000; Zhu et al. 2007; Ao et al. 2013). Most *H. erectus* sites in East Asia were found in the monsoonal areas of China. The monsoonal climate in East Asia apparently offered an environment in which early humans could survive.

Table 5.8 shows the context of important sites which have yielded human fossils and/or paleoliths (see their locations in Fig. 5.12). The currently earliest *H. erectus* site in China was found in the Yuanmou Basin in southwest China, with an age as early as 1.7 Ma (Zhu et al. 2008). The climate of this area is under the influence of

Table 5.8 Selected Hominin or Paleolithic sites in China

Site	Material	Age (Ma)	Dating method	Reference
Yuanmou	*H. erectus*	1.7	Magnetostratigraphy	Zhu et al. (2008)
Nihewan	Artifacts	1.66–0.32	Magnetostratigraphy, ESR, Orbital tuning	Zhu et al. (2007), Liu et al. (2010a), Ao et al. (2010b)
Xihoudu	Artifacts	1.27	Magnetostratigraphy	Zhu et al. (2003)
Lantian Gongwangling	*H. erectus*	1.15	Magnetostratigraphy	An and Ho (1989)
Bose	Artifacts	0.8	$^{40}Ar/^{39}Ar$	Hou et al. (2000)
Lantian Chenjiawo	*H. erectus*	0.65	Magnetostratigraphy	An and Ho (1989)
Yunxian	*H. erectus*	0.87–0.83 0.5	Magnetostratigraphy, ESR	Yan (1993), Chen et al. (1997)
Nanjing	*H. erectus*	0.58	U-series	Zhao et al. (2001)
Zhoukoudian	*H. erectus*	0.5–0.23	Magnetostratigraphy, ESR, U-series	Wu et al. (1985)
Hexian	*H. erectus*	0.4	U-series	Etler (1996)
Dali	Archaic *H. sapiens*	0.27	Sedimentary cycle	Xiao et al. (2002)
Jinniushan	Archaic *H. sapiens*	0.2	U-series	Chen et al. (1994)
Chaoxian	Archaic *H. sapiens*	0.2–0.16	U-series	Shen et al. (2010)
Maba	Archaic *H. sapiens*	0.13	U-series	Chen and Yuan (1988)
Zhiren Cave	Modern *H. sapiens*	0.1	U-series	Liu et al. (2010b)
Liujiang	Modern *H. sapiens*	0.07	U-series	Chen and Yuan (1988)
Ziyang	Modern *H. sapiens*	0.03	U-series	Chen and Yuan (1988)
Salawusu	Modern *H. sapiens*	0.07–0.04	U-series	Chen and Yuan (1988)
Tianyuan Cave	Modern *H. sapiens*	0.04	AMS	Shang et al. (2007)
Upper Cave	Modern *H. sapiens*	0.03	AMS	Chen et al. (1992); Hedges et al. (1992)

the Indian monsoon. The evidence of *H. erectus* at Yuanmou is slightly later than the age of *H. erectus* at Dmanisi (1.85–1.78 Ma) in Georgia (Ferring et al. 2011) and at Java (1.8–1.5 Ma) in southeastern Asia (Larick et al. 2001; Swisher et al. 1994). This might imply that *H. erectus*, once it dispersed beyond Africa, spread rapidly across Asia during the earliest Pleistocene.

In addition to the Indian monsoonal Yuanmou Basin, China has a large number of Pleistocene hominin or Paleolithic sites under the influence of the seasonal reversal of East Asian summer and winter monsoons, such as the Nihewan Basin (North China). In the Nihewan Basin, a series of Paleolithic sites have been found (Table 5.9). In particular, the MJG-III layer (1.66 Ma) documents the oldest coexistence of stone tools and man-made bone modifications in East Asia, indicating possible continuity with the oldest stone tools and artificial bone modifications re-

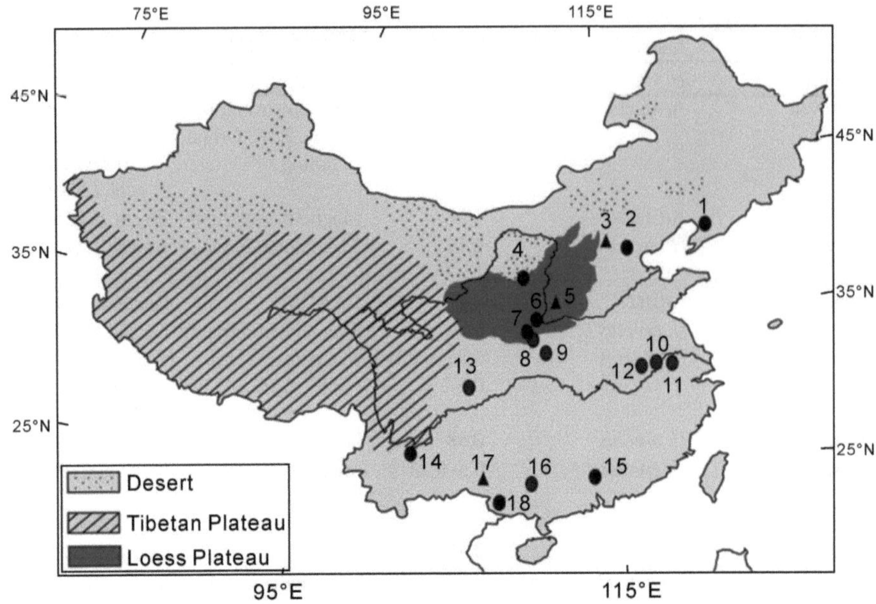

Fig. 5.12 Map of China with locations of Paleolithic or hominid sites. Solid circles represent the hominid sites. Solid triangles represent the Paleolithic sites. 1, Jinniushan; 2, Zhoukoudian-Upper Cave-Tianyuan Cave; 3, Nihewan; 4, Salawusu; 5, Xihoudu; 6, Dali; 7, Lantian; 8, Chenjiawo; 9, Yunxian; 10, Hexian; 11, Nanjing; 12, Chaoxian; 13, Ziyang; 14, Yuanmou; 15, Maba; 16, Liujiang; 17, Bose; 18, Zhiren Cave

Table 5.9 Paleolithic sites in the Nihewan Basin

Site	Age (Ma)	Dating method	References
Majuangou	1.66–1.55	Magnetostratigraphy	Zhu et al. (2004)
Xiaochangliang Dachangliang	1.49	Orbital tuning	Ao et al. (2010b)
Banshan	1.32	Magnetostratigraphy	Zhu et al. (2004)
Feiliang	1.2	Magnetostratigraphy	Deng et al. (2007)
Donggutuo	1.12–1.06	Orbital tuning	Ao et al. (2010b)
Huojiadi Chenjiawan	1.1	Magnetostratigraphy	Liu et al. (2010c), Wang et al. (2006)
Maliang	0.79	Orbital tuning	Ao et al. (2010b)
Dongpo	0.32 Ma	ESR	Liu et al. (2010a)

ported in eastern Africa (Zhu et al., 2004). Combing the ages of these stone tools, we speculate that most or all of these stone tools may be manufactured and used by *H. erectus*. This implies that the *H. erectus* inhabited the Nihewan Basin for a very long time during the Pleistocene, although the vegetation in this area shifted from typical subtropical species, via a well-developed forest vegetation, to grasses and herbs during this period (Yuan et al. 1996). In addition, most of these Paleolithic sites contain a large number of mammalian fossils, which are known as Nihewan Fauna (Qiu 2000; Zhu et al. 2007).

In central China, the Xihoudu Paleolithic site (1.27 Ma) (Zhu et al. 2003) was found in the Sanmenxia Basin; Lantian (1.15 Ma) and Chenjiawo (0.65 Ma) *H. erectus* sites (An and Ho 1989) were found along the southern margin of the LP. The cave site of Zhoukoudian Locality 1 in North China is the largest single source of *H. erectus* in the world. Its 40 m-thick depositional sequence was divided into 17 layers. Stone artifacts and *H. erectus* fossils have been recovered from layers 1–10, from a lower level in layers 8/9 and an upper level in layers 3–4. The hominin-fossil-bearing layers probably have a time range between 230 (based on $^{230}Th/^{234}U$ dating) and 500 (based on fission-track dating of sphene grains) kyr (Wu et al. 1985). The lowest fossiliferous layer 13 is dated to be ca. 0.78 Ma as suggested by palaeomagnetic stratigraphy (Qian et al. 1980).

In South China, several Mid-Pleistocene *H. erectus* fossils have been found at Yunxian (Hubei Province), Nanjing (Jiangshu Province) and Hexian (Anhui Province). Two *H. erectus* skulls from Yunxian show a mixture of features associated with both *H. erectus* and archaic *H. sapiens*, which comprise a crucial basis for studying the Mid-Pleistocene hominid diversity and the relationships among regionally disparate Mid-Pleistocene hominids (Chen et al. 1997). Two *H. erectus* skulls were found in the Nanjing site, with Skull No.1 being dated to ca. 0.58 Ma based on U-series dating (Zhao et al. 2001) and Skull No.2 being dated to 183–127 ka based on faunal correlation (Xu 1999). At Hexian site, *H. erectus* skull, teeth and occipital bone have been found (Etler 1996). In addition, a large number of Mid-Pleistocene stone tools have been found in the Bose Basin. Acheulean-like handaxes have appeared in this basin as early as ca. 0.73 Ma (Hou et al. 2000). This may imply that a spread of bifacial technologies across East Asia from Africa occurred earlier than previously thought.

A large number of archaic *H. sapiens* fossils have been found in China, such as a skull from Dali, a skull and postcranial bones from Jinniushan, fragments of maxilla and occipital from Chaoxian and a skull-cap from Maba. Dali hominid has a morphological mosaic of eastern and western Eurasia, which occurred at the transition from S_3 paleosol formation to L_3 loess accumulation, close to the transition from glacial marine isotope stage (MIS) 8 to interglacial MIS 7 (Xiao et al. 2002). Both Chaoxian and Maba hominids show similar characteristics with the *H. neanderthalensis* in Europe. All of these morphological mosaics indicate probable gene flow between European and Asian hominids during this period or before.

The period after ca. 0.1 Ma is crucial for hominid evolution, during which modern *H. sapiens* became established throughout the Old World and the trend of incremental accretion of increasing modern features in Chinese fossil human record continues unabated. During this period, recognizable differences between northern and southern Chinese also become apparent, although how far back in time this division may go is unclear at present. The earliest modern *H. sapiens* in Asia come from Zhirendong (Zhiren Cave) in South China, with an age as old as 0.1 Ma (Liu et al. 2010b). The hominid specimens include an anterior fragment of mandible and two molars. The age and morphology of the Zhiren Cave human remains support a modern human emergence scenario for East Asia. It also places the Late Pleistocene Asian emergence of modern humans in a pre-Upper Paleolithic context and raises issues concerning the long-term Late Pleistocene coexistence of late archaic and early modern humans across Eurasia. In addition, there are several early modern *H. sapiens* sites, such as the Ziyang and Liujiang. The date of the cranium from the Ziyang is problematical. A Paleolithic site in the immediate vicinity was dated to ca. 0.03 Ma (Chen and Yuan 1988). A well-perserved Liujiang cranium from Guangxi was discovered together with a nearly complete axial skeleton (Wu et al. 1985). The ages of both Ziyang and Liujiang hominids are problematical. In addition to South China, some modern *H. sapiens* sites were also found in North China, such as Salawusu, Tianyuan Cave and Upper cave. Cranial and postcranial bones and fragments are found from a few sites of Salawusu region, northern China. The human fossil-bearing layer has been dated between 65 and 35 ka by U-series assays (Wu et al. 1985). The Upper Cave Men were discovered above the Locality 1 of Zhoukoudian. It was dated to ca. 0.03 Ma using the accelerator mass spectrometry radiocarbon technique (Chen et al. 1992; Hedges et al. 1992). The Tianyuan Cave is only about 6 km southwest of the famous Locality 1 of Zhoukoudian. It has an age of ca. 0.04 Ma based on using direct accelerator mass spectrometry radiocarbon (Shang et al. 2007). Morphological comparison shows that modern *H. sapiens* from Tianyuan exhibits several late archaic human features, such as its anterior to posterior dental proportions, a large hamulus length, and a broad and rounded distal phalangeal tuberosity. This morphological pattern implies that a simple spread of modern humans from Africa is less likely to occur.

In addition to the aforementioned Paleolithic or hominid sites, there are a large number of other sites yielding human fossils in China. The morphological features of *Homo erectus* and *Homo sapiens* in China show a mosaic between these chronospecies, and indicate the changes are transitional. No clear-cut morphological demarcation line could be drawn between them. Human fossils in China have some common characteristics, such as the anterolateral surface of the fronto-sphenoidal process of zygomatic bone facing more forward, the contour of the lower margin of the zygomatic process of maxilla is curved, the high position of the joining of this process with the maxillary body, more obtuse fronto-malar angle and zygomaxillary angles, flatness of nasal saddle, low upper face, the contour of the suture between frontal, nasal and maxilla approximating to a horizontal curve, quadrangular orbit with blunt infero-latera1 orbital margin, presence of median sagittal keeling, the most protruding point of the median sagittal profile of frontal bone locates at its lower half. It is worth emphasizing that all upper incisor fossils found in China are shovel shaped without exception. Among Chinese human fossils there are also a few specimens with morphological features inharmonious with other fossils found in China. This phenomenon indicates the result of gene flow from outside. The inharmonious features include very high nasal saddle in skull No.1 of Nanjing, bulging between pyriform orifice and special shape of brow ridge of Dali cranium, spherical orbit and its sharp inferolateral margin in the Maba specimen, suprainiac fossa in Chaoxian occipital, chignon in crania from Liujiang, Ziyang, Lijiang and Chuandong. All of these features are rare in China and much more frequently seen in Europe, so the gene flow is probably from Europe. Comparisons between the fossils from China and those from areas east and south to China indicate that there are communications among the populations of these areas during Late Pleistocene. Therefore the scenario in China should be considered as continuous evolution with hybridization. This hypothesis is also supported by the evidence from Paleolithic archeology and mammalian faunas. Evidence from all of these aspects strongly supports the Multiregional Evolution Hypothesis for the origin of modern humans which has different submodels in different regions. For example, continuity might exist in much less extent in Europe than in East Asia. In addition, there are distinct differences between the recent human populations of North and South China. The difference could be roughly in certain extent traced to Neolithic and late Paleolithic stage. There is no evidence showing that it could be extended to archaic *H. sapiens* and *H. erectus*. There are distinct interpopulational differences in the latter two stages. These differences might be only the difference between local populations instead of representing the difference between North and South China.

The human evolution in China looks like neither a ladder nor a bush. Species multiplicity of hominids in the Pleistocene could not be extended into early and Middle Pleistocene at least in the case of China, because after the appearance of man-made tools the communication between human populations increased and the possibility of formation of new species by isolation is unlikely. However there were still distinct intra-species and interpopulational differences among hominids. The ability to resist disaster and diseases of ancient humans was rather low. Extinction of small populations is prone to happen. So the human evolution in China could be likened to a river network. There are branches of the river. Some branches could be blind (corresponding to extinction of local populations). Some branches could unite and receive small branches from other river networks (small population of immigrants from outside China). Some could flow to other rivers (human migration to other areas). In a river the increase of water volume in the lower reaches results from the confluence of many branches. In human evolution in China, the expansion of the population in later times is mainly due to reproduction of the populations in China themselves, where the extinction of some small local populations might be followed by replacement in some cases. So the scenario includes continuous evolution with extinction of some small local populations and replacement by other populations in certain areas as well as supplementary hybridization with immigrants from other regions. In short, for human evolution in China, the continuity may be the main process, while the others pathways are subsidiary.

During the Quaternary, China experienced significant climatic and environmental variability of global and regional significance. Specifically, early humans met with long-term aridification and cooling of the Asian continental interior over the last 2.6 Ma (Ao 2010; Ao et al. 2009, 2010a) punctuated by alternations of glacial and interglacial climate (Ao et al. 2010b) and a series of regional environmental shifts, such as stepwise southerly migrations of the Mu Us desert lying to the north of the LP (Ding et al. 2005), C_4 plant expansions in East Asia (An et al. 2005), and $\delta^{13}C$ maxima events in South China (Wang et al. 2004). In addition, early humans were able to survive the increasing climatic contrast between glacial and interglacial climate during and after the Mid-Pleistocene transition (MPT) (Deng et al. 2007). In particular, early humans adjusted to prolonged glacial climate in North China (40°N) during the MPT (Deng et al. 2007; Ao et al. 2010b). The overwintering problem of the 40°N temperate zone during and after the MPT may have been significantly solved by hunting and eating animals. For example, archeological evidence at the Majuangou Paleolithic site suggested that several diaphysis fragments of deer- and horse-sized mammalian long bones show tool percussion damage indicative of marrow extraction (Zhu et al. 2004). It is thus reasonable to hypothesize that subsequent early humans that lived in East Asia were hunter-gatherers. Some Early Pleistocene settlers of the Nihewan Basin, e.g., those of the Feiliang and Xiantai sites, lived in dry paleoclimates associated with steppe paleoenvironments indicated from palynological features (Deng et al. 2006b). These indicate that early humans were quite adapted to the wide environmental changes in East Asia during the Quaternary.

References

An ZS, Ho CK (1989) New magnetostratigraphic dates of Lantian *Homo erectus*. Quatern Res 32:213–221

An ZS, Kutzbach JE, Prell WL, Porter SC (2001) Evolution of Asian monsoons and phased uplift of the Himalayan Tibetan plateau since Late Miocene times. Nature 411(6833):62–66

An ZS, Wang SM, Wu XH, Chen MY, Sun DH, Liu XM, Wang FB, Li L, Sun YB, Zhou WJ, Zhou J, Liu XD, Lu HY, Zhang YX, Dong GG, Qiang XK (1999) Eolian evidence from the Chinese Loess Plateau: the onset of the Late Cenozoic Great Glaciation in the Northern Hemisphere and Qinghai-Xizang Plateau uplift forcing. Science in China 42(3):258–271

An ZS, Huang YS, Liu WG, Guo ZT, Clemens S, Li L, Prell W, Ning YF, Cai YJ, Zhou WJ, Lin BH, Zhang QL, Cao YN, Qiang XK, Chang H, Wu ZK (2005) Multiple expansions of C4 plant biomass in East Asia since 7 Ma coupled with strengthened monsoon circulation. Geology 33:705–708

Ao H (2010) Mineral-magnetic signal of long-term climatic variation in Pleistocene fluvio-lacustrine sediments, Nihewan Basin (North China). J Asian Earth Sci 39:692–700

Ao H, Dekkers MJ, Deng CL, Zhu RX (2009) Paleoclimatic significance of the Xiantai fluvio-lacustrine sequence in the Nihewan Basin (North China), based on rock magnetic properties and clay mineralogy. Geophysical J Intern 177(3):913–924

Ao H, Deng CL, Dekkers MJ, Sun YB, Liu QS, Zhu RX (2010a) Pleistocene environmental evolution in the Nihewan Basin and implication for early human colonization of North China. Quatern Int 223–224:472–478

Ao H, Deng CL, Dekkers MJ, Liu QS, Qin L, Xiao GQ, Chang H (2010b) Astronomical dating of the Xiantai, Donggutuo and Maliang Paleolithic sites in the Nihewan Basin (North China) and implications for early human evolution in East Asia. Palaeogeograp Palaeoclimatol Palaeoecol 297(1):129–137

Ao H, Dekkers MJ, Wei Q, Qiang XK, Xiao GQ (2013) New evidence for early presence of hominids in North China. Scientific Reports 3:2403, Doi 10.1038/Srep02403

Barbour GB (1925) The deposits of the Sankanho Valley. Bulletin of the Geological Society of China 4, 53–55

Brunet M, Guy F, Pilbeam D, Mackaya HT, Likius A, Ahounta D, Beauvilain A, Blondel C, Bocherens H, Boisserie JR, DeBonis L, Coppens Y, Dejax J, Denys C, Duringer P, Eisenmann V, Fanone G, Fronty P, Geraads D, Lehmann T, Lihoreau F, Louchart A, Mahamat A, Merceron G, Mouchelin G, Otero O, Campomanes PP, PonceDeLeon M, Rage J, Sapanet M, Schuster M,

Sudre J, Tassy P, Valentin X, Vignaud P, Viriot L, Zazzo A, Zollikofer C (2002) A new hominid from the Upper Miocene of Chad, Central Africa. Nature 418:145–151

Cande SC, Kent DV (1995) Revised calibration of the geomagnetic polarity timescale for the late Cretaceous and Cenozoic. J Geophys Res 100:6093–6095

Cerdeño E, Sánchez B (2000) Intraspecific variation and evolutionary trends of *Alicornops simorrense* (Rhinocerotidae) in Spain. Zool Scr 29(4):275–305

Chen TM, Yuan S. (1988) Uranium-series dating of bone and teeth from Chinese palaeolithicsites. Archaeometry 30:59–76

Chen TM, Hedges REM, Yuan ZX. (1992) The second batch of accelerator radiocarbon dates for upper cave site of Zhoukoudian. Acta Anthropologica Sinica 11:112–116 (in Chinese with English abstract)

Chen TM, Yang Q, Wu E. (1994) Antiquity of Homo sapiens in China. Nature 368:55–56

Chen TM, Yang Q, Hu YQ, Bao WB, Li TY. (1997) ESR dating of tooth enamel from Yunxian Homo erectus site, China. Quaternary Science Reviews 16:455–458

deMenocal PB (1995) Pliopleistocene African climate. Science 270(5233):53–59

Deng CL, Shaw J, Liu QS, Pan YX, Zhu RX (2006a) Mineral magnetic variation of the Jingbian loess/paleosol sequence in the northern Loess Plateau of China: implications for Quaternary development of Asian aridification and cooling. Earth Planet Sci Lett 241:248–259

Deng CL, Wei Q, Zhu RX, Wang HQ, Zhang R, Ao H, Chang L, Pan YX (2006b) Magnetostratigraphic age of the Xiantai Paleolithic site in the Nihewan Basin and implications for early human colonization of Northeast Asia. Earth Planet Sci Lett 244(1–2):336–348

Deng CL, Xie F, Liu CC, Ao H, Pan YX, Zhu RX (2007) Magnetochronology of the Feiliang Paleolithic site in the Nihewan Basin and implications for early human adaptability to high northern latitudes in East Asia. Geophysical Research Letters 34:L14301, doi:10.1029/2007GL030335

Deng CL, Zhu RX, Zhang R, Ao H, Pan YX (2008) Timing of the Nihewan formation and faunas. Quatern Res 69:77–90

Deng T (1999) Historical distribution of *Equus przewalskii* and its control by climate fluctuation. Acta Theriologica Sinica 19:10–16 (in Chinese)

Deng T, Downs W (2002) Evolution of Chinese Neogene Rhinocerotidae and its response to climatic variations. Acta Geol Sin 76:139–145

Deng T, Wang XM, Ni XJ, Liu LP (2004a) Sequence of the Cenozoic mammalian faunas of the Linxia Basin in Gansu, China. Acta Geol Sin 78(1):8–14

Deng T, Wang XM, Ni XJ, Liu LP, Liang Z (2004b) Cenozoic stratigraphic sequence of the Linxia Basin in Gansu, China and its evidence from mammal fossils. Vertebrata Palasiatica 42:45–66 (in Chinese)

Ding ZL, Derbyshire E, Yang SL, Sun JM, Liu TS (2005) Stepwise expansion of desert environment across northern China in the past 3.5 Ma and implications for monsoon evolution. Earth Planet Sci Lett 237:45–55

Ding ZL, Derbyshire E, Yang SL, Yu ZW, Xiong SF, Liu TS (2002) Stacked 2.6-Ma grain size record from the Chinese loess based on five sections and correlation with the deep-sea $\delta^{18}O$ record. Paleoceanography 17:1033. doi:1010.1029/2001PA000725

Etler DA (1996) The fossil evidence for human evolution in Asia. Annual Review of Anthropology 25:275–301

Fan MJ, Dettman DL, Song CH, Fang XM, Garzione CN. (2007) Climatic variation in the Linxia basin, NE Tibetan Plateau, from 13.1 to 4.3 Ma: The stable isotope record. Palaeogeography Palaeoclimatology Palaeoecology 247:313–28

Fang XM, Garzione C, Van der Voo R, Li JJ, Fan MJ (2003) Flexural subsidence by 29 Ma on the NE edge of Tibet from the magnetostratigraphy of Linxia Basin, China. Earth Planet Sci Lett 210(3–4):545–560.

Ferring R, Oms O, Agusti J, Berna F, Nioradze M, Shelia T, Tappen M, Vekua A, Zhvania D, Lordkipanidze D (2011) Earliest human occupations at Dmanisi (Georgian Caucasus) dated to 1.85–1.78 Ma. Proceedings of the National Academy of Sciences of the United States of America 108(26):10432–10436

Flynn LJ, Tedford RH, Qiu ZX (1991) Enrichment and stability in the pliocene mammalian fauna of north China. Paleobiology 17(3):246–265

Flynn LJ, Wu W, Downs WR (1997) Dating vertebrate microfaunas in the late Neogene record of Northern China. Palaeogeograp PalaeoclimatoL PalaeoecoL 133(3-4):227–242

Flynn LJ, Qiu ZX, Opdyke ND, Tedford RH (1995) Ages of key fossil assemblages in the late Neogene terrestrial record of northern China, in Geochronology, time scales, and global stratigraphic correlation, W. Berggren, D. Kent, M. P. Aubry and J. Hardenbol (eds.), pp. 365-373, Society for Economic Paleontology and Mineralogy Special Publication

Gabunia L, Vekua A, Lordkipanidze D, Swisher CC, Ferring R, Justus A, Nioradze M, Tvalchrelidze M, Anton SC, Bosinski G, Joris O, de Lumley MA, Majsuradze G, Mouskhelishvili A (2000) Earliest Pleistocene hominid cranial remains from Dmanisi, Republic of Georgia: Taxonomy, geological setting, and age. Science 288:1019–1025

Harris N (2006) The elevation history of the Tibetan Plateau and its implications for the Asian monsoon. Palaeogeograp Palaeoclimatol Palaeoecol 241(1):4–15

Hedges REM, Housley RA, Bronk CR, Van Klinken GJ (1992) Radiocarbon dates from the Oxford AMS system: Archaeometry Datelist 14. Archaeometry 34:141–159

Hou YM, Potts R, Yuan BY, Guo ZT, Deino A, Wang W, Clark J, Xie GM, Huang WW (2000) Mid-Pleistocene Acheulean-like stone technology of the Bose basin, South China. Science 287:1622–1626

Howell FC, Petter G (1985) Comparative observations on some Middle and Upper Miocene Hyaenids. Genera: *Percrocuta* Kretzoi, *Allohyaena* Kretzoi, *Adcrocuta* Kretzoi (Mammalia, Carnivora, Hyaenidae). Geobios-Lyon 18(4):419–476

Huang PH (1995) Evolutional process and depositional cycles of Peking man cave in relation to the living environments of Peking Man. Acta Anthropologica Sinica 14:101–109 (in Chinese)

Huang WP, Tang YJ (1974) Observation on the later Cenozoic of Nihewan Basin. Vertebrata Palasiatica 12:99–110 (in Chinese)

Huang WW, Dong GR, Hou YM (2004) Stratigraphic, chronological and ecological contexts of Pleistocene *Homo sapiens* of Sjara-Osso-Gol site, Ordos Plateau of North China. Acta Anthropologica Sinica 23:258–271 (in Chinese)

Ji HX (1975) The lower Pleistoecne mammalian fossils of Lantian district, Shaanxi. Vertebrata Palasiatica 13:169–177 (in Chinese)

Jia L, Wei Q, Li C (1979) Report on the excavation of Hsuchiayao Man Site in 1976. Vertebrata Palasiatica 17:277–293 (in Chinese)

Jia LP, Wei Q (1976) Xujiayao Palaeolithic Site, in Yanggao County, Shanxi Province. Archaeol Bull 2:97–114 (in Chinese)

Løvlie R, Su P, Fan XZ, Zhao ZJ, Liu C (2001) A revised paleomagnetic age of the Nihewan Group at the Xujiayao Palaeolithic Site, China. Quatern Sci Rev 20:1341–1353

Larick R, Ciochon RL, Zaim Y, Suminto S, Rizal Y, Aziz F, Reagan M, Heizler M (2001) Early Pleistocene 40Ar/39Ar ages for Bapang Formation hominins, Central Jawa, Indonesia. Proceedings of the National Academy of Sciences of the United States of America 98(9), 4866–4871

Li CK (1978) Two new lagomorphs from the Miocene of Lantian, Shensi. Profess Paper Stratigraph Paleontol 7:143–148 (in Chinese)

Li Q, Zheng SH, Cai BQ (2008) Pliocene biostratigraphic sequence in the Nihewan Basin, Hebei, China. Vertebrata Palasiatica 46:210–232 (in Chinese)

Li Y (1984) The early Pleistocene mammalian fossils of Danangou, Yuxian, Hebei. Vertebrata Palasiatica 22:60–68 (in Chinese)

Li YH, Han DF (1959) Beast bones in the Neolithic relic of Banpo, Xi'an, Shaanxi. Paleovertebrata et Paleoanthropologia 1959:166–172

Li YX, Ji HX (1981) Environmental change in Peking Mans Time. Vertebrata Palasiatica 19:337–347 (in Chinese)

Li YX, Xue XX (2008) The microtine distribution and its environmental significance since the Middle Pleistocene in Zhangping Caves of the Qinling Mountains. Science in China 51(2):294–299

Li YX, Zhang YX, Xue XX (2012) The composition of three mammal faunas and environmental evolution in the last glacial maximum, Guanzhong area, Shaanxi Province, China. Quaternary International 248:86-91

Lisiecki LE, Raymo ME (2005) A Pliocene-Pleistocene stack of 57 globally distributed benthic $\delta^{18}O$ records. Paleoceanography 20, PA1003, doi:1010.1029/2004PA001071
Liu CR, Yin GM, Gao L, Bahain JJ, Li JP, et al (2010a) ESR dating of Pleistocene archaeological localities of the Nihewan Basin, North China - Preliminary results. Quaternary Geochronology 5:385–390
Liu P, Deng CL, Li SH, Zhu RX (2010c) Magnetostratigraphic dating of the Huojiadi Paleolithic site in the Nihewan Basin, North China. Palaeogeography Palaeoclimatology Palaeoecology 298:399–408
Liu P, Deng CL, Li SF, Cai SH, Cheng HJ, Yuan BY, Wei Q, Zhu RX (2012) Magnetostratigraphic dating of the Xiashagou Fauna and implication for sequencing the mammalian faunas in the Nihewan Basin, North China. Palaeogeogr Palaeoclimatol Palaeoecol 315–316:75–85
Liu TS, Li CK, Zhai RJ (1978) Pliocene vertebrates of Lantian, Shensi. Profess Paper Stratigraphical Paleontology 7:149–200 (in Chinese)
Liu W, Jin C-Z, Zhang Y-Q, Cai Y-J, Xing S, et al. (2010b) Human remains from Zhirendong, South China, and modern human emergence in East Asia. Proceedings of the National Academy of Sciences 107:19201–6
Nie JS, King JW, Fang XM (2008) Link between benthic oxygen isotopes and magnetic susceptibility in the red-clay sequence on the Chinese Loess Plateau. Geophysical Research Letters 35(3), L03703, doi:03710.01029/02007GL032817
Potts R (1996) Evolution and climate variability. Science 273:922–923
Potts R (1998) Environmental hypotheses of hominin evolution. Yearbook of Physical Anthropology 41:93–136
Qi GQ, Dong W (2006) *Lufengpithecus hudienensis* Site, Science Press, Beijing (in Chinese)
Qian F, Zhang JX, Yin WD (1980) Magnetostratigraphic study on the cave deposits containing fossil Peking Man at Zhoukoudian. Chinese Science Bulletin 25:359
Qiu ZD, Li CK (2005) Evolution of Chinese mammalian faunal regions and elevation of the Qinghai-Xizang (Tibet) Plateau. Science in China 48(8):1246–1258
Qiu ZD, Li CK, Wang SJ (1981) Miocene mammalian fossils from Xining Basin, Qinghai. Vertebrata Palasiatica 19:156–173 (in Chinese)
Qiu ZX (1973) A new genus of giant rhinoceros from Oligocene of Dzungaria, Sinkiang. Vertebrata Palasiatica 11:182–191 (in Chinese)
Qiu ZX (2000) Nihewan fauna and Q/N boundary in China. Quat Sci 20:142–154 (in Chinese)
Qiu ZX, Guan J (1986) A lower molar of Pliopithecus from Tongxin, Ningxia Hui Autonomous Region. Acta Anthropologica Sinica 5:201–207 (in Chinese)
Qiu ZX, Qiu ZD (1990) Neogene local mammalian faunas: succession and ages. J Stratigrap 14:241–260
Qiu ZX, Qiu ZD (1995) Chronological sequence and subdivision of Chinese Neogene mammalian faunas. Palaeogeograp Palaeoclimatol Palaeoecol 116(1–2):41–70
Qiu ZX, Xie JY (1997) A new species of Aprotodon (Perissodactyla, Rhinocerot idae) from Lanzhou Basin, Gansu, China. Vertebrata Palasiatica 35:250–267 (in Chinese)
Qiu ZX, Wang BY (1999) Allacerops (Rhinocerotoidea, Perissodactyla), its discovery in China and its systematic position. Vertebrata Palasiatica 37:48–61 (in Chinese)
Qiu ZX, Xie JY, Yan DF (1988) A New Chilothere skull from Hezheng, Gansu, China, with special reference to the Chinese "Diceratherium". Science in China 31(4):493–504
Qiu ZX, Wang BY, Xie JY (1998) Mid-Tertiary chalicothere (Perissodactyla) fossils from Lanzhou, Gansu, China. Vertebrata Palasiatica 36:297–318 (in Chinese)
Qiu ZX, Deng T, Wang BY (2004) Early Pleistocene Mammalian Fauna from Longdan, Dongxiang, Gansu, China, Science Press, Beijing (in Chinese)
Raymo ME (1994) The initiation of Northern Hemisphere glaciation. Annu Rev Earth Planet Sci 22:353–383
Shang H, Tong HW, Zhang SQ, Chen FY, Trinkaus E (2007) An early modern human from tianyuan cave, zhoukoudian, china. Proceedings of the National Academy of Sciences of the United States of America 104:6573–8

Shen GJ, Fang YS, Bischoff JL, Feng YX, Zhao JX (2010) Mass spectrometric U-series dating of the Chaoxian hominin site at Yinshan, eastern China. Quaternary International 211:24–8

Swisher CC, Curtis GH, Jacob T, Getty AG, Suprijo A (1994) Age of the earliest known hominids in Java, Indonesia. Science 263(5150):1118–1121

Tang YJ (1980) Note on a small collection of early Pleistocene mammalian fossils from northern Hebei. Vertebrata Palasiatica 18:314–323 (in Chinese)

Tang YJ, Ji HX (1983) A Pliocene-Pleistocene transitional fauna from Yuxian, northern Hebei. Vertebrata Palasiatica 21:245–254 (in Chinese)

Tang YJ, Li Y, Chen WY (1995) Mammalian fossils and the age of Xiaochangliang paleolithic site of Yangyuan, Hebei. Vertebrata Palasiatica 33:74–83 (in Chinese)

Teford RH, Flynn LL, Qiu ZX, Opdyke ND, Downs WR (1991) Yushe Basin, China: Paleomagnetically calibrated mammalian biostratigraphic standard for the Late Neogene of Easterm Asia. J Vertebrate Paleontol 11:519–526

Teilhard de CP, Licent E (1924) On the discovery of a palaeolithic in north China. Bull Geol Soci China 3:45–50

Teilhard de CP, Piveteau J (1930) Les mammiferes fossils de Nihowan (Chine). Annales de Paleontologie 19, 1–154

Tong HW, Li H, Xie JY (2008) Revisions of some taxa of the Salawusu fauna from Sjara-Osso-Gol area, Neimongol, China. Quat Sci 28:1106–1113

Wang AD (1982) Discovery of the Pliocene mammalian faunas from the Nihewan region and its significance. Chin Sci Bull 27:227–229

Wang BY, Qiu ZX (2003) Aepyosciurinae: a new subfamily of Sciuridae (Rodentia, Mammalia) from basal loess deposits at the northeastern border of Tibetan Plateau. Chin Sci Bull 48(7):691–695

Wang HQ, Deng CL, Zhu RX, Wei Q, Hou YM, Boeda E (2005) Magnetostratigraphic dating of the Donggutuo and Maliang Paleolithic sites in the Nihewan Basin, North China. Quatern Res 64(1):1–11

Wang HQ, Deng CL, Zhu RX, Xie F (2006) Paleomagnetic dating of the Cenjiawan Paleolithic site in the Nihewan Basin, northern China. Science in China 49:295–303

Wang PX, Tian J, Cheng XR, Liu CL, Xu J (2004) Major Pleistocene stages in a carbon perspective: The South China Sea record and its global comparison. Paleoceanography 19, PA4005, doi:4010.1029/2003PA000991

Wang YY, Xue XX, He RC (1985) Division of the Quaternary strata in the Loess district of north Shaanxi and east Kansu. Acta Geol Pol 46:102–117 (in Chinese)

Wei, Q (1991) Geologic sequence of the archaeological sites in the Nihewan Basin, North China. In: IVPP (Institute of Vertebrate Paleontology and Paleoanthropology) (Ed.), Contributions to the XIII INQUA. Beijing Scientific and Technological Publishing House, Beijing, pp. 61–73 (in Chinese)

Wei Q (1994) Banshan Paleolithic site from the Lower Pleistocene in the Nihewan Basin in northern China. Acta Anthropologica Sinica 13:223–238 (in Chinese)

Wei Q, Meng H, Cheng SQ (1985) New Paleolithic site from the Nihewan beds. Acta Anthropologica Sinica 4:223–232 (in Chinese)

Wu RK, Ren ME, Zhu XM, Yang ZG, Hu CK, Kong SH, Xie YY, Zhao SS (1985) Multi-disciplinary Study of the Peking Man Site at Zhoukoudian, Science Press, Beijing (in Chinese)

Wu WY, Ye J, Zhu BC (1991) On Alloptox (Lagomorpha, Ochotonidae) from the Middle Miocene of Tongxin, Ningxia Hui Autonomous Region, China. Vertebrata Palasiatica 29:204–229 (in Chinese)

Wu WY, Meng J, Ye J (2003) The discovery of Pliopithecus from northern Junggar Basin, Xinjiang. Vertebrata Palasiatica 41:76–86 (in Chinese)

Xiao J, Jin CZ, Zhu YZ (2002) Age of the fossil Dali Man in north-central China deduced from chronostratigraphy of the loess-paleosol sequence. Quatern Sci Rev 21:2191–2198

Xie F (2006) Nihewan, Cultural Relics Publishing House, pp. 330, Beijing, China (in Chinese)

Xie F, Li J, Liu LQ (2006) Paleolithic archeology in the Nihewan Basin, pp. 278, Huashan Literature & Arts Presss, Shijiazhuang, China (in Chinese)

Xie JY, Xu JC (1988) Some mammalian fossils from Malan Loess, Huachi, Gansu. Vertebrata Palasiatica 26:149–152 (in Chinese)

Xie JY, Gao SY, Dong GR, Li BS (1995) Zoocoenosium in Sara Wusu. J Desert Resear 15:313–322

Xu QQ (1999) Mammals, geological age and nautral environment, in *Homo erectus* from Nanjing. R. K. Wu, X. X. Li, X. Z. Wu and X. N. Mu (eds.), pp. 280–285, Jiangsu Science and Technology Publishing House, Nanjing (in Chinese)

Xu QQ, Ou YL (1982) Climatic changes during Peking Mans Time. Acta Anthropologca Sinica 1:80–90 (in Chinese)

Xue XX, Zhang YX, Yue LP (1995) Discovery and chronological division of the Hipparion Fauna in Laogaochuan village, Fugu County, Shaanxi. Chin Sci Bull 40(11):926–929

Xue XX, Zhang YX, Yue LP (2006) Paleoenvironments indicated by the fossil mammalian assemblages from red clay-loess sequence in the Chinese Loess Plateau since 8.0 Ma BP. Science in China 49(5):518–530

Xue XX, Zhou WJ, Zhou J, Head J, Jull AJT (2000) Biological records of paleoclimate and paleoenvironment changes from Guanzhong area, Shaanxi Province during the last glacial maximum. Chin Sci Bull 45(9):853–857

Yan GL (1993) A prelimineary study on magnetic stratigraphy of the geological section with the fossil bed of Yunxian Homo of Hubei. Journal of China University of Geosciences 18:221–6 (in Chinese)

You YZ, Xu QQ (1981) The late Pleistocene mammalian faunas of northern China and correlation with deep-sea sediments. Vertebrata Palasiatica 19:77–86

You YZ, Tang YJ, Li Y (1980) Discovery of the Palaeoliths from the Nihewan Formation. Quaternaria Sinica 5:1–11 (in Chinese)

Young CC (1932) On a new ochot onid from north Suiyuan. Bull Geol Soci China 11:255–258

Yuan BY (1978) Sedimentary environment and stratigraphical subdivision of Sjara-Osso-Goll Formation. Chinese J Geology 3:220–234 (in Chinese)

Yuan BY, Zhu RX, Tian WL, Cui JX, Li RQ, Wang Q, Yan FH (1996) The age, subdivision and correlation of Nihean Group. Science in China 26:67–76

Yuan FL, Do HJ (1984) Cenozoic Biostratigraphy in China, Geological Publishing House, Beijing (in Chinese)

Yue LP, Xue XX (1996) Paleaeomagnetism of Chinese Loess, Geological Publishing House, Beijing (in Chinese)

Zachos J, Pagani M, Sloan L, Thomas E, Billups K (2001) Trends, rhythms, and aberrations in global climate 65 Ma to present. Science 292(5517):686–693

Zhang RZ (1997) Distribution of mammalian species in China, Chinese Forestry Publishing House, Beijing (in Chinese)

Zhang YX, Xue XX, Yue LP (1995) Age and division of Neogene “red bed” of Laogaochuan, Fugu County, Shaanxi. J Stratigraphy 19:214–219

Zhang YX, Yue LP, Cao HX (2002) The ecological sequence of Hipparion fauna in the middle reaches of the Yellow River during Neogene. Chin Sci Bull 47(1):79–82

Zhang YX, Sun DH, An ZS, Xue XX (1999) Mammalian fossils from Late Pliocene (Lower MN 16) of Lingtai, Gansu Province. Vertebrata Palasiatica 37:190–199 (in Chinese)

Zhang ZQ, Zheng SH, Liu JB (2003) Pliocene micromammalian biostratigraphy of Nihewan Basin, with comments on the stratigraphic division. Vertebrata Palasiatica 41:306–313 (in Chinese)

Zhao JX, Hu K, Collerson KD, Xu HK (2001) Thermal ionization mass spectrometry U-series dating of a hominid site near Nanjing, China. Geology 29(1):27–30

Zheng SH (1982) Some Pliocene mammalian fossils from Songshan-2 and -3 (Tianzhu, Gansu) and the Songshan Fauna. Vertebrata Palasiatica 20:216–227 (in Chinese)

Zheng SH (1983) The Middle Pleistocene micromammalian fauna from Hexian man locality and its significance. Chin Sci Bull 28:237–239

Zheng SH, 1997. Evolution of the Mesosiphneinae (Siphneidae, Rodentia) and environmental change. In: Tong Y. S., Zhang YY, Wu WY et al. (eds.) Evidence for evolution-Essays in honor of Prof. Chungchien Young on the hundredth anniversary of his birth. Bejing: China Ocean Press. 137–150 (in Chinese)

Zheng SH, Zhang ZQ (2000) Late Miocene-early Pleistocene micromammals from Wenwanggou of Lingtai, Gansu, China. Vertebrata Palasiatica 38:58–71 (in Chinese)

Zheng SH, Zhang ZQ (2001) Late Miocene-early Pleistocene biostratigraphy of the Leijiahe area, Lingtai, Gansu. Vertebrata Palasiatica 39:215–228 (in Chinese)

Zhou MZ, Li CK (1965) Mammalian fossils in association with the mandible of Lantian Man at Chen-Chia-Wo, in Lantian, Shaanxi. Vertebrata Palasiatica 9:377–393 (in Chinese)

Zhou TR, Li HZ, Liu QS, Li RQ, Sun XP (1991) Cenozoic Paleogeography of the Nihewan Basin, pp. 1–162, China Science Press, Beijing (in Chinese)

Zhu RX, Deng CL, Pan YX (2007) Magnetochronology of the fluvio-lacustrine sequences in the Nihewan basin and its implications for early human colonization of Northeast Asia. Quat Sci 27:922–944 (in Chinese)

Zhu RX, An ZS, Potts R, Hoffman KA (2003) Magnetostratigraphic dating of early humans in China. Earth-Science Reviews 61:341–359

Zhu RX, Potts R, Pan YX, Yao HT, Lu LQ, Zhao X, Gao X, Chen LW, Gao F, Deng CL (2008) Early evidence of the genus *Homo* in East Asia. J Hum Evol 55(6):1075–1085

Zhu RX, Hoffman KA, Potts R, Deng CL, Pan YX, Guo B, Shi CD, Guo ZT, Yuan BY, Hou YM, Huang WW (2001) Earliest presence of humans in northeast Asia. Nature 413:413–417

Zhu RX, Potts R, Xie F, Hoffman KA, Deng CL, Shi CD, Pan YX, Wang HQ, Shi RP, Wang YC, Shi GH, Wu NQ (2004) New evidence on the earliest human presence at high northern latitudes in northeast Asia. Nature 431:559–562

Chapter 6
Late Cenozoic Climate Change in Monsoon-Arid Asia and Global Changes

Zhisheng An, Youbin Sun, Hong Chang, Peizhen Zhang, Xiaodong Liu, Yanjun Cai, Zhangdong Jin, Xiaoke Qiang, Weijian Zhou, Li Li, Zhengguo Shi, Liangcheng Tan, Xiaoqiang Li, Xinbao Zhang, and Zhao Jin

Abstract Based on the geological and biological findings, and climate model simulations, current understanding of the Asian monsoon-arid environment has been synthesized here. This chapter addresses the evolution of the Asian monsoon-arid environment and how it is related to the growth of the Tibetan Plateau (TP), and global environmental change since the Cenozoic. Asian monsoon appearance may begin in the Eocene. The basic structure of the monsoon-arid environment was established by late Oligocene, and the modern pattern of the monsoon-arid environment has developed since the late Pliocene. Conceptual models for Asian monsoon-arid environmental change mechanisms at various time scales are proposed here. The occurrence of great Northern hemisphere glaciation has been discussed. Differentiation of the natural background and anthropogenic signals is examined. Policies and countermeasures for sustainable development on the Loess Plateau and arid areas are proposed for future reference.

Keywords Cenozoic · Monsoon-arid environment evolution · Tibetan plateau · Monsoon initiation · Numerical Simulations · Conceptual models · Global Changes · Tectonic-orbital-Millennial Monsoon dynamics · Sustainable development

---

Z. An (✉) · Y. Sun · H. Chang · X. Liu · Y. Cai · Z. Jin · X. Qiang · W. Zhou · L. Li · Z. Shi · L. Tan · X. Li · Z. Jin
State Key Laboratory of Loess and Quaternary Geology,
Institute of Earth Environment, Chinese Academy of Sciences,
No. 10 Fenghui S. Rd, High-Tech Zone,
Xi'an PO Box 17, Xi'an, China
e-mail: anzs@loess.llqg.ac.cn

Y. Sun
e-mail: sunyb@ieecas.cn

H. Chang
e-mail: changh@loess.llqg.ac.cn

P. Zhang
Institute of Geology, China Seismological Bureau, 100029 Beijing, China
e-mail: peizhen@ies.ac.cn

X. Liu
e-mail: liuxd@loess.llqg.ac.cn

Z. An (ed.), *Late Cenozoic Climate Change in Asia*,
Developments in Paleoenvironmental Research 16,
DOI 10.1007/978-94-007-7817-7_6,

Y. Cai
e-mail: caiyj@loess.llqg.ac.cn

Z. Jin
e-mail: zhdjin@ieecas.cn

X. Qiang
e-mail: qiangxk@loess.llqg.ac.cn

W. Zhou
e-mail: weijian@loess.llqg.ac.cn

L. Li
e-mail: lili@ieecas.cn

Z. Shi
e-mail: shizg@ieecas.cn

L. Tan
e-mail: tanlch@ieecas.cn

X. Li
e-mail: lixiaoqiang@ivpp.ac.cn

X. Zhang
Institute of Mountain Hazards and Environment, Chinese Academy of Sciences, No.9, Block 4, Renminnan Road, 610041 Chengdu, China
e-mail: zxbao@imde.ac.cn

Z. Jin
e-mail: jinzhao@ieecas.cn

6.1 Introduction

In the previous chapters, the current understanding of the Asian monsoon-arid environment has been presented, and the relationship between the Asian monsoon and loess, and a variety of bio-geological archives has been discussed on orbital and millennial time scales. The progressive drying of the Asian interior, and its regional and global climatic significance, have been described in terms of dust and aerosol distributions through time. Mammalian evolution and monsoon evolution have been reviewed, including the record of human evolution as it relates to climate change. The results presented indicate that climate change in the eastern humid zone was controlled by Asian monsoon circulation changes across the Asian monsoon-arid environmental gradient, while climate changes in the western arid zone were controlled by the Northern Hemisphere westerlies, and East Asian winter monsoon circulation. Based on the sum of these geological and biological findings, and climate model simulations, this chapter addresses the evolution of the Asian monsoon-arid environment and how it is relates to the growth of the Tibetan Plateau (TP), and global environmental change since the Cenozoic. Asian monsoon-arid environment differentiation began in the Eocene. The basic structure of the monsoon-arid environment was established by the late or Oligocene, and the modern pattern of the monsoon-arid environment has developed since the late Pliocene. Conceptual models for Asian monsoon-arid environmental change mechanisms at various time scales are proposed here. Differentiation of the natural background and human signals are examined. Policies and countermeasures for sustainable development on the Loess Plateau and arid areas are proposed for future reference.

6.2 Tectonics and Climate—The Making of the Monsoon

Tectonics is ultimately responsible for the Asian monsoon as we know it today, with the massive TP playing a central role in monsoon dynamics. Equally important to its size and height is the position of the plateau, attached to the world's largest landmass (Eurasian continent), and juxtaposed to the world's largest ocean basin (the Indo-Pacific Ocean). It is this combination of an extensive land-sea boundary, backed by dramatic topographic relief, which makes the Asian monsoon possible (Raymo and Ruddiman 1992; Kutzbach et al. 1993; Ramstein et al. 1997; An et al. 1999, 2001, 2006b; Wang et al. 2006b; Zhang et al. 2007).

The TP affects both global and regional climate more profoundly than other continental topographic features. The continuous uplift of the massive TP has enhanced continental weathering on a global scale, a process that affects ocean chemistry, atmospheric CO_2, and global climate (Walker et al. 1981; Berner et al. 1983; Garzione 2008). The TP uplift is reflected in seawater Sr records during the past 40 Ma, as they are sensitive to changes in continental weathering. This and other evidence led researchers to suggest that the uplift and erosion of the Himalayan-Tibetan orogen caused a reduction of atmospheric CO_2 and global cooling, leading to the glacial climate that persists today (e.g. Raymo and Ruddiman 1992; Edmond 1992; Zachos and Kump 2005; Dupont-Nivet et al. 2008; Garzione 2008).

During the same period in Asia, uplift of the TP and maybe the retreat of the Paratethys Ocean, an epicontinental sea formerly extending over Eurasia, triggered dramatic aridification and cooling of continental Asia and the onset of the Asian monsoon (Kutzbach et al. 1989; Ruddiman and Kutzbach 1989; Raymo and Ruddiman 1992; An et al. 2001; Guo et al. 2002; Liu et al. 2003a; Harris 2006; Zhang et al. 2007; Dupont-Nivet et al. 2008; Molnar et al. 2010). The plateau causes a standing wave to form in the jet stream (e.g. Held 1983; Manabe and Terpstra 1974), resulting in an essential ingredient of Indian monsoon dynamics (Manabe and Terpstra 1974; Held 1983). On a regional scale, the TP also causes the Siberian high pressure center to move northward, where a high-pressure zone forms in winter over most of the Northern Hemisphere (e.g. Trenberth and Chen 1988). The TP also influences the East Asian summer and winter monsoons aridity of inland Asia (An et al. 2001). Although the history of the TP's upward and outward growth through the Cenozoic remains a topic of debate and active research, we now understand enough to trace its development from a tectonic viewpoint, and infer its effects on the Asian monsoon.

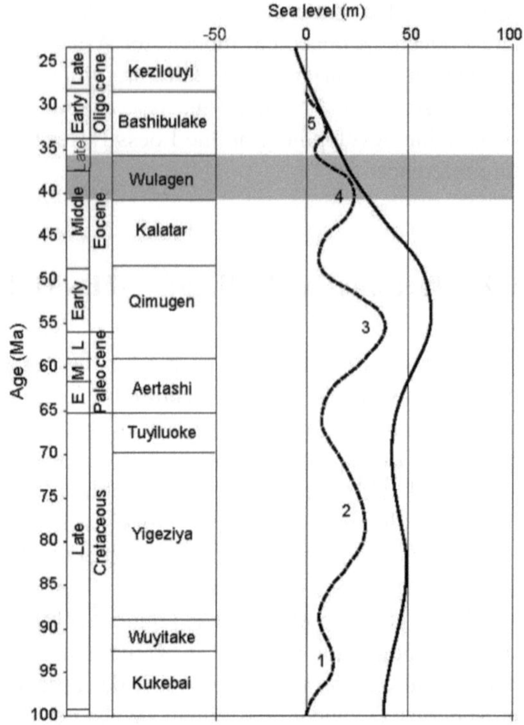

Fig. 6.1 Schematic regional framework of the five marine incursions in the Tarim Basin based on a review of historical literature (Bosboom et al. 2011). The *dotted line* indicates the estimated relative sea level change. The *bold line* shows the long-term sea level curve corresponding to the upper scale, relative to modern sea level

6.2.1 Retreat of the Paratethys Sea

The most important factor that shaped today's land-sea configuration was the retreat of the Paratethys Sea from the Tarim Basin, Ferghana Basin and Afghan-Tadjik Basin. A total of five major Late Cretaceous to Paleogene marine transgressions and subsequent retreat events have been recognized in the sedimentary records of the westernmost Tarim Basin (Fig. 6.1). These correspond to two Late Cretaceous groups: the Kukebai-Wuyitake Formation and the Yigeziya-Tuyiluoke Formation; and three Paleogene formations: the Qimugen Formation, Wulagen Formation, and Bashibuluke Formation (Fig. 6.1; He 1991; Tang et al. 1992; Lan and Wei 1995; Burtman 2000; Bosboom et al. 2011). With the last retreat of the Paratethys from the Tarim Basin in the Oligocene, the Ferghana-Tajik Sea also disappeared, despite the fact that it had persisted throughout the Paleocene and Eocene. This gave rise to the Mediterranean Sea. At this time, the East Asian monsoon had not yet developed, and a "planetary circulation system" controlled climate in the Asian inland (Liu et al. 1998; Sun and Wang 2005; Guo et al. 2008). As the moisture source area for a large region in Asia, the withdrawal of the Paratethys influenced climate change downwind (Dupont-Nivet et al. 2008; Bosboom et al. 2011). The land-sea distribu-

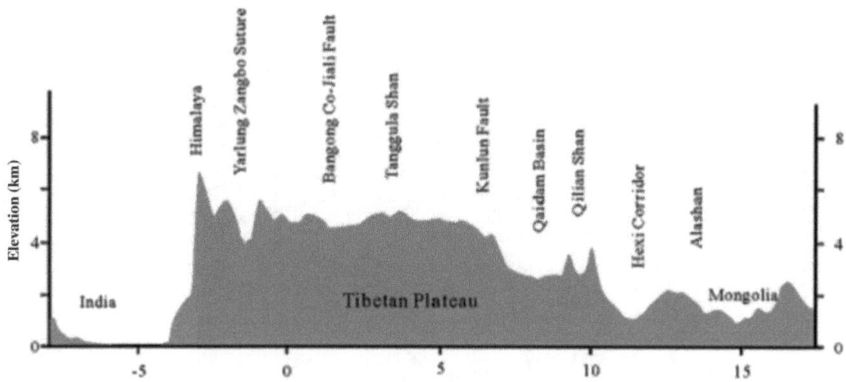

Fig. 6.2 Distance from Lhasa along a direction of N 21°E/degree

tion in Eurasia did not change immediately after the Mediterranean formed. By late Miocene, the closures of marine gateways with the Atlantic left the Mediterranean Sea isolated, at which time thick evaporates formed on the sea floor. The cumulative effect of this salt removal process lowered global ocean salinity by 6 % (Adams et al. 1977). At about the same time global ice volume increased, possibly through climatic feedbacks associated with the isolation of the Mediterranean (Hodell et al. 1986).

6.2.2 *Tectonic Evolution of the Tibetan Plateau*

The topographic profile of the TP is shown diagrammatically in Fig. 6.2. The formation of the TP is one of the most important tectonic events in Earth's history. In the past 30 years, many studies have attempted to understand the tectonic and geodynamic processes responsible for the evolution of the TP (Molnar and Tapponnier 1975; Besse et al. 1984; Patriat and Achache 1984; Dewey et al. 1988; Harrison et al. 1992; Molnar et al. 1993; Rowley 1996; Chung et al. 1998; Tapponnier et al. 2001; Zhang et al. 2004; Rowley and Currie 2006; Wang et al. 2008a; Molnar and Stock 2009). The TP can be divided into five primary structural components on the basis of sutures, fault belts and geological mapping. From south to north these include: the Himalayan terrane, Lhasa terrane, Qiangtang terrane, Songpan-Ganzi terrane, and Kunlun-Qaidam-Qilian terrane. These terranes are separated by the following sutures: the Yalong-Zangbo suture, Bangong-Nujiang suture, Jinshajiang suture and Animaqing-Kunlun-Muztagh suture, respectively (Fig. 6.3; Kapp et al. 2003). Most of the surface of the TP is covered by Cenozoic sediments, thrust over bedrock that only crops out along the Gandese Mountains, Tanggula Mountains, Kunlun Mountains and Altyn Tagh. These ranges are sub-parallel to one another.

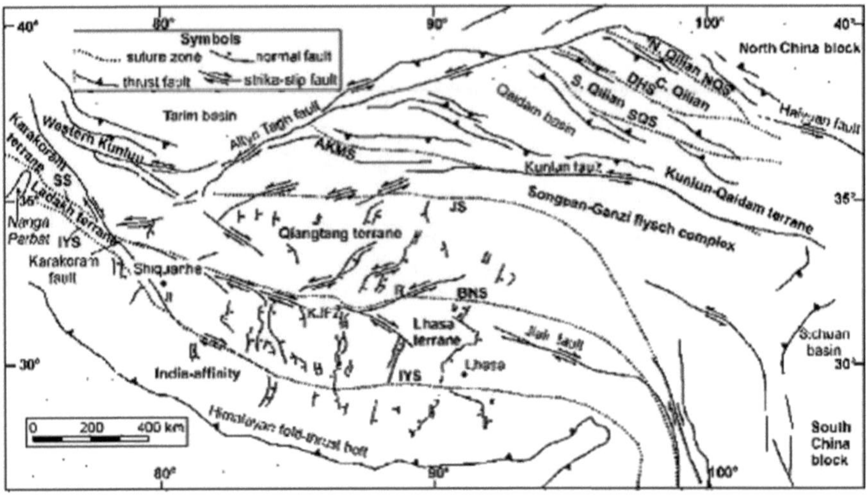

Fig. 6.3 Tectonic map of Tibet showing major terranes, suture zones and active faults. From north to south the sutures are: the *NQS* North Qilian, *DHS* Danghe Nan Shan, *SQS* South Qilian, *AKMS* Anymaqin-Kulun-Muztagh, *JS* Jinsha, *BNS* Bangong-Nuning, *IYS* Indus-Yalu. The Shyok suture (SS), west of the Karakoram fault, is also shown. The system of en echelon right-slip faults along the Bangong-Nujiang suture constitute the Karakorum-Jiali zone. (After Kapp et al. 2003)

6.2.3 Stepwise Growth of the Tibetan Plateau

The stepwise growth of the TP is summarized in this section. Important growth processes in the TP occurred during the periods 55–45 Ma BP, 42–34 Ma BP, 28–22 Ma BP, 15–7 Ma BP, and since 3.6 Ma, based on the evolution of the Paratethys Sea (He 1991; Tang et al. 1992; Lan and Wei 1995; Burtman 2000; Bosboom et al. 2011),

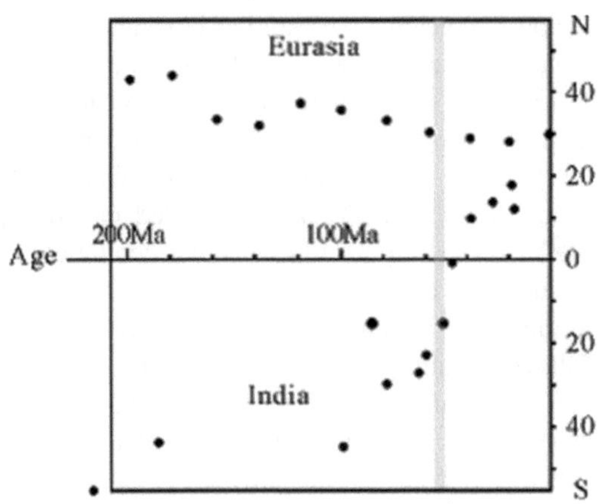

Fig. 6.4 Paleolatitude of a point now in Lhasa as it was bound to Eurasia (*upper points curve*) or India (*lower points curve*). (After Pozzi et al. 1984)

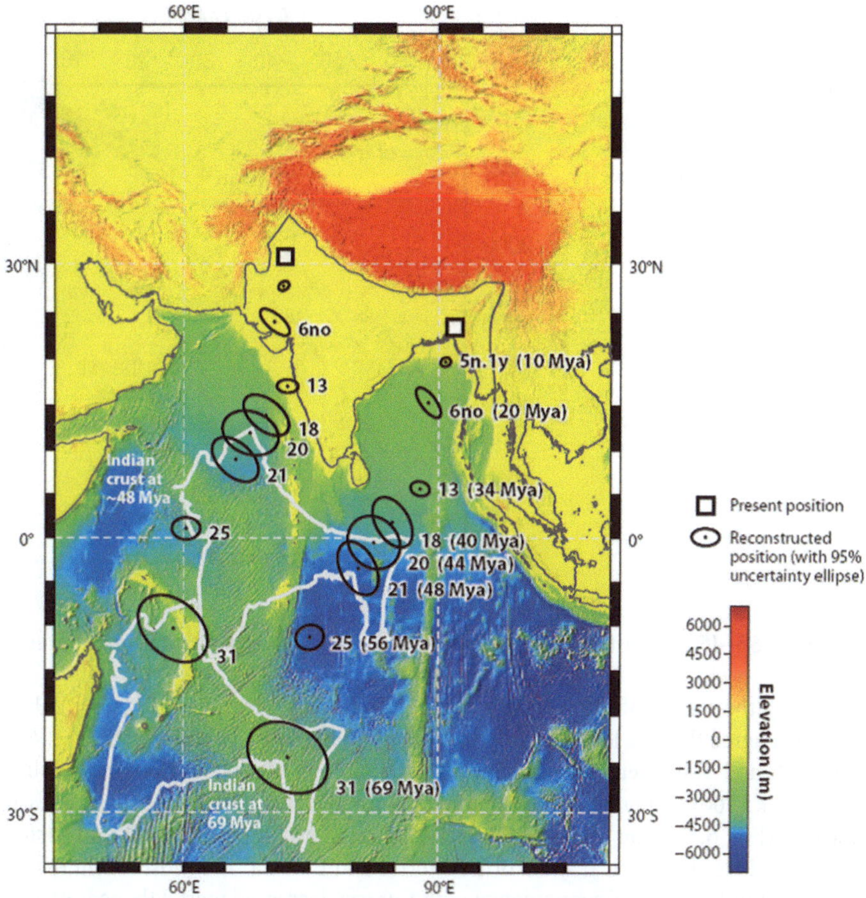

Fig. 6.5 Map showing present and reconstructed positions of two points on the India plate with respect to the Eurasia plate at different times in the past. Numbers next to points identify chrons and corresponding ages. An outline of the Indian coastline and the northern edge of present-day, intact Indian crust is also reconstructed to its positions at 48 Ma, close to the time of collision, and 69 Ma, before the collision. (Modified from Molnar and Stock 2009)

analysis of individual structural units (Dewey et al. 1988; Kapp et al. 2003), reconstructed geological motion by paleomagnetic methods (Pozzi et al. 1984; Besse et al. 1984; Patriat and Achache 1984; Molnar and Stock 2009), evolution of sedimentary environments (Tapponnier et al. 2001; Wang et al. 2008a), estimated paleoaltimetry (Rowley and Currie 2006), analysis of volcanic rocks (Turner et al. 1993; Chung et al. 1998; Roger et al. 2000; Mo et al. 2008), and rapid exhumation events (Jolivet et al. 1999; George et al. 2001; Wan et al. 2001; Wang et al. 2002b, 2003a; Yuan et al. 2006; Sobel et al. 2008; Chen et al. 2008; Zheng et al. 2008a).

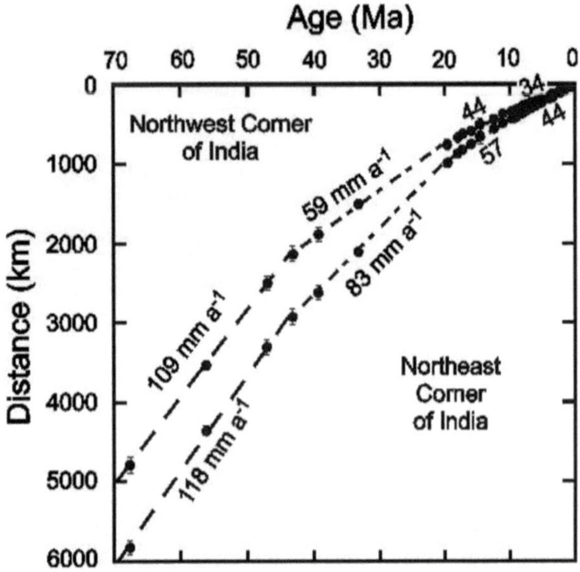

Fig. 6.6 Distances of points in the northwestern and northeastern corners of India at different times in the past, showing average convergence rates during different intervals and integral convergence curves

6.2.3.1 55–45 Ma Initial Collision Between the Indian and Eurasian Plates

Although the timing of the collision between India and Eurasia is still being debated, most agree that the last oceanic crust vanished between 55 and 45 Ma (Fig. 6.4; Garzanti and Vanhaver 1988; Zhu et al. 2005; Jain et al. 2009). Recently Molnar and Stock (2009) calculated the relative positions of India and Eurasia at times in the past (Fig. 6.5) using the known histories of separation between India and Africa, Africa and North America, and North America and Eurasia. The most recent results show that India's rate of convergence with Eurasia decreased markedly near 45 Ma, consistent with the buoyancy of Indian crust resisting continued subduction of the Indian lithosphere. The proposed decreases in convergence rate are 30–38 % (from 118 to 83 mm/yr in northeastern India and from 109 to 59 mm/yr in northwestern India) at 40–45 Ma (Fig. 6.6), and this decrease in convergence rate can be interpreted as an indicator of the onset of collision between the Indian and Eurasian continents.

Before India collided with Eurasia, widespread folding and thrust faulting in southern Tibet suggest that Eurasia's southern margin was elevated, in the manner of the present-day Central Andes (e.g. Burg and Chen 1984; England and Searle 1986; Murphy et al. 1997; Kapp et al. 2003; Volkmer et al. 2007). India's northward movement into Eurasia has been absorbed by the north-south shortening of both the Indian crust to build the Himalaya, and Eurasian crust to build Tibet and high terrain farther north. Thus, at 45–55 Ma, an Andean margin of high terrain, thousands of kilometers long and perhaps 200 to 400 km wide (Wang et al. 2008a), lay at 10°N–20°N (± 5°) along the southern margin of Eurasia (Fig. 6.13 Phase 0), when northern India made contact with that margin (Molnar et al 2010). Volcanic activity in the southern part of Tibet took place in the early Paleocene as marked by volcanic rocks

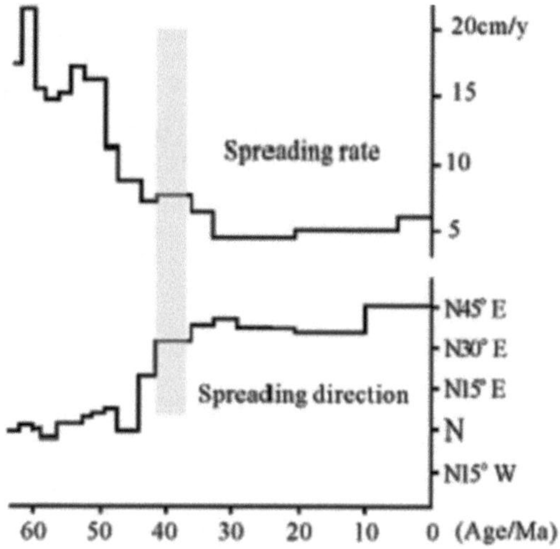

Fig. 6.7 Rate of spreading at the accreting boundary of the India plate during the Cenozoic on the eastern Indian ridge. (Patriat and Achache 1984)

of the Linzizong Formation. The transition from andesitic volcanics (64.4–60.6 Ma BP) to dacitic volcanics (~54 Ma BP; Mo et al. 2008) in the Linzizong formation might indicate the termination of Andes-type subduction of the Indian lithosphere beneath the Eurasian continent and the inception of continental collision.

Subsequent penetration of India into Eurasia caused wide-spread deformation immediately after the collision. Dates of deformation in northern Tibet suggest that deformation began there shortly after collision and continued to the present-day. Slip along the Altyn Tagh fault has occurred since at least the Oligocene (Ritts et al. 2004; Yue et al. 2005). Yin et al. (2002) related the onset of Cenozoic left-lateral strike-slip motion on the Altyn Tagh fault to the inception of Cenozoic sedimentation in the Qaidam basin about 49 Ma ago, and they further suggested that both slip on the Altyn Tagh fault and sedimentation in the Qaidam basin were tectonic responses at the northern margin of the TP to the collision in the plate boundary a few thousand kilometers away to the south. Sediment deposition in at least two basins (Hoh Xil and Qaidam) in north central Tibet does appear to progress northward, but began shortly after collision, at about 49 Ma (Yin et al. 2002; Zhu et al. 2006).

6.2.3.2 42–34 Ma Uplift of the Proto-Tibetan Plateau

The first pulse of tectonic deformation or plateau uplift appeared to occur during the time span after ~42 Ma and before the Eocene-Oligocene Transition, ~34 Ma (Fig. 6.7; Patriat and Achache 1984). Wang et al. (2008a) proposed that the Lhasa and southern Qiangtang terranes have been uplifted to near their modern elevation since 40 Ma and they named the uplifted portion of the plateau as “the proto-TP” based on magnetostratigraphy, sedimentology, paleocurrent measurements, and

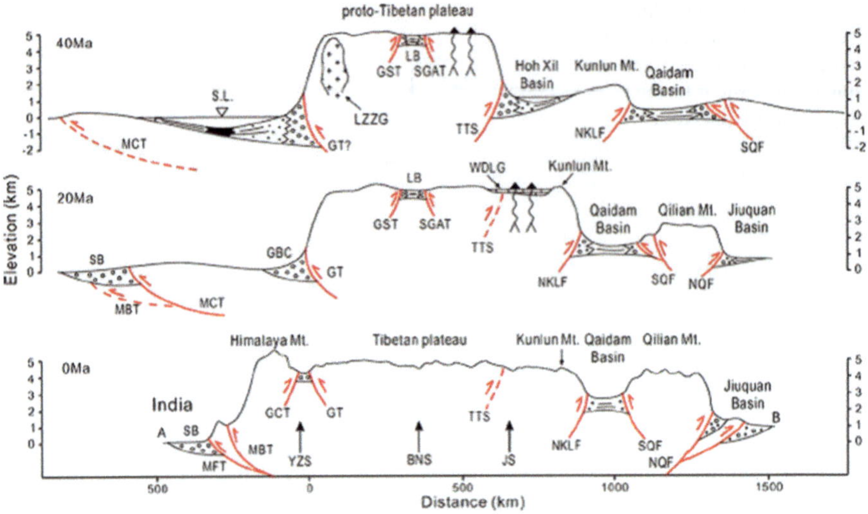

Fig. 6.8 Schematic paleogeographic cross sections of the Himalaya and Tibetan Plateau, showing our proposed surface-uplift history for the Tibetan Plateau, in which the plateau grows northward and southward from an elevated proto-Tibetan Plateau (Lhasa and Qiangtang terranes) beginning in the Late Paleogene. *BT* main boundary thrust, *GCT* great counter thrust, *LZZG* Lingzizhong group volcanics, *WDLG* continental Wudaoliang Group, *TTS* Tanggula thrust system, *NKLF* north Kunlun fault, *NQF* north Qilian Mountain fault, *GBC* Gangrinboche conglomerates, *YZS* Yarlung Tsangpo suture zone, *BNS* Bangong–Nujiang suture zone, *JS* Jinsha River suture zone, *MCT* main central thrust, *SL* sea level, *GT* Gangdese thrust, *GST* Gaize–Siling Tso thrust, *LB* Lunpola basin, *SGAT* Shiquanhe–Gaize–Amdo thrust, *SQF* the south Qilian Mountain fault, *SB* Siwalik foreland basin, *MFT* main frontal thrust. The Yaluzangbu River has been fixed as a reference point. (After Wang et al. 2008a)

$^{40}Ar/^{39}Ar$ and fission-track studies (Fig. 6.8). The Tanggula Shan probably formed the northern margin of the "proto-TP" and the Gangdese arc formed the southern boundary (Kapp et al. 2005, 2007; Spurlin et al. 2005).

This phase of uplifting (Fig. 6.13 Phase I) also registered its signature in sedimentary basins north of the "proto-TP". In the Paleogene records, the most striking first-order paleoenvironmental change is the combined regional appearance of conifers (and in particular the taxon *Picea*), along with a marked change in the geographic distribution of arid versus tropical and subtropical taxa. These regional features are widely recognized as diagnostic for Tibetan uplift with aridification north of the plateau and monsoonal intensification to the south and east (Harris 2006). In the Xining basin near the northern margin of the TP, Dupont-Nivet et al. (2008) found that the most conspicuous palynological event within the analyzed sequence is the sudden appearance of the Pinaceae family, and in particular that of taxon *Picea* (Piceapollenites). Their high resolution magnetostratigaphic dating indicates that this event is recorded in sections bracketed between 38.3 and 37.3 Ma. The 38 Ma conifer appearances in Xining basin is associated with coeval orographic uplift in the TP region.

Evidence for Paleogene tectonism and uplift exists for the central and northern part of the TP.

1. In the center of the plateau, estimates of the paleoaltimetry of late Eocene and younger deposits of the Lunpola and Nima basins (spanning the Bangong-Nujiang suture between the Lhasa and Qiangtang terranes; Fig. 6.3) indicate an elevation of > 4 km (Rowley and Currie 2006; DeCelles et al. 2007).
2. In the east-central regions of the TP, Paleocene– Eocene (older than 37 Ma) shortening is also documented in the Yushu-Nangqiang basins neighboring the Jinsha suture in the eastern part of the Qiangtang terrane (Horton et al. 2002; Spurlin et al. 2005).
3. In the Hoh Xil basin just north of the "proto-TP", sedimentation rates averaged ~150 m/Ma from ~55 to 40 Ma ago, and increased significantly to ~1,500 m/Ma around ~40 Ma ago, and then decreased to ~400 m/Ma from ~39 to 30 Ma ago (Wang et al. 2008a). The rapid acceleration in sedimentation rates around ~40 Ma ago coincides with the deposition of a coarsening-upward boulder-cobble conglomerate that has been interpreted to be syntectonic, in response to uplift or tectonic deformation in the "proto-TP" to the south.
4. In the Kunlun Mountains, immediately north of the Hoh Xil basin, Clark et al. (2010) collected apatite for (U-Th)/He dating along four elevation transects to study early deformation in the region. The general pattern of the age/depth curve for all transects plotted together shows an increase in the age/depth gradient at about 35 Ma, suggesting increased exhumation and erosion that has been inferred to represent initiation of reverse fault deformation in northern Tibet along the eastern Kunlun Range. This initiation is consistent with an independent study that shows exhumation of the central Kunlun Shan between 41 and 21 Ma from accelerated cooling, determined from feldspar $^{40}Ar/^{39}Ar$ multi-domain diffusion modeling (Mock et al. 1999).
5. The land-sea distribution has obviously changed during this period (Fig. 6.9 and 6.10). Although global long-term sea level changes controlled this regression, growth of the TP is one of the reasons for the absence of large-scale marine sediments in the Tarim basin. Climatic factors, especially the planetary wind system in the Eocene, controlled moisture sources, the disappearance of the Tarim sea and the enhanced aridity of the Asian interior (Dupont-Nivet et al. 2008; Abels et al. 2011; Bosboom et al. 2011).

6.2.3.3 28–22 Ma Expansion of the Tibetan Plateau

A major phase of tectonic deformation occurred on the TP during the time period between 28 and 22 Ma (Fig. 6.13 Phase II). The best example of this phase of deformation is manifested in the Hoh Xil Basin of the north-central TP which is a part of the high plateau today. The basin consists of 5,000 m of Cenozoic nonmarine strata deposited in a foreland basin system that developed in concert with the rise and erosion of adjacent high mountain belts, the "proto-TP". According to studies by Liu

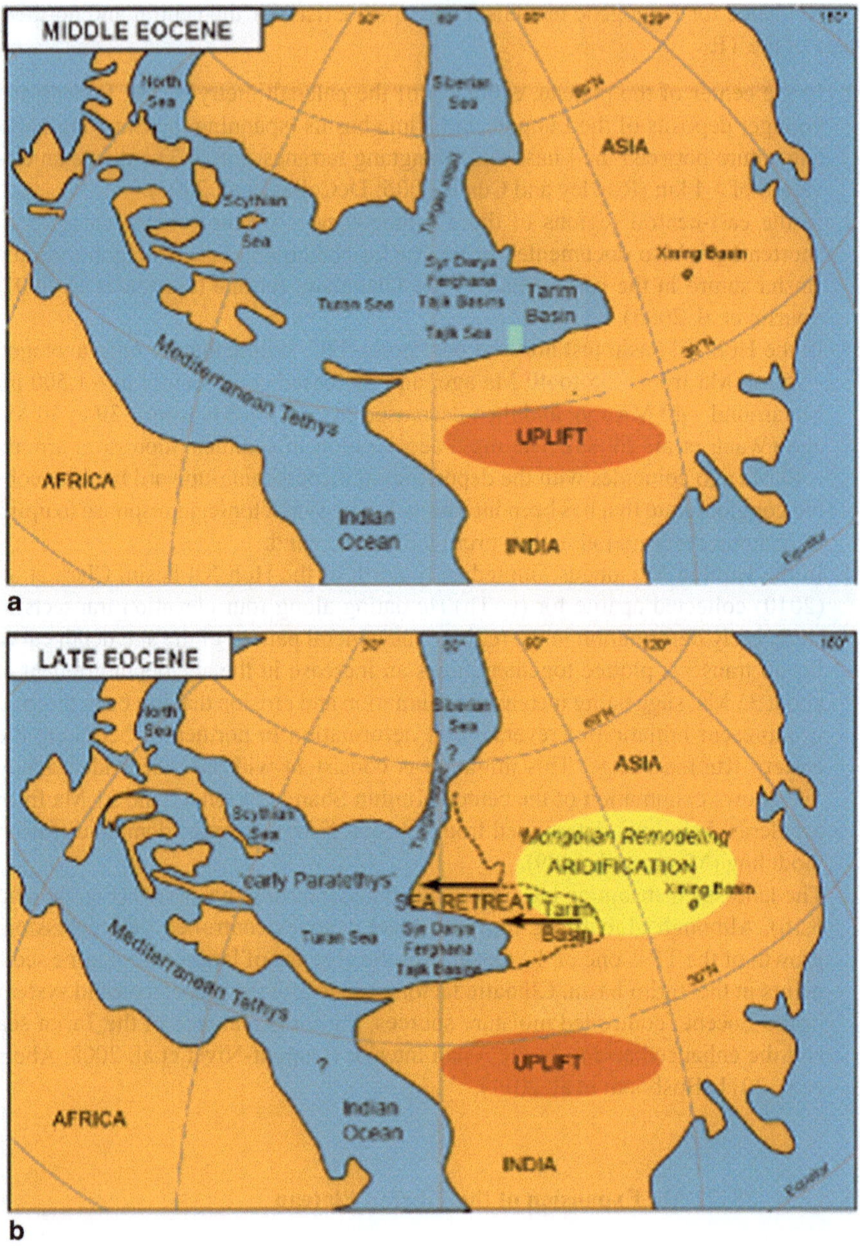

Fig. 6.9 Paleogeographic maps showing the extent of marine conditions in the Middle Eocene (**a**) and Late Eocene (**b**), before and after the last major retreat from the Tarim Basin. (After Bosboom et al. 2011)

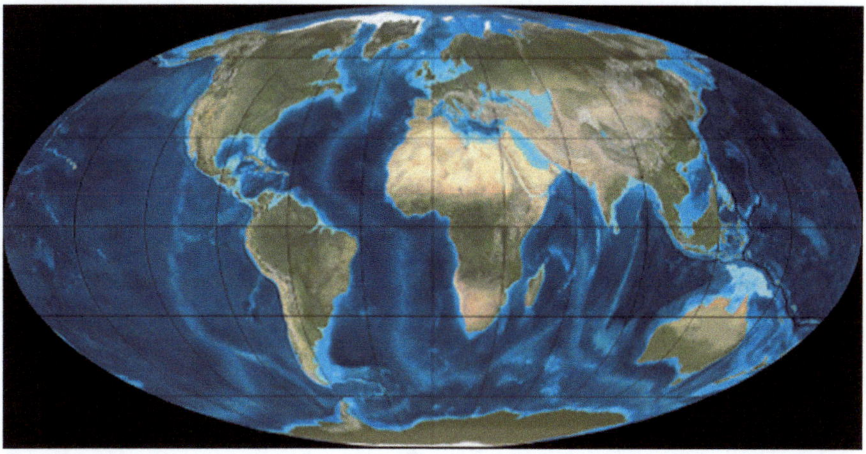

Fig. 6.10 Global land-sea configuration in the Early Miocene. (From http://jan.ucc.nau.edu/~rcb7/mollglobe.html)

and Wang (2001); Wang et al. (2002a) and Liu et al. (2003c), the sediments in the Hoh Xil basin can be divided into three lithostratigraphic units: 1) the Fenghuoshan formation, 2) Yaxicuo formation and 3) Wudaoliang formation. The Fenghuoshan formation consists of a cobble-pebble conglomerate, red sandstone, and bioclastic limestone of fluvial, fan-delta, and lacustrine origin. Magnetostratigraphic dating indicates its age ranges from 52 to 31 Ma. The overlying Yaxicuo formation is distinguished by sandstone, mudstone, marl, and gypsum deposited in fluvial and playa environments with magnetostratigraphic ages from ~31 to ~24 Ma. The Wudaoliang formation lies unconformablly on top of the Fenghuoshan and Yaxicuo formations, and consists of lacustrine marl and minor amounts of black oil shale, whose basal age is measured to be ~22 Ma (Liu et al. 2003c).

A pronounced angular unconformity separates the Yaxicuo formation from the overlying Wudaoliang Group (Liu and Wang 2001; Wang et al. 2002a; 2008a; Liu et al. 2003c). The Fenghuoshan and the Yaxicuo formations have been intensively deformed and are cut by numerous thrust faults. In contrast, the overlaying Wudaoliang formation is only gently tilted. This unconformity indicates that regional tectonic deformation occurred before about 24–22 Ma.

Wang et al. (2008a) also report that apatite fission track samples from 40 to 35 Ma-old Fenghuoshan formation suggest rapid exhumation of the central TP from about 30–22 Ma ago. The northern Qaidam thrust fault (southern Qilian Shan Mountains) became active at 28–24 Ma as determined from lithofacies, magnetostratigraphy, biostratigraphy and detrital apatite fission-track ages (Yin et al. 2002, 2007, 2008). The Linxia Basin since at least 29 Ma is interpreted as a flexural foredeep basin whose structure is likely due to thrust loading of the West Qinling range (Fang et al. 2003).

The end of marine sedimentation and the onset of terrestrial deposition throughout the Tarim basin began in the Late Oligocene (He 1991; Hao et al. 2002). The Fergana-Tajiki Sea also vanished at the same time. As the moisture sources disap-

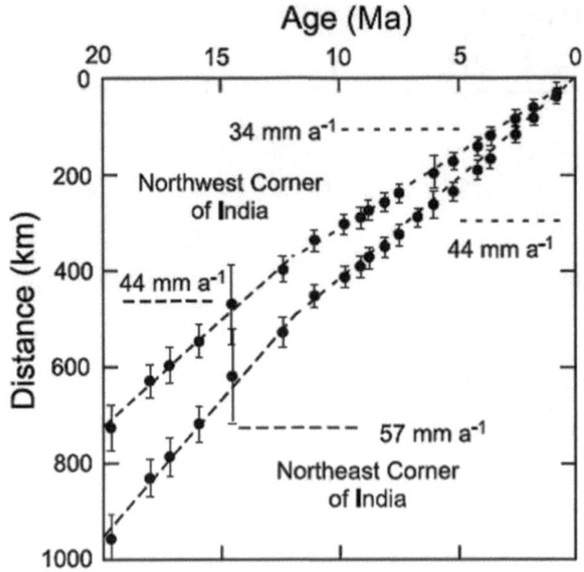

Fig. 6.11 Distance plot of points in the northeast and northwest corners of India from their reconstructed positions relative to Eurasia at selected times in the past. Uncertainties are shown at 1sigma. Note the marked change near 11 Ma and the ~2 Ma uncertainty in the precise time of change. (After Molnar and Stock 2009)

peared, the Tethys dried up, except for the Meterranean (Fig. 6.10;http://jan.ucc.nau.edu/~rcb7/ mollglobe.html). Erosion of the Asian interior, previously covered by the Tarim Sea was initiated by the regression and aridification trend that began in the Early Oligocene (Akhmetyev et al. 2005). This trend eventually resulted in the deserts of northern China which have acted as eolian sources since that time (Guo et al. 2008). Atmospheric circulation changes attributed to the land-sea distribution in Eurasia, and large-scale uplift of the TP and retreat of the Paratethys shifted the climate towards a monsoon system (Ramstein et al. 1997; Liu et al. 1998; Zhang et al. 2007; Guo et al. 2008). The formation of eolian deposits and the change in atmospheric circulation also resulted in the formation of loess in western China in the Late Oligocene (Qiang et al. 2011).

6.2.3.4 15–7 Ma Eastward and Northeastward Growth of the Tibetan Plateau

Widespread deformation in the eastern and northern margin of the TP took place during this time period and probably established the fundamental pattern of modern geomorphology of the TP and northwestern China. One of most significant events that occurred was a major slowing down of the convergence rate between the Indian and Eurasian plates (Molnar and Stock 2009). Reconstructed positions of the northwest and northeast corners of India since 20 Ma indicate a 25% decrease in the rate of convergence between India and Eurasia ~11 Ma (Fig. 6.11). Not only did the direction that India moved relative to Eurasia change (Fig. 6.11), but also the rate for the northwest corner of India dropped from 44 to 34 mm/yr, and that for the northeast corner of India decreased from 57 to 44 mm/yr (Fig. 6.11). As we will show below, the change in the convergence rate occurred at essentially the same time that deformation within and surrounding Tibet seems to have increased. One possibility is that somehow the crust and upper mantle beneath Tibet suddenly exerted an increased outward force per unit length on Tibet's surroundings, including the Indian lithosphere (Molnar and Lyon-caen 1989; Molnar and Stock 2009).

A spectrum of observations suggests changes in the tectonic development of the TP and its surroundings in Late Miocene time. Some such events can be dated precisely at 7–15 Ma, and although dating of others remains less precise, many seem to have occurred around that time. Analysis of Cenozoic stratigraphy in the western Hexi Corridor and adjacent north Qilian Shan range identifies facies changes in the Miocene related to range growth (Bovet et al. 2009). Growth strata within the dated section of the Niugetao formation (~9 Ma) of the Jiuxi Basin (central Hexi Corridor) suggests fault activity of the North Qilian Shan thrust during Late Miocene time (Fang et al. 2005), but does not constrain fault initiation or range growth. New apatite (U-Th)/He dates from the northeastern margin of the TP (north Qilian Shan) indicate rapid cooling began at ~10 Ma, which is attributed to the onset of faulting and topographic growth (Zheng et al. 2010). Preservation of the paleo-PRZ in the hanging wall and growth strata in the footwall allow for the calculation of vertical and horizontal fault slip rates, averaged over the last 10 Myr, of ~0.5 mm/yr and ~1 mm/yr respectively, which are within a factor of two of established Holocene slip rates and geodetic data. Low fault slip rates since the initiation of the northern Qilian Shan fault suggest that the total horizontal offset did not exceed 10 km.

Zheng et al. (2003) measured detrital fission-track ages from the Linxia Basin, on the northeastern margin of the TP. They showed an abrupt change near 8 Ma, though possibly as early as 13 Ma, in the difference between the mean age of detritus and its depositional age (Fang et al. 2003). Zheng et al. (2003) interpret this change to document the emergence of the range just west of the basin, Laji Shan, where Proterozoic rocks outcrop over a large area. In supporting results from detrital fission-track ages from the Linxia Basin, Lease et al. (2011) discovered that the Jishi Shan bounding the Linxia basin on the west has undergone rapid cooling since ~10.6 Ma based on apatite (U-Th)/He dating. This result is consistent with previous work of detrital zircon U/Pb dating in nearby mountains of the Laji Shan. There it was found that 500–1,000 Ma zircons, derived from Silurian strata, was only significant in the 8 Ma stratum of the Guide basin adjacent to the Laji Shan. This distinctive zircon population was otherwise absent in strata deposited both before and after 8 Ma (Lease et al. 2007). They attribute the appearance of this zircon population to the emergence of the Laji Shan as an actively eroding terrane at 8 Ma along the northern and eastern margins of the TP during the Late Miocene.

The Liupan Shan, which lies ~200 km northeast of the Linxia Basin and is marked by a single ridge of high terrain, forms the northeasternmost zone of crustal shortening and thickening associated with the formation and outward growth of Tibet (Zhang et al. 1991). Using reset fission-track ages on detrital material deposited since Cretaceous time, Zheng et al. (2008a) found that the older material had been heated sufficiently to anneal tracks and to reset ages, and then subsequently cooled since ~ 8 Ma. They inferred that thrust faulting, growth of the mountain range, subsequent erosion, and the resulting cooling of buried sediment began at that time.

Using both fission- track and (U-Th)/He dates from different elevations in these regions, Kirby et al. (2002) found slow cooling, at < 1 °C/Myr, from Jurassic to mid-Miocene time, and then rapid cooling at 30–50 °C/Myr. For the Longmen Shan, rapid cooling began no earlier than 12–13 Ma and perhaps as recently as 5–6 Ma. For the Min Shan, it began no earlier than 6–7 Ma and perhaps as recently as 4–5 Ma. Kirby et al. (2002) suggested that mountain building began when cooling accelerated.

Clark (2003) and Clark et al. (2005) studied fission track and (U-Th)/He dates from elevation transects into deep gorges cut into southeastern Tibet. Their results indicated slow cooling of < 1 °C/Myr in Cretaceous and early Cenozoic times, followed by rapid cooling beginning between 9 and 13 Ma. They reported an average erosion rate during the period of slow cooling of < 0.02 mm/yr (< 20 m/Myr) on the interfluves between deep gorges, and they concluded that the present-day gentle surface of low relief almost surely was flat and low when cooling was slow. Thus, since 9–13 Ma, the surface defined by this low-relief terrain rose ~ 2,000 m. They buttressed this inference with a discussion of the present-day drainage and the possibility that rapidly rising terrain led to river capture (Clark et al. 2004).

6.2.3.5 Growth of the Northeastern Tibetan Plateau since 3.6 Ma

Many geologic lines of evidence support the idea that the TP rose from a relatively low plateau of ~ 1,000 m to its present elevation during the last 3.6 Ma (Fig. 6.13 Phase IV). These include inferred sedimentation rates and grain size data of material deposited near the plateau (Li et al. 1997a, b; Li and Fang 1999; Zheng et al. 2000; Pares et al. 2003; Chang et al. 2012), as well as geomorphic features such as recent down-cutting from rivers like the Yellow River (Li et al. 1981, 1996). Li et al. (1979) and Li (1995) have considered the protracted processes of tectonic deformation from 3.6 Ma to the present as an orogenic movement, and have named it the "Qinghai-Xizang movement". Important evidence supporting this view is the widespread increase in deposition rate and grain size of late Pliocene and Quaternary sediments around the TP. Rapid exhumation and uplift occurred in the Kunlun Mountains (Zheng et al. 2000; Wang et al. 2003a) and Himalayan region (Amidon et al. 2005) in the Late Pliocene. The sedimentation rate increased (Métivier et al. 1998) and the depocenter shifted to the southeast in the Qaidam basin (Wang et al. 2006a), and sedimentation rates increased over a large area on the northern TP during the period 3.6–2.6 Ma BP (Fig. 6.12). The diachronous angular unconformity assigned to the late Pliocene in the Qaidam basin (Sun et al. 2005, 2006c) and increased sedimentation rates from the same period indicate tectonic movement of the northern region. An alternative view is that these changes in erosion rates and grain sizes result from climate change (Molnar and England 1990; Zhang et al. 2001; Molnar 2004).

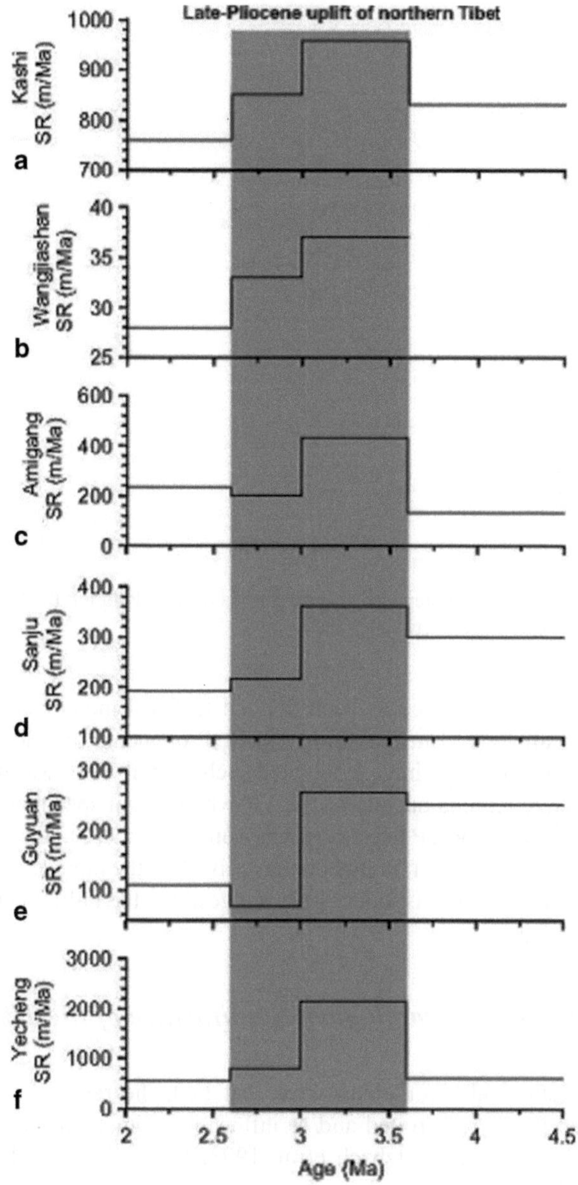

Fig. 6.12 Changes in sedimentation rate (SR) along the TP between 4 and 2 Ma BP. Sedimentation rates from **a** Kashi (Heermance et al. 2007), **b** Wangjiashan (Li et al. 1997a), **c** Amigang (Fang et al. 2005), **d** Sanju (Sun et al. 2006b), **e** Guyuan (Jiang et al. 2007), and **f** Yecheng (Zheng et al. 2000)

Paleo-elevations in southern TP can be estimated by associating nearest living relatives of fossil plants and pollen (Hsü et al. 1973; Hsü 1981). Hsü (1981) reported a rise of 3,000 m since the Late Pliocene. Mercier et al. (1987) subdivided these and similar data regionally to suggest that different parts of the plateau rose at different times, but they too found a rapid Late Cenozoic rise for most of the plateau.

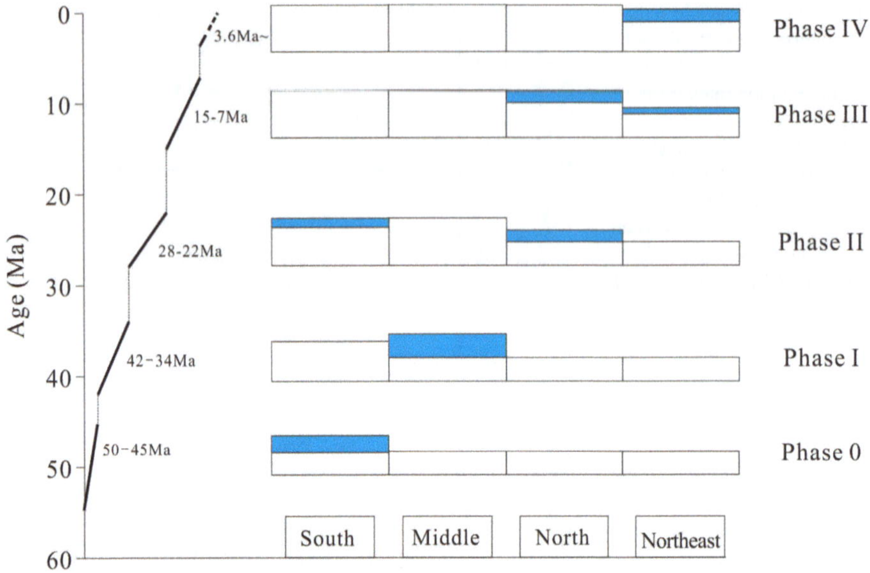

Fig. 6.13 Four stepwise growth periods of the TP during the Cenozoic. *White* (*Blue*) rectangles indicate stable (active) terraces

Similar logic has been applied to fossil mammals, such as Hipparion (e.g. Li et al. 1981; Liu and Ding 1984). Proponents of a low TP in Late Miocene and Pliocene times have associated such animals with warm environments that characterize regions outside of the TP, where fossil specimens have been found. Having excluded the TP from those environments, these authors then deduced that for Hipparion and other animals to have lived on the plateau, the TP must have been much warmer than it is today. They suggest that the uplift occurred in the Late Pliocene.

6.2.4 Climatological Significance of the Tibetan Plateau Growth

Numerical simulations show that both the retreat of the Paratethys and growth of the TP controlled and/or influenced climate change in Asia (Raymo and Ruddiman 1992; Kutzbach et al. 1993; Ramstein et al. 1997; An et al. 1999, 2001; Zhang et al. 2007). This change was episodic, with synchronous inland drought and enhanced Asian monsoons at about 10 Ma and 3.6 Ma BP, at times when the TP experienced significant uplift.

Retreat of the Paratethys was followed by the growth of the TP and resulted in aridification of the Asian interior where climate was controlled by the westerly atmospheric circulation (Dupont-Nivet et al. 2008; Abels et al. 2011). The processes of the TP's growth can be divided into four phases (Fig. 6.13), including formation of a middle proto-plateau during 42–34 Ma, northern and southern expansion of the plateau during 28–22 Ma, east-northeastward growth during 15–7 Ma and northeastward growth since 3.6 Ma. These phases may have influenced the development

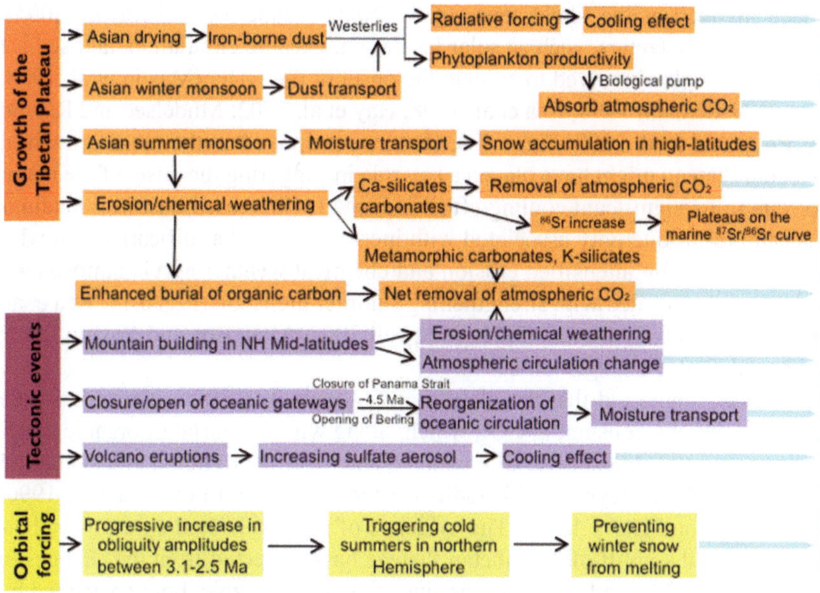

Fig. 6.14 Onset of major northern Hemisphere glaciations during the Pliocene triggered by Tibetan growth and other tectonic and orbital forcings

and evolution of the East Asian and Indian monsoons and the contemporary Asian monsoon-arid zones. The growth of the TP divided the westerlies into north and south branches, and diverted moisture to the south and east parts of the plateau. High elevation also kept moisture from the Indian monsoon off the plateau (Molnar et al. 2010), further enhancing aridification in the interior. Meanwhile, changes in the land-sea configuration and climate induced the deposition of loess sequences in the Late Oligocene-Early Miocene (Qiang et al. 2011). Through time, these tectonic changes account for today's Asian climate, characterized by seasonal monsoons and a vast, high elevation arid interior (An et al. 2006b).

6.2.5 *Implication for the Onset of Major Northern Hemisphere Glaciation*

Initiation of major Northern Hemisphere glaciations (NHG) was first identified by large amplitude $\delta^{18}O$ and IRD fluctuations around 2.4 Ma (Shackleton et al. 1984), and was confirmed by a gradual increase in the benthic $\delta^{18}O$ and IRD values during the interval of ~3.6–2.6 Ma (Raymo et al. 1988; Johnsen et al. 1992). However, mechanisms driving the onset of the NHG are still elusive. Tectonic forcing, especially the closure of the Panama Seaway (4.6–3.6 Ma; Haug and Tiedemann 1998), the opening of the Bering strait (7.4~7.3–4.8 Ma; Marincovich and Gladenkov 1999), and the closing of the Indonesian seaway (9.9–7.5 Ma; Linthout et al. 1997), has been invoked as the likely cause of the NHG (Mudelsee and Raymo 2005). Meanwhile, other factors, such as solar insolation, oceanic circulation, and sea-ice occurrence, are also involved to explain the onset of the NHG (Maslin et al. 1998; Haug and Tiedemann 1998; Rea et al. 1998; Hay et al. 2002; Mudelsee and Raymo 2005; Reynolds et al. 2008; Fig. 6.14). Here, we propose that the uplift/growth of the Tibetan Plateau might have played a key role in triggering the onset of the major NHG, due to its significant cooling effect induced by increased dust/iron loadings and ocean bio-productivity associated with increased inland aridification and winter monsoon, and by intensified erosion and chemical weathering via atmospheric CO_2 decrease associated with strengthening summer monsoon (Fig. 6.14; Rea et al. 1998; Arimoto 2001; An et al. 2001; Ridgwell 2002; Mudelsee and Raymo 2005; Jickells et al. 2005; Zhang et al. 2009).

It has been suggested that the plateau and mountain uplift accelerated silicate weathering, thereby consuming CO_2, which, along with the burial of organic materials, decreased the atmospheric CO_2 content, and thereby promoted global cooling (Raymo et al. 1988; Raymo and Ruddiman 1992; Derry and FranceLanord 1996; France-Lanord and Derry 1997). The evidence of the accelerated silicate weathering induced by the uplift of the Tibetan Plateau have reflected from the changes in the seawater $^{87}Sr/^{86}Sr$ and δ^7Li_{SW} ratios since the late Cenozoic, both controlled by the fluxes and isotopic composition of rivers (Palmer and Edmond 1992; Bickle 1994; Misra and Froelich 2012). However, when seawater $^{87}Sr/^{86}Sr$ ratios (Hodell et al. 1990, 1991) showed its steady variation during the 4–2 Ma period, the sediment flux increased rapidly in the Indian Ocean (Rea 1992). This temporal correspondence between $^{87}Sr/^{86}Sr$ flat plateau and high sediment flux peaks during 4-2 Ma could be associated with rapid chemical weathering of carbonates along with anomalously high erosion rate under a strengthened summer monsoon both induced by the uplift of the Tibetan Plateau at around 3.6 Ma (Farrell et al. 1995). The high mountain denudation at around 3.6 Ma is also recently indicated by an increased δ^7Li_{SW} ratio. High-relief terrains (uplifted mountains and down cutting high plateaus) with weathering-limited regimes such as the Himalayas exhibit strong physical and chemical weathering and high denudation rates with extreme incongruent weathering. Large rates of secondary clay formation that carry most of the weathered Li down rivers, and hence dissolved Li isotopes ratios that are ^7Li-enriched relative to their source rocks (Misra and Froelich 2012).

Along with the growth of the northeastern Tibetan Plateau in the Pliocene, there were increased dust/iron fluxes over the central Asia (An et al. 2001; Cao and An 2011; Shi et al. 2011). Records of atmospheric Fe loadings and the dust fluxes reconstructed from Al concentrations in a loess-paleosol sequence from the Chinese Loess Plateau showed an increasing trend from 6.0 to 2.7 Ma. The air/sea exchange of eolian iron transported from arid lands in central Asia may have a significant impact on marine bioproductivity via an iron fertilization mechanism in the North Pacific. A strong coherence between the evolution of iron loadings with variations of bioproductivity in the North Pacific and global atmospheric CO_2 from 3.6 to 2.7 Ma further implied that the increased dust/iron flux may have accelerated the onset of the NHG.

6.3 Onset and Evolution of the Asian Monsoon-Arid Climate System

The Asian monsoon-arid environment system is an important component of the global environment system and it covers a large portion of Asia. It leads to a humid monsoonal environment in southeastern Asia and a corresponding arid environment in northwestern Asia. Monsoon-associated precipitation is a profound human environmental concern, and socially sustainable development in populous regions of East Asia requires a thorough understanding of the origin and development of the Asian monsoon-arid environment system. During the past decades, the monsoon-arid environment coupling system in Asia has become a focus for climatologists, geographers and geologists, and their studies have substantially increased our understanding of this climate system. The question of exactly when the Asian monsoon climate began remains open. Some researchers proposed an Early Miocene onset for the Asian monsoon (Sun and Wang 2005; Guo et al. 2008), while others suggested an Oligocene, or even Eocene onset (Gu and Renaut 1994; Clemens 2006; Huber and Goldner 2012). Knowledge of this question is crucial for recognition of its forcing mechanisms, understanding of the Asian and global climate evolution and prediction of future climate change. Here we review previously reported data including lithology, pollen and mammalian fossils to reconstruct the Cenozoic evolution of the monsoon-arid system since the Paleocene. We examine the changes of environmental patterns in China through time in the context of a climatic transformation from a uniformly distributed, subtropical planetary climate system, to the present day monsoon-arid environment system, with a pattern of dry western, and humid eastern regions. Our analysis suggests that the Asian monsoon was initiated possibly as early as the Eocene. The Asian monsoon-arid environment system began to develop during the Oligocene and became established during the early Miocene, and reached its current form during the Late Pliocene (Fig. 6.15).

6.3.1 Paleocene (~65–56 Ma; Fig. 6.15a)

In the Paleocene, China was characterized by lower western and higher eastern regions, without the existence of the present high-elevation TP. Under the influence of the planetary climate system, there was a broad, roughly east-west dry belt across China, which resulted in thick red continental deposits associated with halites in most of China. For example, a halite intercalated with thin layers of mudstone of ca. 400 m in thickness occurred in the Qingjiang Basin in southern China (Sun and Wang 2005). All sites with evaporite deposits were characterized by arid pollen taxa, associated with high proportions of *Ephedripites*. Even in southeastern China, there was a high percentage of *Ephedripites* in the halite deposits. Paleocene mammalian fossils are rather sparse, but they are abundant enough to show a similarity in faunal composition across China, without obvious topographic differences

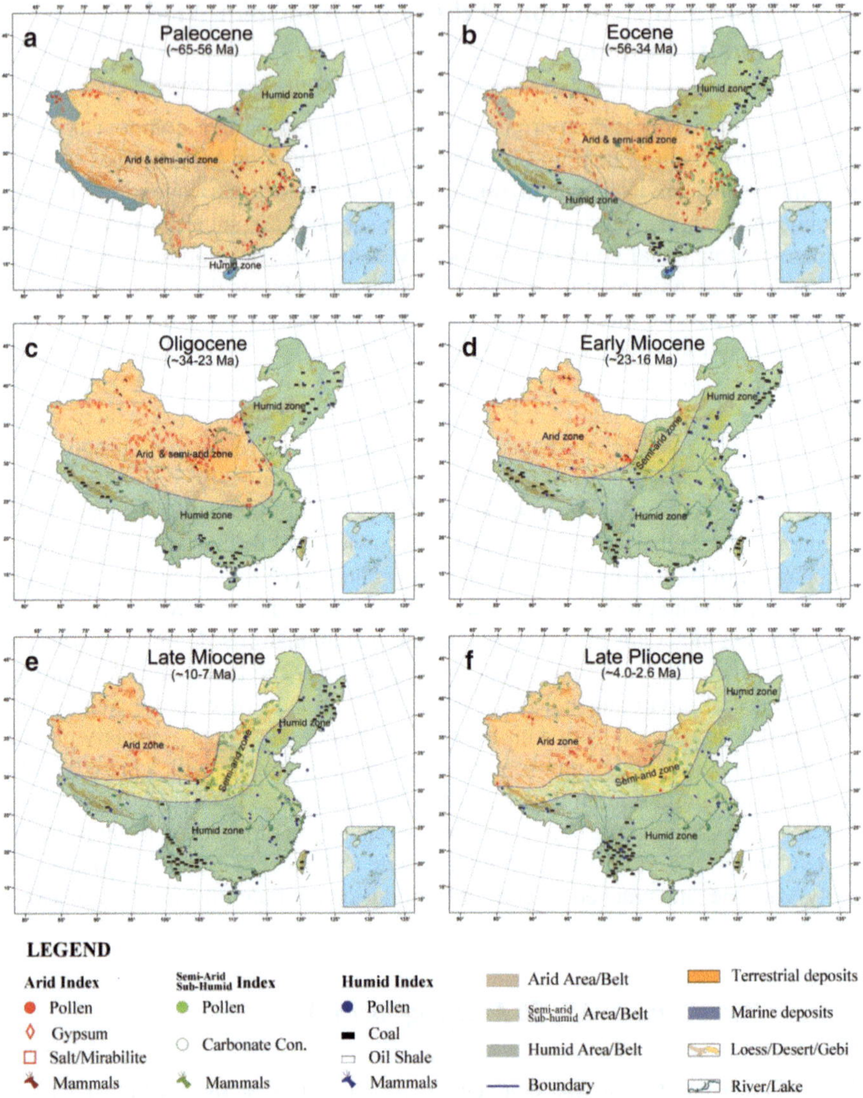

Fig. 6.15 Inferred Cenozoic environments in China based on depositional facies, pollen and mammalian fossil data. **a** Paleocene (~65-56 Ma). **b** Eocene (~56-34 Ma). **c** Oligocene (~34-23 Ma). **d** Early Miocene (~23-16 Ma). **e** Late Miocene (~10-7 Ma) and **f** Late Pliocene (~4-2.6 Ma). The base map is from China Geological Survey (2001). Data on the distribution of the Cenozoic terrestrial deposits are from Chinese Stratum Thesaurus Editorial Board (1999)

between east and west. This is consistent with a planetary climate system. During the Paleocene, a humid climate was only found in northeastern China (>38°N), the Leizhou Peninsula and Hainan Island. Mammals adapted to lakeshore environments were common in Inner Mongolia at that time (Yuan and Du 1984). These areas were possibly beyond the influence of the planetary wind system, with the occurrence of coal strata and oil shale but little halite. Both fossil pollen and leaves from sites in northeastern China point to a warm and humid environment (Sun and Wang 2005).

6.3.2 Eocene (~56–34 Ma; Fig. 6.15b)

During the Eocene, the broad east to west dry belt still crossed China, but a northward migration of its southern boundary is clear (Fig. 6.15b). This implies a northward retreat of the westerly planetary wind system and possible initiation of the Asian monsoon in the southern tropical zone (Gu and Renaut 1994; Clemens 2006; Huber and Goldner 2012). The Eocene was a period of maximum subsidence of the terrestrial basins and large lakes in China. Because of the Eocene uplift of the TP (Wang et al. 2008a), most of the Paratethys Sea had disappeared by the Late Eocene (Dercourt et al. 1993; Ramstein et al. 1997) and a series of tectonism-induced downfaulted basins were formed. Most basins in East and West China are still dominated by thick red continental and halite deposits, which indicate an arid environment (Liu et al. 1998; Sun and Wang 2005). Rich gypsum deposits from the western Tarim Basin serve as a typical example. In these arid areas, the pollen records are generally dominated by arid *Ephedripites* and *Nitraria*. Both West and East China had similarly dominant arid-adapted mammalian faunas at this time (Qiu and Li 2005). There is no evidence for a substantial topographic barrier for the spread of most mammals at this time, nor is there a distinct regional differentiation of faunal elements between east and west.

The sedimentary, pollen and mammalian distributions also indicate a northward migration of the humid zone toward the east-west dry belt. Whereas there are many evaporites in the east-west dry belt, coal and oil shale were widely deposited along the northern and southern margins of the subtropical arid zone, especially along the southern margin of the TP and coastal areas of Southeast China. Even in the arid zone, thin layers of oil shale occurred in thick evaporite deposits. Some evergreen trees such as *Cinnamomum lanceolata*, *Citrus* and *Sabalites* started to appear in southern China, which indicates a tropical humid environment. In northeastern China, however, tropical and subtropical plants were widely distributed, including *Cycas anomostoma* and *Sabalites*, together with four species of ferns that are common to tropical and subtropical environments today. In addition, mammals preferring moist, forested habitats were also common along the eastern coast and on the northern margin of the subtropical arid zone. Examples include: the Huadian of Jilin, Benxi of Liaoning, Changle of Shandong, Liyang of Jiangsu to Maoming of Guangdong and Baise of Guangxi (Yuan and Du 1984; Qiu and Li 2005). These areas contained forest dwellers such as primates and insectivores in their faunas and are often associated with abundant organic sediments such as lignite and oil shale. One exception is the fissure-fillings at Liyang south of Jiangsu Province. Taken as a whole, faunal evidence supports the idea that the Asian monsoon was initiated in tropical South China and the eastern coast during the Eocene.

6.3.3 Oligocene (~34–23 Ma; Fig. 6.15c)

During the Oligocene, in particular the Late Oligocene, a significant retreat of the arid zone from eastern China and further expansion of the humid zone was observed, which was the precursor to the present-day Asian monsoon-arid climate system pattern. West China was further elevated by the continued growth of the TP (Tapponnier et al. 2001), and the Paratethys Sea retreated completely from the western Tarim Basin during the Early Oligocene. In the Oligocene, western China was still dominated by red clastic rocks with gypsum, trona and halite, but a large amount of eolian dust appeared in lacustrine sediments from the Linxia Basin at about 29 Ma (Fang et al. 2003). Continuous eolian red clay deposition occurred on the western part of the Chinese Loess Plateau since ca. 25 Ma (Qiang et al. 2011). The Oligocene pollen of western China is dominated by arid plants, but the proportion of Chenopodiaceae increased considerably, which is consistent with intensified aridification in western China. Two persistent and obvious features mark the evolution of Oligocene mammals: 1) the disappearance of a great quantity of archaic taxa, and 2) the appearance of higher-crowned and steppe-adapted small mammals (tsaganomyids and ochotonids) in the Mongol-Xinjiang region adjacent to the Qinghai-Xizang Plateau. Tsaganomyids and ochotonids are adapted to temperate grasslands today, and their rise is consistent with northward movement of the arid zone. Oligocene pollen records from the Qaidam, Junggar, Kuqa, Xining and Jiuquan basins show typical vertical zonation with herbs, deciduous forests and coniferous forests, in addition to a large number of arid taxa. This indicates not only an arid climate in western China, but also significant differences in topography, possibly due to the uplift of the TP.

During the Oligocene, the humid zone migrated significantly towards western China. In most of southern and eastern China, coal and oil shale were widely deposited. Pollen from northeastern China was dominated by humid Taxodiaceae and fern spores, while in southern China broad-leaved evergreen plants were common, along with some deciduous coniferous plants. Dominant fern spores were found in mudstones and sandstones of the northern South China Sea. High proportions of hygrophilous plants and broad-leaved evergreen plants occurred in eastern China. In contrast to the arid-adaptable mammals of western China, abundant hygrophilous mammals such as *Amynodont* and *Anthracotherium* appeared in southern and eastern China (Yuan and Du 1984). These differences in pollen, sedimentology and mammalian fossils between northwestern China, and eastern-southern China indicate that the Asian monsoon was probably strong enough to affect most areas of eastern-southern China by the Late Oligocene, possibly a little earlier.

6.3.4 Miocene (~23–5.3 Ma; Fig. 6.15d, e)

During the Miocene, a further retreat of the arid zone and expansion of the humid zone were observed, and an Asian monsoon-arid climate system became well-established. In a tectonic and palaeogeographic context, the Miocene of China differed substantially from the Oligocene. During the Miocene, most of the modern TP and its structural configuration had formed. The arid zone retreated further from eastern China and aridity in the west became more intense. The eastern and southern boundaries of the latitudinal arid belt retreated to ca. 105°E and 30°N, respectively. The middle reaches of the Yellow River were dominated by semi-arid conditions. This spatial pattern of climate and environment was similar to that of the present.

Purple-red and grayish green mudstones with gypsum and halite, and arid pollen assemblages of *Ephedripites*, *Nitraria* and Chenopodiaceae were common in northwestern China during the Miocene, for example in the Qaidam, Tarim, Jungaar and Jiuquan basins. Because of a strengthened arid climate in northwestern China and strong monsoon climate in eastern Asia, thick eolian loess-soil sequences were deposited on the Loess Plateau (Liu et al. 1985; Guo et al. 2002). In the northeast, micromammals (including Castoridae, Aplodontidae, Gliridae, Zapodidae, Dipodidae, Gerbillidae, Ochotonidae) of the Xiejia Fauna (Early Miocene), Tunggur-Fauna (Middle Miocene) and Ertemte Fauna (Late Miocene) were distributed over the Holarctic or Palearctic regions, but none are found in the present Oriental Region at the family level. These mammals generally adapted to arid and cold grasslands.

Most of southern and eastern China was characterized by humid climate during the Miocene. Lignite indicative of humid climate was widely deposited in southwestern China and the southern TP, with thicknesses of more than 1,000 m at some locations. During the mid-Miocene climatic optimum (~14–18 Ma), the humid summer monsoon might have expanded to the Lanzhou and Linxia basins and eastern TP as suggested by pollen and sedimentary evidence. For example, fluvial-lacustrine sediments from the Linxia Basin contain a relatively high proportion of forest pollen and less herb pollen during the mid-Miocene climatic optimum (Ma et al. 1998). Eolian red clay sequences on the western CLP were characterized by high dust accumulation rates and magnetic susceptibility during the mid-Miocene climatic optimum (Guo et al. 2002; Qiang et al. 2011), which indicates enhanced aridity in western China and intensified Asian monsoon circulation (An et al. 2006b). Pollen data show that the Shanwang flora in eastern China was roughly comparable to modern vegetation in the summer wet mountain area of the middle and lower Changjiang River Valley, which requires a moist summer climate (Sun and Wang 2005). Since the Miocene, the continental shelf of the South China Sea has preserved a large number of mangrove pollen, marine dinoflagellates and pollen of tropical and subtropical plants, which indicates an expansion of the marine environment and a hot and humid climate. Abundant tropical and subtropical plants in southwestern China during the mid-Miocene climatic optimum also indicate high temperatures and substantial precipitation in this area (Tao and Du 1982; Guo and Chen 1989). During the Late Miocene, possibly because of the growth of the TP

(George et al. 2001; Wan et al. 2001; Wang et al. 2002a; Clark et al. 2005; Molnar 2005), the aridity in western China and monsoon climate were further intensified, which resulted in the salinification of the Qaidam Basin to the west, and an increase in vegetation adapted to humid conditions in eastern China.

During the Miocene, the distribution of mammals differentiated between North and South China. The Yuanmou and Lufeng faunas in southern China contain extant families which live nowadays in the Oriental Region or mainly in tropical-subtropical zones, except for Castoridae and Cricetidae. Unlike the faunas of northern China, that reflect a temperate, relatively arid steppe environment similar to the present Palearctic Region; faunas from southern China are indicative of a tropical or subtropical forest environment characteristic of the Oriental Region. This implies the initiation of the Oriental Region in the Miocene. During the Miocene, there was a "transitional zone" of zoogeographic regions in central China, which was characterized by fauna derived from adjacent faunal regions. The Early Miocene Sihong Fauna (Northwestern Jiangsu Province) suggests the existence of such a transitional zone, which contains either taxa from the modern Holarctic and Palearctic Regions, such as Castoridae, Gliridae and Ochotonidae, or members that are restricted to the Oriental Region, such as Echinosoricinae, Rhizomyidae and Platacantomyidae. The taxonomic composition of the Chinese Miocene mammals is characterized not only by latitudinal ecological variations, but also by longitudinal variations. Take the earlier Miocene faunas for example; the Sihong Fauna and the Shanwang Fauna consist mainly of mammals preferring warm and moist habitats, while the Xiejia Fauna of Qinghai Province, and the Quantougou Fauna of Gansu, are composed of taxa that can withstand dry and cold climate. The humidification of southeast and southwest China since the Early Miocene implies a strong influence of the summer monsoon on southern and eastern China. The combination of pollen, sedimentary and mammalian evidence indicates that the Asian monsoon-arid climate system was well established during the Miocene. It supports the notion of synchronous intensification of the monsoon and arid environment rather than largely diachronous developments.

In addition, the Early and Late Miocene in China had similar distributions of arid zones, but the former had a relatively larger humid zone with a smaller and more humid semi-arid zone (Fig. 6.15). During the Early Miocene, the humid monsoon may have expanded to the eastern margin of the TP (Guo and Chen 1989). There were increased woodlands but decreased grasslands in the Linxia Basin during the Early Miocene (Ma et al. 1998). A slight decreasing trend in the percentage of arid plants occurred in the Qaidam Basin after ca. 23 Ma (Wang et al. 1999).

6.3.5 *Pliocene (~4–2.6 Ma; Fig. 6.15f)*

The distribution patterns of environment and climate in the Pliocene were similar to those of the Miocene, with the arid zone in the northwest part of China and humid zone in southern and eastern China. The modern pattern of the Asian monsoon-arid climate system was established during the Pliocene. By that time, most of the Miocene paleolakes in northwestern China had gradually disappeared. In their place, thick conglomerate layers were deposited. The accumulation rate of eolian dust on the CLP also increased significantly at this time (An et al. 2001). These changes indicate further enhanced aridification in northwestern China and tectonic activity along the northern margin of the TP. The percentage of herb plants also increased significantly in northwestern China in the Pliocene, for example in the Jiuxi, Qaidam and Dunhuang basins (Ma et al. 2004). Fossil mammal localities are also consistent with arid conditions in northwestern China at this time. These arid-adapted mammals included Bovidae, Cricetidae, Siphneidae, Avicolidae, Dipodidae, and Ochotonidae.

In contrast to the disappearance of lakes in northwestern China, a large number of lakes formed during the Pliocene in southern and eastern China, such as those in the Nihewan, Shanmengxia, Yushe and Yuanmou basins. Pollen records also indicate a humid climate in southern and eastern China. Abundant subtropical broad leaf plants and needle forest plants are known from eastern China. Pliocene fossil plants from Yuyao of Zhejiang Province, representative of montane areas of eastern China, have a mix of broadleaved trees (*Machilus*, *Myrica*, *Liquidambar* and *Ormosia*), subtropical montane conifers (*Cedrus* and *Tsuga*) and boreal conifers (*Abies* and *Picea*). The fossil leaves are of moderate size, thin or in some cases coriaceous in texture, with toothed or entire margins indicating subtropical montane forest vegetation and a comparatively warm and humid climate. In southwest China, Pliocene lignite-bearing lacustrine and swampy sediments in several small basins yield plentiful plant fossils. At Eryuan in Yunnan Province, fossil leaves in lignite-bearing sediments consist mainly of alpine oaks such as *Quercus semicarpifolia*, and *Q. spathulata*, and are accompanied by some deciduous trees such as *Acer paxii*, *Celties bungeana* and *Populus*. Cones of *Pinus yunnanensis*, seeds of *Abies* and fruits of *Trapa* are also found. The vegetation recognized through the fossil plants is that of a mixed conifer and broadleaved forest in a subtropical mountain setting at ca. 2,500 m a.s.l., that is, close to the present altitude. Two plant fossil-bearing sites, Lanping of Yunnan Province and Dechang of Sichuan Province, are found to be very similar to Eryuan. These also contain fossil leaves from alpine oaks with entire margins (modern alpine oaks in the region have toothed margins), suggesting a warmer and moister climate, as compared with today (Sun and Wang 2005). Differentiation in the mammalian composition between the south and north is more distinct during the Pliocene than before. A so-called "*Ailuropoda-Stegodon* Fauna" composed mainly of mammals adapted to warm and humid environments, such as *Tapirus*, *Rhinoceros*, *Rhizomys* and *Hystrix*, prevailed throughout South China. These lines of sedimentary, pollen and mammalian evidence indicate that the modern pattern of Asian monsoon-arid climate was established during the Late Pliocene (Qiu and Li 2005).

The combined lithological, palaeobotanical and mammalian fossil data reveal a transition of the Asian climate from a zonal westerly planetary wind system to monsoon-westerly coupled wind system. This climatic transition occurred as early as the Eocene, which was marked by the joint onset and strengthening of the Asian summer and winter monsoon circulations and intensified aridification of the Asian interior. As a result of the moisture-bearing summer monsoon, southern and eastern

China gradually became humid, while northwest China remained arid due to the predominance of the westerlies and the Asian winter monsoon, with a negligible influence from the summer monsoon. The Paleocene had the most extreme typical aridity in the broad east-west belt across China. There were predominant Xerophilous plants like *Ephedripites* and huge deposits of halite and gypsum in southwestern China, particularly in the middle and lower reaches of the Changjiang during the Paleocene, which is in striking contrast to the humid climate in these areas today. The initiation of the Asian monsoon in the Eocene saw the arid zone gradually retreat from these areas. More significant expansion and intensification of the Asian monsoon occurred during the Oligocene, especially in its late stage. The similarity of the climatic patterns from the Late Oligocene to Quaternary indicates that the Asian monsoon-arid environment system have been maintained at least since the Late Oligocene. From about 25 Ma, eolian red clays have been deposited continuously on the western CLP (Qiang et al. 2011).

Despite numerous studies, the causes of the Asian monsoon and inland desertification are still debated. Different interpretations involve land-sea distribution, uplift of the TP, development of the Paratethys Sea, global cooling and expansion of the South China Sea. Actually, all of these processes change significantly as seen in Fig. 6.16. Hence, they may all contribute to the evolution of the monsoon-arid environment in Asia. At about 55 Ma, the Indian Ocean plate collided with the Eurasia continental plate (Garzanti and Vanhaver 1988; Zhu et al. 2005; Jain et al. 2009). Marine benthic $\delta^{18}O$ values indicate a warm global climate at that time (Zachos et al. 2001), while seawater $^{87}Sr/^{86}Sr$ ratios indicate relatively stable continental weathering (Hess et al. 1986; McArthur et al. 2001). There was a significant transgression of the Paratethys Sea (Burtman 2000; Bosboom et al. 2011), and during this period, lithological, palaeobotanical and mammalian fossil evidence point to a dominant east-west belt across China under the influence of a westerly planetary wind system (Yuan and Du 1984; Qiu and Li 2005; Sun and Wang 2005; Guo et al. 2008). During the Late Eocene (~42–34 Ma), gradually increasing benthic $\delta^{18}O$ values suggest increasing global ice volume, especially in the Antarctic (Zachos et al. 2001). Since then, global climate transformed from ice-free to icehouse periods. A decrease in atmospheric CO_2 and increased in $^{87}Sr/^{86}Sr$ ratios in sea water indicate gradually intensified continental silicate weathering (Hess et al. 1986; Raymo et al. 1988; Hodell et al. 1990; Krishnaswami et al. 1992; McArthur et al. 2001). This is synchronous with the uplift of the middle TP (uplifted to $>4{,}000$ m elevation; Tapponnier et al. 2001; Rowley and Currie 2006; Wang et al. 2008a) and large-scale retreat of the Paratethys Sea (Burtman 2000; Bosboom et al. 2011). These lines of evidence indicate that the Late Eocene was characterized by an important transition in both global and Asian climate. The factors involved in this transition include: 1) the uplift of the middle TP, 2) subsequent increased intensity of continental silicate weathering, 3) enhanced land-sea thermal contrast due to the retreat of the Paratethys Sea, and 4) strengthened cross-equatorial atmospheric meridional circulation due to Antarctic glaciations. Taken together, these can account for the onset and intensification of the Asian monsoon, along with the northward and westward retreat of the east-west arid zone (Raymo et al. 1988; Raymo and Ruddiman 1992; Ramstein et al. 1997; An et al. 2011). This is consistent with the climatic evolution

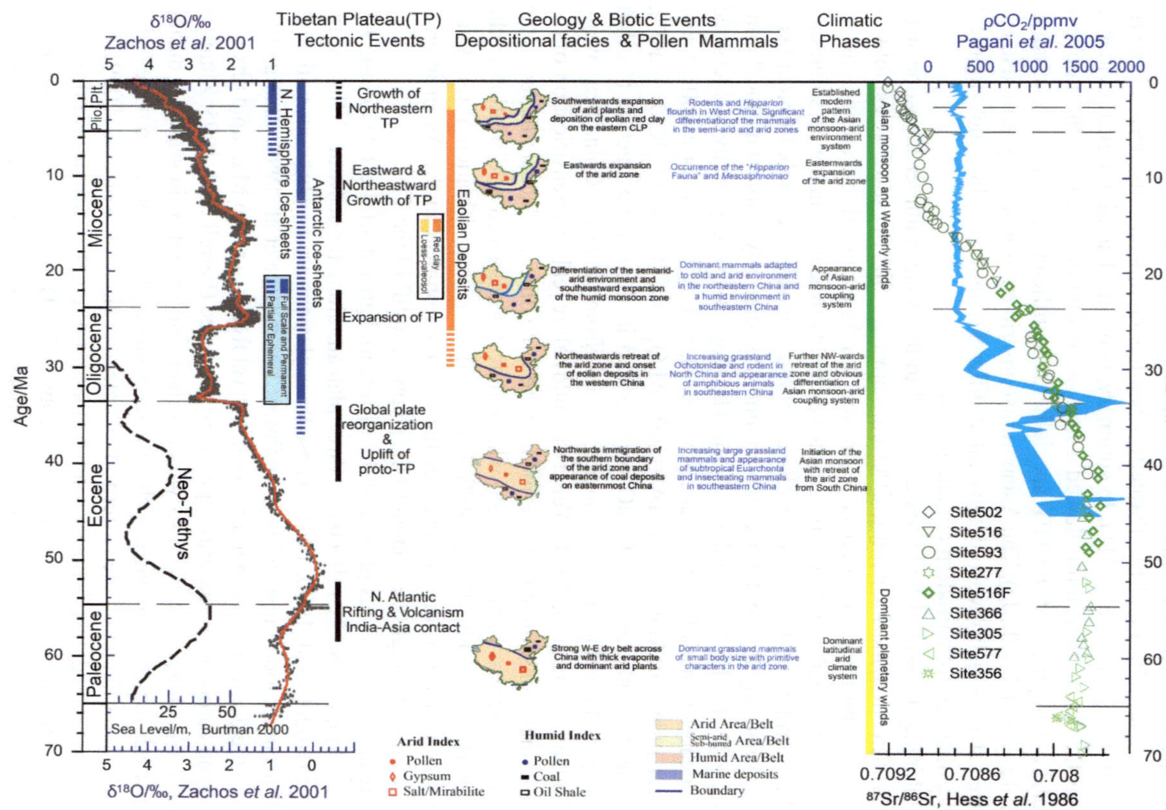

Fig. 6.16 A synthesis of the Asian and global climate, environmental and tectonic changes during the Cenozoic

revealed by lithological, palaeobotanical and mammalian fossil records. Since the Oligocene, icehouse climate has remained (Zachos et al. 2001), the concentration of the atmospheric CO_2 has significantly decreased (Pearson and Palmer 2000; Pagani et al. 2005, 2011), seawater $^{87}Sr/^{86}Sr$ ratios (indicative of enhanced continental silicate weathering) have continued to a relatively steady increase (Hess et al. 1986; McArthur et al. 2001) and the Paratethys Sea has completely disappeared (Burtman 2000; Bosboom et al. 2011). In summary, geological and biological evidence indicates the Asian monsoon followed a stepwise evolution much like the stepwise structural development of the TP (especially the significant uplift periods at about 28–22, 15–7 and 3.6–0 Ma). The Asian monsoon-arid environment system also experienced steps which involved substantial differentiation, coupling, development and establishment of a modern climate pattern.

Combining the established proxies related to the Asian summer monsoon with Asian aridification, marine benthic $\delta^{18}O$ and seawater $^{87}Sr/^{86}Sr$ ratios, we establish a continuous, long-term evolutionary framework for Asian and global climate during the last 26 Ma (Fig. 6.17). In addition to the variability of long-term trends, some specific short-term climatic shifts were also observed. For example, global atmospheric CO_2 was relatively low and global ice volume was relatively stable with local minima at 25–22, 17–14, 10–7 and 4–2.6 Ma. The magnetic susceptibility of loess-red clay sequences (An et al. 2001; Qiang et al. 2011), $\delta^{13}C$ of benthic foraminifers, and Chemical Index of Alteration ($CIA = 100 \times Al_2O_3 / (Al_2O_3 + Na_2O + K_2O + CaO^*)$, where CaO* is the amount of CaO incorporated in the silicate fraction (Nesbitt and Young 1982) values from sediments, records from the South China Sea (Wang et al. 2003b; Cheng et al. 2004; Wei et al. 2006) all reveal an intensification of the Asian summer monsoon in these periods. At the same time, dust accumulation in the northern Pacific was relatively high, which indicates enhanced aridification in the Asian interior (Rea et al. 1985, 1998). It seems that the intensification of the Asian summer monsoon and the presence of the arid environment in the Asian interior were coeval, both might be related to the stepwise uplift of the TP.

There are three obvious phases of the differential growth of the TP (17–14, 9–6 and 4–2.6 Ma, Fig. 6.17), reflected in three plateaus along the seawater $^{87}Sr/^{86}Sr$ ratio curve (Hodell et al. 1990, 1991). These three plateaus are synchronous with high $CaCO_3$ accumulation rates and high sediment fluxes (Rea et al. 1998; Zhang et al. 2001). In the dominant summer monsoon interval from ~ 4 to 2.6 Ma, rapid physical erosion and carbonate weathering apparently occurred (Farrell et al. 1995; Zhang et al. 2001). We suggest that the relatively constant $^{87}Sr/^{86}Sr$ ratios in seawater during those periods may be associated with rapid dissolution of carbonates from the Asian continent, in particular on the eastern TP (Chetelat et al. 2008; Noh et al. 2009), and from Ca-dominated dust deposits, such as loess and red clay, which also produces lower $^{87}Sr/^{86}Sr$ ratios to the ocean (e.g. Jacobson 2004), both induced by strengthened summer monsoon precipitation, rather than with the Himalayan source waters. If so, the plateaus of seawater Sr curve correlating with unusually high erosion rates could be related with the uplift of the Himalayan-Tibetan Plateau and associated increases in summer monsoon precipitation.

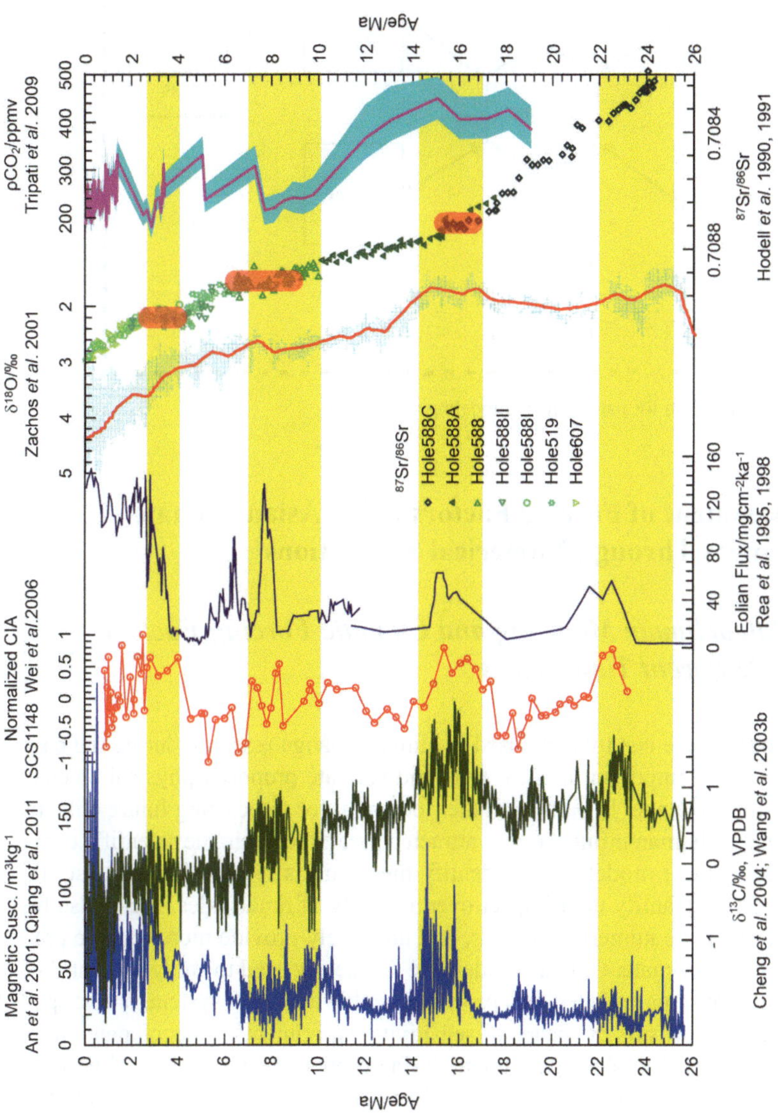

Fig. 6.17 A synthesis of long-term evolutionary framework of the Asian and global climate during the last 26 Ma, based on the summer monsoon proxies of Chinese loess magnetic susceptibility (An et al. 2001; Qiang et al. 2011), $\delta^{13}C$ of benthic foraminifers, CIA records from the South China Sea (Cheng et al. 2004; Wang et al. 2003b; Wei et al. 2006), Asian aridification proxy of dust accumulation in the northern Pacific (Rea et al. 1985, 1998), global climate proxies of benthic $\delta^{18}O$ (Zachos et al. 2001), $^{87}Sr/^{86}Sr$ ratios of sea water (Hodell et al. 1990, 1991) and atmospheric CO_2 concentration (Tripati et al. 2009)

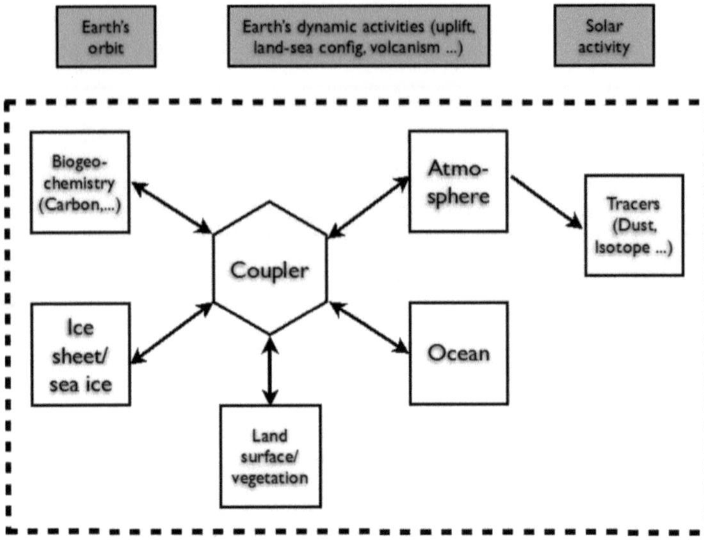

Fig. 6.18 An illustration for paleoclimate modeling

6.4 Assessment of Forcing Factors in the Asian Climate Changes Through Numerical Simulations

6.4.1 Paleoclimate Modeling and Climatic Forcing Factors at Different Time Scales

Climate models have been widely used in climate change research during the past few decades as an important tool to study and validate proposed physical mechanisms of climate change. In particular, their potential for forecasting future climate under increasing human influences has attracted great interest in the scientific community. Paleoclimate modeling can identify mechanisms responsible for past climate change and quantify their impact over a variety of time scales. It can also be employed to validate numerical models, and ultimately provide more reliable predictions of future climate. Climate models are frequently used to study present-day nonlinear relationships and feedback processes within a climate system, or test specific forcing mechanisms by means of sensitivity experiments. Paleoclimate models are typically used to simulate climate during representative geological periods or across abrupt climate events, to explore force-response mechanisms or validate theories of climate change. In this way climate modeling overcomes the temporal and spatial limitations of observational studies.

A climate model is a mathematical model consisting of a system of equations that stem from fluid dynamics and thermodynamics, and simulates the climate system that governs interactions between the atmosphere, oceans, land surface and ice. Numerical simulation methods based on climate models have been extensively applied in paleoclimate research. Paleoclimate modeling often deals with long time spans, on centennial to millenial scales. Different climate models with increasing complexity, from simple energy-balance models (EBM), to Earth system models of intermediate complexity (EMIC) and three-dimensional coupled general circulation models (CGCM), can be chosen for specific scientific requirements so as to balance increasing simulation details while conserving computational resources. An EBM is mainly used to carry out sensitivity experiments of a specified process or feedback, in the absence of atmospheric dynamic processes. A CGCM can better describe spatial-temporal characteristics and feedback mechanisms of climate systems (atmosphere, ocean, land, cryosphere, biosphere; Fig. 6.18). CGCMs have been widely used in snapshot experiments or limited-long transient simulations, although it is not practical to run a CGCM for a very long time at present due to limited computer resources. An EMIC is somewhere between an EBM and CGCM, whose increased computational efficiency is achieved by simplifying dynamical processes to meet the requirement of modeling on long time-scales. Ice-sheet models, as an important but yet uncoupled component of earth system models of long-term climatic changes, such as glacial-interglacial cycles (GIC), are also employed in current paleoclimate modeling (Fig. 6.18).

Whereas modern climate studies focus on internal fluctuations within climate systems, paleoclimate research is focused on external forcing factors, like solar radiation and underlying surface conditions imposed by land, oceans, and ice sheets (Fig. 6.18). Our understanding of past climate through modeling developed from the superposition of climate system responses and various forcing factors at different temporal and spatial scales (Ruddiman 2001).

On the tectonic scale (millions of years), major driving factors for the Asian paleo- environmental variation are changes in the underlying surface conditions including mountain uplift and seaway closing or opening, associated with continental drift. The dominant forcing factor among them is undoubtedly the uplift of the TP, which was the most significant tectonic event in the Asian continent since the Cenozoic. The phased uplift of the TP since the Cenozoic regulated the thermal transport and dynamic processes of Asian climate, with feedbacks through the whole earth's climate system. This event directly contributed to the formation and development of the Asian monsoon-arid climate pattern. Changes in ocean conditions also have a great effect on Asian paleoclimate evolution. For example, a series of tectonic events including the Paratethys Sea retreat and Indonesian seaway closure have significantly changed the land-sea distribution, ocean circulation and global thermal conditions, producing a distinct influence on Asian climate.

Since the Quaternary, climate change has followed a broad pattern of periodic fluctuations. Besides significant glacial boundary conditions, the global climate evolution during the Quaternary is primarily impacted by orbitally-induced insolation changes. Similarly, the Asian monsoon-arid climate system also has quasi-periodic variations, under the influence of changes in orbital forcing, global ice volume and atmospheric greenhouse gas concentration. Based on classical Milankovitch theory, the direct driving factor of climate change at the orbital scale is believed to be the alteration of solar radiation induced by varying earth's orbital parameters including eccentricity, obliquity and precession. After the LGM, Asian climate experienced several rapid cooling or warming events, for example, the Younger Dryas event and Heinrich events. Presently, the explanation for these climatic events mainly focuses on internal feedback mechanisms of the coupled atmosphere-ocean system, particularly on variations of the thermohaline circulation.

During the past thousand years, the major factor impacting solar radiation was solar activity itself. Except for the direct observation with an astronomical telescope, scientists have already reconstructed several varying sequences of solar activity intensity in the past thousand years by using cosmic-ray isotope records such as ^{14}C and ^{10}Be from ice cores and tree-rings. The reconstructed sequences indicate that there are periods of 11, 90 and 200 years in long-term variations of solar activities. It has also been shown that the sun was extremely active in the most recent 70 years, probably the most active period during the past 8,000 years (Solanki et al. 2004). The solar radiation of the Maunder Minimum period, which appeared in the seventeenth century, decreased by 0.15–0.3 % compared to modern times. However, the intensity of solar activity during the Medieval Warm Period (MWP) in the twelfth century was close to the present day (Rigozo et al. 2001).

Volcanic eruptions can emit large amounts of sulfate aerosols into the atmosphere and therefore influence climate in various ways, including: affecting the atmospheric radiation balance and temperature depending on the optical properties of the aerosols, and modulating horizontal and vertical thermodynamic gradients, eventually influencing atmospheric circulation. Volcanic aerosols transported poleward in the atmosphere and deposited in glaciers give us a record of the earth's volcanic activity during the past. However, there is considerable uncertainty in inverting the aerosol settling rate due to imprecise information on their latitudinal distribution and other factors. Up to now, the most detailed and reliable observational record of a volcanic explosion is from the Pinatubo eruption in 1991, providing a good foundation for understanding its radiation effects (Sato et al. 1993).

Since the industrial revolution, climate change has resulted from the dual effects of natural and anthropogenic factors with increasing human activities and greenhouse gas emissions. In the past thousand years, the primary factors driving climate change include both natural forcing from solar and volcanic activities and anthropogenic forcing from enhancing greenhouse gases and changes in land use. The significant increases in greenhouse gases like CO_2, CH_4 and N_2O from anthropogenic emissions are considered to be the principal factors controlling recent global warming. The concentration of major greenhouse gases increased rapidly and continuously during the last 250 years. For example, the concentration of CO_2 increased from 280 ppm before the industrial revolution (1000–1750 AD) to 379 ppm in 2005, a much sharper increase than in the previous ten thousand years; CH_4 increased to 1,774 ppm in 2005, a concentration never achieved during the past 650 ka (IPCC AR4, 2007). In the past hundred years, the human effect on climate change has been close to, or has exceeded the natural effects from solar and volcanic activities.

6.4.2 Influence of Uplift/Growth of the Tibetan Plateau on the Evolution of the Asian Monsoon-Arid Environment in the Late Cenozoic

The uplift of the TP is the most remarkable tectonic event on Earth during the Cenozoic. The TP uplift plays a substantial role in Asian and global climate change and has always been an essential component of paleoclimate modeling studies. Due to the difficulty in reconstructing boundary conditions, climate simulations on a tectonic timescale usually focus on the sensitivity of the climate system to ideal or semi-ideal orographic forcing rather than on the history of climate evolution.

Various ideal sensitivity experiments with or without mountains have revealed more and more deeply the importance of the TP uplift on regulating global climate, particularly on the regional climate over Asia. Manabe and colleagues carried out contrasting simulations, with or without mountains, using early GCMs that showed the significance of the TP on the development of the Asian monsoon (Manabe and Terpstra 1974; Hahn and Manabe 1975). Their results show that the TP not only determines the position and intensity of the Siberian High in winter, but also controls the formation and evolution of the Asian summer monsoon. Afterwards, Kutzbach et al. (1989) used more sophisticated climate models to conduct simulations under no-mountain, half-mountain and whole-mountain conditions. Their results also showed that the TP uplift significantly reinforces the intensity of the South Asian monsoon and results in climate change across vast areas in Asia and beyond. Meanwhile, the appearance of the plateau and mountains reorganizes atmospheric circulation, splitting the westerlies into two branches and diverting airflow around the uplifted orographic barrier. The TP uplift can also cause global climate cooling and regional differentiation of climate. For instance, the uplift promotes wetter conditions east and south of the TP, and drier conditions to the west and north. Earlier studies mainly focused on climatic effects related to the South Asian monsoon. Since then it has become clear that the evolving mechanism of the East Asian paleomonsoon is more complex in many aspects.

In order to evaluate the evolution of Asian climate during the period of phased TP uplift in a manner that is more consistent with geological records, Liu and Yin (2002) used a GCM to carry out a series of experiments with systematic changes in the plateau topography. The elevations of all land grids in the modern large orographic area (10°N –60°N, 50°E–140°E) in the Eurasian continent were set to 100%, 90%, 80%, ... 20% and 10% of their modern terrain elevations, corresponding to the ten experiments named M10, M09, M08, ... M02 and M01. An extra experiment without mountains named M00 was also generated. The land grids on the Eurasian continent where modern elevations are greater than 200 m were set to 200 m in M00. The numerical simulations of the phased TP uplift indicate that the response of the East Asian monsoon is more sensitive to uplift than the South Asian monsoon. Moreover, the change in the East Asian winter monsoon is more remarkable than that of the summer monsoon during the course of the TP uplift.

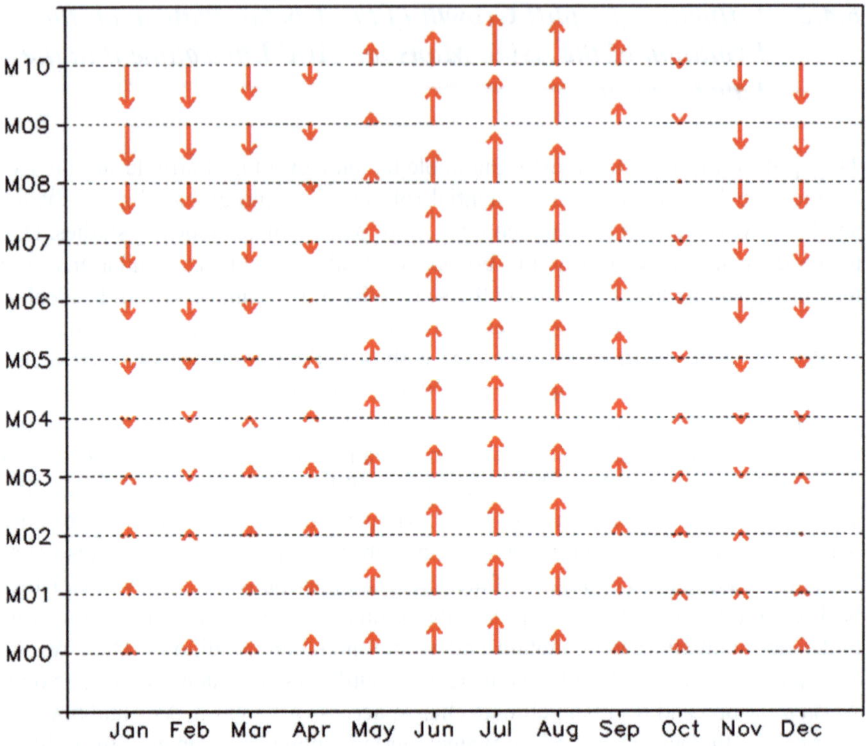

Fig. 6.19 Monthly mean variations of meridional surface wind velocity in the northern part of East Asia (105°E–120°E, 34°N–42°N) during the course of TP uplift. The M00, M01, M02 … M10 along the Y-axis represent numerical experiments with 0 %, 10 %, 20 % … 100 % of present terrain elevations, respectively, approximating a process of phased TP uplift. The upward (downward) wind vector arrows represent southerly (northerly) wind. The length of the July wind vector in M10 represents a wind velocity of 5.5 m/s, and the rest can be reasoned by analogy. (Redrawn from Liu and Yin 2002)

Figure 6.19 shows monthly mean variations of meridional surface wind velocity in the northern part of East Asia (105°E–120°E, 34°N–42°N) during the process of phased TP uplift. It can be seen that the southerly wind is predominant during the whole year in the experiments M00, M01, M02 and M03. Weak northerly winds appear in winter in M04, but they only lasts for 3 months, from November to January. Northerly winds obviously strengthen in winter in M05 and M06, and their duration reaches 5 months. In M07, M08, M09 and M10, the northerly wind is further reinforced, lasting for 6 months, from November to April. In the whole process of TP uplift, the southerly wind is slightly enhanced in summer, though the duration of southerly winds in a year is reduced with gradual increases in northerly winds in winter. The monsoon becomes established over northern East Asia along with the appearance of northerly winds in winter. Thus, there should be

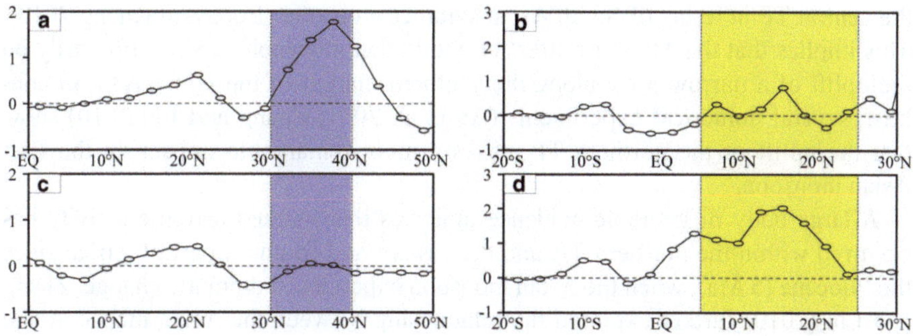

Fig. 6.20 Summer precipitation variation (mm/day) in East-Asian longitude zone (100°E–120°E; a, c) and South-Asian longitude zone (65°E–90°E; b, d) induced by the uplift of the northern TP (**a**, **c**), and uplift of the southeastern African plateau (**b**, **d**). The *violet* and *yellow* shaded areas represent the latitudes of northern East Asia and South Asia, respectively. (Redrawn from Zhang and Liu 2010)

no monsoon activity over northern East Asia in terms of the seasonal reversal of prevailing wind directions, before the TP uplifted to half of its present elevation. On the contrary, the formation of monsoon climate over southern East Asia could be relatively earlier than that over northern East Asia. The above modeling experiments further prove the importance of TP uplift in the development of aridification north of the plateau. The response of Asian inland aridification to TP uplift is nonlinear. Particularly in the later stages of uplift, where a relatively greater contribution to aridification may be expected for inland northwest China. Subsequent simulations with a coupled atmosphere-ocean model show a similar but intricate impact of the TP uplift on Asian climate, when ocean feedback mechanisms are included (Abe et al. 2003).

With progress in reconstructing the history of the TP uplift, modeling studies on the climatic effect of TP uplift are turning from performing simple sensitivity experiments of entire-plateau uplift to emphasizing regional differences, and the effects of various forms of uplift on the Asian paleo-environment. Based on geological facts supporting significant regional differences in the TP uplift, relatively complicated experiments of TP uplift are designed to examine the roles of different forms of uplift in the paleo-environmental evolution in different regions of the Asian continent by comparing modeling results to reconstructed climatic proxies. Various lines of geological evidence show that the TP terrain actually uplifted asynchronously, with uplifts earlier in the south, followed by uplift in the north (Molnar et al. 2010). The Himalayas and nearby mountains of southern Tibet reached more than 4,000 m before 40 Ma, while major tectonic activities occurred later in the northern TP. Recently, a number of modeling studies have explored the influence of sub-regional uplift of the TP on the Asian monsoon and arid environments (see the review by Liu and Dong (2013) for the further details). A numerical simulation merely considering TP uplift or Himalayan uplift shows that the dynamic effect of the lofty Himalayas is much greater than the thermal effect of the central TP in terms of South Asian summer monsoon (Boos and Kuang 2010). This implies that the TP uplift effect on the Indian monsoon depends primarily on the uplift of a narrow zone along the southern margin of the Himalayas. In contrast, several numerical experiments (An et al. 2001; Zhang and Liu 2010) show that the uplift of the northern TP makes a more remarkable impact on the East Asian monsoon.

A large body of geologic evidence indicates that marked tectonic activity has occurred within the northern TP, and the eastern and southern part of Africa since the Pliocene (5 Ma), when the Asian monsoon experienced dramatic change. Zhang and Liu (2010) further explored the relationship between the Asian monsoon and plateau uplift using numerical simulations. They completed a few experiments in which only the elevation of the northern TP or African plateaus was reduced, with a maximum reduction of 2,000 m in the northern TP or a decrease in elevation of up to 20 % of the present height in eastern and central southern Africa. Figure 6.20 shows the variation of summer precipitation averaged for the longitudinal zone of East Asia (100°E–120°E) or South Asia (65°E–90°E). The summer precipitation decreases mainly in northern East Asia (30N°–40°N; Fig. 6.20a) and changes little in South Asia (Fig. 6.20c) with the uplift of the northern TP. On the contrary, summer precipitation increases mainly in South Asia (between 5°N–25°N, Fig. 6.20c) and changes little in East Asia (Fig. 6.20d) as the African plateau becomes uplifted. These experiments provide a good example of how tectonic uplift in two different regions can induce very different effects on the Asian monsoon system and the inland arid environment.

A number of modeling experiments have been performed to study the effect of TP uplift on climate over inland areas. Earlier studies conducted ideal experiments with and without mountains and demonstrated that the TP can significantly enhance inland aridification (Ruddiman and Kutzbach 1989; Manabe and Broccoli 1990; Kutzbach et al. 1993). Intensified inland aridification could subsequently induce a significant increase in the atmospheric dust level, a phenomenon that has been well documented in eolian deposits downwind, e.g., the Chinese Loess Plateau and marine sediments from the North Pacific (Rea et al. 1998; Sun and An 2005). Shi et al. (2011) simulated the variation of eolian dust from Asian deserts during the Pliocene with a global climate-dust model, based on reconstructed dust source maps. Simulated dust mass accumulation rates (MAR) are in good agreement with marine and terrestrial observations. The results show that both regional tectonic uplift and global climate change are responsible for inland desert areas (Fig. 6.21) and MAR increases during the mid-Pliocene. Their relative contributions depend on the location of the depositional basins affected. For example, during the Late Cenozoic climatic cooling mainly controlled the sedimentation flux over the Loess Plateau, while TP uplift was more important in the North Pacific.

Other tectonic events in the Cenozoic, such as the Paratethys retreat, the expansion of the South China Sea and the closure of the Indonesian seaway must also be considered for their role in the development of the Asian monsoon-arid system. Simulations performed by Ramstein et al. (1997) showed that the retreat of the Pa-

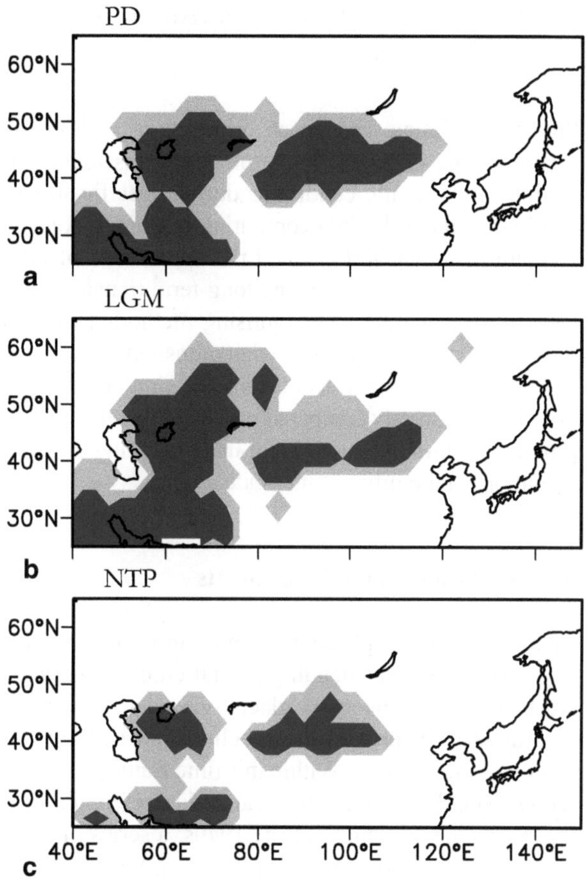

Fig. 6.21 Simulated inland arid regions during the present-day (**a**), last glacial maximum (**b**), and mid-Pliocene with a lower northern Tibet Plateau scenario (**c**). *Dark gray* shows regions with annual precipitation less than 200 mm and *light gray* shows those less than 300 mm. Differences between (b)/(c) and (a) generally indicate the contributions of climatic cooling/northern Tibetan uplift on the inland desert areas. (Redrawn from Shi et al. 2011)

ratethys promoted monsoon precipitation over South Asia before the TP uplift. The Paratethys retreat and expansion of the South China Sea changed both the regional land-ocean thermal contrast and corresponding atmospheric circulation. These factors worked together with the TP uplift to intensify aridity/humidity contrasts between East China and Central Asia and promote monsoon dynamics (Zhang et al. 2007). The closure of the Indonesian seaway is another important tectonic process that affected circulation in the tropical Indian Ocean and Pacific Ocean, and induced a large-scale redistribution of thermal conditions (Yu et al. 2003).

Although sensitivity studies focus on particular forcing factors that influence the Asian monsoon system, it should be kept in mind that they are not independent. It has been shown for example that the uplift of the TP can amplify the response of the East Asian paleomonsoon to orbital insolation forcing. It could be that the remarkable enhancement of orbital-scale variability of the Asian monsoon since the Pliocene has been enhanced by uplift of the TP (Liu et al. 2003a).

6.4.3 *Progress of Asian Monsoon Simulations on the Orbital Timescale*

Studies of climate change on orbital time scales, especially since the LGM, are a cornerstone of paleoclimate modeling research. Most early studies focused on equilibrium climatic conditions during specific stages, such as the mid-Holocene and LGM, due to limited computing resources. Recently, with the improvement of computational capabilities and the application of orbital acceleration technology, researchers have begun testing long-term transient simulations. These have rapidly become one of the most promising methods to investigate detailed relationships between orbital forcing and Asian monsoon fluctuations. Long-term climate evolutionary series obtained from transient simulations rather than equilibrium simulations can be directly compared with geological climate proxies. Thus, this technique provides an in-depth method for understanding the dynamic mechanisms of climate evolution on the orbital time scale.

6.4.3.1 Equilibrium Experiments

As the most recent glacial maximum in Earth's history, the LGM often serves as a key period for understanding glacial climate, particularly with regard to evolving mechanisms of Quaternary glacial-interglacial cycles. Existing plentiful geological records serve as proxies of past climate since the LGM and these have provided a wealth of information. Within this time frame, the LGM and Holocene warm period (HWP) stand out as reliable validation reference points for paleoclimate modeling. Since the beginning of the 1970s, a series of international research programs, e.g., "Climate: Long range Investigation, Mapping, and Prediction" (CLIMAP), the "Cooperative Holocene Mapping Project" (COHMAP) and the "Paleoclimate Modelling Intercomparison Project" (PMIP), have been proposed to reconstruct climatic changes over the past 20,000 years, and to investigate the driving mechanism of glacial cycles and understand the formation processes of ice-age environments. Until now, the most successful and influential projects among them are the two phases of PMIP (Braconnot et al. 2007).

PMIP1 began in the 1990s and includes 18 prominent atmospheric GCMs from around the world. Reconstructed boundary conditions (including distribution of ice sheets, solar radiation and greenhouse gas concentrations) during the LGM (21 ka BP) and middle Holocene (6 ka BP) were used in simulating the mean climate states for these two periods. The ability of AGCMs to simulate paleoclimate is assessed by analyzing similarities and differences in the models when responding to common boundary conditions. The modeling results reflect typical characteristics of global climate and the reinforcing trend of the East Asian paleomonsoon from the LGM to the Holocene, which is consistent with relevant geological evidence. However, the simulated variations are generally smaller than geological records indicate, perhaps because some potential feedbacks in the climate models are neglected. To refine the work of PMIP1, a second PMIP phase has been initiated. Climate models used for PMIP2 include several widely-used CGCMs and EMICs with feedbacks from the ocean and from vegetation. These are being run under improved boundary conditions to simulate LGM and middle Holocene climate.

Results of the PMIP2 model simulations show that the annual mean temperature is lower by about 2–5 °C in tropical regions and by up to 30 °C over ice sheets in the Northern Hemisphere (NH) during the LGM, as compared to the present day. The cooling in the NH is greater than in the Southern Hemisphere (SH) and the cooling over continents is often stronger than over the oceans due to albedo effects of ice sheets and the smaller thermal capacity of oceans (Braconnot et al. 2007). Characteristics of Asian climate change from PMIP2 simulations are also analyzed. Applications of regional climate models allow us to reproduce high-resolution climatic features in Asia (Zheng et al. 2004). A quantitative evaluation of climatic factors shows that the weakened summer monsoon and enhanced winter monsoon over East Asia during the LGM are mainly attributed to changes in sea surface temperature (SST), sea ice, ice sheets and terrain. Moreover, climatic feedbacks involving regional vegetation, soil conditions, dust, and snow-ice cover on the TP during the LGM have also been shown to be significant (Jiang et al. 2003). Climate change during the HWP is much subtler than that during the LGM. Compared to the present day, the most obvious change is a strengthened temperature difference between winter and summer in the NH and corresponding weakened seasonal temperature differences in the SH. Despite these differences, annual mean temperatures do not vary significantly. During this time, warming on land in boreal summer enhances the thermal low and promotes vapor transportation from tropical oceans. This enhances the tropical monsoon system (Braconnot et al. 2007). A coupled climate-vegetation model to simulate mid-Holocene climate has been used to show that the incorporation of vegetation feedback significantly improves climate modeling for the HWP, particularly for North Africa (Wang 1999a). A strengthened Asian monsoon system can intensify the Walker circulation and so increase upwelling associated with the easterlies of the tropical East Pacific Ocean. This can produce weakened El Niño-Southern Oscillation (ENSO) intensity during the mid-Holocene (Zheng et al. 2008b). It is interesting to note that climatic characteristics in East Asia during the HWP are likely to be similar to conditions following future global warming and could serve as a useful analog to evaluate future climate change. For example, a comparative analysis using simulations for the middle Holocene and late twenty-first century show that the patterns of summer precipitation are similar for the two warming periods in some regions of East Asia (Zhang and Liu 2009). This similarity is particularly true in two key regions: 1) the southern TP, which shows increased precipitation and 2) southwestern Xinjiang Province which shows reduced precipitation. The above changes in simulated summer precipitation also coincide qualitatively with geological climate records for the middle Holocene. Therefore, the pattern of East Asian summer precipitation change in the mid-Holocene can be taken as a historical analog for the future in some regions and this will be valuable for future climate change forecasting efforts.

6.4.3.2 Transient Experiments

Even with modern computing resources, it is still impractical to conduct paleoclimate experiments covering tens of thousands of years using coupled climate models. An equilibrium simulation can only retrieve a mean climate state for a specific period, but it cannot provide a long-term evolutionary history of a climate system to compare to geologic records, or analyze driving mechanisms responsible for climate change. While an equilibrium experiment is an expedient way of approaching the problem at present, long-term transient simulations are the goal for the future. Recently, with the improvement of computer resources and the appearance of orbital acceleration technology, scientists in the paleoclimate modeling community have attempted some transient experiments with GCMs on the orbital scale. For instance, the Max Planck Institute for meteorology used a three-dimensional coupled atmosphere-ocean GCM to carry out a climate integration covering the past thousand years (von Storch et al. 2004). Liu et al. (2009d) used the NCAR CCSM to conduct a transient simulation of up to 10,000 years from the LGM to the Bølling-Allerød (BA) warming period. The transient response of the Atlantic meridional overturning circulation induced by the sudden termination of the North Atlantic freshwater discharge and the related initiation of the BA warming were successfully simulated. Tuenter et al. (2005) produced a 130,000-year transient simulation using an earth system model of intermediate complexity and explored the response of the Asian and African monsoons to orbital forcing and their phase relationships. The problem with this approach is how to carry out continuous integrations over tens of thousands of years to understand the transient response of climate to orbital forcing. One solution is to apply orbital forcing acceleration technology (Lorenz and Lohmann 2004).

Kutzbach et al. (2008) made a study of global monsoon evolution using the coupled Fast Ocean Atmosphere Model (FOAM). They integrated for 284 ka with an orbital acceleration scheme and time-evolving forcing from orbital insolation. The results show that the phase of the Asian monsoon is close to the NH summer insolation, which indicates that the Asian monsoon during the Quaternary may be controlled by boreal summer insolation variability. Shi and Liu (2009) used these modeling outputs to further validate the phase relationship between the Asian summer monsoon response and orbital forcing at the precession cycle. Their results show that June-August (JJA) precipitation in the South Asian monsoon region is correlated with NH insolation and land temperatures in central Asia (25°N–40°N, 65°E–105°E), and with SST in the northern tropical Indian Ocean (0–15°N, 65°E–105°E) in April-June (AMJ) and May-August (MJJA; Fig. 6.22). The land temperature response is immediate whereas the SST response exhibits a phase lag of one month. In this case, it is the spring insolation that determines the thermal state on land, and regulates the development of the Asian summer monsoon. Summer precipitation in the South Asian monsoon region is actually controlled by spring insolation in the antecedent period rather than during the summer insolation months.

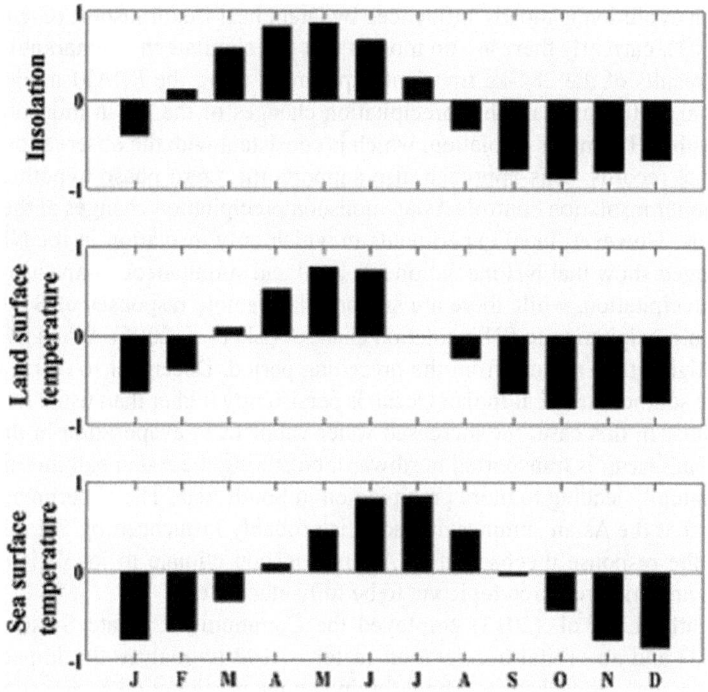

Fig. 6.22 Correlation coefficients of FOAM-simulated South Asian summer (June-July-August) precipitation with northern Insolation (**a**), central Asian land surface temperature (**b**), and northern tropical Indian Ocean surface temperature (**c**) from January to December

Long-term transient simulations are still in their infancy. At present, many attempts are based simply on orbital insolation forcing and greenhouse gas concentrations, due to the immature technology, and these studies often neglect boundary conditions, such as ice sheets. The question of how to design global ice sheet distributions and put them into transient experiments is still a difficult problem. Meanwhile, high-resolution geologic evidence for longer-timescales is relatively sparse. Therefore, it is unrealistic to expect a direct comparison between sequences of model outputs and climate proxies. Nevertheless, comparing model results with observations is still a meaningful approach at the present stage. For example, many geologic climate records and numerical simulations show a precession cycle in the evolution of the Asian monsoon, but what remains unclear is a dynamic understanding of how insolation changes induced by precession control climate (Liu and Shi 2009). The "zero phase hypothesis" stresses that the Asian monsoon evolution is completely controlled by NH summer insolation, while the "latent heat hypothesis" deems that it is controlled by latent heat flux transport from the South Indian Ocean induced by SH insolation changes, except for direct driving of NH summer insolation. Although marine deposits from the Arabian Sea indicate that the Asian monsoon evolution is mostly influenced by latent heat flux transport (Clemens and Prell 2003), currently there are no models that can simulate this remarkable effect.

The results of the 284-ka transient experiment using the FOAM model (Kutzbach et al. 2008) indicate that precipitation changes of the Asian monsoon are in phase with NH summer insolation, which is consistent with the observation of cave stalagmite records. This approach also supports the "zero phase hypothesis" that NH summer insolation controls Asian monsoon precipitation changes at the precession scale. However, ideal experiments in which only insolation in the NH or SH are changed show that NH insolation has local and simultaneous impacts on South Asian precipitation, while there are seasonal-lag remote responses of South Asian monsoon precipitation to SH insolation changes (Liu et al. 2006). When SH insolation is higher than normal from the preceding period, December to April, the SST from the southern tropical Indian Ocean is persistently higher than usual from January to July. In this case, the increased water vapor from evaporation in the southern Indian Ocean is transported northward, bolstering the Asian summer monsoon and eventually leading to more precipitation in South Asia. The experiments above suggest that the Asian summer monsoon is probably influenced by SH insolation. Hence, the response mechanism of Asian monsoon climate to insolation forcing remains an active research topic yet to be fully elucidated.

Recently, Li et al. (2013) employed the Community Climate System Model (CCSM3) and an orbital acceleration factor of 100 to analyze the impact of the insolation changes caused by orbital forcing on the precipitation in monsoonal East Asia (EA) and arid Central Asia (CA) over the past 150ka. Their results suggest that annual precipitation in both EA and CA has strong signals of the 20ka precession cycles and is in-phase with NH mid-latitude summer insolation. Composite analyses based on seven precession cycles suggest that the anti-phase relationship between summer insolation and winter insolation of the Northern Hemisphere induced by the procession cycle controls the corresponding summer precipitation in EA and winter precipitation in CA, respectively. When summer insolation increases, there is a greater warming over the land mass of continental Asia than over the surrounding ocean, and the summer monsoon is enhanced with northward extension due to the amplification of the ocean-land thermal contrast. As a result, precipitation increases significantly in the monsoonal EA region. Meanwhile, the increase in the summer insolation is accompanied by a decrease in the winter insolation that causes dramatic cooling of the troposphere in low latitude regions from North Africa to the Mediterranean Sea and CA, while changes in high latitude regions are minimal. Consequently, the temperature gradient in mid- to high-latitudes is weakened and so is the intensity of the westerly circulation. This circulation pattern tends to induce more frequent cold air outbreaks and also enhances westerly flow in lower latitudes, transporting water vapor from the Atlantic to CA via the Mediterranean, and thereby increasing winter precipitation in CA. Therefore, the responses of the Asian monsoon and westerly circulation to summer and winter insolation induced by the precession cycle determine precipitation in the respective rainy seasons, and are the primary causes leading to the synchronous variation pattern of precipitation in the monsoonal EA and the arid CA at the orbital scale.

6.4.4 Overview of Modeling Climate Change during the Past Millennium, for the Last Century and in the Future

The past millennium is a period when the human population on Earth rose dramatically and human activities became an increasingly important factor for climate. Studying climate change during the past thousand years, particularly the last 100 years, is important not only for understanding dynamic mechanisms within the climate system, but also for providing useful information to predict future climate change. Although there are still disputes about the reconstructions of NH and global mean temperature sequences during the past millennium, the current major view is that climate evolution is characterized by three typical stages, including: 1) the MWP, 2) the LIA, and 3) twentieth century warming. The present abnormal warming is unprecedented, at least for the past 1,300 years (Mann et al. 2008). Hansen et al. (2006) argued that present global mean temperature might be at, or very close to, the maximum value since the Holocene, and it is less than 1 °C from the maximum mean temperature for the past million years. The reconstructed temperature sequence in China shows that the warmest 50 years for the last millennium occurred in the second half of the twentieth century (Wang et al. 2007a). A key question for climate change research is to distinguish relative contributions of natural and human factors. Natural climatic changes include fluctuations like the ocean thermohaline circulation caused by intricate interactions in the coupled ocean-land-atmosphere-ice system, and the variability induced by external forcing factors such as solar and volcanic activities. Human effects include greenhouse gas emissions, aerosol releases, and land use.

For the mechanism of climate change in the past millennium, studies have concentrated mainly on temperature changes for several specific periods including the LIA, MWP and twentieth century warming. Climate models are used to quantitatively analyze the relative contributions of various climatic forcing factors such as solar activity, volcanic eruptions, greenhouse gas emissions and land cover changes. The IPCC Fourth Assessment Report, which summarized available climate simulations on the past thousand years, indicates that all climate models can reproduce basic characteristics of these three typical periods, including the warming during the twelfth to fourteenth centuries, the cooling in the mid-fifteenth, seventeenth and early nineteenth centuries, and the subsequent unprecedented temperature increase of the twentieth century, if the comprehensive impacts of natural external factors (solar and volcanic activities) and human influences (emissions of greenhouse gases and aerosols) are considered. Warming in the MWP can be ascribed to increased solar activity and decreased volcanicity, while the cooling in the LIA is basically due to the opposite conditions. If only the natural forcing is considered, climate models cannot reproduce the global warming of the twentieth century. Only when anthropogenic forcing is included can the models reproduce well the major characteristics of the twentieth century warming. Therefore, the recent global warming, especially in the late twentieth century, is considered to be closely associated with anthropogenic greenhouse gas emissions.

Modeling studies on regional climate evolution in China for the last thousand years have also recently advanced. Solar activity and volcanic eruptions are deemed to be major reasons for the appearance of the LIA and MWP in China. Based on the long transient simulation from the three-dimensional coupled atmosphere-ocean model of the Max Planck Institute for meteorology, Liu et al. (2009c) observed that the simulated LIA (1450–1850 AD) and MWP are stages with reduced and increased global monsoon precipitation, respectively. The modeling results are consistent with reconstructed temperature sequences in East Asia, which demonstrate that climate change is mainly modulated by the above factors during the past thousand years. The CCSM2-simulated summer precipitation over East Asia further shows that the century-timescale change is actually linked to a cyclical fluctuation in solar activity (Shen et al. 2009) and inter-decadal climate variability is probably related to feedbacks within the climate system. In addition, several simulations have dealt with the formation of rapid climate change on timescales of hundreds to thousands of years. For example, experiments using a coupled atmosphere-ocean model (Manabe and Stouffer 1995) show that fresh water injected continuously into the North Atlantic Ocean leads to a rapid climatic cooling in Greenland, Scandinavia and western Europe during subsequent centuries. A numerical experiment with EMIC indicates that East Asian climate change during the LGM is closely related to D/O and Heinrich events in the North Atlantic Ocean (Jin et al. 2007).

Asian climate in the twenty-first century will be changing under the background of global warming induced by increasing greenhouse gas emissions. It is also influenced by other kinds of human natural factors such as aerosol emissions, land cover changes and solar activities. The analyses of the IPCC multi-model outputs show that the Asian monsoon circulation will weaken, although a rise in the Asian continental temperature will cause monsoonal precipitation to increase. Meanwhile, drought frequency in central Asia will increase. Another interesting phenomenon is that the amplitude of surface warming may increase with elevation in the TP and neighboring regions under global warming conditions (Liu et al. 2009a). This is consistent with observations in the twentieth century (Liu and Chen 2000). Liu et al. (2009a) produced a detailed model analysis using outputs simulated by the NCAR CCSM3 model, with T85-resolution (~ 140 km) and steadily increasing atmospheric CO_2forcing, with 1% increases per year. They divided the model grids for the TP and neighboring regions (25°N–42°N, 70°E–110°E) into 11 grades according to their elevation, with 0.5 km increments (0–0.5 km, 0.5–1.0 km, … 5.0–5.5 km). They then calculated the warming rates of annual mean, maximum, minimum temperatures and the mean elevation of grids in each grade (Fig. 6.23a). Fig. 6.23b is the same as Fig. 6.23a but for the winter half of the year (November to March). The X-axis in the figure also shows the numbers of model grid points in each elevation range. During the course of global warming induced by increasing atmospheric CO_2, climate warming presents a significant elevation dependency and the trend of minimum temperature is particularly remarkable in cold seasons. For instance, the warming rates of annual averaged (and minimum) temperature are 3.0 (3.3), 3.5 (3.7) and 4.6 (5.4) °C/100 a for three topographic zones with elevations in the range of 1–1.5 km, 3–3.5 km, and 5–5.5 km, respectively, and they reach 3.1 (3.7), 3.8

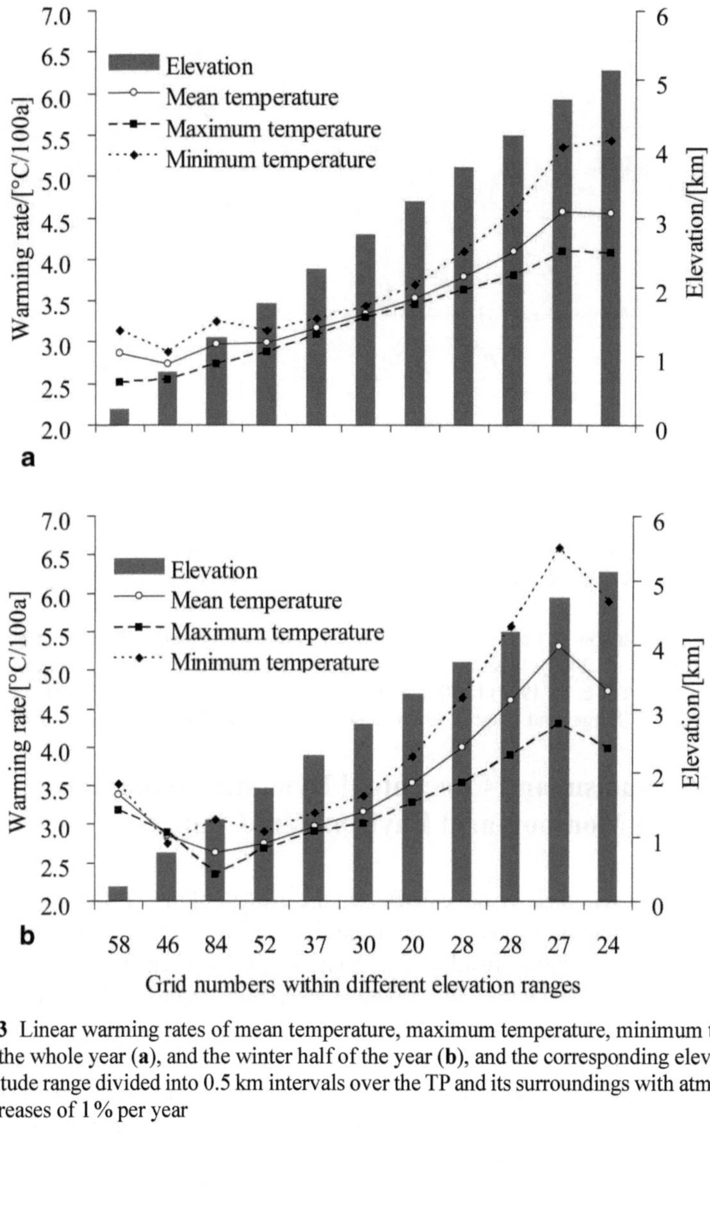

Fig. 6.23 Linear warming rates of mean temperature, maximum temperature, minimum temperature for the whole year (**a**), and the winter half of the year (**b**), and the corresponding elevations in each altitude range divided into 0.5 km intervals over the TP and its surroundings with atmospheric CO_2 increases of 1 % per year

(4.2) and 4.9 (5.9) °C/100 a in the winter half of the year. This strong warming over the TP and its elevation dependency may be caused mainly by increasing greenhouse gases and may also be influenced by feedbacks of ice-snow albedo, clouds and water vapor, and heating effects of dust and absorbed aerosols from South Asia. It should be noted that great uncertainties still exist in the projection of future climate change due to unquantified forcing factors that may be expected to develop in the coming decades.

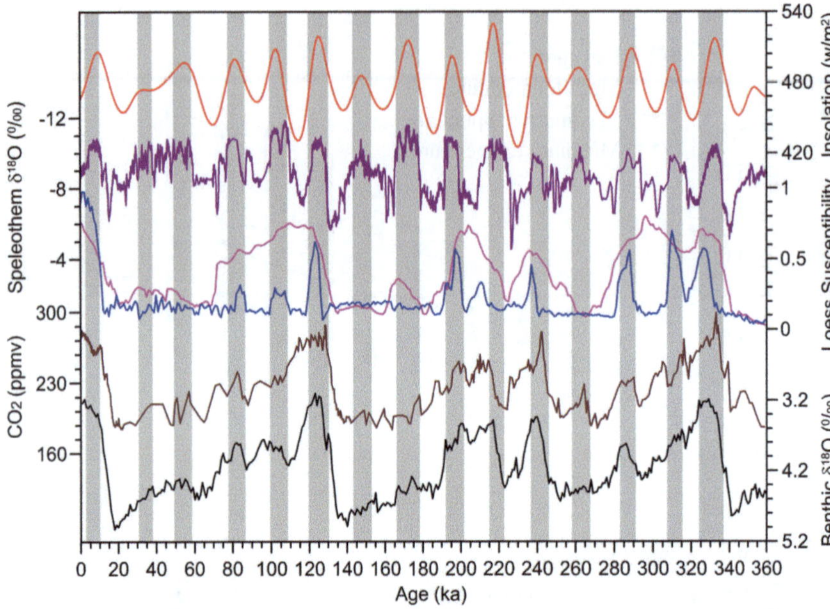

Fig. 6.24 Comparison of East Asian monsoon variation (speleothem $\delta^{18}O$ (Wang et al. 2001, 2008c; Cheng et al. 2009), loess magnetic susceptibility (Sun et al. 2006a, 2006b)) with CO_2 concentration, (Petit et al. 1999) benthic marine $\delta^{18}O$ (Lisiecki and Raymo 2005) and NH July insolation at 65° (Berger and Loutre 1991) changes over the last three glacial-interglacial cycles

6.5 Mechanism and Conceptual Dynamic Models for Asian Monsoon-arid Environment Changes

In the previous sections, we have discussed the growth of the TP, the onset and evolution of the Asian monsoon-arid environment during the Cenozoic, and relevant climate model simulation results. The relationship between the Asian monsoon-arid environment and TP growth, as well as its connection to global environmental change has been discussed on a tectonic time scale. In this section, emphasis will be placed on environmental changes at orbital, glacial-interglacial, millennial and shorter time scales. We will focus on Asian monsoon variations and their dynamic driving forces (e.g. solar radiation, atmospheric CO_2 and lower boundary conditions), as well as conceptual models at different time scales.

6.5.1 Mechanisms of Orbital-Scale Monsoon Variability

The evolution and underlying mechanisms that control the Asian monsoon variability on orbital time scales have received considerable attention from both geological and modeling researchers. Orbital time scale monsoon variations have been investigated extensively from loess (Kukla 1987; An et al. 1991a; Ding et al. 1995;

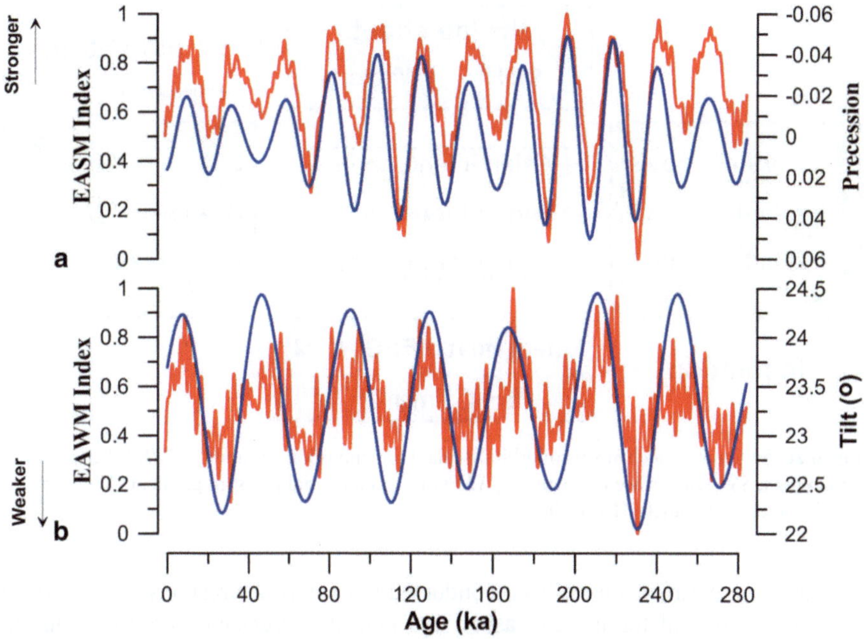

Fig. 6.25 Simulated index of East Asian summer and winter monsoon (*red*) for the past 280 ka: **a** summer monsoon, **b** winter monsoon. The high-frequency components (<5 ka) are filtered. The orbital parameters (*blue*) are shown for comparison. (Recalculated and redrawn from Fig. 3 of Shi et al. 2011)

Liu and Ding 1998; An et al. 2000), marine sediments (Clemens et al. 1991, 1996; Clemens and Prell 2003; Wang et al. 2003b), and speleothems (Wang et al. 2008c; Cheng et al. 2009). Loess-based proxies (e.g., magnetic susceptibility and grain size) display glacial-interglacial variations similar to those observed in deep-sea benthic $\delta^{18}O$ records, with a dominant 100-ka cycle and weak 41- and 23-ka cycles over the past 800 ka, implying that the East Asian monsoon variation was strongly coupled with NH ice volume changes (e.g. Ding et al. 1995; Liu et al. 1999).

In the last decade, high-resolution, absolutely dated speleothem $\delta^{18}O$ records indicate that the East Asian summer monsoon was characterized by dominant precession cycles and nearly in-phase response to NH summer insolation forcing (Wang et al. 2008c; Cheng et al. 2009). Integration of monsoon records from Chinese loess, speleothems and marine sediments from the South China Sea and Arabian Sea, however, indicate that both Northern and Southern Hemisphere (SH) climates have influenced the Asian monsoon on orbital time scales (Clemens et al. 2010). A comparison of Asian monsoon proxy variations with changes in insolation, CO_2 concentration and global ice volume suggests that solar insolation likely plays an important role in driving summer monsoon variations, whereas changing lower boundary conditions (i.e., ice volume and CO_2) may be significant factors in modulating glacial-interglacial monsoon variations (Fig. 6.24).

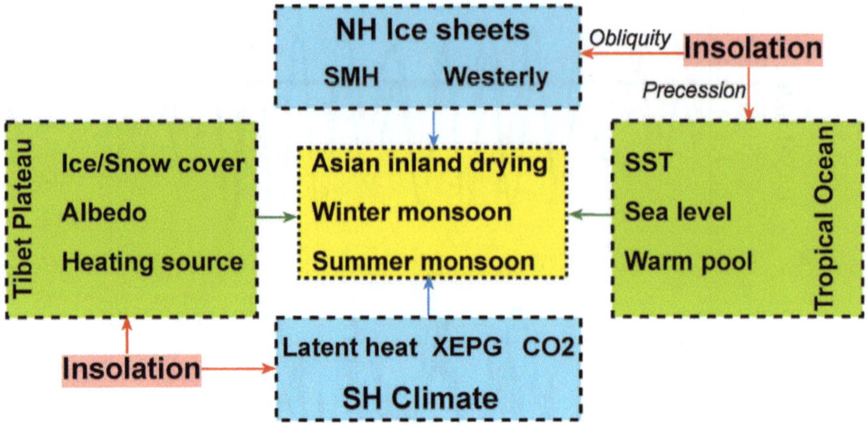

Fig. 6.26 Schematic mechanisms of orbital-scale Asian monsoon variability. *SMH* Siberian-Mongolian high, *SST* sea surface temperature, *XEPG* cross equatorial pressure gradient, *NH* Northern Hemisphere, *SH* Southern Hemisphere

Various experiments have been conducted to explore the response of the Asian monsoon to orbital forcing and associated glacial-interglacial variations. Due to computational resource limitations, the former paleoclimate experiments were performed as "time-slices" or "snapshots". These experiments focused on particular periods, in order to obtain an equilibrium state for the climate system under prescribed boundary conditions. With orbital parameters set for several thousand years ago, monsoon climatology during the Holocene has been obtained (e.g. Kutzbach 1981; Liu et al. 2003b) and precession was proposed as a dominant orbital parameter controlling the Indian monsoon. The developing orbital acceleration technique has recently provided some long-term transient simulations covering several orbital cycles (Tuenter et al. 2005; Weber and Tuenter 2011). These studies offer new insights into the evolving monsoon. For example, a 280 ka transient simulation (Kutzbach et al. 2008), that only takes orbital forcing into account shows that an index of East Asian summer monsoon intensity is nearly synchronous with northern summer insolation changes, supporting the view that the Asian summer monsoon is primarily controlled by precession. However, the East Asian winter monsoon more likely follows obliquity, with a period of approximately 40 ka (Fig. 6.25; Shi et al. 2011). The distinct responses between summer and winter monsoon highlights the complexity of monsoon responses to orbital forcing and also the importance of obliquity, which was unexpected.

The effect of global ice sheets on the Asian monsoon is also very significant. Results from the Paleoclimate Modeling Intercomparison Project Phase II (Braconnot et al. 2007) show that the glacial effect on the intensity of the East Asian monsoon is not as simple as expected. The East Asian summer monsoon consistently shows a significant weakening during the Last Glacial Maximum (LGM; Jiang and Lang 2010), which is also the case for the Indian monsoon (Prell and Kutzbach 1992). The winter monsoon, as measured by meridional wind speed at 850 hPa, strengthened in 10 out of the 21 models but weakened in the other 11 models, with an average weakening of 4% (Jiang and Lang 2010).

Variations in the volume and extent of ice sheets can also modulate the orbital phase of monsoon response, which varies with the distance to the Eurasian ice sheet (Weber and Tuenter 2011). Similarly, aridity in the climate over inland Asia is more or less strengthened during the LGM. Decreased moisture transport can increase the aridity of the dust source areas, especially over the Middle East and Central Asia, which contributes significantly to elevated LGM dust fluxes reconstructed from geological records (Mahowald et al. 2006; Shi et al. 2011). Furthermore, SH climate (e.g., Antarctica temperature, CO_2 emission from the Southern Ocean) can influence the glacial-interglacial Indian monsoon by enhancing the cross-equatorial pressure gradient between the northern Indian low and Mascarene high over the southern Indian Ocean (An et al. 2011).

In summary, changes in the Asian arid-monsoon environmental system were directly driven by external solar radiation and strongly modulated by changing lower boundary conditions. These changes involved interactions among the atmosphere, ocean, land and ice under different solar radiation conditions in East Asia (Fig. 6.26). External orbital forcing is dominated by precessional-scale summer insolation changes, which have a profound influence on the Asian summer monsoon (Kutzbach et al. 2008; Wang et al. 2008c), but also result in a distinct obliquity-scale meridional thermal gradient associated with winter monsoon variability (Shi et al. 2011). SH insolation-induced latent heat transport and cross-equatorial currents also drive Asian monsoon variability on orbital time scales (Clemens et al. 2010). In addition, changes in the tropical ocean (e.g., SST, warm pool, and sea level) can also enhance the summer monsoon but have limited influence on the winter monsoon (Wang et al. 2005). The TP amplifies monsoon variability on orbital time scales as a source of heating during summer and cooling in winter (Liu et al. 2003a).

6.5.2 Glacial-Interglacial Time Scale

The variation of Asian monsoon circulation has been suggested to be a response to solar-radiation on the earth's surface as a result of orbital changes and seasonal cycles. Since the strength of the monsoon depends upon the pressure gradient, it is also suggested that the variation is related to underlying surface conditions on a glacial-interglacial time scale (An et al. 1991b, 2011). Within this time scale there are obvious marine transgressions and regressions, implying that the pressure gradient associated with the inland-sea distance could influence the intensity and range of the monsoon. Under modern climate conditions, the monsoon front rarely penetrates as far as the Qilian and Liupan Mountains, to the northwest. During glacial times, the distance from the oceanic moisture source to inland Asia increased, as the coastline moved southeastward. There are three major large-scale marine transgressions (125–73, 40–25, and 11–6 ka) and two major regressions (50–45 and 18–15 ka) that occurred in the last 130 ka along the eastern and southern coasts of China. During the last glacial cycle, the well-developed marginal seas and continental shelves of

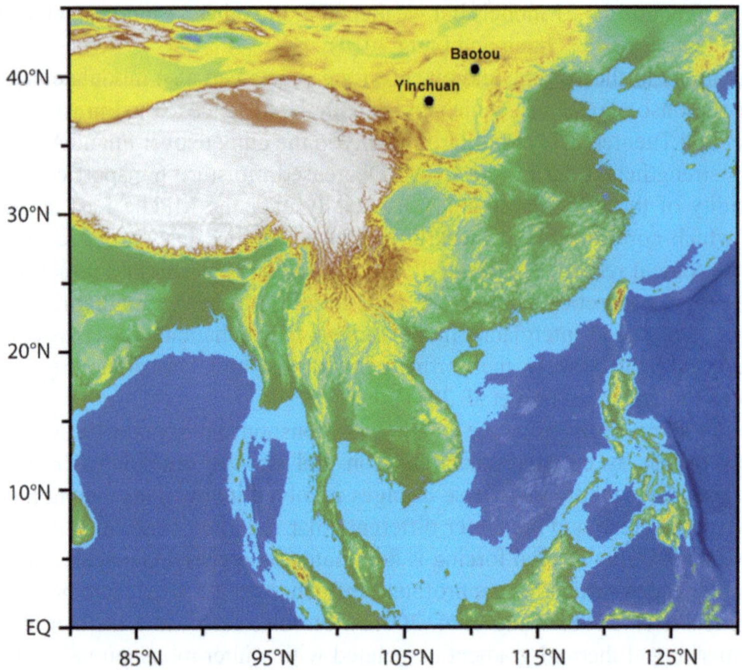

Fig. 6.27 Regional relief of southeastern Asia (GTOPO30 data distributed by U.S. Geological Survey's EROS Data Center). *Light blue* denotes areas of marginal seas emergent during the LGM when the sea level was ~ 130 m lower than present. The cities of Yinchuan and Baotou are marked. It has been shown that the coastline in eastern China shifted eastward by ~ 1,000 km during the LGM. (An et al. 1991a)

East Asia and the West Pacific Ocean experienced significant changes in area extent due to changes in sea level. Lower sea levels during the last glacial would have reduced the marginal seas and led to the exposure of the continental shelves. For instance, the marginal sea from East Asia to Australia is estimated to have exposed about 4×10^6 km^2 during the LGM (An 2000). The distance from inland Yinchuan and Baotou areas to the coast of the Yellow Sea is now 1,100 km, whereas during 18–15 ka of the LGM, when sea level dropped by at least 130 m, the distance was about 2,100–2,200 km, twice as far as at present. During the LGM, the exposure of the continental shelf that resulted from the drop in sea level drastically decreased the area of the Yellow sea north of the Taiwan Strait, to about 1/3 of its current area, and reduced the area of the South China Sea to about 1/5 of its current area (Fig. 6.27). This increase of distance further reduced the weak summer monsoon during the LGM and greatly decreased the amount of summer precipitation in the Yinchuan and Baotou areas (An et al. 1991a). Decreased moisture and heat transported to the atmosphere resulting from the shrinkage of the low latitude oceans and decreased sea surface temperatures (SST) contributed to weak monsoon intensity during the last glaciation. The winter SST decrease in the Chinese marginal sea was greater than that in the West Pacific, and thus strengthened climatic seasonality in the South China Sea. During the Holocene, the Asian Summer Monsoon inten-

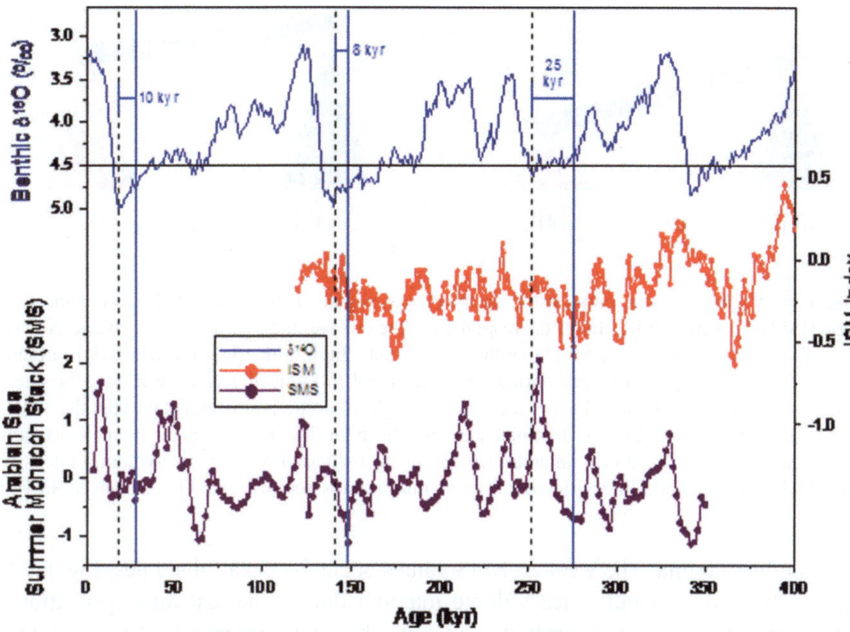

Fig. 6.28 The 350 ka Arabian Sea Summer Monsoon Stack (SMS; Clemens and Prell 2003) and the Heqing ISM index relative to the global ice volume (benthic $\delta^{18}O$, LR04 age scale; Lisiecki and Raymo 2005). The SMS also indicates that the weakest monsoons (*blue lines*) lead ice volume maxima (*black dashed lines*) in glacial times. (cf Fig. 3.14)

sity increased markedly as recorded in many lake and peat sediments in semiarid zones, as the monsoon front penetrated further inland, resulting in relatively high lake levels in the early Holocene, such as at Lake Qinghai (Lister et al. 1991). The decreased distance from coastline to inland likely contributed to the strengthening of the Asian Summer Monsoon through increased moisture supply. Therefore, the change in sea level and associated translocation of the coastline in East and Southeast Asia play important roles in regulating the summer monsoon over inland Asia on glacial-interglacial time scales.

The importance of NH ice volume and Antarctic temperatures in driving the Indian Summer Monsoon (ISM) on glacial-interglacial time scale has been demonstrated with an ISM record from Heqing paleolake, southwestern China, by comparison with marine ISM and global climate records (An et al. 2011). It was shown that the influence of changing lower boundary conditions on ISM circulation is focused, especially with regard to the timing of ISM maxima and minima relative to changes in NH ice volume, and Antarctic temperatures. Transitioning from interglacials into glacials, ISM strength reaches a minimum when benthic $\delta^{18}O$ reaches ~4.5 ‰, corresponding to low, often minimal Antarctic temperatures (δD minima). These ISM minima occur 14–35 ka earlier than the timing of NH glaciation maxima. Similar leads are found in the 350-ka Summer Monsoon Stack (SMS) from five proxy records of summer monsoon variability at three sites in the Northern Arabian Sea (Clemens and Prell 2003). The SMS also shows early monsoon minima relative to

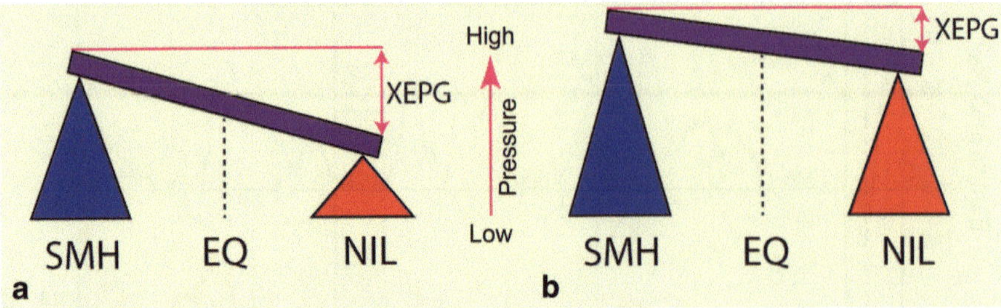

Fig. 6.29 A schematic diagram illustrating the proposed glacial-interglacial Indian summer monsoon (ISM) dynamics. It illustrates the responses of the cross-equatorial pressure gradient (XEPG, body force of the ISM) to the SMH (southern Mascarene high, blue triangles) and NIL (northern Indian low, *red* triangles) changes during interglacial and full glacial periods. **a** During the interglacial, less NH ice volume enhances the NIL and high Antarctic temperatures weaken the SMH. The large XEPG (strong ISM) is dominated by the NIL. **b** During the full glacial, high NH ice volume weakens the NIL and low Antarctic temperatures (increased Antarctic ice volume) intensify the SMH. The early ISM intensification from its minimum is determined by the enhanced SMH

ice volume maxima. The weakest SMS values occur 8–25 ka (blue lines, Fig. 6.28) earlier than corresponding ice volume maxima during the last three glaciations. These two independent records demonstrate that ISM minima lead ice maxima, although the lead timings are not the same. The difference may be related to various proxies and/or time resolution (An et al. 2011). Subsequently, the ISM strengthens gradually (black arrows, Fig. 3.14 of Sect. 3.3) even as NH ice volume continues to increase. The early ISM minima, and the slowly strengthening ISM that follow, are hypothesized to be linked to low SH temperatures as depicted by Antarctic δD (Jouzel et al. 2007), increasing Antarctic ice volume (Pollard and DeConto 2009) and low sea surface temperatures (SST; Mashiotta et al. 1999; pink curve in Fig. 3.14 in Sect. 3.3). To explain this link mechanistically, we note that polar temperatures vary with greater amplitude than do tropical temperatures (Braconnot et al. 2007).

During interglacials, ISM maxima are driven by an enhanced Indian low associated with both global ice volume minima and abundant moisture supply. Moving from peak interglacials into glacials, increasing NH glaciation decreases the strength of the Indian low. This is consistent with previous proxy studies (Clemens and Prell 2003; Clemens et al. 2008) and numerical simulations (Prell and Kutzbach 1992; Kim et al. 2008), indicating weakening of the ISM with growing NH ice sheets. However, ISM minima occur when NH glaciation reaches a level equivalent to ~4.5 ‰ in the benthic $\delta^{18}O$ record and Antarctic temperatures are near their lowest. Subsequently, ISM strength begins to slowly increase even as glaciation continues to build, and the moisture supply is relatively low (Kiefer et al. 2006). This glacial ISM strengthening can be attributed to an increasing cross-equatorial pressure gradient (XEPG) resulting from low Antarctic temperatures (Jouzel et al. 2007) and increasing Antarctic ice volume (Pollard and DeConto 2009). This combination leads to enhanced SH atmospheric meridional circulation and a strengthened Mascarene high. This dynamical SH forcing more than compensates for the weakening of the Indian low associated with increased NH glaciation. SH forcing has been suggested previously to explain Asian monsoon climate events during the last glaciation (An 2000; Cai et al. 2006; Rohling et al. 2009). Modern observations also show that abnormally cold Antarctic temperatures lead to a stronger Mascarene high that tends to strengthen the cross-equatorial Somali jet (Xue et al. 2004) and the ISM.

When global ice volume decreases and passes the threshold of ~4.5‰ into interglacial time, the continental Indian low rapidly intensifies and thus enhances the XEPG and ISM, together with an increased moisture supply from the warming tropical Indian Ocean.

Interglacial ISM peaks that are in phase with NH ice volume minima are interpreted as responses to a dominant enhanced Indian low, in spite of simultaneously increased Antarctic temperatures that act to weaken the Southern Ocean high. Analyses of simulated sea level pressure show a strengthened Indian low and ISM in the mid-Holocene (Liu et al. 2003b), which illustrates the remarkable northern influence on the interglacial XEPG. In this way, the lower boundary condition changes of both hemispheres are indispensable for understanding glacial-interglacial monsoon dynamics.

Meanwhile, the decrease in ISM strength from peak interglacials into glacials is not linearly related to an increase in ice volume. Secondary ISM peaks occur during the slow transition from interglacials into glacials and generally correlate with low values of Antarctic temperatures (peaks a through f, Fig. 3.14 in Sect. 3.3). It is plausible that the secondary peaks result from the combined influence of NH ice volume and SH temperature change, consistent with large concentrations of variance at non-primary orbital periods (e.g. 29, 55, and 73 ka) in the ISM spectrum (An et al. 2011).

In summary, at the glacial-interglacial timescale, ISM intensity is generally enhanced primarily by northern low pressure during interglacials and by southern high pressure during glacials (Fig. 6.29).

6.5.3 Monsoon Variability on Millennial to Centennial Time-Scales and Driving Forces

The cognition of climate change on millennial time-scales was intimately associated with the study of abrupt climate change. In 1950s, as one of few scientists who firstly noticed and promoted the study of abrupt climate change, Wallace Broecker pointed out that there were severe changes around the world all dated to about the same time and suggested that "two stable states exist, the glacial and the interglacial state, and that the system changes quite rapidly from one to the other" (Broecker 1957). Although later studies showed that the coincidence of these changes was an artifact of inaccurate dating, this idea persisted. In the 1970s, using the ^{14}C dating technique, the Younger Dryas event was demonstrated to be a rapid climatic shift near the end of the last deglaciation (Webb and Bryson 1972), and other studies showed cases of other abrupt climate change, e.g. Kukla's study of snail shells and pollen in layers of loess in Czechoslovakia (Kukla and Koci 1972). The instability of climate was then strongly reinforced by ice core studies from Greenland and Antarctic and by deep-sea cores that cover much longer periods. Ice core records from

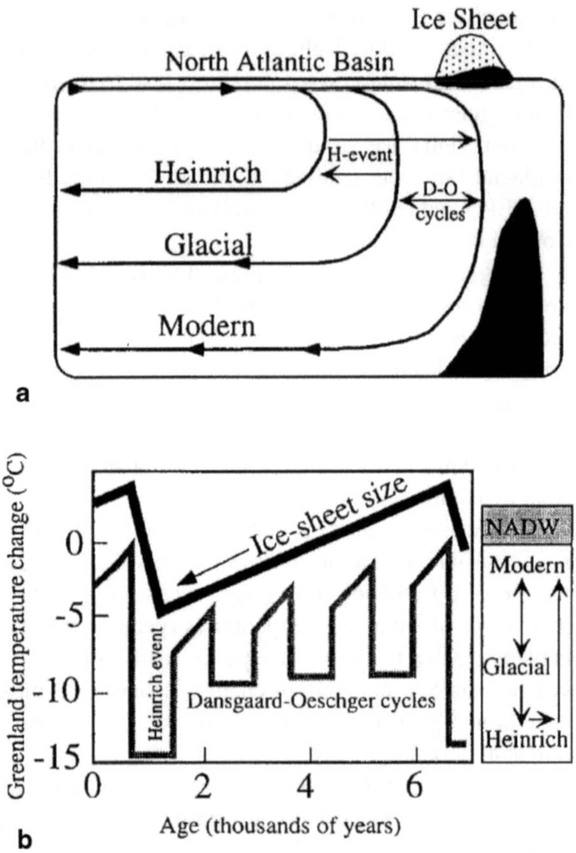

Fig. 6.30 **a** Conceptual schematic of three modes of North Atlantic circulation (modern, glacial, and Heinrich). Oceanic jumps between modern and glacial modes caused the Dansgaard/Oeschger (D/O) oscillations in which the cold, dry and windy signals of the glacial mode were transmitted elsewhere largely through the atmosphere. **b** An idealized time series of changes in climate between Heinrich events, during which several D/O oscillations occur. (From Alley et al. 1999)

Greenland clearly show multiple abrupt climate events over the last 110 ka BP. Each of these millennial-scale Dansgaard–Oeschger cycles were characterized by abrupt temperature shifts in Greenland, and consisted of a cold stadial and a warm interstadial pair (Johnsen et al. 1992; Dansgaard et al. 1993). Some of the cold stadials were also characterized by ice-rafted debris (or Heinrich) events in North Atlantic sediments (Heinrich 1988). Soon after, similar millennial-scale climate events have been found in numerous records from Northern Hemisphere (Porter and An 1995; Guo et al. 1996; An and Porter 1997; Schulz et al. 1998; Altabet et al. 2002; Spötl and Mangini 2002) and Southern Hemisphere (Lowell et al. 1995; Charles et al. 1996; Blunier et al. 1998).

Dansgaard's group once suggested that these abrupt swings of temperature in Greenland may reflect a "shift between different quasi-stationary modes of atmospheric circulation" (Dansgaard et al. 1982), as conveyed by the "dishpan" experiments in 1950s. However, it was hard to see how the atmosphere could settle into an entirely new state unless something drastic happened in the oceans, because seawater holds most of the heat energy and most of moisture and CO_2 of the climate system. The idea advocated by Wallace Breocker, i.e., the pattern of North Atlantic

Ocean circulation, which is responsible for the redistribution of heat between the tropics and the polar regions, could change on a short timescale (Broecker et al. 1985). This view was became accepted as the most likely processes controlling abrupt climate change (Broecker et al. 1989; Broecker and Denton 1989; Alley 2007; Denton and Broecker 2008; Denton et al. 2010).

Both observations and model simulations suggest that North Atlantic ocean circulation can be categorized into three modes: 1) a modern mode associated with warm times in the North Atlantic in which deep-water formed in both the Nordic Seas and further south in the North Atlantic; 2) a glacial mode in which deep-water formation largely stopped in the Nordic Seas, but where vigorous deep-intermediate water formation continued further south; 3) a Heinrich mode in which deep-intermediate water formation was greatly reduced in both locations (Alley et al. 1999). Shifts in ocean circulation between modern and glacial modes have been suggested as the cause of the Dansgaard/Oeschger oscillations. During the glacial mode, a cold, dry and windy signal, resulting from weakened ocean circulation, may have been transmitted through the atmosphere on a hemispheric or even global scale (Zhang and Delworth 2005) and sea ice changes over the North Atlantic regions may have played an important role in amplifying and transmitting the signal (Chiang and Bitz 2005). Extreme Heinrich events caused some similar, but stronger and more-widespread atmospheric anomalies than those of the cold phase of the Dansgaard/Oeschger oscillations (Fig. 6.30; Alley et al. 1999).

Various hypotheses have been proposed to explain the forcing behind these abrupt events. Internal oscillations in the ocean-ice-climate system were initially suggested as causes for these abrupt climate events, because these events operate on timescales that are close to ocean turnover times (Broecker et al. 1985; Ghil et al. 1987; Maasch and Saltzman 1990). Later, Mayewski et al. (1997) suggested that changes in insolation and associated positive feedbacks related to ice sheets were a major factor in explaining favorable time periods and controls on the amplitude of large and rapid climate change events. Recently, by using an intermediate-complexity climate model with glacial boundary conditions, Braun et al. (2005) generated rapid climate shifts similar to the Dansgaard–Oeschger events with a spacing of 1,470 years when adding periodic inputs of freshwater to the North Atlantic Ocean in cycles of ~87 and ~210 years. They attributed the robust 1,470-year response time to the superposition of these two shorter solar cycles, together with strongly nonlinear dynamics and the long characteristic timescale of the thermohaline circulation. They also suggested the D/O cycles were driven by external solar activity. In any case, the key to the climate dynamics of these abrupt events is likely to depend on circulation changes in the North Atlantic (Alley 2007; Broecker 2010).

These abrupt events have their impacts on the East Asian climate indeed. Porter and An (1995) found that the grain-size maxima in the central loess plateau correlated with Heinrich events during the last glaciation and also during the last interglacial time (An and Porter 1997). They proposed that the climate signals from the North Atlantic influenced the East Asian winter monsoon through changes in the westerly wind. Since then, these abrupt events have been identified in a wide number of different records from the LP and other East Asian monsoon regions

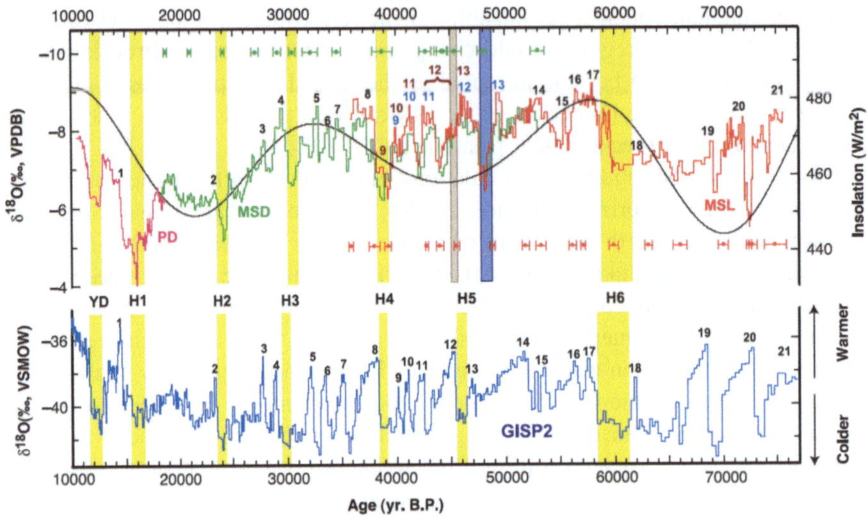

Fig. 6.31 Comparison of the $\delta^{18}O$ record of Hulu cave, East Asian summer monsoon record, and $\delta^{18}O$ of Greenland ice core, North Atlantic climate record during the last glaciation. (From Wang et al. 2001)

(Zhang et al. 1997; Chen et al. 1997; Ding et al. 1998; 1999; Fang et al. 1999; Wu et al. 2006), further confirming the teleconnection between the East Asian winter monsoon and the climate of high northern latitudes.

The East Asian summer monsoon and South Asian monsoon also demonstrate abrupt changes that are synchronous with temperature changes recorded in Greenland ice cores. Guo et al. (1996) found intervals of lower weathering intensity in the last glacial and late penultimate glacial loess, indicating significant weakening of the summer monsoon and/or strengthening of the winter monsoon. These corresponded to those of the coarse-grained Heinrich layers in the North Atlantic Ocean and it was suggested that both summer and winter monsoons have experienced high frequency changes approximately synchronous with Heinrich events if we also take the aforementioned loess studies into consideration. Sediment records from the northeastern Arabian Sea also show that Dansgaard–Oeschger and Heinrich events are strongly expressed in South Asian monsoon variability (Schulz et al. 1998; Altabet et al. 2002). They reveal that mild interstadials found in the Greenland ice cores reflect periods of global warmth, associated with high levels of tropospheric moisture and other greenhouse gases. This suggests that large-scale climate elements in the ocean–atmosphere system were interacting rapidly at both high and low latitudes, and points to the importance of common forcing agents such as atmospheric moisture and other greenhouse gases (Schulz et al. 1998). With high-precision U-series ages, high-resolution speleothem records from southern China further confirmed the correlation of the Asian summer monsoon and North Atlantic climate (Fig. 6.31)(Wang et al. 2001, 2008c; Yuan et al. 2004; Cai et al. 2006; Kelly et al. 2006). The speleothem study also demonstrates that the duration of millennial climatic events decreases, and their frequency increases, during glacial build-up in both the last and penultimate glacial periods, indicating that ice sheet size affects their character and pacing (Wang et al. 2008c), highlighting the effect of mean climate conditions on abrupt events (Sima et al. 2004; Wang 2006). Recently, An et al. (2012) found that the variations of Asian summer monsoon and the Westerlies show an anti-phase relationship both on glacial and millennial time scales, indicating the interplay of the Asian summer monsoon and the Westerlies in this region. This further confirmed the cold air excursions of high-latitude, mostly originated from North Atlantic, may transmit through the Westerlies and the Siberia High (Porter and An 1995).

Modeling studies show that ice-rafted events in the North Atlantic may result in the weakening of North Atlantic Deep Water (NADW) production, and increase sea ice during the winter season. This combination would produce a decrease in the temperature of the Northern Hemisphere (Zhang and Delworth 2005; Zhang et al. 2010). A decrease of temperature at high northern latitudes can increase the temperature gradient over the Northern Hemisphere, resulting in a southward shift of the Intertropical Convergence Zone (ITCZ) and weakening of the Asian summer monsoon (Chiang and Bitz 2005; Deplazes et al. 2013). More recently, high-resolution loess records from the western Loess Plateau dated by optical stimulated luminescence have shown that the strength of the East Asian winter monsoon over the past 60,000 years was broadly correlated with temperature variations over Greenland and Asian summer monsoon speleothem records (Sun et al. 2012). A coupled climate model simulation with added freshwater flux into the northern North Atlantic produces a strengthened winter monsoon circulation over the regions that supply dust to the Loess Plateau and a reduction in summer monsoon precipitation over East Asia. This study further confirmed that Atlantic meridional overturning circulation is a driver of abrupt change in the East Asian winter and summer monsoon systems, and that the northern westerlies play a role in transmitting this signal from the North Atlantic to the Asian monsoon regions (Porter and An 1995).

It is worth noting that these abrupt events have their own characteristics within the Asian monsoon regions, because not only cold air from high northern latitudes, but also moisture transport from the tropical ocean, trans-equatorial air streams from the Southern Hemisphere, and possibly ENSO influences, affect Asian monsoon variability. By synthesizing terrestrial records from different regions, Zhou et al. (2001) found that the Younger Dryas chronozone manifest itself globally in different ways and suggested that ENSO and other tropical factors may be superimposed on the high-latitude cold events through moist air transportation patterns from the tropics to higher latitudes. A comparison of East Asian monsoon records and north Australian eolian sediments during MIS3 demonstrates that the strengthening of the southern trade winds and the associated Southern Oscillation (Dedecker et al. 1991) led to an increase in precipitation in East Asia by cross-equatorial flow (An 2000). Moreover, the speleothem records from southern China, which reflect variations of the Asian monsoon, are not only highly correlated with Greenland ice core records, but also generally resemble Antarctic ice core records. Therefore, it has been suggested that the climate of the Southern Hemisphere contributes

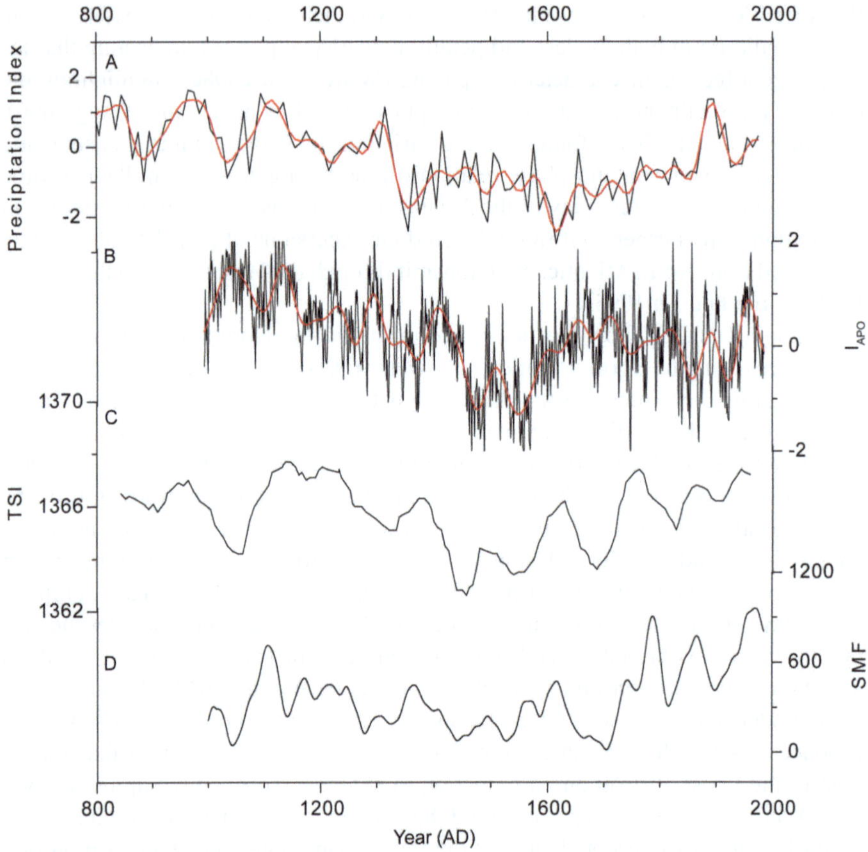

Fig. 6.32 Comparison of the precipitation variations in north central China, the Asian Pacific Oscillation and solar activity. *A* is a synthesized precipitation index record. *B* is an Asian Pacific Oscillation index record (Zhou et al. 2009). *C* is a reconstructed total solar irradiance (*TSI*) record (Bard et al. 2000). *D* is a reconstructed solar modulation function (*SMF*) record (Muscheler et al. 2007). The *red lines* in *A* and *B* have been smoothed with a 50-year low pass FFT filter. (from Tan et al. 2011a)

significantly to variations of the Indian summer monsoon and is an indispensable component of the Asian monsoon system on millennial time-scales (Cai et al. 2006; Rohling et al. 2009).

The Asian monsoon exhibits significant tectonic-, orbital- and millennial-scale variability (e.g., An 2000; Wang et al. 2008c), but also centennial- to decadal-scale variability (Wang 2006; Tan et al. 2011b). Centennial- to decadal-scale abnormal climate is often associated with severe disasters and significant loss of life and property (Liu et al. 2009b; Tan et al. 2011b), so these changes are the subject of increasing concern. Different proxy records such as tree rings (Liu et al. 2009b), stalagmites (Tan et al. 2011b) and historical documents (see Sect. 3.11 of this book) from different regions in monsoonal China show significant climate changes on a

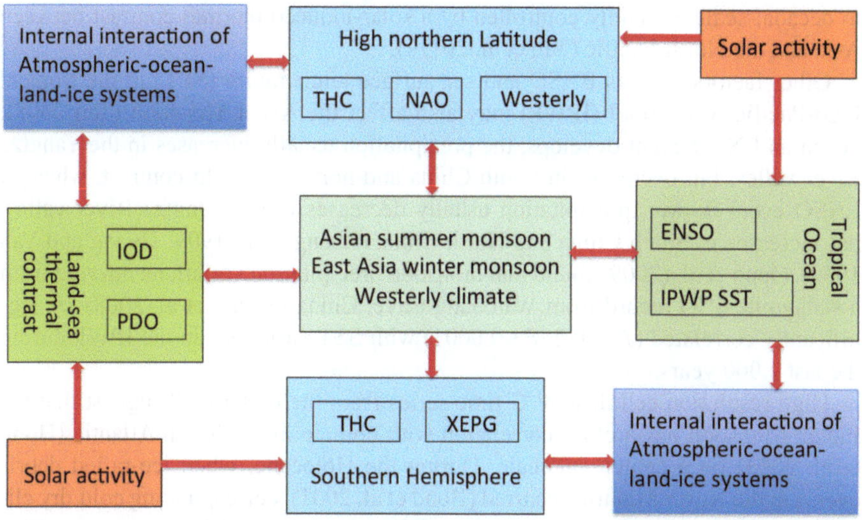

Fig. 6.33 Schematic mechanisms of Asian monsoon variability on millennial to centennial time scales. *THC* thermohaline circulation, *NAO* North Atlantic oscillation, *IOD* Indian Ocean dipole, *PDO* Pacific decadal oscillation, *XEPG* cross-equatorial pressure gradient, *IPWP SST* represents Indo–Pacific warm pool sea surface temperature

centennial scale during historical times. Although there are regional differences in monsoon precipitation variability in China (Tan et al. 2009), they show consistent 80–120 year, ~70 year, 27–35 year, ~20 year, ~15 year and ~11 year periodicities during the last 2,000 years (Tan 2008), indicating that these are important characteristics of the natural variability of the Asian summer monsoon. The quasi-100-year, approximately 35-year, and 10.3-year periodicities may be linked to the Gleissburg periodicity (Xu et al. 2006), the Brückner periodicity (Ku and Li 1998), and the sunspot periodicity (Stuiver and Braziunas 1993; Grootes and Stuiver 1997) of solar activity, respectively. These analyses suggest that solar activity is an important external factor that drives monsoon variability on centennial- to decadal-scales.

Precipitation in north central China is positively correlated with the intensity of the East Asian summer monsoon (EASM; Guo 1983; Huang and Yan 1999; Zhang et al. 2003). The synthesized precipitation record from this region correlates well with a reconstructed Asian Pacific Oscillation index record I_{APO} (Zhao et al. 2007; Zhou et al. 2009) during the last 1,000 years ($R=0.40$, $P<0.001$, $N=99$). On the centennial- to decadal-scale, there is good correspondence between the peaks of the two series (Fig. 6.32). When the synthesized precipitation record and the I_{APO} record (Zhou et al. 2009) are compared with solar activity records (Bard et al. 2000; Muscheler et al. 2007), strong similarities can be observed. During the last millennium, strengthened (weakened) solar activity corresponds to increased (decreased) I_{APO} and precipitation in north central China. This observation suggests that the precipitation variability in north central China and the East Asian monsoon on centennial-to decadal-scale is mainly controlled by a solar-induced thermal contrast between Asia and the North Pacific (Tan et al. 2011a).

Other factors such as ENSO and sea surface temperature (SST) in the tropical Indo-Pacific warm pool (IPWP) may also affect the Asian Monsoon (Tan 2008). When an ENSO event develops, the precipitation usually increases in the Yangtze River valley, but decreases in South China and north China. In contrast, when an ENSO event decays, precipitation usually decreases in the Yangtze River valley, but increases in South China and North China (Zhang et al. 1999; Huang and Yan 1999). Oppo et al. (2009) found that monsoon precipitation variations inferred from a stalagmite $\delta^{18}O$ record from Wanxiang Cave, China (Zhang et al. 2008) are significantly correlated ($R^2=0.2$, $P<0.0001$) with SST variations in the IPWP during the last 2,000 years.

High resolution cellulose $\delta^{13}C$ time series from the eastern TP suggest that the Indian Monsoon has a close correlation with climate in the North Atlantic (Hong et al. 2003) on a centennial scale. During the Holocene, when ice-rafted debris events in the North Atlantic occurred (Bond et al. 2001), corresponding cold/dry climate events were observed on the eastern TP, as weak Indian Monsoon events. This correlation was further confirmed by Holocene stalagmite records from Dongge Cave in southern China (Wang et al. 2005) and from Qunf Cave in southern Oman (Fleitmann et al. 2007). Recently, a stalagmite $\delta^{18}O$ record from Wanxiang Cave (Zhang et al. 2008) also showed similarities with an ice-rafted debris record from the North Atlantic (Bond et al. 2001) and the North Atlantic Oscillation (NAO) record (Proctor et al. 2000) for the last 1,800 years.

In summary, variations of the Asian arid-monsoon environmental system on millennial to centennial time scales were driven by both external solar activity and internal interactions within the atmosphere-ocean-land-ice system (Fig. 6.33). A number of factors contribute to the Asian monsoon variability on millennial to centennial time scales. These include: variations in solar activity, oceanic and atmospheric circulation changes (Wang et al. 2005). Changes in the circulation of the North Atlantic may amplify solar output changes (Bond et al. 2001) and affect NADW production, with feedbacks to atmospheric circulation (Denton and Broecker 2008; Broecker 2010) as well as the influx of fresh melt water from glacial sources (Keigwin and Lehman 1994; Curry and Oppo 1997; Vidal et al. 1997; Curry et al. 1999). It's worth noting that the anthropogenic forcing from greenhouse gases and aerosols released through human activities, has at least partly changed the natural variability of Asian monsoon rainfall since the late twentieth century (Zhang et al. 2008; Liu et al. 2013).

6.6 Sustainable Development of the Asian Monsoon-Arid Areas

6.6.1 Differentiation of Natural Background and Human Signals

During the last 200 years, human activities have greatly changed the environment of the earth. Global warming has resulted from anthropogenic emissions of greenhouse gases, and dramatic changes have occurred with respect to Earth's land surface, oceans, coastlines, biodiversity, water cycle and biogeochemical cycles. Modern industry and agriculture consumes vast quantities of fossil fuels through combustion, and introduces fine particles, pesticides, persistent organic pollutants (POPs) and a variety of toxic chemicals into the environment. The appearance of POPs in sediments, such as polycyclic aromatic hydrocarbons (PAHs) and polychlorinated biphenyls (PCBs) is a clear signal of human activities.

Early human activities such as agricultural production and the utilization of metals have impacted regional environments (Hong et al. 1996a, b; Lee et al. 2008). A 7,000 year-long sediment core from Liangzhi Lake in Hubei province in central China show increasing trends in Cu, Ni, Pb, and Zn from 3000 BC onward, indicating the start of the Bronze Age in ancient China and its anthropogenic impacts on the regional environment (Lee et al. 2008). Heavy metals produced for coinage and weapons during historical times may have increased trace element concentrations in seabirds from island in South China Sea, by atmospheric deposition and bioaccumulation (Yan et al. 2010). Human activities significantly impacted the environment on the LP. During the Qin (221–206 BC) and Han (206 BC–220 AD) Dynasties, forests and grasslands were widely distributed on the plateau. Since then, forests in valleys have almost disappeared, and forests in the mountains were destroyed during the Tang (618–907 AD) and Song (960–1276 AD) Dynasties. Further deterioration of the LP environment followed the Ming (1368–1644 AD) Dynasty, as a result of a dramatically increased human population and dertrimental public policies such as wasteland development (Wang 1999b; Dai et al. 2000).

Ruddiman proposed the "early human influence on climate" hypothesis, and suggested that early human activities in the Holocene led to changes in atmospheric greenhouse gas concentrations (Ruddiman 2003; Ruddiman et al. 2008). He suggested that deforestation for agricultural production by ancestors in the Eurasian continent caused a massive release of carbon stored in trees, resulting in increased CO_2 concentrations since 8,000 year BP (Ruddiman 2003). The abnormal increase in atmospheric CH_4 concentration since 5,000 years BP has been attributed to population expansion and rice cultivation (Ruddiman et al. 2008; Li et al. 2009). East Asia has a long history of civilization where human activities were evident everywhere during historical times. This raises the following questions which merit further scrutiny: 1) did human activities significantly affect the Asian monsoon-arid environment? and 2) how can we differentiate between human environmental changes and natural ones?

In order to differentiate the natural background from human environmental effects we can consider historical climate patterns. The latitudinal shift of the desert/loess transition zone in northern China depends on the position of the northern limit of the East Asian summer monsoon. The northern boundary of the modern summer monsoon is close to the estimated southern margin of desert during the Holocene Optimum period. However, modern desert margin have moved southward about three degrees, which suggested that human activity had affected the natural environment (Zhou et al. 2002). Under natural conditions, when precipitation increases in the upper section of the Heihe River Basin, runoff and groundwater levels are high in the middle section, and *vice versa* in times of low precipitation. However, this relationship was broken after AD 1940, because of human activities, such as the construction of dams and reservoirs (Liu et al. 2010). Tan et al. (2011a) compared the precipitation and temperature variations in the central part of North China, and found that the climate was warm-humid/cool-dry in this region over the past 1,800 years. If the modern warm-dry pattern in North China plain (Ma and Fu 2006) never existed in historical times, it may indicate that human activities have induced climate change in the late twentieth century.

Some geochemical proxies in natural archives reflect natural climatic changes, and others are sensitive indicators of human activities. Comprehensive analyses of multiple geochemical proxies can provide an effective means of differentiating natural background from human signals. Variations of Pb, Zn, Fe, Mn, Cu, Co, Ni, Ti, V, and Al in the peat profile from Hongyuan, China, correlate well with each other and are internally consistent with dust variations before 6 cal ka BP. (Yu et al. 2010). However, Cu, Pb and Zn are significantly elevated above background levels and have an opposing trend to the other elements during some periods after 6 cal ka BP. Abnormal peaks of Cu, Pb and Zn correspond to prosperous times of late Neolithic and Bronze Age cultures in the windward regions, implying anthropogenic sources. Speleothem $\delta^{13}C$ values may also be related to anthropogenic impacts during historical periods. Severe degradation of vegetation and soils in karst areas caused by excessive land utilization can result in marked increases in speleothem $\delta^{13}C$ values (Liu 2008).

Biological proxies such as pollen, seeds, phytoliths and charcoal can also be used to differentiate natural climate conditions from anthropogenic impacts on the environment (Li et al. 2007, 2008). Pollen signals that reflect human impacts include: 1) a decline in certain tree pollen, 2) flourishing pioneer plant pollens, 3) concomitant occurrences of cereal-type pollens and cropland weed pollens; and 4) abrupt changes in pollen concentrations and richness (Li et al. 2008). By comparison of contemporary pollen diagrams between cultural layers and natural sediments in the same region, we can differentiate climatic and anthropogenic impacts on the environment. In the vegetation of Xishanping, Gansu, northwest China, a sudden disappearance of conifer trees and expansion of broadleaved trees (dominated by *Castenea*) occurred in 4,600 cal a BP (Li et al. 2007). However, the climate was relatively cool at that time (An et al. 2005) and cannot explain a decrease of *Picea* and increase of *Castenea.* In addition, wheat, barley, buckwheat, foxtail millet, broomcorn millet, oat, soybean, and rice were found in the Xishanping section, indicating

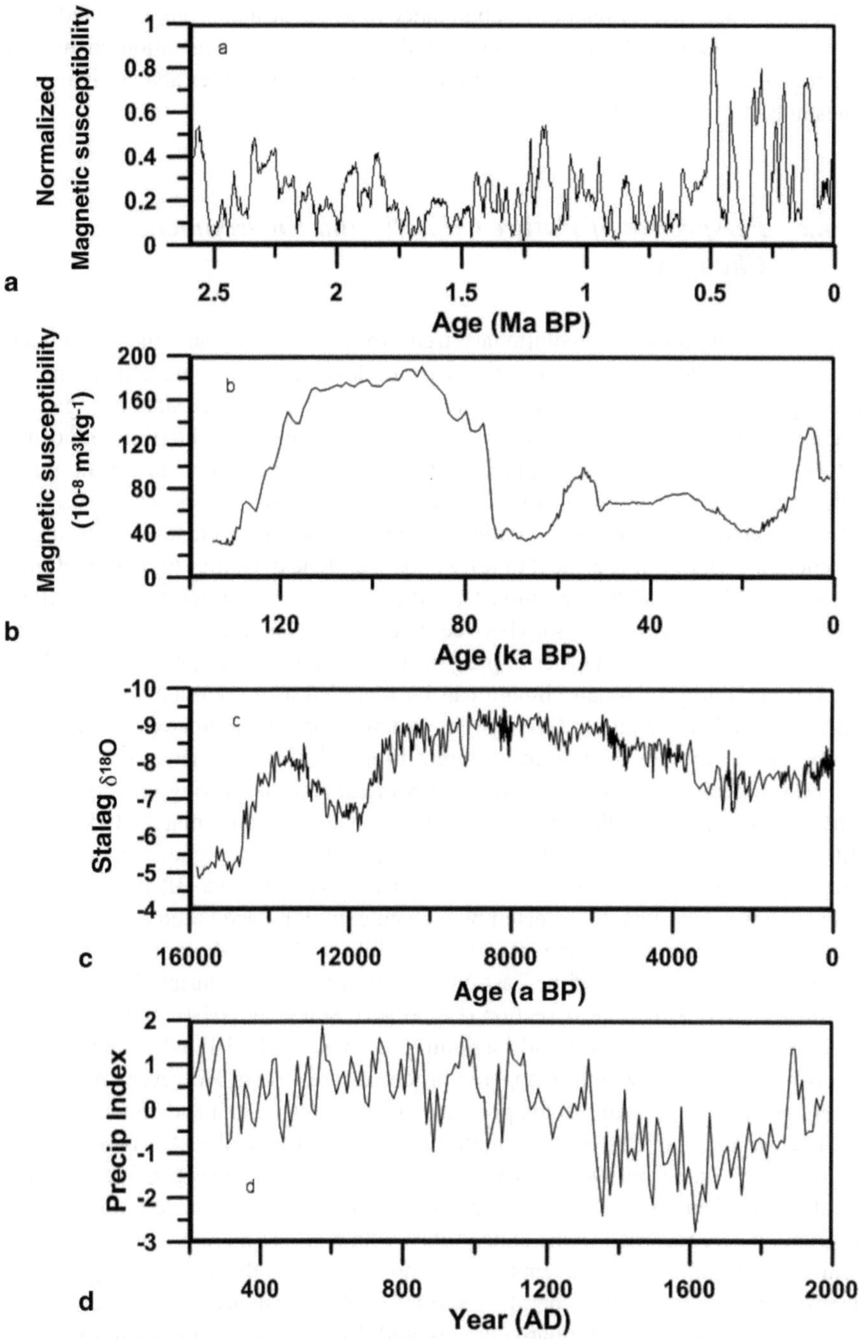

Fig. 6.34 Precipitation proxies over an East Asian monsoonal region on different timescales. **a** Magnetic susceptibility record of Lingtai and Zhaojiachuan loess sections during the past 2.6 Ma (Sun et al. 2006a). **b** Magnetic susceptibility record of Xifeng loess section during the past 128 ka (Sun et al. 2010). **c** Stalagmite $\delta^{18}O$ record from Dongge cave during the past 16 ka (Dykoski et al. 2005). **d** Synthesized precipitation index in central China during the past 1,800 a (Tan et al. 2011a)

agricultural activity. A high pollen abundance of *Poaceae* and *Fagopyrum* indicates intensive agriculture. It has been suggested that the abrupt vegetation changes in the region could be the result of selective hewing of conifer trees and cultivation of chestnuts (Li et al. 2007).

6.6.2 Perspective of Future Climatic and Environmental Changes

After following a natural evolutionary trend for millions of years, the climate and environment in Asia has entered a new stage of development, where human impacts are significant. Human activities began to significantly affect climate change with the introduction of fossil fuel burning and land use changes. In the twentieth century, a strong global warming of approximately 0.74 °C is most likely the result of rapidly increasing concentrations of anthropogenic greenhouse gases. The assessment of future environmental and climatic impacts currently poses a severe and urgent problem for society. It is critical to fully understand the dynamics of natural climate system variability and human impacts, and to quantify their relative contributions. Here, we discuss the relationship between present-day climate and paleoclimate, on a variety of timescales, in order to provide a baseline for future predictions.

The variability of Asian climate can be regarded as a combination of effects, each with different climatic forcings, for example, orbital parameters and solar activity. These forcings operate on different timescales, and evidence for them can be seen in geological proxies as well as historical records. During the Quaternary, East Asian monsoonal climate oscillated with a cycle of approximately 40 ka before 1,000 ka BP and followed a 100 ka cycle after that, based on magnetic susceptibility analyses in loess (Fig. 6.34a; Sun et al. 2006a). After 1,000 ka BP, the East Asian monsoon was significantly affected by a change in global ice volume. Warmer and wetter climate occurred during interglacials while colder and drier climate occurred during glacials. Variation of the East Asian summer monsoon during the last 128 ka is shown in the Xifeng loess section (Fig. 6.34b; Sun et al. 2010). Today's climate follows a broad Holocene trend beginning at about 11 ka BP, where eolian dust deposition is quite low and pedogenic processes are strong. However, the present interglacial is not as warm as the previous one between 128 ka and 73 ka ago.

The variability of East Asian climate since the last deglacial is clearly seen in stalagmite $\delta^{18}O$ data from Dongge cave, in southwestern China (Fig. 6.34c; Dykoski et al. 2005). During the last deglacial, several millennial timescale rapid climatic oscillations are recorded in the stalagmites, within which there are numerous centennial- and multi-decadal events that are not as large as the millennial ones. Millennial climatic events are generally regarded as changes in ocean state. Our present climate is in a period with relatively cool and dry conditions, following the Holocene Optimum (9–5 ka BP).

The last two millennia are important in that they connect the past and present climate change. Stalagmite and historical document records have been combined to reconstruct a high-resolution precipitation proxy record for northern central China during the last 1,800 years (Tan et al. 2011a). Strong quasi-centennial and multi-decadal oscillations are detected in the reconstruction, which are related to solar activity. The Little Ice Age (LIA) is the driest stage over this region during the past 1,800 years. After the LIA precipitation gradually starts to recover. Despite some rapid changes, modern precipitation is still increasing, on a centennial timescale.

Understanding past climate change not only provides a framework to understand the frequency and amplitude of past climate change, it also provides analogues for present day global warming. During Earth's history, there are many times when the planet experienced global warming. One example is the last deglacial period. After the Bølling-Allerød warm period, deglaciation was characterized by intense and rapid climate oscillations, including the Younger Dryas (YD) event. These are well-known from a range of proxies (e.g. Mangerud et al. 1974; Dansgaard et al. 1989). The YD event, which is known as a rapid cooling event after initial warming, is thought to be related to a breakdown in thermohaline circulation due to the injection of fresh water over high-latitude regions (Bond et al. 1993; Liu et al. 2009c).

The YD event exerted an effect on climate around the North Atlantic region and in other parts of the world, including the Asian monsoon region (e.g., Zhou et al. 1996, 1999). In East Asia, climate oscillated rapidly, from a cold-dry climate to a cool-humid condition, then returned to a cold-dry climate. During the YD, significant variability in precipitation occurred in the East Asian monsoon region (Zhou et al. 1996). The YD serves as an example of a climatic teleconnection between the North Atlantic and Asian monsoon-arid environment. Cold air masses probably provide the link necessary to connect the two regions (Zhou et al. 1999). The climatic reversal and associated precipitation variations that occurred during the YD provide an example of a possible direction the earth could follow with continued global warming.

At present, the assessment of future climate change on a shorter timescale is mainly based on projections of climate models under different scenarios of greenhouse gas emissions. During the twenty-first century, a warming trend is predicted for Southeast Asia, with temperatures close to the global mean. Temperatures exceeding the global mean are predicted for South Asia and East Asia. Significant warming is predicted over the inland Asian continent (IPCC 2007). These patterns of temperature increase will no doubt lead to changes in the coupled monsoon-arid system over Asia. Satellite precipitation data show that the areal extent of the Asian monsoon and total precipitation have increased over the past 30 years, which is consistent with other monsoon regions in the world (Hsu et al. 2011).

From a traditional viewpoint, future warming will be larger over the interior of continents than over the oceans, thus inducing a greater thermal contrast between land and sea. This enhanced land-sea thermal contrast is expected to result in stronger summer monsoons over Asia. However, from modeling results, these anticipated changes in the monsoon do not simply follow. Some climate models predict a reduced Asian monsoon circulation due to the weakening of the meridional thermal

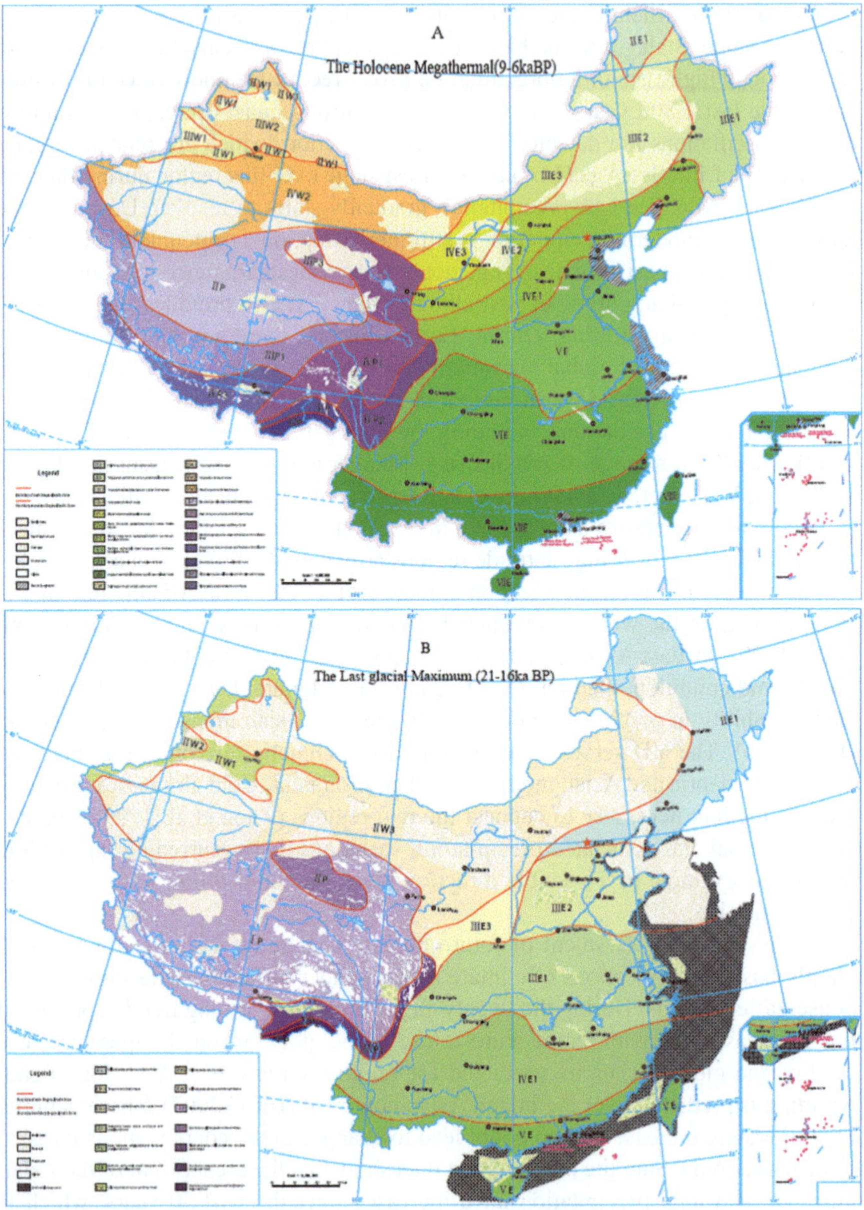

Fig. 6.35 The distribution of bio-climatic zones in China during the Last Glacial Maximum (21,000–16,000 years BP) and the Holocene Optimum (9,000–6,000 years BP). (An et al. 2006a)

gradient (Ueda et al. 2006). Another possibility is that enhanced moisture transport from a warmer Indian Ocean could override the effect of weakened circulation, resulting in an intensification in rainfall (Douville et al. 2000; Ashrit et al. 2003, 2005; Meehl and Arblaster 2003; May 2004). For the winter monsoon, a majority of nine coupled atmosphere-ocean general circulation models simulated a relatively weak winter monsoon, under a global warming scenario (Hori and Ueda 2006). The weakened winter monsoon is closely associated with a strong anticyclonic anomaly over the North Pacific in the models.

6.6.3 Policy and Countermeasures for Sustainable Development in the Arid Zone and Loess Plateau

The arid and semi-arid zones in China meet along a climatic transition zone whose boundary has changed through time. According to the definition proposed by Huang (1958) and Zhao (1983), areas with annual precipitation from 200 to 400 mm are semiarid zones, and areas with annual precipitation lower than 200 mm are arid zones. The arid and semi-arid zones have experienced a long history of natural climatic and environmental change. In the Paleocene, the distribution of arid and humid zones were controlled by planetary wind system. This changed with the growth of the TP. During 3.6–2.6 Ma BP, the modern pattern of western arid and eastern humid monsoon zones was established in China. Since 2.6 Ma BP, the drying tendency in western China continued, punctuated by dry-wet climatic oscillations. In recent times, global warming has led to increased evapotranspiration, decreased humidity and frequent sandstorms in China's arid zones (Academic Division of CAS 2000). Understanding the natural background and development of aridity in western China is critical for policy making.

During the Last Glacial Maximum and Holocene Optimum, significant changes of the bio-climatic zones in China demonstrated their reliance on global and Asian climate conditions (Fig. 6.34; An et al. 2006a). For the Last Glacial Maximum, solar radiation was low in the NH, glaciers significantly expanded, sea level dropped and the coastal continental shelf of eastern China was almost entirely exposed as the coastline moved 600–1,000 km eastward. The desert and semiarid steppe boundary shifted to the south and east, reaching the Greater Khingan Mountains, the Taihang Mountains and north of the Tsinling Mountains. The boundary of the semiarid steppe moved southward by several degrees latitude. The bio-climatic zones of eastern China moved southward accordingly, and tropical forests disappeared in the mainland of China (Fig. 6.35a). During the Holocene Optimum (9,000–5,000 years BP), NH solar radiation significantly increased, glaciers rapidly melted and sea level rose. The paleo-coastline during the Holocene Optimum moved inland by about 200 km in the Yangtze River Delta and about 80 km in the North China Plain, as compared to today. The desert shrank into the inland basin of northwestern China, and the transition belt between the desert and the semiarid steppe drew back close to the Ruoshui River in Gansu and Inner Mongolia. The southern boundary of

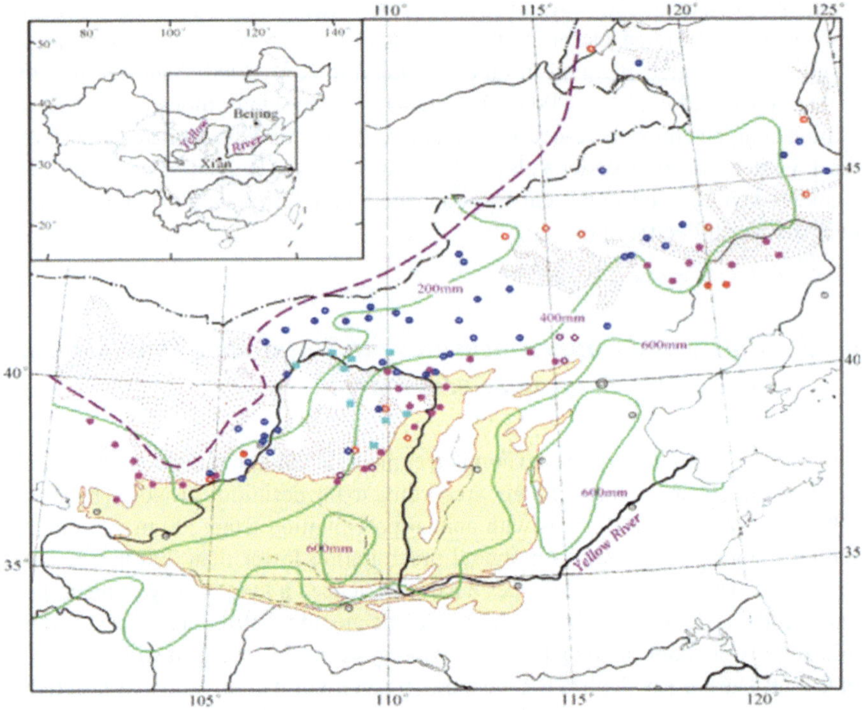

Fig. 6.36 The distribution of archaeological sites in the Mu Us sand land and surrounding areas. Desert and sand land, Loess Plateau region, Microlithic culture (*split blue circles*, 9,000–4,000 ^{14}C yr BP), Yangshao culture (*full red circle*, 7,000–4,000 ^{14}C yr BP), other cultures (*opened red circle*, 4,000 ^{14}C yr BP), recent habitation locations (*blue squares*, since 3,000 ^{14}C yr BP). The distribution of archaeological sites to the south of the dashed line is generally consistent with the distribution of semiarid steppe marked by 200–400 mm annual precipitation. Currently some deserts and sand lands have developed in this zone, suggesting human impacts related to desertification. (Zhou et al. 2002)

the semiarid steppe moved to the Huanglongshan Mountains and the Dai county of Shanxi province (Fig. 6.34b). Present day temperatures are 2–3 °C lower, and annual precipitation is ~100 mm lower than during the Holocene Optimum. Under present climate conditions, the transition belt between the desert and semiarid steppe follows the Helan Mountains, Yinshan Mountains, Daqingshan Mountains and Greater Khingan Mountains. Where the annual precipitation is about 200–400 mm, the natural vegetation is generally semiarid steppe, yet some deserts and sandy lands also develop in these regions. The modern landscape preserves evidence of historical human activities that significantly shaped landforms and altered the environment.

Since the appearance of Chinese culture (about 7,000 years BP, Yangshao culture as representative), the population in China has continuously increased. The semiarid zone and the LP are extremely sensitive to human activities. A large number of archaeological sites indicate the presence of ancient human occupations in the present Mu Us sand land and Northeast China, where the climate used to be warm and humid (Fig. 6.36). Zhou et al. (2002) pointed out that historical human habitation, cultivation and wars were major factors in the change from semiarid steppe-grasslands to desert vegetation. For example, the vegetation cover of the LP was more extensive in the past, with widely distributed forests in gully areas south of Pingliang city, Yan'an city and Suide city, as well as grasslands on the plateau. Much of the native vegetation was destroyed by human activities and soil erosion on the LP became a serious issue.

The arid zone of western China currently faces a range of ecological and environmental threats, such as aridity, desertification, water shortages, and vegetation degradation. These problems have seriously affected the survival and development of local communities (Wang et al. 2008b; Piao et al. 2010). In order to improve the economic, social and environmental conditions of Western China, the Chinese Central Government initiated the "China Western Development" strategy in 1999. The primary goal of this program for the first 10 years was to ameliorate ecological conditions, initiate infrastructure improvements, and increase the efficiency of human resources in western China. As early as 1978, the Chinese Central Government implemented the "Three Norths Forest Shelterbelt Program", whose objective was to construct a "Green Great Wall" in the Chinese arid zone and mitigate sand storms and desertification (Wang et al. 2010). Over the last 30 years, this program has invested several billion Yuan to plant trees. The shelterbelt program has not worked as well as hoped however, as a tendency towards desertification still continues (Wang et al. 2008b; Cao et al. 2011).

In 1999, the Chinese Central Government implemented another huge program in western China, the "Grain for Green Project", whose objective was to increase the vegetation coverage and improve the natural environment. According to the schedule of the Grain for Green Project reported by the Chinese State Administration of Forestry, the total area of reforested farmland will amount to 14.67 million ha and the total budget will reach 337 billion Yuan by the end of 2010 (Xu et al. 2004). The scope and investment of this project are rare in the world. Fortunately, the Grain for Green Project has been very successful over the last 10 years (Wang et al. 2007b; Cao 2011) although some problems remain, such as planting pine trees on thick loess cover, a practice that is not consistent with natural vegetation. More careful plans will be implemented in the future (Zhang 2003; Zhang and Qi 2004).

6.6.3.1 Focus on Sustainability

Sustainability, the capacity to endure, is the potential which maintains the healthy development of human society with emphasis on an optimal harmonious state encompassing the economy, society and environment (Adams 2006). Sustainability is currently a primary research focus in global change studies. The endeavors here on the CLP should closely combine with the guidelines of Future Earth (http://www.

icsu.org/future-earth/) international research initiative proposed by ICSU (International Council for Science). It will develop the knowledge for responding effectively to the risks and opportunities of global environmental change and for supporting transformation towards global sustainability in the coming decades. In order to achieve sustainable development, the natural evolutionary pattern of the Asian monsoon-arid environment should be followed. Given what we have learned, we can offer the following suggestions and countermeasures to achieve sustainability:

The Implementation of Large Scale Ecological Projects Should Comply with Natural Environment Patterns. The distribution pattern of the arid zone in China is the outcome of long-term evolution of the natural environment. In a small-scale geographic unit human beings may change the landscape but broader environmental changes generally result from natural processes. Therefore, ecological projects should not seek to alter long-term natural trends in environmental evolution. The scope of projects should be tailored to maximize the benefits from labor and resources invested in the work.

The Fragile Semiarid Zone Where Annual Precipitation Is 200–400 mm Should Be the Focus of Desertification Prevention Efforts. The natural vegetation for the semiarid zone is grassland, which is especially sensitive to human impacts. Therefore, the semiarid zone should be made a focus of desertification prevention. As the semiarid zone is a transition zone between grazing and farming, effective treatment of desertification can significantly improve the living conditions of local communities.

Reforestation on the Loess Plateau Should Comply with the Climatic Zonal Distribution of Vegetation and Local Landscape. Reforestation is an important measure to mitigate soil and water loss on the LP. As large scale climate conditions determine the zonal distribution of deserts, grasslands and forests from the north to the south on the LP, reforestation should follow established climatic vegetation distribution zones, and selectively plant trees, shrubs, and grasses where it is appropriate. (Academic Division of CAS, 2000). In addition, the local landscape should be considered. For example, rocky mountain and gully areas are fit for planting trees, while terraced land is suitable for grasses (Zhang 2003).

Rapid Development of Water-Saving Technologies and Optimal Use of Water Resources. At present, an effective countermeasure in mitigating water shortages in the arid zone of western China is to promote water-saving technologies and optimal use of water resources among industry, agriculture, ecosystems and inhabitants (Academic Division of CAS, 2000). For agriculture, water-saving irrigation, plastic film and/or straw mulching should be promoted. Moreover, rainfall collection engineering, water-saving technologies and dry farming technologies should be combined and popularized. The regulations on water prices should be supervised and water savings should be greatly encouraged. Ecological water use practices should guide allocation within the limits of available water resources.

Implementation of Effective Management for City Growth and Optimization of Economic Structure. Environmental capacity is limited in the arid zone of western China, especially for water resources. The peak size of cities should be carefully calculated according to water resource capacity, and industrial development should be well-planned. The unique culture of multi-ethnic groups should be protected and promoted for the regional economy. Solar and wind energy are abundant in western China and therefore green energy industries should be promoted through economic development in the arid zone.

Future work should focus on: 1) solution-orientated research for sustainability, linking environmental change and development challenges; 2) effective interdisciplinary collaboration across natural and social sciences, economics, and technology development to find the best scientific solutions to deal with the environment deterioration in the semi-arid areas; 3) timely information for policy-makers through research and consulting reports using the knowledge that integrates global and regional assessments; and 4) increased capacity building in science, technology and innovation, especially for developing countries and a new generation of scientists.

References

Abe M, Kitoh A, Yasunari T (2003) An evolution of the Asian Summer monsoon associated with mountain uplift-simulation with the MRI atmosphere-ocean coupled GCM. J Meteorol Soc Jpn 81(5):909–933

Abels HA, Dupont-Nivet G, Xiao GQ, Bosboom R, Krijgsman W (2011) Step-wise change of Asian interior climate preceding the Eocene-Oligocene Transition (EOT). Palaeogeogr Palaeoclimatol Palaeoecol 299(3–4):399–412. doi:10.1016/j.palaeo.2010.11.028

Academic Division of CAS (2000) Some suggestions for ecology and environment restoration and industrial structure adjustment in the China Western Development. Bull Chin Acad Sci 6:403–406 (in Chinese)

Adams WM (2006) The future of sustainability: re-thinking environment and development in the twenty-first century. IUCN Renowned Thinkers Meeting

Adams CG, Benson RH, Kidd RB, Ryan WBF, Wright RC (1977) The Messinian salinity crisis and evidence of late Miocene eustatic changes in the world ocean. Nature 269(5627):383–386. doi:10.1038/269383a0

Akhmetyev MA, Dodoniv AE, Somikova MV, Spasskaya II, Kremenetsky KV, Klimanov VA (2005) Chapter 8: Kazakhstan and Central Asia (plains and foothills). Geol Soc Am Spec Pap 382:139–161. doi:10.1130/0-8137-2382-5.139

Alley RB (2007) Wally was right: predictive ability of the North Atlantic "Conveyor belt" hypothesis for abrupt climate change. Annu Rev Earth Planet Sci 35:241–272. doi:10.1146/annurev.earth.35.081006.131524

Alley RB, Clark PU, Keigwin LD, Webb RS (1999) Making sense of millennial-scale climate change. In: Clark PU, Webb RS, Keigwin LD (eds) Mechanisms of global climate change at millennial time scales, geophysical monograph. American Geophysical Union, Washington, vol 112, pp 385–394

Altabet MA, Higginson MJ, Murray DW (2002) The effect of millennial-scale changes in Arabian Sea denitrification on atmospheric CO2. Nature 415(6868):159–162. doi:10.1038/415159a

Amidon WH, Burbank DW, Gehrels GE (2005) U-Pb zircon ages as a sediment mixing tracer in the Nepal Himalaya. Earth Planet Sc Lett 235(1–2):244–260. doi:10.1016/j.epsl.2005.03.019

An ZS (2000) The history and variability of the East Asian Paleomonsoon climate. Quat Sci Rev 19(1–5):171–187

An ZS, Porter SC (1997) Millennial-scale climatic oscillations during the last interglaciation in central China. Geology 25(7):603–606

An ZS, Kukla G, Porter SC, Xiao JL (1991a) Late quaternary dust flow on the Chinese Loess Plateau. Catena 18(2):125–132

An ZS, Kukla GJ, Porter SC, Xiao JL (1991b) Magnetic-susceptibility evidence of monsoon variation on the Loess Plateau of Central China during the last 130,000 years. Quat Res 36(1):29–36. doi:10.1016/0033-5894(91)90015-W

An ZS, Wang SM, Wu XH, Chen MY, Sun DH, Liu XM, Wang FB, Li L, Sun YB, Zhou WJ, Zhou J, Liu XD, Lu HY, Zhang YX, Dong GG, Qiang XK (1999) Eolian evidence from the Chinese Loess Plateau: the onset of the Late Cenozoic Great Glaciation in the Northern Hemisphere and Qinghai-Xizang Plateau uplift forcing. Sci China Ser D 42(3):258–271

An ZS, Porter SC, Kutzbach JE, Wu XH, Wang SM, Liu XD, Li XQ, Zhou WJ (2000) Asynchronous Holocene optimum of the East Asian monsoon. Quat Sci Rev 19(8):743–762

An ZS, Kutzbach JE, Prell WL, Porter SC (2001) Evolution of Asian monsoons and phased uplift of the Himalaya-Tibetan plateau since Late Miocene times. Nature 411(6833):62–66. doi:10.1038/35075035

An CB, Tang LY, Barton L, Chen FH (2005) Climate change and cultural response around 4000 cal yr BP in the western part of Chinese Loess Plateau. Quat Res 63(3):347–352. doi:10.1016/j.yqres.2005.02.004

An Z, Zhou J, Wang S (2006a) Paleo-environmental map for typical Periods. SinoMaps, Beijing (in Chinese)

An ZS, Zhang PZ, Wang EQ, Wang SM, Qiang XK, Li L, Song YG, Chang H, Liu XD, Zhou WJ, Liu WG, Cao JJ, Li XQ, Shen J, Liu Y, Ai L (2006b) Changes of the monsoon-arid environment in China and geowth of the Tibetan Plateau since the Miocene. Quat Sci 26 (5):678–693 (in Chinese with English abstract)

An ZS, Clemens SC, Shen J, Qiang XK, Jin ZD, Sun YB, Prell WL, Luo JJ, Wang SM, Xu H, Cai YJ, Zhou WJ, Liu XD, Liu WG, Shi ZG, Yan LB, Xiao XY, Chang H, Wu F, Ai L, Lu FY (2011) Glacial-interglacial Indian summer monsoon dynamics. Science 333(6043):719–723. doi:10.1126/science.1203752

An ZS, Colman SM, Zhou WJ, Li XQ, Brown ET, Jull AJT, Cai YJ, Huang YS, Lu XF, Chang H, Song YG, Sun YB, Xu H, Liu WG, Jin ZD, Liu XD, Cheng P, Liu Y, Ai L, Li XZ, Liu XJ, Yan LB, Shi ZG, Wang XL, Wu F, Qiang XK, Dong JB, Lu FY, Xu XW (2012) Interplay between the Westerlies and Asian monsoon recorded in Lake Qinghai sediments since 32 ka. Sci Rep UK 2:619. doi:10.1038/Srep00619

Arimoto R (2001) Eolian dust and climate: relationships to sources, tropospheric chemistry, transport and deposition. Earth-Sci Rev 54(1–3):29–42

Ashrit RG, Douville H, Kumar KR (2003) Response of the Indian monsoon and ENSO-monsoon teleconnection to enhanced greenhouse effect in the CNRM coupled model. J Meteorol Soc Jpn 81(4):779–803. doi:10.2151/Jmsj.81.779

Ashrit RG, Kitoh A, Yukimoto S (2005) Transient response of ENSO-monsoon teleconnection in MRI-CGCM2.2 climate change simulations. J Meteorol Soc Jpn 83(3):273–291. doi:10.2151/Jmsj.83.273

Bard E, Raisbeck G, Yiou F, Jouzel J (2000) Solar irradiance during the last 1200 years based on cosmogenic nuclides. Tellus B 52(3):985–992

Berger A, Loutre MF (1991) Insolation values for the climate of the last 10 million years. Quat Sci Rev 10 (4):297–317. doi:10.1016/0277-3791(91)90033-q

Berner RA, Lasaga AC, Garrels RM (1983) The Carbonate-silicate geochemical cycle and its effect on atmospheric carbon-dioxide over the past 100 million years. Am J Sci 283(7):641–683

Besse J, Courtillot V, Pozzi JP, Westphal M, Zhou YX (1984) Palaeomagnetic estimates of crustal shortening in the Himalayan thrusts and Zangbo suture. Nature 311(5987):621–626

Bickle MJ (1994) The role of metamorphic decarbonation reactions in returning strontium to the silicate sediment mass. Nature 367 (6465):699–704. doi:10.1038/367699a0

Blunier T, Chappellaz J, Schwander J, Dallenbach A, Stauffer B, Stocker TF, Raynaud D, Jouzel J, Clausen HB, Hammer CU, Johnsen SJ (1998) Asynchrony of Antarctic and Greenland climate change during the last glacial period. Nature 394(6695):739–743. doi:10.1038/29447

Bond G, Broecker W, Johnsen S, McManus J, Labeyrie L, Jouzel J, Bonani G (1993) Correlations between climate records from North-Atlantic Sediments and Greenland Ice. Nature 365(6442):143–147

Bond G, Kromer B, Beer J, Muscheler R, Evans MN, Showers W, Hoffmann S, Lotti-Bond R, Hajdas I, Bonani G (2001) Persistent solar influence on North Atlantic climate during the Holocene. Science 294(5549):2130–2136

Boos WR, Kuang Z (2010) Dominant control of the South Asian monsoon by orographic insulation versus plateau heating. Nature 463(7278):218–222

Bosboom RE, Dupont-Nivet G, Houben AJP, Brinkhuis H, Villa G, Mandic O, Stoica M, Zachariasse WJ, Guo Z, Li C, Krijgsman W (2011) Late Eocene sea retreat from the Tarim Basin (west China) and concomitant Asian paleoenvironmental change. Palaeogeogr Palaeoclimatol Palaeoecol 299(3–4):385–398. doi:10.1016/j.palaeo.2010.11.019

Bovet PM, Ritts BD, Gehrels G, Abbink AO, Darby B, Hourigan J (2009) Evidence of Miocene Crustal Shortening in the North Qilian Shan from Cenozoic Stratigraphy of the Western Hexi Corridor, Gansu Province, China. Am J Sci 309(4):290–329

Braconnot P, Otto-Bliesner B, Harrison S, Joussaume S, Peterchmitt JY, Abe-Ouchi A, Crucifix M, Driesschaert E, Fichefet T, Hewitt C (2007) Results of PMIP2 coupled simulations of the Mid-Holocene and last glacial maximum-part 1: experiments and large-scale features. Clim Past 3(2):261–277

Braun H, Christl M, Rahmstorf S, Ganopolski A, Mangini A, Kubatzki C, Roth K, Kromer B (2005) Possible solar origin of the 1,470-year glacial climate cycle demonstrated in a coupled model. Nature 438(7065):208–211. doi:10.1038/Nature04121

Broecker WS (1957) Application of radiocarbon to oceanography and climate chronology. Columbia University, New York

Broecker WS (2010) The great ocean conveyor: discovering the trigger for abrupt climate change. Princeton University Press, Princeton

Broecker WS, Denton GH (1989) The role of ocean-atmosphere reorganizations in glacial cycles. Geochim Cosmochim Ac 53(10):2465–2501

Broecker WS, Peteet DM, Rind D (1985) Does the ocean-atmosphere system have more than one stable mode of operation. Nature 315(6014):21–26

Broecker WS, Kennett JP, Flower BP, Teller JT, Trumbore S, Bonani G, Wolfli W (1989) Routing of meltwater from the Laurentide ice-sheet during the Younger Dryas cold episode. Nature 341(6240):318–321

Burg JP, Chen GM (1984) Tectonics and structural zonation of Southern Tibet, China. Nature 311(5983):219–223. doi:10.1038/311219a0

Burtman VS (2000) Cenozoic crustal shortening between the Pamir and Tien Shan and a reconstruction of the Pamir-Tien Shan transition zone for the Cretaceous and Palaeogene. Tectonophysics 319(2):69–92

Cai YJ, An ZS, Cheng H, Edwards RL, Kelly MJ, Liu WG, Wang XF, Shen CC (2006) High-resolution absolute-dated Indian Monsoon record between 53 and 36 ka from Xiaobailong Cave, southwestern China. Geology 34(8):621–624. doi:10.1130/G22567.1

Cao S (2011) Impact of China's large-scale ecological restoration program on the environment and society in arid and semiarid areas of China: achievements, problems, synthesis, and applications. Crit Rev Env Sci Tec 41(4):317–335. doi:10.1080/10643380902800034

Cao JJ, An ZS (2011) Atmospheric iron loading in Asia during 6.0-2.0 Myr and implications for the onset of Northern Hemisphere Glaciation. J Earth Envron 1(1):52–59

Cao SX, Chen L, Shankman D, Wang CM, Wang XB, Zhang H (2011) Excessive reliance on afforestation in China's arid and semi-arid regions: lessons in ecological restoration. Earth-Sci Rev 104(4):240–245. doi:10.1016/j.earscirev.2010.11.002

Chang H, Ao H, An ZS, Fang XM, Song YG, Qiang XK (2012) Magnetostratigraphy of the Suerkuli Basin indicates Pliocene (3.2 Ma) activity of the middle Altyn Tagh Fault, northern Tibetan Plateau. J Asian Earth Sci 44:169–175

Charles CD, Lynch-Stieglitz J, Ninnemann US, Fairbanks RG (1996) Climate connections between the hemisphere revealed by deep sea sediment core ice core correlations. Earth Planet Sci Lett 142(1–2):19–27

Chen FH, Bloemendal J, Wang JM, Li JJ, Oldfield F (1997) High-resolution multi-proxy climate records from Chinese loess, evidence for rapid climatic changes over the last 75 kyr. Palaeogeogr Palaeoclimatol Palaeoecol 130(1–4):323–335

Chen Z, Li L, Liu J, Gong H, Jiang R, Li S, Han X, Li X, Wang C, Wang G, Wang G, Lu K (2008) Preliminary study on the uplifting-exhumation process of the western Tianshan range, northwestern China. Acta Petrol Sin 24:625–636

Cheng X, Zhao Q, Wang J, Jian Z, Xia P, Huang B, Fang D, Xu J, Zhou Z, Wang P (2004) Data report: Stable isotopes from Sites 1147 and 1148. In: Wang P, Prell WL, Blum P (eds) Proceedings of the Ocean Drilling Program, Scientific Results, vol 184. College Station, TX, pp 1–12. doi:10.2973/odp.proc.ir.184.2000

Cheng H, Edwards RL, Broecker WS, Denton GH, Kong XG, Wang YJ, Zhang R, Wang XF (2009) Ice age terminations. Science 326(5950):248–252. doi:10.1126/science.1177840

Chetelat B, Liu CQ, Zhao ZQ, Wang QL, Li SL, Li J, Wang BL (2008) Geochemistry of the dissolved load of the Changjiang Basin rivers: anthropogenic impacts and chemical weathering. Geochim Cosmochim Acta 72(17):4254–4277. doi:10.1016/j.gca.2008.06.013

Chiang JCH, Bitz CM (2005) Influence of high latitude ice cover on the marine intertropical convergence zone. Clim Dynam 25(5):477–496. doi:10.1007/s00382-005-0040-5

China Geological Survey (2001) Chinese 1,500,000 Geological Map Database. 1.0 ed

Chinese Stratum Thesaurus Editorial Board (1999) Tertiary Stratum Thesaurus. Geological Publishing House, Beijing

Chung S, Lo C, Lee T, Zhang Y, Xie Y, Li X, Wang K, Wang P (1998) Diachronous uplift of the Tibetan plateau starting 40 Ma ago. Nature 394:769–773

Clark M (2003) Late Cenozoic uplift of southeastern Tibet. Massachusetts Institute of Technology, Cambridge, Massachusetts

Clark M, Schoenbohm L, Royden L, Whipple K, Burchfield B, Zhang X, Tang W, Chen L (2004) Surface uplift, tectonics, and erosion of eastern Tibet. Tectonics 23:1006. doi:10.1029/2002TC001402

Clark M, Bush J, Royden L (2005) Dynamic topography produced by lower crustal flow against rheological strength heterogeneities bordering the Tibetan Plateau. Geophy J Int 162(2):575–590

Clark MK, Farley KA, Zheng DW, Wang ZC, Duvall AR (2010) Early Cenozoic faulting of the northern Tibetan Plateau margin from apatite (U-Th)/He ages. Earth Planet Sci Lett 296(1–2):78–88. doi:10.1016/j.epsl.2010.04.051

Clemens S (2006) Extending the historical record by proxy. In: The Asian monsoon. Springer Praxis Books, Berlin, pp 615–630. doi:10.1007/3-540-37722-0_16

Clemens SC, Prell WL (2003) A 350,000 year summer-monsoon multi-proxy stack from the Owen ridge, Northern Arabian sea. Mar Geol 201(1–3):35–51. doi:10.1016/S0025-3227(03)00207-X

Clemens S, Prell W, Murray D, Shimmield G, Weedon G (1991) Forcing mechanisms of the Indian-Ocean monsoon. Nature 353(6346):720–725. doi:10.1038/353720a0

Clemens SC, Murray DW, Prell WL (1996) Nonstationary phase of the plio-pleistocene Asian monsoon. Science 274(5289):943–948

Clemens SC, Prell WL, Sun Y, Liu Z, Chen G (2008) Southern hemisphere forcing of Pliocene δ18O and the evolution of Indo-Asian monsoons. Paleoceanography 23. doi:10.1029/2008PA1638

Clemens SC, Prell WL, Sun Y (2010) Orbital-scale timing and mechanisms driving Late Pleistocene Indo-Asian summer monsoons: reinterpreting cave speleothem δ18 O. Paleoceanography 25(4):PA4207. doi:10.1029/2010pa001926

Curry WB, Oppo DW (1997) Synchronous, high-frequency oscillations in tropical sea surface temperatures and North Atlantic Deep Water production during the last glacial cycle. Paleoceanography 12(1):1–14. doi:10.1029/96pa02413

Curry WB, Marchitto TM, McManus JF, Oppo DW, Laarkamp KL (1999) Millennial-scale changes in ventilation of the thermocline, intermediate, and deep waters of the glacial North Atlantic.

In: Clark U, Webb S, Keigwin D (eds) Mechanisms of global climate change at millennial time scales. AGU, Washington DC. (Monogr Ser 112:59–76. doi:10.1029/GM112)

Dai Y, Cai J, Wang H (2000) The position and role of vegetation construction in ecological environment construction in Loess Plateau. J Northwest Agric Univ 28(6):130–134 (in Chinese)

Dansgaard W, Clausen HB, Gundestrup N, Hammer CU, Johnsen SF, Kristinsdottir PM, Reeh N (1982) A new greenland deep ice core. Science 218(4579):1273–1277. doi:10.1126/science.218.4579.1273

Dansgaard W, White JWC, Johnsen SJ (1989) The abrupt termination of the younger dryas climate event. Nature 339(6225):532–534. doi:10.1038/339532a0

Dansgaard W, Johnsen SJ, Clausen HB, Dahljensen D, Gundestrup NS, Hammer CU, Hvidberg CS, Steffensen JP, Sveinbjornsdottir AE, Jouzel J, Bond G (1993) Evidence for general instability of past climate from a 250-Kyr Ice-Core Record. Nature 364(6434):218–220. doi:10.1038/364218a0

DeCelles PG, Kapp P, Ding L, Gehrels GE (2007) Late Cretaceous to middle Tertiary basin evolution in the central Tibetan Plateau: changing environments in response to tectonic partitioning, aridification, and regional elevation gain. Geol Soc Am Bull 119(5–6):654–680. doi:10.1130/B26074.1

Dedeckker P, Correge T, Head J (1991) Late Pleistocene record of cyclic eolian activity from tropical Australia suggesting the younger dryas is not an unusual climatic event. Geology 19(6):602–605. doi:10.1130/0091-7613 (1991) 019 <0602:Lproce> 2.3.Co;2

Denton GH, Broecker WS (2008) Wobbly ocean conveyor circulation during the Holocene? Quat Sci Rev 27(21–22):1939–1950. doi:10.1016/j.quascirev.2008.08.008

Denton GH, Anderson RF, Toggweiler JR, Edwards RL, Schaefer JM, Putnam AE (2010) The last glacial termination. Science 328(5986):1652–1656. doi:10.1126/science.1184119

Deplazes G, Luckge A, Peterson LC, Timmermann A, Hamann Y, Hughen KA, Rohl U, Laj C, Cane MA, Sigman DM, Haug GH (2013) Links between tropical rainfall and North Atlantic climate during the last glacial period. Nat Geosci 6(3):213–217. doi:10.1038/ngeo1712

Dercourt J, Ricou L, Vrielinck B (1993) Atlas Tethys palaeoenvironmental maps. Gauthier-Villars, Paris

Derry LA, FranceLanord C (1996) Neogene Himalayan weathering history and river 87Sr/86Sr: impact on the marine Sr record. Earth Planet Sci Lett 142(1–2):59–74. doi:10.1016/0012-821×(96) 00091–X

Dewey J, Shackleton R, Chang C, Sun Y (1988) The tectonic evolution of the Tibetan Plateau. Philos Trans R Soc London 327:379–413

Ding Z, Liu T, Rutter NW, Yu Z, Guo Z, Zhu R (1995) Ice-volume forcing of East Asian winter monsoon variations in the past 800,000 Years. Quat Res 44(2):149–159. doi:10.1006/qres.1995.1059

Ding ZL, Rutter NW, Liu TS, Sun JM, Ren JZ, Rokosh D, Xiong SF (1998) Correlation of Dansgaard-Oeschger cycles between Greenland ice and Chinese loess. Paleoclimates 2:281–291

Ding ZL, Ren JZ, Yang SL, Liu TS (1999) Climate instability during the penultimate glaciation: evidence from two high-resolution loess records, China. J Geophys Res-Sol Ea 104(B9):20123–20132. doi:10.1029/1999jb900183

Douville H, Royer JF, Polcher J, Cox P, Gedney N, Stephenson DB, Valdes PJ (2000) Impact of CO_2 doubling on the Asian summer monsoon: robust versus model-dependent responses. J Meteorol Soc Jpn 78(4):421–439

Dupont-Nivet G, Hoorn C, Konert M (2008) Tibetan uplift prior to the Eocene-Oligocene climate transition: evidence from pollen analysis of the Xining Basin. Geology 36(12):987–990. doi:10.1130/G25063a.1

Dykoski CA, Edwards RL, Cheng H, Yuan DX, Cai YJ, Zhang ML, Lin YS, Qing JM, An ZS, Revenaugh J (2005) A high-resolution, absolute-dated Holocene and deglacial Asian monsoon record from Dongge Cave, China. Earth Planet Sci Lett 233(1–2):71–86. doi:10.1016/j.epsl.2005.01.036

Edmond JM (1992) Himalayan tectonics, weathering processes, and the strontium isotope record in marine limestones. Science 258(5088):1594–1597. doi:10.1126/science.258.5088.1594

England P, Searle M (1986) The Cretaceous-Tertiary deformation of the Lhasa Block and its implications for crustal thickening in Tibet. Tectonics 5(1):1–14

Fang XM, Ono Y, Fukuksawa H, Pan BT, Li JJ, Guan DH, Oi KC, Tsukamoto S, Torii M, Mishima T (1999) Asian summer monsoon instability during the past 60,000 years: magnetic susceptibility and pedogenic evidence from the western Chinese Loess Plateau. Earth Planet Sci Lett 168(3–4):219–232

Fang XM, Garzione C, Van der Voo R, Li JJ, Fan MJ (2003) Flexural subsidence by 29 Ma on the NE edge of Tibet from the magnetostratigraphy of Linxia Basin, China. Earth Planet Sci Lett 210(3–4):545–560. doi:10.1016/S0012-821x(03)00142-0

Fang X, Zhao Z, LI J, Yan M, Pan B, Song C, Dai S (2005) Magnetostratigraphy of the late Cenozoic Laojunmiao anticline in the northern Qilian Mountains and its implications for the northern Tibetan Plateau uplift. Sci China Earth Sci 48(7):1040–1051. doi:10.1360/03yd0188

Farrell JW, Clemens SC, Gromet LP (1995) Improved chronostratigraphic reference curve of late Neogene Seawater 87Sr/86Sr. Geology 23(5):403–406. doi:10.1130/0091-7613

Fleitmann D, Burns SJ, Mangini A, Mudelsee M, Kramers J, Villa I, Neff U, Al-Subbary AA, Buettner A, Hippler D, Matter A (2007) Holocene ITCZ and Indian monsoon dynamics recorded in stalagmites from Oman and Yemen (Socotra). Quat Sci Rev 26(1–2):170–188. doi:10.1016/j.quascirev.2006.04.012

France-Lanord C, Derry LA (1997) Organic carbon burial forcing of the carbon cycle from Himalayan erosion. Nature 390(6655):65–67. doi:10.1038/36324

Garzanti E, VanHaver T (1988) The Indus clastics—fore-arc basin sedimentation in the Ladakh Himalaya (India). Sedi Geol 59 (3–4):237–249. doi:10.1016/0037-0738(88)90078-4

Garzione CN (2008) Research focus surface uplift of Tibet and Cenozoic global cooling. Geology 36(12):1003–1004. doi:10.1130/focus122008.1

George A, Marshallsea S, Wyrwoll K, Chen J, Lu Y (2001) Miocene cooling in the northern Qilian Shan, northeastern margin of the Tibetan Plateau, revealed by apatite fission-track and vitrinite-reflectance analysis. Geology 29(10):939–942

Ghil M, Mullhaupt A, Pestiaux P (1987) Deep water formation and Quaternary glaciations. Clim Dynam 2(1):1–10. doi:10.1007/Bf01088850

Grootes PM, Stuiver M (1997) Oxygen 18/16 variability in Greenland snow and ice with 103 to 105 year time resolution. J Geophys Res 102(C12):26455–26470

Gu CG, Renaut RW (1994) The effect of Tibetan uplift on the formation and preservation of Tertiary lacustrine source-rocks in eastern China. J Paleolimnol 11(1):31–40

Guo QY (1983) The summer monsoon intensity index in East Asia. Acta Geographica Sinica 38(3):207–216

Guo S, Chen J (1989) Cenozoic floras and coal-accumulating environment in Himalayas and hengduan mountains area. Acta Palaeonologica Sinica 28:512–521

Guo ZT, Liu TS, Guiot J, Wu NQ, Lu H, Han J, Liu J, Gu Z (1996) High frequency pulses of East Asian monsoon climate in the last two glaciations: link with the North Atlantic. Clim Dynam 12(10):701–709. doi:10.1007/s003820050137

Guo ZT, Ruddiman WF, Hao QZ, Wu HB, Qiao YS, Zhu RX, Peng SZ, Wei JJ, Yuan BY, Liu TS (2002) Onset of Asian desertification by 22 Myr ago inferred from loess deposits in China. Nature 416(6877):159–163

Guo ZT, Sun B, Zhang ZS, Peng SZ, Xiao GQ, Ge JY, Hao QZ, Qiao YS, Liang MY, Liu JF, Yin QZ, Wei JJ (2008) A major reorganization of Asian climate by the early Miocene. Clim Past 4:153–174

Hahn DG, Manabe S (1975) The role of mountains in the south Asian monsoon circulation. J Atmos Sci 32(8):1515–1541

Hansen J, Sato M, Ruedy R, Lo K, Lea DW, Medina-Elizade M (2006) Global temperature change. Proc Natl Acad Sci 103(39):14288

Hao Y, Guan S, Ye L, Huang Y, Zhou Y, Guan S (2002) Neogene stratigraphy and palaeogeography in the western Tarim Basin. Acta Geol Sin-Engl 76:289–298

Harris N (2006) The elevation history of the Tibetan Plateau and its implications for the Asian monsoon. Palaeogeogr Palaeoclimatol Palaeoecol 241(1):4–15

Harrison T, Copeland P, Kidd W, Yin A (1992) Raising Tibet. Science 255:1663–1670
Haug GH, Tiedemann R (1998) Effect of the formation of the Isthmus of Panama on Atlantic Ocean thermohaline circulation. Nature 393(6686):673–676. doi:10.1038/31447
Hay WW, Soeding E, DeConto RM, Wold CN (2002) The Late Cenozoic uplift—climate change paradox. Int J Earth Sci 91(5):746–774. doi:10.1007/s00531-002-0263-1
He CQ (1991) Late Cretaceous-Early Tertiary microphytoplankton from the western Tarim Basin in southern Xinjiang. Science Press, Beijing
Heermance RV, Chen J, Burbank DW, Wang C (2007) Chronology and tectonic controls of late Tertiary deposition in the southwestern Tian Shan foreland, NW China. Basin Res 19(4):599–632. doi:10.1111/j.1365-2117.2007.00339.x
Heinrich H (1988) Origin and consequences of cyclic ice rafting in the Northeast Atlantic-Ocean during the past 130,000 years. Quat Res 29(2):142–152. doi:10.1016/0033-5894(88)90057-9
Held IM (1983) Stationary and quasi-stationary eddies in the extratropical troposphere: theory. In: Hoskins BJ, Pearce RP (eds) Large-scale dynamical processes in the atmosphere. Academic Press, New York, pp 127–168
Hess J, Bender ML, Schilling JG (1986) Evolution of the ratio of strontium-87 to strontium-86 in seawater from cretaceous to present. Science 231(4741):979–984. doi:10.1126/science.231.4741.979
Hodell D, Elmstrom M, Kennett J (1986) Latest Miocene benthic δ18O changes, global ice volume, sea level and the 'Messinian salinity crisis'. Nature 320:411–414
Hodell DA, Mead GA, Mueller PA (1990) Variation in the strontium isotopic composition of seawater (8 Ma to Present)—implications for chemical-weathering rates and dissolved fluxes to the oceans. Chem Geol 80(4):291–307. doi:10.1016/0168-9622(90)90011-Z
Hodell DA, Mueller PA, Garrido JR (1991) Variations in the strontium isotopic composition of seawater during the Neogene. Geology 19(1):24–27. doi:10.1130/0091-7613
Hong SM, Candelone JP, Patterson CC, Boutron CF (1996a) History of ancient copper smelting pollution during Roman and medieval times recorded in Greenland ice. Science 272(5259):246–249. doi:10.1126/science.272.5259.246
Hong SM, Candelone JP, Soutif M, Boutron CF (1996b) A reconstruction of changes in copper production and copper emissions to the atmosphere during the past 7000 years. Sci Total Environ 188(2–3):183–193. doi:10.1016/0048-9697(96)05171-6
Hong YT, Hong B, Lin QH, Zhu YX, Shibata Y, Hirota M, Uchida M, Leng XT, Jiang HB, Xu H, Wang H, Yi L (2003) Correlation between Indian Ocean summer monsoon and North Atlantic climate during the Holocene. Earth Planet Sci Lett 211(3–4):371–380. doi:10.1016/s0012-821x(03)00207-3
Hori ME, Ueda H (2006) Impact of global warming on the East Asian winter monsoon as revealed by nine coupled atmosphere-ocean GCMs. Geophys Res Lett 33(3):L03713. doi:10.1029/2005gl024961
Horton BK, Yin A, Spurlin MS, Zhou JY, Wang JH (2002) Paleocene-Eocene syncontractional sedimentation in narrow, lacustrine-dominated basins of east-central Tibet. Geol Soc Am Bull 114(7):771–786. doi:10.1130/0016-7606 (2002) 114 <0771:Pessin> 2.0.Co;2
Hsü J (1981) On palaeobotanical evidence for continental drift and the Himalayan uplift. The comprehensive scientific expedition to the Qinghai- Xizang Plateau, Academia Sinica. Studies on the period, amplitude and type of the uplift of the Qinghai-Xizang Plateau. Science Press, Beijing
Hsü J, Kong Z, Sun X, Tao J, Du N (1973) Research on Quaternary paleoplant in Himalaya and its implication for uplift of the Himalaya. Chin Sci Bull 18:274
Hsu P, Li T, Wang B (2011) Trends in global monsoon area and precipitation over the past 30 years. Geophys Res Lett 38(8):L08701. doi:10.1029/2011gl046893
Huang B (1958) A scheme on Physio-geographic regionalization of China. Acta Geographica Sinica 24(4):348–363 (in Chinese)
Huang G, Yan ZW (1999) The East Asian summer monsoon circulation anomaly index and its interannual variations. Chin Sci Bull 44(14):1325–1329
Huber M, Goldner A (2012) Eocene monsoons. J Asian Earth Sci 44:3–23

IPCC FARA (2007) Climate change 2007: the physical science basis. In: Solomon S, Qin D, Manning M et al (eds) Contribution of working group I to the fourth assessment report of the intergovernmental panel on climate change. Cambridge University Press, Cambridge, p 996

Jacobson AD (2004) Has the atmospheric supply of dissolved calcite dust to seawater influenced the evolution of marine 87Sr/86Sr ratios over the past 2.5 million years? Geochem Geophy Geosy 5(12):Q12002. doi:10.1029/2004gc000750

Jain AK, Lal N, Sulemani B, Awasthi AK, Singh S, Kumar R, Kumar D (2009) Detrital-zircon fission-track ages from the Lower Cenozoic sediments, NW Himalayan foreland basin: clues for exhumation and denudation of the Himalaya during the India-Asia collision. Geol Soc Am Bull 121(3–4):519–535

Jiang DB, Lang XM (2010) Last glacial maximum East Asian monsoon: results of PMIP simulations. J Climate 23(18):5030–5038. doi:10.1175/2010jcli3526.1

Jiang D, Wang H, Drange H, Lang X (2003) Last glacial maximum over China: sensitivities of climate to paleovegetation and Tibetan ice sheet. J Geophys Res 108(D3):4102

Jiang HC, Ding ZL, Xiong SF (2007) Magnetostratigraphy of the Neogene Sikouzi section at Guyuan, Ningxia, China. Palaeogeogr Palaeoclimatol Palaeoecol 243:223–234

Jickells TD, An ZS, Andersen KK, Baker AR, Bergametti G, Brooks N, Cao JJ, Boyd PW, Duce RA, Hunter KA, Kawahata H, Kubilay N, laRoche J, Liss PS, Mahowald N, Prospero JM, Ridgwell AJ, Tegen I, Torres R (2005) Global iron connections between desert dust, ocean biogeochemistry, and climate. Science 308(5718):67–71

Jin L, Chen F, Ganopolski A, Claussen M (2007) Response of East Asian climate to Dansgaard/Oeschger and Heinrich events in a coupled model of intermediate complexity. J Geophys Res Atmospheres 112(D6):D06117. doi:10.1029/2006jd007316

Johnsen SJ, Clausen HB, Dansgaard W, Fuhrer K, Gundestrup N, Hammer CU, Iversen P, Jouzel J, Stauffer B, Steffensen JP (1992) Irregular glacial interstadials recorded in a New Greenland ice core. Nature 359(6393):311–313. doi:10.1038/359311a0

Jolivet M, Roger F, Arnaud N, Frunel M, Tapponnier P, Seward D (1999) Exhumation history of the Altun Shan with evidence for timing of the subduction of the Tarim block beneath the Altyn Tagh system, north Tibet. Earth Planet Sci Lett 329:749–755

Jouzel J, Masson-Delmotte V, Cattani O, Dreyfus G, Falourd S, Hoffmann G, Minster B, Nouet J, Barnola JM, Chappellaz J, Fischer H, Gallet JC, Johnsen S, Leuenberger M, Loulergue L, Luethi D, Oerter H, Parrenin F, Raisbeck G, Raynaud D, Schilt A, Schwander J, Selmo E, Souchez R, Spahni R, Stauffer B, Steffensen JP, Stenni B, Stocker TF, Tison JL, Werner M, Wolff EW (2007) Orbital and millennial Antarctic climate variability over the past 800,000 years. Science 317(5839):793–796. doi:10.1126/science.1141038

Kapp P, Murphy M, Yin A, Harrison T, Ding L, Guo J (2003) Mesozoic and Cenozoic tectonic evolution of the Shiquanhe area of western Tibet. Tectonics 22(4):1029. doi:10.1029/2001Tcoo1332

Kapp P, Yin A, Harrison TM, Ding L (2005) Cretaceous-Tertiary shortening, basin development, and volcanism in central Tibet. Geol Soc Am Bull 117(7–8):865–878. doi:10.1130/1325595.1

Kapp P, DeCelles PG, Gehrels GE, Heizier M, Ding L (2007) Geological records of the Lhasa-Qiangtang and Indo-Asian collisions in the Nima area of central Tibet. Geol Soc Am Bull 119(7–8):917–932. doi:10.1130/B26033.1

Keigwin LD, Lehman SJ (1994) Deep circulation change linked to HEINRICH Event 1 and Younger Dryas in a middepth North Atlantic Core. Paleoceanography 9(2):185–194. doi:10.1029/94pa00032

Kelly MJ, Edwards RL, Cheng H, Yuan DX, Cai YJ, Zhang ML, Lin YS, An ZS (2006) High resolution characterization of the Asian Monsoon between 146,000 and 99,000 years BP from Dongge Cave, China and global correlation of events surrounding Termination II. Palaeogeogr Palaeoclimatol Palaeoecol 236(1–2):20–38. doi:10.1016/j.palaeo.2005.11.042

Kiefer T, McCave IN, Elderfield H (2006) Antarctic control on tropical Indian Ocean sea surface temperature and hydrography. Geophys Res Lett 33:L24612

Kim SJ, Crowley TJ, Erickson DJ, Govindasamy B, Duffy PB, Lee BY (2008) High-resolution climate simulation of the last glacial maximum. Clim Dynam 31(1):1–16

Kirby E, Reiners P, Frol M, Whipple K, Hodges K, Farley K, Tang W, Chen Z (2002) Late Cenozoic evolution of the eastern margin of the Tibetan Plateau: inferences from 40Ar/39Ar and (U-Th)/He thermochronology. Tectonics 21(1):1001. doi:10.1029/2000TC001246

Krishnaswami S, Trivedi JR, Sarin MM, Ramesh R, Sharma KK (1992) Strontium isotopes and Rubidium in the Ganga Brahmaputra river system - weathering in the Himalaya, fluxes to the Bay of Bengal and contributions to the evolution of oceanic 87Sr/86Sr. Earth Planet Sci Lett 109(1–2):243–253. doi:10.1016/0012-821x(92)90087-C

Ku T, Li H (1998) Speleothems as high-resolution paleoenvironment archives: records from northeastern China. J Earth Syst Sci 107(4):321–330

Kukla G (1987) Loess stratigraphy in Central China. Quat Sci Rev 6(3–4):191–219. doi:10.1016/0277-3791(87) 90004-7

Kukla G, Koci A (1972) End of the last interglacial in the loess record. Quat Res 2(3):374–383

Kutzbach JE (1981) Monsoon climate of the early Holocene - climate experiment with the Earths orbital parameters for 9000 Years Ago. Science 214(4516):59–61. doi:10.1126/science.214.4516.59

Kutzbach JE, Guetter PJ, Ruddiman WF, Prell WL (1989) Sensitivity of climate to late Cenozoic uplift in Southern Asia and the American West: numerical experiments. J Geophys Res 94(D15):18393–18407

Kutzbach JE, Prell WL, Ruddiman WF (1993) Sensitivity of Eurasian climate to surface uplift of the Tibetan Plateau. J Geol 101(2):177–190

Kutzbach JE, Liu XD, Liu ZY, Chen GS (2008) Simulation of the evolutionary response of global summer monsoons to orbital forcing over the past 280,000 years. Clim Dynam 30(6):567–579. doi:10.1007/s00382-007-0308-z

Lan X, Wei J (1995) Late Cretaceous-early tertiary marine bivalve fauna from the Western Tarim Basin. Chinese Science Pubilishing House, Beijing

Lease RO, Burbank DW, Gehrels GE, Wang ZC, Yuan DY (2007) Signatures of mountain building: detrital zircon U/Pb ages from northeastern Tibet. Geology 35(3):239–242. doi:10.1130/G23057a.1

Lease RO, Burbank DW, Clark MK, Farley KA, Zheng DW, Zhang HP (2011) Middle Miocene reorganization of deformation along the northeastern Tibetan Plateau. Geology 39(4):359–362

Lee CSL, Qi SH, Zhang G, Luo CL, Zhao LYL, Li XD (2008) Seven thousand years of records on the mining and utilization of metals from lake sediments in central China. Environ Sci Technol 42(13):4732–4738. doi:10.1021/Es702990n

Li J (1995) Uplift of Qinghai-Xizang (Tibet) Plateau and global change. Lanzhou University Press, Lanzhou

Li J, Fang X (1999) Uplift of Tibetan Plateau and environmental changes. Chin Sci Bull 44(23):2117–2124

Li J, Wen S, Zhang Q, Wang F, Zheng B, Li B (1979) Discussion on the period, amplitude and type of the uplift of the Qinghai-Xizang Plateau. Sci Sinica 22(11):1314–1328

Li J, Li B, Wang F, Zhang Q, Wen S, Zheng B (1981) The process of the uplift of the Qinghai-Xizang Plateau. In: Liu D (ed) Geological and ecological studies in Qinghai-Xizang Plateau. Science Press, Beijing, pp 111–118

Li JJ, Fang XM, Ma HZ, Zhu JJ, Pan BT, Chen HL (1996) Geomorphological and environmental evolution in the upper reaches of the Yellow River during the late Cenozoic. Sci China Ser D 39(4):380–390

Li JJ, Fang XM, Van der Voo R, Zhu JJ, Mac Niocaill C, Cao JX, Zhong W, Chen HL, Wang JL, Wang JM, Zhang YC (1997a) Late Cenozoic magnetostratigraphy (11-0 Ma) of the Dongshanding and Wangjiashan sections in the Longzhong Basin, western China. Geologie En Mijnbouw 76(1–2):121–134

Li JJ, Fang XM, VanderVoo R, Zhu JJ, MacNiocaill C, Ono Y, Pan BT, Zhong W, Wang JL, Sasaki T, Zhang YT, Cao JX, Kang SC, Wang JM (1997b) Magnetostratigraphic dating of river terraces: rapid and intermittent incision by the Yellow River of the northeastern margin of the Tibetan Plateau during the Quaternary. J Geophys Res-Sol Ea 102(B5):10121–10132

Li XQ, Zhou XY, Zhou J, Dodson J, Zhang HB, Shang X (2007) The earliest archaeobiological evidence of the broadening agriculture in China recorded at Xishanping site in Gansu Province. Sci China Ser D 50(11):1707–1714. doi:10.1007/s11430-007-0066-0

Li YY, Zhou LP, Cui HT (2008) Pollen indicators of human activity. Chin Sci Bull 53(9):1281–1293. doi:10.1007/s11434-008-0181-0

Li XQ, Dodson J, Zhou J, Zhou X (2009) Increases of population and expansion of rice agriculture in Asia, and anthropogenic methane emissions since 5000BP. Quat Int 202(1):41–50

Li XZ, Liu XD, Qiu LJ, An ZS and Yin ZY (2013) Transient simulation of orbital-scale precipitation variation in monsoonal East Asia and arid Central Asia during the last 150ka. J Geophys Res 118:7481–7488. doi:10.1002/jgrd.50611

Linthout K, Helmers H, Sopaheluwakan J (1997) Late Miocene obduction and microplate migration around the southern Banda Sea and the closure of the Indonesian Seaway. Tectonophysics 281(1–2):17–30. doi:10.1016/S0040-1951(97)00156-X

Lisiecki LE, Raymo ME (2005) A Pliocene-Pleistocene stack of 57 globally distributed benthic delta O-18 records. Paleoceanography 20(1):Pa1003. doi:10.1029/2004pa001071

Lister GS, Kelts K, Zao CK, Yu JQ, Niessen F (1991) Lake Qinghai, China: closed-basin lake levels and the oxygen isotope record for ostracoda since the latest Pleistocene. Palaeogeogr Palaeoclimatol Palaeoecol 84(1–4):141–162

Liu Z (2008) Applications of geochemical proxies in speleothem to the study on evolution and impact factor of Karst-desertification in Central Western Guizhou during morden and contemporary period. Southwestern University

Liu XD, Chen BD (2000) Climatic warming in the Tibetan Plateau during recent decades. Int J Climatol 20(14):1729–1742. doi:10.1002/1097-0088(20001130)20:14 <1729::Aid-Joc556>3.0.Co;2-Y

Liu TS, Ding ML (1984) Comparison of stratum including early human fossils with paleoclimatic cycle in deep-sea sediments. J Anthropol 3:93–101

Liu TS, Ding ZL (1998) Chinese loess and the paleomonsoon. Annu Rev Earth Planet Sci 26:111–145

Liu XD, Dong BW (2013) Influence of the Tibetan Plateau uplift on the Asian monsoon-arid environment evolution. Chin Sci Bull 58(34):4277–4291. doi: 10.1007/s11434-013-5987-8

Liu XD, Shi ZG (2009) Effect of precession on the Asian summer monsoon evolution: a systematic review. Chin Sci Bull 54(20):3720–3730

Liu ZF, Wang CS (2001) Facies analysis and depositional systems of Cenozoic sediments in the Hoh Xil basin, northern Tibet. Sediment Geol 140(3–4):251–270. doi:10.1016/S0037-0738(00)00188-3

Liu X, Yin ZY (2002) Sensitivity of East Asian monsoon climate to the uplift of the Tibetan Plateau. Palaeogeogr Palaeoclimatol Palaeoecol 183(3):223–245

Liu TS et al (1985) Loess and the environment. China Ocean Press, Beijing

Liu TS, Zheng M, Guo Z (1998) Initiation and evolution of the Asian monsoon system timely coupled with the ice-sheet growth and the Tectonic movements in Asia. Quat Sci 18(3):194–204 (in Chinese with English abstract)

Liu TS, Ding ZL, Rutter N (1999) Comparison of Milankovitch periods between continental loess and deep sea records over the last 2.5 Ma. Quat Sci Rev 18(10–11):1205–1212

Liu XD, Kutzbach JE, Liu ZY, An ZS, Li L (2003a) The Tibetan Plateau as amplifier of orbital-scale variability of the East Asian monsoon. Geophys Res Lett 30(16):1839. doi:10.1029/2003gl017510

Liu Z, Otto-Bliesner B, Kutzbach J, Li L, Shields C (2003b) Coupled climate simulation of the evolution of global monsoons in the Holocene. J Climate 16(15):2472–2490. doi:10.1175/1520-0442 (2003) 016 <2472:Ccsote> 2.0.Co;2

Liu ZF, Zhao XX, Wang CS, Liu S, Yi HS (2003c) Magnetostratigraphy of Tertiary sediments from the Hoh Xil Basin: implications for the Cenozoic tectonic history of the Tibetan Plateau. Geophys J Int 154(2):233–252. doi:10.1046/j.1365-246X.2003.01986.x

Liu X, Liu Z, Kutzbach JE, Clemens SC, Prell WL (2006) Hemispheric insolation forcing of the Indian Ocean and Asian monsoon: local versus remote impacts. J Clim 19(23):6195–6208

Liu XD, Cheng ZG, Yan LB, Yin ZY (2009a) Elevation dependency of recent and future minimum surface air temperature trends in the Tibetan Plateau and its surroundings. Global Planet Change 68(3):164–174. doi:10.1016/j.gloplacha.2009.03.017

Liu Y, An ZS, Linderholm HW, Chen DL, Song HM, Cai QF, Sun JY, Tian H (2009b) Annual temperatures during the last 2485 years in the mid-eastern Tibetan Plateau inferred from tree rings. Sci China Ser D 52(3):348–359. doi:10.1007/s11430-009-0025-z

Liu J, Wang B, Ding Q, Kuang X, Soon W, Zorita E (2009c) Centennial variations of the global monsoon precipitation in the last millennium: results from ECHO-G model. J Clim 22:2356-2371

Liu Z, Otto-Bliesner B, He F, Brady E, Clark P, Lynch-Steiglitz J, Carlson A, Curry W, Brook E, Jacob R, Erickson D, Kutzbach J, Cheng J (2009d) Transient simulation of last deglaciation with a new mechanism for Bølling-Allerød warming. Science 325:310–314

Liu Y, Sun JY, Song HM, Cai QF, Bao G, Li XX (2010) Tree-ring hydrologic reconstructions for the Heihe River watershed, western China since AD 1430. Water Res 44(9):2781–2792. doi:10.1016/j.watres.2010.02.013

Liu J, Wang B, Cane MA, Yim SY, Lee JY (2013) Divergent global precipitation changes induced by natural versus anthropogenic forcing. Nature 493(7434):656–659. doi:10.1038/Nature11784

Lorenz SJ, Lohmann G (2004) Acceleration technique for Milankovitch type forcing in a coupled atmosphere-ocean circulation model: method and application for the Holocene. Clim Dynam 23(7):727–743

Lowell TV, Heusser CJ, Andersen BG, Moreno PI, Hauser A, Heusser LE, Schluchter C, Marchant DR, Denton GH (1995) Interhemispheric correlation of late pleistocene glacial events. Science 269(5230):1541–1549

Ma ZG, Fu CB (2006) Some evidence of drying trend over northern China from 1951 to 2004. Chin Sci Bull 51(23):2913–2925. doi:10.1007/s11434-006-2159-0

Ma Y, Li J, Fan X (1998) Pollen-based vegetational and climatic records during 30.6 to 5.0 My from Linxia area, Gansu. Chin Sci Bull 43(3):301–304

Ma YZ, Fang XM, Li JJ, Wu FL, Zhang J (2004) Vegetational and environmental changes during late Tertiary-early Quaternary in Jiuxi Basin. Sci China Ser D Earth Sci 34:107–116

Maasch KA, Saltzman B (1990) A low-order dynamic-model of global climatic variability over the full Pleistocene. J Geophys Res-Atmos 95(D2):1955–1963. doi:10.1029/Jd095id02p01955

Mahowald NM, Yoshioka M, Collins WD, Conley AJ, Fillmore DW, Coleman DB (2006) Climate response and radiative forcing from mineral aerosols during the last glacial maximum, pre-industrial, current and doubled-carbon dioxide climates. Geophys Res Lett 33(20). doi:10.1029/2006gl026126

Manabe S, Broccoli A (1990) Mountains and arid climates of middle latitudes. Science 247(4939):192

Manabe S, Stouffer RJ (1995) Simulation of abrupt climate-change induced by fresh-water input to the North-Atlantic ocean. Nature 378(6553):165–167. doi:10.1038/378165a0

Manabe S, Terpstra TB (1974) The effects of mountains on the general circulation of the atmosphere as identified by numerical experiments. J Atmos Sci 31(1):3–42

Mangerud JA, Andersen ST, Berglund BE, Donner JJ (1974) Quaternary stratigraphy of Norden, a proposal for terminology and classification. Boreas 3(3):109–128. doi:10.1111/j.1502-3885.1974.tb00669.x

Mann ME, Zhang Z, Hughes MK, Bradley RS, Miller SK, Rutherford S, Ni F (2008) Proxy-based reconstructions of hemispheric and global surface temperature variations over the past two millennia. Proc Natl Acad Sci 105(36):13252

Marincovich L, Gladenkov AY (1999) Evidence for an early opening of the Bering Strait. Nature 397(6715):149–151

Mashiotta TA, Lea DW, Spero HJ (1999) Glacial-interglacial changes in Subantarctic sea surface temperature and δ18O-water using foraminiferal Mg. Earth Planet Sc Lett 170(4):417–432. doi:10.1016/s0012-821x(99)00116-8

Maslin MA, Li XS, Loutre MF, Berger A (1998) The contribution of orbital forcing to the progressive intensification of Northern Hemisphere glaciation. Quat Sci Rev 17(4–5):411–426. doi:10.1016/S0277-3791(97)00047-4

May W (2004) Potential future changes in the Indian summer monsoon due to greenhouse warming: analysis of mechanisms in a global time-slice experiment. Clim Dynam 22(4):389–414. doi:10.1007/s00382-003-0389-2

Mayewski PA, Meeker LD, Twickler MS, Whitlow S, Yang QZ, Lyons WB, Prentice M (1997) Major features and forcing of high-latitude northern hemisphere atmospheric circulation using a 110,000-year-long glaciochemical series. J Geophys Res-Oceans 102(C12):26345–26366. doi:10.1029/96jc03365

McArthur JM, Howarth RJ, Bailey TR (2001) Strontium isotope stratigraphy: LOWESS version 3: best fit to the marine Sr-isotope curve for 0-509 Ma and accompanying look-up table for deriving numerical age. J Geol 109(2):155–170. doi:10.1086/319243

Meehl GA, Arblaster JM (2003) Mechanisms for projected future changes in south Asian monsoon precipitation. Clim Dynam 21(7–8):659–675. doi:10.1007/s00382-003-0343-3

Mercier JL, Armijo R, Tapponnier P, Careygailhardis E, Lin HT (1987) Change from Late Tertiary compression to Quaternary extension in Southern Tibet during the India-Asia collision. Tectonics 6(3):275–304. doi:10.1029/Tc006i003p00275

Métivier F, Gaudemer Y, Tapponnier P, Meyer B (1998) Northeastward growth of the Tibet plateau deduced from balanced reconstruction of two depositional area: the Qaidam basin and Hexi Corridor basins, China. Tectonics 17:823–842

Misra S, Froelich PN (2012) Lithium isotope history of Cenozoic seawater: changes in silicate weathering and reverse weathering. Science 335(6070):818–823

Mo X, Niu Y, Dong G, Zhao Z, Hou Z, Zhou S, Ke S (2008) Contribution of syncollisional felsic magnatism to continental crust growth: a case study of the Paleogene Linzizong volcanic. Chem Geol 250:49–67

Mock C, Arnaud N, Cantagrel J (1999) An early unroofing in northeastern Tibet? Constraints from 40Ar/39Ar thermochronology on granitoids from the eastern Kunlun range (Qinghai, NW China). Earth Planet Sci Lett 171:107–122

Molnar P (2004) Late cenozoic increase in accumulation rates of terrestrial sediment: how might climate change have affected erosion rates? Annu Rev Earth Planet Sci 32:67–89. doi:10.1146/annurev.earth.32.091003.143456

Molnar P (2005) Mio-Pliocene growth of the Tibetan Plateau and evolution of East Asian climate. Palaeontologia Electronica 8(1):1–23

Molnar P, England P (1990) Late Cenozoic uplift of mountain-ranges and global climate change—chicken or egg. Nature 346(6279):29–34. doi:10.1038/346029a0

Molnar P, Lyon-caen H (1989) Fault plane solutions of earthquakes and active tectonics of the Tibetan Plateau and its margins. Geophys J Int 99(1):123–153

Molnar P, Stock JM (2009) Slowing of India's convergence with Eurasia since 20 Ma and its implications for Tibetan mantle dynamics. Tectonics 28(3). doi:10.1029/2008tc002271

Molnar P, Tapponnier P (1975) Cenozoic tectonics of Asia: effects of a continental collision. Science 189:419–426

Molnar P, England P, Martinod J (1993) Mantle dynamics, uplift of the Tibetan Plateau, and the India monsoon. Rev Geophys 31:357–396

Molnar P, Boos WR, Battisti DS (2010) Orographic controls on climate and paleoclimate of Asia: thermal and mechanical roles for the Tibetan Plateau. Annu Rev Earth Planet Sci 38:77–102

Mudelsee M, Raymo ME (2005) Slow dynamics of the Northern Hemisphere glaciation. Paleoceanography 20(4):Pa4022. doi:10.1029/2005pa001153

Murphy MA, Yin A, Harrison TM, Durr SB, Chen Z, Ryerson FJ, Kidd WSF, Wang X, Zhou X (1997) Did the Indo-Asian collision alone create the Tibetan plateau? Geology 25(8):719–722

Muscheler R, Joos F, Beer J, Muller SA, Vonmoos M, Snowball I (2007) Solar activity during the last 1000 year inferred from radionuclide records. Quat Sci Rev 26(1–2):82–97

Nesbitt HW, Young GM (1982) Early Proterozoic climates and plate motions inferred from major element chemistry of lutites. Nature 299(5885):715–717. doi:10.1038/299715a0

Noh H, Huh Y, Qin JH, Ellis A (2009) Chemical weathering in the Three Rivers region of Eastern Tibet. Geochim Cosmochim Acta 73(7):1857–1877. doi:10.1016/j.gca.2009.01.005

Oppo DW, Rosenthal Y, Linsley BK (2009) 2,000-year-long temperature and hydrology reconstructions from the Indo-Pacific warm pool. Nature 460(7259):1113–1116. doi:10.1038/Nature08233

Pagani M, Zachos JC, Freeman KH, Tipple B, Bohaty S (2005) Marked decline in atmospheric carbon dioxide concentrations during the Paleogene. Science 309(5734):600–603. doi:10.1126/science.1110063

Pagani M, Huber M, Liu ZH, Bohaty SM, Henderiks J, Sijp W, Krishnan S, DeConto RM (2011) The role of carbon dioxide during the onset of Antarctic glaciation. Science 334(6060):1261–1264. doi:10.1126/science.1203909

Palmer MR, Edmond JM (1992) Controls over the strontium isotope composition of river water. Geochim Cosmochim Acta 56(5):2099–2111. doi:10.1016/0016-7037(92)90332-D

Pares JM, Van der Voo R, Downs WR, Yan MD, Fang XM (2003) Northeastward growth and uplift of the Tibetan Plateau: magnetostratigraphic insights from the Guide Basin (vol 108, Art no. 2017, 2003). J Geophys Res-Sol Ea 108(B8):2400. doi:10.1029/2003jb002624

Patriat P, Achache J (1984) India-Eurasia collision chronology has implications for crustal shortening and driving mechanism of plates. Nature 311:615–621

Pearson PN, Palmer MR (2000) Atmospheric carbon dioxide concentrations over the past 60 million years. Nature 406(6797):695–699

Petit JR, Jouzel J, Raynaud D, Barkov NI, Barnola J-M, Basile I, Bender M, Chappellaz J, Davis M, Delaygue G, Delmotte M, Kotlyakov VM, Legrand M, Lipenkov VY, C. Lorius C, PÉpin L, Ritz C, Saltzman E, Stievenard M (1999) Climate and atmospheric history of the past 420,000 years from the Vostok ice core, Antarctica. Nature 399:429–436

Piao S, Ciais P, Huang Y, Shen Z, Peng S, Li J, Zhou L, Liu H, Ma Y, Ding Y (2010) The impacts of climate change on water resources and agriculture in China. Nature 467(7311):43–51

Pollard D, DeConto RM (2009) Modelling West Antarctic ice sheet growth and collapse through the past five million years. Nature 458(7236):329–332

Porter SC, An ZS (1995) Correlation between climate events in the North Atlantic and China during the last glaciation. Nature 375(6529):305–308. doi:10.1038/375305a0

Pozzi J, Westphal M, Girardeau J, Besse J, Zhou Y, Chen X, Li S (1984) Paleomagnetism of the Xigaze ophiolite and flysch (Yarlung Zangbo suture zone, southern Tibet): latitude and direction of spreading. Earth Planet Sci Lett 70:383–394

Prell WL, Kutzbach JE (1992) Sensitivity of the Indian monsoon to forcing parameters and implications for its evolution. Nature 360(6405):647–652

Proctor CJ, Baker A, Barnes WL, Gilmour WA (2000) A thousand year speleothem proxy record of North Atlantic climate from Scotland. Clim Dynam 16(10–11):815–820. doi:10.1007/s003820000077

Qiang XK, An ZS, Song YG, Chang H, Sun YB, Liu WG, Ao H, Dong JB, Fu CF, Wu F, Lu FY, Cai YJ, Zhou WJ, Cao JJ, Xu XW, Ai L (2011) New eolian red clay sequence on the western Chinese Loess Plateau linked to onset of Asian desertification about 25 Ma ago. Sci China Earth Sci 54(1):136–144. doi:10.1007/s11430-010-4126-5

Qiu Z, Li C (2005) Evolution of Chinese mammalian faunal regions and elevation of the Qinghai-Xizang (Tibet) Plateau. Sci China Ser D Earth Sci 48(8):1246–1258. doi:10.1360/03yd0523

Ramstein G, Fluteau F, Besse J, Joussaume S (1997) Effect of orogeny, plate motion and land-sea distribution on Eurasian climate change over the past 30 million years. Nature 386(6627):788–795

Raymo ME, Ruddiman WF (1992) Tectonic forcing of late Cenozoic climate. Nature 359(6391):117–122

Raymo ME, Ruddiman WF, Froelich PN (1988) Influence of late Cenozoic mountain building on ocean geochemical cycles. Geology 16(7):649–653. doi:10.1130/0091-7613

Rea DK (1992) Delivery of Himalayan sediment to northern Indian Ocean and its relation to global climate, sea level, uplift and seawater strontium. In: Duncan RA (ed) Synthesis of results from scientific drilling in the Indian Ocean. AGU, Washington. (Geophys Monogr Ser 70:387–402. doi:10.1029/GM070p0387)

Rea DK, Leinen M, Janecek TR (1985) Geologic approach to the long-term history of atmospheric circulation. Science 227(4688):721–725

Rea DK, Snoeckx H, Joseph LH (1998) Late Cenozoic eolian deposition in the North Pacific: Asian drying, Tibetan uplift, and cooling of the northern hemisphere. Paleoceanography 13(3):215–224

Reynolds BC, Frank M, Halliday AN (2008) Evidence for a major change in silicon cycling in the subarctic North Pacific at 2.73 Ma. Paleoceanography 23(4):Pa4219. doi:10.1029/2007pa001563

Ridgwell AJ (2002) Dust in the Earth system: the biogeochemical linking of land, air and sea. Philos Trans A Math Phys Eng Sci 360(1801):2905–2924. doi:10.1098/rsta.2002.1096

Rigozo NR, Echer E, Vieira L, Nordemann DJR (2001) Reconstruction of Wolf sunspot numbers on the basis of spectral characteristics and estimates of associated radio flux and solar wind parameters for the last millennium. Solar Phys 203(1):179–191

Ritts BD, Yue Y, Graham SA (2004) Oligocene-Miocene tectonics and sedimentation along the Altyn Tagh Fault, Northern Tibetan Plateau: analysis of the Xorkol, Subei, and Aksay Basins. J Geol 112(2):207–229. doi:10.1086/381658

Roger F, Tapponnier P, Arnaud N, Schärer U, Brunel M, Xu Z, Yang J (2000) An Eocene magmatic belt across central Tibet: mantle subduction triggered by the India collision. Terra Nova 12:102–108

Rohling EJ, Liu QS, Roberts AP, Stanford JD, Rasmussen SO, Langen PL, Siddall M (2009) Controls on the East Asian monsoon during the last glacial cycle, based on comparison between Hulu Cave and polar ice-core records. Quat Sci Rev 28(27–28):3291–3302. doi:10.1016/j.quascirev. 2009.09.007

Rowley D (1996) Age of initiation of collision between India and Asia: a review of stratigraphic data. Earth Planet Sci Lett 145:1–13

Rowley D, Currie B (2006) Palaeo-altimetry of the late Eocene to Miocene Lunpola basin, central Tibet. Nature 439(7077):677–681

Ruddiman WF (2001) Earth's climate: past and future. WH Freeman, New York

Ruddiman WF (2003) The anthropogenic greenhouse era began thousands of years ago. Clim Chang 61(3):261–293

Ruddiman WF, Kutzbach JE (1989) Forcing of Late Cenozoic northern hemisphere climate by plateau uplift in Southern Asia and the American West. J Geophys Res-Atmos 94(D15):18409–18427

Ruddiman WF, Guo ZT, Zhou X, Wu HB, Yu YY (2008) Early rice farming and anomalous methane trends. Quat Sci Rev 27(13–14):1291–1295. doi:10.1016/j.quascirev.2008.03.007

Sato M, Hansen JE, McCormick MP, Pollack JB (1993) Stratospheric aerosol optical depths, 1850–1990. J Geophys Res 98(D12):22987–22994

Schulz H, von Rad U, Erlenkeuser H (1998) Correlation between Arabian Sea and Greenland climate oscillations of the past 110,000 years. Nature 393(6680):54–57

Shackleton NJ, Backman J, Zimmerman H, Kent DV, Hall MA, Roberts DG, Schnitker D, Baldauf JG, Desprairies A, Homrighausen R, Huddlestun P, Keene JB, Kaltenback AJ, Krumsiek KAO, Morton AC, Murray JW, Westbergsmith J (1984) Oxygen isotope calibration of the onset of ice-rafting and history of glaciation in the North-Atlantic region. Nature 307(5952):620–623. doi:10.1038/307620a0

Shen C, Wang W, Peng Y, Xu Y, Zheng J (2009) Variability of summer precipitation over Eastern China during the last millennium. Clim Past 5:129–141

Shi Z, Liu X (2009) What is driving the Asian monsoon evolution: inspirations from a transient simulation. Quat Sci 29(6):1025–1032. doi:1001-7410 (2009) 29:6<1025:sqdtyz>2.0.tx;2–3

Shi ZG, Liu XD, An ZS, Yi BQ, Yang P, Mahowald N (2011) Simulated variations of eolian dust from inner Asian deserts at the mid-Pliocene, last glacial maximum, and present day: contributions from the regional tectonic uplift and global climate change. Clim Dynam 37(11–12):2289–2301. doi:10.1007/s00382-011-1078-1

Sima A, Paul A, Schulz M (2004) The Younger Dryas—an intrinsic feature of late Pleistocene climate change at millennial timescales. Earth Planet Sci Lett 222(3-4):741–750. doi:10.1016/j.epsl.2004.03.026

Sobel E, Oskin M, Burbank G, Mikolaichuk A (2008) Exhumation of basement-cored uplifts: example of the Kyrgyz Range quantified with apatite fission track thermochronology. Textonics 25:2008. doi:10.1029/2005TC001809

Solanki SK, Usoskin IG, Kromer B, Schussler M, Beer J (2004) Unusual activity of the sun during recent decades compared to the previous 11,000 years. Nature 431(7012):1084–1087. doi:10.1038/Nature02995

Spotl C, Mangini A (2002) Stalagmite from the Austrian Alps reveals Dansgaard–Oeschger events during isotope stage 3: implications for the absolute chronology of Greenland ice cores. Earth Planet Sci Lett 203(1):507–518. doi:10.1016/S0012-821x(02)00837-3

Spurlin MS, Yin A, Horton BK, Zhou J, Wang J (2005) Structural evolution of the Yushu-Nangqian region and its relationship to syncollisional igneous activity, east-central Tibet. Geol Soc Am Bull 117(9):1293–1317. doi:10.1130/b25572.1

Stuiver M, Braziunas TF (1993) Modeling atmospheric 14C influences and 14C ages of marine samples to 10 000 BC. Radiocarbon 35(1):137–189

Sun YB, An ZS (2005) Late Pliocene-Pleistocene changes in mass accumulation rates of eolian deposits on the central Chinese Loess Plateau. J Geophys Res-Atmos 110:D23101. doi:10.1029/2005jd006064

Sun X, Wang P (2005) How old is the Asian monsoon system?—Palaeobotanical records from China. Palaeogeogr Palaeoclimatol Palaeoecol 222(3):181–222

Sun Z, Yang Z, Pei J, Ge X, Wang X, Yang T, Li W, Yuan S (2005) Magnetostratigraphy of Paleogene sediments from northern Qaidam Basin, China: implications for tectonic uplift and block rotation in northern Tibetan plateau. Earth Planet Sci Lett 237:635–646

Sun YB, Clemens SC, An ZS, Yu ZW (2006a) Astronomical timescale and palaeoclimatic implication of stacked 3.6-Myr monsoon records from the Chinese Loess Plateau. Quaternary Sci Rev 25(1–2):33–48. doi:10.1016/j.quascirev.2005.07.005

Sun YB, Chen J, Clemens SC, Liu QS, Ji JF, Tada R (2006b) East Asian monsoon variability over the last seven glacial cycles recorded by a loess sequence from the northwestern Chinese Loess Plateau. Geochem Geophy Geosy 7:Q12Q02. doi:10.1029/2006GC001287

Sun YB, Wang XL, Liu QS, Clemens SC (2010) Impacts of post-depositional processes on rapid monsoon signals recorded by the last glacial loess deposits of northern China. Earth Planet Sci Lett 289(1–2):171–179. doi:10.1016/j.epsl.2009.10.038

Sun YB, Clemens SC, Morrill C, Lin XP, Wang XL, An ZS (2012) Influence of Atlantic meridional overturning circulation on the East Asian winter monsoon. Nat Geosci 5(1):46–49. doi:10.1038/ngeo1326

Tan LC (2008) High-resolution monsoon precipitation variations in China during the last 2000 years. PhD. Dissertation, Chinese Academy of Sciences, Xi'an

Tang T, Xue Y, Yu C (1992) Characteristics and sedimentary environments of the Late Cretaceous to early Tertiary marine strata in the western Tarim. Science Press, Beijing

Tan LC, Cai YJ, Cheng H, An ZS, Edwards RL (2009) Summer monsoon precipitation variations in central China over the past 750 years derived from a high-resolution absolute-dated stalagmite. Palaeogeogr Palaeoclimatol Palaeoecol 280(3–4):432–439. doi:10.1016/j.palaeo.2009.06.030

Tan LC, Cai YJ, An ZS, Yi L, Zhang HW, Qin SJ (2011a) Climate patterns in north central China during the last 1800 year and their possible driving force. Clim Past 7(3):685–692. doi:10.5194/cp-7-685-2011

Tan LC, Cai YJ, An ZS, Edwards RL, Cheng H, Shen CC, Zhang HW (2011b) Centennial-to decadal-scale monsoon precipitation variability in the semi-humid region, northern China during the last 1860 years: records from stalagmites in Huangye Cave. Holocene 21(2):287–296. doi:10.1177/0959683610378880

Tao JR, Du NQ (1982) Neogene flora of Tengchong Basin in western Yunnan, China. Acta Botanica Sinica 24:273–281

Tapponnier P, Xu Z, Roger F, Meyer B, Arnaud N, Wittlinger G, Yang J (2001) Oblique stepwise rise and growth of the Tibet Plateau. Science 294(5547):1671–1677

Trenberth KE, Chen S-C (1988) Rotational and divergent geopotential components. J Atmos Sci 45(20):2949–2960. doi:10.1175/1520-0469 (1988) 045<2949:radgc>2.0.co;2

Tripati AK, Roberts CD, Eagle RA (2009) Coupling of CO2 and ice sheet stability over major climate transitions of the last 20 million years. Science 326:1394–1397

Tuenter E, Weber S, Hilgen F, Lourens L, Ganopolski A (2005) Simulation of climate phase lags in response to precession and obliquity forcing and the role of vegetation. Clim Dynam 24(2):279–295

Turner S, Hawkesworth C, Liu J, Rogers N, Kelley S, van Calsteren P (1993) Timing of Tibetan uplift constrained by analysis of volcanic rocks. Nature 364:50–54

Ueda H, Iwai A, Kuwako K, Hori ME (2006) Impact of anthropogenic forcing on the Asian summer monsoon as simulated by eight GCMs. Geophys Res Lett 33(6):L06703. doi:10.1029/2005gl025336

Vidal L, Labeyrie L, Cortijo E, Arnold M, Duplessy JC, Michel E, Becqué S, van Weering TCE (1997) Evidence for changes in the North Atlantic Deep Water linked to meltwater surges during the Heinrich events. Earth Planet Sci Lett 146(1–2):13–27. doi:10.1016/s0012-821x(96)00192-6

Volkmer JE, Kapp P, Guynn JH, Lai Q (2007) Cretaceous-Tertiary structural evolution of the north central Lhasa terrane, Tibet. Tectonics 26(6):1–18. doi:10.1029/2005tc001832

von Storch H, Zorita E, Jones JM, Dimitriev Y, Gonzalez-Rouco F, Tett SFB (2004) Reconstructing past climate from noisy data. Science 306(5696):679–682. doi:10.1126/science.1096109

Walker JCG, Hays PB, Kasting JF (1981) A negative feedback mechanism for the long-term stabilization of Earth's surface temperature. J Geophys Res 86(C10):9776. doi:10.1029/JC086iC10p09776

Wan J, Wang Y, Li Q, Wang F, Wang E (2001) FT evidence of Northern Altyn uplift in Late-Cenozoic. Bull Mineral Petrol Geochem 20:222–224

Wang HJ (1999a) Role of vegetation and soil in the Holocene megathermal climate over China. J Geophys Res 104(D8):9

Wang X (1999b) Some points on management of soil and water loss in the Loess Plateau. Soil Water Conserv China 12:17–19 (In Chinese with English abstract)

Wang B (2006) The Asian monsoon. Springer, Berlin

Wang J, Wang YJ, Liu ZC, Li JQ, Xi P (1999) Cenozoic environmental evolution of the Qaidam Basin and its implications for the uplift of the Tibetan Plateau and the drying of central Asia. Palaeogeogr Palaeoclimatol Palaeoecol 152(1):37–47

Wang YJ, Cheng H, Edwards RL, An ZS, Wu JY, Shen CC, Dorale JA (2001) A high-resolution absolute-dated Late Pleistocene monsoon record from Hulu Cave, China. Science 294(5550):2345–2348

Wang CS, Liu ZF, Yi HS, Liu S, Zhao XX (2002a) Tertiary crustal shortenings and peneplanation in the Hoh Xil region: implications for the tectonic history of the northern Tibetan Plateau. J Asian Earth Sci 20(3):211–223

Wang Y, Wan J, Li Q, Wang F, Wang E (2002b) Fission-track evidence for the Cenozoic uplift and Erosion of the northern segment of the Altyn Tagh fault zone at the Aksay-Dangjin pass. Acta Geological Sinica 76:191–198

Wang E, Wan J, Liu J (2003a) Late Cenozoic geological evolution of the foreland basin bordering the West Kunlun range in Pulu area: constraints on timing of uplift of northern margin of the Tibetan Plateau. J Geophys Res Solid Earth 108(B8):2401. doi:10.1029/2002JB001877

Wang P, Zhao Q, Jian Z, Cheng X, Huang W, Tian J, Wang J, Li B, Sun X (2003b) Thirty million year deep-sea records in the South China Sea. Chin Sci Bull 48(23):2524–2535

Wang YJ, Cheng H, Edwards RL, He YQ, Kong XG, An ZS, Wu JY, Kelly MJ, Dykoski CA, Li XD (2005) The Holocene Asian monsoon: links to solar changes and North Atlantic climate. Science 308(5723):854–857. doi:10.1126/science.1106296

Wang E, Xu F, Zhou J, Wan J, Burchfiel B (2006a) Eastward migration of the Qaidam basin and its implications for Cenozoic evolution of the Altyn Tagh fault and associated river system. Geol Soc Am Bull 118:349–365

Wang X, Auler AS, Edwards RL, Cheng H, Ito E, Solheid M (2006b) Interhemispheric anti-phasing of rainfall during the last glacial period. Quat Sci Rev 25(23–24):3391–3403. doi:10.1016/j.quascirev.2006.02.009

Wang G, Innes JL, Lei J, Dai S, Wu SW (2007a) Ecology. China's forestry reforms. Science 318(5856):1556–1557. doi:10.1126/science.1147247

Wang SW, Wen XY, Luo Y, Dong WJ, Zhao ZC, Yang B (2007b) Reconstruction of temperature series of China for the last 1000 years. Chin Sci Bull 52(23):3272–3280

Wang C, Zhao X, Liu Z, Lippert PC, Graham SA, Coe RS, Yi H, Zhu L, Liu S, Li Y (2008a) Constraints on the early uplift history of the Tibetan Plateau. Proc Natl Acad Sci 105(13):4987–4992. doi:10.1073/pnas.0703595105

Wang XM, Chen F, Hasi E, Li JC (2008b) Desertification in China: an assessment. Earth-Sci Rev 88(3–4):188–206

Wang YJ, Cheng H, Edwards RL, Kong XG, Shao XH, Chen ST, Wu JY, Jiang XY, Wang XF, An ZS (2008c) Millennial- and orbital-scale changes in the East Asian monsoon over the past 224,000 years. Nature 451(7182):1090–1093. doi:10.1038/nature06692

Wang XM, Zhang CX, Hasi E, Dong ZB (2010) Has the Three Norths Forest Shelterbelt Program solved the desertification and dust storm problems in arid and semiarid China? J Arid Environ 74(1):13–22. doi:10.1016/j.jaridenv.2009.08.001

Webb TI, Bryson RA (1972) Late- and postglacial climatic change in the northern midwest, USA: quantitative estimates derived from fossil pollen spectra by multivariate statistical analysis. Quat Res 2:70–115

Weber SL, Tuenter E (2011) The impact of varying ice sheets and greenhouse gases on the intensity and timing of boreal summer monsoons. Quat Sci Rev 30(3–4):469–479. doi:10.1016/j.quascirev.2010.12.009

Wei G, Li X-H, Liu Y, Shao L, Liang X (2006) Geochemical record of chemical weathering and monsoon climate change since the early Miocene in the South China Sea. Paleoceanography 21(4):1–11. doi:10.1029/2006PA001300

Wu GJ, Pan BT, Gao HS, Guan QY, Xia DS (2006) Climatic signals in the Chinese loess record for the Last Glacial: the influence of northern high latitudes and the tropical Pacific. Quatern Int 154–155:128–135. doi:10.1016/j.quaint.2006.02.007

Xu J, Tao R, Xu Z (2004) Sloping Land Conversion Program: cost-effectiveness, structural effect and economic sustainability. China Econ Q 4(1):139–162 (in Chinese)

Xu H, Hong YT, Lin QH, Zhu YY, Hong B, Jiang HB (2006) Temperature responses to quasi-100-yr solar variability during the past 6000 years based on δ18O of peat cellulose in Hongyuan, eastern Qinghai-Tibet plateau, China. Palaeogeogr Palaeoclimatol Palaeoecol 230(1–2):155–164. doi:10.1016/j.palaeo.2005.07.012

Xue F, Wang H, He J (2004) Interannual variability of Mascarene high and Australian high and their influences on East Asian summer monsoon. J Meteorol Soc Jpn 82(4):1173–1186. doi:10.2151/jmsj.2004.1173

Yan H, Sun L, Wang Y, Liu X, Qiu S, Cheng W (2010) A 2000-year record of copper pollution in South China Sea derived from seabird excrements: a potential indicator for copper production and civilization of China. J Paleolimnol 44(2):431–442. doi:10.1007/s10933-010-9413-9

Yin A, Rumelhart P, Butler R, Cowgill E, Harrison T, Foster D, Ingersoll R, Zhang Q, Zhou X, Wang X, Hanson A, Raza A (2002) Tectonic history of the Altyn Tagh fault system in northern Tibet inferred fron Cenozoic sedimentation. Geol Soc Am Bull 114:1257–1295

Yin A, Dang Y, Zhang M, McRivette MW, Burgess WP, Chen X (2007) Cenozoic tectonic evolution of Qaidam basin and its surrounding regions (part 2): wedge tectonics in southern Qaidam basin and the Eastern Kunlun Range. Geol Soc Am Special Pap 433:369–390. doi:10.1130/2007.2433(18)

Yin A, Dang YQ, Zhang M, Chen XH, McRivette MW (2008) Cenozoic tectonic evolution of the Qaidam basin and its surrounding regions (Part 3): structural geology, sedimentation, and regional tectonic reconstruction. Geol Soc Am Bull 120(7–8):847–876

Yu Y, Zhou Z, Zhang X (2003) Impact of the closure of Indonesian seaway on climate: a numerical modeling study. Chin Sci Bull 48:83–93

Yu X, Zhou W, Liu X, Xian F, Liu Z, Zheng Y, An Z (2010) Peat records of human impacts on the atmosphere in Northwest China during the late Neolithic and Bronze Ages. Palaeogeogr Palaeoclimatol Palaeoecol 286(1–2):17–22. doi:10.1016/j.palaeo.2009.11.034

Yuan F, Du H (1984) Chinese Cenozoic biostratigraphy. Geological Publishing House, Beijing
Yuan DX, Cheng H, Edwards RL, Dykoski CA, Kelly MJ, Zhang ML, Qing JM, Lin YS, Wang YJ, Wu JY, Dorale JA, An ZS, Cai YJ (2004) Timing, duration, and transitions of the last interglacial Asian monsoon. Science 304(5670):575–578
Yuan W, Dong J, Wang S, Carter A (2006) Apatite fission track evidence for Neogene uplift in the eastern Kunlun Mountains, northern Qinghai-Tibet Plateau, China. J Asian Earth Sci 27:847–856
Yue Y, Graham SA, Ritts BD, Wooden JL (2005) Detrital zircon provenance evidence for largescale extrusion along the Altyn Tagh fault. Tectonophysics 406(3–4):165–178. doi:10.1016/j.tecto.2005.05.023
Zachos J, Kump L (2005) Carbon cycle feedbacks and the initiation of Antarctic glaciation in the earliest Oligocene. Glob Planet Chang 47(1):51–66. doi:10.1016/j.gloplacha.2005.01.001
Zachos J, Pagani M, Sloan L, Thomas E, Billups K (2001) Trends, rhythms, and aberrations in global climate 65 Ma to present. Science 292(5517):686–693
Zhang X (2003) Scientific approaches and suggestions on vegetation construction on the Loess Plateau. Soil Water Conserv China 1:17 (in Chinese)
Zhang R, Delworth TL (2005) Simulated tropical response to a substantial weakening of the Atlantic thermohaline circulation. J Clim 18(12):1853–1860. doi:10.1175/jcli3460.1
Zhang R, Liu XD (2009) An analogy analysis of summer precipitation change patterns between mid-holocene and future climatic warming scenarios over East Asia. Scientia Geographica Sinica 29:679–683
Zhang R, Liu XD (2010) The effects of tectonic uplift on the evolution of Asian summer monsoon climate since Pliocene. Chin J Geophys 53(6):948–960
Zhang X, Qi Y (2004) Scientific review and suggestions on vegetation recovery of hard regions for afforestation in non-arid zone. Soil Water Conserv China 10:5–7 (in Chinese)
Zhang PZ, Burchfiel BC, Molnar P, Zhang WQ, Jiao DC, Deng QD, Wang YP, Royden L, Song FM (1991) Amount and style of Late Cenozoic deformation in the Liupan Shan Area, Ningxia Autonomous Region, China. Tectonics 10(6):1111–1129
Zhang XY, Arimoto R, An ZS (1997) Dust emission from Chinese desert sources linked to variations in atmospheric circulation. J Geophys Res 102:28041–28047
Zhang RH, Sumi A, Kimoto M (1999) A diagnostic study of the impact of El Niño on the precipitation in China. Adv Atmos Sci 16(2):229–241. doi:10.1007/bf02973084
Zhang PZ, Molnar P, Downs WR (2001) Increased sedimentation rates and grain sizes 2–4 Myr ago due to the influence of climate change on erosion rates. Nature 410(6831):891–897
Zhang Q, Tao S, Chen L (2003) The interannual variability of East Asian summer monsoon indices and its association with the pattern of general circulation over East Asia. Acta Meteorologica Sinica 61(4):559–568 (in Chinese)
Zhang P, Shen Z, Wang M, Gan W, Bürgmann R, Molnar P, Wang Q, Niu Z, Sun J, Wu J, Sun H, You X (2004) Continuous deformation of the Tibetan Plateau from global positioning system data. Geology 32:809–812
Zhang Z, Wang H, Guo Z, Jiang D (2007) Impacts of tectonic changes on the reorganization of the Cenozoic paleoclimate patterns in China. Earth Planet Sci Lett 257:622–634
Zhang PZ, Cheng H, Edwards RL, Chen FH, Wang YJ, Yang XL, Liu J, Tan M, Wang X, An C, Dai Z, Zhou J, Zhang D, Jia J, Jin L, Johnson KR (2008) A test of climate, sun, and culture relationships from an 1810-year Chinese cave record. Science 322(5903):940–942. doi:10.1126/science.1163965
Zhang YG, Ji JF, Balsam W, Liu LW, Chen J (2009) Mid-Pliocene Asian monsoon intensification and the onset of Northern Hemisphere glaciation. Geology 37(7):599–602. doi:10.1130/G25670a.1
Zhang R, Kang SM, Held IM (2010) Sensitivity of climate change induced by the weakening of the Atlantic meridional overturning circulation to cloud feedback. J Clim 23(2):378–389. doi:10.1175/2009jcli3118.1
Zhao S (1983) A new scheme for comprehensive physical regionalization in China. J Geogr Sci 1:1–10

Zhao P, Zhu YN, Zhang RH (2007) An Asian-Pacific teleconnection in summer tropospheric temperature and associated Asian climate variability. Clim Dynam 29(2–3):293–303

Zheng HB, Powell CM, An ZS, Zhou J, Dong GR (2000) Pliocene uplift of the northern Tibetan Plateau. Geology 28(8):715–718

Zheng DW, Zhang PZ, Wan JL, Li CY, Cao JX (2003) Late Cenozoic deformation subsequence in northeastern margin of Tibet—detrital AFT records from Linxia Basin. Sci China Ser D 46:266–275. doi:10.1360/03dz0021

Zheng Y, Yu G, Wang S, Xue B, Zhuo D, Zeng X, Liu H (2004) Simulation of paleoclimate over East Asia at 6 ka BP and 21 ka BP by a regional climate model. Clim Dynam 23(5):513–529

Zheng D, Zhang P, Wan J, Yuan D, Li C, Yin G, Zhang G, Wang Z, Wei M, Chen J (2008a) Rapid exhumation at ~ 8 Ma on the Liupan Shan thrust fault from apatite fission-track thermochronology: implications for growth of the northeastern Tibetan Plateau margin. Earth Planet Sci Lett 248:198–208

Zheng W, Braconnot P, Guilyardi E, Merkel U, Yu Y (2008b) ENSO at 6 ka and 21 ka from ocean-atmosphere coupled model simulations. Clim Dynam 30(7):745–762

Zheng D, Clark MK, Zhang P, Zheng W, Farley KA (2010) Erosion, fault initiation and topographic growth of the North Qilian Shan (northern Tibetan Plateau). Geosphere 6(6):937–941. doi:10.1130/ges00523.1

Zhou WJ, Donahue DJ, Porter SC, Jull TA, Li XQ, Stuiver M, An ZS, Matsumoto E, Dong GG (1996) Variability of monsoon climate in East Asia at the end of the last glaciation. Quat Res 46(3):219–229

Zhou WJ, Head MJ, Lu XF, An ZS, Jull AJT, Donahue D (1999) Teleconnection of climatic events between East Asia and polar, high latitude areas during the last deglaciation. Palaeogeogr Palaeocl 152(1–2):163–172

Zhou WJ, Head MJ, An ZS, De Deckker P, Liu ZY, Liu XD, Lu XF, Donahue D, Jull AJT, Beck JW (2001) Terrestrial evidence for a spatial structure of tropical-polar interconnections during the Younger Dryas episode. Earth Planet Sci Lett 191(3–4):231–239

Zhou WJ, Dodson J, Head MJ, Li BS, Hou YJ, Lu XF, Donahue DJ, Jull AJT (2002) Environmental variability within the Chinese desert-loess transition zone over the last 20,000 years. Holocene 12(1):107–112

Zhou XJ, Zhao P, Liu G (2009) Asian-Pacific Oscillation index and variation of East Asian summer monsoon over the past millennium. Chin Sci Bull 54(20):3768–3771. doi:10.1007/s11434-009-0619-z

Zhu B, Kidd WSF, Rowley DB, Currie BS, Shafique N (2005) Age of initiation of the India-Asia collision in the East-Central Himalaya. J Geol 113(3):265–285. doi:10.1086/428805

Zhu L, Wang C, Zheng H, Xiang F, Yi H, Liu D (2006) Tectonic and sedimentary evolution of basins in the northeast of Qinghai-Tibet Plateau and their implication for the northward growth of the Plateau. Palaeogeogr Palaeoclimatol Palaeoecol 241(1):49–60. doi:10.1016/j.palaeo.2006.06.019

Index

Z. An (ed.), *Late Cenozoic Climate Change in Asia*,
Developments in Paleoenvironmental Research 16,
DOI 10.1007/978-94-007-7817-7, © Springer Science+Business Media Dordrecht 2014

MIX
Papier aus verantwortungsvollen Quellen
Paper from responsible sources
FSC® C105338

If you have any concerns about our products,
you can contact us on
ProductSafety@springernature.com

In case Publisher is established outside the EU,
the EU authorized representative is:
Springer Nature Customer Service Center GmbH
Europaplatz 3, 69115 Heidelberg, Germany

Printed by Libri Plureos GmbH
in Hamburg, Germany